Y0-CMF-758

DISCARDED

MODELING OF RIVERS

MODELING OF RIVERS

GB
1201.72
M35
M6

Edited by

HSIEH WEN SHEN
Professor of Civil Engineering
Colorado State University

LAMAR UNIVERSITY LIBRARY

647902

A WILEY-INTERSCIENCE PUBLICATION

JOHN WILEY & SONS, New York • Chichester • Brisbane • Toronto

Published by John Wiley & Sons, Inc.

Copyright© 1979 by H. W. Shen

All rights reserved. Published simultaneously in Canada

Reproduction or translation of any part of this work beyond that permitted by Sections 107 or 108 of the 1976 United States Copyright Act without the permission of the copyright owner is unlawful. Requests for permission or further information should be addressed to H. W. Shen.

Library of Congress Cataloging in Publication Data

Main entry under title:
Modeling of rivers.

Based on notes delivered at the Institute on River Mechanics: Modeling of Rivers, Colorado State University, July 1977.
1. Rivers—Mathematical models—Congresses.
2. Hydraulic models—Congresses. I. Shen, Hsieh Wen, 1931- II. Institute on River Mechanics—Modeling of Rivers, Colorado State University, 1977.

GB1201.72.M35M6 551.4'83'0184 79-3913
ISBN 0-471-05474-7

Printed in the United States of America

10 9 8 7 6 5 4 3 2 1

TABLE OF CONTENTS

Chapter

1 Introduction, Flow Resistance, Sediment Transport
H. W. Shen, Colorado State University

2 Flow Variability and Modeling of River Processes
V. Yevjevich, Colorado State University

3 Forecasting of Flows-Flood Frequency Analysis
H. J. Morel-Seytoux, Colorado State University

4 Flow Forecasting Based on Pre-season Conditions
H. J. Morel-Seytoux, Colorado State University

5 River and Canal Morphology
D. B. Simons, Colorado State University

6 Principles of Physical Modeling
H. W. Shen, Colorado State University

7 Sediment Sources and Impacts in the Fluvial System
D. B. Simons, Colorado State University
T. J. Ward, Colorado State University
R. M. Li, Colorado State University

8 Computer Modeling of Rivers: HEC 1-6
W. A. Thomas, U.S. Army Corps of Engineers, Vicksburg, Mississippi

9 Water and Sediment Routing from Watersheds
R. M. Li, Colorado State University

10 Water and Sediment Routing in Rivers
Y. H. Chen, Colorado State University

11 Analysis of Watershed and River System
D. B. Simons, Colorado State University
R. M. Li, Colorado State University

12 Water Quality Model for Agricultural Runoff
A. S. Donigian, Hydrocomp, Inc., Palo Alto, California
N. H. Crawford, Hydrocomp, Inc., Palo Alto, California

13 Unsteady Flow Mathematical Modeling Techniques
B. C. Yen, University of Illinois at Urbana-Champaign

TABLE OF CONTENTS (continued)

Chapter

14 Modeling of Ice in Rivers
G. D. Ashton, U.S. Army Corps of Engineers, Hanover, New Hampshire

15 Shore-Attached Thermal Plumes in Rivers
W. W. Sayre, The University of Iowa

16 Data Collection Planning and Survey of Water Quality Models
T. G. Sanders, Colorado State University

17 River Water Quality Model Construction
P. A. Krenkel, The University of Nevada
V. Novotny, Marquette University

18 Basic Approach to Water Quality Modeling
P. A. Krenkel, The University of Nevada
R. J. Ruane, Tennessee Valley Authority, Chattanooga, Tennessee

19 Application of Water Quality Models
D. W. Hendricks, Colorado State University

20 Additional Remarks on Extremal Floods, Basic Equations, River Channel Patterns, Modeling Techniques and Research Needs
H. W. Shen, Colorado State University

FOREWORD

This book is based essentially on the lecture notes delivered at the Institute on River Mechanics - Modeling of Rivers, Colorado State University, July 1977. Some modifications have been made from the original notes in several chapters.

One of the recent major advancements in river behavior analysis is the construction of mathematical models. Despite the fact that many uncertainties are still involved in establishing basic relationships, mathematical models are effective tools in providing us with quantitative estimates. Several popular mathematical models are presented here by their respective authors, with emphasis on the presentation of basic approaches and assumptions. For a complete description of the models, the reader should refer to the original reports.

The purpose of this book is to present current knowledge related to various types of river modeling. The chapters in general are interrelated, yet stand by themselves with minimum dependence on other chapters. Chapter 1 gives a brief introduction to the material covered in this book plus some of the current knowledge on alluvial river resistances, sediment production and sediment transport rates.

Special acknowledgements are due the following persons: Stephanie Allen and Barbara Gibbens for their capable assistance in language editing and typing of the manuscript, and Arlene Nelson and her typing pool for their excellent typing service.

Hsieh Wen Shen
Colorado State University, 1978

Chapter 1

INTRODUCTION, FLOW RESISTANCE, SEDIMENT TRANSPORT

by

H. W. Shen, Professor of Civil Engineering, Colorado State University, Fort Collins, Colorado

Chapter 1

INTRODUCTION, FLOW RESISTANCE, SEDIMENT TRANSPORT

1.1 INTRODUCTION

An intimate relationship has existed between man and rivers since the beginning of civilization. Water has always been an integral part of man's development. Most early civilizations developed near river valleys, such as Mesopotamia, which means, "the land between the rivers" (namely the Euphrates and the Tigris in western Asia); in the Nile Valley of Egypt; in the valley of the Yellow River in China; and along the Indus River in western Asia. Rivers provide water for human consumption, agricultural needs and industrial demands, navigation, hydropower, recreation, etc. Rivers also cause severe damage to human lives and property through floods, droughts and erosion.

Investigations of river behavior have been conducted since the beginning of time. These investigations have, perhaps, followed these steps:

a) field observation and data collection
b) analysis of field results
c) field experimentation
d) conclusions from field investigations
e) theoretical analysis based on fluid mechanics principles
f) laboratory physical modeling
g) analog models
h) computer-assisted mathematical models
i) combination of field data collection and investigation, theoretical analysis, physical modeling, and mathematical modeling.

Mathematical modeling usually deals with the techniques of solving basic equations. However, the reliability of the results cannot exceed that of the equations used. The continuity equations for flow and sediment as well as the momentum equations are essentially correct. The major uncertainties are in the friction used in the momentum equation and in the sediment transport rate used in the sediment continuity equation (Sections 1.6 and 1.7 of this chapter are devoted to discussions of these two topics, respectively).

1.2 HYDROLOGIC INPUT

The study of the properties of inflow is usually the first step in analyzing possible river response. There may be several flow characteristics of interest: 1) peak flow for estimation of dam spillway capacity, bank erosion and protection, scour near piers, formation of channel regime and channel geometry, and water surface profiles for flood control, etc.; 2) maximum volume of flow in a continuous period for reservoir volume, spillway capacity, water surface profiles, aggradation and degradation, etc.,; 3) rate of change of flow for bank erosion and stability, dam spillway capacity, etc.; and 4) low flows for estimation of minimum water supply, river water quality, and the effects of both of these on aquatic biota.

The term "dominant river discharge" has been used rather loosely. Actually, there are many dominant river discharges--one for each distinct purpose. For instance, one may determine a "dominant river discharge" which is the constant discharge that will shape a particular channel geometry. As a rule of thumb, the bank-full discharge is about equivalent to a flood with an average return period of 2.3 years. The variance of this return period is rather great. One can also have a dominant river discharge for sediment transport; a large river discharge carries a large sediment load, and thus the "dominant river discharge" is somehow greater than the mean river discharges. One can perhaps define a dominant river discharge by the following method:

If a quantity of interest A (such as sediment discharge) is a function of Q so that

$$A = f(Q) \tag{1-1}$$

the dominant discharge Q_1 for this quantity A is thus

$$f(Q_1)T = \int_0^T f(Q)dt \tag{1-2}$$

It is clear that Q_1 is a function of time and f.

The next chapter describes the effects of rainfall and runoff on river modeling. Chapter 3 discusses the forecast of flow from long-term flood frequency analysis and Chapter 4 presents flow forecasting based on pre-season conditions.

1.3 RIVER AND CANAL MORPHOLOGY

The morphology or behavior of the river is the result of interaction between the flow, the quantity and character of the sediment discharge, the local topography, and the local soil and geological conditions.

Chapter 5 discusses the characteristics of rivers, the variables affecting channel geometry and bed roughness, relationships defining river geometry at a cross section, alluvial bed form characteristics, prediction of bed forms, alluvial river patterns (meandering, braided and straight), flow characteristics at river bends, variation of size of bed material and longitudinal channel slope with distance, and qualitative response of river systems to disturbances.

1.4 PHYSICAL MODELING

Physical modeling has contributed significantly to our understanding of flow phenomena, especially in the cases for which no entirely satisfactory theoretical solution is available. This is particularly true for three-dimensional flow studies. For an undistorted rigid boundary model with one dominant similarity criterion, the design, testing and interpretation of model results are well defined. However, the situation is much more complicated for any of a combination of the following cases: 1) more than one similarity criterion, 2) distorted model scales, and 3) movable bed. Chapter 6 discusses the basic principles in dimensional analysis, selection of similarity criteria (through physical reasoning and inspectional analysis of the basic flow equation), use of similarity criteria to determine model-prototype ratios for various flow properties, and how to deal with distorted river models with movable beds. Although a thorough knowledge is required to deal with complex model problems, the design of models and interpretation of model results still require knowledge in the particular field, plus experience in modeling. Calibration of the model with field data is a must for complex situations.

1.5 MATHEMATICAL MODELING

The basic equations used to study movement of water and sediment through rivers are:

a) Flow continuity equation
b) Flow momentum equation
c) Sediment continuity equation
d) Sediment transport equation

An equation to describe flow resistance, such as the Manning equation, is needed to calculate the energy loss, and the variation of the Manning n value with flow equations is also required. When either aggradation or degradation processes are encountered, the area of channel cross section changes, and one needs to know the relationship between the change in channel width and the change in channel depth.

Dispersion of pollutants in rivers or canals is another important area of study for which a dispersion equation is needed. The general dynamic equations of motion governing one-dimensional, gradually varied unsteady flow in open channels are the St. Venant equations. These are first-order quasi-linear hyperbolic equations and do not have explicit solutions, except for simple cases (for example, slow transients).

With the recent great advancement of large-sized electronic computers for high-speed calculations, numerous investigations have introduced numerical techniques for solving these governing equations. A main theme (Chapters 7 through 13) of this book is to present the current status of knowledge of mathematical modeling. The term "mathematical modeling" is used to cover gradually varied unsteady and steady flow problems (with and without consideration of sediment movements).

Chapter 7 discusses the sources and inputs of sediment in a fluvial system. Both watershed and channel sources are discussed, with examples. These include surface erosion, landslides, channel bottom erosion, stream bank erosion, man-induced disturbances, raindrop splash, overland flow, vegetation effects, etc. Impacts of sediment on water quality and aquatic life are also briefly mentioned.

HEC Models

Currently, the most comprehensive set of computer models of rivers is the HEC series produced by the Hydrologic Engineering Center, U.S. Army Corps of Engineers, Davis, California, discussed in Chapter 8. HEC-1 is for the calculation of a runoff hydrograph from a watershed with specified precipitation pattern and precipitation rates for a single storm event. In addition, hydrographs at any number of intermediate points in the watershed may be produced.

The most popular program of this series is the HEC-2, which calculates the water surface profile for a steady flow discharge. Pertinent hydraulic parameters are determined in the process: depth of flow, water surface width, elevation of the total energy line, friction slope, flow velocity, critical depth and volume of water beneath the computed profile. A secondary purpose of HEC-2 is to calculate the permissible encroachment width for a specified rise in the water surface profiles.

The purpose of HEC-3 is to perform a routing of flows through a system of multipurpose reservoirs. Inflows, water demands and reservoir operating rules are prescribed. System performance is measured in terms of storage requirements and in terms of dollars, with shortages declared during any period that a demand is not met.

HEC-4 is used to generate a sequence of monthly flow volumes of any desired length. Statistical characteristics used in the generation are calculated from observed monthly stream flows or from regionalized statistics. Missing records are filled in based on concurrent flows at other stations. Maximum and minimum monthly flows are flagged in the observed, reconstituted and generated flow records.

The purpose of the HEC-5 program is to determine the release from each reservoir project in a system of reservoirs; both conservation and flood control purposes may be included. The system will be operated automatically according to prescribed criteria. Inflow may be deterministic, as in planning studies, or forecasted, with an assigned contingency factor for error.

HEC-6 calculates water surface profiles and stream bed profiles. The water velocity, water depth, energy slope, sediment load, gradation of the sediment load and gradation of the bed surface are also computed. Because a large number of water surface profiles are calculated, a completely different algorithm to HEC-2 was developed.

Water and Sediment Routing from Watersheds

At upstream watersheds where the channel slope is steep and the flow depth is small, kinematic wave modeling is particularly attractive. The channel bottom slope is assumed to be the same as the energy slope. Chapter 9 describes a comprehensive effort to develop models for water and sediment routing from watersheds. Vegetation cover, infiltration rates and sediment production by raindrops are usually rather important in the study of flow and sediment movement in watersheds; current knowledge in these areas is presented in this chapter. Both "analytical solutions" (method of characteristics) and numerical solutions are discussed. The flow equations are solved first; the sediment equation is solved independently of flow routing. Finally, suggestions for simplification of procedures to obtain approximate solutions are given.

River Modeling

Dynamic wave modeling is usually used to solve for flow and sediment movements in large rivers. There are several approaches to a numerical solution of the set of differential equations: One can solve all equations simultaneously; one can solve the flow equation first and then the sediment equation; one can solve all equations simultaneously with the assumption that flow rate does not change rapidly, and thus the flow discharge can be introduced as a step function, neglecting the effect of unsteadiness; or one can solve the flow equation first and then the sediment equation with the assumption that flow rate does not change rapidly. Chapter 10 presents a rather comprehensive investigation in this area. Both one- and two-dimensional approaches and their related problems are discussed.

Analysis of Watershed and River Systems

General problems related to the analysis of watershed and river systems are discussed in Chapter 11. This chapter reviews the current state of the art in watershed and river system modeling; it describes physical processes important to watershed and river systems modeling, provides criteria for a useful model, and discusses the stages and planning of model development. Actual examples from the modeling of the Beaver Creek Watershed, Upper Mississippi River and Yazoo River System are given.

Modeling Agricultural Runoff Quality and Modeling Applications in Environmental Planning: "Hydrocomp Model"

The most comprehensive study on the modeling of agricultural runoff quality and environmental planning is by Hydrocomp, Inc. Chapter 12 discusses the model development philosophy components of practical modeling systems and applications of the

Hydrocomp models. The main purposes of these models are to investigate 1) agricultural runoff and movements of pesticides, nutrients and sediments; and 2) alternatives in planning comprehensive watershed management, including frequency analysis in environmental planning, uses of time series in aquatic ecology evaluations, and costs and benefits from continuous simulation results.

This chapter also discusses calibration and investigated methods, tasks in model applications, software requirements, software concepts for models, input devices, common language, data space and organization, output devices, file management systems for time series, storage and retrieval of watershed characteristics, data utilities and graphics, and physical process models. Process models include the effects of interception, infiltration, surface runoff, and groundwater.

Analysis of Various Models and Research Needs

Chapter 13 presents the derivation of the basic equations used for mathematical modeling. Due to the various simplification schemes, the following four well-known types of mathematical models are available: a) the dynamic wave model is used to solve the complete momentum equation, b) the diffusive wave model is used to assume that the energy slope is equal to the channel slope minus the change of water surface profile by neglecting the convective accumulation terms due to spatial variation and the flow velocity and local accumulation terms, c) the kinematic wave model is used to assume that the energy slope is equal to the channel slope (this is valid for steep channel slope), and d) the storage routing model is used to assume that the momentum equation can be neglected completely and routing of flow is controlled entirely by the flow continuity equation.

A brief discussion is given on the current status of knowledge of these models. Suggested research needs are given, also.

1.6 FRICTIONAL RESISTANCE

As stated previously, the reliability of mathematical model results is very dependent on the reliability of the equations used. One major problem is how to describe the hydraulic resistance of frictions in the momentum equations. Certain current knowledge on this subject is reviewed here.

Frictional resistance is divided into the following four sections: 1) rigid boundary without rainfall, 2) rigid boundary with rainfall, 3) alluvial river boundary, and 4) boundary with vegetation.

Rigid Boundary Without Rainfall

The two most well-known equations developed and used by hydraulic engineers are the Manning and Chezy equations. Manning's equations is

$$V = \frac{1.49}{n} R^{2/3} S^{1/2} \tag{1-3}$$

and Chezy's equations is

$$V = C\sqrt{RS} \tag{1-4}$$

In the above two equations, V is the average flow velocity, R is the hydraulic radius, S is the energy slope, n is Manning's roughness n value, and C is Chezy's roughness C value. Many textbooks (Chow, 1959; Henderson, 1966) provide pictures and detailed tables on the approximate n and C values for difficult rigid boundaries. Table 1-1 provides a few commonly used roughness values.

Table 1-1. Resistance Parameters for Overland Flow (after Woolhiser, 1975)

Surface	Laminar Flow k_o	Turbulent Flow Manning's n	Turbulent Flow Chezy's C ($ft^{1/2}$/sec)
Concrete or Asphalt	24- 108	.01 -.013	73 -38
Bare Sand	30- 120	.01 -.016	65 -33
Graveled Surface	90- 400	.012-.03	38 -18
Bare Clay-Loam Soil (eroded)	100- 500	.012-.033	36 -16
Sparse Vegetation	1000- 4,000	.053-.13	11 - 5
Short Grass Prairie	3000-10,000	.10 -.20	6.5- 3.6
Bluegrass Sod	7000-40,000	.17 -.48	4.2-1.8

Strickler (1923) proposed the following formula to relate Manning's n value to median size of the roughness element on the boundary.

$$n = 0.0342\, d_{50}^{1/6} \tag{1-5}$$

where d_{50} is the median size of the roughness element on the boundary in feet.

In laboratory studies, the most popular approach is to use the Darcy-Weisbach friction coefficient and the Moody diagram. The Darcy-Weisbach frictional resistance f is expressed as follows:

$$S = \frac{f}{4R}\frac{V^2}{2g} \qquad (1\text{-}6)$$

For laminar flow over a smooth surface, the theoretical relationship between the friction factor and Reynolds number ($R_e = VR/\nu$) is

$$f = 24/R_e \qquad (1\text{-}7)$$

where ν is the kinematic viscosity of the fluid.

For laminar flow over a rough surface, a similar relationship has been observed:

$$f = k_o/R_e \quad ; \quad k_o > 24 \qquad (1\text{-}8)$$

The values of k_o are given in Table 1-1.

The relationships between f and n, and between f and C can be derived respectively from Eqs. (1-3) and (1-6) and from Eqs. (1-4) and (1-6).

$$f = \frac{8gn^2}{1.49^2 R^{1/3}} \qquad (1\text{-}9)$$

and

$$f = \frac{8g}{C^2} \qquad (1\text{-}10)$$

Laboratory researchers would prefer to use the Darcy-Weisbach friction coefficient because the variation of f with flow conditions is well tested by Moody's diagram (available in all basic fluid mechanics texts). Moody's diagram was derived from closed conduit studies. However, as shown in Chow (1959) (Fig. 1-1 and Fig. 1-2), the f and R_e relationship also applies reasonably well for flow through open channel.

River engineers usually prefer to use either Manning's n values or Chezy's C values because they have more experience with these coefficients. In any case, it must be remembered that normally, for a rigid boundary, if the bed is rougher than the banks and if the channel is reasonably wide, both Manning's n value and Chezy's C value decrease with increasing flow depth.

Rigid Boundary With Rainfall

The effect of raindrops on flow can be very significant if the flow depth is not too large as compared with the raindrop size. Shen and Li (1973a) found that when R_e is smaller than 900, f is a function of both R_e and rainfall intensity; and when R_e is greater than 2000, f is only a function of R_e and is independent of the rainfall intensity. Figure 1-1 shows this effect. For R_e between 900 and 2000, insufficient data is available to define the trend precisely.

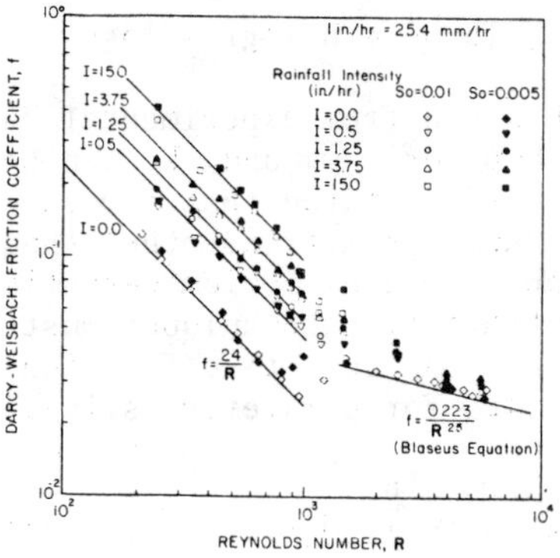

Fig. 1-1. Relationship among f, R, I, S_o (after Shen and Li, 1973a; experimental data after Yoon, 1970).

From data collected by Yoon (1970) and Li (1972) from a smooth channel bottom with slope between 0.005 and 0.0108, rainfall intensity between 0.5 and 17.5 in./hr., and Reynolds number between 126 and 12,600, Shen and Li (1973a) found that

$$k = k_o + AI^b \tag{1-11}$$

where $k = fR_e$ and is the parameter including the effect of rainfall, I is rainfall intensity in inches per hour, and A as well as b is an empirical coefficient. For R_e less than 900, A = 27.162 and b = 0.407; for R_e greater than 2000, A = 0.

Izzard (1944) found that A = 5.67 and b = 1.333, and Fawkes (1972) found that A = 10.0 and b = 1.0. These were derived from less experimental data than used by Shen and Li (1973a).

Woolhiser (1975) also presented some suggestions on how to derive a given relationship based on experimental data. For laminar flow, the equation for the rising hydrograph is:

$$Q = \frac{8gS_o}{k\nu} i^m t^m \tag{1-12}$$

where S_o is the bottom channel slope, Q is the flow discharge, t is the time, and m and k are parameters to be determined. Let

$$Q^1 = \frac{Q\nu}{8gS_o i^m} \quad (1\text{-}13)$$

thus

$$\log Q^1 = n \log t - \log k \quad (1\text{-}14)$$

If Q^1 and A, obtained from experimental data, are plotted on logarithmic graph paper, m and k can be obtained graphically. Woolhiser (1975) stated that for experimental watersheds or plots that have geometrics other than a plane, the rising hydrograph and the equilibrium detention are very complex; therefore, optimization techniques must be utilized for parameter estimation. One objective function that is frequently used and is appropriate for parameter estimates for a single event is

$$F(k, A, b) = \sum_{i=1}^{n} [Q_o(i\Delta t) - Q_c(i\Delta t)]^2 \quad (1\text{-}15)$$

where the subscripts o and c refer to observed and computed discharges, respectively, and Δt is a fixed time increment. If a set of n runoff events is available for analysis, the objective function could incorporate the sum of squares of deviations between observed and computed peak rates.

Alluvial River Boundary

When fluid flows over a movable river bed, the bed surface normally is deformed into various configurations. Some discussion on bed forms is given in Chapter 5. Einstein (1950) was perhaps the first to separate the total resistance into skin resistance and form resistance. He assumed that

> "On such a bed, friction develops in two distinctly different ways: (1) along the sediment grains of the surface as a rough wall with the representative grain diameter equal to k_s; and, in addition, (2) by separation of the flow from the surface at characteristic points of the ripples or bars. This separation causes wakes to develop on the lee side of the bars, characterized by rollers or permanent eddies of basically stagnant water such as those observed behind most submerged bodies of sufficient size. This flow pattern causes a pressure difference to develop between the front and rear sides of each bar so that part of the flow resistance is transmitted to the wall by this shape resistance, i.e. by normal pressure ..."

He then proceeded to divide the cross-sectional flow area into two parts: one area attributed to the grain resistance and the other area to the shape or form resistance. Through the results of Einstein and Banks (1950), Einstein felt that these two resistances could be added together to obtain the total resistance.

Einstein and Barbarossa (1952) proposed the use of the universal logarithmic velocity relationship, developed for a plane rigid boundary, to estimate the skin resistance. They also found from laboratory and field data that their form resistance, obtained by subtracting skin roughness from total roughness, varied with ψ', an inverse of a Froude number based on flow shear velocity and sediment grain sizes. According to Einstein (1950), ψ' is also an indication of sediment transport rate.

Subsequently, Vanoni and Brooks (1957) developed a graphical solution for the estimation of the skin resistance to facilitate the use of Einstein and Barbarossa's procedure. Shen (1962) found that: (1) Einstein and Barbarossa's form resistance varied with both ψ' and sediment size; and (2) the Reynolds number (based on the fall velocity of the particle and the particle diameter) seems to correlate well with sand, coal, and plastic pellets. Since then, many researchers such as Simons and Richardson (1966), Haynie and Simons (1968), Engelund and Hansen (1967), Lovera and Kennedy (1969), Alam and Kennedy (1969), and Mahmood (1971) have proposed various methods to estimate the skin roughness and the form roughness. Sentark also developed a procedure to estimate the alluvial bed form roughness, as presented in Chapter 5.

As discussed by Shen (1975), there are many difficulties in using some of the above approaches:

<u>Rigid Boundary Versus Movable Boundary</u>: When fluid flows over a flat rigid boundary, the total resistance contributed by the boundary to the flow occurs at the interface between the two. However, when fluid flows over a flat movable boundary, the total resistance consists of the resistance between the fluid and the top moving layer as well as the resistance existing among the various sediment layers moving at different speeds. Lovera and Kennedy (1969) found that the skin resistance for a movable flat boundary was quite different from a flat rigid boundary.

<u>Flat Boundary Versus Curved Boundary</u>: The total skin resistance on a dune or bar with flow separations occurring at its downstream end can be very different than that of a plane rigid boundary. As indicated by Raudkivi (1963) and Mahmood (1971), the flow velocity distribution above ripples and dunes is rather complicated. There is still insufficient information on this aspect.

Mutual Interference of Various Bed Forms: Vanoni (1946) and Vanoni and Nomicos (1959) found that a suspended load concentration of 1.2-3.3 g/l can reduce the Darcy-Weisbach friction factor for flat beds by about 18%. Simons, Richardson and Haushild (1963) concluded from their experiments that the resistance to flow and bed material transport decreased in the lower flow regime and increased in the upper flow regime when fine sediment was added to the flow.

The effect of suspended sediment load has not been specifically included in the many resistance formulas, although it has been taken care of at least partially by the use of some flow parameters such as ψ'. The presence or the absence of wash load has not been considered in the friction formula.

Effect of Temperature: Straub, Anderson, and Flammer (1958) found from their laboratory experiments that the energy slope (or the frictional resistance) increased by more than 50% when flow temperature decreased from 86°F to 35°F for a mean flow velocity of about 2.2 fps and a flow depth of about 0.23 ft. Franco (1968) determined from his laboratory experiments that the energy gradient increased by about 20% when the temperature decreased from 80°F to 40°F for the same flow depths and flow velocities. Hubbell and Al-Shaikh (1961) and Cunha (1973) presented separate evidence that increasing temperature could either increase or decrease flow resistances for different flow conditions.

Actually, the temperature could affect the flow resistance differently depending on the range of flow conditions that one is in. It was concluded by Taylor and Vanoni (1972a,b) from their experiments that: (1) In a stream with a rippled bed of fine sand, an increase in water temperature (discharge and depth held constant) may effect either an increase or a reduction in bed roughness and sediment discharge; (2) under certain conditions (unspecified by Taylor), an increase in water temperature alone can accomplish a change in bed form; and (3) in flows where the Reynolds number (based on the shear velocity and sediment size) is less than 8, the transition from ripples to dunes is accomplished at a lower velocity in the warmer water flow.

As shown in Fig. 1-2, Shen, Mellema and Harrison (1977) developed criteria to show the variation of bed forms with water temperature for the same flow discharge in the Missouri River near Omaha. For instance, in a river flow discharge of 32,500 cubic feet per second, the river bed form changes from dunes to plane bed at a water temperature of about 60°F.

Porosity of Bed: Munoz and Gelhar (1968) have shown that the friction factor for a porous pipe boundary is much greater than for a rigid boundary. However, for a case with fine sediment, this effect is perhaps small.

Effect of Three-Dimensional Flow: Secondary currents and other variations in the transverse direction may have significant effects on the development of bed configurations.

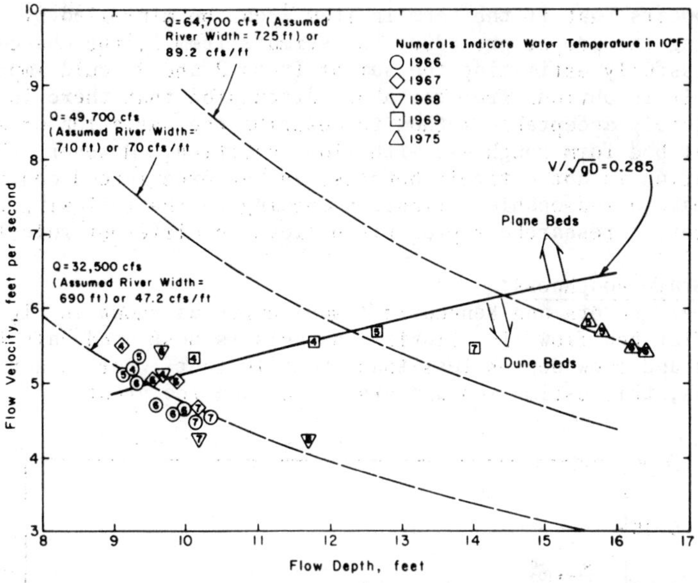

Fig. 1-2. Prediction of Bed Forms of Missouri River near Omaha (after Shen, Mellema and Harrison, 1977).

Characteristics of Time Fluctuation of Boundary Shear Stress: Blinco, Mahmood, and Simons (1973) presented measurements for the time fluctuation of shear stress over a flat rigid boundary. Conceivably, the time variation of shear stress over dunes can be quite different than that for a flat rigid boundary.

Limitation of Flow Depth in Laboratory Flumes: This is an extremely important problem and no conclusive evidence is presently available. Annambhotla, Sayre, and Livesey (1972) and Cheong and Shen (1973) have shown that some statistical properties of alluvial bed forms in field data (collected by Annambhotla, et al.) are quite similar to those obtained from laboratory flumes. If the development of bed forms should be limited by small flow depths or narrow flow depths, or both, the usefulness of laboratory experimental results in the field would be questionable.

There are many other difficulties in addition to the aforementioned. Since there is no known way to correctly estimate the skin resistance of an alluvial channel bed, perhaps the total resistance can be separated into the following three items: (1) "skin resistance" according to whatever procedure is selected; (2) correction of skin resistance due to incorrect procedure used in the first item; and (3) form resistance. The main problem will then be to find a procedure for the estimation of the "skin resistance" so that one can find a way to estimate the sum of the second and third items. Of course, at this stage it appears that if the term in Item 2 can be minimized by selecting the proper procedure to estimate Item 1, the chance of successfully estimating the sum of Items 2 and 3 would improve.

It is obvious from the above discussion that there is no completely acceptable method to describe the variation of alluvial bed form roughness with flow condition. However, the situation is not entirely hopeless and an experienced engineer may make a reasonable estimate according to the following rules (different researchers probably believe in different rules):

<u>For Grain Roughness</u>:

a. Lovera and Kennedy's (1969) curve as shown in Fig. 1-3 for flat-bed flows in alluvial channels is best used only for dunes and flow depths less than about 10 feet. For larger rivers, this estimation appears to be much too great.

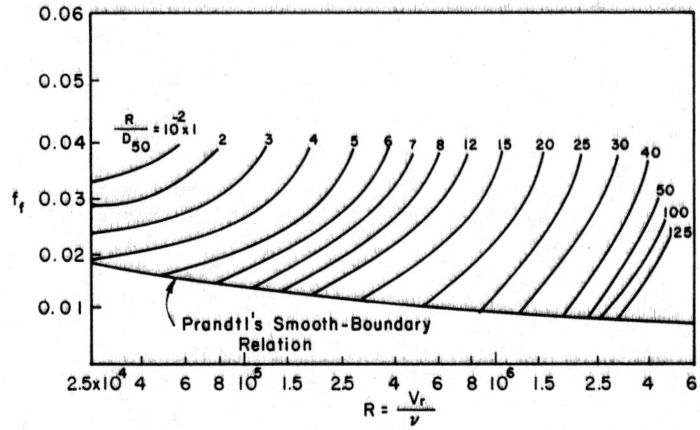

Fig. 1-3. Flat-Bed Friction Factor Diagram for Alluvial Streams (after Lovera and Kennedy, 1969).

b. Einstein's (1950) procedure for calculating grain roughness by the logarithm's velocity distribution is reasonably valid in all cases. For river depths less than 10 feet and for dune beds, Lovera and Kennedy's (1969) curve appears to be slightly better.

Einstein and Barbarossa's (1952) recommended equation for grain roughness is

$$\frac{V}{\sqrt{U_*'}} = 5.75 \log_{10}(12.2) \frac{R_b' x}{k_s} \tag{1-16}$$

where U_*' is the boundary shear stress pertaining to grain roughness, R_b' is the hydraulic radius related to grain roughness, k_s is the representative grain roughness, and x is a parameter depending on the ratio between the representative grain roughness and the viscous sublayer thickness. U'_* and R'_b are related by

$$U'_* = \sqrt{gR'_bS_f} \qquad (1\text{-}17)$$

With known flow conditions, i.e., V, S_f, k_s (which, according to Einstein is approximately the grain size at which 65% of the sediment is finer) and the viscous sublayer thickness, one can calculate the value of either R'_b or U'_* through a trial and error procedure. Vanoni and Brooks (1957) provided a graphical solution of Eq. (1-16) without trial and error.

c. For small relative roughness (i.e., the ratio between the flow depth and the k_s is small, say less than 10.6), grain roughness cannot be calculated by a normal grain roughness equation. In this situation, significant eddies are formed behind the large grains. Grain roughness must be treated as a series of low baffles. The following references should be consulted: Judd and Petersen (1969), Robinson and Albertson (1952), Waterways Experiment Station (1953), Sayre and Albertson (1961), Koloseus and Davidian (1966), and Shen and Li (1973b). Figure 1-4 shows the curve presented by Shen and Li (1973b) for analyzing some selected previous data. For Fig. 1-4, the dimensions of the blocks are given by Fig. 1-5. Kao and Shen (1977) found that the resistances of baffles in flow from the above articles can probably be applied to flow resistance over large rocks in their data collected in Taiwan, China.

For Form Roughness:

a. Form roughness in a laboratory flume is usually less than the grain roughness if either Einstein and Barbarossa's (1952) procedure as shown in Fig. 1-6 or Lovera and Kennedy's (1969) procedure is used to calculate grain roughness.

b. Form roughness in a river can be greater (up to twice as large) than the grain roughness.

c. For sediment particles less than 0.6 mm (or Wd/ν less than 100, where W is the fall velocity of the particle, d is the grain size and ν is the kinematic viscosity of fluid) form roughness always decreases with the increase in sediment transport rate.

d. For sediment particles not much greater in diameter than 0.6 mm, form roughness first decreases, then increases with the increase of sediment transport rate. Minimum form roughness occurs when the shear stress is about two and one-half times the critical shear stress for incipient motion.

e. For very large sediment particles, so that R/R_s is less than 10, the alluvial bed form is almost flat (perhaps with some large and low amplitude bars).

f. For sediment particles less than 0.6 mm in diameter, there is a greater tendency to form larger dunes as suggested by Shen (1962).

g. Einstein and Barbarossa's (1952) procedure is particularly applicable to rivers less than 20 feet deep, with

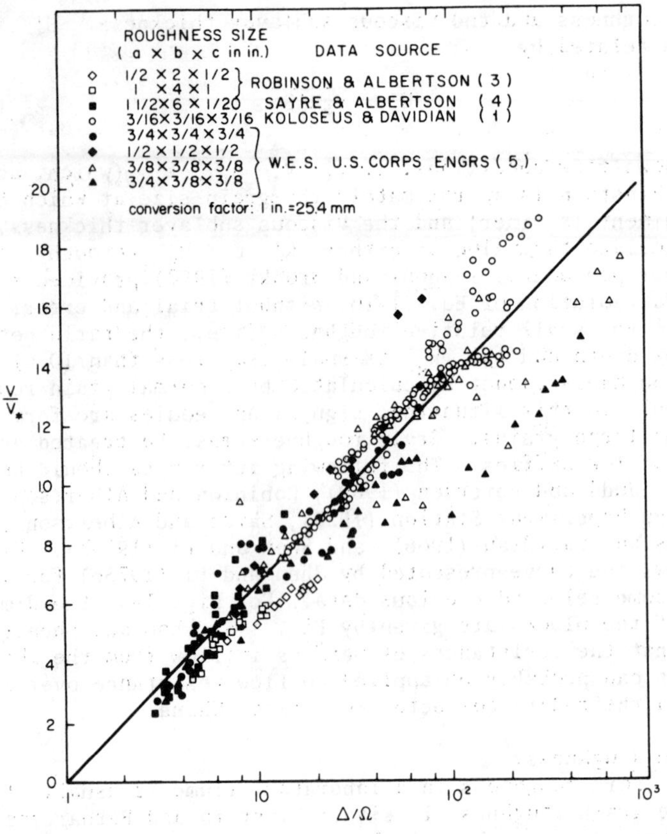

Fig. 1-4. V/V_* versus Δ/Ω for Rigid Roughness Elements (after Shen and Li, 1973b).

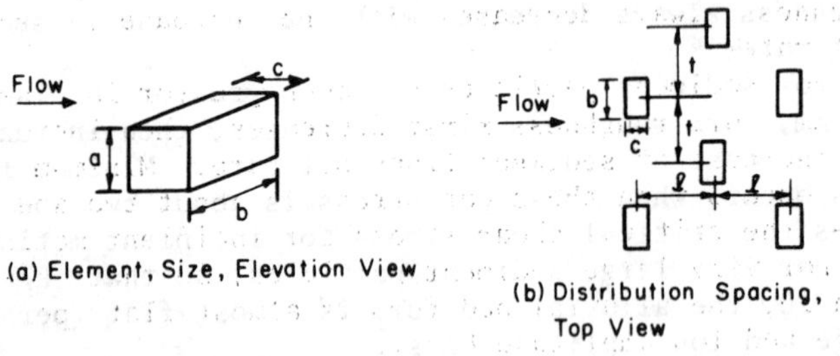

Fig. 1-5. Roughness Elements (after Shen and Li, 1973b).

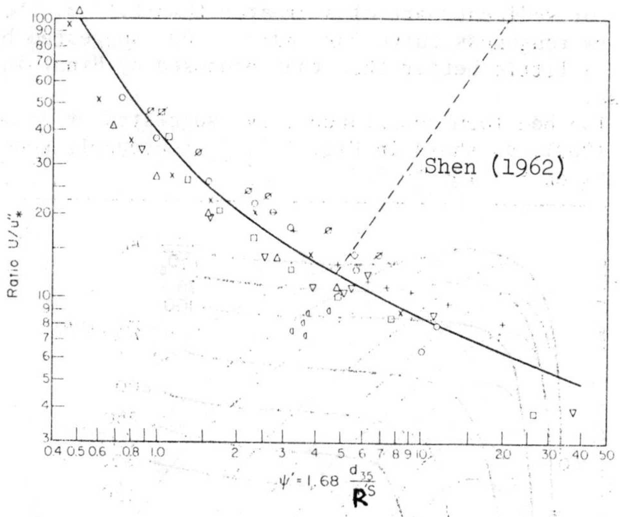

Fig. 1-6. Friction Loss Due to Channel Irregularities as a Function of Sediment Transport (after Einstein and Barbarossa, 1952).

sediment sizes of about 0.2 to 0.3 mm. A recent study by Shen, Mellema and Harrison (1978) indicates this, as shown in Fig. 1-7.

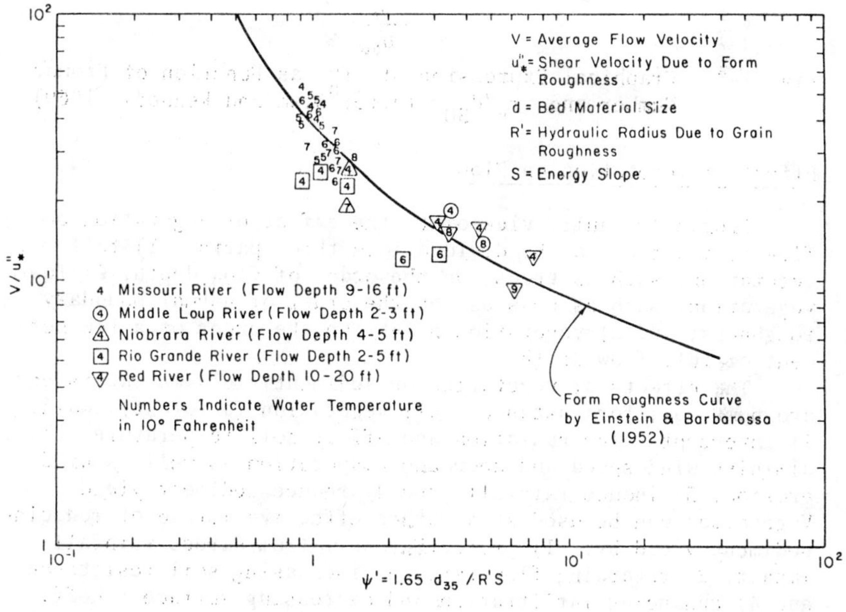

Fig. 1-7. Variation of Form Resistance (after Shen, Mellema and Harrison, 1978).

h. For sediment particles greater than 0.6 mm, the branch of the form roughness curve for $Wd/\nu > 100$ suggested by Shen (1962) is a little better than that proposed by Einstein and Barbarossa.

i. The bed form resistance curve suggested by Alam and Kennedy (1969), as shown in Fig. 1-8, is reasonably good for R/d_{50} between 5×10^2 to 10^4.

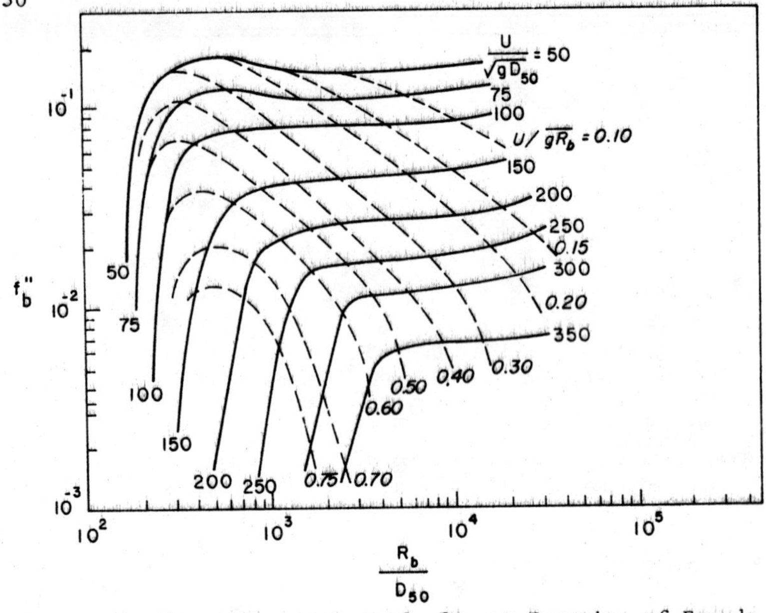

Fig. 1-8. Graphical Expression of f_b'' as Function of Froude Number and r_b/d_{50} (after Alam and Kennedy, 1969).

Effect of Vegetation on Flows

From a hydraulic viewpoint, the effect of vegetation on flow retardation can be divided into three parts: 1) tall vegetation, such as trees, on the order of flow depth; 2) low vegetation, such as grasses, on the order of normal boundary roughness; and 3) vegetation height on the order of a few percent of full flow depth.

The effects of vegetation on sediment yield in watersheds are numerous; for instance, tall vegetation (a forest) could: 1) intercept solar radiation and affect soil temperature, 2) diminish wind speed and decrease evaporation as well as wind erosion, 3) induce rainfall, and 4) reduce sediment yield. Vegetation can be used as a rather effective method of reducing sediment yield by: 1) protecting soil from direct rainfall impact, 2) retarding flow rate, 3) increasing soil resistance, and 4) enhancing infiltration and decreasing surface runoff.

Very low vegetation has the same effect as surface roughness, and Manning's n values given by various textbooks such as Chow (1959) would provide reasonable solutions. For slightly taller vegetation (with height varying from zero to full flow depth) there are three comprehensive studies.

Ree and Palmer (1949) summarized their results of the estimates of Manning's n value for various vegetation and hydraulic conditions in Table 1-2 and Fig. 1-9.

Table 1-2. Classification of Degree of Retardance for Various Kinds of Grass (after Ree and Palmer, 1949).

Retardance	Cover	Condition
A Very high	Weeping love grass	Excellent stand, tall (av 30 in.)
	Yellow bluestem ischaemum	Excellent stand, tall (av 36 in.)
B High	Kudzu	Very dense growth, uncut
	Bermuda grass	Good stand, tall (av 12 in.)
	Native grass mixture (little bluestem, blue grama, and other long and short Midwest grasses)	Good stand, unmowed
	Weeping love grass	Good stand, tall (av 24 in.)
	Lespedeza sericea	Good stand, not woody, tall (av. 19 in.)
	Alfalfa	Good stand, uncut (av 11 in.)
	Weeping love grass	Good stand, mowed (av 13 in.)
	Kudzu	Dense growth, uncut
	Blue grama	Good stand, uncut (av 13 in.)
C Moderate	Crab grass	Fair stand, uncut (10 to 48 in.)
	Bermuda grass	Good stand, mowed (av 6 in.)
	Common lespedeza	Good stand, uncut (av 11 in.)
	Grass-legume mixture--summer (orchard grass, redtop, Italian rye grass, and common lespedeza)	Good stand, uncut (6 to 8 in.)
	Centipede grass	Very dense cover (av 6 in.)
	Kentucky bluegrass	Good stand, headed (6 to 12 in.)
D Low	Bermuda grass	Good stand, cut to 2.5 in. height
	Common lespedeza	Excellent stand, uncut (av 15 in.)
	Buffalo grass	Good stand, uncut (3 to 6 in.)
	Grass-legume mixture--fall, spring (orchard grass, redtop, Italian rye grass, and common lespedeza)	Good stand, uncut (4 to 5 in.)
	Lespedeza sericea	After cutting to 2 in. height, very good stand before cutting
E Very Low	Bermuda grass	Good stand, cut to 1.5 in. height
	Bermuda grass	Burned stubble

Kouwen and Unny (1973) found that deflection by vegetation can be described approximately by a simple combination of m (number of roughness limits per square centimeter), E (the modulus of elasticity of the simulated vegetation), I (the second moment of area of its cross section), ρ (the density of the fluid), and u_* (the density shear stress).

The relationship between $(mEI/\rho u_*^2)^{1/4}/h$ and k/h is given in Fig. 1-10, where k is the deflected roughness height and h is the original roughness height. Series A, B, C, D,

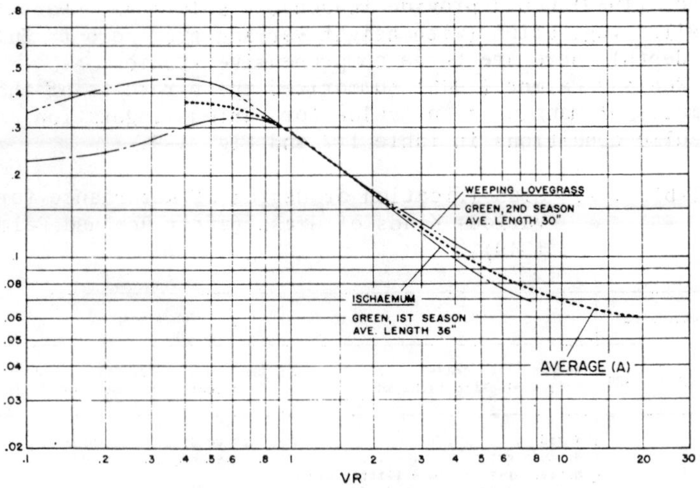

(a) Curves for A or very high vegetal retardance.

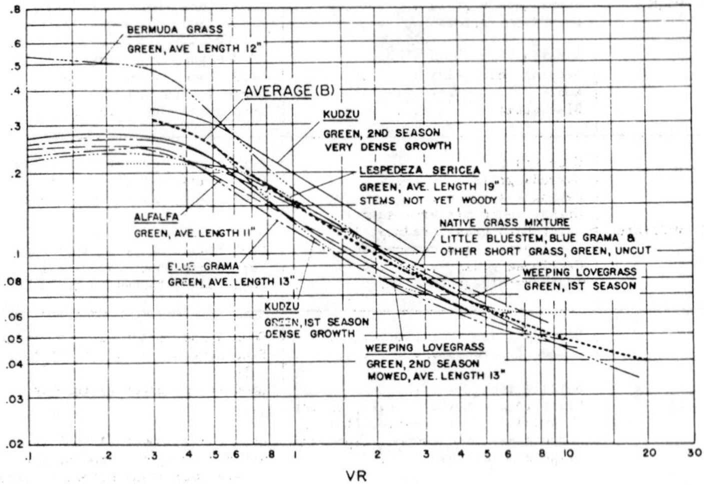

(b) Curves for B or high vegetal retardance.

Fig. 1-9. Experimental n-VR Curves (after Ree and Palmer, 1949).

and E were conducted with difficult simulated vegetation spacings.

Kouwen and Unny (1973) also presented two figures (Fig. 1-11 and 1-12) to indicate the change in Manning's n value and the Darcy-Weisbach f value with the flow conduits, respectively. In these two figures, y_n is the flow depth.

Chen (1976) conducted a series of laboratory experiments in a flume 20 ft wide, 20 ft long and 2 ft deep with Kentucky Blue

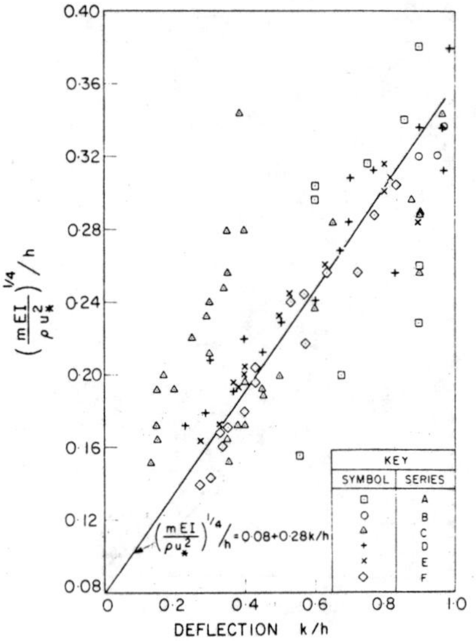

Fig. 1-10. Plot of $[mEI/(\rho u_*^2)]^{1/4}/h$ Versus Deflection k/h (after Kouwen and Unny, 1973).

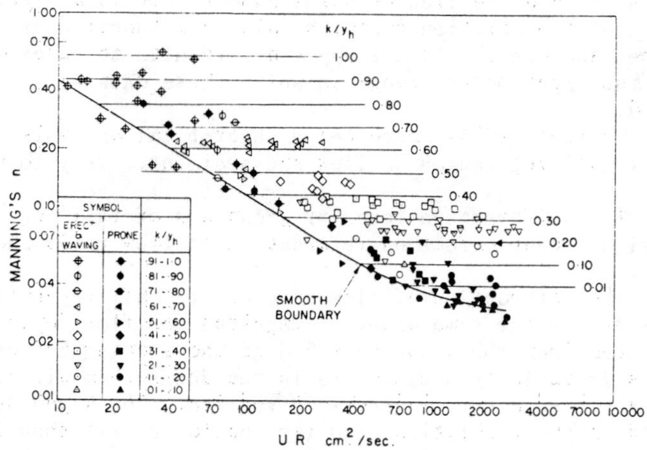

Fig. 1-11. Variation of Manning's n as Function of UR and k/y_n (after Kouwen and Unny, 1973).

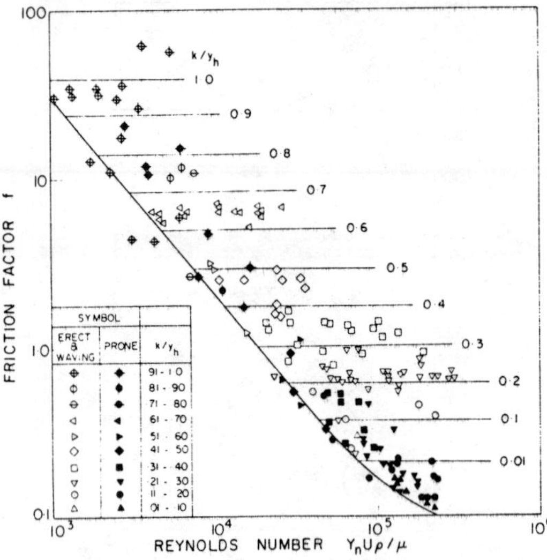

Fig. 1-12. Variation of Friction Factor with Relative Roughness (after Kouwen and Unny, 1973).

Grass and (hybrid) Bermuda Grass. Several channel slopes were used. His results are given in Fig. 1-13. He found that the Darcy-Weisbach friction factor f is a function of both the channel slope and the flow Reynolds number. It is a little strange that the friction factor should be a function of channel slope, and one should not apply his results to cases outside of the experimental range in which these data were collected.

Li and Shen (1973) presented a theoretical analysis of the effect of tall vegetation on flow and sediment. They found that:

1. The different patterns or groupings of tall vegetation have a significant effect on retardation of flow rates and sediment yields.

2. The retardation of flow rate due to tall vegetation with heights of the same order of magnitude as flow depth can be estimated from the present method if the spacing between the cylinders is at least 6 diameters in the downstream direction and 3 diameters in the transverse direction. The Reynolds number based on the vegetation diameter should be less than 2×10^5, although there is no evidence to indicate that this method will not be applicable for other ranges of Reynolds numbers.

3. Tall vegetation in staggered patterns is much more effective in reducing flow rate than any other pattern for the same amount of tall vegetation. Thus, if the main aim of planting tall vegetation is to reduce flow rates and sediment yield, it should be planted in a staggered pattern.

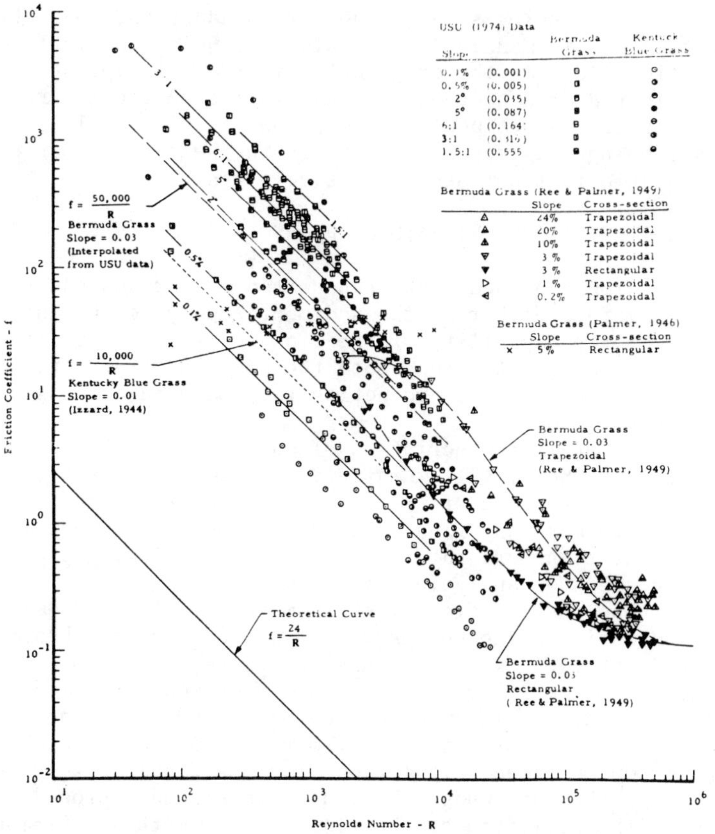

Fig. 1-13. Relationships between Darcy-Weisbach Friction Coefficient and Reynolds Number for Flow on Natural Turf Surfaces (after Chen, 1976).

4. Since it is difficult to harvest tall vegetation in a staggered pattern, the next best thing is to harvest it in rows perpendicular to the flow direction.

5. The average boundary shear stress on the bed is much more sensitive to a change of bottom slope and the size of vegetation than the variation of flow discharge and sediment size. The retardation of boundary shear stress on the bed will increase significantly as trees grow larger. All these effects can be estimated quantitatively by the following method.

Petryk and Bosmajian (1975) found that the composite n value for channel roughness and vegetation roughness can be expressed as

$$n = n_p \sqrt{1 + \frac{C_d}{2g} \frac{\Sigma A_i}{AL} \left(\frac{1.49}{n_b}\right)^2 \left(\frac{A}{P}\right)^{4/3}} \qquad (1\text{-}17)$$

where n is Manning's n value for boundary and vegetation effects, n_b is Manning's n value excluding the influence of vegetation, C_d is the drag coefficient of each plant, A_i is the projected area of the ith plant in the upstream direction, A is the cross-sectional area of flow, L is the length of channel being considered, and p is the wetted perimeter. This equation has not been extensively tested.

1.7 SEDIMENT TRANSPORT RATE BY FLOWS

The amount of sediment transported by streams and rivers can be governed either by the sediment supply (or production) rate or by the capability of the flow to transport sediment. Therefore, one has to study both the sediment transport capability and the sediment production rate, and take the smaller value of the two.

For a detailed description of current knowledge on this subject, see Shen (1971), Graf (1971), Raudkivi (1976), and many others. Because of the tremendous uncertainties, it is difficult to make any final recommendations on the estimation of sediment transport rate. However, the following procedures are suggested for analyzing field data:

1. Use the modified Einstein method (Colby and Hembree, 1955) to estimate the unmeasured suspended load and bed load on measured data. In applying this procedure, there is a possible difficulty in relating suspended sediment concentration profile to particle size. If a regression method is used, one may have to decide whether or not to rely more on suspended load calculations or on bed load calculations for the final answer. At this stage Shen and Hung (1977) have developed a procedure to minimize the deviation of the final result with measurements.

2. Separate bed material load from wash load and analyze them separately. See Shen (1973) for several suggested methods of separating wash load from bed material load.

3. Decide which available sediment transport equations best agree with the measured data and use them where actual measurement is not available to estimate the sediment transport load for the design flow.

When no measured data is available, the following methods are suggested:

1. Einstein's (1950) procedure, if bed load is a significant portion of the total bed material load and the suspended sediment concentration is not too great.

2. Colby's (1964) method for rivers with flow depth up to about 10 feet; also see Item 4 below.

3. Toffaleti's (1969) method for large rivers.

4. Shen and Hung's (1971) method for flume data and small rivers, as shown in Fig. 1-14, gives too low a sediment concentration for large rivers with bed material load sizes assumed to be greater than 0.0625 mm. Yang (1973) also recommended another regression equation based on his stream power concept.

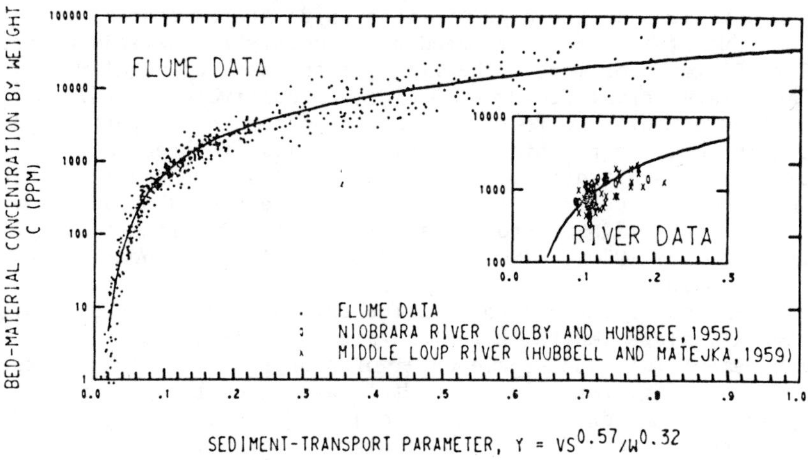

Fig. 1-14. Bed-Material Loads versus Flows (All Primary Data) where V is the Average Flow Velocity in Feet per Second, S is the Energy Slope and W is the Terminal Fall Velocity of the Mean Particle Size in Feet per Second (after Shen and Hung, 1971).

According to an analysis by Yang (1977), the equation by Shen and Hung (1971) and Yang (1973) gave approximately the same accuracy based on all data available to him. Shen and Hung's equation is preferred because it is much simpler and involves fewer regression coefficients. Actually, no one really knows the total sediment bed material load in streams because suspended sediment samplers can only measure suspended load in the measured zone. The lower unmeasured zone may contain at least 10% and can reach up to 120% of the measured value in the measured zone. The ratio between the suspended load plus the bed load in the unmeasured zone and the suspended load in the measured zone is a function of flow and sediment characteristics. One should always remember the fact that with more adjustable coefficients one can also adjust the empirical curve to better fit the data. The main question is how well the final curve can be applied to new data. A split sampling technique is very useful in establishing an empirically determined curve.

1.8 SEDIMENT PRODUCTION

The most well-known equation is the "Universal Soil Loss Equation" derived by the U.S. Department of Agriculture (see Wischmeier and Smith, 1965; and Soil Conservation Society of America, 1977).

The U.S. Department of Agriculture began to study soil loss about 1930 when the first ten Federal-State Cooperative Stations began operation. Thirty-two additional stations were established in the next 25 years. Measurements of precipitation, runoff and soil loss at these 42 stations in 23 states east of the Rocky Mountains were collected continuously for periods of from 5 to 30 years or more. The field plots were rectangular to facilitate normal flow row spacing for cultivated units. A "unit plot" of 72.6 feet long on a nine percent uniform slope, continuously in bare fallow soil and tilled to break surface crusts was arbitrarily selected to serve as a common reference point for evaluation. These dimensions were used because most of the plots in U.S. erosion studies by the Department of Agriculture from 1930 to 1960 were 72.6 feet long and near nine percent slope.

Six major factors to describe rainfall (R), soil erodibility (K), slope length (L), slope steepness (S), cropping and management (C), and supplemental erosion control practices such as contouring, terracing, etc. (P), were used to develop the following "Universal Soil Loss Equation" based on the field data described above and rainfall simulation data by a multiplicative model:

$$\underline{E} = R\ K\ L\ S\ C\ P \tag{1-18}$$

where $\underline{E}$ is the soil loss with the time period R and soil dimension of K, and A is the average soil loss for the time interval represented by factor R, expressed in the dimensions of factor K (usually tons/acre). If the value of R is obtained from the iso-erodent map, A will be the average annual soil loss.

R is a measure of the erosive force of rainfall and runoff. It usually equals the local value of the erosion index (EI) rainfall parameter, but two exceptions will be pointed out.

K reflects the inherent erodibility of a particular soil. For a given soil, it equals the average soil loss per unit of factor R from a 72.6-foot length of nine percent slope in clean-tilled continuous fallow. It can be measured on other slope gradients and adjusted to the nine percent standard with a slope-effect formula. It can also be computed as a function of the soil's physical and chemical properties.

L and S are dimensionless factors that adjust the soil loss estimate for effects of length, steepness, and shape of the field slope.

C introduces the effects of the cropping system and management variables on soil loss.

P reflects the benefits of supporting practices, such as contouring and stripcropping.

According to Wischmeier (see Soil Conservation Society of America, 1977), "the equation computes long-term average annual soil losses for specific combinations of physical and management conditions. It was designed for field use by soil conservation planners and technicians. This required that it be easy to solve and include only factors whose values at a particular site can be determined from available data. Some potential details and refinements were sacrificed in the interests of utility.

"Each of the six erosion factors in the equation is a function of numerous secondary variables and interaction effects that must be considered when computing local values of the factors. For field application, local values for the equation's factors are obtained from published charts and tables. These chart values were developed from research data collected at 48 stations in 26 states. They average the effects of random fluctuations in many of the variables, which can bias short-term soil losses.

"The relation of a particular parameter to soil loss is often appreciably influenced by the levels at which other parameters are present. To the extent that these interaction effects could be evaluated from existing data, they are reflected in the equation through the established procedures for computing local factor values. Factor R reflects the interaction of storm size and rain intensities. The soil loss ratios and EI distribution curves used to evaluate C account for the interactions of the crop system with management and rainstorm distribution. The relation of slope steepness to effectiveness of contouring and stripcropping is reflected in the table of P values. There are, however, indications of interactions between soil, topography, and surface conditions that require further research before their effects can be incorporated in the factor-evaluation procedures."

Wischmeier (1973) and the Soil Conservation Society of America (1977) provide detailed discussions of these factors. Several other equations are presented; Shen and Li (1976) summarized some selected formulas in Table 1-3. This table gives a brief summary of some available soil loss equations, where L_o is the plot length in feet, S_o is the bed slope, and S_c and L_c are values of S_o and L_o at which erosion begins.

Table 1-3. Soil Loss Equations (after Shen and Li, 1976).

Investigator	Relationship Presented
Zingg (1940)	$\underline{E} \alpha L_o^{0.66} S_o^{1.37}$
Musgrave (1947)	$\underline{E} \alpha RKL_o^{0.37} S_o^{1.35} C$
Wischmeier and Smith (1965)	$\underline{E} \alpha RKL_o^{0.5} (0.00076 S_o^2 + 0.0053 S_o + 0.0076) CP$
Meyer and Monke (1965)	$\underline{E} \alpha L^{0.9} S_o^{3.5}$
Meyer (1965)	$\underline{E} \alpha (L_o - L_c)^{1.-1.5}$, $E \alpha (S_o - S_c)^{1.5}$
Meyer and Kramer (1968)	$\underline{E} \alpha L^{0.5} (S_o - S_c)^{1.4}$
Young and Mutchler (1969)	$\underline{E} \alpha L^{1.24} S_o^{0.74}$
Kilinc (1972)	$\underline{E} \alpha L^{1.035} S_o^{1.664}$

Soil loss is a very complicated process and hence no purely theoretical treatment of this problem has yet been developed.

It is believed that simplifying assumptions are needed to obtain an analytical solution.

Li, Shen and Simons (1973) made the following analysis. They first assumed that:

1. The soil layer is loose, and uniformly consists of soil with fine sediment size.
2. The sediment concentration is small so that the equation of motion for sediment-laden water can be approximated by the equation of motion for water only.
3. The variation in bottom slope, S_o, is negligible.
4. The rainfall intensity and the infiltration rate are constant; the approximate solutions to the mechanics of steady sheet flow can be developed.

The continuity equation for sediment can be expressed as

$$\frac{d\,q_s}{dx} = p_s \tag{1-19}$$

where q_s is the sediment discharge per unit width of channel and p_s is the fine sediment pick-up rate per unit area.

The raindrop splash loosens the soil layer, but its transporting capability is generally limited. With the assumption of a loose soil layer, the raindrop splash erosion is eliminated in this analysis. Then, the rate of fine sediment pick-up is assumed to be a power function of the τ_o boundary shear stress of surface runoff. The relation is

$$p_s = a_* \tau_o^{b_*} \tag{1-20}$$

where a_* is some constant describing the erodibility of a specific soil and b_* is some exponent which, judging from the existing sediment transport equations, is assumed to be 2.0. However, a_* and b_* can be determined by an optimization scheme.

With no base flow from the upstream end of an overland flow plot $(q_o = 0)$, the entire erosion rate resulting from a plot length L_o can be estimated as follows (for $N_r < 900$):

$$\underline{E} = \frac{1}{L_o} q_s(L_o) = \frac{1}{L_o}\int_0^{L_o} p_s\, d_x = \frac{a}{L_o}\int_0^{L_o} \tau_o^2 dx$$

$$= \frac{a\gamma^2 (k_1 + 27.162\, i^{0.407})^{2/3} \nu^{2/3} i^{2/3}}{(8g)^{2/3}} S_o^{4/3} \frac{1}{L_o}\int_0^{L_o} x^{2/3} dx$$

$$= R_k\, L_o^{2/3} S_o^{4/3}$$

$$= R_k \; L_o^{0.666} S_o^{1.333} \qquad (1\text{-}21)$$

and for $2000 < N_r < 12600$

$$\underline{E} = \frac{a\gamma^2 (k_2 + 0.012)^{2/3} \; \nu^{1/6} i^{7/6}}{(8g)^{2/3}} \; S_o^{4/3} \; \frac{1}{L_o} \int_0^{L_o} x^{7/6} dx$$

$$= R_k \; L_o^{7/6} S_o^{4/3}$$

$$= R_k \; L_o^{1.166} S_o^{1.333} \qquad (1\text{-}22)$$

where $q_s(L_o)$ is the sediment discharge at the end of the plot, i is the rainfall intensity in inches per hour and R_k is a factor describing the rainfall characteristics, soil erodibility, and fluid properties. For $900 < N_r < 2000$, the results may be approximated by linear interpolation.

The exponents of slope length and percent slope in the above equations are consistent with the regression equations given in Table 1-3. However, the analytical results indicate that different exponents of slope length should be used for different ranges of Reynolds number.

Soil erosion by raindrops can be a rather significant factor. This is especially true for the flow and sediment routing model for watersheds described in Chapter 9. Soil erosion by raindrops should be studied in three aspects: the characteristics of raindrops, the characteristics of soil, and the interaction between the two. The mechanics of soil erosion involve three distinct purposes: detachment, transportation, and disposition. Mutchler (1967, 1971) and Carter, Greer, Braud and Floyd (1974) analyzed raindrop characteristics and splash droplets produced by water drop impact. Laws (1940), Ellison (1944, 1947), Woodburn (1948), McIntyre (1958), Wischmeier and Smith (1958), Free (1960), Moldenhauer and Koswara (1968), Meyer and Wischmeier (1969), Young and Mutchler (1969), Foster and Martin (1969), Mazurak and Mosher (1970), Mutchler and Larson (1971), Epstein and Grant (1971), Bubenzer and Jones (1971), Barnett, Carreker, Abruna and Dooley (1971), Farmer (1973), and Young and Weirsma (1973) all introduced various aspects of soil erosion by raindrops. It is difficult to summarize all their investigations because many different types of soil were used. The erodibility of soil is difficult to study if cohesive material is involved.

1.9 MODELING OF ICE

Ice problems are increasingly being analyzed by a variety of modeling techniques. Chapter 14 summarizes the nature of river ice phenomena and describes various means by which ice behavior is modeled. Attention is not only directed to physical models, but includes analytical and numerical models that have been used to make predictions of ice behavior. Topics covered in this chapter are background literature, formation evolution and break up of ice covers, hydraulics associated with the presence of ice, thermal effects, and ice forces.

1.10 THERMAL PLUMES

Chapter 15 describes in detail a modeling of shore-attached thermal plumes in the Missouri River. This model is applicable to relatively shallow water and reasonably fast flow so that the vertical mixing is complete. It discusses the formulation of a theoretical model and analysis of field data for the verification of models. Special attention is given to the incorporation of the effects of difficult channel geometry and flow distribution on the thermal plumes. The limitation and applicability of this model are discussed. Brief discussions are also given on research needs in this area.

1.11 DATA COLLECTION PLANNING

Data collection planning is a rather important phase of modeling. Chapter 16 describes various data collection practices, data applicability and limitations, sampling techniques, sampling location criteria, sampling frequency, statistical data analysis, and data collection for parameters estimated in water quality models.

1.12 WATER QUALITY MODELING

Chapters 17, 18, and 19 discuss the construction and application of a water quality model. A water quality model can be formulated based on mass balance equations. Difficult boundary conditions and calibration and verification of models, as well as coefficients, are described.

1.13 RESEARCH NEEDS

Some additional comments and future research needs are discussed in Chapter 20.

1.14 REFERENCES

Alam, A. M. Z., and Kennedy, J. F., 1969. Friction factors for flow in sand bed channels. Journal of the Hydraulics Division, ASCE, Vol. 95, No. HY6, Proc. Paper 6900, November, pp. 1973-1992.

Annambhotla, V. S. S., Sayre, W. W., and Livesey, R. H., 1972. Statistical properties of Missouri River bed forms. Journal of the Waterways, Harbors and Coastal Engineering Division,

ASCE, Vol. 98, No. WW4, Proc. Paper 9358, November, pp. 489-510.

Barnett, A. P., Carreker, J. R., Abruna, F., and Dooley, A. E., 1971. Erodibility of selected tropical soils. Transactions of the ASAE.

Bazilewich, V. A., 1974. Incipient motion and sediment transport. Discussion, Journal of the Hydraulics Division, Proceedings of the ASCE, Vol. 100, No. HY9, September.

Blinco, P. H., Mahmood, K., and Simons, D. B., 1973. Stochastic structure of the turbulent boundary shear stress process. Proceedings of the 15th Congress, Vol. 1, International Association for Hydraulic Research, Turkey, pp. 371-380.

Bubenzer, G. D., and Jones, B. A., Jr., 1971. Drop size and impact velocity effects on the detachment of soils under simulated rainfall. Transactions of the ASAE.

Carter, C. E., Greer, J. D., Braud, H. J., and Floyd, J. M., 1974. Raindrop characteristics in South Central United States. Transactions of the ASAE.

Chen, C. L., 1976. Flow resistance in broad shallow grassed channels. Journal of the Hydraulics Division, Proceedings of the ASCE, Vol. 102, No. HY3, March.

Cheong, H. F., and Shen, H. W., 1973. Spectral properties of alluvial bed forms in rivers and flumes. First International Symposium on River Mechanics, Asian Institute of Technology, Bangkok, Thailand, January.

Chow, V. T., 1959. Open-channel hydraulics. McGraw-Hill Book Company.

Colby, B. R., 1964. Discharge of sands and mean-velocity relationships in sand-bed streams. U.S. Geol. Survey Prof. Paper 462-A.

Colby, B. R., and Hembree, C. H., 1955. Computations of total sediment discharge Niobrara River near Cody, Neb. U.S. Geol. Survey Water Supply Paper 1357.

Cunha, L. V., 1973. The influence of water temperature on the roughness of alluvial forms. Proceedings of the 15th Congress, Vol. 1, International Association for Hydraulic Research, Turkey, pp. 15-20.

Einstein, H. A., 1950. The bed-load function for sediment transportation in open channel flows. Technical Bulletin 1026, Soil Conservation Service, United States Department of Agriculture, September.

Einstein, H. A., and Banks, R. B., 1950. Fluid resistance of composite roughness. Transactions, American Geophysical Union, Vol. 31, No. 4, August, pp. 603-610.

Einstein, H. A., and Barbarossa, N. L., 1952. River channel roughness. Transactions, ASCE, Vol. 117, Paper No. 2528, pp. 1121-1132.

Ellison, W. D., 1944. Studies of raindrop erosion. Agricultural Engineering, April.

Ellison, W. D., 1947. Soil erosion studies. Parts I-V, Parts VI and VII with Ellison, O. T., Agricultural Engineering, April-October.

Engelund, F., and Hansen, E. A., 1967. Monograph on sediment transport in alluvial streams. Teknisk Forlag, Copenhagen, Denmark.

Epstein, E., and Grant, J. W., 1971. Soil erodibility as affected by soil surface properties. Transactions of the ASAE.

Farmer, E. E., 1973. Relative detachability of soil particles by simulated rainfall. Soil Science Society of America Proceedings, Vol. 37, No. 4, Madison, Wisconsin, July-August.

Fawkes, P. E., 1972. Roughness in a model of overland flow. M.S. Thesis, Colorado State University, Fort Collins.

Foster, R. L., and Martin, G. L., 1969. Effect of unit weight and slope erosion. Journal of the Irrigation and Drainage Division, Proceedings of the ASCE, Vol. 95, No. IR4, December.

Franco, J. J., 1968. Effects of water temperature on bed load movement. Journal of the Waterways and Harbors Division, ASCE, Vo. 94, No. WW3, Proc. Paper 6083, August, pp. 343-352.

Free, G. R., 1960. Erosion characteristics of rainfall. Agricultural Engineering, July.

Graf, W. H., 1971. Hydraulics of sediment transport, McGraw-Hill Book Company.

Haynie, R. B., and Simons, D. B., 1968. Design of stable channels in alluvial materials. Journal of the Hydraulics Division, ASCE, Vol. 94, No. HY6, Proc. Paper 6217, November, pp. 1399-1420.

Henderson, F. M., 1966. Open-channel flow. The MacMillan Co., New York.

Hubbell, D. W., and Al-Shaikh Ali, K. S., 1961. Qualitative effects of temperature on flow phenomena in alluvial channels. Professional Paper 424-D, United States Geological Survey.

Izzard, C. F., 1944. The surface profile of overland flow. Transactions, A.G.U., pp. 959-968.

Judd, H. E., and Peterson, D. F., 1969. Hydraulics of large bed element channels. Utah Water Research Laboratory, College of Engineering, Utah State University, Logan, Utah, August.

Kao, C. S., and Shen, H. W., 1977. Unpublished report.

Koloseus, H. J., and Davidian, J., 1966. Free surface instability corrections and roughness-concentration effects on flow over hydrodynamically rough surfaces, laboratory in open channel flows. U.S. Geological Survey Water-Supply Paper 1592-C, D, U.S. Government Printing Office, Washington, D.C.

Kouwen, N., and Unny, T. E., 1973. Flexible roughness in open channels. Journal of the Hydraulics Division, Proceedings of the ASCE, Vol. 99, No. HY5, May.

Laws, J. O., 1940. Recent studies in raindrops and erosion. Agricultural Engineering, Vol. 21, No. 11, November.

Li, R. M., 1972. Sheet flow under simulated rainfall. M.S. Thesis, Colorado State University, Fort Collins.

Li, R. M., and Shen, H. W., 1973. Effect of tall vegetation on flow and sediment. Journal of Hydraulics Division, ASCE, Proc. Paper 9748, Vol. 99, No. HY5, May.

Li, R. M., Shen, H. W., and Simons, D. B., 1973. Mechanics of soil erosion by overland flow. Proceedings XVth IAHR Congress, State Hydraulic Works of Turkey, September.

Lovera, F., and Kennedy, J. F., 1969. Friction factors for flat-bed flows in sand channels. Journal of the Hydraulic Division, ASCE, Vol. 95, No. HY4, Proc. Paper 6678, July, pp. 1227-1234.

Mahmood, K., 1971. Flow in sand-bed channels. Water Management Technical Report No. 11, Colorado State University, Fort Collins, Colorado.

Mazurak, A. P., and Mosher, P. N., 1970. Detachment of soil aggregates by simulated rainfall. Soil Sci. Soc. Amer. Proc., Vol. 34.

McIntyre, D. S., 1958. Soil splash and the formation of surface crusts by raindrop impact. Commonwealth Scientific and Industrial Research Organization.

Meyer, L. D., and Wischmeier, W. H., 1969. Mathematical simulation of the process of soil erosion. Transaction of ASAE.

Moldenhauer, W. C., and Koswara, J., 1968. Effect of initial clod size on characteristics of splash and wash erosion. Soil Sci. Soc. Amer. Proc., Vol. 32.

Munoz, G. R. J., and Gelahar, L. W., 1968. Turbulent pipe flow with rough and porous walls. Hydrodynamic Laboratory Report 109, Massachusetts Institute of Technology, Cambridge, Massachusetts.

Mutchler, C. K., 1967. Parameters for describing raindrop splash. Journal of Soil and Water Conservation.

Mutchler, C. K., 1971. Splash droplet production by waterdrop impact. Water Resources Research.

Mutchler, C. K., and Larson, C. L., 1971. Splash amounts from waterdrop impact on a smooth surface. Water Resources Research.

Petryk, S., and Bosmajian, G., III, 1975. Analysis of flow through vegetation. Journal of the Hydraulics Division, Proceedings of the ASCE, Vol. 101, No. HY7, July.

Raudkivi, A. J., 1963. Study of sediment ripple formation. Journal of the Hydraulics Division, ASCE, Vo. 89, No. HY6, Proc. Paper 3692, November, pp. 15-33.

Raudkivi, A. J., 1974. Incipient motion and sediment transport. Discussion, Journal of the Hydraulics Division, Proceedings of the ASCE, Vol. 100, No. HY9, September.

Raudkivi, A. J., 1976. Loose boundary hydraulics. 2nd Edition, Pergamon Press.

Ree, W. O., and Palmer, V. J., 1949. Flow of water in channels protected by vegetative linings. U.S. Soil Conservation Bulletin No. 967, February, 115 p.

Robinson, A. R., and Albertson, M. L., 1952. Artificial roughness standard for open channels. Transactions, American Geophysical Union, Vol. 33, No. 6, December, pp. 881-888.

Sayre, W. W., and Albertson, M. L., 1961. Roughness spacing in rigid open channels. Journal of the Hydraulics Division, ASCE, Vol. 87, No. HY3, Proc. Paper 2823, May, pp. 121-150.

Shen, H. W., 1962. Development of bed roughness in alluvial channels. Journal of Hydraulics Division, ASCE, Proc. Paper 3113, Vol. 88, No. HY3.

Shen, H. W., 1971. River Mechanics. Volume I, P.O. Box 606, Fort Collins, Colorado 80521.

Shen, H. W., 1973. Flow over alluvial bed. Chapter 2, Environmental Impact on Rivers, H. W. Shen, ed., P.O. Box 606, Fort Collins, Colorado.

Shen, H. W., 1975. Hans A. Einstein's contributions in sedimentation. Journal of Hydraulics Division, ASCE, Proc. Paper 11290, Vol. 101, No. HY5, May.

Shen, H. W., and Hung, C. S., 1971. An engineering approach to total bed material load by regression analysis. Proc. Sedimentation Sym., June 16-19, Berkeley.

Shen, H. W., and Hung, C. S., 1977. Unpublished report.

Shen, H. W., and Li, R. M., 1973a. Rainfall effect on sheet flow over smooth surface. Journal of Hydraulics Division, ASCE, Proc. Paper 9733, Vol. 99, No. HY5, May.

Shen, H. W., and Li, R. M., 1973b. Analysis of resistance over staggered roughness. Journal of the Hydraulics Division, ASCE, Technical Note, Proc. Paper 10147, No. HY11, November.

Shen, H. W., and Li, R. M., 1976. Watershed sediment yield. Chapter 21, Stochastic Approaches to Water Resources, H. W. Shen, ed., P.O. Box 606, Fort Collins, Colorado.

Shen, H. W., Mellema, W., and Harrison, A. S., 1977. Analysis of temperature effects on stage-discharge relationship in a Missouri River reach near Omaha. U.S. Army Engineering Division, Missouri River, Omaha, Nebraska, MRD Sediment Series, No. 15, February.

Simons, D. B., and Richardson, E. V., 1966. Resistance to flow in alluvial channels. Professional Paper 422-J, United States Geological Survey.

Simons, D. B., Richardson, E. V., and Haushild, W. H., 1963. Some effects of fine sediment on flow phenomena. Water Supply Paper 1498-G, United States Geological Survey.

Soil Conservation Society of America, 1977. Soil erosion prediction and control. 7515 Northeast Ankeny Road, Ankeny, Iowa.

Straub, L. B., Anderson, A. G., and Flammer, G. H., 1958. Experiments on the influence of temperature on the sediment load. MRD Sediment Series No. 10, United States Army Engineer Division, Missouri River, Omaha, Nebraska.

Strickler, A., 1923. Beiträge zur Frage der Geschwindigkeitsformel und der Rauhigkeitszahlen für Ströme, Kanäle und geschlossene Leitungen (some contributions to the problem of velocity formula and roughness coefficients for rivers, canals, and closed conduits), Mitteilungen des eidgenössischen Amtes für Wasserwirtschaft, Bern, Switzerland, No. 16

Taylor, B. D., and Vanoni, V. A., 1972a. Temperature effects in low-transport, flat-bed flows. Journal of the Hydraulics Division, ASCE, Vol. 98, No. HY8, Proc. Paper 9105, August, pp. 1427-1445.

Taylor, B. D., and Vanoni, V. A., 1972b. Temperature effects in high-transport, flat-bed flows. Journal of the Hydraulics Division, ASCE, Vol. 98, No. HY12, Proc. Paper 9456, December, pp. 2191-2206.

Toffaleti, F. B., 1969. Definitive computations of sand discharge in rivers. Proc. Paper 6350, Vol. 95, No. HY1, Jour. of Hydraulics Div., ASCE, January.

Vanoni, V. A., 1946. Transportation of suspended sediment by water. Transactions, ASCE, Vol. 111, Paper No. 2267, pp. 67-102.

Vanoni, V. A., and Brooks, N. H., 1957. Laboratory studies of the roughness and suspended load of alluvial streams. Sedimentation Laboratory No. E-68, California Institute of Technology, Pasadena, California, December.

Vanoni, V. A., and Nomicos, G. N., 1959. Resistance properties of sediment-laden streams. Journal of the Hydraulics Division, ASCE, Vol. 85, No. HY5, Proc. Paper 2020, May, pp. 77-107.

Waterways Experiment Station, 1953. Roughness standards for hydraulic models, report no. 4, study of finite boundary roughness in rectangular flumes. U.S. Army Corps of Engineers, Vicksburg, Mississippi, Tech. Memorandum No. 2-364.

Wischmeier, W. H., 1973. Upslope erosion analysis. Chapter 15, Environmental Impact on Rivers, edited and published by H. W. Shen, Colorado State University, Fort Collins, Colorado.

Wischmeier, W. H., and Smith, D. D., 1958. Rainfall energy and its relationship to soil loss transactions. American Geophysical Union.

Wischmeier, W. H., and Smith, D. D., 1965. Predicting rainfall-erosion losses from cropland east of the Rocky Mountains. Agriculture Handbook, No. 282, USDA.

Woodburn, R., 1948. The effect of structural condition on soil detachment by raindrop action. Agricultural Engineering.

Woolhiser, D. A., 1975. Simulation of unsteady overland flow. In Unsteady Flow in Open Channels, edited by K. Mahmood and V. Yevjevich, Chapter 12, Vol. II, Water Resources Publications, P.O. Box 303, Fort Collins, Colorado 80522.

Yang, C. T., 1973. Incipient motion and sediment transport. Journal of the Hydraulics Division, Proceedings of ASCE, Vol. 99, No. HY10, October.

Yang, C. T., 1977. The movement of sediment in rivers, Geological Surveys, Vol. 3, D. Reidel Publishing Co., Dordrecht, Holland, pp. 39-68.

Yoon, N. Y., 1970. The effect of rainfall on the mechanics of steady spatially varied sheet flow on a hydraulically smooth boundary. Ph.D. Thesis, Department of Civil Engineering, University of Illinois, Urbana, Illinois.

Young, R. A., and Mutchler, C. K., 1969. Soil movement on irregular slopes. Water Resources Research.

Young, R. A., and Wiersma, J. L., 1973. The role of rainfall impact in soil detachment and transport. Water Resources Research, Vol. 9, No. 6, December.

Chapter 2

FLOW VARIABILITY AND MODELING OF RIVER PROCESSES

by

V. Yevjevich, Professor of Civil Engineering, Colorado State University, Fort Collins, Colorado

Chapter 2

FLOW VARIABILITY AND MODELING OF RIVER PROCESSES

2.1 INTRODUCTORY REMARKS

Looking at the variability of water, sediment, and water quality variables of a stream system from the viewpoint of modeling of river processes, it is often feasible to clearly distinguish between stream system inputs, responses, and outputs. Important inputs for modeling are water, sediment, and water quality. Because of a dominance of sediment inputs on river processes, the sediment is separated here from the other water quality variables.

Water Inputs

Currently, the information on water inputs into a stream system is obtained by one or more of the following approaches:

1. By direct measurements of runoff at points at which the information is needed.
2. By transferring information on runoff from points of observation of runoff to points of needed runoff inputs.
3. By transferring information on precipitation into information on runoff by using either the physical or statistical methods of information transfer. The character of precipitation (rainfall, snowfall and snowmelt) and of information transfer methods are relevant to modeling of processes in river mechanics. The accuracy of this approach in securing the water input information, regardless whether the physical or statistical methods are used, is most often smaller than the input information of direct runoff measurements or of their judicious transfer along a stream system.
4. By transferring simultaneously both the available runoff and the available precipitation information at observational points to points of needed runoff information for modeling of river processes. Figure 2-1 schematically illustrates these information procurement approaches.

Sediment Inputs

Characteristics of sediment inputs are determined in one of the following approaches:

1. By direct measurements of sediment transported at the selected points (usually but not necessarily at points of runoff measurements).
2. By establishing the relationship between the water flow and the sediment transport at a runoff gauging station and the transfer of information on runoff to information on sediment.
3. By using precipitation data in modeling of the erosion and sediment inflow into streams. The accuracy of the

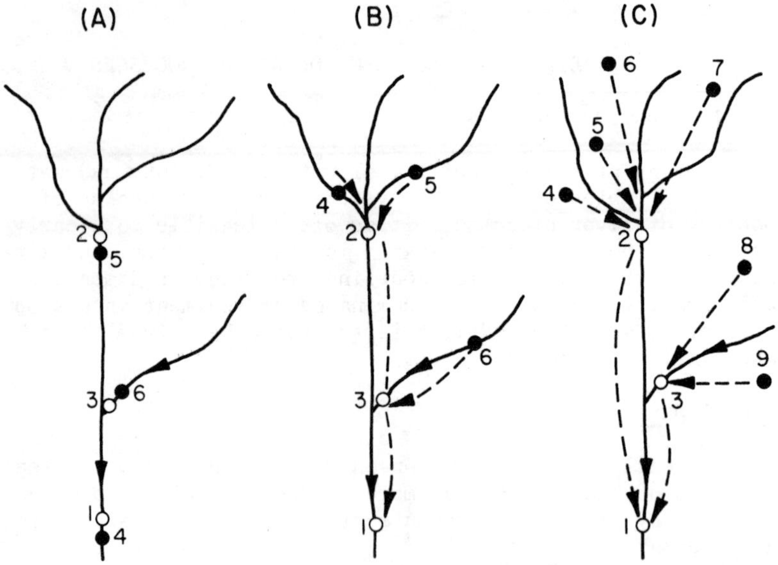

Fig. 2-1. Methods of Transferring Information on Runoff: (A) Points of Available Information (4, 5, 6) Coincide with Points (1, 2, 3) of Needed Inputs for Modeling of River Processes; (B) Transfer of Runoff Information from Points of Available Data (4, 5, 6) to Points of Needed Information (1, 2, 3); (C) Transfer of Precipitation Information from Points of Available Data (4 through 9) to Points of Needed Runoff Information (1, 2, 3); and (D) Various Combinations of Cases A, B and C

information procurement usually decreases in the order of these three approaches.

4. By using simultaneously the combinations of the above three approaches.

Water Quality Inputs

The information on water quality input variables is composed of a large number of physical, chemical and biological random variables, generally designated by the term "water quality." The procurement of this information for modeling of stream system responses to water quality inputs is much more complicated and less accurate than is the case either for water flow or sediment transport inputs. Time series of water quality random variables, multivariate statistical relationships either among these variables and relationships between these variables and water flow and sediment transport variables, belong the the most complex time series and relationships not only in hydrology and hydraulics but also in the entire geophysics.

The complexity of water quality inputs, and their interrelationships with water flow and sediment transport, require a special treatment. A concept which will be inevitably implemented in the future is that of multivariate-multipoint inputs and outputs to the stream system, as illustrated by Fig. 2-2. The character of inputs and outputs for water quali-

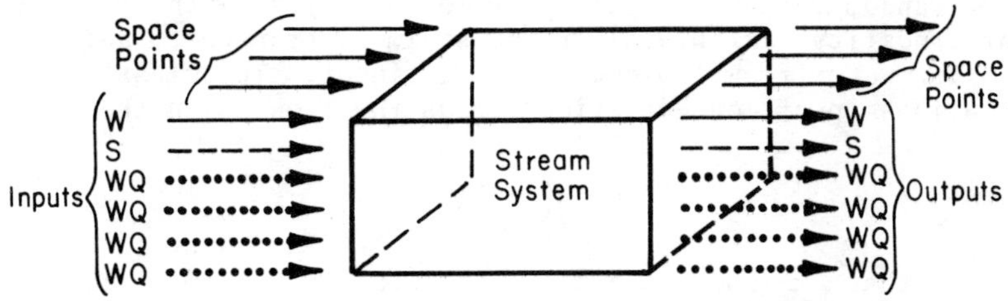

Fig. 2-2. A Conceptual Representation of Multivariate-Multi-Point Inputs and Outputs of a Stream System (W = Water Variable, S = Sediment Variable, WQ = Water Quality Variables)

ty variables, each being a component of the multivariate at a given point, is that they are time-dependent, periodic-stochastic processes, often also intermittent processes, each component being dependent on all the other components at a given point, and as multivariates for a number of points they are also highly space dependent. Besides, the system is changeable both in space and time, defined here as the changing processes of river mechanics.

Three expected practical outcomes of this multivariable-multipoint input-response-output processes are: (1) Identification of system properties and response characteristics, if not provided by observations; (2) Mathematical modeling of inputs, responses, and outputs; and (3) Finding the best management techniques of such systems under given optimization objectives.

Comparison of Information Procurement Methods

A comparison of the available methods for information procurement on water, sediment, and water quality input and output variables for modeling of processes in river mechanics is of utmost importance for producing the results of a high reliability. No model could give more accurate results than the accuracy of input information permits. This comparison and the proper selection of methods to be used in providing the input and output data on water, sediment and water quality are prerequisites for a reliable modeling of processes in river mechanics.

A word of caution is warranted in some of the current practices in procurement of input information for modeling of processes in river mechanics, because they are subject to spurious correlation. When it is not identified and avoided, this spurious correlation provides an illusion of good relationships between the water flow and the sediment transport, on one side, or between the water quality random variables and the water flow and sediment transport variables, on the other side. The spuriousness comes from using the concentration type random variables in the form of $C = y/x$, with y = the water quality variable(s) or the sediment transport, and x = the water flow, whichever is pertinent. The classical relationship of concentration versus the flow is in the form

$$C = \frac{y}{x} = F(x) \tag{2-1}$$

Assume that x and y are two independent random variables. Even in this case of using $(y/x) = f(x)$, a high correlation is produced though the relationship $y = f(x)$ has no significant correlation because x and y are independent. The observed points (C,x) may have smaller deviations from the curve $C = f(x)$ than are the deviations of observed points (x,y) from the curve $y = f(x)$, or vice versa, depending on how the spuriousness has been introduced.

Types of Information Transfer

In transferring information from one variable to another, or from one point to another for the same variable, or in transferring information from several variables to one variable, or from several points to one point for the same variable, three basic approaches can be used:

1. Transfer of every value of a discrete series of historic data so as to produce the same number of values for a variable or at a point;
2. Transfer of the information only by using the parameters, from the parameters of variables or points with the relevant information to the parameters of variables or points which need the information; and
3. Transfer of information through a relationship between the mathematical models.

The predominate approach in practice is the use of historic data by applying the relationships between variables having information and the variables needing information, or between points having information and the points needing information, for the same variable. In recent times, trends are developed for the transfer of information by using the relationships between the parameters or the relationships between the mathematical models.

When the information is transferred by the parameter and model methods, the inputs on water, sediment, and water quality variables to a system of streams or stream channels are produced by using the Monte Carlo method as the method of generating the new samples. It then produces new samples, small or long, of input variables by simulating properties of their processes. In such an approach, the response of a river or a system of channels, and modeling of their processes, can be investigated for many of these generated input time series. The modeling of inputs and the generation of new samples by these models is becoming more and more of interest in hydrology and in modeling of processes in river mechanics.

2.2 MODELING OF INTERMITTENT PROCESSES

In studying the precipitation and other intermittent processes, various methods in presenting the information at a point or regionalizing the parameters from data on several points are available in literature. Two problems need appropriate solutions in an eventual generation of new samples of intermittent processes over a region or in a river basin: (i) intermittency in processes and (ii) problems arising in generating new samples at several points by preserving the periodicity in the series and the interstation correlation. The largest number of stations for which the new samples have been generated by preserving the time structure and the space dependence of intermittent station series has been up to three, by using the combinatorial analysis. The most recent investigations enable the solutions of both problems, namely the intermittency and the preservation of time structure and space dependence. They are presented herein.

Solution of the Problem of Intermittency

Hydrologic time processes have been classified for practical purposes as perennial (non-intermittent) and intermittent. Most climatologic and hydrologic time processes are non-intermittent series, meaning that there is a non-zero value of that variable at any time. Instantaneous precipitation, evaporation, some runoff (usually for small rivers with negligible underground or surface water storage), sediment transport of rivers, and others, represent the typical hydrologic intermittent time series. For some time the observed values are zeros; for other time values are greater than zero. Though there may be a continuous flux of water molecules through the liquid-gaseous or solid-gaseous inter-phases on the continental areas, with a non-zero difference in the number of molecules passing continuously in two directions, the original concept of precipitation variable was designed in such a way that the process of instantaneous or short-interval precipitation is intermittent, and similarly for some other intermittent processes.

In practice, many intermittent processes, with positive series values for some time periods and zero values for the other time periods, are observed as totals for given time intervals (Δt of discrete series), usually counted in minutes, hours, days, or a longer interval is used. Therefore, a sequence (period) of one or more Δt intervals with values greater than zero is followed by a sequence (period) of one or more Δt intervals of zero values, and vice versa, as schematically shown in Fig. 2-3. Most of the observed or computed

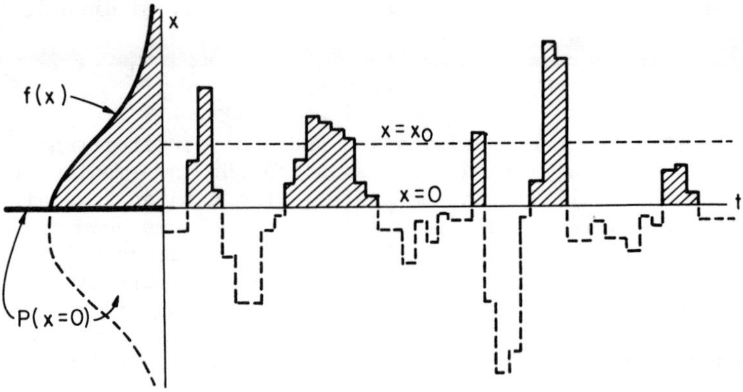

Fig. 2-3. Conceptual Treatment of an Intermittent Series, as a Truncated Part of a Non-Intermittent Series

time series are processed and their data published in such a way as to be intermittent. A large amount of available data of this type makes it necessary to design the most feasible methods for their investigation and mathematical description, that permit the simulation of these intermittent series by the data generation method. Figure 2-3 shows also the mixed distribution of values of intermittent series, namely a given probability, $P(x = 0)$, of zero values, and the probability densities, $f(x)$, for values greater than zero. In this case the classical probability of non-exceedence of the value x_o is

$$F(x_o) = P(x = 0) + \int_0^{x_o} f(x)dx . \qquad (2\text{-}2)$$

Because of spatial interrelationship for most of climatologic variables, the resulting hydrologic variables such as precipitation, evaporation, runoff of small rivers, sediment transport, and similar variables, may all have intermittent series. They are all spatially dependent also. When a point series is studied independently of the time series at the other points, methods are already available for a description of these intermittent series in the form of mathematical models and the estimation of their parameters. One approach to the description of the univariate (or point) intermittent time series is to first describe the random process of the time sequence of zero and non-zero periods, each period consisting of one or more intervals of a discrete time series. The difficulty in this approach arises from the fact that nearly all the parameters, especially the interval mean, interval standard deviation and interval autocorrelation coefficients (and sometimes also the interval skewness and kurtosis coefficients) are periodic. To avoid the difficulty of this combination of periodicities in series parameters and intermittency in time series, an approach to the analysis of such processes starts by dividing the annual cycle into the seasons (months, or weeks, or otherwise), and the daily cycle into its parts. The basic assumption is then that all the parameters are constants within these seasons or parts. This assumption requires a division of three two-dimensional cycles into a relatively large number of seasons or parts, each with the constant parameters, in order to justify its use whenever the amplitudes of harmonics fitting the periodic parameter curves are large. This approach of using the seasons or parts of cycles, with the constant parameters within each season or part, requires the estimation of a large number of parameters. A sudden change in parameters occurs at the borderlines of seasons or parts of these cycles.

Because of various difficulties, an alternative approach is needed to the often used or already classical approach for the description of intermittent time series. This classical approach first uses a process of zero and non-zero periods, and then the stochastic structure of non-zero values within the non-zero periods of a given season or cycle part. Often a non-zero period covers the borderlines between the seasons or the parts, so that parameters change by a jump within a non-zero period. A feasible replacement alternative to this classical approach is to consider the intermittent series of Fig. 2-3 as a truncated series of non-intermittent series. This is equivalent to replacing the negative (unknown) values of the non-intermittent series by zeros in order to produce the truncated, intermittent series. In this case, $x = 0$ is the truncation level for the non-intermittent series.

The probability of negative, unknown values of the non-intermittent series must be equal to the probability of zero values, $P(x = 0)$ of Fig. 2-3, for the intermittent series. The problem then becomes one of estimation of properties of the non-intermittent series from the available sample of data on the intermittent series, and particularly:

1. The estimation of probability distribution functions, $F(x)$, at each time interval of a cycle, or equivalently the estimation of periodic parameters of assumed or inferred, but same probability distribution function at all time intervals of the cycle, with the parameters of the function changing periodically along the cycle of the year or the day; and

2. The estimation of stochastic structure of time dependence of the non-intermittent series.

To estimate the character of a reconstructured non-intermittent series, the information of the observed intermittent series must be used and particularly:

1. The frequency curves of non-zero values at each time interval of the cycle of a discrete series, as well as frequencies of zero value occurrence at each time interval as estimates of $P(x = 0)$;

2. The estimates of the periodic parameters from the frequency curves of non-zero values as well as from the autocorrelation structure of intermittent series; and

3. The estimates of the stochastic dependence structure of stochastic component in the non-intermittent series from the corresponding dependence of the intermittent series.

Probability Distribution Functions of Non-Intermittent Series

The probability density functions such as that of Fig. 2-3, for the reconstructed non-intermittent series, may be different from a normal distribution function, and its truncated part, such as $f(x)$ of Fig. 2-3, may not be the truncated normal function. Therefore, there is a problem of inferring the best probability and truncated probability distribution functions for the case of each intermittent series and its cycle intervals.

The experience shows that the easiest and most accurate analysis of time series is for the case of the normal (Gaussian) probability distribution function when it can well fit the frequency distribution of truncated intermittent, and reconstructed non-intermittent processes. However, their parameters vary periodically along the time series intervals of the basic astronomical cycles. The advantages of using the normal function will be evident in the further discussion, especially in the analysis of time structure and space dependence of intermittent time series, as well as in the generation of new samples of multipoint series simultaneously by preserving all their properties.

One approach (Kelman, 1977) is to fit a normal function at each interval of the annual cycle (or daily cycle) by using the transformation

$$y = x^{\alpha} \qquad (2\text{-}3)$$

with x = the non-normal distribution of intermittent series, y = the normal distribution of the reconstructed non-intermittent series, or the truncated normal distribution of transformed intermittent series, and α = the parameter to be estimated (usually one should expect α to be a periodic parameter).

Another, simplified approach (Richardson, 1977) is to use either the square root, or $\alpha = 1/2$ (or the cubic root or $\alpha = 1/3$) transformation so that $y = \sqrt{x}$ (or $y = \sqrt[3]{x}$) becomes close to normal. Or, one can use the gamma distribution to fit the untransformed values, and then the normal distribution for the transformed values, as Figs. 2-4 through 2-6 demonstrate. Kelman (1977) used an analytical method to best estimate the needed transformation for a normal distribution fit.

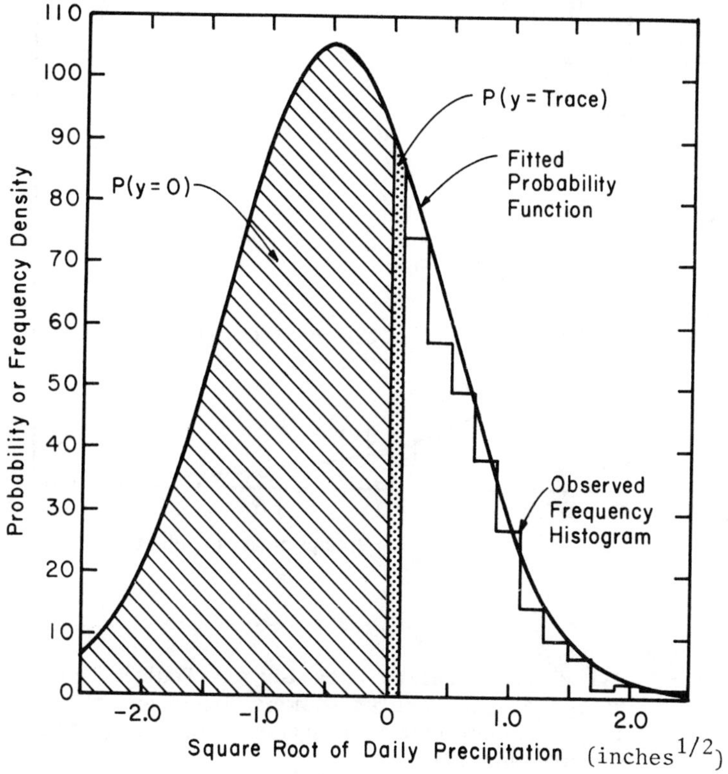

Fig. 2-4. The Truncated Normal Distribution of Daily Precipitation with a Square Root Transformation (after Richardson, 1977)

Periodicity in Parameters and Time Dependence

The estimation of parameters of the normal distribution of transformed, non-intermittent series automatically enables the fitting of the periodic functions by the Fourier series analysis to the interval mean and interval standard deviation. Similarly, it is feasible to infer the autocorrelation structure of the non-intermittent series, with the periodicity in parameters removed, from the corresponding intermittent time series. This is particularly feasible in the case the normal distribution function is used to fit the transformed intermittent and non-intermittent series.

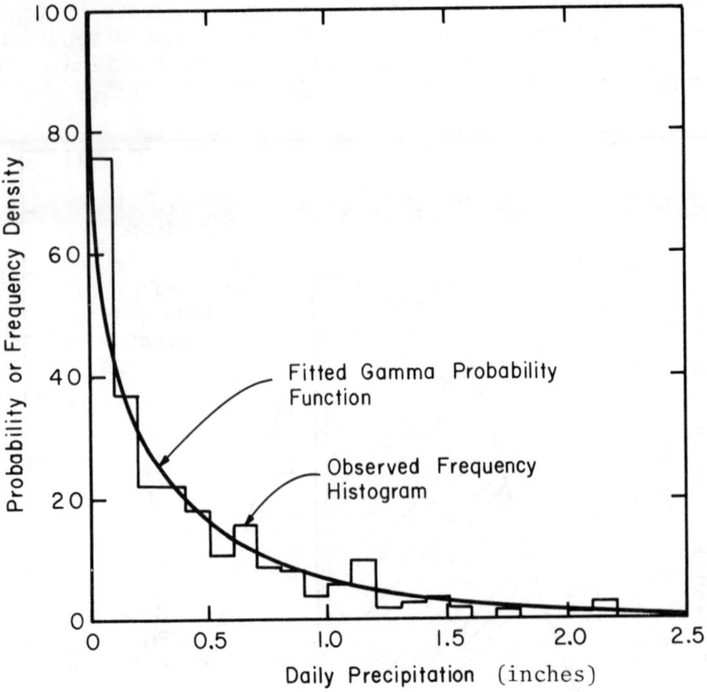

Fig. 2-5. Daily Precipitation Data for the Fourth 28-Day Period of the Year at Gilmer, Texas and the Fitted Two-Parameter Gamma Frequency Function (after Richardson, 1977)

The basic approach is then in postulating that an intermittent time series, with the short time interval, is only a truncated process of a non-intermittent, discrete time series. Basically, it is assumed that the probability distribution of non-zero values of an intermittent time series is only a tail, or a part of, either a truncated normal distribution, or a truncated other distribution, such as gamma, lognormal and similar. Therefore, techniques are needed for the estimation of properties of a non-intermittent process from a periodic-stochastic, intermittent process. Techniques are further needed for the transformation of original variables or of their stochastic residuals in such a way that the periodic-stochastic, intermittent process of an asymmetric variable becomes only the truncated part of a normal distribution in case of the non-normal distribution of variables.

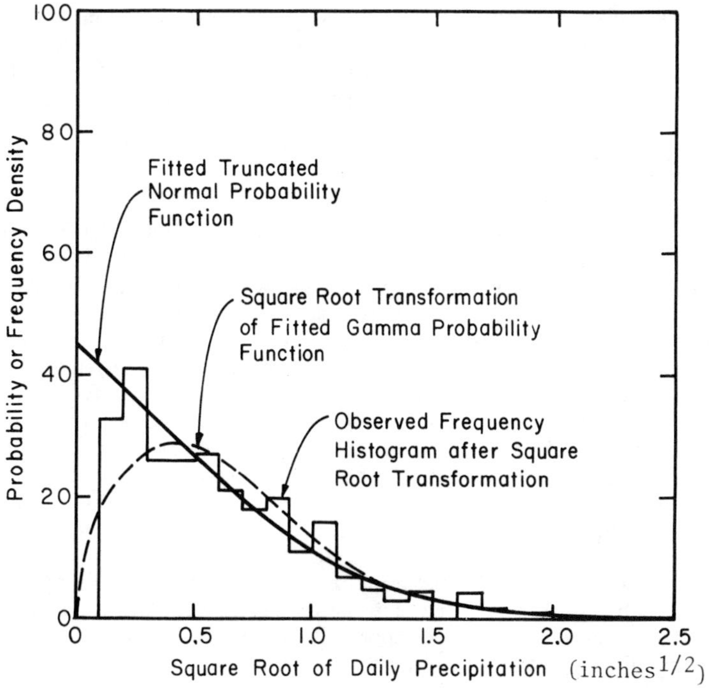

Fig. 2-6. Square Root of Daily Precipitation Data for the Fourth 28-Day Period of the Year at Gilmer, Texas, the Square Root Transformation of the Fitted Gamma Frequency Function, and the Fitted Truncated Normal Frequency Function (after Richardson, 1977)

Space Dependence Problem

When the problem of generating new samples by using the Monte Carlo (experimental statistical) method is posed, with the requirement that the generated samples of intermittent processes preserve both the time and space properties of random variables involved, this problem becomes that of a mathematical description and that of the generation of new samples in case of time dependent, periodic-stochastic, intermittent and space dependent time series. Both the periodicity in parameters, and the fact that the non-zero values occur only at some points, with the zero values not observed simultaneously at all the points, create difficulties in generating the new samples of the multipoint intermittent time series.

Attempts have been undertaken to apply the combinatorial analysis and Markov chains in order to generate simultaneously the series of 2-3 stations. First, their zero and non-zero periods of one or more intervals are generated. Then both the space and time dependence within the non-zero periods are preserved by modeling. Researchers following this approach have been able to simulate jointly the samples of only 2-3 station series. For more than three stations, the combinatorial approach becomes so complex that it is unfeasible to extend to cases of four, five, six and more intermittent time series.

2.3 GENERATION OF NEW SAMPLES OF MULTIPOINT, INTERMITTENT TIME SERIES

The generation of multivariate time series, which are periodic-stochastic, intermittent and also stochastically dependent both in time and space, is best performed by using the approach of the multivariate normal distribution and the principal component analysis. It seems logical to proceed in that direction also for variables that have the asymmetric probability distributions and are periodic-stochastic, intermittent time series. When a multivariable process is found to be periodic-stochastic, intermittent, non-normal stochastic process, difficulties arise both in mathematical description and in generation of new multivariate samples. When it becomes feasible to study intermittency by assuming it to be a truncated process of a non-intermittent time series, then removing periodicities in parameters, and transforming the original variables or their random residuals into the normal variables, the principal component analysis becomes a feasible and very desirable approach for the generation of new samples.

Method for Generation of Multivariate Series

Once the approach of constructing the non-intermittent discrete time series from intermittent series is selected for each point of a multipoint set of series, it then becomes feasible to approximate closely their transformed multivariate non-normal distribution by a multivariate normal distribution. From it then the periodic parameters can be estimated by fitting a set of harmonics in the Fourier analysis of a sequence of estimated interval means and interval standard deviations. The periodicity in parameters is then appropriately removed from the series. The remaining stationary stochastic components may come out to be either dependent or independent time processes. For a dependent process, linear dependence models can be inferred and their parameters estimated. This permits the computation of the independent identically distributed residuals, as the time independent stochastic components (called herein the TISC-variables). Once the series have been reduced to a set of normal, time independent, identically distributed stochastic processes, their spatial lag-zero correlation matrix enables a transformation of this set of series to their principal components, as a new set of space and time independent multivariate normal process.

To generate the new samples of multipoint series, the normal independent samples of principal components are generated for each point, and the reverse procedure applied on these time and space normal independent processes. Further transformation of the reverse order produces the periodic-stochastic, non-intermittent process at each point. They preserve the space dependence, periodicity and time dependence. By equating each negative value with zero, the multivariate, periodic-stochastic, truncated or intermittent, normal process is simulated by a set of new samples. Variables are then transformed from the normal to the corresponding non-normal distribution. This approach for the generation of new multi-variate samples by using the Monte Carlo (or experimental statistical) sample generation method is a feasible, practical method to model a set of periodic-stochastic, intermittent, time and space dependent processes by using the multivariate normal distributions and the principal components analysis processes.

Example of Multivariate Generation of Series at a Set of Regular Grid Points

The case of monthly precipitation over the Great Plains of the United States of America is presented here to show these techniques: (i) How the information on monthly precipitation can be condensed in the form of mathematical models for a given area; (ii) How the generation of new multivariate samples can be carried out for a systematic grid of points instead of generating them at points of available information; and (iii) What procedure should be followed in order to use the simple multinormal distribution and the principal components analysis in generating the new multivariate samples. Similarly as for the monthly precipitation of Great Plains, the new multivariate samples of daily precipitation, hourly precipitation, or similar short-interval time processes over a river basin, an urban area, an irrigation area, etc., can be generated.

For the real cases of generating precipitation samples, all the available information over an area should be condensed in form of mathematical models and their estimated parameters, and not only in form of a limited number of station series of a given, same sample size. To obtain best estimates of models and their parameters, all observations over that region should be included in practical cases. Models should represent the time structure of precipitation series, and their estimated parameters presented in the form of their changes or isolines over the region.

Figure 2-7 presents the location of 79 precipitation stations in the Upper Great Plains. Only a fraction of all the available precipitation stations has been used (Tase, 1976) for the purpose of minimizing the computer time in generating the new multivariate samples of monthly precipitation, in this particular case for the study of droughts in the Great Plains.

Figure 2-8 presents the systematic grid of 80 points over the same area of the Great Plains, at which new samples are generated for the study of droughts in comparison what would be

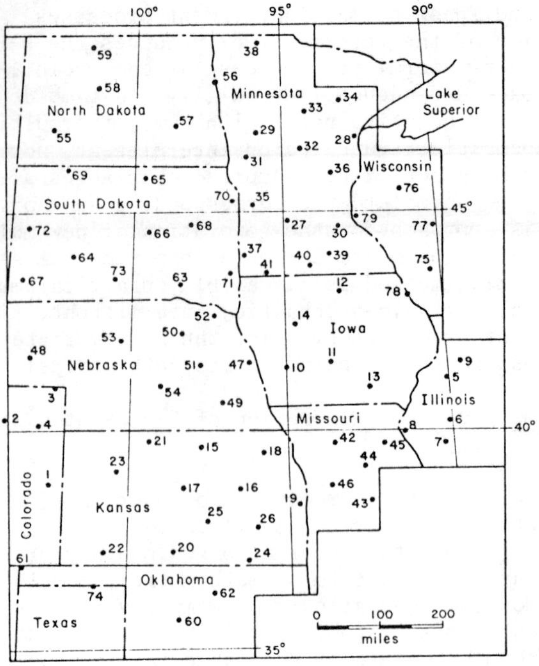

Fig. 2-7. The Study Area and Location of the 79 Stations (after Tase, 1976)

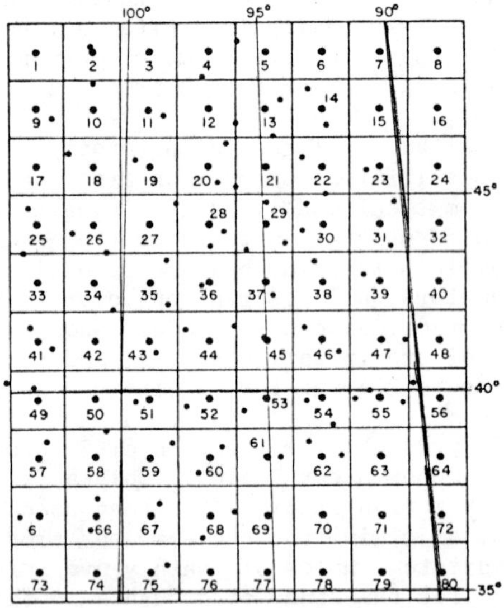

Fig. 2-8. Grid System (Big Points) with 100-Mile Grid Interval over the Upper Great Plains and Observed Stations (Small Points) with Their Series Used for Objectives of Modeling (after Tase, 1976)

the case if droughts were studied by using the observational points of Fig. 2-7. In the grid points, every point has the same area weight in studying the areal characteristics of precipitation process. To be able to generate the new multivariate samples of precipitation series at a set of grid points of Fig. 2-8, the precipitation process must be mathematically modeled, and the model parameters regionalized by fitting the areal functions to parameter estimates over the observed points. Then the general models are used for the new grid of points, and parameters computed for these points from the regional equations of each parameter.

Figures 2-9 and 2-10 give the isolines of general monthly means and general monthly standard deviations, respectively.

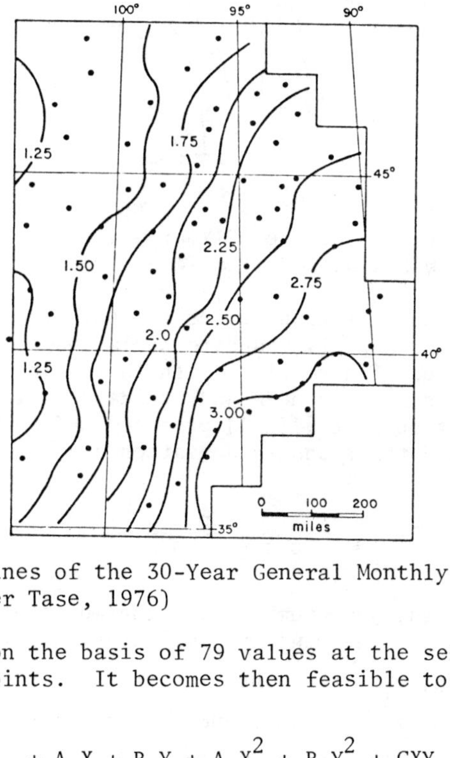

Fig. 2-9. Isolines of the 30-Year General Monthly Mean, $\bar{m}$ (after Tase, 1976)

They are drawn on the basis of 79 values at the selected observational points. It becomes then feasible to fit a curve of the type

$$\nu = A_0 + A_1X + B_1Y + A_2X^2 + B_2Y^2 + CXY \qquad (2\text{-}4)$$

where ν = the symbol for any regionalized parameter, X = the longitude and Y = the latitude of station position, and A_0, A_1, B_1,..., = the coefficients of the regional parameter equations. These equations then permit the computation of parameter estimates at each point of a systematic grid of points, such as one shown in Fig. 2-8. Similarly as for the general monthly means and the general monthly standard deviations, the fit of

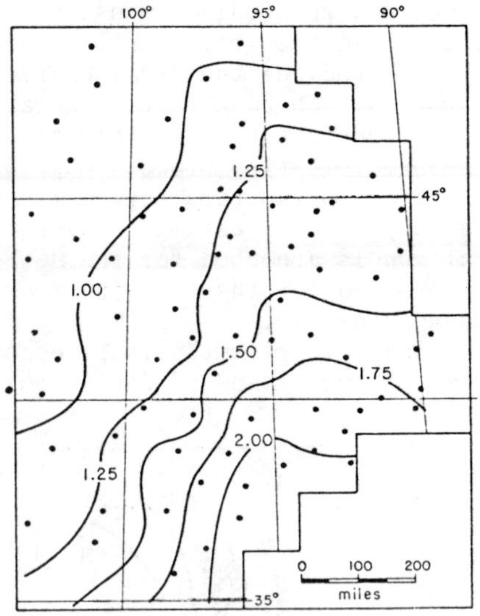

Fig. 2-10. Isolines of the 30-Year General Monthly Standard Deviation, $\bar{s}$ (after Tase, 1976)

harmonics for each station series for the time fluctuation of monthly mean and monthly standard deviation produces the amplitudes and phases of harmonics, with these parameters similarly regionalized as for the mean and standard deviation, and equations of the type of Eq. (2-4) produced. Then the periodicities in the mean and the standard deviation are removed by

$$\varepsilon_{p,\tau} = \frac{x_{p,\tau} - m_{\tau}}{s_{\tau}} \qquad (2\text{-}5)$$

in which p = the year number, τ = the month of the year (1, 2,...,12), $x_{p,\tau}$ = the monthly precipitation values at points of observation, m_{τ} = the monthly mean, s_{τ} = the monthly standard deviation, and $\varepsilon_{p,\tau}$ = the standardized values with the removed periodicities in the mean and standard deviation. If $\varepsilon_{p,\tau}$ is dependent in time sequence, a linear model can be found for the dependence, and a time independent stochastic component, $\xi_{p,\tau}$, found at each observational point. For a good fit of the second-order autoregressive model, then

$$\xi_{p,\tau} = \frac{1}{\sigma_{\xi}}(\varepsilon_{p,\tau} - \alpha_1 \varepsilon_{p,\tau-1} - \alpha_2 \varepsilon_{p,\tau-2}), \qquad (2\text{-}6)$$

in which σ_ξ = the standard deviation of $\xi_{p,\tau}$, and α_1 and α_2 = the autoregressive coefficients (constant or periodic).

Once the time independent stochastic components, $\xi_{p,\tau}$, or (TISC)-variables, of precipitation have been determined for all stations in a region, a river basin, an urban area, their inter-station dependence in the form of the lag-zero cross-correlation coefficients can be determined as a model relating these coefficients to station position, distance and orientation. Figure 2-11 shows for the station no. 52 of the selected precipitation stations in the Great Plains, how the lag-zero cross-correlation coefficient of that station monthly precipitation

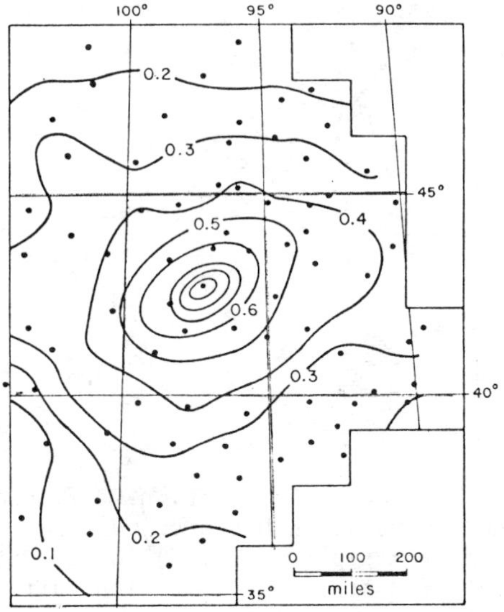

Fig. 2-11. Isocorrelation Patterns Based on Relationship Between Station 52 and All Other Stations for $(\xi - \xi_{min})^{1/3}$ (after Tase, 1976)

series to all the other station series in the region changes with the change of latitude and longitude. This coefficient decreases with an increase of the distance between the correlated stations. Though there is a slight ellipticity of isolines of this coefficient, it is valid to approximate the model of $r = f(x,y)$ only by its relationship to the distance, d, between the stations. The cross-correlation coefficients are computed for the transformed TISC variable $\xi_{p,\tau}$, namely for $(\xi - \xi_{min})^{1/3}$.

Figure 2-12 gives the relationship of the lag-zero cross-correlation coefficient, r, to the distance, d, in the form

$$r = e^{-\alpha d} , \tag{2-7}$$

with $\alpha = 0.00418$ in this particular case.

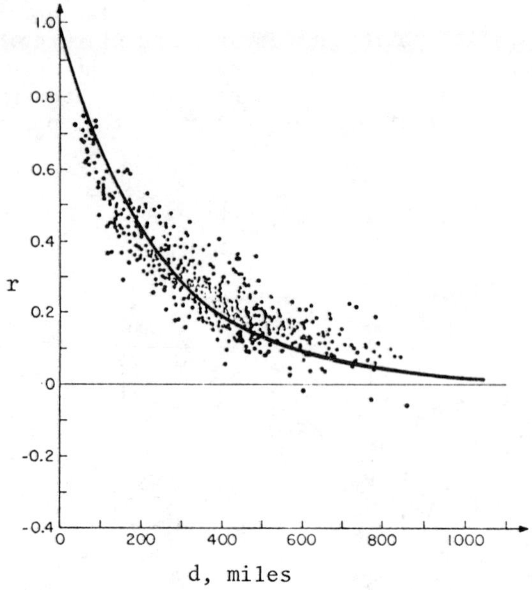

Fig. 2-12. Lag Zero Cross Correlation Coefficient, r, Versus the Interstation Distance, d, and the Fitted Function r = exp (-0.00418d) for the Stochastic Component of Monthly Precipitation (after Tase, 1976)

By condensing all the information on monthly precipitation series over a large region in the form of a set of mathematical equations, the generation of new samples of monthly precipitation process over that region becomes feasible. It can then be independent of the observational points. Similarly, this approach can be used for any variable and any interval Δt of a time series. To simplify this generation, it is feasible to cover the region, basin or area of investigation by any square or rectangular grid of points, with each point being associated with a well-defined unit area. It can be applied also to any other distribution of points over an area. In other words, the use of sample generation method for the investigation of periodic-stochastic processes can be separated from the observational points. This is important because the observational points were selected in the past basically by two criteria: as points at which the observations could be easily organized, and by the constraint of available funds for observations.

2.4 CHARACTERISTICS OF RIVER PROCESSES

To model the various aspects of processes in river mechanics, it is beneficial to assess the various characteristics of river processes, mainly the general properties of its major variables and the important features of environments in which rivers are created.

Variability of River Processes

The variability problem in modeling of processes in river mechanics results basically from the variability in climatologic and hydrologic variables. Temperature, atmospheric pressure, wind velocity, humidity, precipitation, evaporation, effective precipitation (difference between precipitation and evaporation), runoff, erosion, sediment transport, water quality variables, and similar quantities, all have time processes that are composed of within-the-year (and/or daily) astronomical periodicities in parameters, with a stochastic variation superimposed on these periodic parameters. All geophysical time processes vary in space as trend processes. Usually, the parameters vary by trends, while the fluctuating sample values about these population parameters undergo a stochastic variation. Variables which affect the modeling of processes in river mechanics, as well as the river mechanics variables, are in general the four-dimensional space-time random processes. Since functions of random variables are random variables, quantities involved in modeling of river mechanics must be also the random variables. Variability of river flows and of quantities that are functions of river flows and climatologic variables, must be by necessity conceived of and treated as the random variables in any accurate modeling of processes in river mechanics.

Figure 2-13 shows a space photo of the 110 miles of the Mississippi River in the states of Arkansas and Louisiana for the river conditions in March 1969. It does not show only a high variability with time of river bed shapes and alignments, but also that all the geometric river variables are random variables, both in space and time, with the non-negligible variations around the means of geometric variables.

General Characteristics of Rivers and Their Environment

In their present characteristics and forms, the rivers are functions of material in which they flow. The general river characteristics are an integral of all the past regimes of river water and sediment flows and of the river basin characteristics. The detail river forms, however, are functions of the present-day water and sediment regimes, with these detail forms shaped in such a way as to enable the river to carry through the water and sediment with the least quantity of energy. In most valleys rivers flow in their own sediments as Fig. 2-13 clearly indicates. Because the sediment deposits in river valleys are

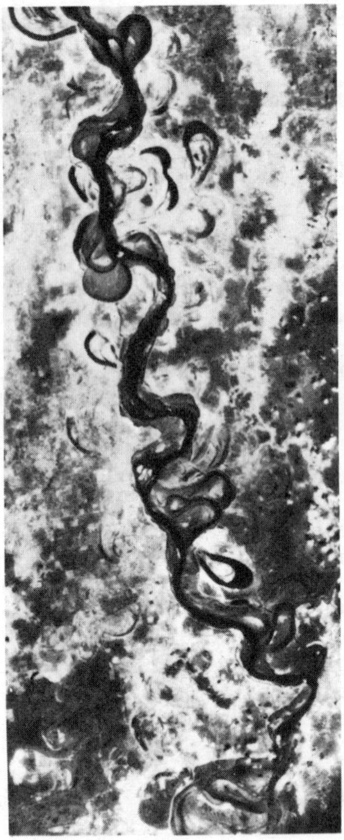

Fig. 2-13. 110 Miles of Mississippi River, USA between Lake Village, Arkansas (top) and Pt. Pleasant, Louisiana (bottom) (NASA Apollo 9 flight, March 1969)

the results of past regimes of precipitation, soil type and erosion, runoff, and sediment transport, rivers have adjusted their general and detail river forms to both the past and the present water flow and sediment transport. The present river environments can be properly understood and described only by using the findings of the fluvial paleo-morphology. The paleo-morphology takes into account the past history of river basin development and runoff and sediment regimes, as much as the historic, chronologic data permit its reconstruction. Since the past fluvial processes which have shaped river valleys have integrated various random processes of those times, it is not feasible to treat the contemporaneous river environments only as the simple effects of the present-day water flow and sediment transport and the resulting river forms. The constraints of the old geomorphologic processes cannot be neglected.

Two examples of the old geomorphologic processes, that still constrain the present day general river forms and processes, are the melting of ice of the last glacial age (about 15,000 to 20,000 years ago), and the draining by erosion, with the filling by sedimentation, of the large continental lakes.

The fluvio-glacial flow of water and moranic sediments in the relatively rapid melting of ice has filled many river valleys by sediment, in which now the rivers either flow or they are bounded by the remaining terraces. Some terraces may be remanents of the previous phases of the Pleostecene glacial age (lasting about two million years).

The example of the large lake or the internal, continental sea is the old Pannonian Lake in Europe. When the Danube River cut the Carpathean Mountains in the Iron Gates region, as well as filled sufficiently the bottom of the lake by its sediment, the present rivers of the Pannonian Plains flow through different sediment compositions. Two extreme examples are the Danube River to the West, and the Tisza River to the East. Figure 2-14 shows the schematic representation of the paleomorphologic constraints on the present general river forms.

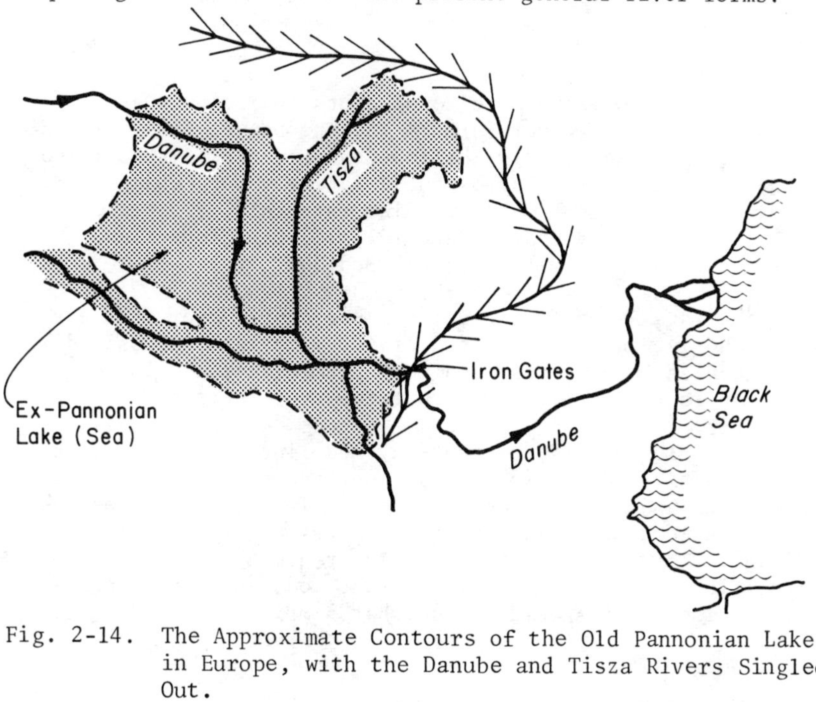

Fig. 2-14. The Approximate Contours of the Old Pannonian Lake in Europe, with the Danube and Tisza Rivers Singled Out.

The Danube River along some of its reaches has the sand and gravel material at the base, which have been deposited by the river at its estuaries into the Pannonian Lake. Later these sediments have been covered by many meters of consolidated, cohesive fine materials. The river easily erodes the banks by undercutting them (Fig. 2-15, A), so that the cross section is relatively wide but shallow for the type of the river.

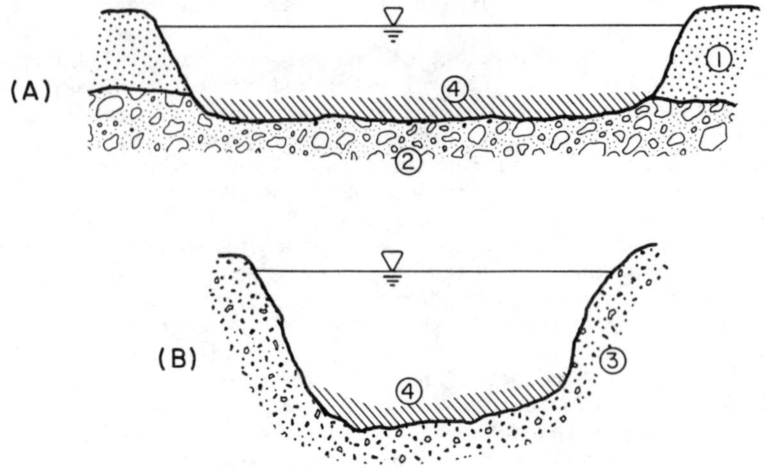

Fig. 2-15. Paleo-Geomorphic Constraints to Present-Day General River Forms: (A) The Danube River of Wide but Relatively Shallow Channel, and (B) the Tisza River of a Narrow but Relatively Deep Channel; (1) Recent Cohesive Fine Materials, (2) Old Coarse, Sand and Gravel Material; (3) Old Compact and Cohesive Material; (4) Bedload.

The Tisza River flows in its lower course over the very fine, cohesive and consolidated clayey materials, deposited in the Pannonian Lake by fine sediment precipitation. Therefore, the river is relatively narrow but deep for the size of that river (Fig. 2-15, B).

In conclusion, the present-day modeling of processes in river mechanics must take into account the paleo-morphologic constraints and composition of river environments besides taking into account the contemporaneous regimes of water flow and sediment. Whenever only these latter regimes predominate in the composition of river environments, the modeling of processes in river mechanics should be somewhat simpler.

Variations in River Forms

Variations in river forms may be conveniently classified into three basic categories with all transitions and combinations:

1. Short-range variations, which are basically the changes related to variations of bed forms, such as ripples, dunes, antidunes and plainbed forms, with various combinations and interactions of these forms. During the passage of a flood, say from a low flow to a peak flow and back to a low flow, the riverbed may pass through all these forms.

2. Middle-range variations, which represent most often the changes resulting from the intervention by man, or some accidents in the nature, with the river phenomena of aggradation and degradation. Some stochastic variations in water and sediment inflows into rivers, or the very rare events in river basins, belong to this type of variations. It is sufficient that an unusually high intensity rainfall occurs over a region of high erodibility for a large quantity of sediment to enter a river. Since the capacity of rivers to carry the given bed-load sediment is limited by river flows and bed characteristics, the river may not be able to carry along the resulting large quantity of sediment of the rare event. The sediment spreads over the riverbed as a long bar, or as a temporarily aggraded river reach. The future flows with less saturated bed-load material than the river could carry along this bar slowly degrade this bar, with all the consequences in changes of the lateral and bed forms that these bars may require.

3. Long-range variations in river forms are functions of various long-range changes in river basins. Such eventual changes are the changes in erodibility of river basin soils, as function of vegetation cover and similar factors. Terraces at different levels along many river valleys help to understand these long-range variations of rivers and their basic forms.

Water and Sediment Carrying Capacities of Rivers as Integrals of River Shaping Processes

A river historically develops a capacity to transport water and sediment. Theoretically, no practical limit exists for a river to transport the finest suspended sediment in water. It is sufficient to investigate the mud flows to ascertain that the concentration of fine materials in a river flow may be very large. The limits exist, however, for any river flow to carry the bed-load material of sand and gravel. Significant differences exist in velocities between floodwater waves and bed-load waves. Because of the large lag of bed-load sediment wave in comparison with the water wave, bed-load sediment travels by shifts as a water wave passes over a river reach. Significant information is available on transport capacities of rivers for water and sediment. Differences in travel velocity of water and sediment have an impact on shaping of riverbeds at short-range and middle-range time scales.

Interactions exist between water flow and sediment transport, periodic-stochastic processes. They create some finite capacities for rivers to carry water and sediment load for given levels and slopes. This can be conceived as a feedback process; namely, the past regimes of water flow and sediment transport have shaped rivers to their present capacities. Whenever an extreme high water flow occurs, these capacities may not be sufficient to carry water and sediment. Then disruptions result in the valley. Either a flood occupies flood plains or sediment fills the river channel and force the riverbed to change the carrying capacities for non-flood flows. Since the average channel flow predominates, a threshold is usually conceived for the bank-full flow, though the bank-full flow may change along the river from one reach to another.

River Response to Water and Sediment Inputs

The river response to floodwater input is mainly reflected in an attenuation of the wave by a decrease in flood peak and an extension of wavelength along the river. The water storage of river channels and flood plains performs in a similar way as any other storage capacity. Differences in celerities of various parts of a water wave, with the wave peak traveling faster than the water of the peak, make the river response to floodwave non-linear.

The river response to the total sediment input depends on sediment characteristics. Sediment is transported by three processes: suspension, saltation, and bed-load transport. The river response to any sudden input of bed-load sediment is to spread the sediment wave along the river as a longitudinal bar. The response is similar to the response of water input. The bar decreases its height at the aggraded bed as the sediment wave travels downstream with the conservation of sediment mass accomplished by the alongation of sediment wave.

Importance of Rates of Change of Water and Sediment Transport Processes, and the Power Relationships of Sediment Variables to Water Variables

Denoting $Q(t)$ as the water flow, $S_i(t)$ as the transport of the sediment quantity, with i designating the type of sediment transport (suspended, saltation, bed-load), the rates of change of these two processes and their interrelationships are important in modelings of processes in river mechanics. The rates of change are given in the form of first derivatives, $dQ(t)/dt$ for the water, and $dS_i(t)/dt$ for the sediment. It can be hypothesized that the selection of models to be applied to river mechanics should be a function of both $Q(t)$ and $|dQ(t)/dt|$, or $S_i(t)$ and $|dS_i(t)/dt|$, respectively or jointly.

It is generally accepted in the treatment of the unsteady free-surface flow phenomena of open channels, that the faster is the rate of change of the flow, or the greater the absolute value of the $dQ(t)/dt$ term, the more important is the dynamic effect of the flow. In that case, it is necessary to use both the conservation of mass and the momentum equations of unsteady free-surface flow, in order to accurately model the flood wave propagation.

Because of the power function relationships of several sediment transport characteristics to flow discharge, or discharge-related variables, in the form $S_i = F(Q^n)$, $n > 1$, the absolute rate of change of bed-load transport is usually much greater than the absolute rate of change of the water discharge. In case of the relatively small absolute rate of change of river flows, some basic approximations to the solution of partial differential equations of flood wave movement along the rivers are introduced by replacing the two basic partial differential equations of unsteady free-surface flow of open channels by the simplified methods of flood routing. Because of the much greater absolute rates of change of bed-load transport in comparison with the corresponding water discharge, the parallel use of simplified routing methods of water flow and bed-load sediment transport should not be used automatically, when the simplified methods of floodwater routing are selected. The greater are the absolute rates of change of bed-load transport in comparison with the absolute rates of change of water discharge for a given water discharge, the less feasible is a parallel use of simplified methods for both floodwater routing and the bed-load sediment routing. When both the water and sediment are considered, their interaction would require a selection of those routing models for floods and sediments, that can treat a relatively large absolute rate of change of the bed-load transport in flood routing, regardless of a smaller absolute rate of change of water flow, regardless that the flood routing alone may be treated sufficiently accurately by a simple routing method.

Several relationships between the sediment transport variables and the water discharge or its related variables are in the power form

$$X_s = aQ^n \tag{2-8}$$

with a and n parameters, and usually $n \gg 1$, and X_s = the sediment transport variable. Often n is of the order three to five. Because the powers Q^3 or Q^5 are very large for flood flows, very small for average flows, and negligible for low flows, the basic premise of modeling of processes in river mechanics is that only large water flows significantly affect the most important processes of shaping the rivers, particularly its aggradations and degradations. Two examples:

(1) One of the relatively slow changes of the daily flow hydrograph (for the year 1960 of the Boise River near Twin Springs, Idaho, USA); and (2) One of the relatively rapid changes of the daily flow hydrograph (for the year 1960 of the Powell River near Arthur, Tennessee, USA), are used here to show the potential effects of the large Q values on river processes.

Figures 2-16 and 2-17 shows these two hydrographs of daily flows, while Figs. 2-18 and 2-19 show the corresponding

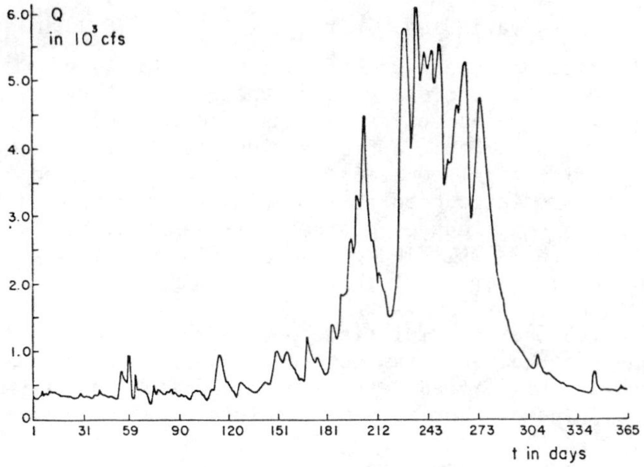

Fig. 2-16. The Hydrograph of Daily Flows for the Year 1960 of the Boise River Near Twin Springs, Idaho, USA

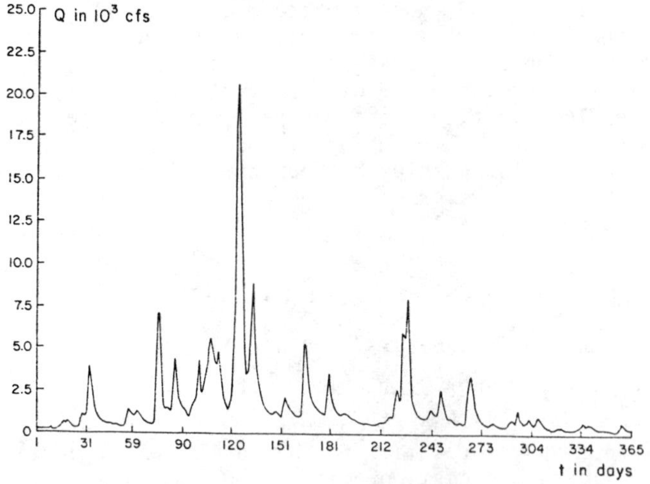

Fig. 2-17. The Hydrograph of Daily Flows for the Year 1960 of the Powell River Near Arthur, Tennessee, USA

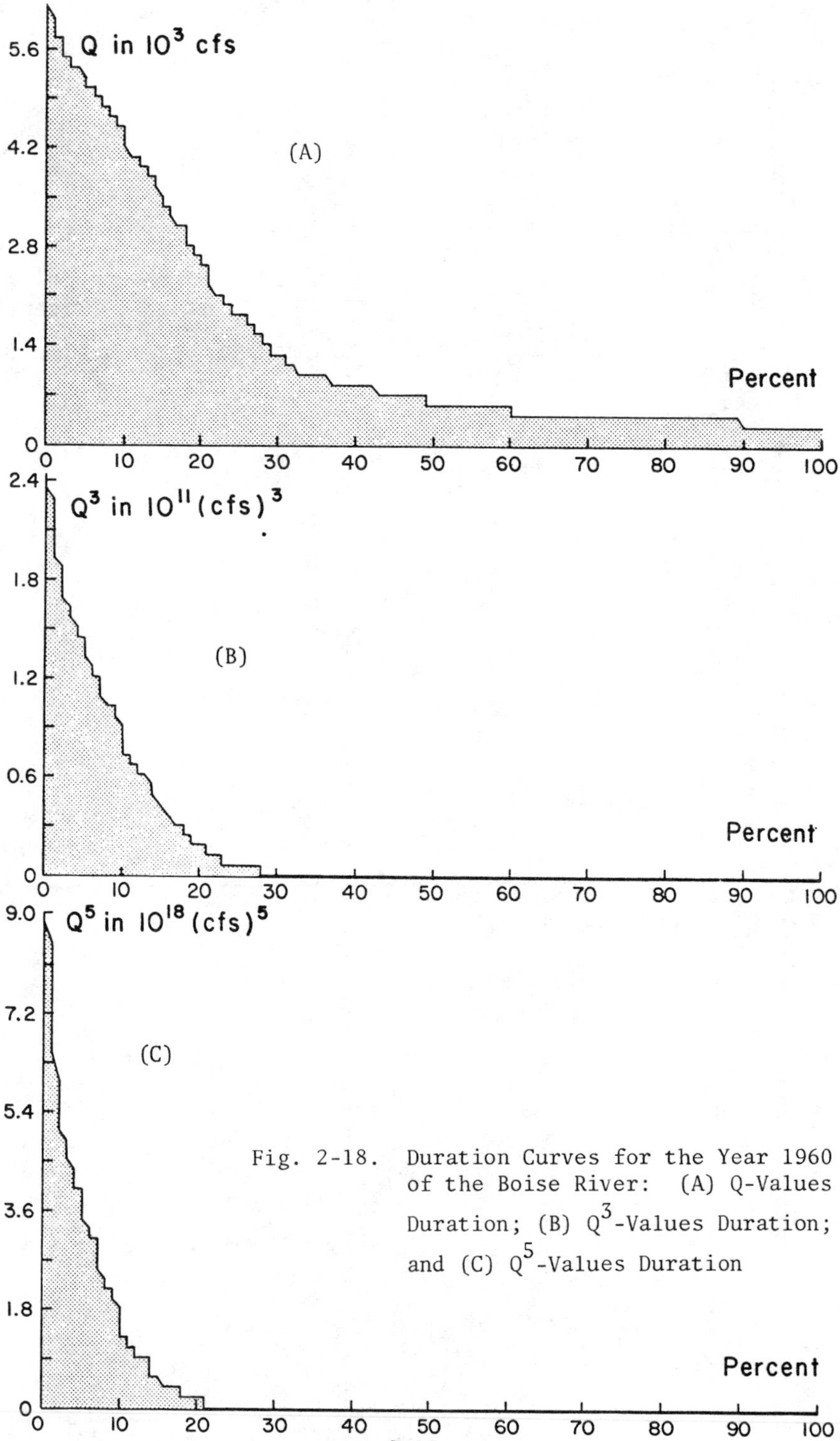

Fig. 2-18. Duration Curves for the Year 1960 of the Boise River: (A) Q-Values Duration; (B) Q^3-Values Duration; and (C) Q^5-Values Duration

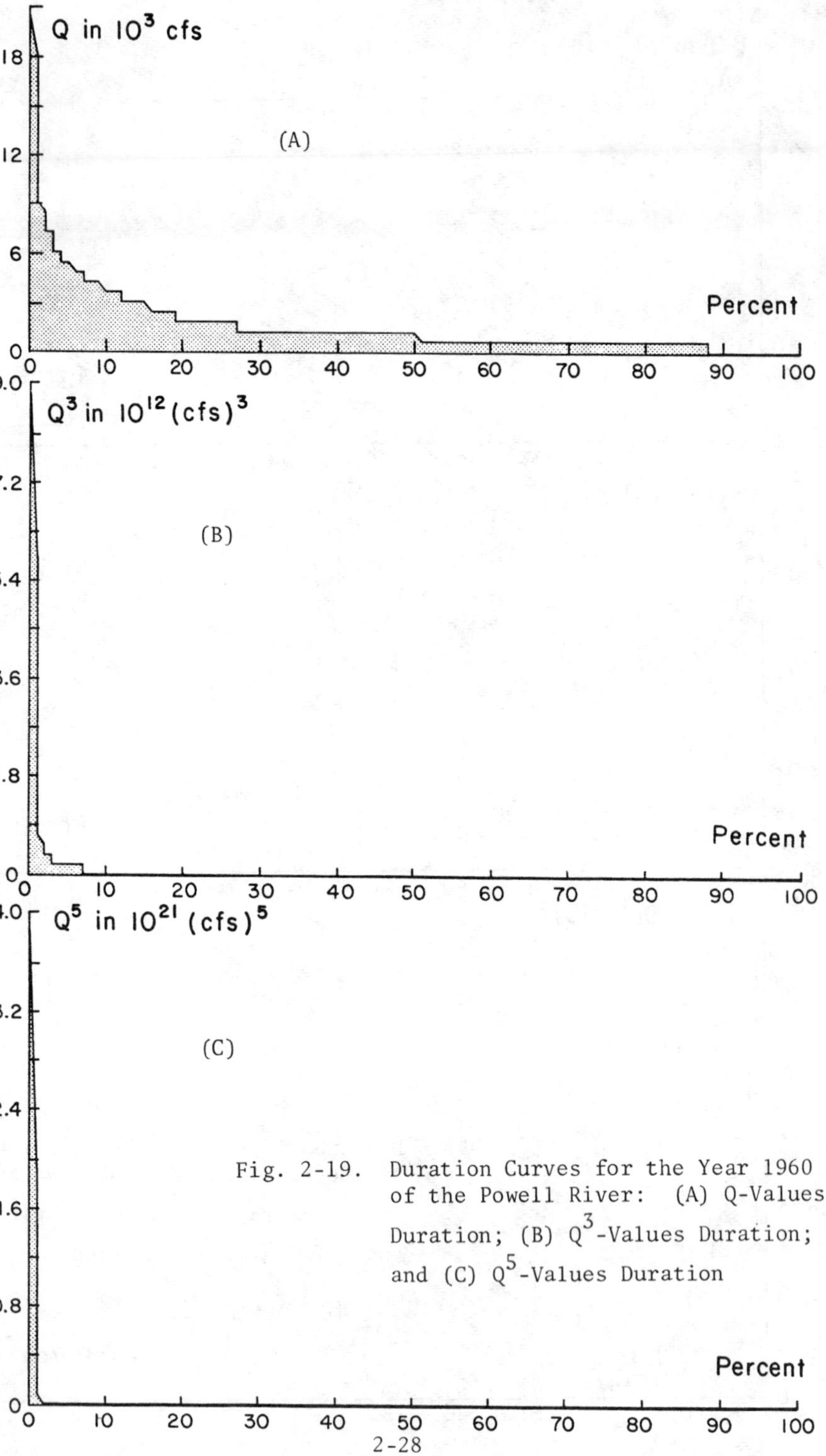

Fig. 2-19. Duration Curves for the Year 1960 of the Powell River: (A) Q-Values Duration; (B) Q^3-Values Duration; and (C) Q^5-Values Duration

duration (cumulative frequency) curves for three cases of each river, respectively: (i) the observed Q-values; (ii) the Q^3-values; and (iii) the Q^5-values.

Figures 2-18 and 2-19 clearly show the effects of flows on processes of river mechanics for which the powers Q^3 and Q^5 play a fundamental role. Some of the most important river shaping and sediment transport processes are very much affected by a small percentage of highest flows. Therefore, the problem of estimating the occurrence, magnitude and duration of the largest river flows is of the fundamental importance in modeling the sediment, bed and channel processes in river mechanics.

2.5 REFERENCES

Kelman, J., 1977. Stochastic modeling of hydrologic, intermittent daily processes. Colorado State University Hydrology Papers Series, No. 89, February 1977, Colorado State University, Fort Collins, Colorado, USA.

Richardson, E. V., and Clarence, W., 1977. A model of stochastic structure of daily precipitation over an area. Colorado State University Hydrology Papers Series, July 1977, Colorado State University, Fort Collins, Colorado, USA.

Tase, N., 1976. Area-deficit characteristics of droughts. Colorado State University Hydrology Papers Series, No. 87, November 1976, Colorado State University, Fort Collins, Colorado, USA.

Chapter 3

FORECASTING OF FLOWS-FLOOD FREQUENCY ANALYSIS

by

H. J. Morel-Seytoux, Professor of Civil Engineering, Colorado State University, Fort Collins, Colorado

Chapter 3

FORECASTING OF FLOWS-FLOOD FREQUENCY ANALYSIS

3.1 INTRODUCTION

Thank goodness that God in *her* wisdom did not choose to imbue man with the power of knowing his destiny! Gone would be the excitement and the anticipation of the future.

Typically though, now as in the days of the Garden of Eden, man cannot resist the temptation of trying to know (Holy Bible, Genesis, p. 3) ... and for some good reasons, because his very existence depends on his ability to react against the vagaries of his surroundings. Life cannot tolerate but minor fluctuations about a mean condition. This requirement for equilibrium (and survival) is constantly contradicted by the variations of the surroundings. To guard against these fluctuations one must anticipate them. Forecasting in hydrology like in many other sciences (meteorology, economics, medecine or politics) is thus a necessity of life.

One must distinguish between *two types* of forecasting, for the *short-run* and for the *long-run*. This distinction corresponds to the two practical problems of *operations* and of *planning*. The two problems call for two distinct methodologies for their solutions. To the question of whether or not an individual will be alive next week, one would answer by taking his temperature, blood pressure, etc... various indices of his present health condition. To the question of whether he would still be alive fifty years from now, one would answer by consulting the life expectancy tables of an insurance company. For operational purposes, one would extend the present condition into the future by deterministic application of the natural laws. For planning purposes, one must be satisfied to express the future in probabilistic terms. This chapter is concerned with this latter preoccupation. The next chapter is concerned with the former.

3.2 FLOW FREQUENCY ANALYSIS

A question one might ask regarding the flow of a river, could be: what will the flow of the Cache la Poudre River be on May 20, 1987? It is not a very interesting question (who cares?) and, besides, nobody knows the answer! A more meaningful question put by the engineer concerned with design of a bridge on the river might be: will the flow exceed 10,000 cfs on that day? The question is more interesting but in view of his interest for the safety of the bridge, the specific date May 20 does not appear to be very relevant. Would not the following question be more to the point: "Will the daily flow in 1987 exceed 10,000 cfs?" without reference to a specific date such as May 20? Proceeding similarly the specific year does not appear of great significance either. The meaningful question is: "what is the *chance* that a given magnitude of flow be *exceeded* in a given year?", or: "what is the chance that an even larger given magnitude of flow be *exceeded over a period* of 10 years? 50 years? 100 years? etc..." Clearly, as our concerns extend over longer horizons, the precise dates of occurrence of the various events become less and less important in comparison to the possible occurrence of events of extreme magnitudes (great or small). How frequently then will an extreme event such as hurricane Hazel of 1954 or the drought of the thirties or, closer to home, the August 1, 1976 Big Thompson flood (Grozier, et al., 1976), hit us?

3.3 RELEVANT PROBABILISTIC CONCEPTS

Given that a hydrologic variable such as river flow or daily precipitation is a *random variable* it is not possible to predict exactly what the value of that variable is going to be on any future date. It is, however, possible to state a probability that the value of the variable will be within a given interval. For example, there might be a 5 percent chance (probability) that the flow of the Poudre River be in the interval 4500-5500 cfs on May 20 of any year. To assess quantitatively the probability, one must know the *law of chance* governing the random hydrologic variable. Probabilists describe the law in a variety of forms. They talk of a *cumulative distribution* function, of a *probability density* function, of *parameters* of the distribution, *moments* of the distribution etc. Loosely and concisely speaking one says that the random variable comes from a *(statistical) distribution* (Mood, et al., 1974, p. 51).

(It is wise to keep in mind *always* that every hydrologic variable wears three hats: (1) its *physical* meaning (e.g. runoff), (2) the *role* it plays (e.g. random variable) and (3) the *numerical* value it takes (observation, realization). When the role of random variable is emphasized, it is traditional to use a *capital* letter, say X. When the numerical value is concerned a lower-case letter is used, say x.)

Cumulative Distribution Function

For a continuous random variable X such as river flow (and most natural hydrologic variables are continuous variables) one defines the *probability* that X be *less than a given* value, x. Naturally, this probability depends on the value of x and one can write symbolically:

$$P\{X \leq x\} = F_X(x) \quad (3\text{-}1)$$

If there is no ambiguity that the cumulative (probability) distribution function (abbreviated c.d.f.) refers to the random variable X and to no other random variable, $F_X(x)$ is then written simply as F(x). From the definition of F(.) it is clear that the function is a monotonically increasing function of its argument and takes values in the range 0, 1. Fig. 3-1 displays two cumulative distribution functions (c.d.f.) for two nonnegative hydrologic variables, X_1 and X_2.

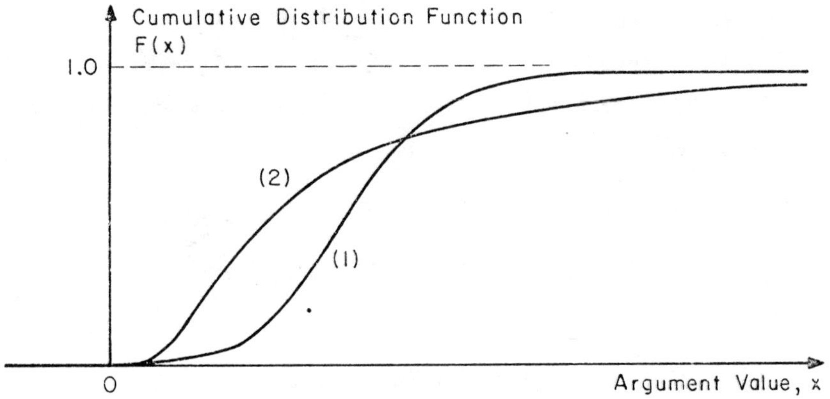

Fig. 3-1. Two Typical Cumulative Distribution Functions

The simplest and most studied c.d.f. is the *normal* distribution, and is defined by the expression:

$$F_X(x) = \int_{-\infty}^{x} \frac{1}{\sigma\sqrt{2\pi}} e^{-\frac{(\xi-\mu)^2}{2\sigma^2}} d\xi \tag{3-2}$$

where μ is the *mean* value of the random variable X, σ^2 is its *variance* (a *parameter* to be defined later), σ is the standard deviation (always *positive*) and ξ is a dummy variable of integration. The range of values of the variable x and of the parameter μ is not limited. It is convenient to define a standardized normal variable, usually denoted U and defined as:

$$U = \frac{X-\mu}{\sigma} \tag{3-3}$$

The random variable (appreviated, r.v.) U has zero mean and a variance of unity. The probability that X be $\leq$ x is then the same as the probability that U $\leq$ u when u and x are related by Eq. (3-3). Thus,

$$P\{X \leq x\} = F_X(x) = P\{U \leq u\} = F_U(u) \tag{3-4}$$

and the c.d.f. $F_U(.)$ is tabulated in many textbooks (e.g.

Brownlee, 1965, p. 559; Mood, et al., 1974, p. 552) and usually denoted $\Phi(.)$.

Probability Density Function

The probability that X falls in a finite interval of width Δx centered about the value x is the probability that $x + \frac{\Delta x}{2}$ not be exceeded while $x - \frac{\Delta x}{2}$ be exceeded. This probability denoted $P\{x - \frac{\Delta x}{2} < X \leq x + \frac{\Delta x}{2}\}$ is from the very definition of the c.d.f. simply:

$$P\{x - \frac{\Delta x}{2} < X \leq x + \frac{\Delta x}{2}\} = F(x + \frac{\Delta x}{2}) - F(x - \frac{\Delta x}{2}) \qquad (3\text{-}5)$$

For an infinitesimal interval dx this probability is:

$$\lim_{\Delta x \to o} \frac{F(x + \frac{\Delta x}{2}) - F(x - \frac{\Delta x}{2})}{\Delta x} \Delta x = \frac{dF}{dx} dx = F'(x)dx \qquad (3\text{-}6)$$

The derivative of the c.d.f. with respect to its argument (here denoted x) is the *probability density function* (p.d.f.) usually denoted f(x), namely:

$$f_X(x) = \frac{dF_X(x)}{dx} \qquad (3\text{-}7)$$

(This derivative usually exists except for c.d.f. which experience discontinuities (jumps) in ordinates for certain values of x. In this chapter we are only concerned with smooth c.d.f. and the derivative will always exist.) With this new notation the probability for X to be in the interval dx, centered about the value x, is:

$$P\{x - \frac{dx}{2} < X \leq x + \frac{dx}{2}\} = f(x)\, dx \qquad (3\text{-}8)$$

The probability density functions for two nonnegative hydrologic variables corresponding to the two c.d.f. of Fig. 3-1 are shown on Fig. 3-2.

From the definition of the p.d.f. and its relation to the c.d.f. (Eq. 3-7) it follows that the p.d.f. for the normal distribution, obtained by differentiation of Eq. (3-2), is:

$$f(x) = \frac{1}{\sigma\sqrt{2\pi}} e^{-\frac{(x-\mu)^2}{2\sigma^2}} \qquad (3\text{-}9)$$

Typical shapes of the normal p.d.f. for a given value of the mean and three values of the variance are shown on

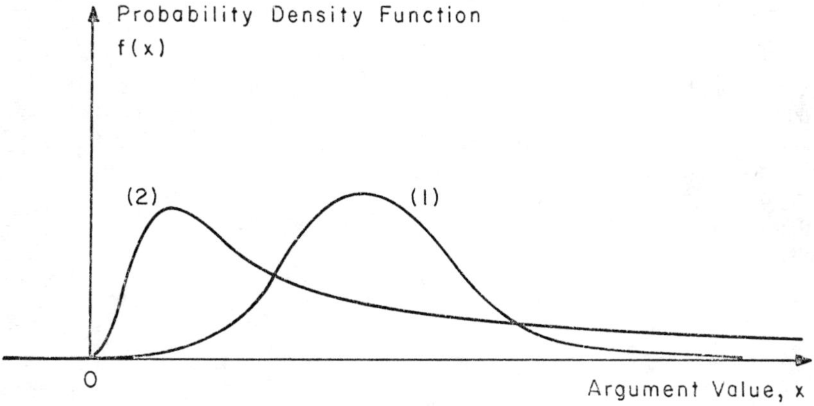

Fig. 3-2. Two Typical Probability Density Functions.

Fig. 3-3. The larger σ the more widespread is the distribution. The smaller σ the more concentrated the distribution or in other words the sharper its peak.

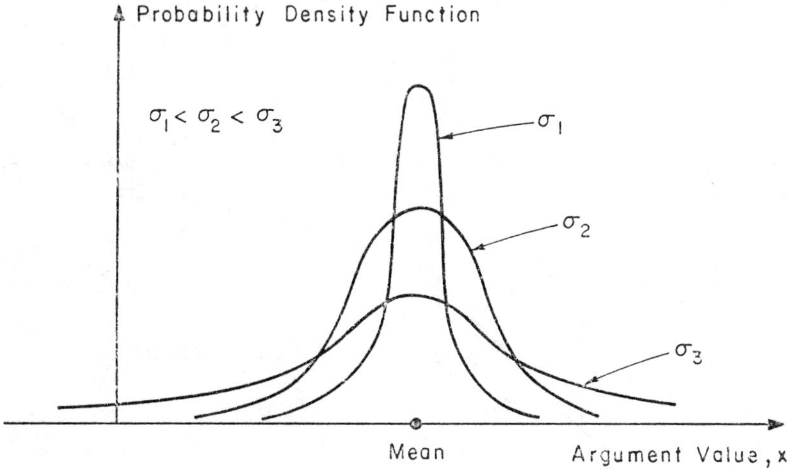

Fig. 3-3. Variation of Shape of Normal Probability Density Function with Variance

The value of the abscissa for which the ordinate is maximum is known as the *mode*. For the normal distribution the mode equals the mean. The points of inflexion are located at a distance σ on either side of the mean.

Exceedance Probability

Traditionally the c.d.f. has been defined to represent the probability of the r.v. X not to exceed a given value x. But in many instances and in particular when concerned with *undesirable* large flows (floods) the probability of interest is not the c.d.f. probability, but its complement, i.e. the *probability* of *exceeding* a given value or the *exceedance* probability, denoted p and by definition calculated by the relation:

$$p = P\{X > x\} = 1 - P\{X \leq x\} = 1 - F(x) \qquad (3\text{-}10)$$

and similarly:

$$q = P\{X \leq x\} = F(x) = 1 - p \qquad (3\text{-}11)$$

where q is the probability (hopefully large) of being on the *safe side* (see Fig. 3-4). For this reason it will be referred to as the *safe* probability for short. Note that for droughts the c.d.f. probability would be the undesirable probability whereas the exceedance probability would be the safe one. In what follows floods not droughts will be discussed and q will be called the *safe* probability for practical mnemonic purpose.

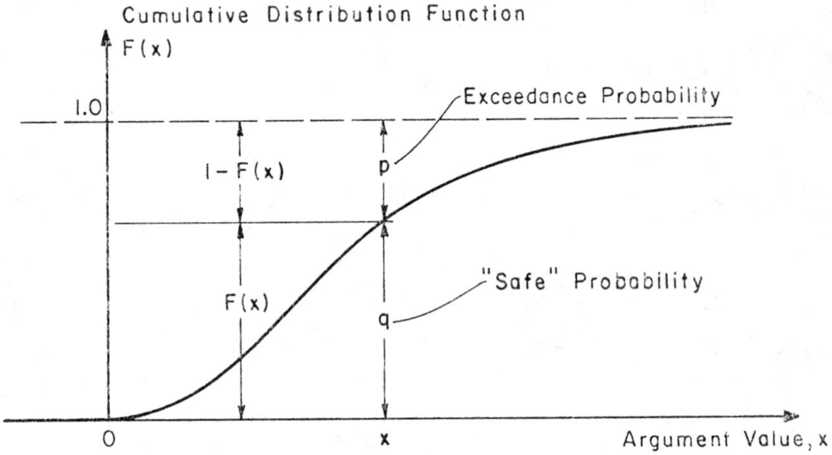

Fig. 3-4. Illustration of the Meaning of the Exceedance and Safe Probabilities

The expression "*exceedance* probability" has appeared in the literature (Gumbel, 1958; Beard, 1962) to replace the more cumbersome one: "probability of exceeding." Though the word exceedance itself does *not* appear (yet) in the dictionary (Websters, 1958) the use of the expression has been adopted by the profession, and the word spelled indifferently exceedence (Beard, 1962; Yevjevich, 1972) or exceedance (U.S. Water Resources Council, 1976).

Expectation Operator

The operator defined mathematically by the integration of any function g(.) of the r.v. X weighted by the p.d.f. of the r.v. X, namely:

$$E\{g(X)\} = \int_{-\infty}^{+\infty} g(x)\ f(x)\ dx \qquad (3\text{-}12)$$

is known as the *expectation operator* and the result of the operation is known as the *expected value* of g(X). The simplest function g(X) is X itself and the expected value of X also known as its *mean*, μ, is:

$$E(X) = \mu = \int_{-\infty}^{+\infty} xf(x)\ dx \qquad (3\text{-}13)$$

For the particular function $g(X) = (X-\mu)^2$ the expected value is known as the *variance* of X and denoted σ^2 and thus:

$$E\{(X-\mu)^2\} = \sigma^2 = \int_{-\infty}^{+\infty} (x-\mu)^2\ f(x)\ dx \qquad (3\text{-}14)$$

As is clear graphically from Fig. 3-3 and mathematically from the definition of Eq. (3-14) the variance is a *measure* of the *spread* of the distribution *about its mean.* In many instances the actual functional form (see Fig. 3-1) of the c.d.f. is not known. However, if the mean and variance (the second centered moment) are known, one can get a fair idea of the shape of the distribution (narrow or broad) (see Fig. 3-3) and the exact form of the distribution is not of paramount importance *if* one is concerned with the *central* range of values of the random variable. On the other hand when concerned with *extreme* events the *shape* of the distribution is very *important.* For example Figs. 3-1 and 3.2 display two distinct distributions. Though the mean of the second distribution is less than that of the first and though the spreads of the two distributions are roughly equal, it is apparent that for high values of x the exceedance probability is larger for the second distribution than for the first.

3.4 RELEVANT STATISTICAL CONCEPTS

Given the distribution of a random variable (r.v.) and its parameters (mean, variance, etc.), probability theory enables us to make statements regarding probability of exceedance of a given value, probability of falling within a given interval, etc... Probability theory does not tell us, however, how to *infer* from data that the distribution is normal or otherwise, how to *estimate* the mean, variance and other parameters. *Inference* and *estimation* are the domain of Statistics.

Choice of a Distribution

The *first serious* question when confronted with observed values (data) of a r.v. is to find the statistical distribution from which it comes. There are several procedures to answer the question and the simplest one is to plot the data in a special way on a special paper. Consider the c.d.f. of a normal r.v. on Fig. 3-5a. The c.d.f. is curved because the probability scale is the regular linear scale. On the other hand, had the probability scale been chosen as shown on Fig. 3-5b, the c.d.f. would be straight. The construction of the *warped* scale is shown on Fig. 3-5. The slope of the straight line on Fig. 3-5b is totally arbitrary. With this construction a normal probability paper was designed. Naturally, it is not necessary to repeat the design of the normal probability paper every time one needs it! It has been done much more precisely and is available commercially. The point is that by a *proper* transformation of the probability scale any statistical c.d.f. can be transformed into a linear function of the argument or, in other words, made to plot as a straight line on the special paper. Such papers can be constructed for the normal distribution or *any other* distribution.

Consider the data of annual (water-year) runoff for the Hinterrhein River at Hinterrhein village in Switzerland for the 19 year period 1951-1969 (see Table 3-1). The data have also been ranked in decreasing order. The data define classes, such as values between 128 and 127, 127 and 122.3, 122.3 and 122.0, etc. down the line to 86 and 85, and two special classes: values greater than 128 and values less than 85, altogether (19-1) + 2 = 20 classes. Presuming from experience (the 19 years of observations) that occurrence of flows in all these classes are *equiprobable* (after all, they all did occur once!) then $P\{X > 128\} = \frac{1}{20}$, $P\{X > 127\} = \frac{2}{20}$, $P\{X > 122.3\} = \frac{3}{20}$, etc. If N is the number of years of record then the estimate of the exceedance probability p of the runoff of rank n is given by the *plotting position* formula:

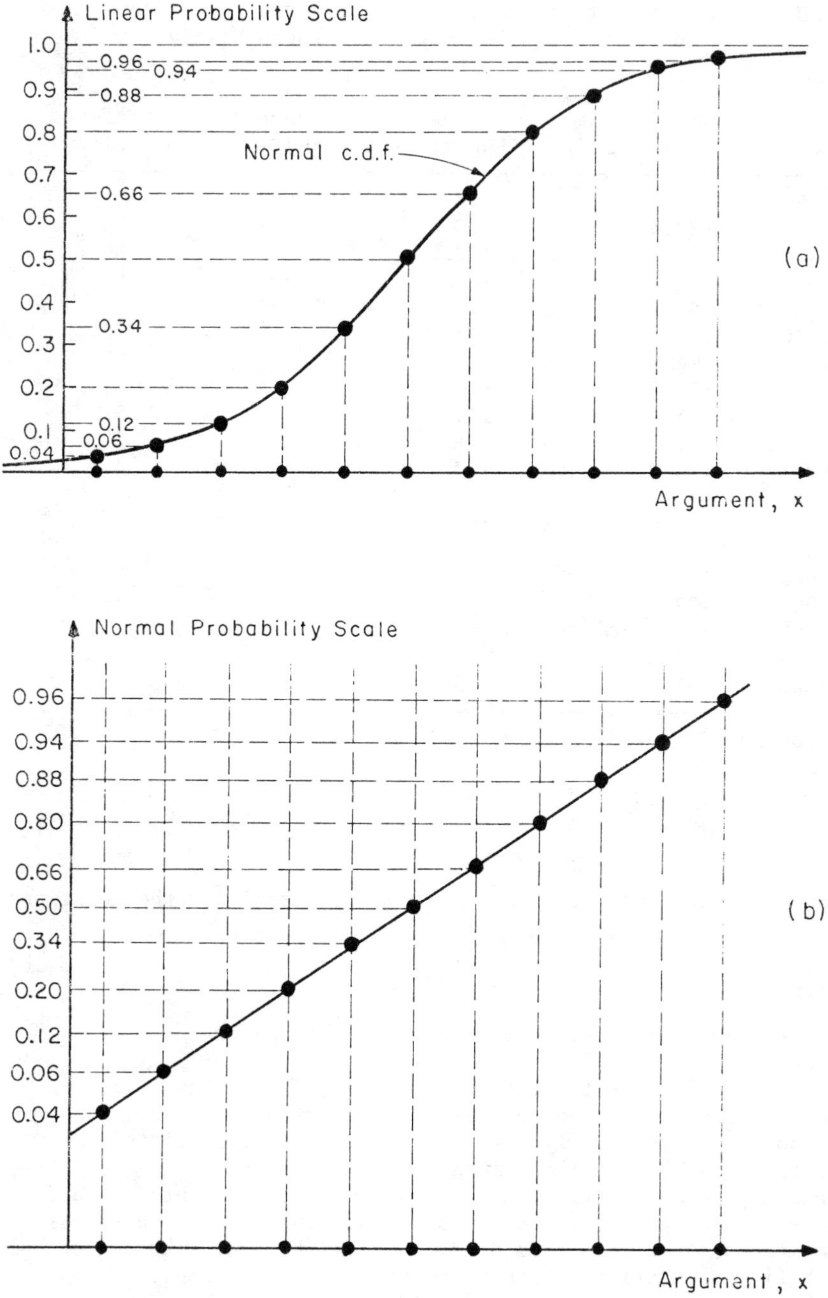

Fig. 3-5. Construction of Normal Probability Paper

Table 3-1. Data of Annual (water-year) Runoff for the Hinterrhein River at Hinterrhein Village, Switzerland

Year	Runoff Millions m^3	Rank	Ranked Runoff	p, Exceedance Probability %	q = 1-p, "Safe" Probability %
1951	127	1	128	5	95
1952	122.3	2	127	10	90
1953	104	3	122.3	15	85
1954	119	4	122.0	20	80
1955	86	5	121.1	25	75
1956	113	6	121.0	30	70
1957	96	7	119	35	65
1958	111	8	113	40	60
1959	93	9	111	45	55
1960	121.1	10	108	50	50
1961	108	11	104	55	45
1962	85	12	102	60	40
1963	121.0	13	100	65	35
1964	102	14	98	70	30
1965	100	15	97	75	25
1966	97	16	96	80	20
1967	128	17	93	85	15
1968	122.0	18	86	90	10
1969	98	19	85	95	5

(rounded data from Laufer, 1977, p. 54)

$$p = \frac{n}{N+1} \tag{3-15}$$

One also defines the *return period*, denoted T_r by the relation:

$$T_r = \frac{1}{p} \tag{3-16}$$

The lower the exceedance probability the higher the return period.

In Table 3-1 the *exceedance* and "*safe*" probabilities were calculated. The data are then plotted on normal probability paper (Fig. 3-6) and a straight line fitted (visually) through the points. In this case it seems that the data do not deviate appreciably from a straight line and one can say with reasonable confidence that the annual flows at the station come from a normal distribution. The degree of fit is relative. From previous experience (Figs. 3-7 and 3-8) it can be said that the fit is rather good. For the entire San Juan mountain area in Colorado, Fig. 3-7 displays the best fit and Fig. 3-8 displays the worst fit.

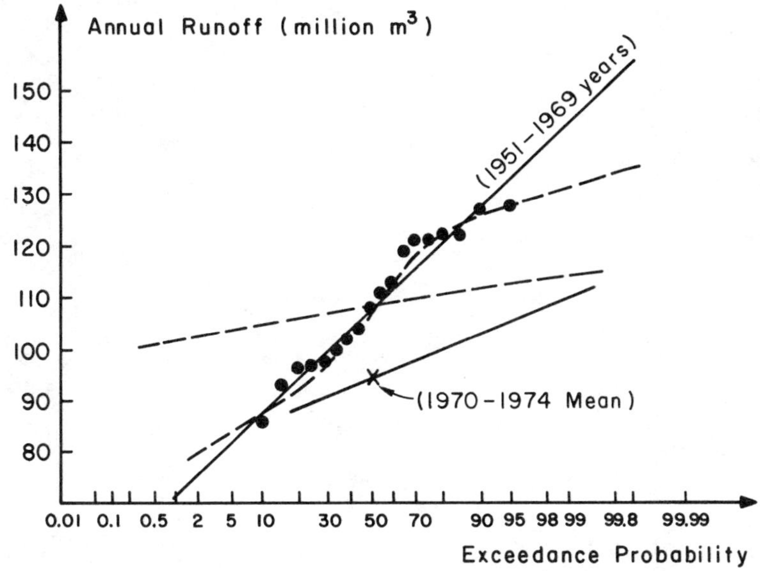

Fig. 3-6. Cumulative Frequency Curve of Annual Flows of Hinterrhein River at Hinterhein Village, Switzerland

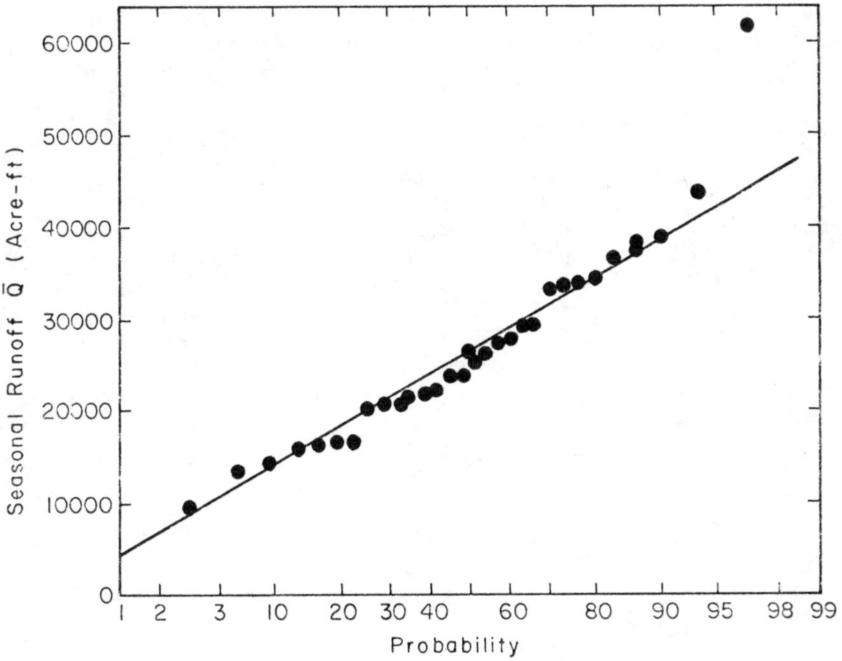

Fig. 3-7. Graphical Test of Normality for Seasonal Flows of Smith Fork Near Crawford, Colorado (after Morel-Seytoux and Saheli, 1972)

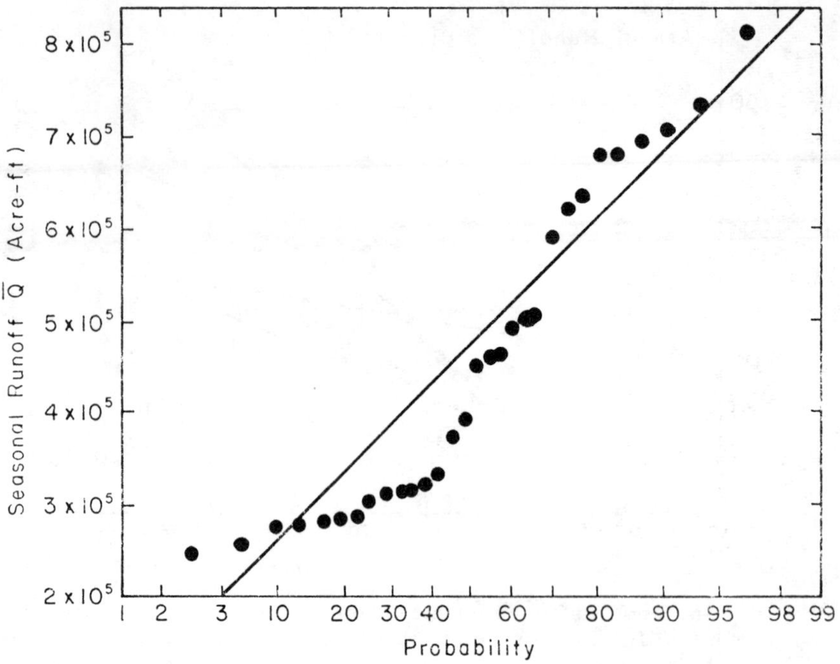

Fig. 3-8. Graphical Test of Normality for Seasonal Flows of Animas River at Durango, Colorado (after Morel-Seytoux and Saheli, 1972)

Estimation of Parameters

Once it has been accepted that the data came from a certain distribution, the *second serious* question is how to estimate the parameters of the distribution in terms of the observations. For example, the annual river flows of the Hinterrhein River being, for practical purposes, normally distributed, the two parameters μ and σ of Eq. (3-2) must be estimated. Staticians (Mood, et al., 1974, p. 109) have shown that these parameters are the expected value and the variance of the flows respectively (therefrom the use of the symbols to denote the parameters) or:

$$E\{X\} = \mu \qquad E\{(X-\mu)^2\} = \sigma^2 \tag{3-17}$$

Given the observations x_n of the r.v. X, *everybody knows* that the estimate of the mean or *sample* mean is given by the expression:

$$\bar{x} = \frac{\sum_{n=1}^{N} x_n}{N} \tag{3-18}$$

and that the *unbiased* sample variance is given by the expression:

$$\hat{S}^2 = \frac{\sum_{n=1}^{N} (x_n - \bar{x})^2}{N-1} \qquad (3\text{-}19)$$

What probably *everyone has forgotten* is how these estimates are obtained (derived)!

There are two (major) methods to derive expressions of sample estimates: (1) the maximum likelihood method (Mood, et al., 1974, p. 276) and (2) the method of moments (Mood, et al., 1974, p. 274).

These are by no means the only methods. For example, a third method is the method of least squares, widely used in regression analysis (Graybill, 1961, p. 114). In this chapter only the method of moments will be used and for this reason only this method will be discussed when needed in a later section to estimate the parameters of the Gumbel distribution.

Distribution of Sample Statistics

In the statistical literature, the estimate $\bar{X}$ is referred to as a "*statistic*." According to an authoritative reference (Mood, et al., 1974, p. 226): "A *statistic* is a function of *observable* random variables, which is itself an observable *random* variable, which does *not* contain any *unknown parameters*." (Italics are author's not Mood's.) Clearly $\bar{X}$ and S^2 are statistics. However, the point to be emphasized here is that these statistics e.g. $\bar{X}$ are *random variables themselves*. $\bar{X}$ may be an unbiased, close, sufficient, consistent, best estimator in this or that way, and all that jazz, etc. (Mood, et al., 1974, p. 288) but nevertheless $\bar{x}$ may be a poor (i.e. rather inaccurate) estimate of the true mean. $\bar{X}$ is a random variable and therefore comes from a statistical distribution. The estimate $\bar{x}$ of μ will be *unreliable* if the distribution of $\bar{X}$ is broad. Thus, it will be necessary to know the c.d.f. of the estimators $\bar{X}$, $\hat{S}^2$, etc... and the parameters of their distribution. Fortunately, many results exist in this regard and they will be called upon as needed in the later sections.

3.5 EXTREME EVENTS

One question that could be asked regarding the annual flows of the Hinterrhein River is: what is the magnitude of an annual high flow with a return period of 100 years? The answer to this question is obtained immediately by extrapolation of the straight line on Fig. 3-6 to the 1 percent exceedance probability. Thus, a large annual runoff volume of 144 million m^3 is to be expected on the average every 100 years, a volume of 156 million m^3 every 200 years, etc. Would these events be extreme? In a sense they are but with present day engineering hydrology terminology they are not, or rather they do not come from an extreme value c.d.f. The 144 or 156 million m^3 are extreme *realizations* (observations) of a random variable which more often than not will take moderate (central) values. On the other hand one could construct a population consisting *only* of extreme value variables. In any one year the maximum daily flow will occur on different dates. If it occurs on May 20 on a given year, it will be an extreme realization for the otherwise unconspicuous May 20 daily flow population. On different years the maximum daily flow will occur on different dates. If the maximum daily flows for every year are pooled together they will constitute a sample from an *extreme value* population. Clearly, made as it is of the crests of all the yearly hydrographs, one can expect that the 100 year high flow (flood) from the extreme value distribution will be of a magnitude far superior to the 100 year high flow among the daily flows for a particular date. If the daily flows on various dates were normally distributed, one can seriously doubt that the extreme flows will be normally distributed. In fact, if the May 18, May 19, May 20, etc... daily flows indeed each came naturally (causally, physically) from a homogeneous (*stationary*) population, one might seriously question (philosophically) whether there exists a stationary population from which the extreme values are drawn. On the other hand, as with snowmelt produced peak flows, if a variety of natural factors combine repetitively every year to create the peak flow (high solar radiation, high temperatures, clear skies, saturated soils from early melt, recharged aquifer, melt spread over the entire slopes, etc.) the peak flows may well come from a homogeneous population whereas the May 20 flows may come from several populations. In this case, the peak flows may be more normally or log-normally distributed than the calendar daily flows (Morel-Seytoux, 1976b, p. 10.10). When a variety of many natural factors act multiplicatively to create a random variable, statistical theory shows that the variable tends to be log-normally distributed.

Annual and Partial Duration Series

The *annual* and *partial duration* series of extreme values at a station are determined from climatological or water supply records for precipitation or streamflow, respectively. The maximum value, of say the 24-hour cumulative precipitation, is determined for each year of record. These maximum amounts, one for each year, comprise the *annual series*. The *partial duration series* includes the maximum values for the year but also includes other high values above a threshold, say 5-cm of 24-hour cumulative precipitation. For each year of record there will be the maximum for the year and one, two, three, or more additional values as well. The partial duration series contains a greater number of values than does the annual series. This series can improve the reliability of the analysis of extreme events for large return periods when the number of years of the record for the station is small.

The major source of difficulty with the partial duration series is the question of *independence* of events. Floods occurring one after another in short succession may not be independent. The time interval between events is therefore critical. Two storms occurring in arid areas will probably be independent. In humid areas, these same two storms will not usually be independent. During the monsoon period in India, hydrograph separation must be a real headache!

In the remainder of this chapter the discussion will be limited to the case of annual series. Methods for converting partial duration series to annual series are discussed in several texts and/or reports (Hershfield, 1961, p. 3; Hjelmfelt and Cassidy, 1975, p. 36; Linsley, et al., 1975, p. 356).

3.6 DISTRIBUTION OF EXTREME VALUES

There are two important questions to ask: Why is it important to know the distribution of extreme values or, more precisely, to know the functional form of the statistical distribution? What is it?

From Fig. 3-6 it was concluded that the runoff was normally distributed. The fit is good but not perfect. Suppose that after all the c.d.f. was not normal! From the data a hypothetical *curved* c.d.f. as indicated by the dashed line on Fig. 3-6 is not totally unrealistic. In fact, it fits the data just as well as the straight line. Within the range of values of the observations probability statements based on one c.d.f. or the other will not be noticeably different. However, the extrapolations to high or low exceedance probabilities will lead to very different flow magnitudes. The importance of the knowledge of the shape of the *tail* of the distribution is graphically illustrated on Figs. 3-1 and 3-2. For reliable extrapolation into values outside the range of observations it must be known.

The Gumbel Distribution

Let there be M observations of a hydrologic series such as daily flows within a year. Suppose that K of these observations can be considered independent. Let m(x) be the average number in a year of observations which exceed x. Then the "safe" probability for one daily flow is $[1 - \frac{m(x)}{K}]$. The joint probability that none of the *independent* daily flows exceed x is $[1 - \frac{m(x)}{K}]^K$ (Feller, 1957, p. 118). Since the maximum daily flow is one of these daily flows, $[1 - \frac{m(x)}{K}]^K$ is also the safe probability for the maximum daily flow (Roche, 1963, p. 27). If K is large and since:

$$\lim_{K \to \infty} [1 - \frac{m(x)}{K}]^K = e^{-m(x)} \tag{3-20}$$

then the safe probability for the extreme values, or in other words the c.d.f. is:

$$F(x) = e^{-m(x)} \tag{3-21}$$

It is convenient (Gumbel, 1958, p. 169) to define the function m(x) in the special form:

$$m(x) = e^{-a(x-x_o)} \tag{3-22}$$

and the Gumbel c.d.f. takes the form:

$$F(x) = e^{-e^{-a(x-x_o)}} \tag{3-23}$$

Given an annual series of extreme values, it remains to estimate the unknown parameters a and x_o in terms of the data.

Estimation of Parameters

The expectation operator was defined generally in Eq. (3-12). In the more special case when g(x) is a power function of x with positive integer exponent r, the expected value is also referred to as the rth *moment* of the distribution (Mood, et al., 1974, p. 72), symbolically:

$$\mu_r^o = \int_{-\infty}^{+\infty} x^r f(x)\, dx \tag{3-24}$$

or the *centered* r^{th} *moment*:

$$\mu_r = \int_{-\infty}^{\infty} (x - \mu)^r f(x)\, dx \tag{3-25}$$

One defines similarly the first sample moment as

$$\overline{X} = \frac{1}{N} \sum_{n=1}^{N} X_n \tag{3-26}$$

and the rth centered sample moments as:

$$M_r = \frac{1}{N} \sum_{n=1}^{N} (X_n - \overline{X})^r \qquad (3\text{-}27)$$

where N is the sample size. Note that the sample moments are a discrete form of the population moments presuming that all realizations $(X_n - \overline{X})$ are equiprobable. The population moments are functions of the distribution parameters. The method of moments to estimate parameters consists of equating enough population and sample moments to have the same number of equations as there are parameters of the distribution.

Since the Gumbel distribution involves only two parameters, it will suffice to equate the first two moments, i.e. the mean and variance. The calculations are straight forward but somewhat tedious. For this reason, they are relegated to an Appendix, for the very eager readers! The derived relations (Appendix 1 or Roche, 1963, p. 41) between the parameters and the sample mean, $\overline{x}$, and standard deviation, $\hat{\sigma}$, are:

$$\overline{x} = \hat{x}_o + \frac{0.577}{a} \qquad (3\text{-}28)$$

$$\hat{a} = \frac{1}{0.78\ \hat{\sigma}} \qquad (3\text{-}29)$$

Substitution of the expressions of $\hat{x}_o$ and $\hat{a}$ in terms of $\overline{x}$ and $\hat{\sigma}$ in Eq. (3-23) yields for the safe probability:

$$q = e^{-e^{-\frac{1}{0.78\ \hat{\sigma}}[x - \overline{x} + 0.577(0.78\ \hat{\sigma})]}}$$

or

$$q = e^{-e^{-\frac{1}{0.78\ \hat{\sigma}}[x - \overline{x} + 0.45\ \hat{\sigma}]}} \qquad (3\text{-}30)$$

Taking the natural logarithm of both sides one obtains:

$$\ell n\ q = -e^{-\frac{1}{0.78\ \hat{\sigma}}(x - \overline{x} + 0.45\ \hat{\sigma})}$$

Taking the natural logarithm of both sides once more, one obtains:

$$\ell n(-\ell n\ q) = -\frac{1}{0.78\ \hat{\sigma}}(x - \overline{x} + 0.45\ \hat{\sigma})$$

and solving for x one obtains:

$$x = \overline{x} + \hat{\sigma}\ [0.78\{-\ell n(-\ell n\ q)\} - 0.45] \qquad (3\text{-}31)$$

It is interesting to note that, as for the normal distribution (see Eq. (3-2)), the c.d.f. or in other words, q is an explicit function of x in Eq. (3-30). In addition with the Gumbel distribution, the inverse of the c.d.f. can be obtained explicitly, namely the function x(q) given in Eq. (3-31) whereas it cannot be done for the normal distribution. In that sense, the Gumbel distribution is a much easier distribution. In fact, there is no need to tabulate it at all!

Graphical Estimation of Floods of a Given Return Period

Defining the *reduced variate*

$$y = -\ell n(-\ell n\ q) \tag{3-32}$$

it is apparent that x(q) given by Eq. (3-31) is a *linear* function of y. This means that the c.d.f. x(q) will plot as a straight line for the special probability scale $-\ell n(-\ell nq)$. In this case, Eq. (3-31) is explicit in q or y and the laborious construction shown in Fig. 3-5 for the normal distribution is not needed. From the relations between p, q and T_r, one can express x in terms of T_r or:

$$x = \bar{x} + 0.78\ \hat{\sigma}\{-\ell n[-\ell n(1 - \frac{1}{T_r})] - 0.577\} \tag{3-33}$$

Given data of peak discharges it is possible to estimate the 50, 100, 250 years floods etc... *strictly graphically* using Gumbel extreme value paper in the same way the 100 year annual runoff for the Hinterrhein River was estimated using normal probability paper. A similar table is thus drawn for Fishkill Creek at Beacon, New York (data from U.S. Water Resources Council Bulletin 17) for a period of 9 years (Table 3-2). The seven low value points seem to line up

Table 3-2. Peak Daily Runoffs for Fishkill Creek at Beacon, New York for Period 1951-1959.

Year	Annual Peak Runoff (cfs)	Rank	Ranked Runoff (cfs)	Exceedance Probability (%)	Safe Probability (%)
1951	2490	1	8800	10	90
1952	3170	2	8280	20	80
1953	3220	3	3220	30	70
1954	1760	4	3170	40	60
1955	8800	5	2500	50	50
1956	8280	6	2490	60	40
1957	1310	7	1960	70	30
1958	2500	8	1760	80	20
1959	1960	9	1310	90	10

fairly well but not the two high value points (Fig. 3-9). If a best fit straight line is drawn through the points it should be drawn through *all* the points. The tentative conclusion from the visual inspection of the graph is that either the Gumbel distribution does not fit the data, or that the two high value points are *outliers*. Actually, 24 years of record are available for the creek. The complete set of values is shown in Table 3-3. The plotted positions of the data are shown on Fig. 3-10. It is clear again that a straight line can be fitted well to the 22 low flow points. This line will deviate considerably from one that would attempt to fit *all* the data say in a least-square sense. The next serious question is whether or not the two high flow points should be ignored. Prior to answering this question one may want to try the Gumbel distribution on the *logarithms* of the data rather than on the data themselves. The results are displayed on Fig. 3-11. There is a marked improvement in the alignment of the points and a straight line may be drawn through all the points. The extrapolation of the line gives for the 100 year logarithm of the flood the value 4.15 and consequently a 100 year flood of 14,000 cfs. It is clear that a relatively small error in the slope of the straight line would result in a relatively large error in flow. "Studies by Benson (1952), Ott (1971) and Nasseri (1976) show that the most important single factor in accurate estimates of flood probability is the length of record analyzed. Benson found that to determine the 100 year flood within ± 10 percent accuracy requires 115 years of record while 48 years of record are required for a ± 25 percent accuracy" (Hydrocomp, Simulation Network Newsletter, 1976, p. 2).

Analytical Estimation of Floods of a Given Return Period

The graphical procedure had the advantage of not requiring even the calculation of the sample mean and sample standard deviation. It does appear, however, somewhat subjective. Naturally, the best fit of the line could be obtained by least square analysis but then more work would be required than calculating the mean and the standard deviation of the data. On the other hand, the exploitation of Eq. (3-33) to determine the magnitude of a flood of a given return period T_r requires little effort beyond that of determining the mean and standard deviation.

For the logarithms of the flows the mean is 3.37 and the standard deviation is 0.25. Substitution in Eq. (3-33) yields:

$$\log(x_r) = 3.37 + 0.25(0.78)\{-\ell n[-\ell n(1 - \frac{1}{T_r})] - 0.577\}$$

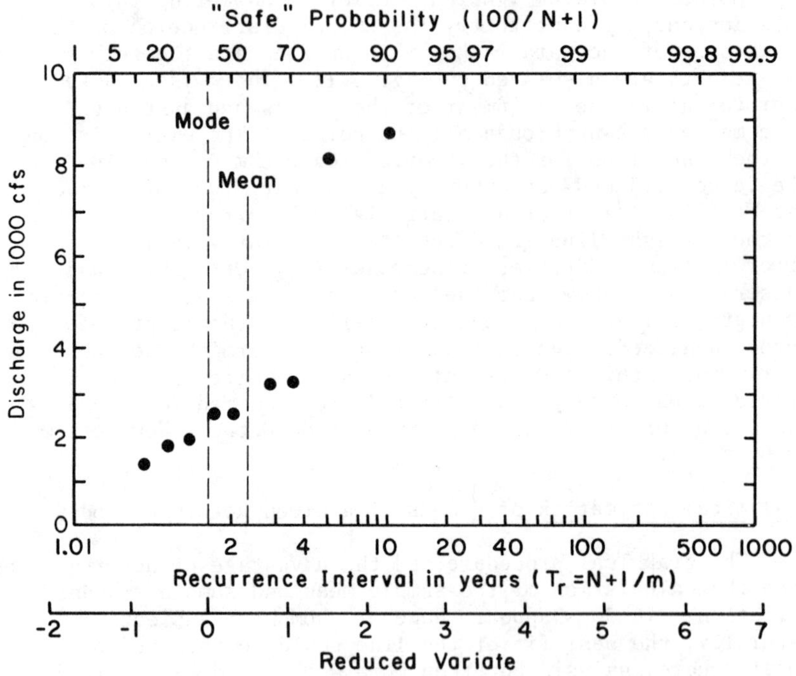

Fig. 3-9. Cumulative Frequency Curve of Annual Maximum Daily Flows for Fishkill Creek at Beacon Using a 9 Years Sample

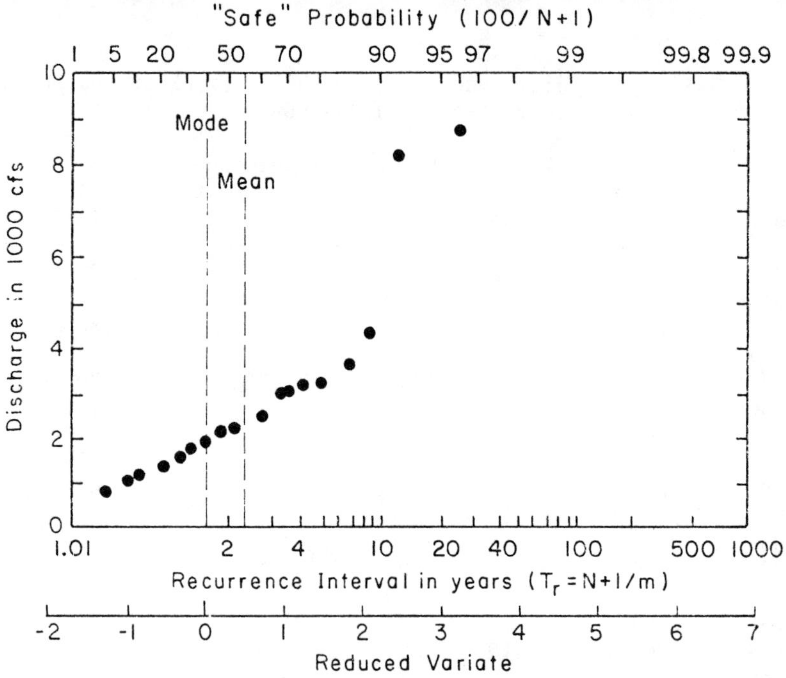

Fig. 3-10. Cumulative Frequency Curve of Peak Daily Flows for Fishkill Creek Using a 24 Years Sample

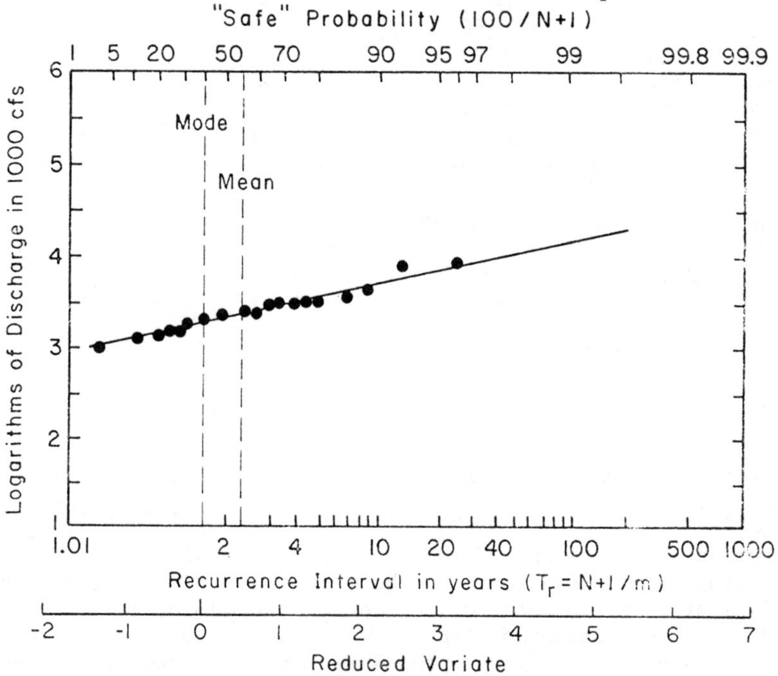

Fig. 3-11. Cumulative Frequency Curve of Logarithms of Peak Daily Flows for Fishkill Creek

Table 3-3. Peak Daily Runoffs for Fishkill Creek at Beacon, New York for Period 1945-1968

Year	Annual Peak Runoff (cfs)	Rank	Ranked Runoff (cfs)	Exceedance Probability (%)	Safe Probability (%)
1945	2290	1	8800	4	96
1946	1470	2	8280	8	92
1947	2220	3	4340	12	88
1948	2970	4	3630	16	84
1949	3020	5	3220	20	80
1950	1210	6	3170	24	76
1951	2490	7	3060	28	72
1952	3170	8	3020	32	68
1953	3220	9	2970	36	64
1954	1760	10	2500	40	60
1955	8800	11	2490	44	56
1956	8280	12	2290	48	52
1957	1310	13	2220	52	48
1958	2500	14	2140	56	44
1959	1960	15	1960	60	40
1960	2140	16	1780	64	36
1961	4340	17	1760	68	32
1962	3060	18	1580	72	28
1963	1780	19	1470	76	24
1964	1380	20	1380	80	20
1965	980	21	1310	84	16
1966	1040	22	1240	88	12
1967	1580	23	1040	92	8
1968	3630	24	980	96	4

and in particular for $T_r = 100$,

$$\log(x_{100}) = 3.37 + 0.25(0.78)\{-\ln[-\ln(0.99)] - 0.577\}$$

$$\log(x_{100}) = 3.37 + 0.25(0.78)\ (4.61 - 0.58)$$

$$\log x_{100} = 3.37 + 0.25(0.78)(4.03) = 3.37 + 0.25(3.14) = 4.16$$

This is the logarithm of x_{100} so that the value of x_{100} is $x_{100} = 14{,}300$ cfs. This value is slightly higher than the graphical estimate one of 14,000 cfs. The implication is that the graphical procedure should be used principally to check the alignment of the points to see whether the data or their logarithms or some other transformation of the data should be used. Then the analytical procedure should be used to estimate the 100 year (or 200 years, etc...) flood. Naturally, the analytical procedure eliminates one element of uncertainty due to the subjective visual fit of a straight line through the data. The estimate remains very uncertain if the sample size is small.

The Log Pearson Type III

The Gumbel distribution depends on two parameters which are estimated by the two statistics $\overline{X}$ and S (also denoted $\hat{\sigma}$). The coefficient of skewness (related to the 3rd centered moment of the distribution) has a fixed value, namely: 1.14 (Natural Environment Research Council, 1975, vol. I, p. 42; Gumbel, 1958, p. 25). Given the observations the coefficient of skewness (which is *dimensionless*) can be estimated according to the formula:

$$G = \frac{N \sum_{n=1}^{N} (x_n - \overline{x})^3}{(N-1)\ (N-2)\ S^3} \tag{3-34}$$

For the data of Fig. 3-11 the coefficient of skewness is 0.73, which is somewhat different from 1.14. It appears that a distribution that could *preserve* the first three sample moments of the observations would be superior to the Gumbel one that preserves only the first two moments.

A frequently used such distribution is the Pearson Type III distribution. Its cumulative distribution function is:

$$F(x) = q = \int_{x_o}^{x} \frac{(\xi - x_o)^{\gamma-1}\ e^{-\frac{(\xi - x_o)}{\beta}}}{\beta^{\gamma}\ \Gamma(\gamma)}\ d\xi \tag{3-35}$$

where β, γ and x_o are parameters, and $\Gamma(.)$ is the *gamma* function, namely:

$$\Gamma(t) = \int_{0}^{\infty} x^{t-1}\ e^{-x}\ dx \qquad t > 0 \tag{3-36}$$

(For estimation of floods of low exceedance probability it is clear that inaccuracies in the estimation of x_o could not have significant consequences. On the other hand, had the annual series of concern been that of extreme low flows rather than that of extreme high flows, the estimation of x_o would be critical.)

The relation between q and x is explicit (though not very convenient) but the relation x(q) can only be obtained *implicitly* for specific values of the parameters. From the form of the x(q) or $x(T_r)$ relation for the Gumbel distribution, Eq. (3-31) or (3-33) one suspects that the standardized variate $\frac{x-\bar{x}}{\hat{\sigma}}$ is only a function of T_r and of the coefficient of skewness or:

$$x_r = \bar{x} + K(T_r, G)\,\hat{\sigma} \qquad (3\text{-}37)$$

In the case of the Gumbel distribution K is only a function of T_r since G is constant equal to 1.14, namely (by identification of terms between Eq. (3-37) and Eq. (3-33)):

$$K(T_r) = 0.78\{-\ln[-\ln(1 - \frac{1}{T_r})] - 0.577\} \qquad (3\text{-}38)$$

In this case K can be evaluated very simply. For the Pearson Type III distribution $K(T_r, G)$ is not known explicitly and depends on two parameters. Fortunately, it has been tabulated extensively (U.S. Water Resources Council, 1976, Appendix 3, pp. 2-27). Shorter tables are available in many textbooks (e.g. Linsley, et al., 1975, p. 344; Hjelmfelt and Cassidy, 1975, p. 30; Schulz, 1974, p. 415). For the data displayed on Fig. 3-11 (that is for the case when the logarithms of the flows are used in the Pearson Type III procedure) then for example:

$$\log x_{100} = \overline{\log x} + \hat{\sigma}_{\log} K(100, G_{\log})$$

or

$$\log x_{100} = 3.37 + 0.25\,K(100, 0.73)$$

From the table (U.S. Water Resources Council, 1976, p. 3-3)* one finds K(100, 0.73) = 2.84, thus:

$$\log x_{100} = 3.37 + 0.25(2.84) = 4.08$$

or

$$x_{100} = 12{,}000 \text{ cfs.}$$

This value is quite a bit lower than the estimated value of 14,000 cfs obtained using the Gumbel extreme value distribution. The difference is probably due to the large deviation of G (0.73 in this case) from the Gumbel value of 1.14.

---

* In the Bulletin, K is tabulated as a function of p rather than T_r. Thus, one must look for K (0.01, 0.73) since $\frac{1}{100} = 0.01$.

3.7 CONFIDENCE LIMITS ON FLOOD MAGNITUDE OF A GIVEN RETURN PERIOD

It is apparent that the estimated value of a flood of a large return period is likely to be in error because the underlying statistical distribution was assumed erroneously of a certain type or because even if the correct distribution was assumed, its parameters are not accurately determined when the sample size is small. The purpose of this section is to place *confidence limits* on the estimated value of floods of given return periods.

Distribution of the Sample Mean When Variance is Known

It is clear from Eq. (3-37) that errors in estimation of x will result from errors in estimation of $\overline{x}$ and $\hat{\sigma}$. Can we place confidence limits on the mean $\overline{x}$ in the first place?

Let X_n be the nth observation of the random variable X in a sequence of N observations. Let us assume that all the observations are independent and come from the same distribution. For maximum annual flows these assumptions are very reasonable. However, even with these assumptions, unless the distribution from which the sample is drawn is simple, the distribution of the mean of the observations, $\overline{X} = \frac{1}{N} \sum_{n=1}^{N} X_n$, cannot be obtained analytically. Fortunately, the statistitians have shown that if X is normally distributed with population mean μ and population variance σ^2, then $\overline{X}$ is also distributed normally with mean μ and variance $\frac{\sigma^2}{N}$. Figure 3-12 displays a comparison of the probability density functions of X and of $\overline{X}$, and Fig. 3-13 displays a comparison of their c.d.f.

Confidence Limits on Mean When Variance is Known

Suppose that the population mean μ is known and that σ is known. In this case then there is no need to estimate them! Regardless, suppose one were to draw one observation from the population with known parameters. Before drawing the observation one could make a probability statement about it taking a value greater than a given value $x_o > \mu$. In this case,

$$p(\mu, \sigma, x_o) = P\{X > x_o\} \qquad (3\text{-}39)$$

Vice versa suppose σ is known, but not μ, and x_o was observed. Then one could ask for what value of μ_ℓ would x_o have had a *specified* probability α_ℓ of exceeding x_o (see Fig. 3-14). The answer is obtained by calculating the solution for μ of the equation:

$$p(\mu, \sigma, x_o) = \alpha_\ell \qquad (3\text{-}40)$$

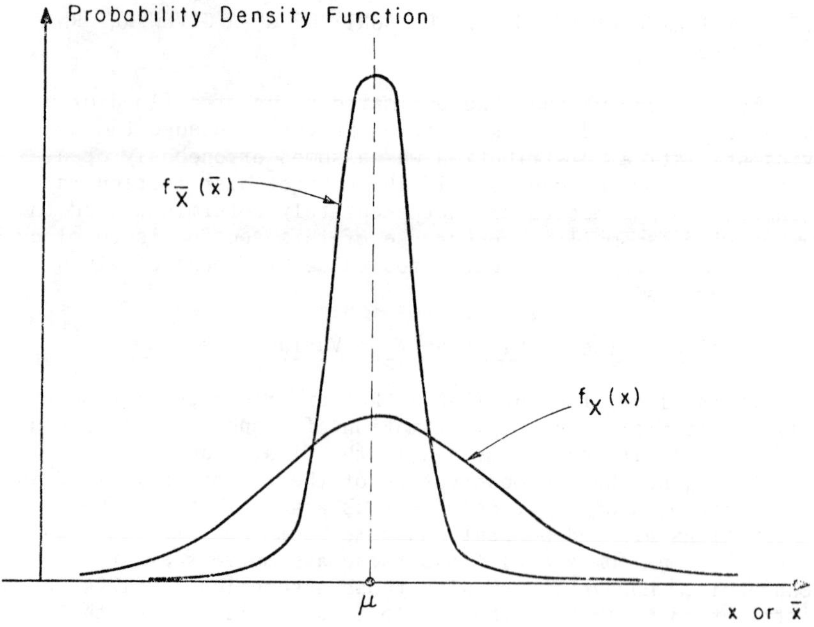

Fig. 3-12. Probability Density Functions for r.v. X and for Its Size N-Sample mean $\overline{X}$

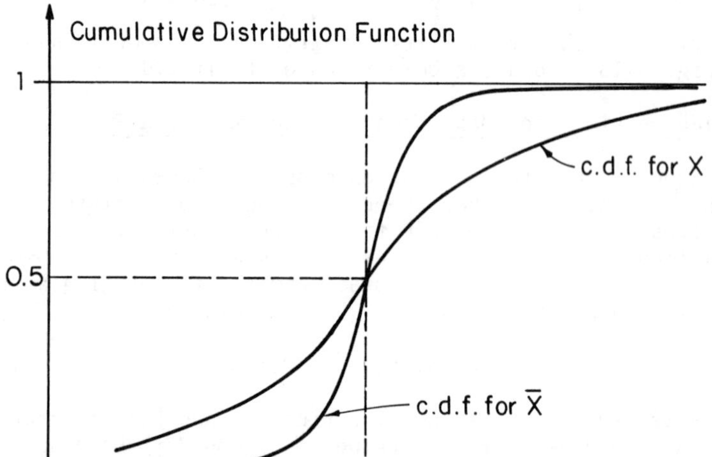

Fig. 3-13. Cumulative Distribution Function for r.v. X and for Its Size N-Sample Mean X

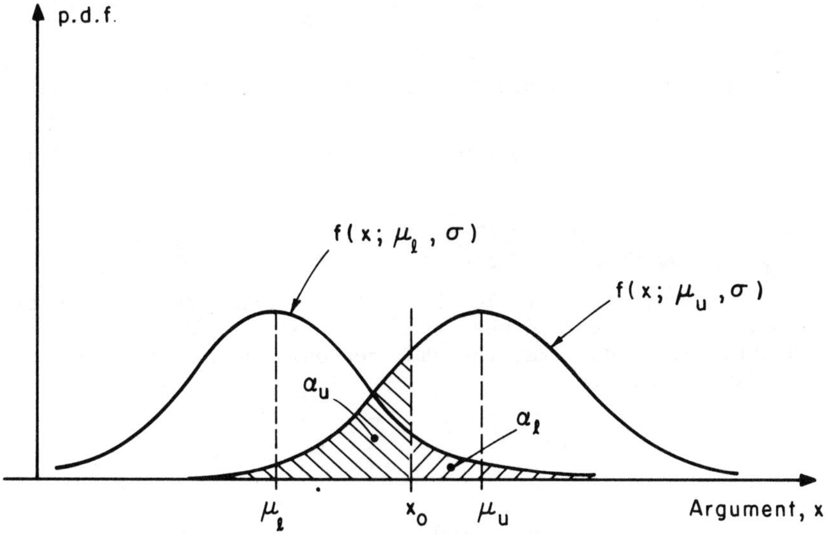

Fig. 3-14. Graphical Illustration of Significance Levels for Confidence Limits

In Eq. (3-39) p is the unknown while μ, σ are known and x_o specified. In Eq. (3-40) μ is the unknown, σ is known, x_o is observed and α is specified. The solution of Eq. (3-40) is μ_ℓ and is known as the *lower confidence limit* for μ. Similarly, one could have asked for what high value of μ say μ_u, would the observation x_o be a realization of a speficied low probability, α_u. The answer is the solution for μ of:

$$q(\mu, \sigma, x_o) = 1 - p(\mu, \sigma, x_o) = \alpha_u \qquad (3\text{-}41)$$

The solution μ_u is called the *upper confidence limit* of μ. The values α_ℓ and α_u are usually selected small (say 10, 5, 2 or 1 percent), and usually selected equal to a common value: α.

For a normal distribution Eq. (3-40) can be solved quite simply, because (see Eqs. (3-3) and (3-4))

$$P\{X \le x_o\} = 1 - \alpha_\ell = F(x_o) = \Phi\left(\frac{x_o - \mu}{\sigma}\right) \qquad (3\text{-}42)$$

From a table of $\Phi(u)$ one would find the value u for which $\Phi = 1 - \alpha_\ell$, denoted $u_{1-\alpha_\ell}$ and known as the $(1-\alpha_\ell)$ *quantile* of the U-variate. Given α_ℓ the value $u_{1-\alpha_\ell}$ is completely determined. Then having found $u_{1-\alpha_\ell}$ from the table then the value of μ_ℓ is simply:

$$\mu_\ell = x_o - \sigma\, u_{1-\alpha_\ell} \tag{3-43}$$

A similar reasoning would lead to the upper confidence value:

$$\mu_u = x_o + \sigma\, u_{1-\alpha_u} \tag{3-44}$$

Loosely speaking, the *probability* is $(1-\alpha_\ell-\alpha_u)$ that the *population value of the mean lies in the range* (μ_ℓ, μ_u).

If instead of a single observation there were N observations and $\bar{x}_o$ was the observed mean then since $\bar{X}$ has the mean μ and standard deviation $\sigma/\sqrt{N}$, the same reasoning would lead to the confidence limits:

$$\mu_\ell = \bar{x}_o - \frac{\sigma}{\sqrt{N}}\, u_{1-\alpha_\ell} \tag{3-45}$$

$$\mu_u = \bar{x}_o + \frac{\sigma}{\sqrt{N}}\, u_{1-\alpha_u} \tag{3-46}$$

Such confidence limits can be evaluated from tables of $\Phi(u)$ but also graphically. Suppose that five more years of data had been available for the Hinterrhein River (Table 3-1, Fig. 3-6) and that the mean of these five years had been 94 million m^3. The exceedance probability for the mean is 50 percent so that the plotting position for the mean of this new sample is as shown on Fig. 3-6 (x symbol). The standard deviation for this new sample is assumed to be the same as for the nineteen other years. Then the line drawn from the point (50 percent, 94 million m^3) with a slope $\frac{1}{\sqrt{5}}$ the slope of the nineteen other data points (shown on Fig. 3-6) will intersect the 10 percent exceedance probability at the 10 percent *upper* confidence limit of μ, namely at a value 103 millions m^3. The mean value for the first 19 years *is* 108 millions m^3! The implications are several: (1) a small sample is very unreliable, (2) a 10 percent *significance* level (another name given to α_u or α_ℓ) is *not* very significant, or (3) the new sample was not drawn from the same population (i.e. after 1969 the river regime changed). Naturally, confidence limits for the population mean based on the 1951-1969 years of record can be obtained graphically by drawing a line through the (50 percent, 108) point with a slope = $\frac{1}{\sqrt{19}}$ times the slope of the 19 points. The intersection of this line with the 90 percent and 10 percent exceedance probability lines give the 10 percent (*one sided*) upper and lower confidence limits for the mean, namely 111.5 and 104.5, respectively. The probability that the true mean falls in that interval is only 100-10-10 = 80 percent. When

$\alpha_\ell = \alpha_u = \alpha$ then α is the one-sided significance level. When $\alpha_\ell = \alpha_u = \alpha/2$ then α is the two-sided significance level because the probability for the true mean to be in that range is $1-\alpha$.

Distribution of the Sample Mean When Variance is Unknown

Naturally the cases when the mean is unknown but the variance is not (i.e. can be estimated with great accuracy) are rare. Fortunately, statisticians have shown (Brownlee, 1965, p. 119; Mood and Graybill, 1963, p. 252; Mood, et al., 1974, p. 381) that $\frac{\bar{X} - \mu}{\hat{\sigma}/\sqrt{N}}$ has a so-called *Student's* t *distribution with (N-1) degrees of freedom.*

Confidence Limits for the Mean When the Variance is Unknown

A reasoning similar to that of the previous sections leads to the formulae for the lower and upper confidence limits, respectively:

$$\mu_\ell = \bar{x} - \frac{\hat{\sigma}}{\sqrt{N}} t_{1-\alpha_\ell}(N-1) \qquad (3\text{-}47)$$

and

$$\mu_u = \bar{x} + \frac{\hat{\sigma}}{\sqrt{N}} t_{1-\alpha_u}(N-1) \qquad (3\text{-}48)$$

In these equations $t_{1-\alpha_\ell}(N-1)$ and $t_{1-\alpha_u}(N-1)$ are the $(1-\alpha_\ell)$ and $(1-\alpha_u)$ quantiles, respectively of the Student t-variate with N-1 degrees of freedom. For the normal distribution a probability paper exists. For the t-distribution a different special paper would be required for each value of the number of degrees of freedom. As a result it is neither used nor available commercially. The graphical construction illustrated previously, assuming that the variance was estimated adequately, cannot be used when the sample size is small (i.e. less than 30).

Confidence Limits for Floods Assuming a Gumbel Distribution

It is apparent from Eq. (3-37) and Eq. (3-38) that the r.v. X_r, magnitude of a flood of return period T_r or, equivalently, magnitude of a flood with exceedance probability $p = 1/T_r$ (also called a (1-p) percent *quantile*) is a linear combination of two random variables, $\bar{X}$ and S. For the Gumbel distribution K_r is a definite number, not a statistic. Thus, the *general* results for expectation and variance of a linear combination of random variables, namely:

$$E\{\sum_{n=1}^{N} a_n X_n\} = \sum_{n=1}^{N} a_n E(X_n) \tag{3-49}$$

$$\text{Var}\{\sum_{n=1}^{N} a_n X_n\} = \sum_{n=1}^{N} \sum_{\nu=1}^{N} a_n a_\nu \text{ cov}(X_n, X_\nu) \tag{3-50}$$

where the X are random variables and the a_n are fixed numbers, apply. It is important to note that Eqs. (3-49) and (3-50) apply *regardless* of the distributions of the X_n. Application of these equations for the variables $\overline{X}$ and S yields:

$$E\{X_r\} = E\{\overline{X}\} + K_r E\{S\}$$

$$E\{X_r\} = \mu + K_r \sigma \tag{3-51}$$

$$\text{Var}\{X_r\} = \sigma_r^2 = \text{Var}\{\overline{X}\} + K_r^2 \text{ Var}[S] + 2K_r \text{ cov}[\overline{X}, S]$$

$$\sigma_r^2 = \frac{\sigma^2}{N} + K_r^2 \text{ Var}[S] + 2K_r \text{ cov}[\overline{X}, S] \tag{3-52}$$

The variance of the sample standard deviation, S, can be expressed in terms of the moments of the distribution approximately and accurately for large N (Natural Environmental Research Council, 1975, p. 101) as:

$$\text{Var}[S] = \frac{\mu_4 - \sigma^4}{4N\sigma^2} = \frac{(\gamma_2 - 1)\sigma^2}{4N} \tag{3-53}$$

where μ_4 is the 4th centered moment of the original distribution from which the X are drawn and γ_2 is the *kurtosis*, ratio of $\frac{\mu_4}{\sigma^4}$. In the case of the Gumbel distribution, this 4th moment has the value 5.4 σ^4 (Natural Environmental Research Council, 1975, p. 103). (The value of the kurtosis for the normal distribution is 3.) Thus, for the Gumbel distribution:

$$\text{Var}[S] = \frac{4.4\,\sigma^2}{4N} = \frac{1.10\,\sigma^2}{N} \tag{3-54}$$

It remains to evaluate cov$[\overline{X}, S]$ in Eq. (3-52). An approximate but reasonably accurate formula for large N is:

$$\text{cov}[\overline{X}, S] = \frac{\mu_3}{2N\sigma} = \frac{\gamma_1\,\sigma^2}{2N} \tag{3-55}$$

where μ_3 is the 3rd centered moment and γ_1 is the coefficient of skewness. For the Gumbel distribution γ_1 is a constant equal to 1.14 and consequently:

$$\text{cov}[\overline{X}, S] = \frac{0.57\ \sigma^2}{N} \tag{3-56}$$

Substitution in Eq. (3-52) yields:

$$\text{Var}[X_r] = \frac{1}{N}[\sigma^2 + 1.10\ K_r^2\ \sigma^2 + 1.14\ K_r\ \sigma^2]$$

or

$$\text{Var}[X_r] = \frac{\sigma^2}{N}[1 + 1.14\ K_r + 1.10\ K_r^2] \tag{3-57}$$

Thus the r.v. X_r has a known expectation given by Eq. (3-51) and known variance given by Eq. (3-57). The distribution of X_r is not known but depending on $\overline{X}$ and S i.e. depending on many observations, one can expect from the Central Limit Theorem which states that $\overline{X}$ will tend to be normally distributed as N increases *regardless* of the original distribution of the X, that the *standardized* X_r variate namely:

$$\frac{X_r - \overline{X} - K_r\ S}{\frac{S}{\sqrt{N}}[1 + 1.14\ K_r + 1.10\ K_r^2]^{1/2}} = X_r^* \tag{3-58}$$

will tend to be Student's t distributed with N-1 degrees of freedom. Thus, the confidence limits for X_r will be respectively, for the upper limit:

$$x_{ru} = \overline{x} + K_r\hat{\sigma} + [1 + 1.14\ K_r + 1.10\ K_r^2]^{1/2}\ \frac{\hat{\sigma}}{\sqrt{N}}\ t_{1-\alpha_u}(N-1) \tag{3-59}$$

and for the lower limit:

$$x_{r\ell} = \overline{x} + K_r\hat{\sigma} - [1 + 1.14\ K_r + 1.10\ K_r^2]^{1/2}\ \frac{\hat{\sigma}}{\sqrt{N}}\ t_{1-\alpha_\ell}(N-1) \tag{3-60}$$

As an example let us calculate the 10 percent one-sided upper confidence limit for the 100 year (logarithm of the) flood for Fishkill Creek (see Fig. 3-11). The value of K_{100} in this case is 3.14 (see p. 3.22). The estimate of log x_{100} is 4.16. The estimate of $\hat{\sigma}_{\log}$ is 0.25. Substitution of these values in Eq. (3-59) yields:

$$(\log x)_{100,u} = 4.16 + \frac{0.25}{\sqrt{24}} [1+1.14(3.14) + 1.1(9.86)]^{1/2} t_{90\%}(23)$$

$$(\log x)_{100,u} = 4.16 + \frac{0.25}{\sqrt{24}} \sqrt{15.43}\ t_{90\%}(23)$$

$$(\log x)_{100,u} = 4.16 + 0.20\ t_{90\%}(23)$$

The value of $t_{90\%}(23)$ is obtained from a Student's t table (e.g. Brownlee, 1965, p. 561; Mood, et al., 1974, p. 556) and has the value 1.319. Finally the upper confidence value of $(\log x)_{100}$ is 4.42 and the corresponding flow is 26,600 cfs whereas the sample estimate was 14,300 cfs.

The derivations leading to this estimation of the confidence limits were *not exact*. Nevertheless, the calculation of the confidence limits are very useful even if the percentage of confidence is not truly 90 percent but 92 or 87 due to the approximations in the derivations. "The standard error of an estimated quantile" (i.e. the standard deviation of X_r) "can usually be expressed in the form $C\sigma/\sqrt{N}$ where σ is the population standard deviation, N is the sample size and C is a factor depending on the return period and on the form of the distribution. If, as is always the case in practice, an *estimated* value of σ is used in this expression, the expression may differ greatly from its true value. In such cases it may not be worthwhile to split hairs over the precise value of C..." (Natural Environmental Research Council, 1975, p. 100).

Confidence Limits for Floods Assuming Log Pearson Type III Distribution

The relation between the logarithm of a flood of given return period T_r is analogously to Eq. (3-37):

$$\log X_r = \overline{\log X} + S_{\log} K(T_r, G_{\log}) \qquad (3\text{-}61)$$

The general results of the section on Confidence Limits for Floods assuming a Gumbel distribution regarding the expected value of $\log X_r$ and its variance apply only if $G_{\log}$ can be estimated accurately. In this case $K(T_r, G_{\log})$ can be considered a constant and not a statistic. When the coefficient of skewness is zero, the log-Pearson distribution reduces to the log-normal distribution. This simpler case will be studied first.

Zero Skewness Coefficient: If the coefficient of skewness of the distribution is assumed to have a population value of zero, then $K(T_r, G_{log})$ in Eq. (3-61) is only a function of T_r. In fact in this case since log X is normally distributed, K_r is precisely the quantile value of Student's t with N-1 degrees of freedom with an exceedance probability of $p = \frac{1}{T_r}$. For example, if it was assumed that the underlying extreme value distribution for the peak flows of Fishkill Creek at Beacon (Table 3-3, Fig. 3-11) was log-normal in spite of the fact that the sample coefficient of skewness is 0.73, then the estimate of the 100 year flood would be:

$$\log x_{100} = \overline{\log x} + S_{log}\, t_{99\%}(N-1)$$

or

$$\log x_{100} = 3.37 + 0.25\, t_{99\%}(23)$$

$$\log x_{100} = 3.37 + 0.25(2.5) = 4.00$$

and consequently x_{100} = 10,000 cfs. This is the 3rd estimate for the same data of the 100 year flood! Assuming a Gumbel distribution (G = 1.14) a value of 14,000 cfs was obtained and using the log-Pearson distribution with the sample coefficient of skewness ($\hat{G}$ = 0.73) a value of 12,000 cfs was obtained. For the *same* mean and standard deviation the greater the skewness the higher the 100 year flood.

If the population value of G is assumed to be known equal to zero, and if the underlying distribution is assumed to be log-normal, then K in Eq. (3-61) is only a function of T_r and it is not a statistic. In particular

$$E\{\log X_r\} = E\{\overline{\log X}\} + K_r\, E\{S_{log}\} \qquad (3\text{-}62)$$

$$Var\{\log X_r\} = \frac{\sigma^2_{log}}{N} + K_r^2\, Var[S_{log}] + 2K_r Cov[\log \overline{X}, S_{log}] \qquad (3\text{-}63)$$

where K_r is a *constant* numerically equal to the $(1-p)^{th}$ quantile of the Student's t distribution with (N-1) degrees of freedom. The results are even simpler than for the Gumbel distribution because for a normal distribution the covariance of the sample mean and variance is zero. (It follows from Eq. (3-55) since $\gamma_1 = 0$). Since for the normal distribution the kurtosis is 3 it follows from Eq. (3-53) that:

$$Var[S_{log}] = \frac{0.5}{N}\,\sigma^2_{log} \qquad (3\text{-}64)$$

Thus Eqs. (3-62) and (3-63) reduce to:

$$E\{\log X_r\} = \mu_{\log} + t_{1-p}(N-1)\,\sigma_{\log} \tag{3-65}$$

$$Var\{\log X_r\} = \{1 + 0.5[t_{1-p}(N-1)]^2\}\,\frac{\sigma^2_{\log}}{N} \tag{3-66}$$

It was argued that the standardized X_r variate defined by Eq. (3-58) would tend to be t-distributed for large N when the underlying distribution has a Gumbel distribution. In the case of a log-normal distribution the standardized variate defined as:

$$(\log X_r)^* = \frac{\log X_r - \overline{\log X} - t_{1-p}(N-1)S_{\log}}{\frac{S_{\log}}{\sqrt{N}}\{1 + 0.5[t_{1-p}(N-1)^2\}^{1/2}} \tag{3-67}$$

is more closely t-distributed. Thus the upper confidence limit for $\log X_r$ is:

$$(\log x_r)_u = \overline{\log x} + t_{1-p}(N-1)S_{\log}$$

$$+ \{1 + 0.5[t_{1-p}(N-1)^2\}^{1/2}\,\frac{S_{\log}}{\sqrt{N}}\,t_{1-\alpha_u}(N-1) \tag{3-68}$$

or

$$(\log X_r)_u = \overline{\log x} + S_{\log}\,K_c \tag{3-69}$$

where K_c is short notation for the expression:

$$K_c = t_{1-p}(N-1) + \{1 + 0.5[t_{1-p}(N-1)]^2\}^{1/2}\,\frac{t_{1-\alpha_u}(N-1)}{\sqrt{N}} \tag{3-70}$$

As an example let us calculate the 10% one-sided upper confidence limit for the 100 year flood. Application of Eq. (3-70) yields:

$$K_c = t_{99\%}(23) + \{1 + 0.5[t_{99\%}(23)]^2\}^{1/2}\,\frac{t_{90\%}(23)}{\sqrt{24}}$$

$$K_c = 2.5 + \{1 + 0.5(2.5)^2\}^{1/2}\,\frac{1.319}{\sqrt{24}}$$

$$K_c = 2.5 + \frac{2.03 \times 1.319}{4.90} = 2.5 + 0.55 = 3.05 \tag{3-71}$$

Consequently the upper confidence limit for $\log x_{100}$ is

$$(\log x_{100})_u = 3.37 + 0.25(2.5 + 0.55)$$

$$= 4.00 + 0.25(0.55) = 4.14$$

and

$$(x_{100})_u = 13{,}725 \text{ cfs.}$$

With the assumption of a Gumbel distribution the upper confidence limit for x_{100} was 26,600 cfs.

Nonzero Skewness Coefficient: Naturally since $\hat{G} = 0.73$ neither the Gumbel distribution nor the log-normal distributions are appropriate. Unfortunately with the log-Pearson distribution there exists no simple explicit formula for the confidence limits such as given by Eq. (3-59) and Eq. (3-68) for the Gumbel and log-normal distributions, respectively. Fortunately values of K_c in Eq. (3-69) applicable to the log-Pearson distribution have been tabulated. Actually what has been tabulated is not K_c but the ratio of K_c/K where K is defined in Eq. (3-61), as a function of level of significance (one-sided), sample size and exceedance probability (U.S. Water Resources Council, 1976, Table 9-2, p. 9-7). For a 100 year flood (p = 0.01), a sample size of 24 and a 10% (one-sided) level of significance the value of K_c/K is 1.27. The value of K was previously found to be 2.84. Thus:

$$(\log x_{100})_u = \overline{\log x} + S_{\log}[K_c]$$

$$(\log x_{100})_u = 3.37 + 0.25[2.84 + 0.27(2.84)]$$

$$(\log x_{100})_u = 4.08 + 0.25(0.27)(2.84) = 4.08 + 0.19 = 4.27$$

and consequently $(x_{100})_u$ = 18,700 cfs, compared to 13,725 cfs assuming a log-normal distribution and 26,600 cfs assuming a Gumbel distribution. The upper confidence width i.e. the difference between $(\log x_{100})_u$ and the sample estimate $\log x_{100}$ is respectively for the log-normal, log-Pearson and Gumbel distribution: 0.14, 0.19, 0.26. The 0.19 value for $\hat{G} = 0.73$ could have been estimated roughly by linear interpolation between the value 0.14 and 0.26 according to the formula:

$$[W_u]_G = \frac{(1.14 - G)(W_u)_o + G(W_u)_{1.14}}{1.14} \qquad (3\text{-}72)$$

Application of this formula for G = 0.73 yields

$$(W_u)_{0.73} = \frac{(1.14 - 0.73)(0.14) + 0.73(0.26)}{1.14} = 0.22$$

which is a little high. However in view of the usual great uncertainty in the estimation of G, Eq. (3-72) can be used particularly if Tables of the K/K_c are not available. For convenience to the user U.S.W.R.C.'s Table 9-2 is reproduced here as Table 3-4.

Table 3.5 summarizes the various numerical results obtained in the previous sections by different methods for Fishkill Creek. Table 3.6 tabulates for convenience the important formulae.

3.8 RELATION BETWEEN RETURN PERIOD AND CRITICAL DESIGN PERIOD

In the previous sections various techniques of estimation of the magnitude of a flood of a given return period were discussed. Techniques for placing confidence limits on these estimates were presented. For pedagogic purposes the Gumbel and the log-normal distributions were discussed at some length though the recommended (accepted) practice is to use the log-Pearson distribution.

The central question remains. For design (safety) purposes what return period should be selected? The answer to this question is mostly economic. In theory the answer is simple. The design return period is the one that satisfies the equilibrium criterion, namely the equality of: (1) the added marginal design cost of guarding against the $(T_r + 1)$ flood rather than the T_r one and of: (2) the corresponding increase in expected future marginal reduction in damage (Massé, 1946; Morel-Seytoux, 1976a, p. 9-26).

In this section we shall continue to address the problem in probabilistic rather than economic terms. It is known from probability theory that there is a definite chance that the T_r-year event (flood) will occur before T_r years have elapsed. In fact there is a smaller but definite chance that it will occur more than once.

In many situations a given structure is particularly susceptible to damage and outright destruction during the construction phase (e.g. earth dam). If T_c designates the duration of the critical (construction) period (in years) one wants to guard against the probability of a damaging flood to occur during the period T_c. Let α be the acceptable risk, presumably determined by economic and safety considerations. This risk is the probability that a T_r year flood be exceeded once or more during the T_c period. The safe probability is q in a given year. The safe probability that the T_r flood not be exceeded in any of the T_c years is q^{T_c}. In other words the risk which is the complement of the safe probability is:

$$\alpha = 1 - q^{T_c} \tag{3-73}$$

or

Table 3-4. Ratios to Compute Standard Deviates above Sample Mean for Confidence Limits of Peaks Computed from Log-Pearson Type III Population with Skew Coefficients between ±0.50.

Level of Significance*	Sample Size N	Ratios K_c/K by which standard deviates in Appendix 3 should be multiplied for given exceedance probability .001	.005	.010	.020	.050	.100	.200	.400
.01	10	2.14	2.16	2.18	2.21	2.27	2.38	2.67	5.01
	15	1.78	1.80	1.82	1.84	1.89	1.97	2.19	3.98
	20	1.62	1.64	1.65	1.67	1.71	1.78	1.96	3.46
	25	1.52	1.54	1.55	1.56	1.60	1.66	1.82	3.14
	30	1.46	1.47	1.48	1.50	1.53	1.58	1.73	2.92
	40	1.38	1.39	1.40	1.41	1.44	1.48	1.61	2.62
	50	1.33	1.34	1.34	1.35	1.38	1.42	1.53	2.44
	70	1.26	1.27	1.28	1.29	1.31	1.34	1.43	2.20
	100	1.21	1.22	1.22	1.23	1.25	1.28	1.36	1.98
.05	10	1.68	1.70	1.71	1.73	1.77	1.84	2.02	3.55
	15	1.49	1.50	1.51	1.53	1.56	1.61	1.76	2.96
	20	1.40	1.41	1.42	1.43	1.46	1.50	1.63	2.65
	30	1.30	1.31	1.32	1.33	1.35	1.39	1.49	2.31
	40	1.25	1.26	1.26	1.27	1.29	1.32	1.41	2.12
	50	1.22	1.23	1.23	1.24	1.26	1.28	1.36	1.99
	70	1.18	1.18	1.19	1.20	1.21	1.23	1.30	1.84
	100	1.15	1.15	1.15	1.16	1.17	1.19	1.25	1.69
.10	10	1.50	1.51	1.52	1.53	1.56	1.61	1.75	2.90
	15	1.36	1.37	1.38	1.39	1.42	1.46	1.57	2.48
	20	1.30	1.31	1.31	1.32	1.34	1.38	1.47	2.26
	25	1.26	1.26	1.27	1.28	1.30	1.33	1.41	2.11
	30	1.23	1.23	1.24	1.25	1.26	1.29	1.37	2.00
	40	1.19	1.20	1.20	1.21	1.22	1.25	1.31	1.86
	50	1.17	1.17	1.18	1.18	1.19	1.22	1.28	1.76
	70	1.14	1.14	1.14	1.15	1.16	1.18	1.23	1.65
	100	1.11	1.12	1.12	1.12	1.13	1.15	1.19	1.54
.25	10	1.25	1.25	1.26	1.26	1.28	1.30	1.37	1.95
	15	1.18	1.19	1.19	1.20	1.21	1.23	1.29	1.76
	20	1.15	1.16	1.16	1.16	1.17	1.19	1.24	1.65
	25	1.13	1.14	1.14	1.14	1.15	1.17	1.21	1.57
	30	1.12	1.12	1.12	1.13	1.14	1.15	1.19	1.52
	40	1.10	1.10	1.10	1.11	1.11	1.13	1.16	1.45
	50	1.09	1.09	1.09	1.09	1.10	1.11	1.14	1.40
	70	1.07	1.07	1.08	1.08	1.08	1.09	1.12	1.34
	100	1.06	1.06	1.06	1.06	1.07	1.08	1.09	1.29

Table 3-4 (continued)

Level of Significance*	Sample Size N	Ratios K_c/K by which standard deviates in Appendix 3 should be multiplied for given exceedance probability .001	.005	.010	.020	.050	.100	.200	.400
.75	10	.87	.87	.86	.86	.84	.81	.74	.16
	15	.89	.89	.88	.88	.86	.84	.79	.31
	20	.90	.90	.90	.89	.88	.86	.81	.41
	25	.91	.91	.91	.90	.89	.87	.83	.47
	30	.92	.92	.91	.91	.90	.88	.84	.51
	40	.93	.93	.92	.92	.91	.90	.86	.58
	50	.93	.93	.93	.93	.92	.91	.88	.62
	70	.94	.94	.94	.94	.93	.92	.89	.68
	100	.95	.95	.95	.95	.94	.93	.91	.73
.90	10	.76	.75	.74	.72	.70	.65	.51	- .64
	15	.80	.79	.79	.78	.76	.72	.62	- .26
	20	.82	.81	.80	.79	.77	.74	.64	- .14
	25	.83	.83	.82	.81	.79	.76	.68	- .02
	30	.85	.84	.83	.83	.81	.78	.70	.07
	40	.86	.86	.85	.85	.83	.81	.74	.20
	50	.88	.87	.87	.86	.85	.83	.76	.28
	70	.89	.89	.89	.88	.87	.85	.80	.40
	100	.91	.91	.90	.90	.89	.87	.83	.50
.95	10	.70	.68	.67	.66	.62	.56	.38	-1.14
	15	.74	.73	.72	.71	.68	.63	.48	- .71
	20	.77	.76	.75	.74	.71	.67	.55	- .47
	25	.79	.78	.77	.76	.74	.70	.59	- .31
	30	.81	.80	.79	.78	.76	.72	.62	- .19
	40	.83	.82	.81	.81	.79	.76	.67	- .03
	50	.84	.84	.83	.82	.81	.78	.70	.07
	70	.86	.86	.86	.85	.84	.81	.75	.22
	100	.89	.88	.88	.87	.86	.84	.79	.35
.99	10	.60	.58	.56	.54	.49	.40	.13	-2.19
	15	.65	.64	.63	.61	.57	.49	.28	-1.49
	20	.69	.68	.67	.65	.61	.55	.37	-1.12
	25	.72	.70	.69	.68	.65	.59	.43	- .88
	30	.74	.72	.72	.70	.67	.62	.48	- .70
	40	.77	.76	.75	.74	.71	.66	.54	- .47
	50	.79	.78	.77	.76	.74	.70	.58	- .27
	70	.82	.81	.80	.79	.77	.74	.65	- .10
	100	.84	.84	.83	.82	.81	.78	.70	.08

*Chance of true value being greater than given by Eq. (3.69)

(after USWRC Table 9-2, Bulletin 17, 1976)

Table 3.5. Summary of Estimates of the 100 Year Flood and Its Upper Confidence Limit for Fishkill Creek at Beacon.

Quantity of Interest	Assumed Distribution		
	log-normal	log Pearson	Gumbel
$\overline{\log x}$	3.37	3.37	3.37
$S_{\log}$	0.25	0.25	0.25
skewness	0	0.73	1.14
$\log x_{100}$	4.00	4.08	4.16
x_{100} (cfs)	10,000	12,000	14,300
K_{100}	2.50	2.84	3.14
90% upper confidence limit of $\log x_{100}$	4.14	4.27* (4.30)	4.42
corresponding flow (cfs)	13,725	18,700 (19,950)*	26,600
K_c	3.05	3.61	4.20

* Obtained by linear interpolation (see Eq. 3.72).

Table 3.6. Summary of Basic Formulae for Estimation of Floods and Upper Confidence Limits

	Eq. No. in text
$(\log x)_r = \overline{\log x} + S_{\log} K(T_r, G_{\log})$	(3.61)
$(\log x)_{ru} = \overline{\log x} + S_{\log} K_c(T_r, G_{\log})$	(3.69)

Gumbel Distribution

$$K_r = 0.78\{-ln[-ln(1 - \frac{1}{T_r})] - 0.577\} \qquad (3.38)$$

$$K_c = K_r + [1 + 1.14\, K_r + 1.10\, K_r^2]^{1/2} \frac{t_{1-\alpha_u}(N-1)}{\sqrt{N}} \qquad (3.59)$$

Log-Normal Distribution

$$p = \frac{1}{T_r}$$

$$K_r = t_{1-p}(N-1) \qquad (3.16)$$

$$K_c = K_r + [1 + 0.5\, K_r^2]^{1/2} \frac{t_{1-\alpha_u}(N-1)}{\sqrt{N}} \qquad (3.70)$$

Log-Pearson Type III Distribution

Tables of K_r (see U.S. Water Resources Council, 1976; or Linsley, et al., 1975; or Hjelmfelt and Cassidy, 1975; or Schulz, 1974).

Table of $\frac{K_c}{K_r}$ (see Table 3.4).

$$\alpha = 1 - (1 - \frac{1}{T_r})^{T_c} \qquad (3\text{-}74)$$

Equation (3-74) provides an estimate of the risk associated with an *a priori* selection of a return period and corresponding flood. Equation (3-74) can be *turned around* to define T_r in term of α, namely:

$$T_r = \frac{1}{1 - (1-\alpha)^{1/T_c}} \qquad (3\text{-}75)$$

For a given *a priori* selected *acceptable* risk α then during the construction period provisions must be made to guard against the T_r flood claculated from Eq. (3-75). Suppose that the construction period is to extend over 4 years, and that an acceptable risk for destruction is 4%, then the return period is:

$$T_r = \frac{1}{1 - (6.94)^{1/4}} = \frac{1}{1 - 0.98} = \frac{1}{0.02} = 50 \text{ years}$$

Had the acceptable risk been 1% then the value of T_r would have been:

$$T_r = \frac{1}{1 - (0.99)^{1/4}} = \frac{1}{1 - 0.9975} = \frac{1}{0.0025} = 400 \text{ years}$$

Associated with each return period is a flood magnitude which must be handled safely by proper design during the construction procedure. Naturally the same analysis can be performed for the design life of the project T_L after construction. In the construction phase a design flood is used to size the coffer dam and diversion channel. In the operation phase a different design flood is used to size the permanent spillway. The same equation (3-75) applies but the risk value α will be different and in one case T_c is used and in the other T_L is used.

3.9 MISCELLANEOUS

The emphasis in this chapter has been on the probabilistic and statistical basis of flood frequency analysis. In addition there are other considerations of a hydrologic nature which must be included in the analysis. For example how does one handle an *outlier*? An outlier is a data point which appears to depart significantly from the trend of the other data (e.g. 2 high flow values on Fig. 3-10). However these 2 points no longer seem to deviate from the general trend on a logarithmic scale (Fig. 3-11). In this case it is clear that a data point may appear to be an outlier because of the wrong choice of a distribution by the analyst. "*All* procedures for treating outliers *ultimately* require *judgment* involving both mathematical and hydrologic considerations. The recommended use of a *generalized* skew coefficient tends to reduce the adverse effect of outliers" (U.S. Water Resources Council, 1976, p. 16).

The estimate of the coefficient of skewness is very *sensitive* to the extreme values and very *unreliable* for *small* samples. If an average value of this coefficient for the region is available from previous studies of flows in the surrounding areas it is better to weigh the sample station value with the regional average value. The weights depend on the sample size at the station of interest and on the *accuracy-equivalent* sample size for the regional mean value (McCain and Jarrett, 1976).

Prior to the systematic data collection at a station, information about past flood is available. How is this information to be included in the analysis? The standard flood frequency analysis requires adjustment (U.S. Water Resources Council, 1976, Appendix 6). The U.S.W.R.C. 1976 Bulletin No. 17 provides a good checklist of factors to be considered (e.g. broken record, incomplete record, zero flood years, mixed populations, etc.).

In some situations at the site of interest no gauge has ever existed and flood information must be transferred to the site from upstream or downstream stations or from stations on a different basin (McCain and Jarrett, 1976, p. 3-5). The watershed conditions may have been drastically altered by man's activities (e.g. urban development) and the past data, systematic or historical may contain little useful information. In this situation one tends to rely on rainfall-runoff models which incorporate the effect of man's activities on the response of the system (Dawdy, Lichty, and Bergmann, 1972). The inputs to the model are natural variables (precipitation, temperature etc...) which are generally and practically not affected by human activities. Flood routing techniques can also be used. These various techniques are discussed in the next chapter.

3.10 CONCLUSIONS

The subject of flood frequency analysis is vast. It is hoped that this chapter has provided a succinct introduction to an understanding of the statistical theories underlying flood frequency analysis. For actual practice it is recommended that the reader would do well to consult the monumental work of the Natural Environmental Research Council (1975) and naturally the USWRC Bulletin 17.

3.11 ACKNOWLEDGEMENTS

A Department Science Development grant from the National Science Foundation for the period 1970-1973 considerably helped the author sharpen up his statistical skills. Much insight was gained from teachers such as Dr. Graybill, Dr. Boes, Dr. Bryson, Dr. Srivastava, and Dr. Siddiqui, all from the *outstanding* staff of Colorado State University, Department of Statistics. Thanks are extended to Dr. Yevjevich who was instrumental in the development of the NSF grant and to other colleagues from the Hydrology and Water Resources Program; Dr. Salas, Dr. Sanders, Dr. Todorovic now at the Institut Polytechnique of Montreal, and visiting scientists such as Dr. Pegram of the University of Natal, South Africa, for many useful discussions.

Much of the material developed in this chapter is based on lectures given by the author in the course Engineering Hydrology during the Spring 1977 semester and lecture notes faithfully taken (and improved) by several student scriptwriters. I want to express my appreciation to these students for their very effective help: Mr. Bradford Price (Corps of Engineers), Jorge Restrepo (Colombia), S. Shanmugan (India), and Ricardo A. Smith (Venezuela).

Finally I must thank Dr. Shen, Director of the 1977 River Mechanics Institute for his *trusting* invitation to participate as a lecturer and an author. Collaboration with Dr. Shen has always proved a pleasurable and worthy endeavor.

3.12 REFERENCES

Beard, L. R., 1962. Statistical methods in hydrology. Civil Works Investigations Project CW-151, U.S. Army Engineer District, Corps of Engineers, Sacramento, California, January 1962, 62 p.

Benson, M. A., 1952. Characteristics of frequency curves based on a theoretical 1000 years record. Geological Survey, Water-Supply Paper 1543-A, Government Printing Office, Washington, D.C., pp. 51-74.

Brownlee, K. A., 1965. Statistical theory and methodology in science and engineering. John Wiley and Sons, New York, 590 p.

Dawdy, D. R., Lichty, R. W., and Bergmann, J. M., 1972. A rainfall-runoff simulation model for estimation of flood peaks for small drainage basins. U.S. Geological Survey Professional Paper 506-B, U.S. Government Printing Office, Washington, D.C., 28 p.

Feller, W., 1957. An introduction to probability theory and its applications. Volume I, John Wiley and Sons, Inc., 461 p.

Graybill, F. A., 1961. An introduction to linear statistical models. McGraw Hill Book Co., Inc., New York, 463 p.

Grozier, R. U., McCain, J. F., Lang, L. F., and Merriman, D.C., 1976. The Big Thompson River flood of July 31-August 1, 1976. Report, Colorado Water Conservation Board, Colorado Department of Natural Resources, 1313 Sherman Street, Denver, Colorado 80203, October 1976, 78 p.

Gumbel, F. J., 1958. Statistics of extremes. Columbia University Press, 375 p.

Hershfield, D. M., 1961. Rainfall frequency atlas of the United States for durations from 30 minutes to 24 hours and return periods from 1 to 100 years. Technical Paper No. 40, U.S. Department of Commerce, Weather Bureau, Washington, D.C., 71 p., May 1961.

Hjelmfelt, A. T., and Cassidy, J. J., 1975. Hydrology for engineers and planners. Iowa State University Press, Ames, Iowa, 210 p.

Hydrocomp, 1976. Flood damage mitigation. Simulation Network Newsletter, Vol. 8, No. 6, October 1976.

Laufer, F., 1977. Weekly control of alpine seasonal reservoir. Engineering Research Center, HYDROWAR Program, Colorado State University, Fort Collins, Colorado 80523, CER76-77FL49, March 1977, 311 p.

Linsley, R. K., Kohler, M. A., and Paulhus, J. L. H., 1975. Hydrology for engineers. 2nd edition, McGraw-Hill Book Co., Inc., 482 p.

Massé, P., 1946. Les réserves et la régulation de l'avenir dans la vie économique: Vol. 1, avenir déterminé. Hermann and Co., Paris, 148 p.

McCain, J. F., and Jarrett, R. D., 1976. Manual for estimating flood characteristics of natural-flow streams in Colorado. Colorado Water Conservation Board Technical Manual No. 1, 1845 Sherman Street, Denver, Colorado 80203, 68 p.

Mood, A. M., and Graybill, F. A., 1963. Introduction to the theory of statistics. McGraw-Hill Book Co., Inc., New York, 443 p.

Mood, A. M., Graybill, F. A., and Boes, D. C., 1974. Introduction to the theory of statistics. 3rd edition, McGraw-Hill Book Co., Inc., 564 p.

Morel-Seytoux, H. J., 1976a. Optimization and uncertainty. Chapter 9 in Stochastic Approaches to Water Resources, Vol. 1, H. W. Shen, editor, P.O. Box 606, Fort Collins, Colorado 80521.

Morel-Seytoux, H. J., 1976b. Water resources planning. Chapter 10 in Stochastic Approaches to Water Resources, Vol. 1, H. W. Shen, editor, P.O. Box 606, Fort Collins, Colorado 80521.

Morel-Seytoux, H. J., and Saheli, F., 1972. A system approach to minimal time detection of changes in watersheds. American Water Resources Association, National Symposium on Watersheds in Transition, Colorado State University, Fort Collins, Colorado.

Nasseri, I., 1976. Regional flow frequency analysis using multi-station stochastic and deterministic models. Technical Report No. 210, Department of Civil Engineering, Stanford University, Stanford, California.

Natural Environment Research Council, 1975. Flood studies report, vol. I, hydrological studies. Whitefriars Press Ltd., London, 550 p.

Old Testament, 1952. Revised standard version. Thomas Nelson and Sons, 997 p.

Ott, R. F., 1971. Streamflow frequency using stochastically generated hourly rainfall. Technical Report No. 151, Department of Civil Engineering, Stanford University, Stanford, California.

Roche, M., 1963. Hydrologie de surface. Gauthier-Villars, Paris, 430 p.

Schulz, E. F., 1974. Problems in applied hydrology. Water Resources Publications, P.O. Box 303, Fort Collins, Colorado, 80521, 1973, revised 1974, 501 pp.

U.S. Water Resources Council, 1976. Guidelines for determining flood flow frequency. Bulletin No. 17 of the Hydrology Committee, 2120 L. Street, N.W., Washington, D.C. 20037, March 1976.

Webster, 1958. Webster's new international dictionary of the english language. G. and C. Merriam Co., Publishers, Springfield, Massachusetts.

Yevjevich, V., 1972. Probability and statistics in hydrology. Water Resources Publications, Fort Collins, Colorado, 302 p.

3.13 APPENDIX

The Gumbel c.d.f. is:

$$F(x) = e^{-e^{-a(x-x_o)}} \tag{1}$$

The p.d.f. is

$$f(x) = \frac{dF}{dx} = a\ e^{-a(x-x_o)}\ e^{-e^{-a(x-x_o)}} \tag{2}$$

The mean, μ, is by definition:

$$\mu = \int_{-\infty}^{+\infty} xf(x)dx \tag{3}$$

Let

$$y = e^{-a(x-x_o)} \tag{4}$$

Then $dy = -a\ e^{-a(x-x_o)}\ dx = -aydx$. Substitution in Eq. (3) yields:

$$\mu = \int_{\infty}^{0} [aye^{-y}]\ \frac{dy}{-ay} = \int_{0}^{\infty} xe^{-y}dy \tag{5}$$

From Eq. (4) one obtains x in terms of y, namely:

$$x = x_o - \frac{1}{a}\ \ell ny \tag{6}$$

Substitution in Eq. (5) yields:

$$\mu = \int_0^\infty [x_o - \frac{1}{a} \ell n y] e^{-y}\, dy = x_o(-e^{-y})\Big|_0^\infty - \frac{1}{a}\int_0^\infty \ell n y e^{-y} dy$$

$$\mu = x_o + \frac{1}{a}\{\int_0^\infty - \ell n y\ e^{-y}\ dy\}$$

The definite integral on the right-hand side is known as Euler's constant (usually denoted γ) and has the value 0.577. Thus the relation between the mean μ and the 2 parameters x_o and a is:

$$\mu = x_o + \frac{\gamma}{a} = x_o + \frac{0.577}{a} \tag{7}$$

With the same change of variable defined by Eq. (4) the expression for the variance has the form:

$$\sigma^2 = \int_0^\infty [x_o - \frac{1}{a} \ell n y - \mu]^2 e^{-y} dy$$

$$\sigma^2 = (x_o - \mu)^2 + 2(x_o - \mu)(\frac{1}{a})0.577 + \frac{1}{a^2}\int_0^\infty [\ell n y]^2 e^{-y} dy$$

$$\sigma^2 = (\frac{0.577}{a})^2 - 2(\frac{0.577}{a})^2 + \frac{1}{a^2}\int_0^\infty [\ell n y]^2 e^{-y} dy$$

$$\sigma^2 = \frac{1}{a^2}\int_0^\infty (\ell n y)^2 e^{-y} dy - (\frac{0.577}{a})^2$$

The definite integral on the right-hand side is available from mathematical tables and turns out to be $\frac{\pi^2}{6} + \gamma^2$. Thus:

$$\sigma^2 = \frac{1}{a^2}[\frac{\pi^2}{6} + \gamma^2 - \gamma^2] = \frac{\pi^2}{6a^2} = \frac{1.644}{a^2}$$

or

$$a^2 = \frac{\pi^2}{6\sigma^2} = \frac{1.644}{\sigma^2}$$

$$a = \frac{1.28}{\sigma} = \frac{1}{0.78\sigma} \tag{8}$$

Chapter 4

FLOW FORECASTING BASED ON PRE-SEASON CONDITIONS

by

H. J. Morel-Seytoux, Professor of Civil Engineering,
Colorado State University, Fort Collins, Colorado
80523

Chapter 4

FLOW FORECASTING BASED ON PRE-SEASON CONDITIONS

4.1 INTRODUCTION

As indicated in Chapter 3 there are two distinct types of forecasting: (1) for the short run and (2) for the long run. For the long run, flow frequency analysis is the appropriate technique. It is not important on the long run, that is, when planning, to know exactly *when* a drought will occur. It is vital to know the *frequency* of its occurrence and the *degree* of *severity* of the occurrence.

On the short term one is interested on the contrary to forecast what is going to happen on a *specific* date. Every bit of available information is utilized to make the forecast as accurate as possible. For flow frequency analysis the recipe calls for probabilistic and statistical ingredients, a cup of empiricism and a touch of engineering judgment. For short-term forecasting the essential ingredients are, increasingly, deterministic mathematical models with various amounts of empiricism. Determinism is a philosophy that presumes that the future is entirely determined from current existing conditions. The problems with this approach are: (1) the appropriate selection of the important initial conditions among an infinitude of factors and (2) the knowledge of the cause-effect relations of these factors to describe the evolution of the phenomena. Depending on the objective of the forecast it may be worthless to try to predict too far into the future as a growing coalition of random errors tend to conceal the deterministic signal behind a heavy smoke screen of noise.

Since there exists an agency of the U.S. government whose mission is to make forecasts of river conditions, it is useful to discuss *first* the practical operational aspects of river forecasting and the types of information and services available from the various *River Forecast Centers*. *Second* the techniques of river forecasting are presented. Many of these techniques were developed or have been adopted by the NWS River Forecast System.

4.2 NATIONAL WEATHER SYSTEM RIVER FORECASTING SYSTEM

> "The primary purpose of the U.S. National Weather Service (NWS) hydrology program is to provide *accurate** and *timely* information to the general public. While flood forecasts and warnings are the most widely known hydrology products, NWS river forecasts are *also* used for water supply, navigation, irrigation, power, reservoir operation, recreation, and water quality interests." (Curtis and Smith, 1976, p. 1)

---

*italics are writer's, not quoted authors'

At the core of the river forecast network is the River Forecast Center (RFC). There are 12 such centers with an areal coverage of services of about 97 percent of the United States. They prepare forecasts and warnings which are transmitted to Weather Service Forecast Offices (WSFO) for dissemination to some 2500 communities. In addition, in the Western U.S., five centers prepare estimates of annual or seasonal runoff. Two centers in the Northeast prepare estimates both of seasonal snowmelt volumes and of monthly flows. Figure 4-1 shows the geographical distribution and the functions of the centers in the United States. Figure 4-2 indicates the agencies and organisms served by the network and the travel of the information through the system.

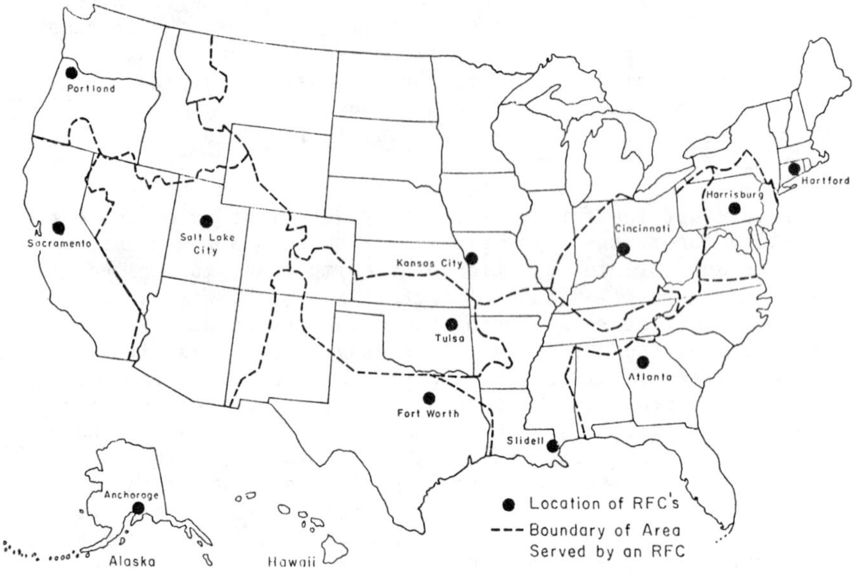

Fig. 4-1. Geographical distribution of NWS river forecast centers (after Curtis and Smith, 1976)

Some of these forecast centers have been in operation for many years. A center was created in Cincinnati, Ohio, in 1870 under the U.S. Signal Corps (Relyea, 1975, p. 43). This center is nowadays responsible for an area of approximately 163,000 square miles. This center, the *Ohio River Forecast Center*, receives some 750 precipitation reports and about 250 stage reports, as frequently as every 6 hours during a storm or a rise in the river. A brief history of the center, a list of services it provides, and a list of needs it serves are contained in a very instructive article for persons not familiar with the operations of such centers (Relyea, 1975, pp. 43-48).

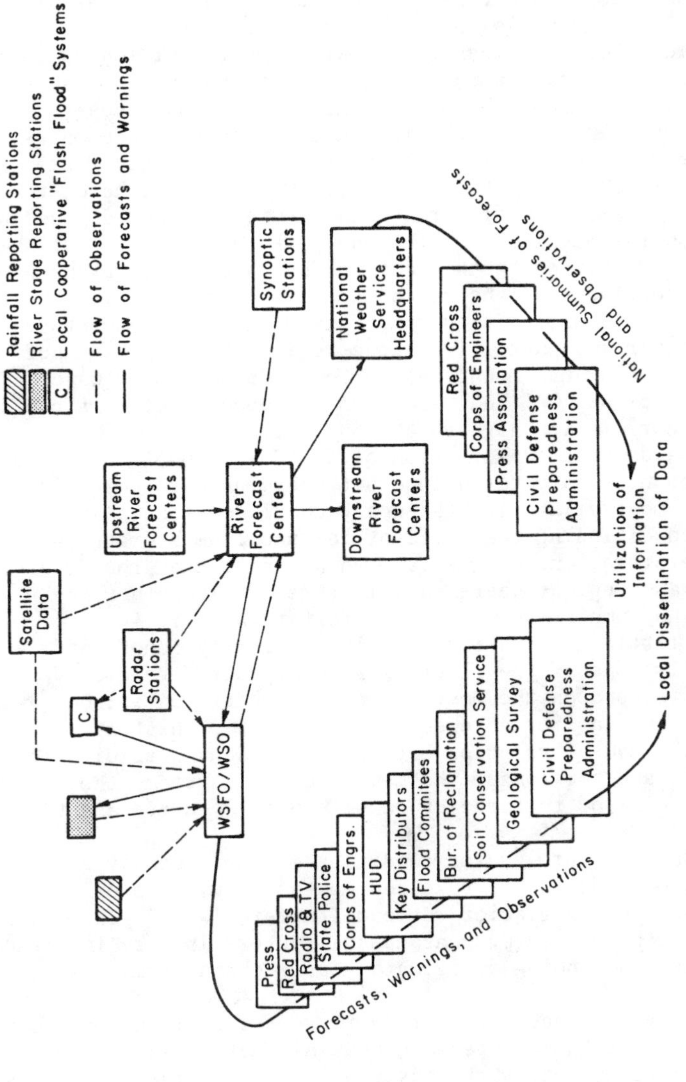

Fig. 4-2. Travel of information within the NWS RFC network (after Curtis and Smith, 1976)

4.3 TYPES OF PRE-SEASON CONDITIONS

All runoff initiates with an anterior precipitation. The information content of the precipitation event depends both upon the nature of the precipitation and its time of occurrence and upon the nature of the runoff forecast to be made and of the time lag between the forecast and the observed precipitation. The water content of the accumulated snowpack as of April 1 in the mountains of Colorado contains a lot of information about the seasonal snowmelt volume of runoff for that year. In fact, on the basis of that information (snowpack as of April 1) and of the March runoff, one can forecast seasonal runoff (April-August) with quite reasonable accuracy by a simple regression equation (correlation coefficient of the order of 90 percent on the average). On the other hand, the forecast of daily runoff during the snowmelt season depends not so much on the magnitude of the snowpack water content at the end of the winter as on the weather (hot and clear versus overcast and cool) during the preceding days. For example, the date of peak snowmelt daily runoff is almost independent of the end of winter snowpack. In fact a regression of peak runoff date in the San Juan mountains of Colorado with snow water content as of April 1 and March runoff, never yielded a coefficient of correlation higher than 60 percent.

For a detailed description of the daily runoff hydrograph, it will be necessary to include not only factors indicative of water volume availability (snowpack condition) but also of factors conditioning the rates of release from storage (such as air temperature, incoming solar radiation at the ground, etc.).

In warm regions where snow is rare or altogether inexistent, the cumulative winter precipitation up to April 1 contains little information relative to spring runoff and practically none if the river is intermittent. Volume of flow will depend on the most recent precipitation episode and little on the earlier antecedent precipitation history. The clearer the causal relation between antecedent precipitation and runoff and the longer the memory of the system, the easier it will be to forecast runoff in an accurate and timely fashion.

In short, a pre-season condition may be a condition which existed from a few hours to a few months or even several decades during a period of glacier retreat (Laufer, 1977). The selection of the proper antecedent causal factors is very important. It will vary depending on climate, basin characteristics, river regime, etc.

In many instances the forecasting of flow is not performed on the basis of known antecedent causal factors such as ground solar radiation, but on the basis of surrogate factors such as air temperature, a much easier variable to observe and measure. Similarly a rise in stage (or a rate of rise) at an upstream river gauging station does not cause a later downstream flood, but it is an indicator of the propagation of a flood wave.

More often than not the pre-season conditions are not described in terms of causal factors but rather in terms of *indicators*. The substitution of indicators for causal factors inevitably leads to loss of accuracy in forecasting, particularly under changing watershed conditions, and requires the expense of a substantial amount of work to calibrate the forecast model.

4.4 ELEMENTS OF RIVER FORECASTING

River forecasting has been traditionally and very practically in the past mostly concerned with the rise of a storm hydrograph and particularly with its crest. The prediction of low flows, however, may become increasingly important as the water demand increases continually while the water availability remains stationary on the average but displays pronounced variations in time. Low flow predictions are important also for navigation. The Ohio River Forecast Center performs a 30-day low flow forecast used to give a weekly outlook for shipment planning on the Mississippi River (Relyea, 1975, p. 46).

According to Richards and Strahl (1969, p. 1), there are three major steps to forecasting the rise of a river hydrograph caused by a rainstorm:

"1. Estimating the *volume* of water that will run directly off the land surface into the stream. This step utilizes the rainfall runoff relation.

"2. Forecasting the *distribution* of this volume of water with *time* as it passes a forecast point. The unitgraph is usually used for this purpose.

"3. Forecasting the *change* in shape of the floodwave as it *moves downstream*. This is known as streamflow routing."

Several processes join forces to prevent a fraction of rainfall volume to appear as runoff volume in the stream. These processes are interception, depression, and infiltration. The vegetal cover of a watershed intercepts some of the initial rainfall but reaches rapidly its limit, estimated in densely forested areas at a maximum of 0.1 inch. The intercepted water ultimately evaporates and is lost to runoff. It is well established that the (net) interception depth is a function of the total precipitation and also of its intensity, as shown on Figure 4-3. Table 4-1 provides equations of interception as a function of precipitation for various species (Blake, 1975, pp. 63-68).

For large storms where runoff forecasts are most important, interception is essentially negligible. Similarly the depression storage capacity of the small land depressions (not the lakes) constitutes a small loss factor for large storms. In final analysis, infiltration, the process of passage of water through the soil surface, is responsible for the greater portion of the loss to runoff (naturally a corresponding gain for soil moisture). The infiltration capacity of the basin is particularly large early in the storm, declining rather regularly with time. For low intensity rainfall the totality of the rainfall infiltrates and would infiltrate indefinitely unless the soil voids completely filled with water. Only then will runoff occur. On the other hand, with intense rainfall rates the infiltration capacity of the soil is quickly reached and runoff occurs even though the soil below the drenched surface may be very dry.

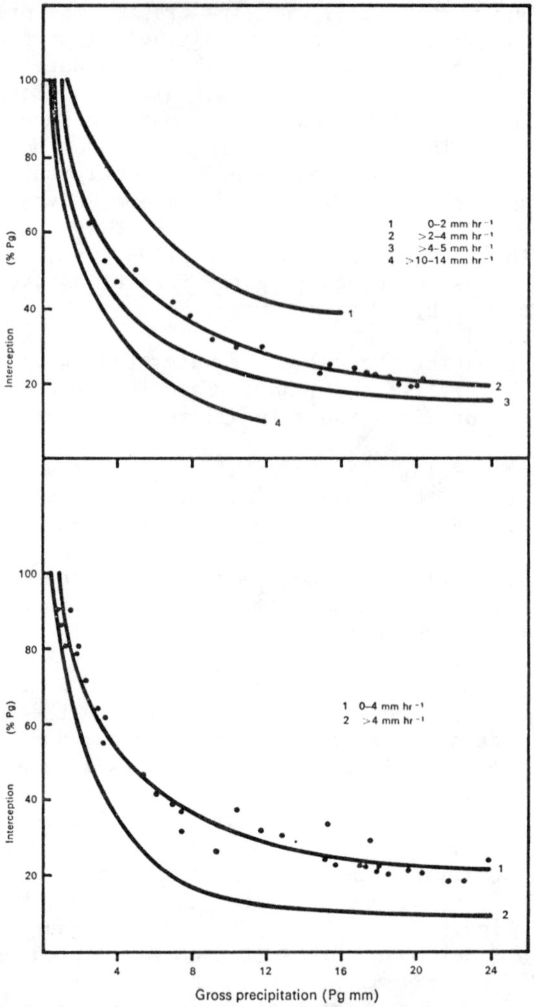

Fig. 4-3. Relationship of interception and gross precipitation for *Pinus Radiata* forest and *Neopanax arboreum* dominant indigenous scrub, New Zealand (after G.J. Blake, 1975, p. 63)

Table 4-1. Linear regression equations comparing interception with gross precipitation for selected New Zealand vegetation communities.

Common Name (see Fig. 2)	Dominant Botanical Name	Site	Reference	Equation	Sample Size	Coefficient of Correlation
Kauri forest	*Agathis australis*	Trounson Park		$V = 0.43\ P_g + 1.0$	10	0.877
Manuka scrub	*Leptospermum scoparium*	Puketurua		$V = 0.18\ P_g + 1.19$	39	0.877
Manuka scrub	*Leptospermum scoparium*	Taita	Aldridge and Jackson 1968	$V = 0.26\ P_g - 1.52$	---	---
Pine 1948 forest	*Pinus radiata*	Whakarewarewa		$V = 0.20\ P_g + 0.94$	183	0.801
Pine 1968 forest	*Pinus radiata*	Whakarewarewa		$V = 0.23\ P_g + 0.05$	25	0.893
Pine	*Pinus radiata*	Silverstream	Fahey 1964	$V = 0.31\ P_g + 1.25$	---	---
Five finger scrub	*Neopanax arboreum*	Otutira		$V = 0.14\ P_g + 0.66$	36	0.857
Gorse scrub	*Ulex europeaus*	Moutere		$V = 0.33\ P_g + 2.57$	19	0.971
Gorse scrub	*Ulex europeaus*	Taita	Aldridge 1968	$V = 0.64\ P_g + 0.95$	39	---
Kamahi scrub	*Weinmannia racemosa*	Taita	Jackson 1973	$V = 0.20\ P_g + 0.81$	141	0.900
Hard beech forest	*Nothofagus truncata*	Taita	Aldridge and Jackson 1973	$V = 0.39\ P_g + 0.0$	11	---
Mountain beech forest	*Nothofagus solandri var. cliffortiodes*	Camp Stream	Rowe 1975	$V = 0.29\ P_g + 1.90$	88	---

(after G.J. Blake, 1975)

All these processes of interception, depression, and infiltration affect the runoff and must be included in the development of a relation between rainfall and runoff. There are several methods to develop such a relation. A most popular one, and still much used today, is the multivariable coaxial graphical correlation described by Kohler and Linsley in 1951 (Linsley, et al., 1975, p. 268).

Coaxial Graphical Method for the Rainfall-Runoff Relation

> "Due to the many physical processes which affect runoff and the complexities of even a small natural basin, any sort of a direct physical or analytical approach to the problem of forecasting runoff is not practical. As a result, the usual solution is to analyze storms covering a wide range of conditions for the drainage area above a point for which forecasts are desired. Rainfall and runoff for these storms are evaluated and procedures developed to correlate them" (Richards and Strahl, 1969, p. 6).

The selection of the proper variables to use in the establishment of the correlation is crucial. Experience has shown that best results for predicting runoff from precipitation resulted from the use of an index which takes into account antecedent precipitation, season of the year, and storm duration. The antecedent precipitation index (API), index to soil moisture deficiency, is defined by the expression

$$I = \sum_{\nu=1}^{n} b_\nu P_\nu \tag{4-1}$$

where b_ν is a constant and P_ν is the basin precipitation which occurred ν days before the storm of interest. It is convenient to assume a decreasing geometric progression for the API from a day to the next during rainless periods or symbolically:

$$I_{\nu+1} = k\, I_\nu \tag{4-2}$$

In other words the index of a given day is the index of the previous day discounted by the factor k, of value in the range 0.85 to 0.95, with 0.90 the most common value. Figure 4-4 illustrates the procedure for the calculation of the API.

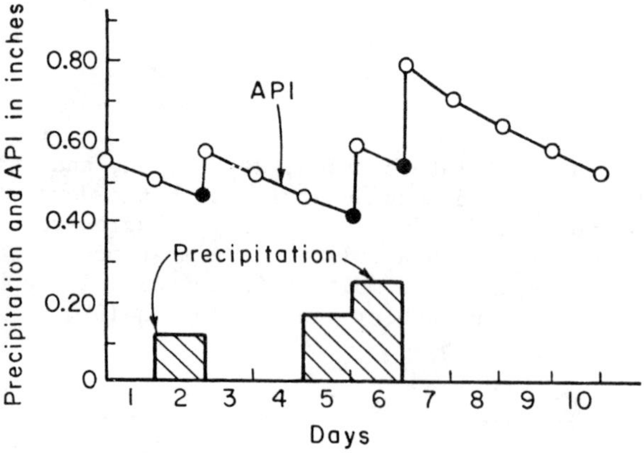

Date	1	2	3	4	5	6	7	8	9	10
0.9 of Yesterday's API	0.50	0.45	0.51	0.46	0.41	0.52	0.69	0.62	0.56	0.50
Precipitation in Past 24 hrs	0	0.12	0	0	0.17	0.25	0	0	0	0
API for Today	0.50	0.57	0.51	0.46	0.58	0.77	0.69	0.62	0.56	0.50

Fig. 4-4. Definition and calculation of antecedent precipitation index (API) (after Richards and Strahl, 1969, p. 45)

The depletion factor k is a measure of the rate of drying of the basin. One would expect that it will vary with the seasons. To incorporate this and other seasonal effects on the rainfall-runoff relation, the time of year is introduced as a third variable. Figure 4-5 displays a typical multivariable coaxial relation where duration of storm, an index of the intensity of the storm, is a fourth variable. The use of the chart is illustrated by the arrows on Figure 4-5. For an API of 2.6 inches, in early May, for a storm lasting a few hours, and a cumulative depth of precipitation of 1 inch, the runoff expressed as a depth of water is 0.3 inches.

The Weather Bureau Technical Memorandum HYDRO 9 (Richards and Strahl, 1969) gives detailed instructions on how to analyze the data for the preparation of a coaxial relation (pp. 9-14) and on the stepwise graphical development of the relation itself (pp. 14-16). Though the relation is initially developed from an understanding of the physical processes, the physical processes are not developed in details and the physical variables are replaced by surrogate indices. The

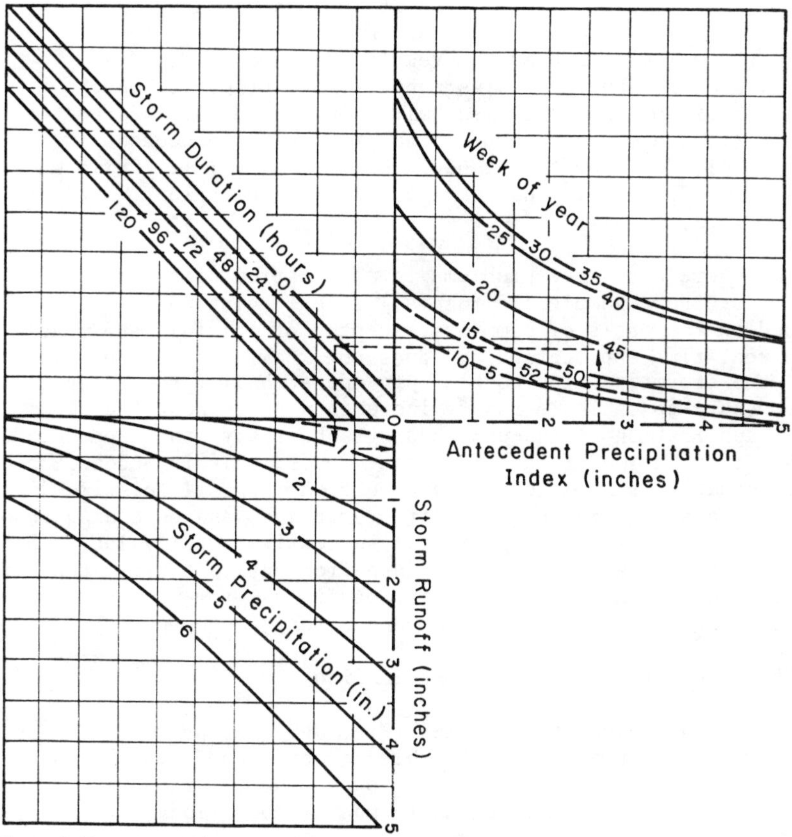

Fig. 4-5. A typical coaxial rainfall-runoff relation (after Richards and Strahl, 1969, p. 47)

relation is "quasi-statistical" (Relyea, 1975, p. 58) since it uses the method of statistics without its jargon or some of its measures of errors such as standard error of estimate, standard error of prediction, etc.

There exist many variations in the selection of variables in the coaxial relation and in the definition of indices. For example in the Ohio River Forecasting Center procedures, two parameters are used to define antecedent conditions:

> "1. Antecedent Precipitation Index (API) to simulate the drying out of the soil. A .90 recession coefficient" (depletion or drying rate factor) "is applied to a running total of rainfall and snowmelt. If no rain, this index will go near zero and remain there, *showing* that the ground is *dry*, but *not how dry*. To show degrees of drought we continue by deducting a small value after the API is below about .05. This decrement has been found by trial and error to be roughly 1/4 lake evaporation. Negative values of the API may result and are taken into account by extension of the season curves.
>
> "2. The season of the year as defined by the week number. Further adjustment is made primarily on the basis of the 2-day mean air temperature .... Parameters (1) and (2) are combined to produce an Antecedent Index (AI) for operational use. Hence RO = f(AI, PCPN), where RO is runoff and PCPN liquid precipitation." (Relyea, 1975, p. 51).

There are a few problems in application of the coaxial method. For example the rainfall-runoff relation is developed on the basis of cumulative depths of rainfall and runoff. In real-time application one does not know what the duration of the storm is going to be. "In operational forecasting it is necessary to estimate runoff increments throughout the storm for application of the unit hydrograph or any other approach to the distribution of runoff. *Mechanically* this is accomplished by computing runoff for the accumulated rainfall for time increments required by the unit hydrograph, and obtaining runoff increments by the subtraction of successive total values. Since the relation was developed on data from entire storms, there is some question as to the validity of its use for forecasts of incremental runoff." (Richards and Strahl, 1969, p. 16).

Alternate Methods of Runoff Estimation

The graphical correlation (Fig. 4-5) has proven to be an "...extremely useful and reliable forecast tool" (Richards and Strahl, 1969, p. 18). Other approaches do exist, however, which may have advantages under special conditions.

The most noteworthy method is the infiltration approach which seems to be the rational approach to the problem. As most other losses are small during heavy rainstorms, the determination of excess rainfall is essentially that of evaluation of the infiltration rates. Hydrologists for a long time have shied away from trying to understand in depth the process of infiltration. Practically all well known textbooks on hydrology have been written by surface hydrologists, few of whom have had extensive personal research experience in the hydrology of the zone of reaeration. Some of these texts are sometime even misleading. Most *infiltration capacity* formulae in the literature deal with the case of *immediate ponding*, i.e. the situation when the water availability at the surface is so great that the soil at the surface saturates instantaneously and remains saturated indefinitely. The definition of the infiltration capacity, credited to Horton (Musgrave and Holtan, 1964, p. 12.11; Richards and Strahl, 1969, p. 19) is "the maximum rate at which water can enter the soil at a particular point under a given set of conditions." (Linsley, et al., 1975, p. 260). The concept is correct but with few exceptions the infiltration capacity formulae, including that of Horton, namely:

$$I_c = I_a + (I_o - I_a)e^{-kt} \tag{4-3}$$

where I_c is the infiltration *capacity*, I_o is the initial (*original*) infiltration capacity, I_a is the time *asymptotic* limit infiltration capacity, k is a *constant* and t is *time*, are *not valid* for *any* given set of conditions. They are usually valid for the limited case of immediate and continued ponding at the soil surface. They will be valid for storms such as shown on Figure 4-6 where the rain is very intense at the very

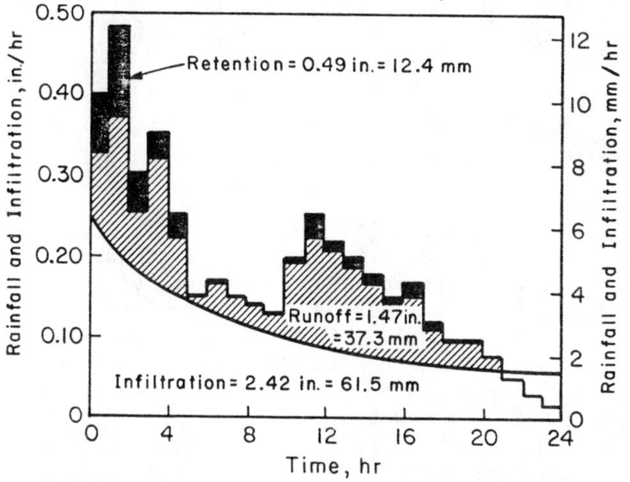

Fig. 4-6. Infiltration curve for intense rain (after Linsley, et al., 1975, p. 261)

beginning of the storm and remains continually above the *special* infiltration capacity curve corresponding to immediate and permanent soil surface saturation. The infiltration component of the hydrological cycle seems to suffer more than other components from a misuse and ignorance of what is actually well known and understood. "I have been depressed to find people cheerfully putting into a grand hydrological model a time-dependence of infiltration capacity relevant to ponded-water and not to the rainstorms they had in mind." (Philip, 1975, p. 25).

The infiltration approach must be able to handle rainfall patterns such as indicated on Figure 4.6. It is rather clear that the usual infiltration capacity curves for the case of immediate and permanent ponding at the soil surface are of little help for this situation. The important problem is to estimate the time(s) of ponding! Most hydrology textbooks are silent on this point and for some good reason. Much work was expanded either on the complex numerical simulation of the soil moisture evolution or on analytical solutions for simplified and stereotyped boundary conditions. The first approach is costly and ultimately leads to little insight beyond what was already known from physical principles. In short it is *silent*. The second approach speaks but is of limited applicability. Fortunately, recently the unsaturated flow specialists have finally turned their attention to the development of simple (approximate) analytical expressions under natural and time-varying boundary conditions. For example, two formulae have appeared recently to predict ponding time under conditions of constant rainfall intensity, r, or equivalently the cumulative infiltration up to ponding time, W_p, expressed as a water depth, namely:

$$rt_p = W_p = (\tilde{\theta}-\theta_i)\, H_c \left(\frac{\tilde{K}}{r\overline{\beta} - \tilde{K}}\right) \qquad (4\text{-}4)$$

$$rt_p = W_p = (\tilde{\theta}-\theta_i)\, H_c \left[e^{\frac{\tilde{K}}{r\overline{\beta} - \tilde{K}}} - 1 \right] \qquad (4\text{-}5)$$

where H_c is a measure of the capillary pull of the soil, $\tilde{K}$ is the hydraulic conductivity of the soil at natural saturation (i.e. saturation except for trapped air content), r is the rainfall rate, $\overline{\beta}$ is an average viscous resistance correction factor in the range 1.0-1.4, $\tilde{\theta}$ is the maximum water content at natural saturation, and θ_i is the initial water content. Eq. (4-4) is a variation on a formula obtained by Mein and Larson (1973, p. 386) and Eq. (4-5) was derived by Morel-Seytoux (1975a, p. 230). Extensions have been developed for variable rainfall (James and Larson, 1974; Morel-Seytoux, 1976; Morel-Seytoux, et al., 1976; Morel-Seytoux, 1977) which are

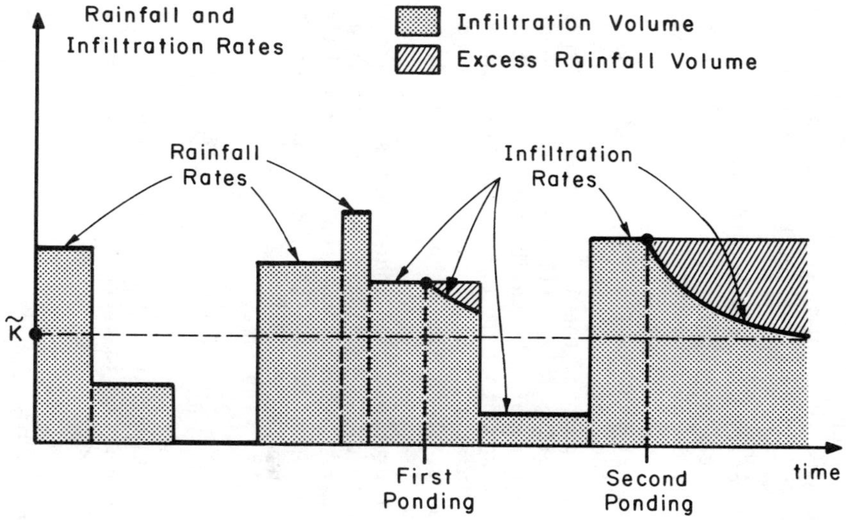

Fig. 4-7. Infiltration curve for low or moderate intensity rain.

quite simple. The good agreement with experimental observations (Figures 4-8, 4-9) is very encouraging in view of the simplicity of the methods and of the approximations in the derivations of the equations. None of these methods have been tested on field data but it is evident that because of their overall correct physical behavior and, if anything else, because of the higher number of degrees of freedom (H_c, $\tilde{K}$, $\overline{\beta}$) for calibration, these methods are bound to be superior, for example, to the so-called Φ-index infiltration method (Hjelmfelt and Cassidy, 1975, p. 87; Linsley, et al., 1975, p. 273; Gray, 1970, p. 7.21; Réméniéras, 1970, p. 346; Musgrave and Holtan, 1964, p. 12-29; Wilson, 1971, p. 44) which has only one degree of freedom. A flow chart of calculations of the ponding time for variable rainfall is illustrated on Figure 4-10. The use of this algorithm is discussed in a recent journal article (Morel-Seytoux, 1978).

The major difficulty with the infiltration approach lies with the estimation of *interflow*. With the coaxial correlation method a runoff value is obtained, 0.3 inches in the example of Figure 4-5, for a rainfall of 1 inch. The 0.3 inches is the total direct runoff that is the sum of overland runoff and *interflow*. In fact the infiltration may have exceeded 1 - 0.3 = 0.7 in. but part of the infiltrated water sprang again as overland runoff. If interflow is a significant portion of the direct runoff the calibrated infiltration curve will not correspond to *true local* infiltration. It will underestimate it. In short infiltration models must be coupled with interflow models. Such simple coupled models do not exist currently.

The antecedent precipitation index is a surrogate variable to describe the state of soil moisture at the time of a new rain. An obvious alternative is to keep a material balance of average soil water content in the upper part of the soil as a function of time, namely:

$$\frac{d\overline{\theta}}{dt} = I - Q_r - E \tag{4-6}$$

where I is infiltration rate, Q_r is deep percolation rate or aquifer recharge and E is evapotranspiration rate. It is easier said than done!

The Unit Hydrograph Concept

The coaxial correlation method provides the *total* (cumulative) depth of watershed runoff for the duration of the storm at a given site (catchment outlet). It provides per se no information regarding the distribution of the runoff past the outlet. If one presumes that a 72 hour *perfectly steady* storm of 6 inches behaves like a 12-hour storm of 1 inch, then like a 24-hour storm of 2 inches, etc., then one can read the chart of Fig. 4-5 to find a 12-hour time distribution of *local*

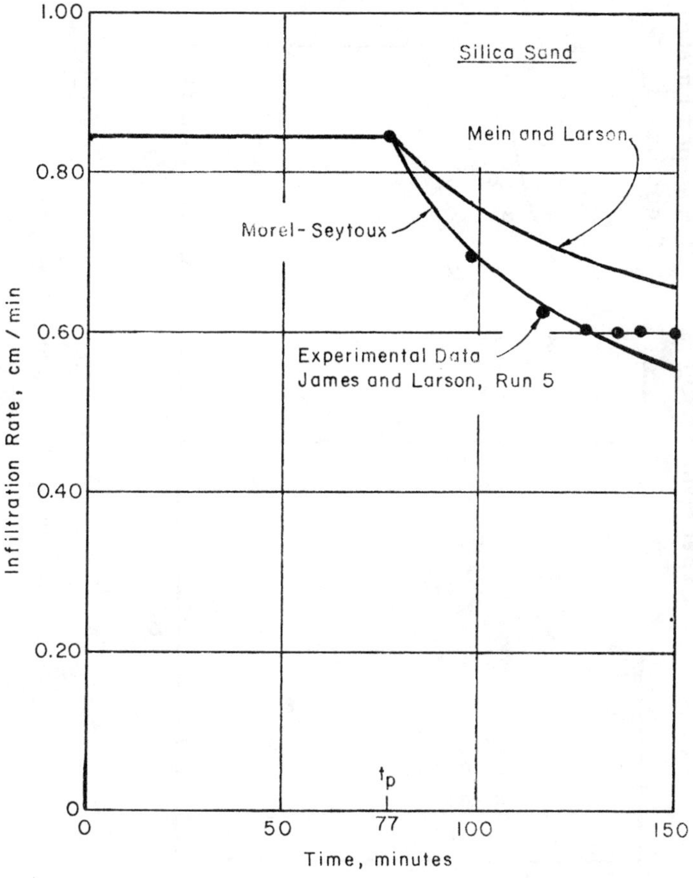

Fig. 4-8. Comparison of observed and predicted infiltration curve for constant rainfall.

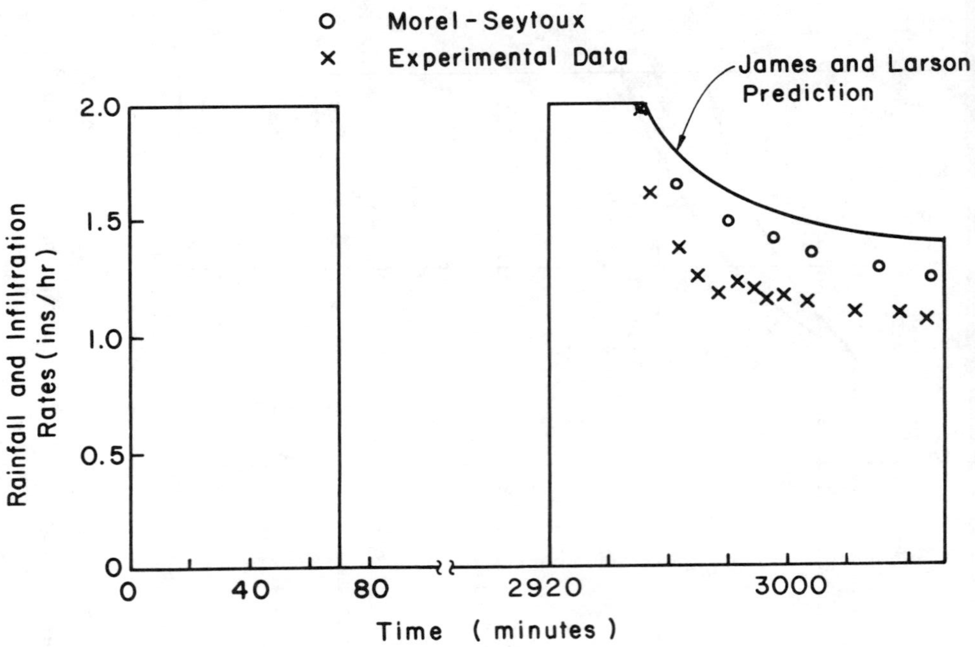

Fig. 4-9. Comparison of observed and predicted infiltration for intermittent rainfall.

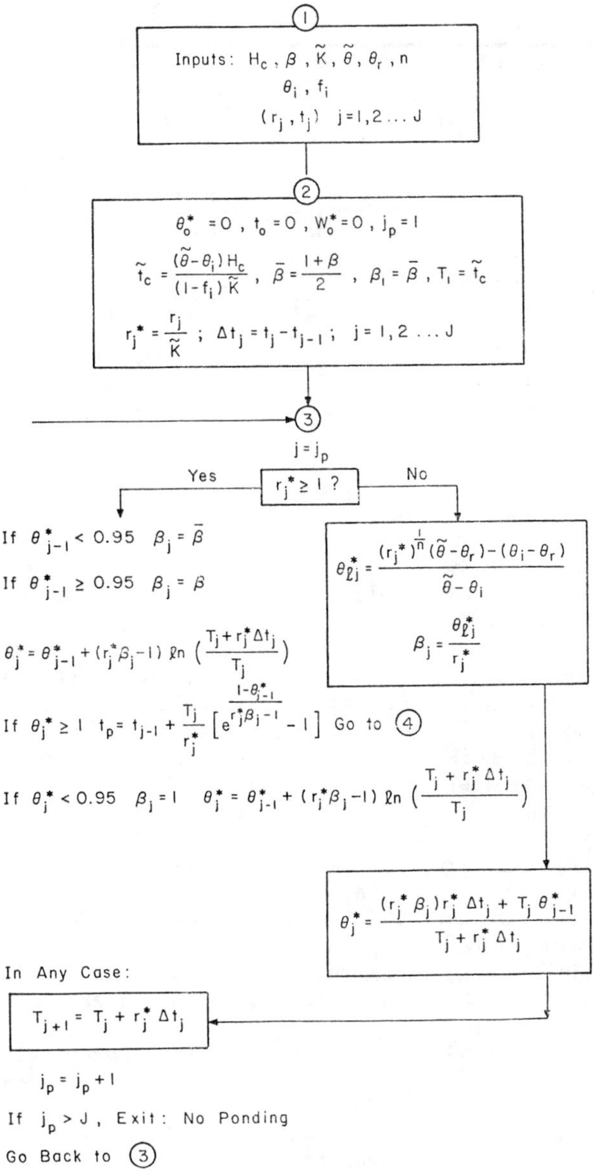

Fig. 4-10. Steps in the calculation of ponding time for variable rainfall pattern.

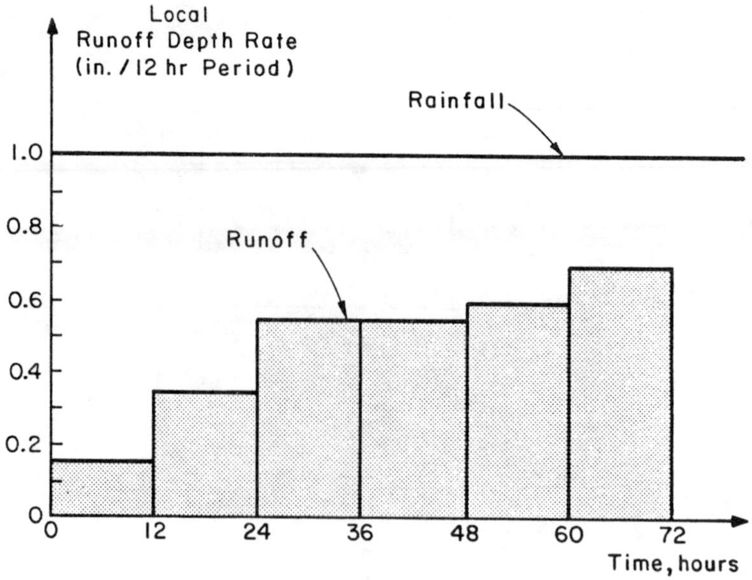

Fig. 4-11. Unit step hydrograph derived from coaxial rainfall-runoff relation.

runoff. Assuming an API of 1 in. and week 20 of the year, the time distribution of runoff for a 72-hour storm of 6 inches total precipitation is as shown on Fig. 4-11. If the storm continued at a rate of 1 in./12 hours indefinitely the runoff per 12 hours would tend to a steady limit somewhat less than but close to 1 in./hour. Fig. 4-11 provides a time distribution of the *local* runoff (the runoff generated at the point where rain falls) which is truly better named *excess* rainfall. It is not the time distribution of the *catchment* runoff, i.e. the runoff past the catchment outlet (site, gauge, etc.). For example, Fig. 4-11 states that 0.15 in. of local runoff was generated by the 1 in. rain during the first 12 hour duration. Possibly the entire 0.15 in. may pass the outlet point within these 12 hours or the 0.15 in. may take several days to pass the catchment outlet. Depending on the position of a point in the watershed, where local runoff is generated, relative to the outlet, it will take more or less time for water to travel to the catchment outlet.

Fig. 4-12 illustrates the nature of the problem. Given a time pattern of excess rainfall falling *uniformly* over the watershed (the *excitation* of the system) what is the time distribution of the catchment outlet runoff (the *response* of the system)? If one assumes that the system (catchment) responds linearly to the excess rainfall pattern, then the direct runoff response $q(t)$ is related to the excess rainfall rate pattern $r(\tau)$ by the relation:

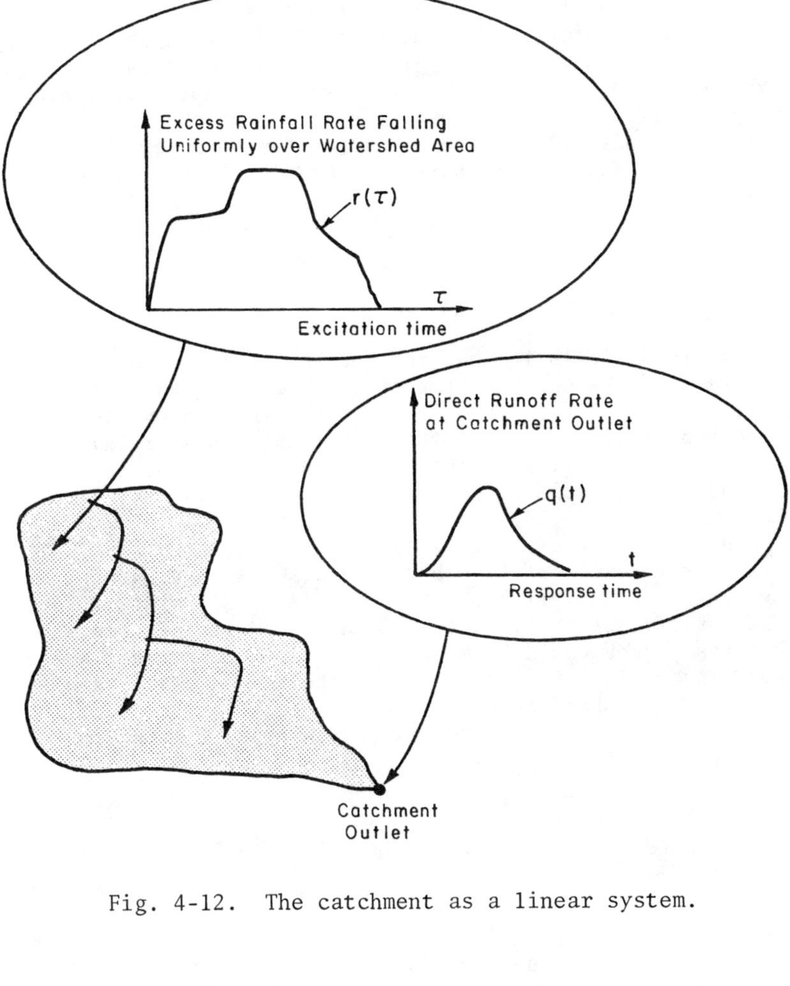

Fig. 4-12. The catchment as a linear system.

$$q(t) = \int_0^t k(t,\tau)\; r(\tau)\; d\tau \tag{4-7}$$

In this equation the function $k(t,\tau)$ the *kernel* function translates the *integrated* effect (in a layman as well as a mathematical sense) of the excitation $r(\tau)$ occurring from the beginning of the rainfall $(\tau=0)$ to response time t, into the response at time t. If in addition one assumes that the part of the response at time t due to the excitation at time τ, does not depend on the dates t and τ but only on their difference (the lag time) then the system is not only linear but in addition *time-invariant*. In this case the kernel function depends only on one argument $(t-\tau)$, rather than on two arguments t and τ. In this case Eq. (4-7) reduces to the simpler form:

$$q(t) = \int_0^t k(t-\tau)\; r(\tau)\; d\tau \tag{4-8}$$

What Eq. (4-8) tells is that the catchment response is *entirely characterized* by one function of one argument, the catchment kernel $k(\cdot)$. If this function is known (by whatever means) then direct runoff can be forecast for *any* excess rainfall pattern using Eq. (4-8). In the more usual terminology of hydrologists, $k(\cdot)$ is known as the *instantaneous unit hydrograph*.

In practice the rainfall pattern is not known as a continuous function of time but as a discrete hyetograph (Fig. 4-13). Thus $r(\nu)$ is the *mean* (excess) rainfall rate between the discrete times $\nu-1$ and ν. Units of time are immaterial; they could be minutes, 10 minute interval, hour, 3 hour interval, day, week, etc. Since the rainfall rate is constant in the time intervals (0,1), (1,2), $(\nu-1,\nu)$ ... (n-1,n), etc., the direct runoff response at time n can be expressed using Eq. (4-8) as:

$$q(n) = \int_0^1 k(n-\tau)\; r(1)d\tau + \int_1^2 k(n-\tau)\; r(2)d\tau$$

$$+ \ldots + \int_{\nu-1}^{\nu} k(n-\tau)\; r(\nu)d\tau + \ldots + \int_{n-1}^{n} k(n-\tau)\; r(\tau)d\tau \tag{4-9}$$

or more concisely as:

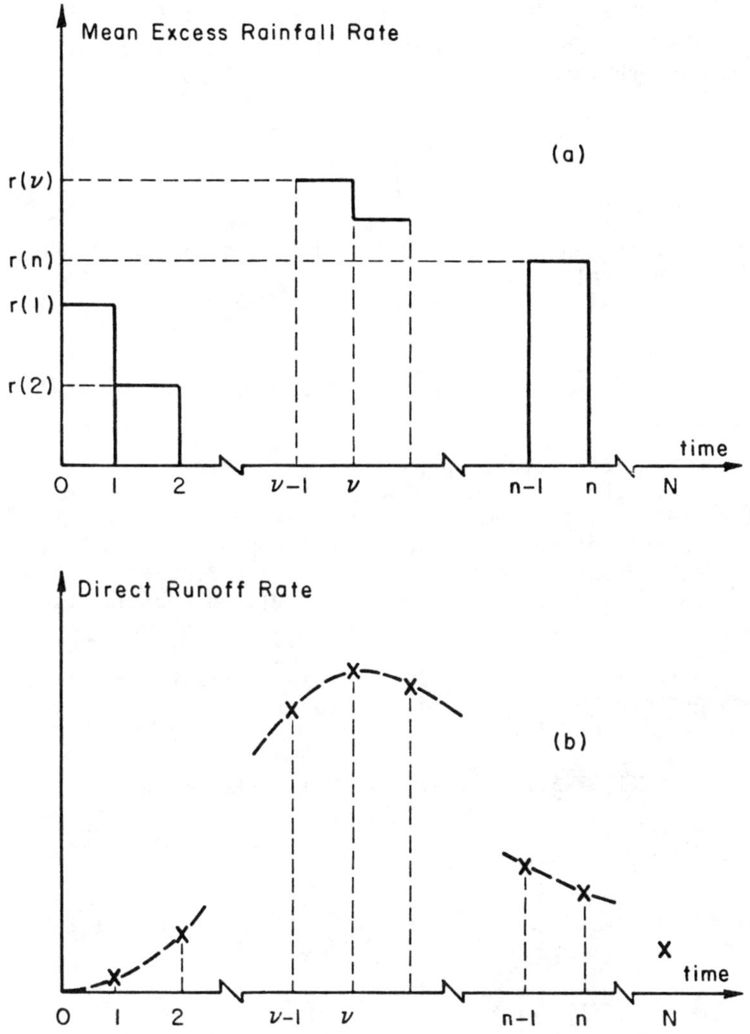

Fig. 4-13. Hyetograph and continuous runoff.

$$q(n) = \sum_{\nu=1}^{n} r(\nu) \int_{\nu-1}^{\nu} k(n-\tau)d\tau \qquad (4\text{-}10)$$

The integral in Eq. (4-10) can be put in a more standard form by defining the *discrete kernel*:

$$\delta(m) = \int_{0}^{1} k(m-\tau)\, d\tau \qquad (4\text{-}11)$$

A change of variable of integration in Eq. (4-10) will show that the integral in Eq. (4-10) is the discrete kernel of argument $(n-\nu+1)$ so that ultimately Eq. (4-10) takes the form:

$$q(n) = \sum_{\nu=1}^{n} \delta(n-\nu+1)\ r(\nu) \qquad (4\text{-}12)$$

The discrete kernel appears at this stage as a mathematical convenience with little hydrological meaning. Its hydrological value will become clear by considering the special case of a rain of unit duration (and stopping indefinitely thereafter) dropping uniformly over the watershed a total excess rainfall depth equal to one unit (inch, centimeter, 2.54 cm, ...). This particular hyetograph is shown on Fig. 4-14.

Application of Eq. (4-12) for this particular rainfall excess pattern to calculate direct runoff at times $n = 1,2 \ldots N$, yields for $n=1$

$$q(1) = \delta(1-1+1)\ r(1) = \delta(1)\ r(1) = \delta(1) \qquad (4\text{-}13)$$

For $n=2$ the summation in Eq. (4-12) includes two terms corresponding to $\nu=1$ and $\nu=2$, thus yielding:

$$q(2) = (2-1+1)\ r(1) + (2-2+1)\ r(2) = (2)\ r(1) + (1)\ r(2) \qquad (4\text{-}14)$$

Since the rainfall depth for the unit duration is one unit depth r(1) equals 1. Since the rain stops after the first period of time, r(2) is zero. Thus Eq. (4-14) reduces to:

$$q(2) = \delta(2) \qquad (4\text{-}15)$$

Successive applications of Eq. (4-12) for all n values yields the general result:

$$q(n) = \delta(n) \qquad n = 1,2 \ldots N \qquad (4\text{-}16)$$

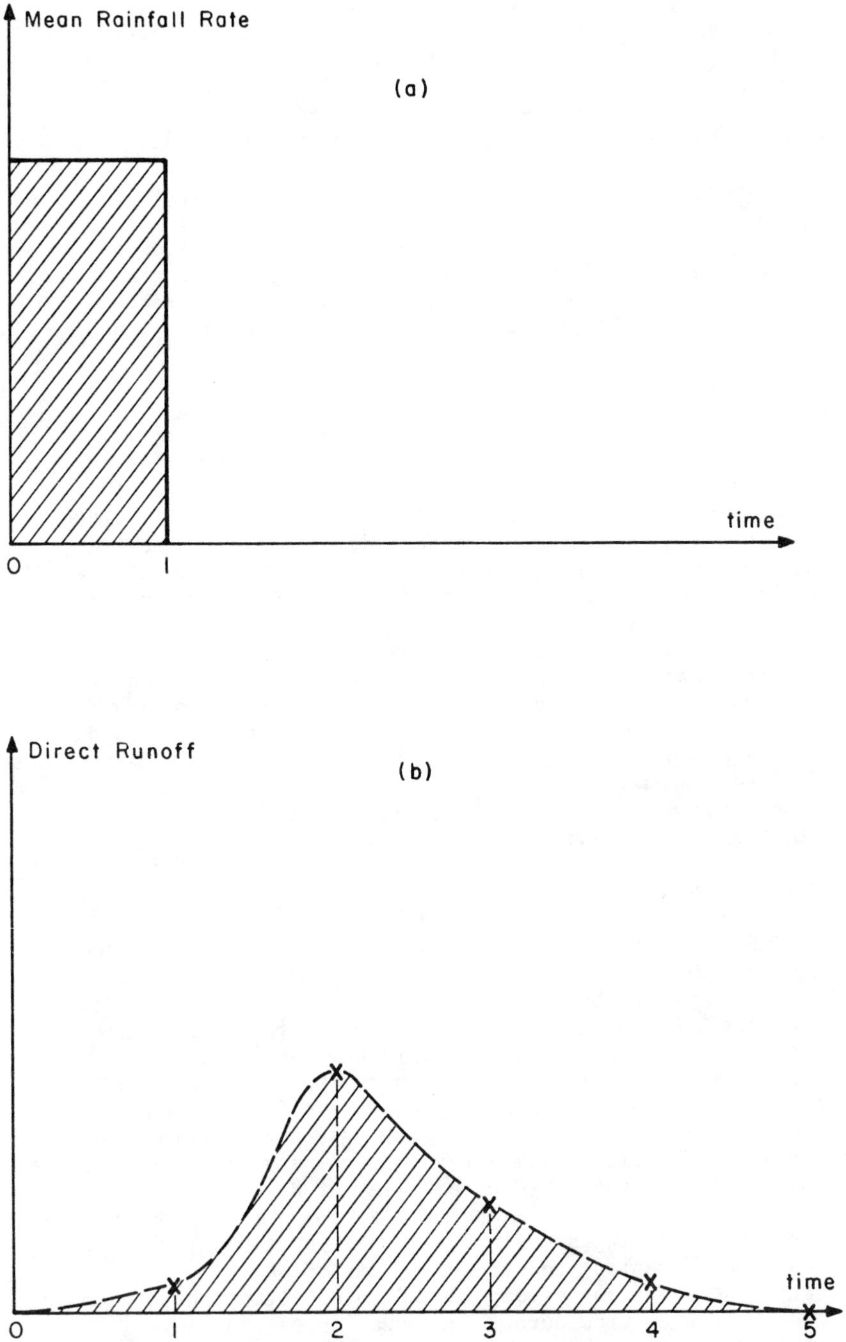

Fig. 4-14. Unit pulse hyetograph and runoff discrete kernel.

Thus the mysterious mathematical animal $\delta(n)$ is simply the direct runoff one *would* observe for the ideal excess rainfall pattern of a unit depth of water in a unit time and no rain thereafter.

The *discrete kernel* $\delta(n)$ is the ordinate of the so-called and well known *unit hydrograph*. In practice the discrete kernel values $\delta(1), \delta(2) \ldots \delta(n)$ are obtained from actual observations of rainfall and of runoff. If a rainfall episode lasts one unit duration with depth per unit time, r, and if the $q(1), q(2) \ldots q(n)$ are the observed values of (direct) runoff after one, two, ... n periods of time then due to the linearity of the assumptions and by direct application of Eq. (4-12), one calculates the $\delta(n)$ as:

$$\delta(n) = \frac{q(n)}{r} \tag{4-17}$$

Naturally uniform rainfall events of the precise desired duration are rare and the direct exploitation of Eq. (4-17) to determine the discrete kernels from the observed rainfall and runoff data is seldom performed. There exist several methods to calculate (*identify*, estimate) the discrete kernels for time-varying rainfall patterns. These are discussed in Appendix 1 of this chapter.

Once the discrete kernels have been identified, then direct runoff can be predicted from Eq. (4-12) for any rainfall forecast. Naturally the validity of the procedure depends on the actual degree of linear behavior of the watershed. Regarding this unit hydrograph theory "we still get many rides from this side-wheeler. Its application in the computer is simple and most of the time nature is not too fussy about the limitations." (Relyea, 1975, p. 51)

Hydrologic Flood Routing

For the relatively small watershed A (Fig. 4-15) the travel time of water on the land to the catchment outlet is insignificantly increased by the travel time in the small well defined channels. On the other hand once water travels no longer predominantly as overland flow but rather as channel and river flow, then speed of travel is controlled by the stream characteristics (slope, roughness, etc.) instead of the watershed characteristics. The question is: given the inflow (upstream flow) distribution in time (hydrograph) say at A, what is the outflow (downstream) hydrograph, say at B?

The fluid mechanics approach to solve this problem would consist of solving simultaneously the equation of continuity (mass conservation) and the equations of motion. These equations being quite complicated, hydrologists have adopted a much simpler procedure. The equation of continuity is retained in a one-dimensional form, namely:

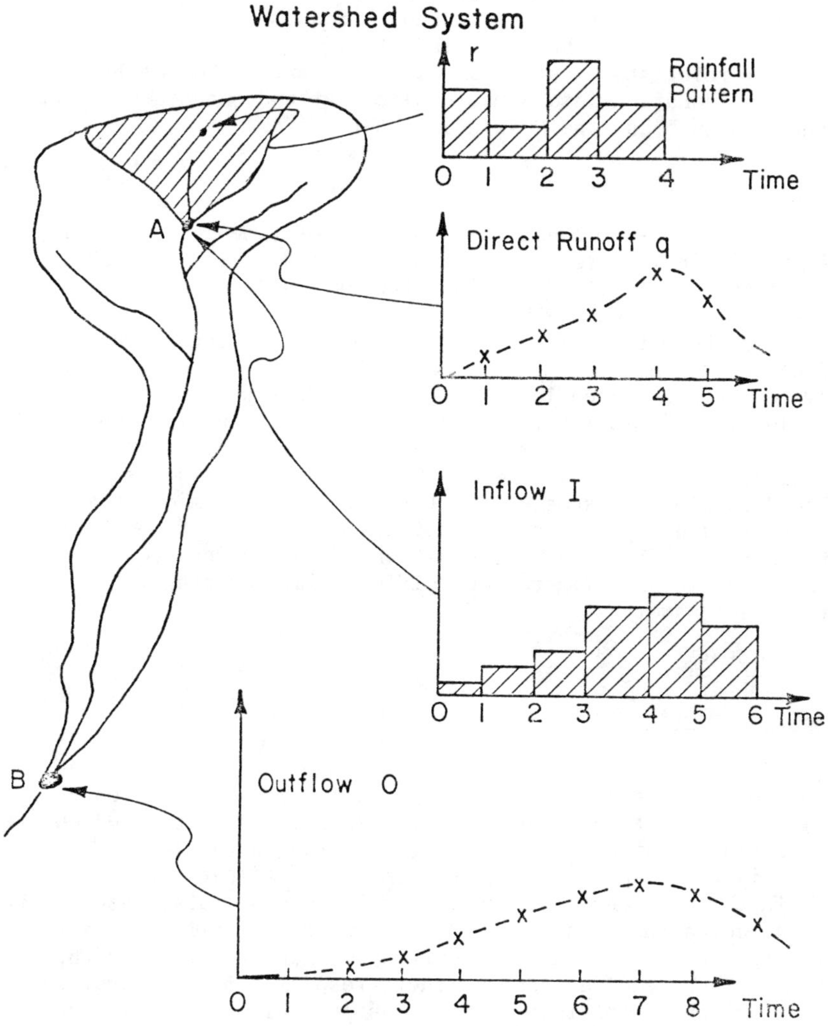

Fig. 4-15. Head basin runoff and river routing.

$$\frac{dS}{dt} = I - O \qquad (4\text{-}18)$$

where S is storage in the reach (or deviation of storage from a reference level), I is instantaneous inflow rate (or deviation from a reference level) and 0 is instantaneous outflow rate (or deviation from a reference level). Since inflows are usually known Eq. 4-18 contains two unknown, S and 0. Thus, one needs *another* equation. Where to find it?

Hydrologists, masters of illusion, enter the complicated momentum equation into a magic black hat and out comes a simple equation such as the Muskingum equation:

$$S = K\ [\xi I + (1-\xi)O] \qquad (4\text{-}19)$$

The parameter K has the dimension of time and ξ, usually denoted x in most textbooks [Linsley, et al., 1975, p. 300; Viessman, et al., 1977, p. 232; Lawler, 1964, p. 25.40], is dimensionless. Eq. 4-19 can only be successful if the kinematics of the motion are as a whole little affected by the dynamics of the system. Naturally Eq. 4-19 implies that the system behaves linearly. Though hydrologists have attempted to justify physically the Muskingum equation talking about *prism* and *wedge* storages [Linsley, et al., 1975, p. 302; Lawler, 1964, p. 25.40; Gray, 1973, p. 8.50] it is just as lucid to simply assume explicitly that the system behaves linearly and depends on two empirical parameters K and ξ without further justification. Elimination of 0 between Eqs. 4-18 and 4-19 yields a first order differential equation in S (the response) namely:

$$\frac{1}{\eta}\frac{dS}{dt} + S = KI \qquad (4\text{-}20)$$

where $\eta = \dfrac{1}{K(1-\xi)}$ (4-21)

During the discussion on the unit hydrograph concept it was stated that the linear system could be characterized entirely by a single function, the kernel function in Eq. 4-8. This function is the response of the system to a *particular* excitation. In fact, it is the response to a unit impulse (also known as the Dirac delta function). Other particular excitations of interest are the unit pulse (used in unit hydrograph theory) and the unit step. These excitations and their responses for a linear catchment are shown in Figure 4-16. The general response (direct runoff in the case of a watershed) can be expressed in several (equivalent) ways in terms of a general excitation (excess rainfall in the case of a watershed), as:

$$q(t) = \int_0^t k(t-\tau)r(\tau)d\tau \qquad (4\text{-}22)$$

or

$$q(t) = K(t)r(o) + \int_o^t K(t-\tau) \frac{\partial r(\tau)}{\partial \tau} d\tau \qquad (4\text{-}23)$$

or

$$q(n) = \sum_{\nu=1}^{n} \delta(n-\nu+1)\, r(\nu) \qquad (4\text{-}24)$$

The functions k( ), K( ) and δ( ) are not independent. They can be deduced one from the other from the relations:

$$\frac{dK(t)}{dt} = k(t) \qquad (4\text{-}25)$$

$$\delta(m) = \int_o^1 k(m-\tau)\, d\tau \qquad (4\text{-}26)$$

$$\delta(m) = K(m) - K(m-1) \qquad (4\text{-}27)$$

and

$$K(m) = \sum_{\nu=1}^{m} \delta(\nu) \qquad (4\text{-}28)$$

It is better to secure the step kernel K(t) and deduce the other functions from it because it is better behaved (see Figure 4-16).

The first step in the general solution of Eq. 4-20 for an arbitrary upstream hydrograph I(t) function is to solve Eq. 4-20 for a unit step excitation, or in other words, solve the equation:

$$\frac{1}{\eta}\frac{dS}{dt} + S = 1 \qquad (4\text{-}29)$$

The solution for the *Muskingum storage step kernel is*:

$$K(t) = 1 - e^{-\eta t} \qquad (4\text{-}30)$$

and using Eq. 4-25, one obtains the *Muskingum storage impulse kernel*:

$$k(t) = \eta e^{-\eta t} \qquad (4\text{-}31)$$

The general solution for S(t) using Eq. 4-22 is:

$$S(t) = \int_o^t K\eta e^{-\eta(t-\tau)} I(\tau)\, d\tau \qquad (4\text{-}32)$$

Using Eq. 4-19 to deduce O(t) from S(t), the general solution for O(t) is:

$$O(t) = \frac{1}{K(1-\xi)^2} \int_o^t e^{-\eta(t-\tau)} I(\tau) d\tau - \frac{\xi}{1-\xi} I(\tau) \qquad (4\text{-}33)$$

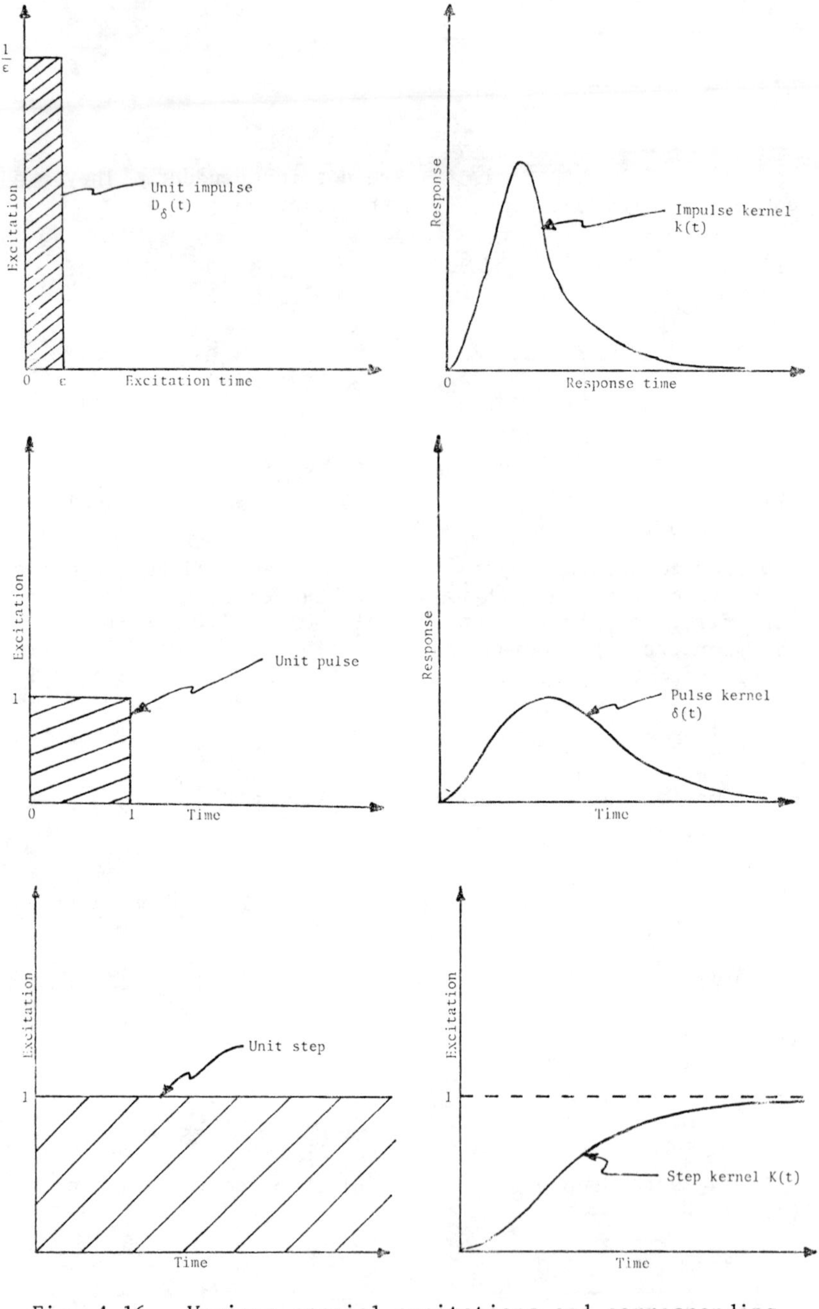

Fig. 4-16. Various special excitations and corresponding response kernels.

By identification of Eq. (4-33) with the general expression Eq. 4-22, one deduces the analytical expression for the Muskingum *outflow* impulse kernel, namely:

$$k_o(u) = \frac{1}{K(1-\xi)^2} e^{-\frac{u}{K(1-\xi)}} - \frac{\xi}{1-\xi} D_\delta(u) \qquad (4\text{-}34)$$

where $D_\delta(\cdot)$ is the Dirac delta function. Application of Eq. 4-26 yields the expression for the Muskingum outflow discrete kernels (for details, see Appendix 2):

$$\delta(1) = 1 - \frac{e^{\frac{-1}{K(1-\xi)}}}{1-\xi} \qquad (4\text{-}35)$$

$$\delta(m) = \frac{e^{\frac{1}{K(1-\xi)}} - 1}{1-\xi} \; e^{-\frac{m}{K(1-\xi)}} \qquad (4\text{-}36)$$

Once the $\delta(\cdot)$ have been evaluated the downstream flows can be predicted in terms of the upstream flows by the relation:

$$O(n) = \sum_{=1}^{n} \delta(n-\nu+1)\, I(\nu) \qquad (4\text{-}37)$$

Note again that in Eq. 4-37 as with the unit hydrograph theory, $I(\nu)$ is the *average* inflow rate during the time interval $(\nu-1,\nu)$ but $O(n)$ is the *instantaneous* outflow rate at time n.

Of course, one must be able to identify the Muskingum parameters from observed data of inflows and outflows, before the above methodology can be implemented. The traditional graphical procedure is described in many textbooks (Linsley, et al., 1975, p. 301; Lawler, 1964, p. 25-41; Viessman, et al., 1977, p. 235; Wilson, 1969, p. 141; Gray, 1973, p. 8.53; Hjelmfelt and Cassidy, 1975, p. 130). Actually the parameters K and ξ can be estimated by a method analogous to the method of moments used in Statistics (see Chapter 3). The mathematics are almost identical, though the interpretation is different since in this case no analysis of random variables or errors is involved. It can be shown (Appendix 2) that:

$$K = \frac{\sum_{n=1}^{N} (n-1/2)\left[\frac{O(n)+O(n-1)}{2} - I(n)\right]}{\sum_{n=1}^{N} I(n)} \qquad (4\text{-}38)$$

Note that the numerator is the difference between the first moment of the outflows and the first moment of the inflows. In Eq. 4-38, K is expressed in the same units of time as the discrete times (n=1,2 . . . N). Once K has been calculated from Eq. 4-38 ξ can be calculated from the equation:

$$2K^2(1-\xi) + 1/12 = \frac{\sum_{n=1}^{N} (n-1/2)^2 \left[\frac{O(n)+O(n-1)}{2} - I(n)\right] - 2K \sum_{n=1}^{N} (n-1/2)\, I(n)}{\sum_{n=1}^{N} I(n)} \tag{4-39}$$

With this method no trial and error repeated calculations are required.

The case of reservoir routing when outflow depends solely on storage and the relation is linear, is a particular case of the Muskingum river routing method for $\xi=0$. All previous results are applicable for this case by setting $\xi=0$ in all the general Muskingum equations.

Finally a warning must be sounded. It is possible for the Muskingum first discrete kernel $\delta(1)$ to be negative. It will happen if

$$\frac{1}{K} < (1-\xi)\, \ell n\left(\frac{1}{1-\xi}\right) \tag{4-40}$$

that is, if K (a *lag* time) is large compared to 1 (the selected unit of time). As a rule of thumb, it is recommended that the chosen unit of time be no smaller than 1/2 to 1/3 of K.

4.5 CONCEPTUAL MODELS FOR CONTINUOUS PREDICTION OF HYDROGRAPH

The first electronic computer to be used in river forecasting by the National Weather Service (NWS) was installed at Forth Worth, Texas in 1961 (Sittner, 1972, p. 1). It was expected that with the computer, newer models of continuous simulation of streamflow would in time replace the traditional three elements: API coaxial rainfall-runoff relation, unit hydrograph, and hydrologic flood routing, and provide more accurate forecasts. As a number of new hydrologic simulation techniques utilizing the computer had been advocated, it was necessary to evaluate them. A project initiated in 1966 for this purpose concluded that two models could meet the NWS specifications for performance: (1) the Stanford Watershed Model IV with minor modifications, and (2) a model developed by the Sacramento River Forecast Center (Burnash, et al., 1973), which is also very similar to the Stanford Watershed Model.

"These are conceptual models. That is, the parameters and variables involved in the mathematical formulations represent actual physical quantities rather than, as in the case of

API, indices to the quantities. This quality is expected to enhance the applicability of forecast models to problems outside the field of pure river forecasting. Examples are land use and groundwater pollution studies." (Sittner, 1972, p. 3).

According to this author, improved accuracy was demonstrated during and after long periods of drought. In particular, the two methods can reproduce situations of little or no river flow even for large rainfalls if soil moisture was very deficient.

The conversion of the River Forecast Centers from the traditional procedures (Richards and Strahl, 1969) to the computerized conceptual models is not complete. The reprinting in 1975 of the 1969 WBTM HYDRO 4 is indirect evidence that many centers still utilize these procedures. The only centers to operate 100% with conceptual models are the Slidell, Louisianna center (created in 1972) covering the Lower Mississippi River Basin and the Sacramento center, covering the State of California, which uses its own Sacramento model. The biggest center in Kansas City, covering the entire Missouri and Mississippi Basins above Saint Louis still uses entirely the traditional API, etc. procedures. It may go hybrid in the near future but is faced with the overwhelming task of calibrating over 1000 basins! The Atlanta, Georgia center uses the conceptual model at 25%, the Tulsa (Arkansas River Basin) center at about 10%, and the other centers in various but still relatively low proportions.

The National Weather Service is simultaneously proceeding with the operational conversion of the centers to the current conceptual models and developing new procedures to remedy the deficiencies of the current models which have become apparent with uses. Improved methodology has been reported in modeling of snow accumulation and ablation (Anderson, 1973) and hydraulic routing (Fread, 1973 and 1974). Research is currently underway to model infiltration into frozen soils (Morel-Seytoux, 1976b).

Though in principle conceptual, the computer models still involve many empirical components. "Conceptual models use an explicit evapotranspiration function. This particular function is probably the most neglected area to date in catchment model research." (Sittner, 1972, p. 6). To this list the writer would like to add the infiltration and interflow components. The treatment of these components in the Stanford Watershed Model and its posterity has *never* been and still is not physical (Chapman and Dunin, 1975).

4.6 EXTENDED STREAM FLOW PREDICTIONS

It is of interest to a variety of water users to have some ideas of the water availability into the near future and of the risks of flood or drought. In Chapter 3, flood frequency analysis was discussed. To be more precise, it should have been called *marginal* flood frequency analysis. The chance of a snowmelt flood exceeding a given magnitude in 1979 can only be predicted today from this marginal extreme value distribution.

On the other hand, given the watershed conditions as of April 1, 1978, one can develop more precise probability statements regarding the May-June, 1978, cumulative flow. If the snowpack is low and the soil moisture very deficient as of April 1, the marginal 100-year flood has an exceedance probability which is much less than 1%. With the added information of the *initial conditions* of the system, one can develop a *conditional* distribution of flows for the near future.

The traditional regression equation of seasonal flow versus snowpack as of April 1 (or some other date) allows one to make an estimate of the conditional expected flow. However, "most of these approaches based on some form of regression/correlation analysis, because of their mathematical structure and discontinuous operation, are capable of providing only seasonal estimates of streamflow volume and do not associate their results with any chance of occurrence." (Twedt, et al. undated, p. 1). The NWS has developed a method to provide more detailed information on flows and their chances of occurrence. The procedure has been developed so far only for headwater basins.

The procedure consists of utilizing the NWS RFS conceptual models to simulate the future starting at a given date with given (known, observed) watershed conditions utilizing for the future the past climatological records. Given the initial conditions as of May 1, 1971 (namely, moisture storage contents, snowpack water-equivalents and other snow cover variables) one simulates a future using the climatological data for May and June of the year 1950. One repeats for the data of 1951, 1952 . . . etc. . ., say for 20 years. Thus 20 *traces* of flows are generated. The sample can then be analyzed by standard statistical techniques. Naturally, the procedure requires that (1) the hydrologic model parameters for the basin be known from previous calibration, (2) the initial basin conditions be known, and (3) that climatological time series be available as inputs. Other conceptual models other than the NWS RFS ones can be used but "such models should contain a soil accounting procedure to describe the movements of water into and through the soil mantle, a method of representing snowpack accumulation and subsequent melt, and a technique for routing the flow of water through river channels" (ibid, p. 4).

The current NWS RFS procedure utilizes a 6-hour computational time step and assumes that the statistical distribution of the flows is log-normal. Typically, total volume of flow, maximum mean daily flow, minimum mean daily flow and average mean daily flow are provided for the period of interest. Table 4-2 (after Twedt, et al.) provides results for the Eagle River, Colorado. Figure 4-17 displays a comparison between the marginal (actual observed) and the conditional (calculated by the ESP model) flow frequency analysis. The value of conditioning the analysis is evident.

A lower level of uncertainty in prediction is attained by the conditional analysis compared to the marginal one due to its significantly flatter slope. Unfortunately, the use of large

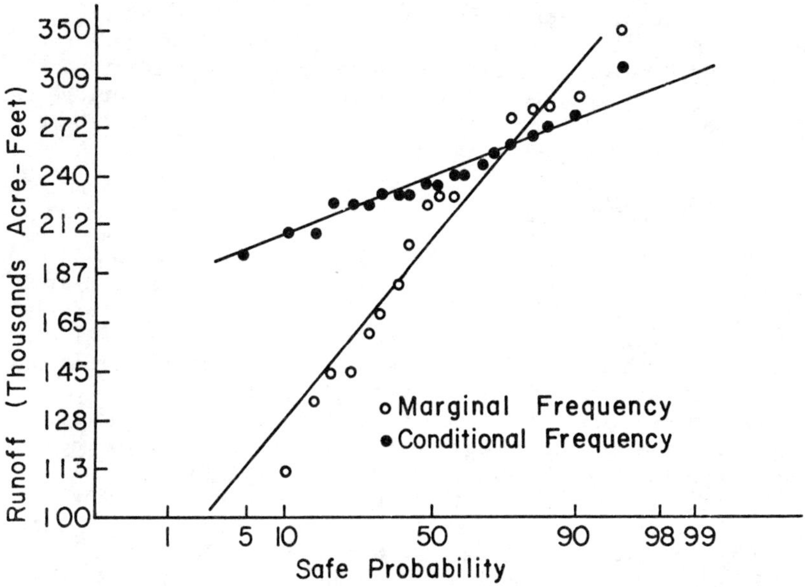

Fig. 4-17. Comparison of marginal and conditional flow frequencies (after Twedt et al., undated, p. 7)

amounts of historical data requires extensive computer storage. To alleviate this problem, the NWS may increase the computational time step from 6 hours to one day.

4.7 CONCLUSIONS

The traditional approaches to river forecasting were described. Limitations with this approach have led to the development of more detailed conceptual models. Though in principle physical, in fact the detailed analysis of many conceptual models indicates that they remain at least in part empirical.

A detailed description of the conceptual models would require a full volume. A recent text provides an adequate introduction to many such models (Viessman et al. 1977, p. 408-478). Several models are discussed in subsequent chapters.

Table 4-2. Estimates of Quantities of Total Volume of Flow for May and June, 1971, as of May 1, 1971 (after Twedt, et al., p. 6)

Classification	Exceedance Probability	Total Volume (Thousands Acre-Feet)	Percent Normal
Reasonable Maximum	0.10	277.5	97.7
Most Reasonable	0.50	239.9	103.2
Reasonable Minimum	0.90	207.3	109.7

Ultimately, river forecasting is dependent on precipitation forecasting which is a subject in itself. The ESP procedure avoids the need of precipitation forecasting to predict not the immediate, but the extended near future. It casts a bridge between short-term deterministic predictions and the long-term statistical flow frequency analysis.

4.8 ACKNOWLEDGEMENTS

Much of the material developed in this chapter is based on lectures given by the author in the course Engineering Hydrology during the Spring, 1977 semester and lecture notes faithfully taken (and improved) by several student scriptwriters. I want to express my appreciation to these students, coming from many different countries, for their very effective help:
Mr. Ken Steele (Corps of Engineers), Silvio Wille (Brazil), Ralph A. Wurbs (Corps of Engineers), Francisco Baquero (Ecuador), Bob Barkau (Corps of Engineers), Hugh Barrett (Australia), and Fred Lux. Finally, I must thank Dr. H. W. Shen, Director of the 1977 River Mechanics Institute, for his trusting invitation to participate as a lecturer and author. Collaboration with Dr. Shen has always proved a pleasurable and worthy endeavor.

4.9 REFERENCES

Anderson, E.A., 1973. NWS RFS-snow accumulation and ablation model. NOAA T.M. NWS HYDRO-17, U. S. Department of Commerce, Silver Spring, Maryland, November 1973.

Blake, G.J., 1975. The Interception Process in Prediction in Catchment Hydrology. Ed. Chapman and Dunin Australian Academy of Sciences, Griffin Press, Netley, South Australia, 482 pages.

Burnash, R.J.C., R.L. Ferral, and R.A. McGuire, 1973. A generalized streamflow simulation system. Conceptual modeling for digital computers. U. S. Department of Commerce, National Weather Service, State of California, Department of Water Resources, Joint Federal-State River Forecast Center, March, 1973, 204 pages.

Butler, S.S., 1957. Engineering hydrology. Prentice-Hall, Inc. Englewood Cliffs, New Jersey, 356 pages.

Chapman, T.G. and F.X. Dunin, 1975. Prediction in catchment hydrology. National Symposium in Hydrology, 25-27 November, 1975, Australian Academy of Science, Griffin Press, Netley, South Australia, 482 pages.

Chow, V.T., editor, 1964. Handbook of applied hydrology. McGraw-Hill Book Company, New York.

Curtis, D.C. and G.F. Smith, 1976. The National Weather Service River Forecasting System-Update 1976. For publication in minutes of International Seminar on Organization and Operation of Hydrological Services in Conjunction with the Fifth Session of the WMO Commossion for Hydrology, Ottawa, Canada, July 15-16, 1976.

Dawdy, D.R., R.W. Lichty, and J.M. Bergmann, 1972. A rainfall-runoff simulation model for estimation of flood peaks for small drainage basins. U.S. Geological Survey Professional Paper 506-B, U.S. Government Printing Office, Washington, D.C., 28 pages.

Dooge, J.C.I., 1973. Linear theory of hydrologic systems. USDA-ARS, Technical Bulletin No. 1468, U.S. Government Printing Office, Washington, D.C. 20402, 327 pages.

Duband, D., 1975. Application des methods d'analyse des donnees a l'hydrologie. Reunion due Group de Travail Hydrologie Stochastique du 12 Juin 1975, Electricite de France, Division Technique Generale, 28 pages.

Eagleson, P.S., 1970. Dynamics hydrology. McGraw-Hill, New York.

Fread, D.L., 1974. Numerical properties of implicit four-point finite difference equations of unsteady flow. NOAA T.M. NWS HYDRO-18.

Fread, D.L., 1973. A dynamics model of stage-discharge relations affected by changing discharge. NOAA T.M. NWS HYDRO-16.

Gray, D.M. and J.M. Wigham, 1973. Runoff-Rainfall-General. Section 7 in Handbook on the Principles of Hydrology, D.M. Gray editor, Water Information Center Publication, PP 7-1/7-24.

Hjelmfelt, A.T. and J.J. Cassidy, 1975. Hydrology for Engineers and Planners. Iowa State University Press, Ames, Iowa, 210 pages.

James, L.G. and C.L. Larson, 1974. Modeling infiltration and redistribution of soil water during intermittent applications. Paper presented at 1974 winter meeting of ASAE, Chicago, Illinois, December 10-13, 1974, 26 pages.

Jønch-Clausen, T. and H.J. Morel-Seytoux, 1976. User's manual for LPTOR, a Fortran IV linear programming routine. HYDROWAR Program, Engineering Research Center, Colorado State University, Fort Collins, Colorado 80523, CER75-76TJ-HJM36, May, 1976, 39 pages.

Jønch-Clausen, T. and H.J. Morel-Seytoux, 1977. User's manual for QPTOR, a Fortran IV quadratic programming routine. HYDROWAR Program, Engineering Research Center, Colorado State University, CER 76-77TJ-HJM48, March, 1977, 72 pages.

Laufer, F., 1977. Weekly control of an alpine seasonal reservoir. HYDROWAR Program, Engineering Research Center, Colorado State University, Fort Collins, Colorado 80523, CER 76-77FL49, March, 1977, 311 pages.

Lawler, E.A., 1964. Hydrology of flow control. Part II. Flood Routing. Section 25 in Handbook of Applied Hydrology. V.T. Chow, ed., McGraw-Hill Book Company.

Linsley, R.K., M.A. Kohler, and J.L.H. Paulhus, 1975. Hydrology for engineers. 2nd edition, McGraw-Hill, 482 pages.

Mein, R.G. and C.L. Larson, 1973. Modeling infiltration during a steady rain. Water Resources Research Journal, Vol. 9, No.2, pp. 384-394.

Mondschein, H.F., 1971. Forecasting the spring 1969 midwest snowmelt floods. NOAA technical memorandum NWS CR-40, U.S. Dept. of Commerce, Central Region, Kansas City, Mo., Feb., 1971.

Monro, J.C., 1971. Direct search optimization in mathematical modeling and a watershed model applciation, NOAA, TM NWS HYDO 12, April, 1971, 52 pages.

Morel-Seytoux, H.J., 1978. Derivation of equations for variable rainfall infiltration. Water Resources Research Journal, Vol. 14, No. 4, August, pp. 561-568.

Morel-Seytoux, H.J., 1977. Derivation of equations for variable rainfall infiltration. HYDROWAR Program, Engineering Research Center, CEP 76-77, HJM47, Colorado State University, Fort Collins, Colorado 80523, May.

Morel-Seytoux, H.J., 1976a. Derivation of equations for rainfall infiltration. Journal of Hydrology, Vol. 31, pp. 203-219.

Morel-Seytoux, H.J., 1976b. Mass and heat flow equations in soils under cold conditions. HYDROWAR Program, Engineering Research Center, Colorado State University, CEP 75-76HJM14, February, 1976, revised April, 1976, 81 pages.

Morel-Seytoux, 1975a. Pour une theorie modifiee de l'infiltration le cas de la pluie. Cah. ORSTOM, ser. Hydrol., Vol. XII, No. 4, pp. 227-233.

Morel-Seytoux, H.J., 1975b. A combined model of water table and river stage evolution. Water Resources Research Journal, Vol. 11, No. 6, December, 1975, pp. 968-972.

Morel-Seytoux, H.J., 1972. Foundations of engineering optimization (optimization in hydrology and water resources). Lecture notes for course CE724, supplemented and reprinted, 1976.

Morel-Seytoux, H.J., T.A. Pick, and T. Jønch-Clausen, 1976. Computation of infiltration for unsteady uninterrupted high rainfall. Engineering Research Center, CEP75-76HJM-TP-TJ33, May, 1976.

Musgrave, G.W. and H.N. Holtan, 1964. Infiltration. Section 12 in Handbook of Applied Hydrology. V.T. Chow, editor, McGraw-Hill Book Company, pp. 12-1/12-30.

Overton, D.E. and M.E. Meadows, 1976. Stormwater modeling. Academic Press, 358 pages.

Philip, J.R., 1975. Some remarks on science and catchment prediction. in Prediction in Catchment Hydrology, Chapman and Dunin, editors, Australian Academy of Science, 482 pages.

Relyea, C.M., 1975. Hydrologic forecasting for the Ohio River waterway. Proceedings of the Symposium on Modeling Techniques, American Society of Civil Engineers, San Francisco, California, Sept. 3-5, 1975, pp. 43-62.

Relyea, C.M., 1967. River forecasting. Proceedings of the IBM Scientific Computing Symposium on Water and Air Resources Management, October 23-25, 1967, Thomas J. Watson Research Center, Yorktown Heights, N.Y., pp. 63-79.

Réméniéras, G., 1970. L'hydrologie de l'ingenieur. Editions Eyrolles, 456 pages.

Richard, M.M. and J.A. Strahl, 1969. Elements of river forecasting (revised). ESSA technical memorandum WBTM HYDRO 9 U. S. Department of Commerce, Weather Bureau, Office of Hydrology, Silver Spring, Maryland, March 1969. Reprinted May 1975, 57 pages.

Roche, M., 1963. Hydrologie de surface. Gauthier-Villars, Paris, 430 pages.

Schulz, E.F., 1974. Problems in applied hydrology. Water Resources Publications, Fort Collins, Colorado 80521, 501 pages.

Sittner, W.T., 1972. Modernization of National Weather Service river forecasting techniques. Paper presented at the Eighth American Water Resources Conference, Saint Louis, Missouri, November 2, 1972, 9 pages.

Staff, Hydrology Research Laboratory, 1972. National Weather Service River Forecast System Forecast Procedures. NOAA Technical memorandum NWS HYDRO-14, December, 1972.

Twedt, T.M., J.C. Schaake, and E.L. Peck, undated publication. National Weather Service extended streamflow prediction. Hydrologic Research Laboratory, NWS, Silver Spring, Maryland, 10 pages.

Viessman, W., J. W. Knapp, G.L. Lewis, and T.E. Harbaugh, 1977. Introduction to hydrology. 2nd edition, IEP a Dun-Donnelley Publisher, New York, 704 pages.

Wilson, E.M., 1969. Engineering hydrology. Macmillan Press, 182 pages.

Yevjevich, V., 1972. Probability and statistics in hydrology. Water Resources Publications, Fort Collins, Colorado, 302 pages.

Appendix 4.1

Identification of Discrete Kernels by Mathematical Programming Techniques

THE INVERSE PROBLEM

Once the discrete kernels $\delta(\cdot)$ have been identified the direct runoffs $q(n)$ can be predicted in terms of the excess rainfalls $r(\nu)$ by the relation:

$$q(n) = \sum_{\nu=1}^{n} \delta\,(n - \nu + 1)\; r(\nu) \qquad (1)$$

Vice versa, given values of excess rainfall and direct runoff, Eq. 1 for $n = 1, 2, \ldots$ can be used to calculate (estimate) the discrete kernels $\delta(1)$, $\delta(2)$, . . etc. . . . However, the problem is not that simple because for a finite memory system, (i.e. a system for which $\delta(m) = 0$ once m exceeds the memory time M) there are more equations (1) than there are unknowns $\delta(\cdot)$. For example on Figure 4-18, the memory time of the watershed is 4 and $\delta(5) = \delta(6) = \ldots = 0$. Since there are 13 observations of rainfall and nonzero runoff, the system of equations has 13 equations for only 4 unknowns: $\delta(1)$, $\delta(2)$, $\delta(3)$, and $\delta(4)$. The linear systems of Equations 1 is overdetermined and cannot be solved exactly.

THE LEAST SQUARES METHOD

Let the difference between the observed values $q°(n)$ and their predictions for given observed values of rainfall $r°(\nu)$ be denoted $e(n)$ (for error), namely:

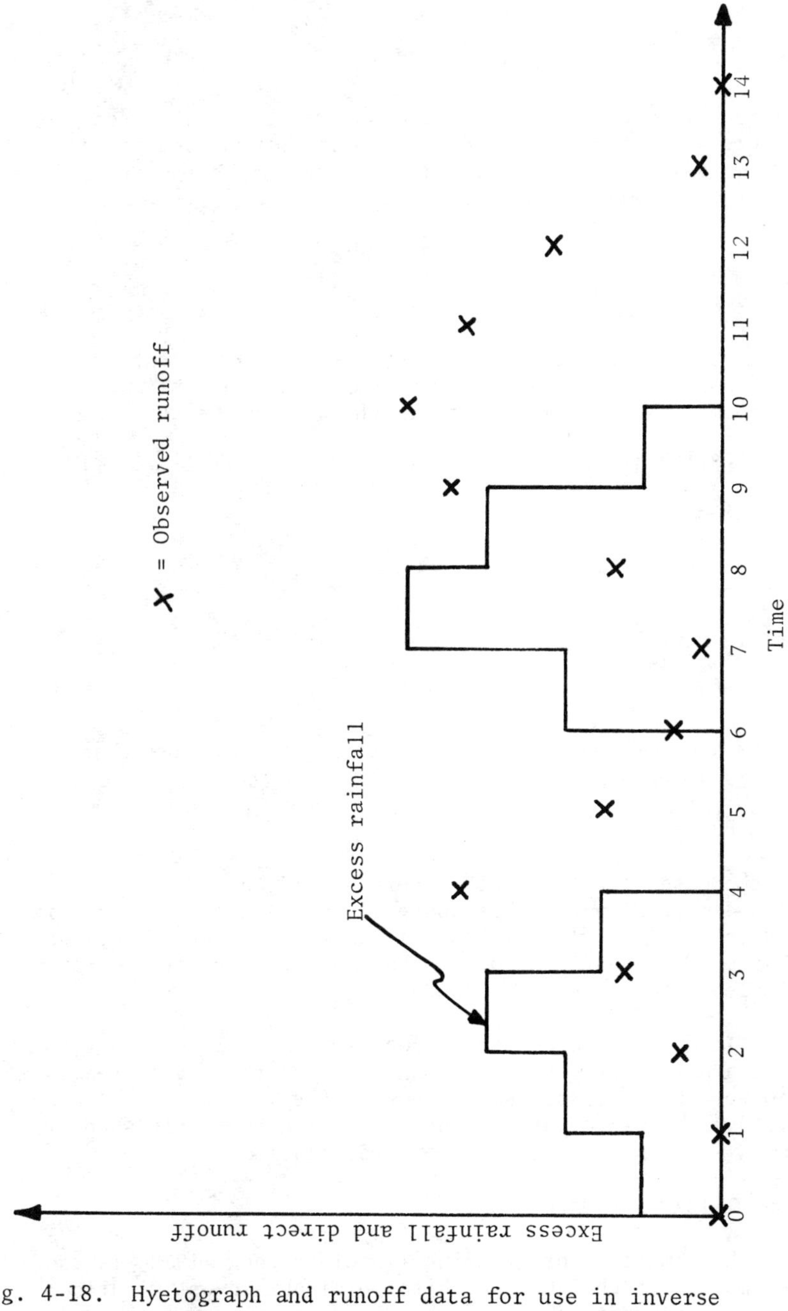

Fig. 4-18. Hyetograph and runoff data for use in inverse procedure to determine the discrete kernels.

$$e(n) = q^\circ(n) - \sum_{\nu=1}^{n} \delta(n - \nu + 1)\, r^\circ(\nu) \quad n = 1,2 \ldots N \tag{2}$$

where N is the total number of observations (in the case of Figure 4-18, N = 13, M = 4). Then it is proposed to choose the M $\delta(\cdot)$ values so as to minimize the sum of the squares of the errors. Formally, the problem is:

$$\text{Minimize} \qquad \{y = \sum_{n=1}^{n} [e(n)]^2\} \tag{3}$$

with respect to the unknown $\delta(\cdot)$. Let $\underline{q}^\circ$ be the column vector of the runoff data, $\underline{\delta}$ be the column vector of the unknowns, and $\underline{R}$ the matrix of coefficients in the linear system:

$$\begin{aligned} r^\circ(1)\delta(1) & = q^\circ(1) \\ r^\circ(2)\delta(1) + r^\circ(1)\delta(2) & = q^\circ(2) \\ r^\circ(3)\delta(1) + r^\circ(2)\delta(2) + r^\circ(1)\delta(3) & = q^\circ(3) \\ \text{----------} & \end{aligned} \tag{4}$$

or, in matrix notation:

$$R^\circ\underline{\delta} = \underline{q}^\circ \tag{5}$$

It is a well-known result (Graybill, 1961, p. 114) that $\underline{\delta}$ is the solution of the linear system:

$$R'R\,\underline{\delta} = R'\underline{q}^\circ \tag{6}$$

In the case of Figure 4-18, where N = 14, and M = 4, R is a 14 X 4 matrix and its transpose R' is a 4 X 14. The matrix R'R is a (4 X 14)(4 X 14) = 4 X 4 matrix. Thus Eq. 6 is a system of 4 equations for 4 unknowns and therefore, a rather simple system to solve.

In many instances this procedure will yield good results, i.e. a unit hydrograph shape which looks physically realistic. However, there is no guarantee that the method will always yield *positive* values for the $\delta(\cdot)$, or a realistic shape. The $\delta(\cdot)$ values may show strong oscillations.

THE MINISAD APPROACH

Mathematical programming (Morel-Seytoux, 1972) is the ensemble of techniques which solve algebraic optimization problems subject to constraints in particular, for example, the constraints that the variables must be non-negative (non-negativity conditions). Mathematical programming reduces to linear programming (LP) when the objective function is a linear function of the variables and when all constraints are linear. Thus a typical linear programming would read, in matrix and vector notation,

$$\underset{\underline{x}}{\text{Min}} \{y = \underline{c}'\underline{x}\}$$

$$\text{subject to:} \quad A\underline{x} \geq \underline{r} \tag{7}$$

The least squares formulation is not linear and cannot be solved by LP techniques. Instead of minimizing the sum of squares of the errors one could minimize the *sum* of the *absolute* values of the errors, (*deviations*, therefore the name MiniSAD), namely:

$$y = \sum_{n=1}^{n} |e(n)| \tag{8}$$

However, in this form the objective function y is still not acceptable for LP and one is led to a *trick*. One defines the error $e(n)$ which *could* be positive or negative to be the difference between two *positive* variables u_n and v_n. In particular, the systems of Eqs. 2 is replaced by the system:

$$u_n - v_n + \sum_{\nu=1}^{n} r^\circ(\nu)\ \delta\ (n - \nu + 1) = q^\circ\ (n) \tag{9}$$

$$n = 1, 2, \ldots N$$

An objective function equivalent to that of Eq. 8 is:

$$y = \sum_{n=1}^{N} (u_n + v_n) \tag{10}$$

In addition, one imposes nonnegativity conditions on the variables $\delta(\cdot)$, namely:

$$\delta(n) \geq 0, \ n = 1, 2, \ldots M \tag{11}$$

If p designates the time of peak of the discrete kernel hydrograph, then one imposes the *ordering* constraints, namely:

$$\delta(1) \leq \ldots \delta(n) \leq \delta(n+1) \leq \ldots \leq \delta(p)$$
$$\delta(p) \geq \delta(p+1) \geq \ldots \delta(p+m) \geq \delta(p+m+1) \geq \ldots \tag{12}$$

NOTE that the number of observations N is defined as the sum of the time index of the last nonzero rainfall rate (on Figure 4-18, this index is 10) plus the memory time (in this case, M = 4), minus 1, or symbolically:

$$N = E_R + M - 1 \tag{13}$$

If one adds all the Eqs. 9, one obtains the equation:

$$\sum_{n=1}^{N} (u_n - v_n) + \left(\sum_{n=1}^{N} r^\circ(n)\right) \sum_{n=1}^{M} \delta(n) = \sum_{n=1}^{N} q^\circ(n) \tag{14}$$

which is different from the equation of conservation of mass which requires that:

$$\sum_{n=1}^{M} \delta(n) = \frac{\sum_{n=1}^{N} q^\circ(n)}{\sum_{n=1}^{N} r^\circ(n)} \tag{15}$$

unless:

$$\sum_{n=1}^{N} u_n - \sum_{n=1}^{N} v_n = 0 \tag{16}$$

Note that Eq. 15 reduces to the usual condition:

$$\sum_{n=1}^{M} \delta(n) = 1 \tag{17}$$

only if runoff and rainfall are expressed in consistent units (i.e., depths of water in same units). The final LP problem is:

$$\underset{\underline{u},\underline{v},\underline{\delta}}{\text{Min.}} \{y = \sum_{n=1}^{N} (u_n + v_n)\} \tag{18}$$

subject to the nonnegativity conditions (11), the ordering constraints (12), the deviation definition constraints (9), and the conservation of mass constraint (17). For the case of Fig. 4-18, the optimization problem involves a total of 30 (i.e. 2N+M) unknowns ($4\delta(\cdot)$, 13 u., 13 v.), 3 (i.e. M-1) ordering inequality constraints, one mass balance equality constraint, 13 (i.e. N) deviation definition equality constraints, or a total of 17 (i.e. N+M) constraints ([M-1] inequalities and [n+1] equalities). A 30 unknowns, 17 constraints problem can be solved without great difficulty with existing LP codes (Jønch-Clausen and Morel-Seytoux, 1976).

THE MINIMAD APPROACH

With the previous approach, the minimum of the sum of the absolute deviations could be made of 13 relatively small absolute deviations or of say 12 very small deviations and a relatively large one. The approach minimizes the error as a whole but not necessarily *uniformly*. Another approach consists of minimizing the *maximum absolute deviation* (therefore the name).

Let e designate this maximum absolute deviation. Clearly e is a nonnegative variable. By definition, the error e(n) is less in absolute value than e, or:

$$\left| \sum_{\nu=1}^{n} r^\circ(\nu)\ \delta(n-\nu+1) - q^\circ(n) \right| \le e \tag{19}$$

If e(n) was negative, than Eq. 19 would be equivalent to:

$$\sum_{\nu=1}^{n} r^\circ(\nu)\ \delta(n-\nu+1) - e \le q^\circ(n) \tag{20}$$

If e(n) was positive, then Eq. 19 would be equivalent to:

$$\sum_{\nu=1}^{n} r^\circ(\nu)\ \delta(n-\nu+1) + e \ge q^\circ(n) \tag{21}$$

Since it is not known *a priori* whether e(n) will be positive or negative, constraints 20 and 21 must be included in the minimization problem. The final formulation of the MiniMAD approach is:

$$\min_{e,\underline{\delta}} \{y = e\} \tag{22}$$

subject to the nonnegativity conditions, the ordering constraints (12), the deviation definition constraints (20) and (21) and the conservation of mass constraint (15). The problem involves (M+1) unknown, one equality constraint and (2N + M-1) inequality constraints. In the case of Figure 4.16, the problem has 5 unknowns (versus 30 with MiniSAD) and 30 constraints (versus 17 with MiniSAD). The number of degrees of freedom (or degrees of difficulty) being that of the number of variables, in principle the MiniMAD solution will require less computer time than MiniSAD, in this case. This will always be true since M+1 is less than 2N+M. However, with MiniMAD the larger number of constraints requires the inversion of bigger matrices and the reduction in number of degrees of difficulty with MiniMAD may be compensated by this computational factor.

THE QUADRATIC PROGRAMMING APPROACH

Another alternative is to minimize the sum of squares of errors subject to the constraints discussed previously. The problem is:

$$\min_{\underline{e},\underline{\delta}} \{y = \sum_{n-1}^{N} [e(n)]^2\} \tag{23}$$

subject to the nonnegativity conditiosn on the $\delta(\cdot)$, but not on the $e(\cdot)$, the ordering constraints (12), the deviation definition constraints (2), and the mass balance equation:

$$\sum_{n=1}^{N} e(n) = 0 \tag{24}$$

The problem involves 17 unknowns (versus 30 with MiniSAD and 5 with MiniMAD) and 17 constraints (versus 17 with MiniSAD and 30 with MiniSAD). However, only 4 of the 17 unknown are nonnegative which eases the computational times (Jønch-Clausen and Morel-Seytoux, 1977). The sizes of the various problems are summarized in Table 4-3.

Table 4-3. Problem Size Summary

Approach	MiniSAD	MiniMAD	QP
Number of Nonnegative Variables	2N+M	M+1	M
Number of Free Variables	0	0	N
Total Number of Variables	2N+M	M+1	N+M
Number of Equality Constraints	N+1	1	N+1
Number of Inequality Constraints	M-1	2N+M-1	M-1
Total Number of Constraints	N+M	2N+M	N+M

Appendix 4.2

Formulae for Calculation of the Muskingum Kernels and Parameters in Terms of Observed Inflow and Outflow Data

The Muskingum Routing Kernel

The kernel of the downstream discharge due to a unit impulse excitation of upstream inflow is:

$$k\ (u) = \frac{e^{\frac{-u}{K(1-\xi)}}}{K(1-\xi)^2} - \frac{\xi}{1-\xi} D_\delta(u) \qquad (1)$$

where K and ξ are the Muskingum parameters and $D_\delta(u)$ is the Dirac delta function.

Moments of the Muskingum Outflow Impulse Kernel

By definition the first moment of the response is:

$$\mu_1 = \int_0^\infty t\ k(t)\ dt \qquad (2)$$

Substitution of Eq. (1) in Eq. (2) yields the expression:

$$\mu_1 = \int_0^\infty \frac{t\ e^{\frac{-t}{K(1-\xi)}}}{K(1-\xi)^2} dt - \frac{\xi}{1-\xi} \int_0^\infty t\ D_\delta(t)\ dt$$

The second integral is identically zero from the very definition of the Dirac delta function. Thus μ_1 reduces to the first integral which can be rewritten as:

$$\mu_1 = \frac{1}{1-\xi} \int_0^\infty \frac{t\ e^{\frac{-t}{K(1-\xi)}}}{K(1-\xi)} dt$$

Let us calculate the definite integral defined by the relation:

$$I(a,b) = \int_a^b \frac{t}{K(1-\xi)} e^{\frac{-t}{K(1-\xi)}} dt$$

Integration by parts leads to:

$$I(a,b) = -t\, e^{\frac{-t}{K(1-\xi)}} \Big|_a^b + \int_a^b e^{\frac{-t}{K(1-\xi)}}\, dt$$

or

$$I(a,b) = -t\, e^{\frac{-t}{K(1-\xi)}} \Big|_a^b - K(1-\xi)\, e^{\frac{-t}{K(1-\xi)}} \Big|_a^b$$

Finally the expression for I(a,b) is obtained:

$$I(a,b) = -\, e^{\frac{-t}{K(1-\xi)}} [t + K(1-\xi)] \Big|_a^b \qquad (3)$$

With these notations the first moment can be written as:

$$\mu_1 = \frac{1}{1-\xi} I(o,\infty)$$

From Eq. (3) it follows that $I(o,\infty) = K(1-\xi)$. Thus, finally one obtains the (simple) result that:

$$\mu_1 = K \qquad (4)$$

By definition the second moment of the response is:

$$\mu_2 = \int_o^\infty t^2\, k(t)dt \qquad (5)$$

Substitution of the Muskingum outflow impulse kernel in Eq. (5) leads to:

$$\mu_2 = \frac{1}{1-\xi} \int_o^\infty \frac{t^2}{K(1-\xi)} e^{\frac{-t}{K(1-\xi)}} dt - \frac{\xi}{1-\xi} \int_o^\infty t^2\, D_\delta(t)dt$$

Again from the definition of the Dirac delta function the second integral is identically zero. The second moment takes the form:

$$\mu_2 = \frac{1}{1-\xi} J(o,\infty)$$

where J(a,b) is the integral:

$$J(a,b) = \int_a^b \frac{t^2}{K(1-\xi)} e^{\frac{-t}{K(1-\xi)}} dt$$

This integral can be evaluated by parts, namely:

$$J(a,b) = -t^2 e^{\frac{-t}{K(1-\xi)}} \Bigg|_a^b + 2 \int_a^b t \, e^{\frac{-t}{K(1-\xi)}} dt$$

or

$$J(a,b) = -t^2 e^{\frac{-t}{K(1-\xi)}} \Bigg|_a^b + 2K(1-\xi) \, I(a,b) \qquad (6)$$

In particular one can evaluate $J(o,\infty)$ namely:

$$J(o,\infty) = 0 + 2K(1-\xi) \, K(1-\xi)$$

Finally one obtains the second moment:

$$\mu_2 = 2K^2 (1-\xi) \qquad (7)$$

<u>The Muskingum Discrete Kernels</u>

By definition the discrete kernel is

$$\delta(t) = \int_o^t k\,(t-\tau)\, d\tau \qquad \text{for } t \leq 1$$

and $$\delta(t) = \int_o^1 k\,(t-\tau)\, d\tau \qquad \text{for } t > 1$$

Basically $\delta(t)$ is the response to a unit pulse. In particular for the Muskingum kernel one obtains for $t \leq 1$:

$$\delta(t) = \int_o^t \frac{e^{\frac{-(t-\tau)}{K(1-\xi)}}}{K(1-\xi)^2} d\tau - \frac{\xi}{1-\xi} \int_o^t D_\delta(t-\tau) d\tau$$

Because τ can take the value t the second integral (by definition of the Dirac delta function) is 1. Thus for $t \leq 1$ the expression for $\delta(t)$ is:

$$\delta(t) = \frac{1}{1-\xi} e^{\frac{-t}{K(1-\xi)}} \int_o^t \frac{e^{\frac{\tau}{K(1-\xi)}}}{K(1-\xi)} d\tau - \frac{\xi}{1-\xi}$$

or

$$\delta(t) = \frac{e^{\frac{-t}{K(1-\xi)}}}{1-\xi}\left[e^{\frac{t}{K(1-\xi)}} - 1\right] - \frac{\xi}{1-\xi}$$

or

$$\delta(t) = 1 - \frac{e^{\frac{-t}{K(1-\xi)}}}{1-\xi} \qquad t \leq 1 \qquad (8)$$

For t > 1 the expression for δ(t) is:

$$\delta(t) = \int_0^1 \frac{e^{\frac{-(t-\tau)}{K(1-\xi)}}}{K(1-\xi)^2}\,d\tau - \frac{\xi}{1-\xi}\int_0^1 D_\delta(t-\tau)\,d\tau$$

By definition of the Dirac celta function the second integral is zero and δ(t) reduces to the relation:

$$\delta(t) = \frac{1}{1-\xi}\, e^{\frac{-t}{K(1-\xi)}} \int_0^1 \frac{e^{\frac{\tau}{K(1-\xi)}}\,d\tau}{K(1-\xi)}$$

$$\delta(t) = \frac{1}{1-\xi}\, e^{\frac{-t}{K(1-\xi)}} \left[e^{\frac{1}{K(1-\xi)}} - 1\right]$$

$$\delta(t) = \frac{[e^{\frac{1}{K(1-\xi)}} - 1\,]}{1-\xi}\, e^{\frac{-t}{K(1-\xi)}} \qquad t > 1 \qquad (9)$$

From Eqs. (8) and (9) one deduces the values of the discrete kernels for various integer time periods:

$$\delta(1) = 1 - \frac{e^{-\frac{1}{K(1-\xi)}}}{1-\xi} \qquad (10)$$

$$\delta(m) = \frac{e^{\frac{1}{K(1-\xi)}} - 1}{1-\xi}\, e^{\frac{-m}{K(1-\xi)}} \qquad m = 2,3. . . . \qquad (11)$$

<u>Moments of the Muskingum Discrete Kernels</u>

By definition then the first moment is:

$$m_1 = \int_0^\infty t\,\delta(t)\,dt$$

Because there are 2 separate expressions for $\delta(t)$ for $t \leq 1$ and $t > 1$, the integral is broken down into 2 integrals:

$$m_1 = \int_0^1 t \left[1 - \frac{e^{\frac{-t}{K(1-\xi)}}}{1-\xi} \right] dt +$$

$$\int_1^\infty \frac{[e^{\frac{1}{K(1-\xi)}} -1]}{1-\xi} \; t \; e^{\frac{-t}{K(1-\xi)}} \; dt$$

which can be rewritten in the form:

$$m_1 = \int_0^1 tdt - K\, I(0,1) + K\, [e^{\frac{1}{K(1-\xi)}} -1]\, I(1,\infty)$$

$$m_1 = \frac{1}{2} - K \left[- e^{\frac{-1}{K(1-\xi)}} [K(1-\xi) + 1] + K(1-\xi) \right]$$

$$+ K\, [e^{\frac{1}{K(1-\xi)}} -1]\, e^{-\frac{1}{K(1-\xi)}} [1 + K(1-\xi)]$$

$$m_1 = \frac{1}{2} + K\, e^{-\frac{1}{K(1-\xi)}} [1 + K(1-\xi)] - K^2(1-\xi)$$

$$+ K\, [1 + K(1-\xi)] - K\, e^{-\frac{1}{K(1-\xi)}} [1 + K(1-\xi)] \quad (12)$$

which simplifies to:

$$m_1 = K + \frac{1}{2}$$

The second moment is:

$$m_2 = \int_0^\infty t^2\, \delta(t)\, dt$$

Again the integral is broken down into two integrals, giving:

$$m_2 = \int_0^1 t^2 \left[1 - \frac{e^{\frac{-t}{K(1-\xi)}}}{1-\xi} \right] dt \quad +$$

$$\int_1^{\infty} \frac{[e^{\frac{1}{K(1-\xi)}} - 1]}{1-\xi}\, t^2\, e^{-\frac{t}{K(1-\xi)}}\, dt$$

In terms of the integrals I(a,b) and J(a,b) previously defined one obtains for m_2 the expression:

$$m_2 = \frac{1}{3} - K\, J(0,1) + [e^{\frac{1}{K(1-\xi)}} - 1]\, J(1,\infty)$$

$$m_2 = \frac{1}{3} - K\, J(0,\infty) + K\, e^{\frac{1}{K(1-\xi)}}\, J(1,\infty)$$

From the expression of J(a,b) in Eq. (6) one finds that:

$$J(0,\infty) = 2K^2(1-\xi)^2$$

and

$$J(1,\infty) = e^{\frac{-1}{K(1-\xi)}} + 2K(1-\xi)\, I(1,\infty)$$

and from Eq. (3) one finds that:

$$I(1,\infty) = e^{\frac{-1}{K(1-\xi)}}\, [1 + K(1-\xi)]$$

Thus m_2 transforms to:

$$m_2 = \frac{1}{3} - 2K^3(1-\xi)^2 + K + 2K^2(1-\xi)\,[1 + K(1-\xi)]$$

Finally:

$$m_2 = \frac{1}{3} + K + 2K^2(1-\xi) \qquad (13)$$

<u>Moments of Response Due to a Unit Pulse Excitation Occurring During the νth Period.</u>

By definition the first moment in this case, denoted m_1^{ν}, is:

$$m_1^{\nu} = \int_0^{\infty} t\, \delta(t - \nu + 1)\, dt \qquad (14)$$

Since the value of $\delta(\ )$ for negative arguments is identically zero, then Eq. (14) is equivalent to:

$$m_1^{\nu} = \int_{\nu-1}^{\infty} t\ \delta(t - \nu + 1)dt$$

Let the new variable $t' = t - \nu + 1$ be defined.
The integral expression for m_1^{ν} becomes:

$$m_1^{\nu} = \int_0^{\infty} (t' + \nu - 1)\ \delta(t')\ dt'$$

or

$$m_1^{\nu} = \int_0^{\infty} t'\ \delta(t')dt' + (\nu - 1) \int_0^{\infty} \delta(t')dt'$$

or

$$m_1^{\nu} = m_1 + (\nu - 1)$$

$$m_1^{\nu} = K + (\nu - \frac{1}{2}) \qquad (15)$$

The second moment is:

$$m_2^{\nu} = \int_0^{\infty} t^2\ \delta(t - \nu + 1)\ dt \qquad (16)$$

Proceeding similarly one obtains:

$$m_2^{\nu} = \int_0^{\infty} t'^2\ \delta(t')dt' + 2(\nu - 1) \int_0^{\infty} t'\ \delta(t')dt' + (\nu - 1)^2 \int_0^{\infty} \delta(t')dt'$$

or:

$$m_2^{\nu} = m_2 + 2(\nu-1)m_1 + (\nu - 1)^2$$

or:

$$m_2^{\nu} = \frac{1}{3} + K + 2K^2\ (1-\xi) + 2(\nu - 1)(K + \frac{1}{2}) + (\nu - 1)^2$$

which noting that $(\nu - \frac{1}{2})^2 = (\nu - 1 + \frac{1}{2})^2 = (\nu - 1)^2 + (\nu - 1) + \frac{1}{4}$

can be rearranged in the form:

$$m_2^{\nu} = \frac{1}{12} + 2K^2 (1-\xi) + 2K(\nu - \frac{1}{2}) + (\nu - \frac{1}{2})^2 \quad (16)$$

Moments of the Response to a General Sequence of Pulses of Magnitudes I (ν).

From the method of superposition (note that the moments are linear in the response) then the general first moment of the response is:

$$M_1 = K \sum_{n=1}^{N} I(n) + \sum_{n=1}^{N} (n - \frac{1}{2}) I(n)$$

where N is large enough to include all time periods for which the response 0(n) is nonzero. The first moment in terms of the 0(n) is

$$M_1 = \sum_{n=1}^{N} (n - \frac{1}{2}) \left[\frac{0(n) + 0(n-1)}{2} \right] \quad (16)$$

Thus, given the inflows I(ν) and the outflows 0(n) the parameter K can be estimated:

$$K = \frac{M_1 - \sum_{n=1}^{N} (n - \frac{1}{2}) I(n)}{\sum_{n=1}^{N} I(n)} \quad (17)$$

Similarly the second moment is:

$$M_2 = [2K^2(1-\xi) + \frac{1}{12}] \sum_{n=1}^{N} I(n) + 2K \sum_{n=1}^{N} (n - \frac{1}{2}) I(n) + \sum_{n=1}^{N} (n - \frac{1}{2})^2 I(n)$$

The second moment can be calculated directly from the outflows 0(n) as:

$$M_2 = \sum_{n=1}^{N} (n - \frac{1}{2})^2 \left[\frac{0(n) + 0(n - 1)}{2} \right] \quad (18)$$

Thus given the inflows and outflows the parameter ξ can also be estimated by the relation:

$$2K^2(1-\xi) + \frac{1}{12} = \frac{M_2 - 2K \sum_{n=1}^{N} (n - \frac{1}{2})\, I(n) - \sum_{n=1}^{N} (n - \frac{1}{2})^2\, I(n)}{\sum_{n=1}^{N} I(n)} \qquad (19)$$

Chapter 5

RIVER AND CANAL MORPHOLOGY

by

D. B. Simons, Associate Dean of Engineering Research and Professor of Civil Engineering, Colorado State University, Fort Collins, Colorado

Chapter 5

RIVER AND CANAL MORPHOLOGY

5.1 INTRODUCTION

Waterways have been continuously studied by man. He has been alternately blessed by those life-giving streams when under control and plagued by their destructive nature when out of control. The purpose of this chapter is to review the morphology of alluvial rivers and canals.

The geometry of alluvial channels is the result of the interaction between the water discharge, the quantity and character of the sediment discharge, and the composition of the bed and bank material. The physical characteristics of channels include the shape of the channel cross section, the configuration of the bed including the forms of bed roughness, the riffles and bars, the longitudinal profile of the river channel, and the channel pattern implying the configuration of the plan of the river--straight, meandering, or braided. Also, the geometry of meanders and braids and the characteristics of the river valleys and river flood plains, which normally confine the flow at flood stage, play an important role in channel geometry.

5.2 CHARACTERISTICS OF RIVERS

River characteristics are important to everyone dealing with water resources, whether from the viewpoint of flood control, navigation, stabilization, environmental problems, irrigation or water resources development for municipalities and industry. A discussion of the diverse and complex facets of river geometry and related river mechanics follows.

5.3 VARIABLES AFFECTING CHANNEL GEOMETRY AND BED ROUGHNESS

Variables affecting channel geometry and bed roughness are numerous and interrelated. Their nature is such that, unlike rigid boundary hydraulic problems, it is not possible to isolate and study the role of an individual variable. For example, if on the average velocity, additional related variables respond to the changing depth, not only will the velocity respond to the change in depth but also the form of bed roughness, the position and shape of alternate, middle, and point bars, the shape of cross section, the magnitude of sediment discharge, and so on.

There are several variables which influence the geometry of river channels. Consider some of the most important variables. The principal ones are:

$$\phi[V, D, S, \rho, \mu, g, d, \sigma, \rho_s, S_p, S_R, S_c, f_s, C_T] = 0 \qquad (5\text{-}1)$$

in which

C_T = concentration of bed-material discharge,
V = velocity,
D = depth,
S = slope of energy grade line,
ρ = density of water-sediment mixture,
μ = apparent dynamic viscosity of the water-sediment mixture,
g = gravitational constant,
d = representative fall diameter of the bed material,
σ = measure of the size distribution of the bed material,
ρ_s = density of sediment,
S_p = shape factor of the particles,
S_R = shape factor of the reach of the stream,
S_c = shape factor of the cross section of the stream, and
f_s = seepage force in the bed of the stream.

The majority of these variables were discussed in the analysis of forms of bed roughness by Simons and Richardson (1962) and by Simons and Sentürk (1976).

Shape Factor

The shape factor for the reach (S_R) in Eq. (5-1) was included to focus attention on the energy losses resulting from the nonuniformity of the flow in a natural stream caused by the bends and the nonuniformity of the banks. The study of these losses in natural channels has long been neglected.

Seepage Force

Slope stability is difficult to analyze because of the engineer's somewhat limited knowledge regarding the shear strength of soils and the heterogeneous nature of most embankments. When the flow of ground water is encountered or created by the impoundment of water, seepage forces and their accompanying pressures complicate the design of stable earthen slopes.

Slope failures in which steady seepage forces have a significant role can be divided into two basic types: a) the massive movement of earth along a slip surface, and b) the progressive sloughing of the slope surface. The massive movement of earth has received the most attention because many engineers associate slope failure with the design of earth dams. In such designs, seepage is generally minimized and confined within the downstream slope by means of impervious cores and rock toes, thereby eliminating the need for consideration of surface failure on the downstream slope. Upstream rip-rap protection limits surface deterioration. However, the instantaneous drawdown case analyzed in determining the upstream slope of a dam represents the most severe seepage condition for the massive failure of earth slopes. Unfortunately, in much engineering construction, seepage must be allowed to penetrate the embankment slope. In these instances, the second type of failure is possible and may result in local failures or trigger considerable material displacement.

In channel bank stability problems, knowledge of the influence of steady seepage is needed. Here the engineer is faced with the two types of failure and needs to know when and to what magnitude the stability of a slope is affected by various seepage conditions. Also, if flow in the channel is controlled by weirs or some other device, information as to the effect of tailwater on slope stability is needed. This latter item is not covered in the literature but an appraisal of the standard computation methods indicates that tailwater could adversely affect stability either by reduction of friction forces through bouyant action or forcing a concentration or change in the potential pattern near the surface.

Also, the seepage force, f_s, which occurs whenever there is inflow or outflow through the bed material and banks of a channel in permeable alluvium, may have a significant effect on bed configuration and resistance to flow (Simons and Richardson, 1962). The inflow or outflow through the interface between water and the bed and the bank depends on the difference in pressure across the interface and the permeability of the bed material. For flow in a natural channel, the pressure difference will depend on the location of the water table in the alluvium and also on the variables in Eq. (5-1); now, seepage force is an independent variable insofar as resistance to flow is concerned.

If there is inflow, the seepage force acts to reduce the effective weight of the sand and, consequently, the stability of the bed material. If there is outflow, the seepage force acts in the direction of gravity and increases the effective weight of the sand and the stability of the bed material. As a direct result of changing the effective weight, the seepage forces can influence the form of bed roughness and the resistance to flow for a given channel slope, channel shape, bed material, and discharge.

A rather common field condition is outflow from the channel during the rising stage; this process builds up bank storage and increases the stability of the bed and bank material. In the falling stage, the situation is reversed; inflow to the channel reduces the effective weight and stability of the bed and bank material and influences the form of bed roughness, the resistance to flow, the stability, and the geometry of the channel.

5.4 RELATIONS DEFINING RIVER GEOMETRY AT A CROSS SECTION

Relations Defining River Geometry at a Cross Section

Relations defining the channel cross section have been proposed by various researchers. One of the most widely accepted sets of relations are probably those presented by Leopold and Maddock (1953). Their relations are based on the analysis of extensive field data. More recently this problem has been treated theoretically at CSU. The theoretical relations are similar to those proposed by Leopold and Maddock. The derived relations at a section are:

$$W \sim Q^{0.24} \tag{5-2}$$

$$y_o \sim Q^{0.46} \tag{5-3}$$

$$S \sim Q^{0.00} \tag{5-4}$$

$$V \sim Q^{0.30} \tag{5-5}$$

Equation (5-4) implies that slope is constant at a cross section. This is not quite true. At low flow the effective channel slope is that of the thalweg that flows from pool through crossing to pool. At higher stages the thalweg straightens, somewhat shortening the path of travel and increasing the local slope. In the extreme case river slope approaches the valley slope at flood stage. It is during high floods that the flow often cuts across the point bars developing chute or flow channels. This path of travel verifies the shorter path the water takes and that a steeper channel prevails under this condition.

Relations Defining River Geometry Along the Channel

In addition to the at-a-station hydraulic geometry, Leopold and Maddock (1953) developed relations for the variations of hydraulic geometry in the downstream direction. The relations were developed using the bankfull geometry and bankfull discharge. These relations have been derived using the basic equations of fluid mechanics at CSU. The derived equations which are almost identical to the empirical relations proposed by Leopold and Maddock are:

$$W_b \sim Q_b^{0.46} \tag{5-6}$$

$$Y_b \sim Q_b^{-0.46} \tag{5-7}$$

$$S \sim Q_b^{-0.46} \tag{5-8}$$

$$V_b \sim Q_b^{0.08} \quad (5\text{-}9)$$

Here the subscript "b" indicates the bankfull condition. These relations indicate how bankfull depth, velocity, slope and width vary as one moves from the headwaters of a channel to its downstream end.

5.5 BED CONFIGURATIONS

Sand bed rivers have well-recognized forms of bed roughness. The accepted forms of bed roughness are ripples, dunes, a transition region, plane or flat bed, standing waves, and antidunes. These forms were studied, identified, and reported by Simons and others. Their characteristics depend on depth of flow, velocity of water, and suspended sediment characteristics including viscosity, size, density and gradation of the bed material.

These forms of bed roughness are subclassified into a lower and upper regime. In the lower regime the bed forms are ripples and dunes; in the upper regime the bed is plane or there are standing waves and antidunes. In the lower regime resistance to flow is large, velocities are small, and bed material transport is small. In the upper regime resistance to flow is small, velocities are relatively large, and transport rates of bed material are large. The shift from lower regime through transition to upper regime occurs at a Froude number of about 0.17-0.18 for fine sands.

It is important to note that as river stage rises much of the river bed may shift from lower regime flow to upper regime flow. In so doing Mannings n values drop from 0.025-0.030 to 0.014-0.018. Hence, the decrease in resistance with rising discharge permits the river to handle large increases in discharge with less increase in stage than one normally expects. For example, the stage at Hardinge bridge on the Ganges rose only 5 ft for a change in discharge from 1,350,000 cfs to 2,585,000 cfs. The increase may have been less except for the partial control of the flow at the bridge.

Sand Bars

Some synonyms of bars are sand waves, banks, sand banks and deltas. Bars are bed forms having lengths of the same order as the channel width or greater, and heights comparable to the mean depth of the generating flow. Several types of bars are defined: (a) Point bars are deposits of sediment that occur on the convex side or inside of channel bends. Their shape may vary with changing flow conditions, but they do not move significantly relative to bends. However, the general magnitude and location of the bar varies with discharge. An example of point bars is shown in Fig. 5-1; (b) Alternate bars are distributed periodically along one and then the other bank of a channel. Their lateral extent is significantly less than

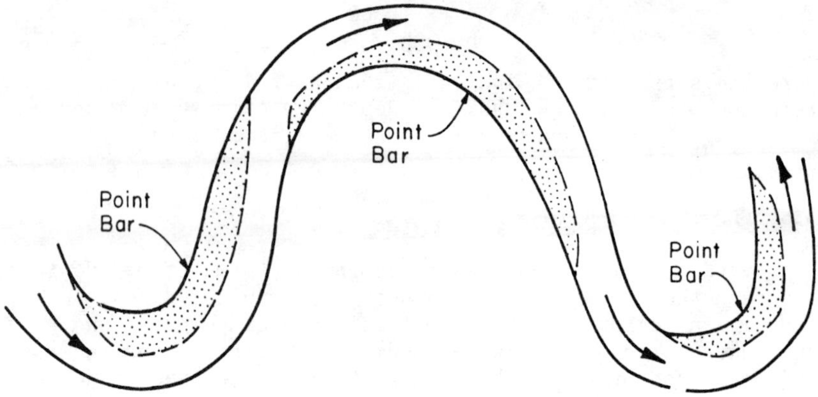

Fig. 5-1. View of Three Point Bars.

the channel width. Alternate bars move slowly downstream. An example of alternate bars is shown in Fig. 5-2. These bars occupy nearly the full channel width. They occur both as isolated and as periodic forms along a channel, and move downstream. Examples of point bars, alternate bars, and tributary bars are illustrated in Fig. 5-3.

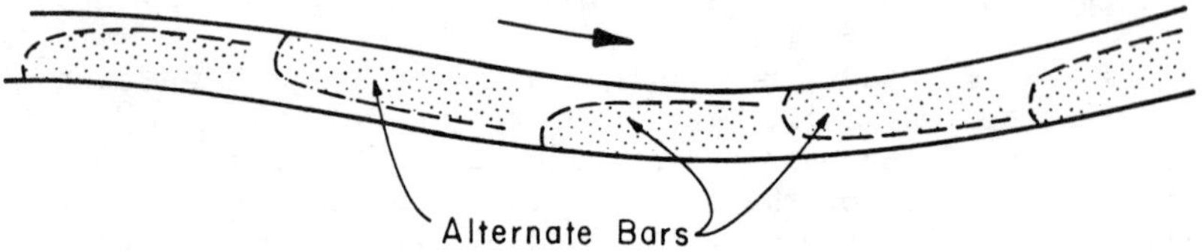

Fig. 5-2. Aerial View of Alternate Bars in a Straight Uniform Reach.

In longitudinal sections tributary bars are approximately triangular, with long gentle upstream slopes and short downstream slopes. These slopes are approximately equal to the angle of repose of the bed material. Bars generated by high flows frequently appear as small islands during low flows. Parts of the upstream slopes of bars are often covered with ripples and dunes.

Mode of Bed Material Transport

The most common modes of bed material transport are (1) for the individual grains to move up the back of the ripple or dune and avalanche down the face, or (2) to advance the face of the ripple or dune in the downstream direction. These two modes of

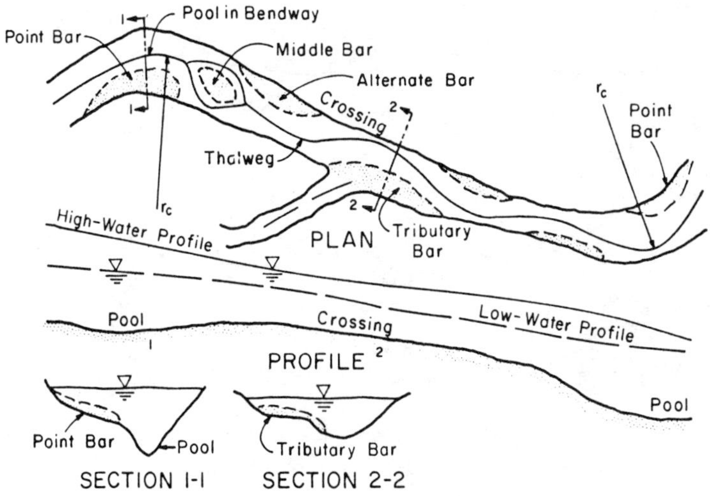

Fig. 5-3. Illustration of Types of Bars that Occur in Alluvial Channels.

transport are illustrated in Figs. 5-4 and 5-5. In Fig. 5-4 the general pattern of the movement can be seen. The return current in the wake retains the downstream slope at an angle a little larger than the angle of repose of the material. The coarser materials are deposited on the toe over a thin deposit of finer materials. The smaller bed material particles are transported towards the summit of the configuration (Fig. 5-4). There is a critical value of the angle of repose after which the toe of the slope slides in the direction of the main flow (Fig. 5-4). The bed material, and particularly the accumulated coarse material, is spread into a layer, forming a permeable zone overlying a less permeable zone. In general, when the bed material is excavated it is possible to see these layers presenting higher and lower permeability than other strata (Fig. 5-4).

The second mode of transport is illustrated in Fig. 5-5. The accumulated material (line 2 on the figure) avalanches down in the direction of the arrows to form a more stable slope (line 3 on the figure). This results in the formation of pockets of material of different size and gradation.

The velocity of the bed forms is not constant. New bed forms can be generated on existing ones. This new bed form proceeds downstream more rapidly until it reaches the crest of the underlying bed form, adding to its amplitude. During this period of time the velocity of the bed form is incessantly variable.

The solid particles travel rapidly on the surface of the bed forms. They jump over the crest of it, following either the path shown by A or by B on Fig. 5-4. If they follow path B

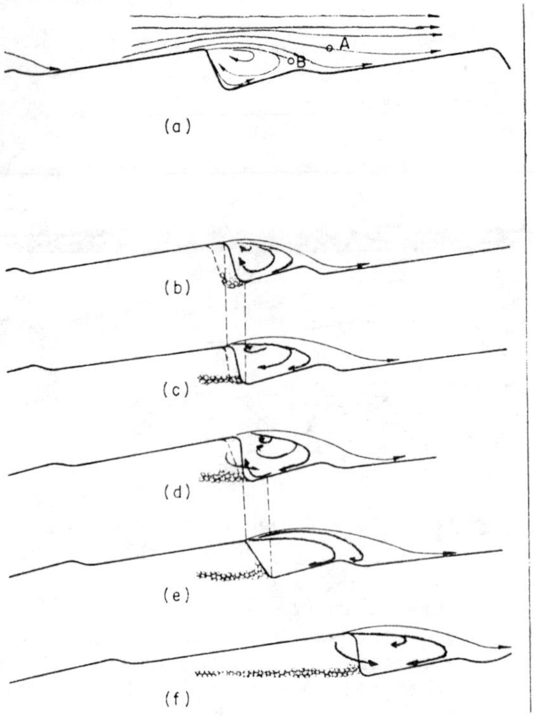

Fig. 5-4. Formation of a Highly Permeable Bottom Layer in Undulating Beds Overriding a Thin Layer of Fine Material (after Simons and Sentürk, 1976).

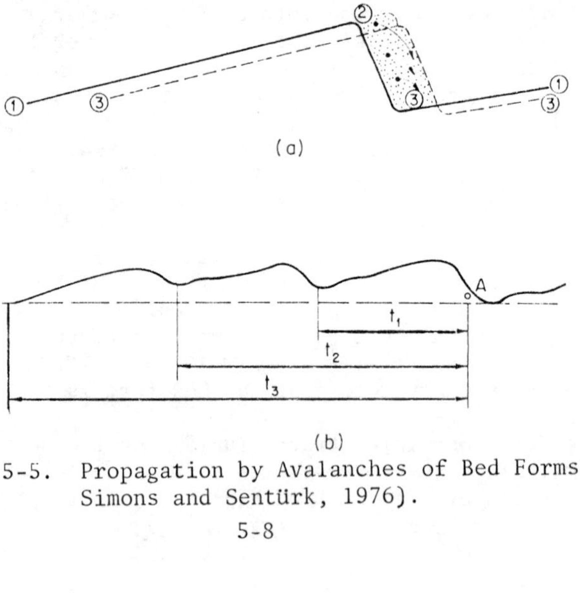

Fig. 5-5. Propagation by Avalanches of Bed Forms (after Simons and Sentürk, 1976).

they are retained in the valley formed between two configurations. The return current might be so high sometimes that new and smaller bed forms can be formed in that region. When the bed material comes to a stop it is buried by the advancing bed forms and rests there until it is liberated by a new valley arriving at that location. The particle A shown on Fig. 5-5 must wait a period of time (t) until it continues its motion again. Thus, most movement of the bed material particles is in steps of the order of magnitude of one or more ripple or dune lengths, separated by rest periods dependent on ripple or dune height and their velocity.

If the velocity of single particles is compared to the velocity of bed forms, it is obvious that the maximum instantaneous velocity of particles greatly exceeds the maximum instantaneous velocity of the bed forms. However, it is possible to see that the average velocity of bed forms is higher than the long term average velocity of single particles. Those two velocities are approximately equal to each other in case of a plane bed with sediment movement. In upper flow regime the velocity of bed forms may be negative if the direction of the main flow is assumed positive. The celerity of bed forms is approximated by many authors.

Upper Flow Regime

In the upper flow regime, resistance to flow is relatively small and sediment transport is large. The usual bed forms are plane bed or antidunes. The water surface is in phase with the bed surface except when an antidune breaks, and normally the fluid does not separate from the boundary. A small separation zone may exist downstream from the crest of an antidune prior to its breaking. Resistance to flow is the result of grain roughness with the grains moving, of sand waves and their subsidence, and of energy dissipation when the antidunes break. The mode of sediment transport is for individual grains to roll almost continuously downstream in sheets one to several grain diameters thick. However, when antidunes break, much of the bed material in the region of the breaking wave is briefly suspended, then most sediment movement stops temporarily and there is return of suspended particles to the bed.

Transitions

The "transition" zone encompasses the bed forms that occur during the passage from lower regime to upper regime. This transition is not unique in loose boundary hydraulics. The passage from plane bed to ripples and the passage from ripples to dunes constitute two more transitions with very interesting properties (Fig. 5-6).

The transition from dunes to plane bed is discussed in this section. The transition from plane bed to ripples and dunes and the transition from ripples to dunes will be considered subsequently.

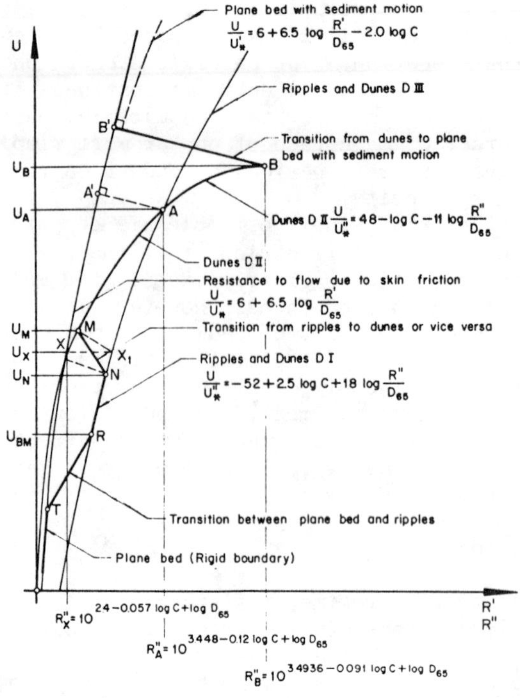

Fig. 5-6. Variation of the Resistance to Flow for a Given Bed Material and Constant Energy Slope (after Simons and Sentürk, 1976). R' and R" are shown Separately on the Abscissa and U is the Ordinate Value. The Black Lines show the Variation of the Resistance for S Constant and Bed Material Constant.

The Transition from Dunes to Plane Bed with Sediment Transport

When certain flow conditions occur dunes are washed out and a plane bed with sediment transport develops and the flow changes from a transitional state to upper regime.

The transition requires a sufficient slope of energy gradient. This critical slope is a function of the material size. Also, it is possible to obtain a dune bed for slopes larger than the critical slope, due to hysteresis phenomena. Hence, the bed configuration in the transition zone is erratic. It may range from that typical of the lower flow regime to that typical of the upper flow regime, depending mainly on antecedent conditions. If the bed configuration is dunes, the depth or slope can be increased to values more consistent with those of the upper flow regime without changing the bed form; or, conversely, if the bed is plane, depth and slope can be decreased to values more consistent with those of the lower flow regime without changing the bed form. Often in the transition from the lower to the upper flow regime, the dunes will decrease in amplitude and increase in length before the bed becomes plane (washed-out dunes). Resistance to flow and sediment transport have the same variability as the bed configuration in the transition.

The magnitude of the stream power ($\tau_o U$) at which the dunes or transition roughness changes to the plane bed is dependent on the fall velocity of the bed material. Dunes of fine sand (low fall velocity) are washed out at lower values of stream power than are dunes of coarser sand. Consequently, in the flume experiments, where the depths were a shallow 15-30 cm (0.5-1.0 ft), the plane bed formed with finer sands at smaller velocities and smaller Froude numbers (F_r). With coarse sands larger energy slopes are required to affect the change from transition to the plane bed; the result is larger velocities and larger Froude numbers. Hence, in a flume containing fine sand, the plane-bed condition commonly exists after the transition and persists over a significant range of Froude numbers ($0.3 \leq F_r \leq 1.0$). However, if the sand is coarse and the depth is shallow, the transition zone may not terminate until the Froude number is so large that the subsequent bed form may be antidunes rather than plane bed. In natural streams, because of their greater depths, the change from transition to plane bed may occur at a much lower Froude number than in flumes.

The resistance to flow for flow over a plane bed results from a moving grain roughness and its magnitude is small. The Manning's n varies from 0.012 to 0.015 and the discharge coefficient $C/\sqrt{g}$ ranges from 14 to 23. For flow over a plane bed with sediment movement, grain roughness is not the usual type because grains roll, hop, and slide along the bed. The resistance to flow is slightly less than for flow over a static plane bed, which is essentially an artificial rigid-boundary condition that exists after molding the bed in that form and when stream power is insufficient to cause significant transport of the bed material.

In many instances when the flow conditions are such that the bed form is in the transition zone, the bed configuration will oscillate between dunes and plane bed. This phenomenon can be explained by the changes in resistance to flow and, consequently, the changes in depth and slope as the bed form changes. The median diameter D_{50} of the sand forming the boundaries is also variable. This phenomena is illustrated in Fig. 5-6.

These intermediate bed forms are not stable. Also, they are different from dunes. They are sometimes called washed-out dunes. In reality they are bed forms that occur between the dune bed and plane bed. They are particular bed forms belonging to the transition region.

The Transition from Ripples to Dunes

The transition from ripples to dunes may encompass a relatively large region and a variety of flow conditions. In general, during this transformation from ripples to dunes the resistance to flow decreases slightly and is greater than the resistance of a dune bed with similar flow conditions for shallow depths.

The geometry of bed forms in this transition includes both ripples and dunes. With an increase in energy input to the channel, the ripples are eliminated and only the dunes remain. The slope, S, and the diameter of the bed material, D_{50}, vary randomly with time and the velocity of a single particle is significantly affected by the formation of troughs between newly-formed dunes and by the apparent random elimination of ripples on the dunes with time.

The Transition from Plane Bed to Ripples

When the flow conditions are such that $\frac{\tau_o}{\gamma_s' D_{50}}$ is smaller than a given critical value the bed material is static and the channel behaves as though rigid. The beginning of sediment transport causes the beginning of ripple development. The development of ripples is rapid. The resistance to flow increases rapidly with the formation of ripples. If the discharge is constant the velocity decreases but the motion of the solid particles may continue due to the increase of the hydraulic radius, R. This is another transition. It occurs before the ripple bed is fully established. The resistance to flow is rather small, but decreases with an increase in the value of Q or increases rapidly with a decrease in the value of Q and stops when the hydraulic radius drops approximately to half of the hydraulic radius corresponding to the beginning of motion. If Q is increased subsequently, the resistance is dictated by a rigid boundary condition until the hydraulic conditions exceed the critical value required for the initiation of motion. If Q is increased while motion of the bed material exists, the resistance to flow is intermediary between the resistance of a plane bed and a rippled bed.

5.6 PREDICTION OF BED FORMS

Work on prediction of bed forms has involved both theoretical and empirical approaches. The latter approach has been more successful than the former.

Theoretical Approach

Because of the complexity of predicting bed forms, all mathematical models to date have essentially made use of the two-dimensional potential flow assumption. This approach is based on the continuity equation for sediment and yields

$$\gamma_s \frac{\partial y}{\partial t} + \frac{\partial q_s}{\partial x} = 0 \qquad (5\text{-}10)$$

Empirical Approach

The variables that affect the flow in alluvial channels are given in Eq. (5-1). From dimensional analysis and eliminating parameters of secondary importance or for which data are not available, the following functional equation can be derived

$$\text{Bed form type} = f_3(S_E, \frac{R}{D_i}, F_r, F_{r_*}, \frac{U_*}{w_{50}}, R_w) \qquad (5\text{-}11)$$

Graphical Analysis: In the absence of universally acceptable analytical solutions for the prediction of bed forms, some research workers have tried to bridge the gap by presenting dimensional and nondimensional plots based primarily on flume data and some data from natural channels.

A relation, Fig. 5-8, was developed by Simons and Richardson (1966) that related stream power, median fall diameter of bed material, and form roughness. If the depth, slope, velocity and fall diameter of bed material are known, one can predict the form of bed roughness by using this relation. Figure 5-7 is based on extensive data from flumes collected by Simons and Richardson and reported by Guy, et al. (1966), and on data from several rivers and canals including: (1) the Elkhorn River near Waterloo, Nebraska (Beckman and Furness, 1962); (2) the Rio Grande, 20 miles above El Paso, Texas; (3) the Middle Loup River at Dunning, Nebraska (Hubbell and Matejka, 1959); (4) the Rio Grande at Cochiti, near Bernalillo, and Angostura heading, New Mexico (Culbertson and Dawdy, 1964); and (5) India and Pakistan data where both lower and upper regime flow have been observed in large irrigation canals with fine-sand beds. The bed forms given by Fig. 5-7 are in good agreement with those observed by Nordin (1964) in the Rio Grande in which the depths were less than 5 ft and the velocity was relatively high. However, in the case of the Mississippi River (Jordan, 1965), with velocities in the same range as for the Rio Grande but with depths up to 50 ft (15 m), the figure predicts flat beds in cases where evidence indicates that the beds are dune-covered.

Other investigators have proposed graphical relations. Some of these are tabulated in Table 5-1.

Athuallah (1968) working with Simons, studied many groups of dimensionless parameters based on Eq. (5-11). All of them exhibited considerable scatter. Of the graphical procedures they attempted, two have the most promise. The first of these,

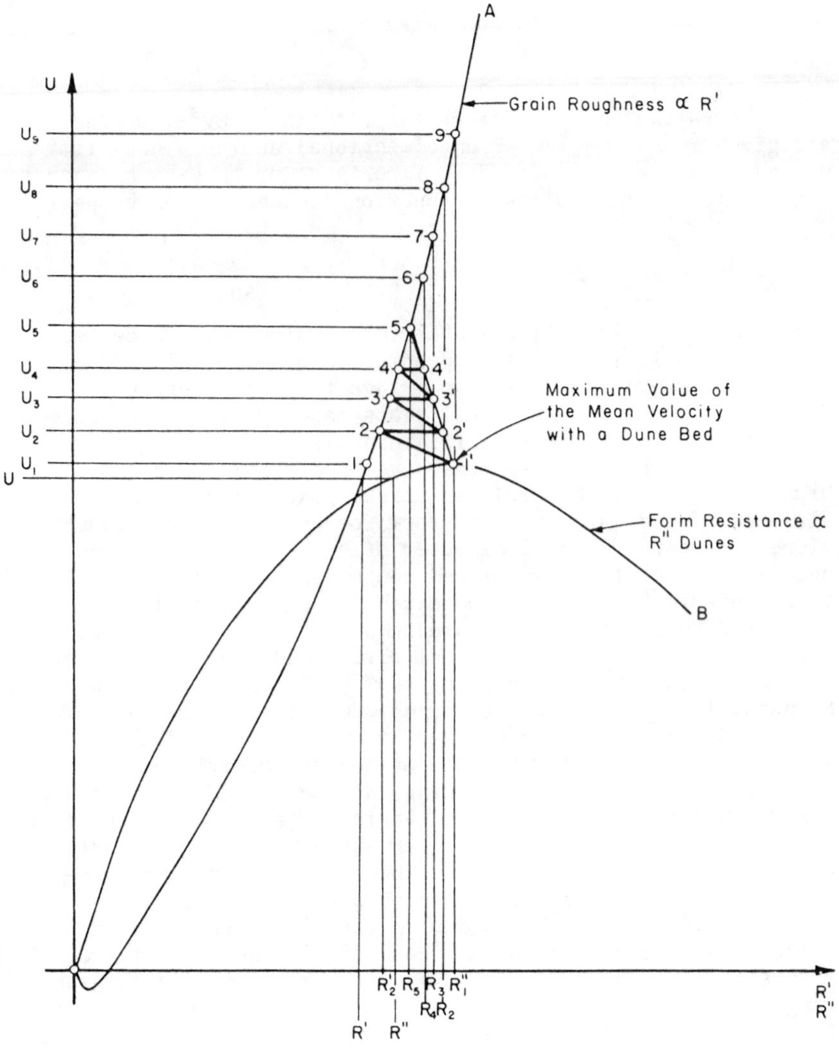

Fig. 5-7. Transitional Resistance and Velocity as the Bed Changes from Dunes to Plane (after Simons and Sentürk, 1976).

Table 5-1. Graphical Analysis

Variables	Investigator	Comments
$\frac{U_* D_{50}}{\nu}, \frac{U_*}{w}$	Liu (1957) Albertson et al. (1958)	Criterion based on flume data did not predict field data well. Most promise appears to be for purpose of prediction of beginning of motion.
$\frac{\tau_o}{(\rho_s - \rho) g D_{50}}, \frac{U}{\sqrt{gR}}$	Garde and Albertson (1959)	Criterion did not predict Rio Grande data well.
$\frac{R}{D_{50}}, \frac{S}{\frac{\rho_s - \rho}{\rho}}$	Garde and Raju (1963)	Considerable scatter evident, especially with the Gilbert and U.S. Geological Survey data.
$\frac{g D_{50}}{(U_*)^2}, D_{50}$	Bogardi (1958)	Difficulty in using the same criterion for flume and field data.
$\tau_o U, D_{50}$	Simons and Richardson (1966)	Difficulty in using the same criterion for flume and large rivers, but relation does fairly well for natural streams.
$\frac{U_*}{w}, S$	Athaullah (1968)	Failed to discriminate between bed forms in natural systems.
$\frac{\tau_o}{\gamma_s' D_{50}}, \frac{Dw}{\nu}$	Sentürk (1973)	Define bed forms according to their resistance to flow. Failure to define antidunes.

shown in Fig. 5-9, uses F_r and R/D_{50} as parameters. The Froude number F_r best describes the effect of inertia and gravitational phenomena on the system whereas R/D_{50} is a measure of relative roughness.

Figure 5-8 can be used to predict bed forms if the mean velocity, hydraulic radius and mean fall diameter of the bed material are known. It should be noted that all graphical methods are limited by their inability to consider all of the variables involved in the problem. No doubt, this limitation is responsible for much of the scatter evident in all relations. The multivariate analysis relying on the statistical or stochastic approach can be applied to consider all of the variables involved. However, this method of analysis is complicated by the new concepts and the advanced mathematical knowledge required. The reader can refer to Nordin and Algert's (1966), Ashida and Tanaka's (1967), Nordin's (1968) and Athaullah's (1968) works if interested.

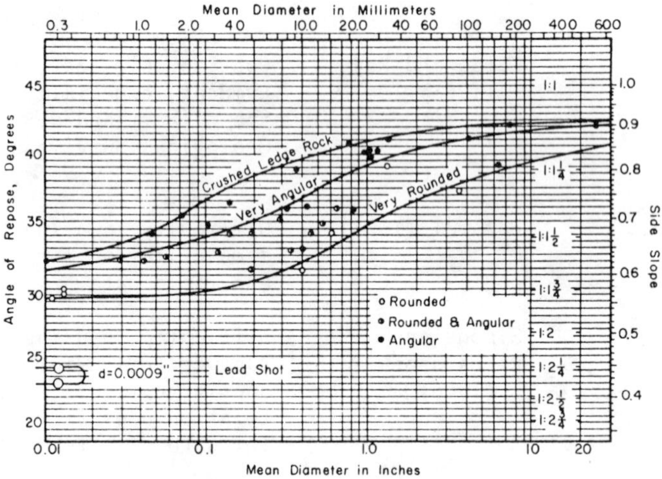

Fig. 5-8. Angle of Repose for Non-Cohesive Material, after Simons and Albertson (1960).

5.7 ROUGHNESS CHARACTERISTICS OF CHANNELS

The roughness of alluvial channels is variable and complex. Roughness is a function of such variables as channel geometry, channel irregularities, type of bed and bank material, response of bed material to flow at the bed-water interface resulting in dunes and bars, the rate of bed material discharge in the channel, the characteristics of channel alignment and slope, the temperature of the water-sediment complex flowing in the channel, the characteristics and quantity of wash load, the intensity of turbulence, and other factors.

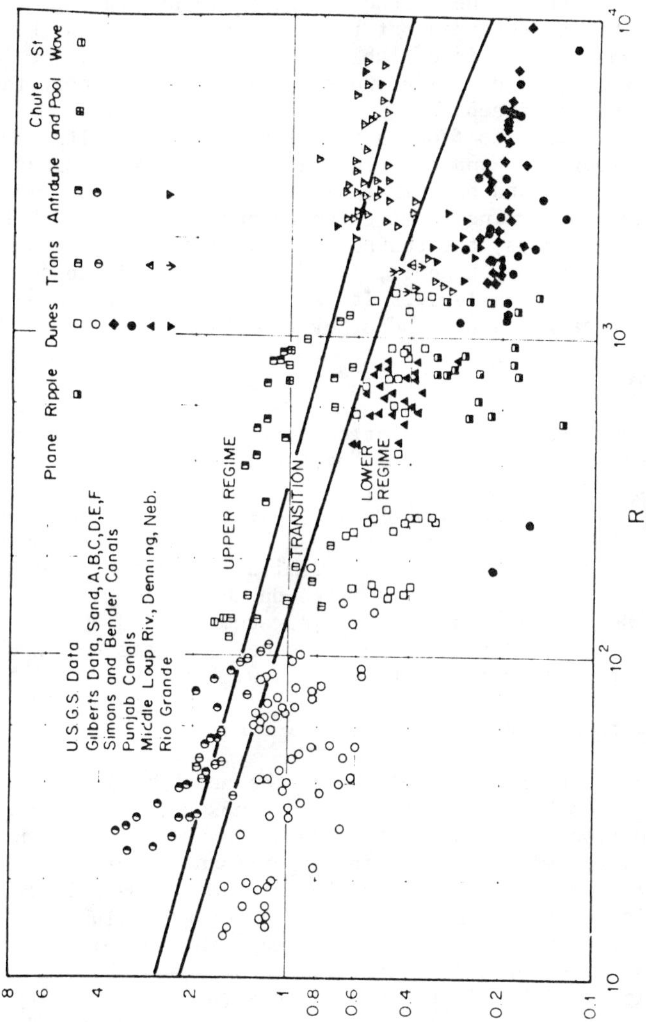

Fig. 5-9. Bed Form Prediction as a Function of F_r and R/D_{50} (after Simons and Sentürk, 1976).

Main Channel

The shape of the main channel depends on the type of boundary material comprising both bed and banks, the channel alignment, and the magnitude of hydraulic variables including shear distribution, τ; stream power, τV; the velocity and velocity distribution. A channel formed in homogeneous coarse material may form a semielliptical cross section as assumed by Lacey. Channels with berms of fine silt and clay materials and beds composed of sand develop steeper banks and a relatively flat bed. In bends, the channel cross sections are more triangular being deep adjacent to the outside bank and shallow near the inner bank. In unstable reaches of a river, the channel may be divided, wide and shallow, and banks near the water line may be steep.

The resistance to flow in the main channel results from both grain roughness and form roughness. The form roughness is a function of bed and bank irregularities significantly larger in size than the grains forming the channel perimeter. With only grain roughness, the channel is relatively smooth; d/D is small. With rougher channels, d/D is relatively large and resistance to flow is large. It is obvious that d/D is also a function of channel slope and other variables.

Flood Plain

The roughness characteristics on the flood plain are complicated by the presence of vegetation, natural and artificial irregularities, buildings, lack of definition of direction of flow, varying slopes and other complexities. Resistance factors reflecting these effects must be selected largely on the basis of past experience with similar conditions. In general, resistance to flow is large on the flood plains. In some instances, conditions are further complicated by deposition of sediment and development of dunes and bars which affect resistance to flow and direction of flow.

Channels with Sand Beds

In channels with erodible beds, the bed material is molded into various combinations of ripples, dunes and bars by the flow. The ripples are small triangular shaped elements which form when water velocity is just larger than the critical velocity required to initiate motion of the bed material. As velocity is increased, dunes develop (often with ripples superposed). With still further increase in the velocity, the ripples and dunes gradually diminish in size and the bed may become quite plain or flat. Beyond this range of velocity standing waves and antidunes may develop. As velocity, depth, and slope of energy gradient vary from point to point in the channel, the bed forms change. It is common to find more than one type of roughness in the cross section. The bed may be plain or flat in the thalweg of the stream where the velocities are largest.

Large bars (sand waves) are also common to sand bed channels. The point bars form on the inside of the bends, the alternate bars in straight reaches and middle bars in the wide, shallow, unstable reaches. These roughness elements significantly increase resistance to flow by increasing form roughness.

Ice Conditions

The presence of ice affects channel roughness and resistance to flow in various unique ways. When an ice cover occurs, the open channel is more nearly comparable to a closed conduit. There is an added shear stress developed between the flowing water and the ice cover. This surface shear is much larger than the normal shear stresses developed at the air-water interface. A study of ice cover by the U.S. Geological Survey has revealed that the ice-water interface is not always smooth. In many instances, the underside of the ice is deformed so that it resembles ripples or dunes observed on the bed of sand bed channels. This may cause overall resistance to flow in the channel to be further increased.

With total or partial ice cover, the drag of the ice retards flow, decreasing the average velocity and increasing the depth. Another serious effect is its influence on bank stability, in and near water structures such as docks, loading ramps, and ships. For example, the ice layer may freeze to bank stabilization materials, and when the ice breaks up, large quantities of rock and other material embedded in the ice may be floated downstream and subsequently thawed loose and dumped randomly leaving banks raw and unprotected.

Resistance Coefficients

Resistance to flow in alluvial channels is usually expressed in terms of resistance coefficients. These coefficients are used in appropriate equations to evaluate loss of energy, average velocity, and related quantities. The most common equations utilizing these resistance coefficients are the Chezy equation:

$$V = C\sqrt{RS} = C/\sqrt{g}\,\sqrt{gRS}\,, \tag{5-12}$$

or

$$V = \sqrt{\frac{8\gamma}{f\rho}}\,\sqrt{RS}\,, \tag{5-13}$$

and the Manning equation

$$V = \frac{1.49}{n} R^{2/3} S^{1/2} = \frac{1.49}{n} R^{1/6} \sqrt{RS}\,. \tag{5-14}$$

In these relations

C is the Chezy resistance coefficient,

f is the Darcy-Weisbach resistance coefficient, and

n is the Manning resistance coefficient.

By grouping V, D, S, ρ, and g into the Darcy-Weisbach resistance coefficient or into the Chezy discharge coefficient, Eq. (5-12) becomes:

$$\frac{8gRS}{V^2} = f = 8/\left(\frac{C}{\sqrt{g}}\right)^2 = \phi[S, D, d, \omega, \sigma, g, \rho]. \tag{5-15}$$

or the bed configuration can be substituted for the dimensionless resistance coefficient to give,

$$\text{Bed configuration} = \phi[S, D, d, \omega, \sigma, g, \rho]. \qquad (5\text{-}16)$$

The variables were not grouped into dimensionless parameters to present more clearly the essential role of each. Because of the interdependency of slope, depth, bed-material characteristics, bed configuration, and resistance to flow, it is difficult to isolate the effect of any one variable. For example, an increase in fall velocity may increase resistance to flow at one slope and decrease it at another slope.

5.8 THE BED MATERIAL OF ALLUVIAL CHANNELS

The most common and most important materials that form the beds of alluvial channels are clay, silt, sand, gravel, and rock. The size of sand ranges from 1/16 to 2 mm, gravel from 2 to 64 mm, and rock is that material larger than 64 mm. The bed material size increases with distance from the mouth of the stream. For example, the bed material ranges from silt and fine sand near the mouth of the Mississippi to large angular rock in the steep mountainous headwaters of many of its numerous tributaries. The significant variables used to describe the bed material are its size, size distribution, shape, and mass density. The fall velocity of sedimentation diameter is often used as a measure of size, particularly within the sand range. The major advantage of this measure is that it combines the effects of several variables into a single parameter; for example, mass denisty, particle shape and particle roughness. In fact, as previously shown, it can be defined in such a way that it includes the effect of suspended sediment concentration and water temperature. The size distributions of silts and clays may be determined by a hydrometer, x-ray diffraction, and other methods. The diameter of sand sizes can be established by standard sieving techniques or by fall diameter methods, Colby and Christensen (1956). Gravel and rock sizes can be established by sieving the smaller sizes and using photographic and grid techniques as well as direct observation to establish the size of the larger more random elements. The greatest drawback when investigating the size of large rock is the large sample that must be analyzed to establish a statistically significant measure of size and size distribution.

The characteristics of the bed material of rivers has an important effect on sediment transport, the forms of bed roughness, the resistance to flow, and, consequently, a very significant effect on the channel geometry.

The basic methods of size analysis have been cited. From a size distribution curve of bed material the percent of the material coarser or finer than any particular size can be easily determined. In general, d_{98} would signify the size of bed material for which 98 percent of the material was finer.

Size distribution can be expressed in various ways and by various parameters. Two of the most common measures of gradation are the standard deviation and the gradation coefficient, σ. The latter parameter is defined as

$$\sigma = 1/2 \left[\frac{d_{84}}{d_{50}} + \frac{d_{50}}{d_{16}} \right] , \qquad (5\text{-}17)$$

in which d_{84} is the size of the bed material for which 84 percent is finer, etc. Hence, σ is the average slope of two segments of the size distribution curve. When the size distribution curve is plotted on log-probability paper most sands from the beds of natural streams, when analyzed, tend to plot as a straight line. For a straight line $d_{84}/d_{50} = d_{50}/d_{16}$ and

$$\sigma = \frac{d_{84}}{d_{50}} = \frac{d_{50}}{d_{16}} .$$

The shape of the particles is most commonly defined by the Corey (1949) shape factor

$$S_p = \frac{c}{\sqrt{ab}} \qquad (5\text{-}18)$$

in which a, b, and c are the dimensions of the three mutually perpendicular diameters of the particle, c being the minor axis. Most quartz sands have a S_p of about 0.7.

There is no widely accepted method of describing the surface roughness of particles. However, from a physical and hydraulic viewpoint e/d should be a good parameter where d is the nominal diameter and e is the average height of particle roughness.

The angle of repose for noncohesive materials such as sand, gravel, and rock increases with both size and angularity. Lane (1955) presented a figure relating angle of repose, median diameter of the material, and shape of particles. Subsequent investigation of this relationship by Simons (1957) verified that additional information on angle of repose was needed. As a result of further study, Fig. 5-10 was developed. The angle of repose, as indicated in this figure, is for dry or saturated material. If the material is saturated and subjected to the action of flowing water, the side slope should be reduced below that indicated for static conditions by five degrees to ten degrees or more depending on design conditions and size and angularity of the rock rip-rap.

The characteristics of the bed are illustrated in Table 5-2 and Fig. 5-11 for the Mississippi River downstream of Cairo, Illinois, the figure is based upon field data and was taken from WES Paper No. 17.

Table 5-2. Variation in Composition of Bed Materials in the Mississippi River.

Miles below Cairo, Illinois	Percent in Size Range								
	0.0-0.008 mm	0.008-.074 mm	.074-.104 mm	.104-.295 mm	.295-.589 mm	.589-1.168 mm	1.168-2.362 mm	2.362-13.33 mm	13.33 and above mm	Total
25	0.63	3.12	1.25	16.82	31.28	18.10	8.20	11.80	8.80	100%
100	--	--	--	8.12	33.38	31.20	10.40	9.60	7.30	100%
200	--	1.87	--	11.63	44.7	26.8	6.00	6.50	2.5	100%
300	--	--	--	17.70	50.40	23.70	3.20	2.50	2.50	100%
400	3.00	2.36	1.26	32.48	35.9	7.50	2.50	6.20	8.80	100%
500	0.38	0.87	1.00	27.15	46.7	9.20	2.20	5.60	6.90	100%
600	1.00	2.37	1.25	34.18	41.20	9.30	--	5.20	5.50	100%
700	--	--	1.0	41.50	43.70	7.40	1.40	2.70	2.30	100%
800	--	--	0.63	52.57	39.30	4.30	--	2.30	0.90	100%
900	--	3.00	--	71.30	25.20	0.50	--	--	--	100%
1000	10.0	10.0	6.5	64.80	8.70	--	--	--	--	100%
1075	25.0	22.50	6.10	45.20	1.20	--	--	--	--	100%

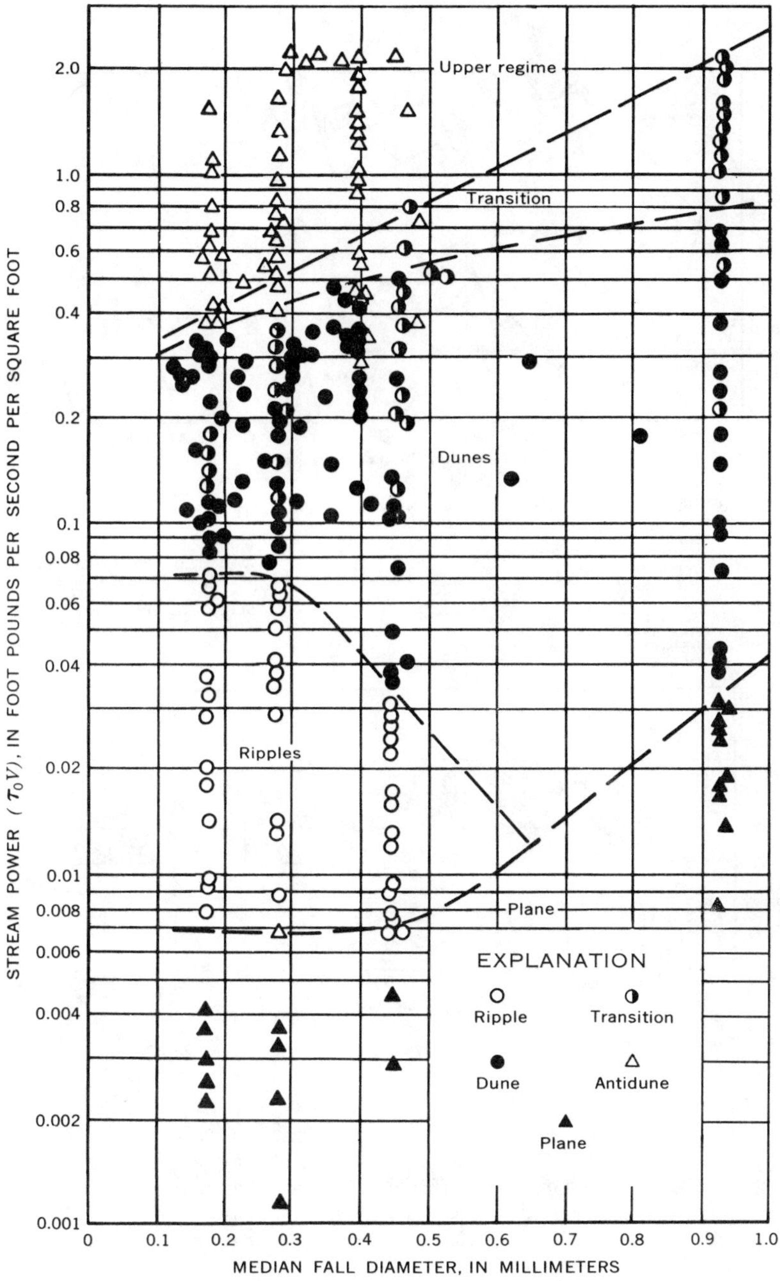

Fig. 5-10. Relation of Bed Form to Stream Power and Median Fall Diameter of Bed Sediment (after Simons and Richardson, 1966).

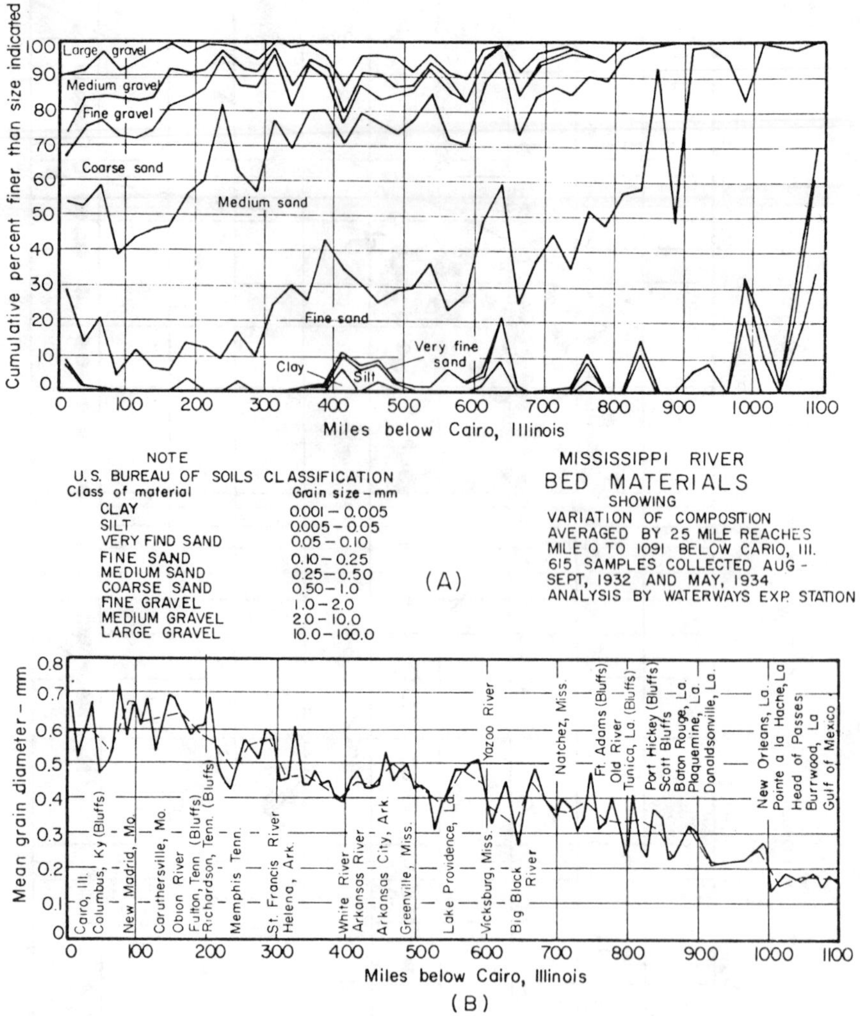

Fig. 5-11. Variation in mean Grain Size of Sand Only, after Waterways Experiment Station Paper 17 (1935) (0.074-1.168 mm). Average by Reaches Other Materials Eliminated from Calculations.
Legend: 10 mile reach averages ——
25 mile reach averages ----.

The Bank Material of Alluvial Channels

The bank material of alluvial channels is even more variable than the bed material. The banks may consist of homogeneous or nonhomogeneous material, fine or coarse material, stratified material, cohesive or noncohesive material, or some heterogeneous combination of many of these materials. In the head reaches of river systems, the banks usually consist of coarse rock, gravel and in many cases outcroppings of solid rock. Down stream the size of the bank material becomes smaller and more uniform. As one follows a river system from the sharp V-shaped canyons where both bed and bank material is largely rock to the wider river valleys, the bed and bank material change from rock to combinations of rock, gravel and limited quantities of sand and soil. Proceeding on downstream to wide open valleys and the plain areas, the banks consist of various combinations of sands, silts and clays. The banks may be highly stratified consisting of layers of sand, silt, and clay sandwiched together in a random pattern consistent with the environmental and geological conditions at the time of deposition.

Still farther downstream, the size of bank material may consist largely of silts and clays deposited at some time in the past by the fluvial system.

To illustrate the complexity of the channel structures refer to Fig. 5-12. This shows the flood plain and indicates the manner in which the channel has changed its alignment, shape, and position in geologic time. The present channel is formed in the old sand bars that are highly stratified, ancient channels cut off at some time in the past and filled with silts and clays and in some instances the banks may be impregnated with varying quantities of vegetal matter. The randomness and nonhomogeneity of the bank materials is typified by Fig. 5-13, which shows the analysis of bank material and types of slide failures along the Mississippi River.

Where the bank material is loose and noncohesive, the alluvial channel tends to be very wide; with coarse rock material and cohesive bank materials the width-depth ratio for the channel is much smaller. Simons and Albertson (1963) clearly illustrated the effect of type of bank materials on the wetted perimeter of small stable channels, see Fig. 5-14. Schumm (1960) has shown somewhat similar effects in river channels, and he has classified alluvial channels as stable, depositing, and eroding and related these classifications to mode of sediment transport.

Those channels carrying water and fine sediment, usually referred to as wash load, may develop berms formed by the deposition of sediment. The process of berming usually begins by a widening of the channel due to excessive erosion during large flows. At smaller flows the velocities adjacent to the banks are low enough that the silts and clays, by sedimentation and chemical processes, deposit on the banks at below the water

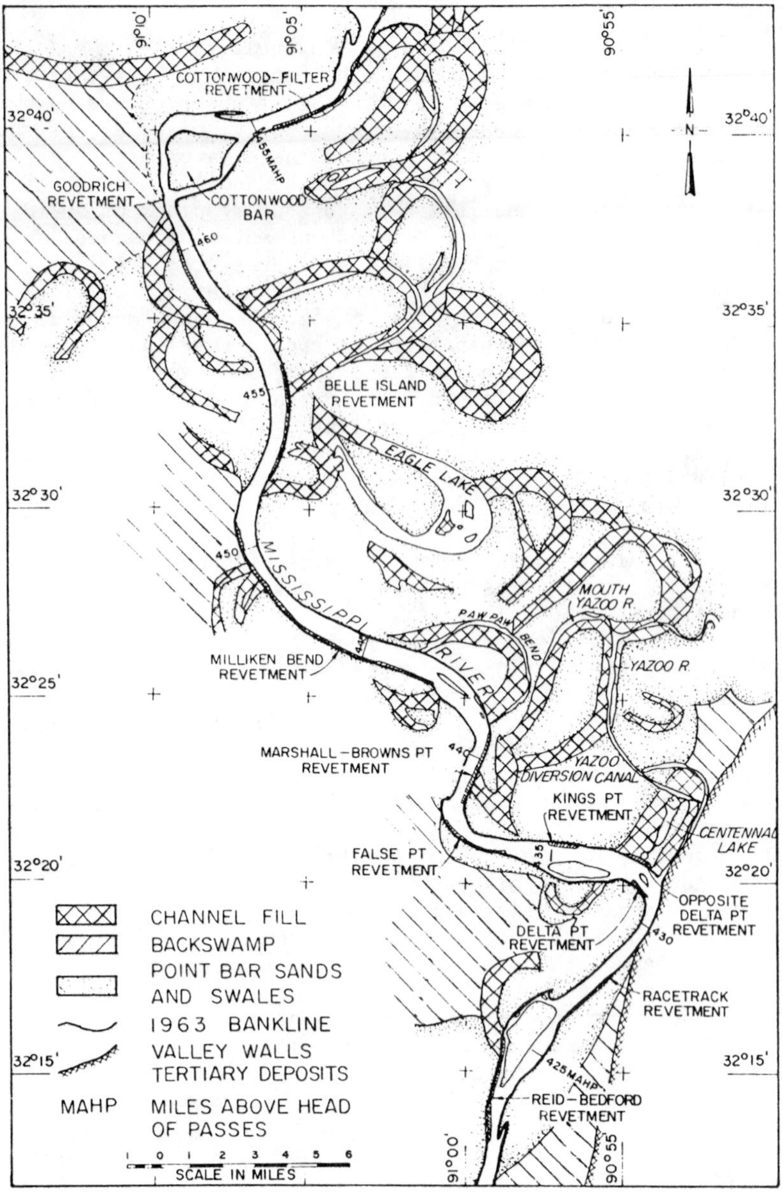

Fig. 5-12. Major Floodplain Deposits in the Meander Belt of the Mississippi River, after Waterways Experiment Station Potamology Investigation Report No. 12-15 (1965). (Vicinity of Goodrich to Reid-Bedford Revetments).

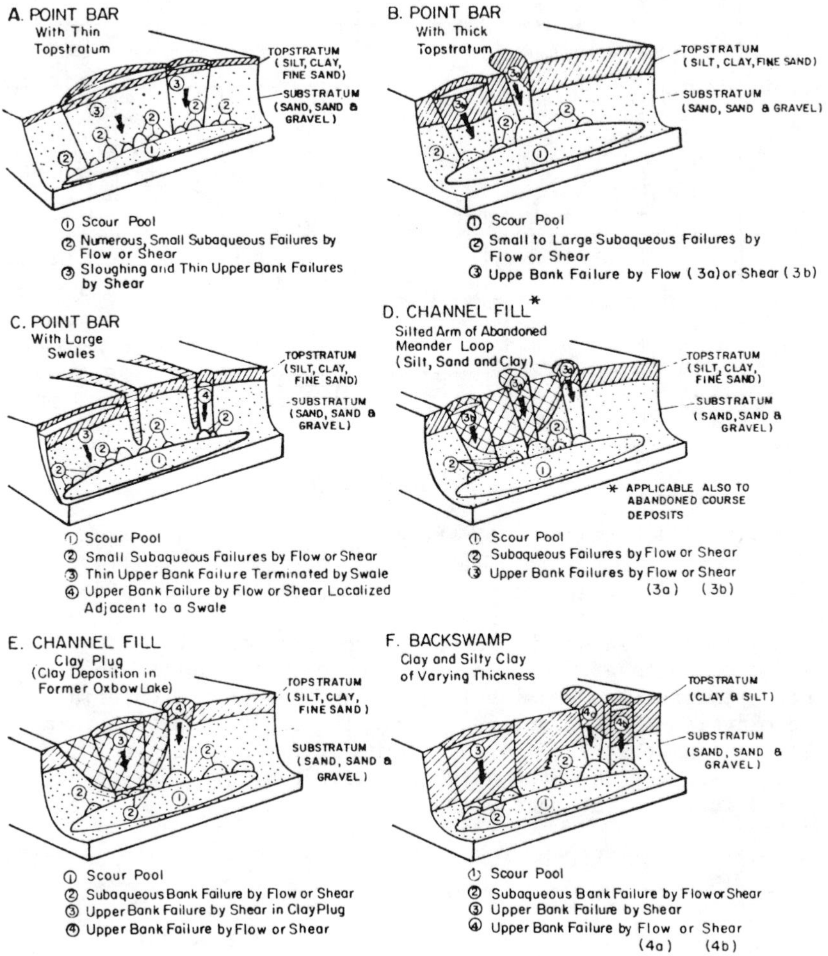

Fig. 5-13. General Influences of the Geology of Riverbank Soils on the Mechanics of Bank Failure, after Waterways Experiment Station Potamology Investigation Report No. 12-15 (1965).

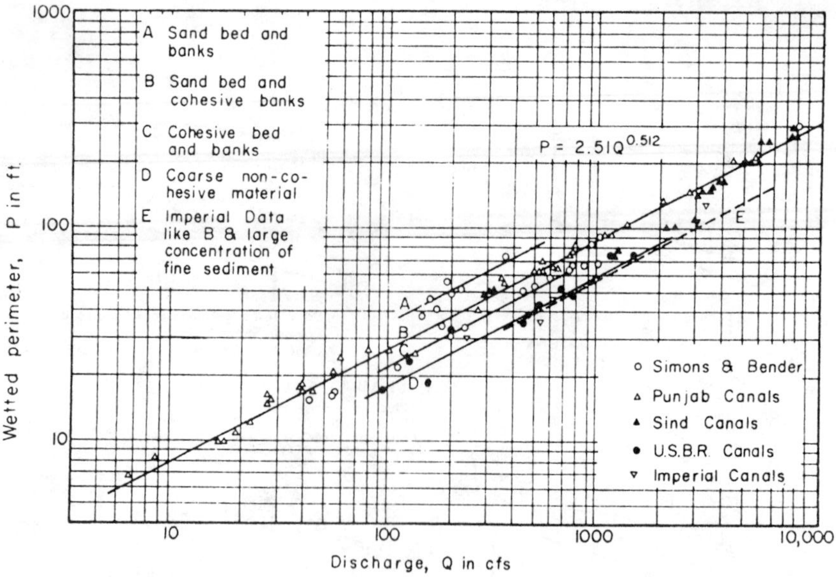

Fig. 5-14. Variation of Wetted Perimeter P with Discharge Q Type of Channel and Concentration of Fine Sediment, after Simons and Albertson (1963).

line, thus building the banks and reducing channel width. These deposits are referred to as berms. The berms are usually more resistant to erosive forces than the natural material. Berming is often encouraged in canals to give a smoother, more stable bank that is also less permeable, reducing water loss due to seepage when the adjacent water table is lower than the water level in the channel.

5.9 RELATIVE INFLUENCE OF VARIABLES ON BED MATERIAL AND WATER DISCHARGE

In predicting the response of alluvial rivers to various changes, it is often helpful to know the relative influence of variables on water and sediment discharge.

The study of the relative influence of viscosity, slope, bed material size and depth on bed material and water discharge is examined in detail using Einstein's bed-load function (1950) and Colby's (1964) relationships. Einstein's bed-load function was chosen because it is the most detailed and comprehensive treatment, from the point of fluid mechanics. Colby's relations were chosen because of the large amount and range of data used in their development.

The data required to compute the total bed material discharge using Einstein's relations are:

S = energy slope

D_{65} = size of bed material for which 65 percent is finer

D_{35} = size of bed material for which 35 percent is finer

ν = kinematic viscosity

n_w = Manning's wall friction coefficient

A = cross-sectional area

P_b = wetted perimeter of the bed

P_w = wetted perimeter of the banks

D_i = size of bed-material fraction i

i_B = percentage of bed material in fraction i

γ_s = specific weight

V = average velocity.

To study the relative influence of variables on bed material and water discharges, the data taken by the U.S. Geological Survey from October 1, 1940 to October 1, 1970 on the Rio Grande near Bernalillo are used. The width of the channel reach was 270 ft. In the analysis the energy slope was varied from 0.7S to 1.5S, in which S is the average bed slope assumed to be equal to the average energy slope. Further, the kinematic viscosity was varied to correspond with variations in temperature from 39.2° to 100°F inclusive. The variation of D_{65}, D_{35}, D_i and i_B was accomplished by using the average bed-material distribution given by Nordin (1964) and shifting the curve representing the average bed-material distribution along a line parallel to the abscissa drawn through D_{50}. The average water temperature was assumed equal to 70°F and the average energy gradient of the channel was assumed equal to 0.00095 ft/ft = 5.0 ft/mi. The water and sediment discharges were computed independently for each variation of the variables and for three subreaches of the Rio Grande of different width near Bernalillo. The applicability of the results depends on the reliability of the modified Einstein bed-load function and Colby's relationships used in the analysis rather than on the choice of data.

The computed water and sediment discharges are plotted in Fig. 5-15, and show the variation of sediment discharge due to changes in bed material size, slope and temperature for any given water discharge. Figure 5-15 shows that when the bed material becomes finer, the sediment discharge increases considerably. The second most important variable affecting sediment discharge is the slope variation (see Fig. 5-16). Temperature is third in importance (Fig. 5-17). The effects of variables on sediment discharge were studied over approximately the same range of variation for each variable.

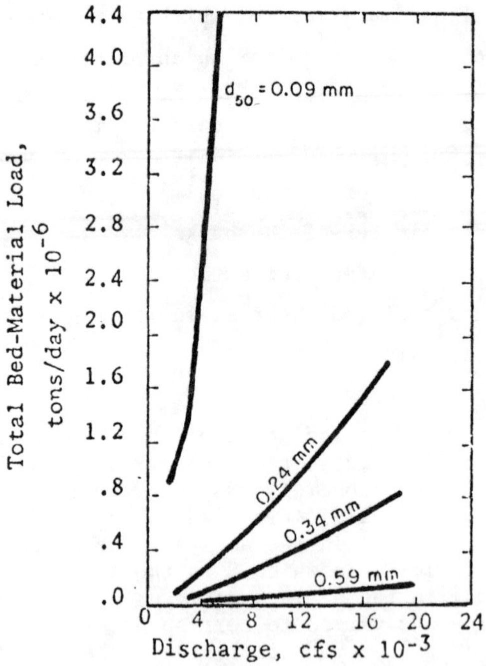

Fig. 5-15. Bed-Material Size Effects on Bed-Material Transport.

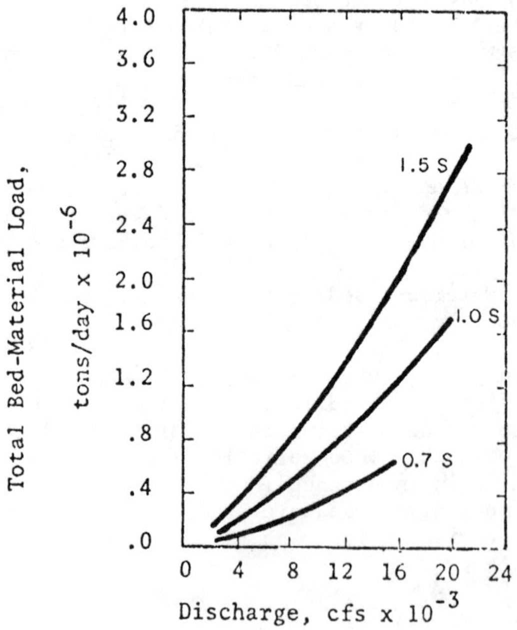

Fig. 5-16. Effect of Energy Gradient on Bed-Material Transport.

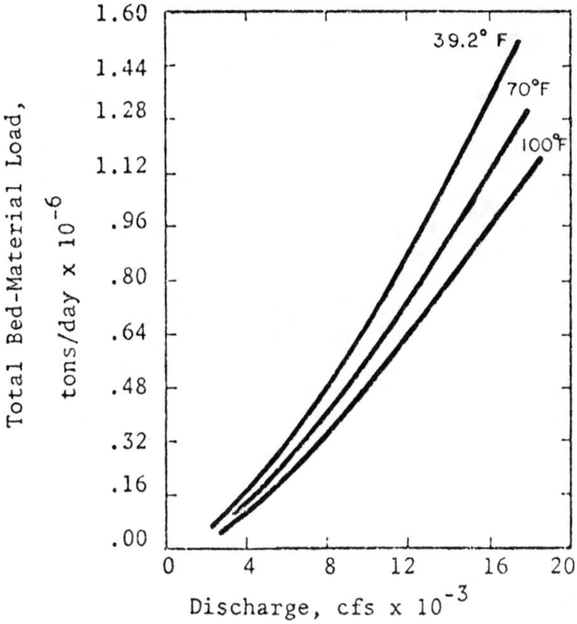

Fig. 5-17. Effect of Kinematic Viscosity on Bed-Material Transport.

5.10 RIVER FORM

Rivers may be classified as meandering, straight, and braided. However, there are many transitional forms sandwiched between these types.

Meander

A meandering river has more or less regular inflections that are sinuous in plan. It consists of a series of bends connected by crossings. In the bends deep pools are carved adjacent to the concave bank by the relatively high velocities. The centrifugal force in the bend causes a transverse water surface slope and helicoidal flow in the bend. These transverse currents, with a magnitude of about 15 percent of the average channel velocity, flow toward the convex bank. In so doing, they sweep the heavier concentrations of bed load toward the convex bank where they are deposited to form the point bar. The bends are connected by crossings (short straight reaches) which are quite shallow compared to the pools in the bendways. At low flow, large sand bars form in the crossings if the channel is not well confined. The scour in the bend causes the bend to migrate downstream and sometimes laterally. Movements as large as 2500 feet per year have been observed in alluvial rivers. Much of the sediment eroded from the outside bank is deposited in the crossing and on the point bar in the next bend downstream. Meandering rivers have relatively flat slopes.

Braiding

The river channel is wide, the banks are poorly defined and unstable. There are two or more main channels that cross one another giving the river bed a braided appearance at low flow. Between sub-channels there are numerous sand bars and islands. These sub-channels and sand bars rapidly change position with time and stage and in an unpredictable manner. At flood stage the flow straightens, most of the sand bars are inundated or destroyed and the river has a canal-like appearance except that the river is much wider and has a higher flow velocity. Such rivers have relatively steep slopes and carry large concentrations of sediment. Much of the Jamuna River is braided.

Straight

The channel has negligible sinuosity at bankfull stage. At low stage the channel develops alternate sand bars and the thalweg meanders around them in a sinuous fashion. Straight channels are often considered as a transitional stage preceeding meandering. If the channel is unconfined, more than one channel develops, creating middle bars as well as point bars, and the river is braided. This way of classifying channels has characteristics common to both meandering and braided rivers.

Subclassification of River Channels

There are subclassifications within the major types of meandering, straight and braided channels that are of use to the geomorphologist and engineer. Low, moderate and high sinuosity are illustrated in Fig. 5-18. A classification based on oxbow lakes is illustrated in Fig. 5-18. In Fig. 5-18 types of meander scroll formations are illustrated. By studying scroll formations in terms of age of vegetation it is possible to quantify rate and direction of channel migration. The bank height classification of rivers is given in Fig. 5-18. Bank height is often an important index to age and activity of the river. Classification based on natural levees is illustrated in Fig. 5-18. As pointed out earlier, well developed levees are associated with older rivers. Typical modern floodplains are illustrated in Fig. 5-18. The floodplain that is broad in relation to the channel width is indicative of an older river. Conversely, when the river valley is narrow and confined by terraces or valley walls the river flowing therein is usually classified as mature. Typical vegetative patterns that are observed along meandering channels are shown in Fig. 5-18. In general, the growth of vegetation is indicative of the presence of silts and clays in the river banks and the

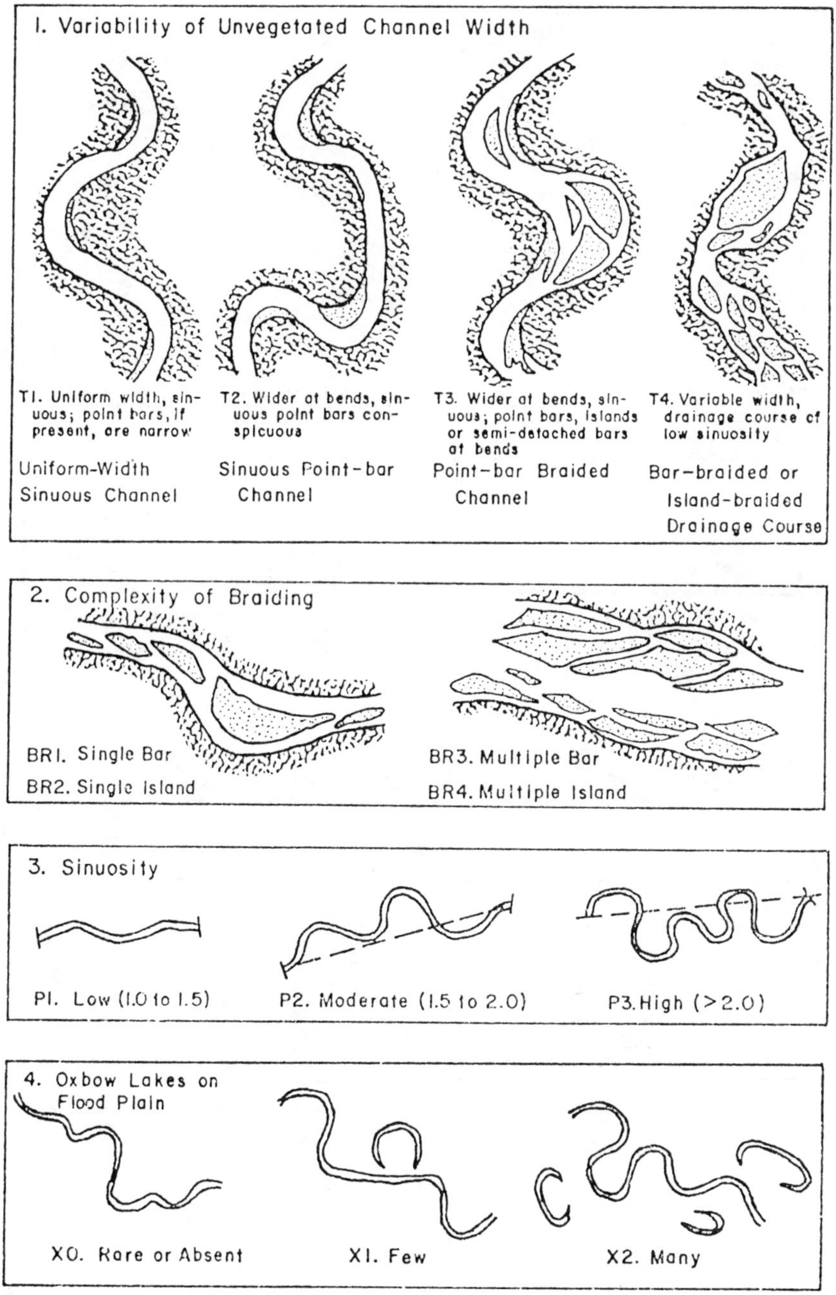

Fig. 5-18. Classification of Alluvial Channels (after Culbertson, et al., 1967).

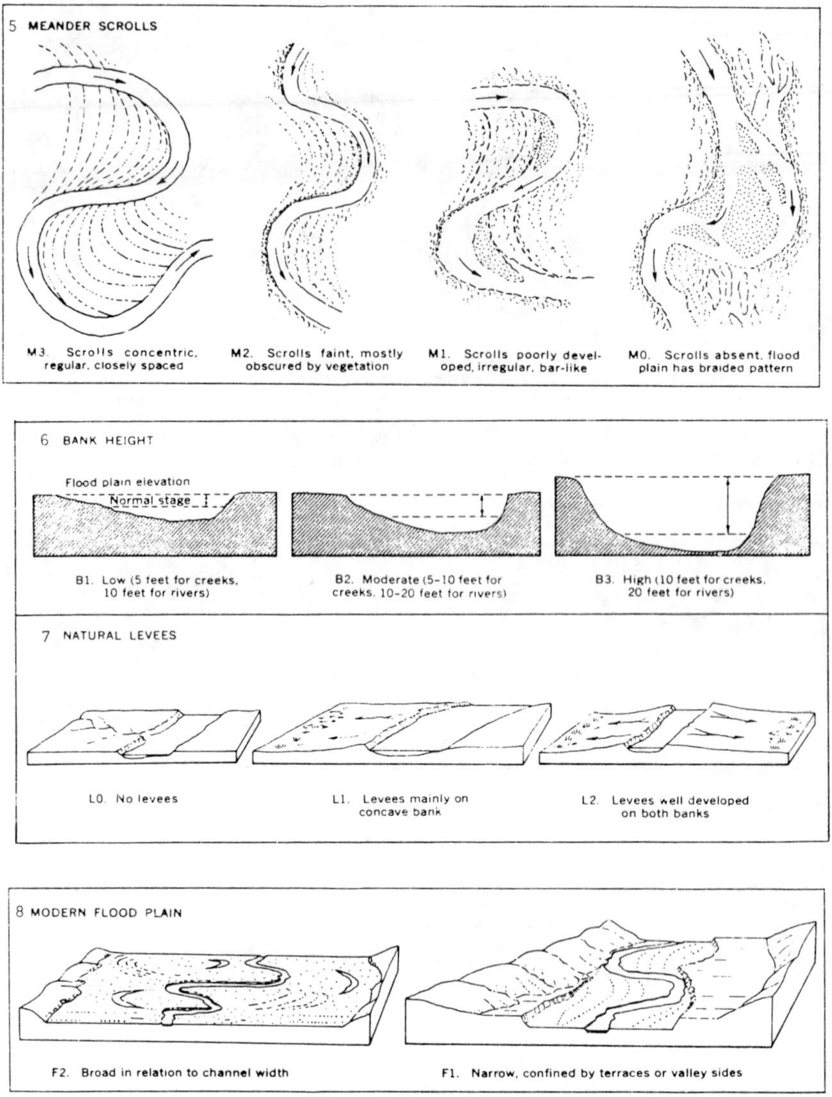

Fig. 5-18. Classification of Alluvial Channels (after Culbertson, et al., 1967) (continued).

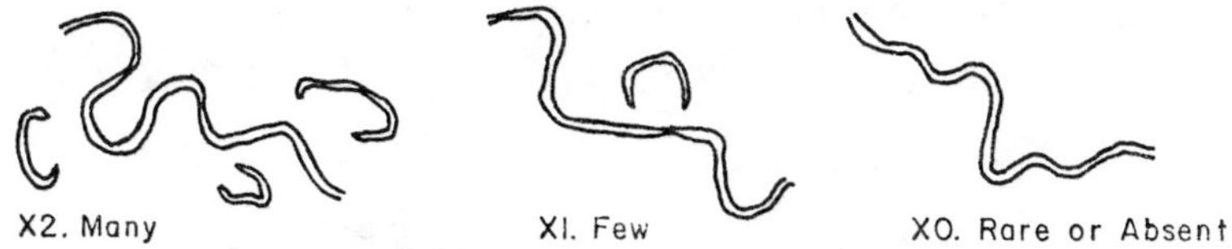

(d) Oxbow Lakes on Floodplain

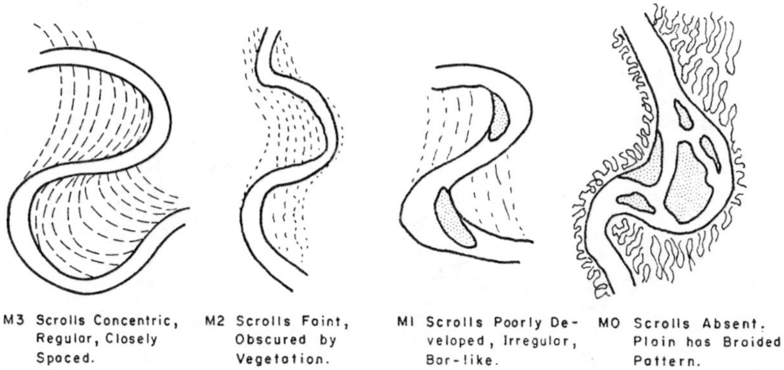

(e) Types of Meander Scroll Formations

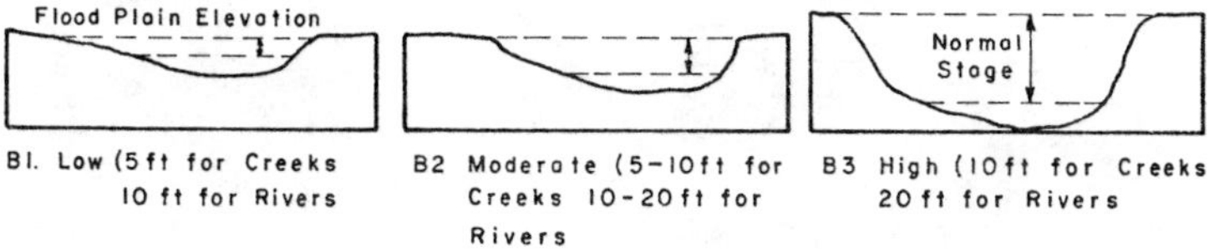

(f) Types of Bank Heights

Fig. 5-18. Classification of River Channels (after Culbertson, et al., 1967) (continued).

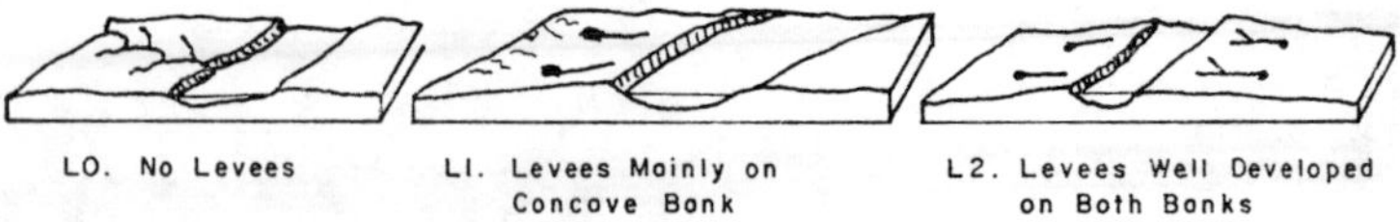

(g) Types of Natural Levee Formations

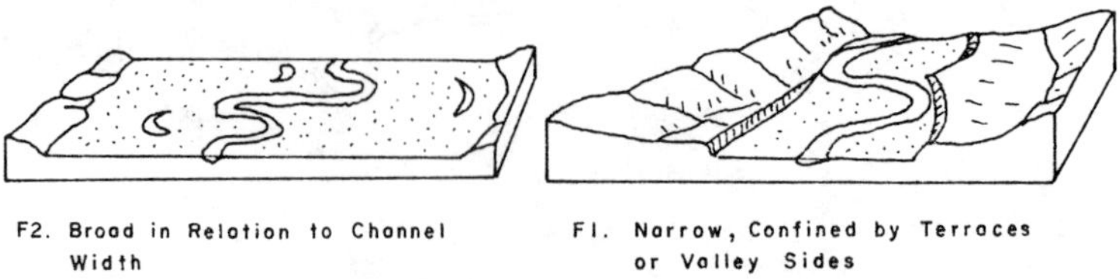

F2. Broad in Relation to Channel Width
FI. Narrow, Confined by Terraces or Valley Sides

(h) Types of Modern Floodplains

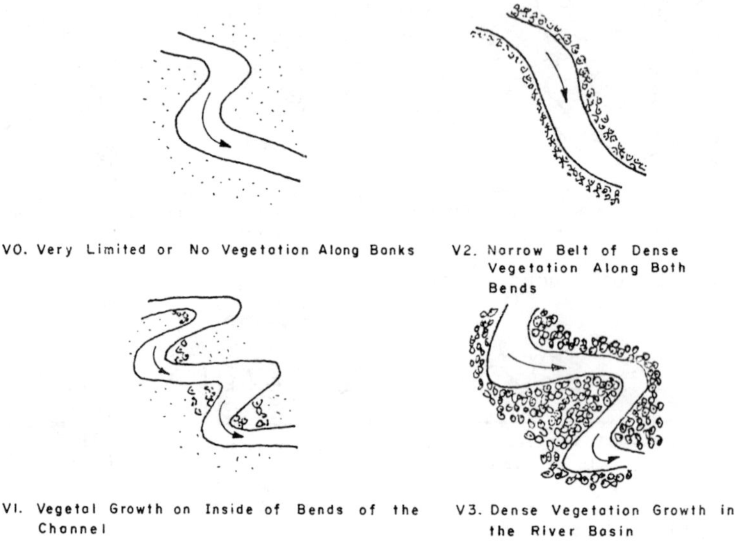

(i) Types of Vegetal Patterns

Fig. 5-18. Classification of River Channels (after Culbertson, et al., 1967) (continued).

floodplain. This is particularly true if the floodplain is well drained. With good drainage the silt and clay are essential to the growth of vegetation because of their water holding capability.

5.11 FORMATION OF BENDS IN ALLUVIAL CHANNELS

Most bends in sand-bed rivers are part of a meander or deformed meander system. Bends are normally formed as a result of the natural tendency for sinuous flow in alluvial channels when the slope of the river is less than $S = 0.0017\ Q^{1/4}$, see Fig. 5-19. This tendency for bends to develop in sand channels which flattens the slope, has been clearly demonstrated by Friedkin (1945), Lane (1955), and many others.

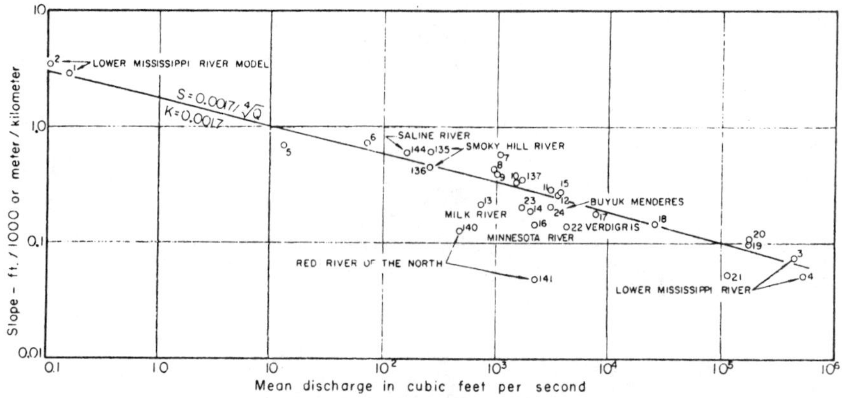

Fig. 5-19. Slope-Discharge Relation--Meandering Sand Streams, after Lane (1955).

The actual shape of the bends varies from beautifully symmetrical patterns such as those pictured in most geomorphology texts to the deformed bends encountered most frequently in nature, particularly on large river systems.

The shapes of two typical bends surveyed on the Mississippi River are illustrated in Figs. 5-20 and 5-21. To further illustrate the random nature of Mississippi River bends refer to Figs. 5-22 and 5-23, which respectively show the percent occurrence of bend radii r_c in feet, and the percent occurrence of bend deflection angles ϕ in radians. Note that the most common radius is about 5,000 feet. This radius was observed on 11 percent of 179 river bends. Similarly, the most common deflection angle is about 1.15 radians. This deflection angle occurred for about 15 percent of these same bends. These distribution curves illustrate the variability of the characteristics of bends in river systems.

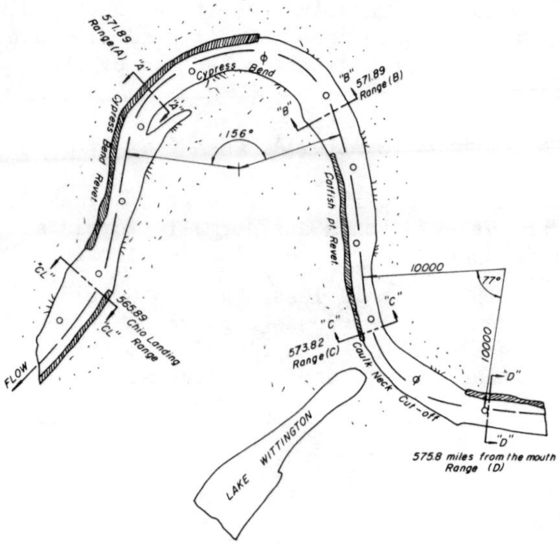

Fig. 5-20. Cypress Bend, Mississippi River -- 565-576 (1962) from the Mouth, after Assifi (1966).

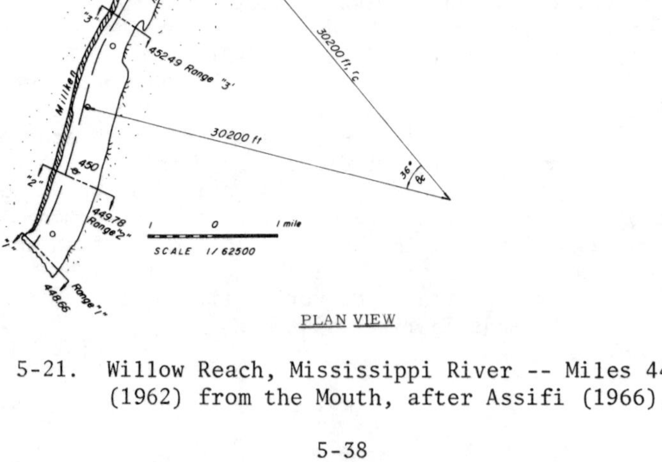

Fig. 5-21. Willow Reach, Mississippi River -- Miles 448-463 (1962) from the Mouth, after Assifi (1966).

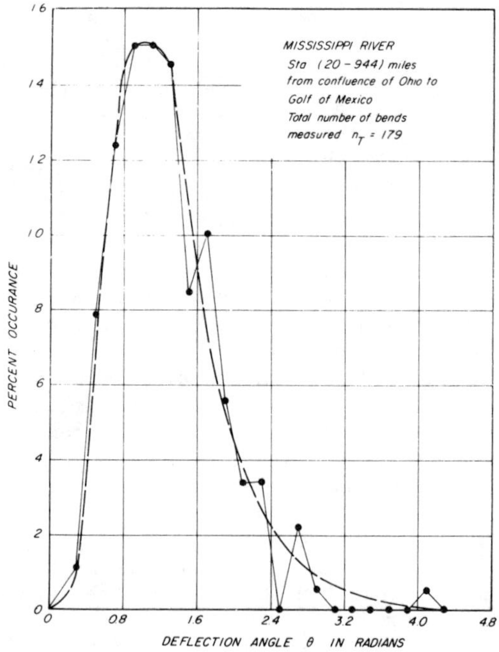

Fig. 5-22. Mississippi River Station (20-944) Miles from Gulf of Mexico to C onfluence of Ohio. Total Number of Bends (N_t) Measured N_t = 179, after Assifi (1966).

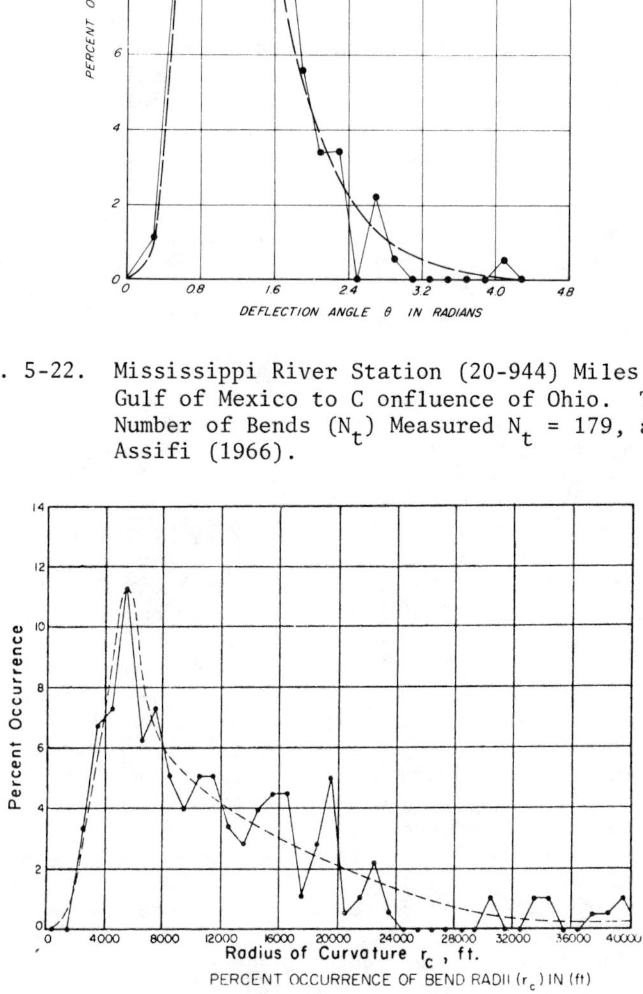

Fig. 5-23. Percent Occurrence -- Bend Deflection Angles (ϕ) in Radians, after Assifi (1966).

Types of Bends

Two principal types of bends are deepened or entrenched bends and meandering surface bends. The first type also includes those in which the river bends follow the curves of the valley so that each river bend includes a promintory of the parent plateau. The second type includes bends which are formed only by the river on a flat, alluvium covered valley floor, and where the slopes of the valley are not involved in the formation of such bends. This division of bends is correct and sufficiently definite with respect to external forms of the relief and the process of formation and development of bends. It is, however, incomplete from the standpoint of the work of the river and of the physical nature of this phenomenon. Both of the morphological types of bends can be put into one category--the category of freely meandering channel, i.e., meandering determined only by the interaction of the stream and the bed material. Such meandering, not disturbed by the influence of external factors, proceeds at an approximately equal rate along the length of the river.

Under natural conditions, there is often encountered a third type of bend. This bend occurs when the stream impinging on a practically noneroding parent bank forms a forced curve which is gradually transformed into a river bend of a more constricted shape.

In all cases the effect of the character (density) of the material composing the banks is important and, to a certain degree, determines the radius of curvature of the channel. In a free bend the radius of curvature increases with the density of the material. The radius of curvature is smallest in a forced bend.

Both from the standpoint of the action of the stream and the interaction between the stream and the channel, as well as from the standpoint of the general laws of their formation, one can distinguish the following three types of bends of a natural river channel.

(a) Free bends - Both banks are composed of alluvial floodplain material which is usually quite mobile; this corresponds to the common concept about a surface bend.

(b) Limited bends - The banks of the stream are composed of consolidated parent material which limits the intensive development of lateral erosion by the stream; this corresponds to entrenched bends (meanders).

(c) Forced bends - The stream impinges onto an almost straight parent bank at a large angle (60°-100°).

Two characteristic features of the different types of bends are as follows:

(1) The first typical feature is a close relationship between the type of stream bend and the radius of curvature. The forced bend has the smallest radius of curvature. Next in size are the radii of free bends. The limited bends have the greatest radii. The average values of the ratios of the radii

of curvature to the width of the stream at a given point (bankfull) for the three types of bends are,

free bends 4.5-5.0
limited bends 7 - 8, and
forced bends 2.5-3.0.

(2) The second characteristic feature is the distribution of depths along the length of the bend.

In the first two types of bends the depth gradually increases and the maximum depth is found some distance below the apex of the bend. In the third type (the forced bend) the depth sharply increases at the beginning of the bend and then gradually diminishes. The greatest depth is located in the middle third of the bend, where there appears to be a concentrated deep scour.

Transverse Slope of Water Surface in Bends

Theory: The development of sinuosity of the flow and the corresponding centrifugal forces causes the free surface of the stream to acquire a transverse slope. This slope can be evaluated quantitatively. In general, the various methods involve isolating the forces acting on fluid elements, introducing them into the equation of transverse equilibrium for streams, and solving the resultant equation. Figure 5-24 illustrates conditions for the transverse equilibirum of the stream and the resultant transverse flow.

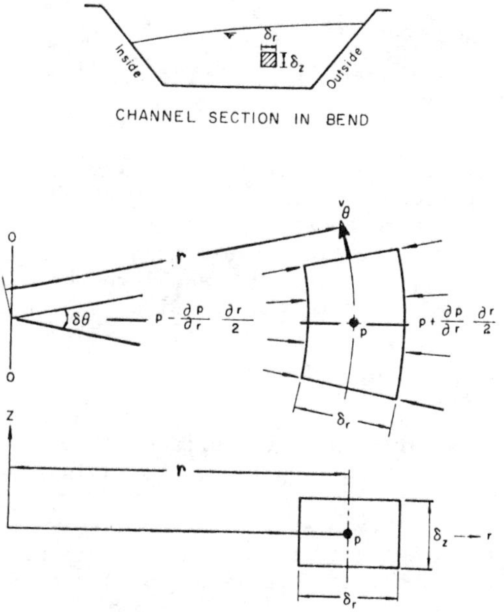

Fig. 5-24. Elemental Fluid Volume in Cylindrical Coordinates Within a River Bend.

Adopting cylindrical coordinates and referring to Fig. 5-24, the dynamics of super elevation may be presented as follows. Total force towards the center equals mass x centrifugal acceleration,

$$\left[\left(P + \frac{\partial p}{\partial r}\frac{\partial r}{2}\right) - \left(P - \frac{\partial p}{\partial r}\frac{\delta r}{2}\right)\right] r\delta\theta\delta_z = (\rho\cdot\delta r\cdot\delta_z\cdot r\delta\theta)\frac{V\theta^2}{r}$$

or

$$\frac{1}{\rho}\frac{\partial p}{\partial r} = \frac{V\theta^2}{r} \tag{5-19}$$

The total super elevation between the outer and inner bank is:

$$\Delta z = \frac{1}{\rho g}\int_{p_{i\omega}}^{p_{o\omega}} dp$$

$$= \frac{1}{g}\int_{r_{i\omega}}^{r_{o\omega}} \frac{V_\theta^2}{r}\, dr. \tag{5-20}$$

Now two assumptions are made:

(i) The radial and vertical velocities are small compared to the tangential velocities such that $V_\theta \simeq V$.

(ii) The pressure distribution in the bend is hydrostatic, i.e., $P = \gamma h$ then,

$$\Delta z = \frac{1}{g}\int_{r_{i\omega}}^{r_{o\omega}} \frac{V^2}{r}\, dr. \tag{5-21}$$

In order to solve Eq. (5-20) it is necessary to assume a velocity distribution along the radius of the bend so that corresponding Δh can be computed. The results obtained by various velocity distribution assumptions are given below:

(1) Woodward (1920) assumed V equal to the average velocity and r equal to the radius to the center of the stream and obtained

$$\Delta z = z - z_i = \alpha_o \frac{V^2}{gr_c}(r_e - r_i), \tag{5-22}$$

in which z_i and r_i are the water surface level and the radius at the inside of the bend, and r_e is the radius to the concave bank of the bend.

(2) If the velocity distribution is assumed to be parabolic with maximum velocity at the centerline and zero velocity at the boundaries, Woodward (1920) showed that the super elevation is,

$$\Delta z = \frac{gV^2}{4g}\left[\frac{20r_c}{3b} - \frac{16r_c^3}{b^3} + \left(\frac{4r_c^2}{b^2} - 1\right)^2 \ell n \frac{2r_c + b}{2r_c - b}\right], \quad (5\text{-}23)$$

in which b = breadth of stream = W = $r_e - r_i$.

(3) Shukry (1950) by assuming the velocity distribution to approximate that of a free vortex, presented the formula,

$$\Delta z = \frac{K_o^2}{2g}\left[\frac{1}{r_i^2} - \frac{1}{r_e^2}\right] \quad (5\text{-}24)$$

in which K_o = free vortex constant. By assuming the depth of flow upstream of the bend is equal to the average depth in the bend, Ippen and Drinker (1962) reduced Eq. (5-24) to,

$$\Delta z = \frac{V^2}{2g}\frac{2W}{r_c}\left[\frac{1}{1 - \left(\frac{W}{2r_c}\right)^2}\right]. \quad (5\text{-}25)$$

(4) For situations where high velocities occur near the outer bank of the channel a forced vortex may approximate the flow pattern. This with the assumption of constant average specific head enabled Ippen and Drinker (1962) to arrive at the equation,

$$\Delta z = \frac{V^2}{2g}\frac{2W}{r_c}\left[\frac{1}{1 + \frac{W^2}{12r_c}}\right]. \quad (5\text{-}26)$$

(5) By assuming that in the middle section of the bend the maximum velocities are close to the centerline of the channel and that the flow pattern inward and outward from the centerline can be represented as forced and free vortices, respectively, then:

$$\Delta z = \frac{1}{g}\int_{r_{i\omega}}^{r_c} \frac{k_i^2 r^2}{r}\, dr + \frac{1}{g}\int_{r_c}^{r_{o\omega}} \frac{k_o^2}{r^3}\, dr, \quad (5\text{-}27)$$

and when $r = r_c$, $V = \overline{V}_{max}$

$$k_i = \frac{\overline{V}_{max}}{r_c} \quad (5\text{-}28)$$

and

$$k_o = \overline{V}_{max}\, r_c \quad (5\text{-}29)$$

Substituting Eq. (5-28) and (5-29) into Eq. (5-27) and simplifying, the new equation for computing the super elevation becomes;

$$\Delta z = \frac{\overline{V}_{max}^{2}}{2g}\left[2 - \left(\frac{r_i}{r_c}\right)^2 + {\frac{r_c}{r_e}}^{2}\right]. \qquad (5\text{-}30)$$

Referring to natural conditions, at least on small streams, the cross channel profile of the water surface has other interesting characteristics. Measurements have shown that the transverse water surface profile is concave upward and it may have a central trough. This trough has been observed on both the rising and falling stages at the same reach. In narrow channels debris tends to accumulate in the troughs. On larger, wider streams more than one trough has been observed by Lane (1957). These troughs may be related to the transverse (secondary) flows in the section.

Transverse Velocity Distribution in Bends

To prove the existence of transverse currents in a bend, let us separate a fluid element with length dx and depth h and let us consider the forces acting on it in the direction of the x-axis. Two forces will act on the element under consideration, the centrifugal force, whose variation over the vertical will correspond to the variations of the square of tangential velocities, and an opposing force, due to the existence of a transverse slope; this force can be expressed by the quantity ydz. A graphical summation of these two curves (Fig. 5-25) indicated the absence of equilibrium in the fluid in the plane of the stream cross section. The curve obtained from the summation of the transverse velocity components indicates the presence of a flow directed toward the concave bank in the upper part of the vertical, and of a reverse flow in its lower part. Since there should be no residual discharge in the transverse direction, the areas of these curves of opposite direction must be equal. Transverse currents, superimposed on the longitudinal flow, form a screwlike type of circulation which can be observed in river bends.

The velocity distribution in open channels is extremely complex in a natural stream. This distribution consists of the longitudinal velocities, which vary with proximity to the bed, banks, and water surface, and many interrelated variables as shown by Eq. (5-1), the transverse or secondary velocities, which relate to the centrifugal force, and other variables and the velocity fluctuations associated with turbulent flow. The most usual way to describe the velocity distribution in alluvial channels is by actual measurements. In this way accurate knowledge of the various velocities and velocity components can be obtained. With less precision, it is possible to use theoretical and experimental equations to describe the velocity

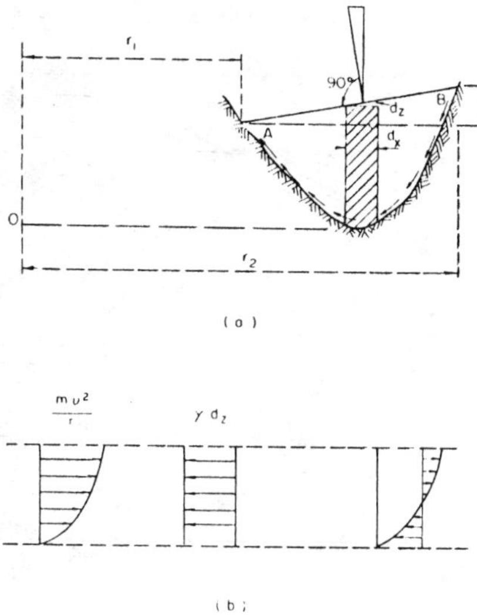

Fig. 5-25. Schematic Representation of Transverse Currents in a Channel Bend.

field. At any vertical in the stream the variation of longitudinal velocity with respect to depth can be described by the von Karman relation which states that the point velocity in a vertical is a function of the logarithm of depth in accordance with Eq. (5-31).

$$\frac{v}{V^*} = \frac{2.303}{K} \log_{10} \frac{y}{K_s} + 8.5 \ , \qquad (5\text{-}31)$$

in which v is the velocity at depth y,

V^* is the shear velocity,

K_s is equivalent to the diameter of the sediment grains that compose the bed, and

K is the universal velocity coefficient.

Consequently, a plot of velocity versus log of depth plots as a straight line on semi-log paper. This line has slope equal to K, and K is the von Karman universal velocity coefficient, approximately equal to 0.39 for rigid boundary, open channel flow.

Extending this concept, if one can describe the longitudinal velocity distribution at several verticals in a cross section, the variation of the longitudinal velocity over the width of the stream is known.

For a gentle bend of a parabolic cross section, Fig. 5-26 was developed. Figure 5-26 shows the curves for velocities across the width of a channel for consecutive sections along a bend. In Fig. 5-26

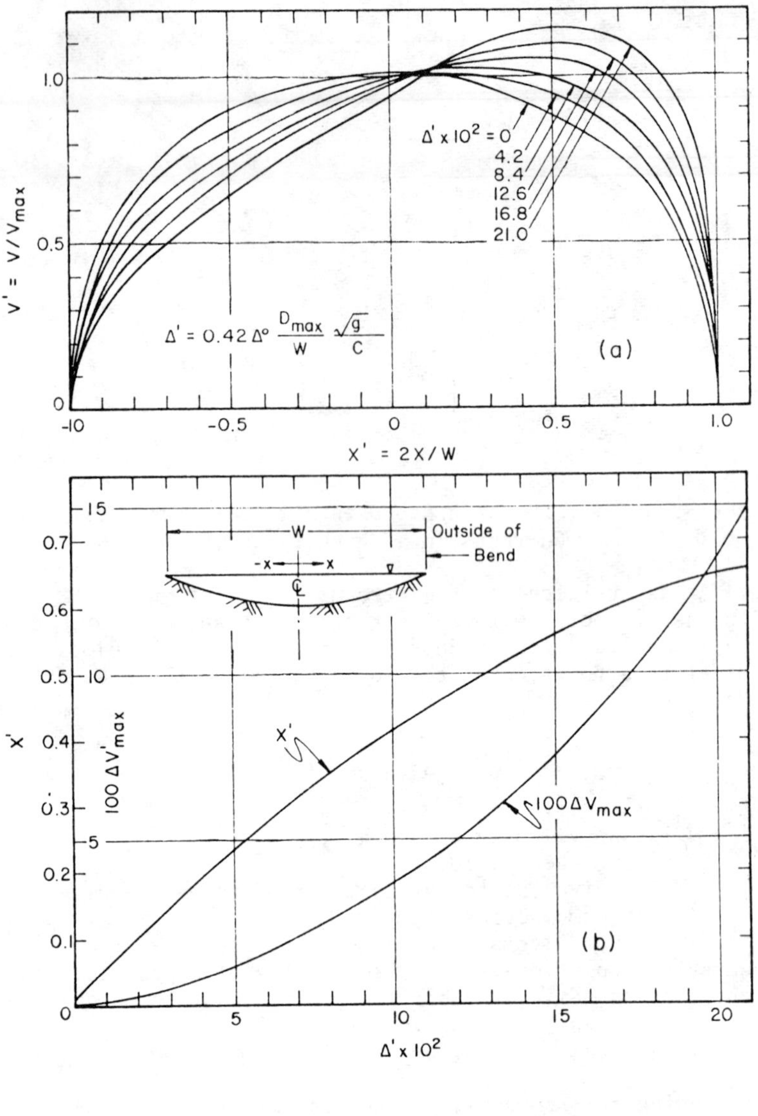

Fig. 5-26. Lateral Distribution of Velocity in an Alluvial Channel, after Rozovskii (1957).

$$V' = \frac{V}{V_{max}} \qquad (5\text{-}32)$$

in which V is the average velocity in any vertical, and V_{max} is the maximum velocity in the straight channel. Continuing,

$$X' \frac{2X}{W} \qquad (5\text{-}33)$$

in which X is measured from the center of the cross section and

$$\Delta' = 0.42\Delta \frac{D_{max}}{W} \frac{\sqrt{g}}{C}, \qquad (5\text{-}34)$$

in which Δ is the angle of the bend in degrees. The distribution of velocity in the straight reach is assumed to follow the form:

$$V = V_{max} \left(\frac{D}{D_{max}}\right)^{0.4} \qquad (5\text{-}35)$$

The V' values for sections within a bend are referenced to V_{max} in the straight reach. The depth across the width of the channel is assumed to vary as,

$$D = D_{max}(1 - X')^2 \qquad (5\text{-}36)$$

Figure 5-26b is constructed from Fig. 5-26a and shows the increase in V_{max} and the distance from the center of the section to the vertical with V_{max}. Equation (5-34) indicates that the transformation of velocities in a bend is greater when C decreases, and when D_{max}/W increases.

The procedure to be followed to determine the approximate nonerosive velocities in a bend, whose cross-section is roughly parabolic, and whose $W > 8D_{max}$ is as follows. Data required: Q, W, D_{max}, cross section, r_c, Δ, $C/\sqrt{g}$, θ. The procedure is:

1. Compute Δ' from Eq. (5-34).
2. From Fig. 5-26b find $\Delta V'max$ and X' and compute K_1, from

$$K_1 = \frac{1}{1 + \Delta V'_{max}} \qquad (5\text{-}37)$$

3. Compute X from Eq. (5-33) and determine D for V'max.
4. Compute K_2:

$$K_2 = \left(\frac{D}{D_{max}}\right)^{1/6}. \qquad (5\text{-}38)$$

5. Compute K_3:

$$K_3 = \left[\left(\frac{\text{Cot}^2\phi - \text{Cot}^2\theta}{1 + \text{Cot}^2\theta}\right)^{1/2} - 11 \frac{D_{max}}{r_c} \frac{\text{Cot}\theta}{+ \text{Cot}^2\theta}\right]^{1/2} \qquad (5\text{-}39)$$

6. Compute K' from Eq. (5-40), assuming

$K_4 = 1$, as

$$K' = K_1 K_2 K_3 K_4, \tag{5-40}$$

in which K_1 is due to the increase in maximum velocity in the bend,

K_2 is due to the displacement of maximum velocity over an area with less depth,

K_3 is due to the affect of bank slope and transverse circulation, and

K_4 is the affect of the increase in bottom velocity.

7. From Fig. 5-26a construct the curves of velocity distribution over the width for the straight approach and the exit section of the bend. For a parabolic section $V_{max} = 1.06\ \overline{V}$ in the straight reach.
8. Having the two velocity distribution curves over the channel width, the conditions for channel stability can be analyzed at any vertical.

The increased velocities usually present near the convex bank at entry into a bend caused no great erosion problem. At that location the transverse velocity is directed up the side slopes, increasing their stability. Several studies have been made of the transverse velocity field in the cross-section of an open channel. Two equations resulting from these studies are:

$$v_r = \frac{1}{3} V \frac{y}{r} \frac{m^2}{g} \left\{ \left(1 - 0.067 \frac{m}{c}\right) \left[\left(2\eta - \eta^2 \right) - \frac{8}{15} \right] + \frac{8}{45} \frac{m}{c} \left[3(\eta - 1)^2 - 1 \right] \right\}, \tag{5-41}$$

and

$$v_r = \frac{1}{k^2} V \frac{y}{r} \left[F_1(\eta) - \frac{\sqrt{g}}{kc} F_2(\eta) \right], \tag{5-42}$$

in which

v_r is the radial velocity of depth y,

m is the Boussinesq constant, equal to 22,

C is the Chezy coefficient,

η is the relative depth,

K is the von Karman coefficient, kappa, and

$F_1(\eta)$ and $F_2(\eta)$ can be determined from Fig. 5-27.

A comparison of the theoretical and observed transverse velocity distributions for a river bend is given in Fig. 5-28. For such sections fairly good results can be obtained. For the more irregular sections the results are less impressive.

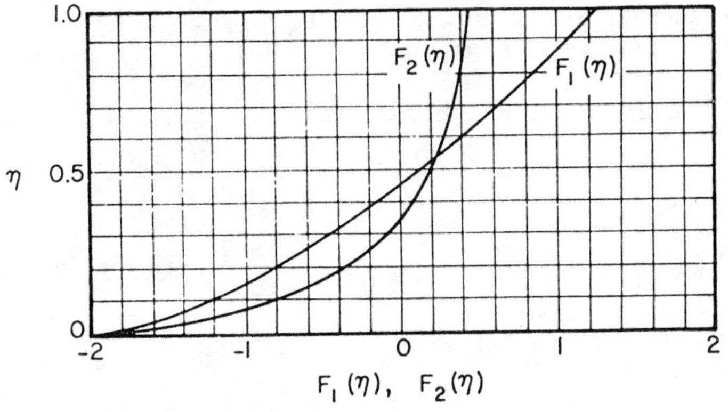

Fig. 5-27. Graph of the Functions $F_1(\eta)$ and $F_2(\eta)$, after Rozovskii (1957).

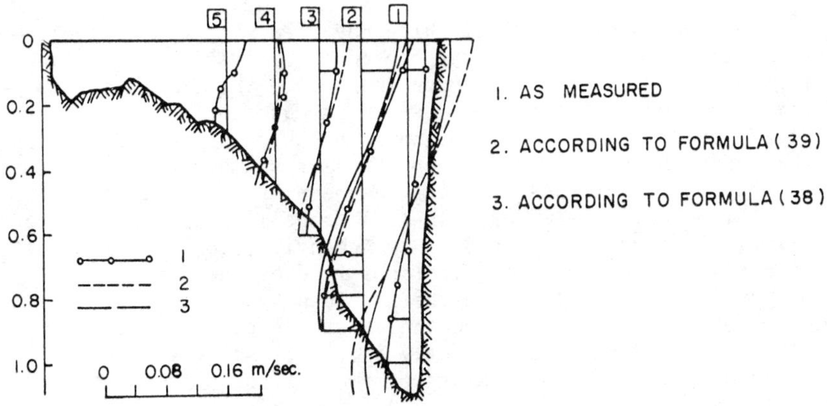

Fig. 5-28. Comparison of Theoretical Formulas with Data Field Observation on the Snov River, after Rozovskii (1957).

Another form of secondary circulation occurs in open channels which consists of cells of flow. In a channel with a large width to depth ratio and approximately uniform depth, transverse flow cells occur, usually in pairs, one rotating clockwise, the other counterclockwise. Between alternate pairs of cells the transverse surface flows come together and dive downward and conversely. This flow phenomena may accumulate debris, ice or other material floating on the surface into distinct parallel lines oriented in the direction of longitudinal flow.

Relations for the Radius of Curvature in Bends: A number of formulas relating the radius of curvature of a bend to the basic parameters of the channel (depth, slope, discharge, and velocity) are given in the following table.

Table 5-3. Radius of Curvature of Bends

Authors	Formulas in Metric System	Formulas in English System
Ripley	$r = 40\sqrt{A}$	$r = 40\sqrt{A}$
Pazin	$r = \frac{100\sqrt{Q_*}}{Q^2}$	$r = \frac{181\sqrt{Q_*}}{Q^2}$
Makaveyev	$r = \frac{0.004\sqrt{Q_*}}{s}$	$r = \frac{0.00726\sqrt{Q_*}}{s}$

in which $Q_* = V^2/gd - 15$ (an index of channel mobility).

Secondary Circulation

Secondary circulation, which may occur in open channels--both straight and curved, has caused considerable speculation regarding its existence in open channels. The extent of its effect on sediment transport, sediment distribution, velocity distribution, and channel roughness has been presented by various authors. A rather comprehensive summary of the beliefs and hypotheses of these writers has been presented in a work by Paul F. Nemenyi (1946, pp. 116-124) as well as his own concepts of this phenomenon. According to Nemenyi, secondary circulation in open channels was first observed by the German Geophysicist Max Moller and the American hydraulic engineer F. P. Sterns. They observed the existence of circulation simultaneously and independently in 1882. According to them, secondary circulation consists of two perfectly symmetrical parts, as illustrated in Fig. 5-29.

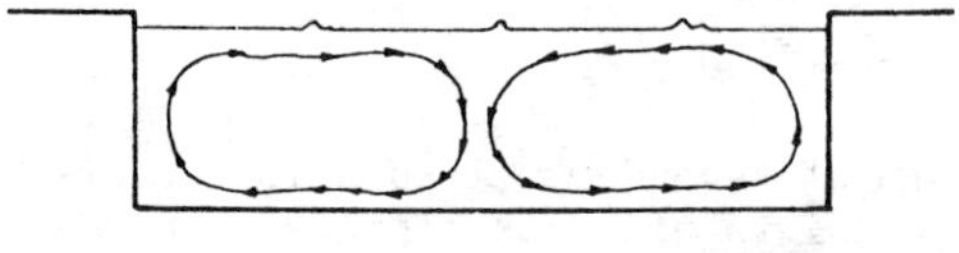

Fig. 5-29. Original Observation of Secondary Currents.

The simple form indicated in this figure is challenged by the results of the more thorough research of L. Prandtl conducted in closed conduits. His research when applied to open channels suggests the presence of several cells as shown in Fig. 5-30.

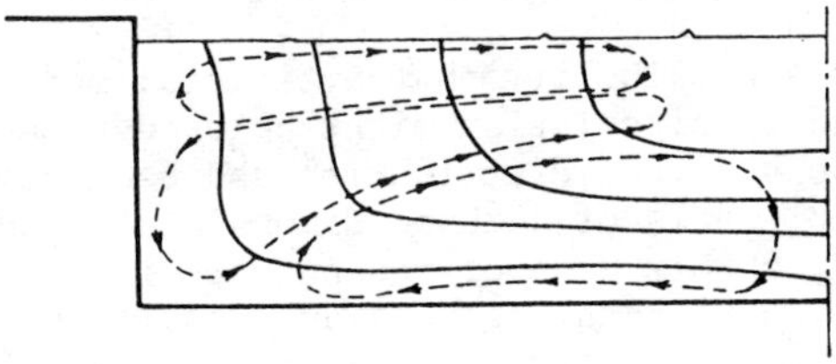

Fig. 5-30. Secondary Circulation, after Nemenyi (1946).

Others contributing to the knowledge of secondary circulation in open channels mentioned by Nemenyi are Terada of Tokyo Imperial University and Hugh Casey. Based on laboratory observations in a wide and steep channel heated from below. Terada found that secondary circulation resulting from temperature differences was of the form indicated in Fig. 5-31. In a paper prepared by Casey, a photograph was presented illustrating secondary circulation for a wide open channel with movable bed as shown in Fig. 5-32. This pattern is very similar to that presented by Terada.

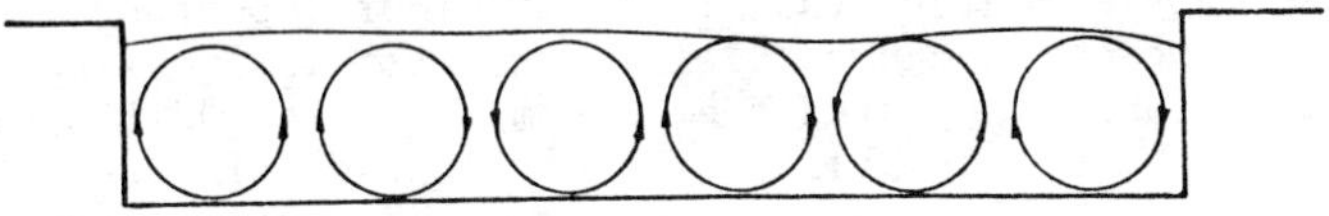

Fig. 5-31. Secondary Circulation in a Wide Open Channel Resulting from Temperature Differences.

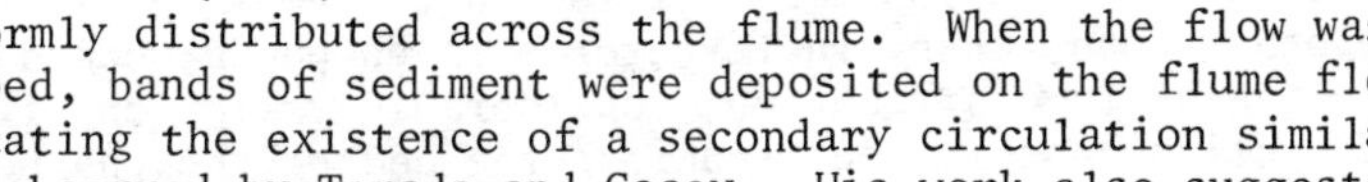

Fig. 5-32. Secondary Circulation.

Vanoni (1946) observed that the sediment was not uniformly distributed across the flume. When the flow was stopped, bands of sediment were deposited on the flume floor indicating the existence of a secondary circulation similar to that observed by Terada and Casey. His work also suggests that the number of cells is probably a function of width to depth ratio, W/D.

In Nemenyi's summary on circulation, he states that:

1. Secondary circulation is a normal occurrence in open channel flow and it can occur without the influence of sediment or temperature.
2. The longitudinal velocity distribution, quantity and distribution of suspended sediment, and the magnitude of secondary circulation are all interrelated in

such a way that if one is varied, the others will be affected also.

3. There is a possibility that artificial modification of secondary circulation might be used to modify or stabilize sediment distribution and bed formation.
4. Turbulent flow is of a three dimensional nature and this is probably a major reason for discrepancies between experimental results and results based on the von Karman theory of velocity distribution and sediment suspension.

The cause of circulation is debatable. In the above summary, it was stated that circulation exists independently of temperature gradient and/or suspended sediment, and the nonuniform sediment distribution across the section is the result of secondary circulation. In contrast to this, Vanoni (1946) believes that secondary circulation is either caused or at least appreciably strengthened by the lateral nonuniform distribution of sediment.

The extent of the effect of secondary circulation on factors related to channel stability is unknown. Based on existing knowledge, its influence may be negligible, of considerable importance, or somewhere in between these limits; however, many speculate that it is of minor importance. Only additional research can completely answer this question.

Lane (1955) suggested that one might gain additional insight to channel stability by studying the existence of secondary circulation and the number of cells generated in channels having different W/D ratios by sprinkling material such as saw dust uniformly across the water surface in a straight reach, then observing to see if the saw dust collects into bands, as one suspects that it would. The number of cells being generated should be directly related to the number of bands observed. In accordance with the foregoing concept, Simons (1957) has observed that secondary circulation may gather the floating surface material into distinct bands indicating the existence of secondary cells.

A study of the data collected by Simons (1957) and others shows that very little variation in lateral sediment distribution occurs that could not be attributed to sampling error. On the other hand, when tractive force is computed, based on vertical velocity distribution, considerable variation of an almost cyclical nature occurs across the bed of the channels. There is a possibility that this may be related directly to secondary circulation. By combining these results with a visual study of the circulation pattern, possibly more could be learned about this subject.

Secondary Motion at Channel Branches

When a channel subdivides into branches, channel geometries, bed roughness, and velocity distributions very different from those in the parent channel may result. Even greater differences exist when comparing the secondary flow in the parent channel with that which finally develops in branch channels. In the region of channel subdivision and for some distance downstream, the incompatibility of the secondary flow superposed by the main channel may cause an increase in turbulence, surface waves, and other instabilities in the head reaches of the branch channels. As flow conditions adjust to the new boundary conditions, these effects diminish.

When the branch channels rejoin, the adjustment in flow conditions causes turbulence, waves, vortices, and a gradual mixing of the flow occurs. When the joining flows differ in sediment discharge, temperature or both, the mixing process may be quite slow. Complete integration of the branch channels may require several miles.

The qualitative effects of the branching and rejoining of channel branches on secondary flow has not been studied and can only be discussed qualitatively in terms of limited field observations.

Effects of Secondary Flow on Channel Geometry

In the straight reaches of a river, little is known of the effects of secondary flow on channel geometry. However, circulating cells have been identified that have a diameter with an order of magnitude equal to the depth of flow. These secondary flows undoubtedly affect bed roughness in sand channels as evidenced by sand ridges roughly paralleling the flow.

In bends of rivers, the secondary flow has a more obvious effect on channel geometry. The transverse slope causing a strong secondary flow contributes to the development of erosive conditions within and downstream of the bend on the outside bank, shapes the channel into a triangular geometry, and causes a large point bar to deposit along the inside of the bend. At large discharges, the main current tends to pull away from the outer bank and may cut a channel across the point bar of such magnitude that a divided channel and, in some cases, a complete channel change may occur.

5.12 SLOPES OF ALLUVIAL RIVERS

Slopes of alluvial rivers vary with discharge and other physiographic factors. In general, slope decreases as discharge increases. In the downstream direction, as discharge increases, slope decreases as shown in Fig. 5-33. For the Yellowstone River basin for meandering alluvial channels, Lane (1955) suggested the slope discharge relation as,

$$S = 0.0017\ Q^{1/4}, \qquad (5\text{-}43)$$

in which Q varies from 0.10 cfs to 500,000 cfs.

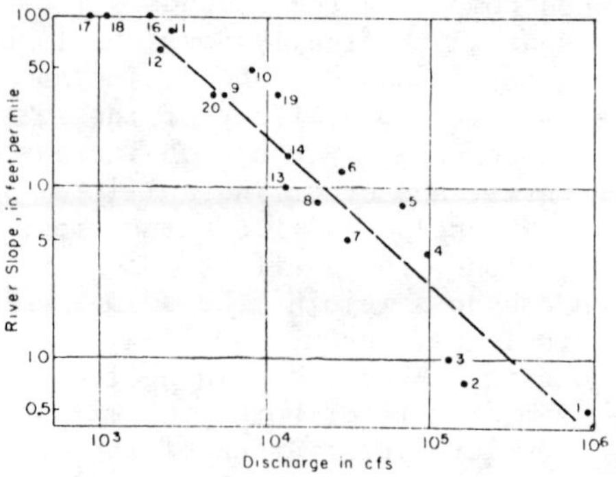

Fig. 5-33. Slope vs. Discharge, after Leopold and Langbein (1962).

Another basic relation often used to describe the variation of channel slope with distance states that

$$S_x = S_o e^{-\theta x} \tag{5-44}$$

in which S_x is channel slope at any distance x downstream of the reference station where the slope is S_o. The θ term is related to the characteristics of the river and its bed and bank material.

5.13 VELOCITIES OF ALLUVIAL RIVERS

The usual reference to velocity in natural streams is not to a velocity at a point but rather to a mean velocity for the channel.

Maximum Velocities

A tabulation was made by the USGS of the largest measured values of velocity at a single point (not the average for the whole cross section). The maximum point velocity usually is on the order of 25 to 50 percent greater than the average velocity for the cross section. Out of 2950 measurements included in the sample, the median value was 4.11 fps, the mean 4.84 fps, and less than 1 percent of the total exceeded 13 fps. One of the highest velocities ever measured by current meter by the USGS was 22 fps in a rock gorge of the Potomac River at Chain Bridge near Washington D.C., during the flood of March, 1936. Velocities up to 30 fps have occasionally been observed, but none have been recorded greater than this value.

Mean Velocities

The mean velocity of river corresponds to the mean or average discharge of a stream. During the flood stages, the mean velocities in the river vary from about 6 to 10 fps. The mean velocity attained in large rivers is generally slightly larger than that in small ones. There are, of course, many local situations where, owing to the constrictions or rapids, velocity attains greater values. The figures cited above would include a large majority of river channels in reaches that have no unusual features. Figure 5-34 shows the variations of mean velocity along Yellowstone-Missouri-Mississippi River system, as discharge increases in the downstream direction. From the plot, it shows that for this particular river system, velocity remains essentially constant in the downstream direction.

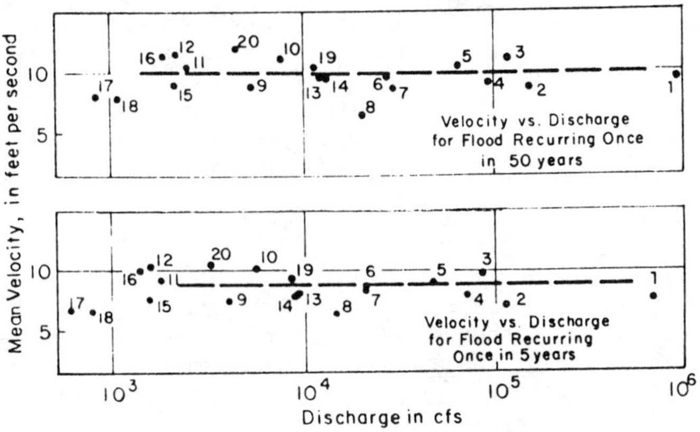

Fig. 5-34. Mean Velocity vs. Discharge, after Leopold and Langbein (1962).

Minimum Velocity

The minimum velocities of alluvial rivers correspond to the average velocity of the river during low flow stage. Under "minimum discharge" it was mentioned that for the 34 streams investigated, the low flows follow a Pearson Type III or Gumbel distribution and that the minimum flows varied from 0 to about 4500 cfs. Therefore, the minimum velocity could also attain a value very close to zero and up to 2-3 fps.

Velocity Fluctuations

Turbulent velocity fluctuations in alluvial rivers may reach 60-70 percent or more than the time average velocity. When the turbulent fluctuating component of velocity is expressed as a ratio of r.m.s. or standard deviation to mean average velocity, the ratio can attain 30-40 percent. For the Mississippi River, Kalinski (1942) found that the maximum fluctuating component of velocity is about 3 times the standard deviation. Figure 5-35 shows a typical plot of fluctuating velocity against time for the Mississippi River. Tiffany collected velocity fluctuations data for the Mississippi River;

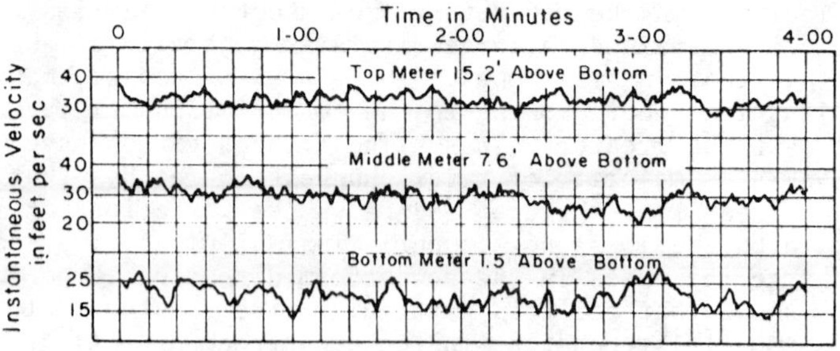

Fig. 5-35. Velocity Fluctuations in Mississippi River as Obtained with Price Current Meter. Total Depth 19.0 ft, after Kalinski (1942).

from his data, the fluctuating component of velocity varied as much as 53 percent of the average velocity.

5.14 VARIATION OF SIZE OF BED MATERIAL AND LONGITUDINAL CHANNEL SLOPE WITH DISTANCE

The reduction of rock material as a consequence of denudation and weathering of the earths crust has been of interest to geologists and geomorphologists from antiquity. Of still greater interest, is the transportation of the movable material--boulders, gravels, sands, and silts by streams. Transportation and depositional phenomena have developed our alluvial plains and river valleys.

Herodotus (about 400 BC) known as the father of history is also remembered for some of his geological observations. He recognized the importance of yearly increments of silt and clay deposited by the Nile. He is attributed the statement "Egypt is the gift of the river." Studies of river bed materials have shown that the size of particles forming the beds of rivers gradually decrease in size in the downstream direction. Knowledge of the size reduction of bed material is interesting and useful because it is closely related to channel geometry, sediment transport, and flow variables such as the velocity.

The principal causes of particle size reduction with distance measured along the channel in a downstream direction are abrasion of the particles, hydraulic sorting, and chemical weathering. In 1875 Sternberg presented a relation sometimes referred to as his abrasion law. This relation is

$$W = W_o e^{-\alpha L}, \tag{5-45}$$

in which W is the weight of the particle at distance L downstream of the reference station where the particle weight is W_o and α is a numerical constant. This relation implies two basic assumptions: (1) that the reduction in weight of particles is due entirely to abrasion, and (2) that the reduction is directly proportional to the distance moved by the flow and to the instantaneous weight of the particle. That is $dW \propto WdL$. Also, the size reduction relation can be expressed as

$$d = d_o e^{-\alpha_2 L} \qquad (5\text{-}46)$$

in which d represents the size of particles at some distance L downstream of the reference station, and d_o is the size of bed material at the reference station.

Paralleling this size reduction phenomonon the channel profiles tend to be concave upward, steep at their source, and gradually flattening exponentially in the downstream direction, similar to the phenomena of particle size and weight reduction. The Lane (1955) equilibrium relation

$$Q_s d \sim Q_w S, \qquad (5\text{-}47)$$

in which Q_s and Q_w are sediment discharge and water discharge, S is slope of energy gradient, and d is the representative size of bed material, implies a decrease in slope with decrease in size of bed material. In fact, this relation is very useful to predict the effect of mans work on river response as illustrated in Fig. 5-36. However, thus far, only qualitative relations between S and d have been found for river systems.

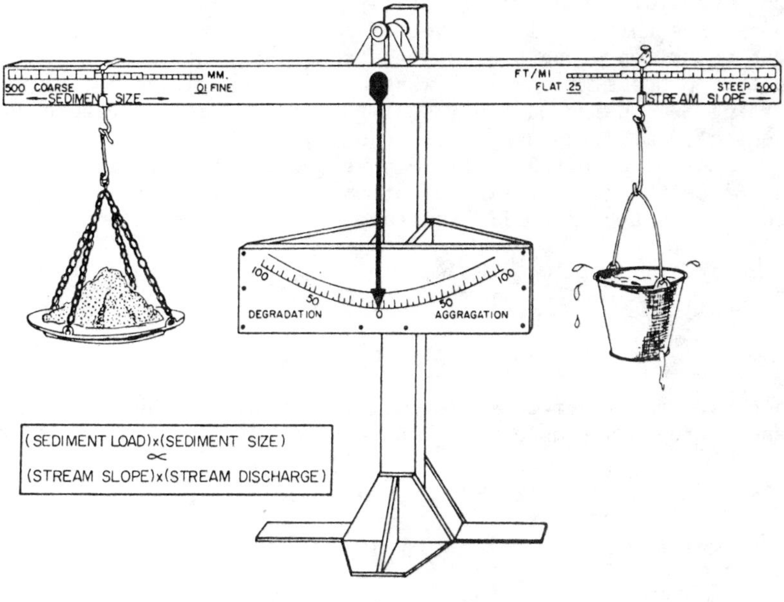

Fig. 5-36. Stable Channel Balance, W. H. Borland, USBR Personal Communication.

A corresponding slope reduction relation similar to Eq. (5-54) has been suggested by Shulits (1936) and others. It states that

$$S = S_o e^{-\delta L} \tag{5-48}$$

in which S_o is the slope at the reference station, S is the slope at distance L downstream, and δ is a numerical value which varies within limits from river to river. A more qualitative relation to describe and relate α and δ would be very useful. The slope equation has been referred to by Shulits as a "rational equation of the river bed profile." On the other hand, some insist that it should be treated as an empirical equation. This caused Shulits to reply "strictly speaking only divinely inspired knowledge will be nonempirical."

Returning to the abrasion concept several objections have been raised:

(1) Laboratory abrasion tests show very little particle size reduction even after particles have been subjected continuously to severe abrasive conditions for long periods.

(2) Values of α based on laboratory tests are quite different from those determined from field data.

(3) Abrasion does not provide a suitable physical explanation of the size reduction of sand size particles.

Studies by Straub (1935), Jopling (1964), Rafay and Simons (1965), and others have illustrated that this phenomena can occur by hydraulic sorting totally independent of abrasion. In most cases this provides a better physical explanation of the size reduction process within the sand range of sizes. Still another process is chemical weathering; even though hydraulic sorting is considered the most important process, abrasion and chemical weathering cannot be ignored.

Hydraulic sorting will be governed by those variables that influence the characteristics of the flow and the transport of sediment in the channel. Consequently, in functional form

$$\alpha = f(d, \rho_s, W, w, V, C_f, \rho, \mu, S, C_T, D) \tag{5-49}$$

in which

$$w = f(d, S_f, \rho_s, \rho, g, \mu). \tag{5-50}$$

To consider further possible relations between α and δ let S be assumed to be a simple power function of d,

$$d \sim S^d \quad \text{or} \quad d = K\,S^d, \tag{5-51}$$

and similarly,

$$d_o \sim S_o^d \quad \text{or} \quad d_o = KS_o^d\ . \tag{5-52}$$

Then in the relation

$$S = S_o e^{-\delta L}, \tag{5-53}$$

we can raise each side of the equation to the power d, and multiply by K. Then substituting d and d_o, respectively, for KS_o^α we obtain

$$d = d_o\, e^{-d\delta L} \,. \tag{5-54}$$

Comparing this relation with

$$d = d_o\, e^{-\alpha L}$$

it can be stated qualitatively that

$$\alpha = d\delta \tag{5-55}$$

The equation $d = d_o\, e^{-\alpha L}$ may be differentiated to yield

$$\alpha = \frac{-dd}{d} \,/\, dL \tag{5-56}$$

in which dL denotes the environmental distance within which the profile size decreases an incremental amount dd. The greater the value of α the greater the rate of decrease in particle size with distance. Similarly, a large value of δ will signify a rapid flattening of channel slope with distance downstream.

Data illustrating particle size and slope reduction along 15 rivers in Europe, Japan, and the United States were obtained from various publications by Rafay (1964). The lengths of rivers varied from 10.6 to 1100 miles and the median diameters of the bed material ranged from 0.12 mm (the Mississippi near its mouth) to 200 mm for the Makita River in Japan.

Particle size reduction is shown for the Mississippi River and the Rio Grande in Figs. 5-37 and 5-38. Slope variation with distance for the Rio Grande is shown in Fig. 5-39.

Further investigations will be required to establish more quantitative relations for α and δ. Both laboratory and field studies are needed. Such information should further improve methods of analysis of rivers and river systems and response of rivers to changing environments.

5.15 QUALITATIVE RESPONSE OF RIVER SYSTEMS

Many rivers have achieved a state of approximate equilibrium throughout long reaches. For practical engineering purposes, these reaches can be considered stable and are known as "graded" streams by geologists and as "poised" streams by engineers. However, this does not preclude significant changes over a short period of time or over a period of years. Conversely, many streams contain long reaches that are actively aggrading or degrading.

Regardless of the degree of channel stability, man's local activities may produce major changes in river characteristics both locally and throughout an entire reach. All too frequently the net result of a river improvement is a greater departure

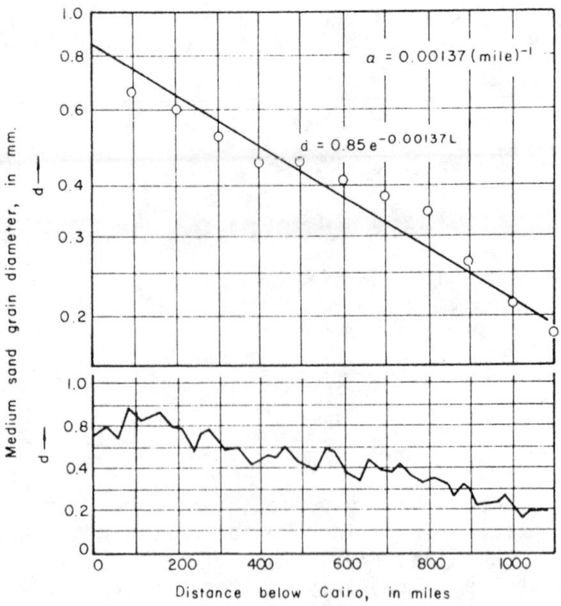

Fig. 5-37. Particle-Size Reduction Along Mississippi River, after Waterways Experiment Station Paper 17 (1935).

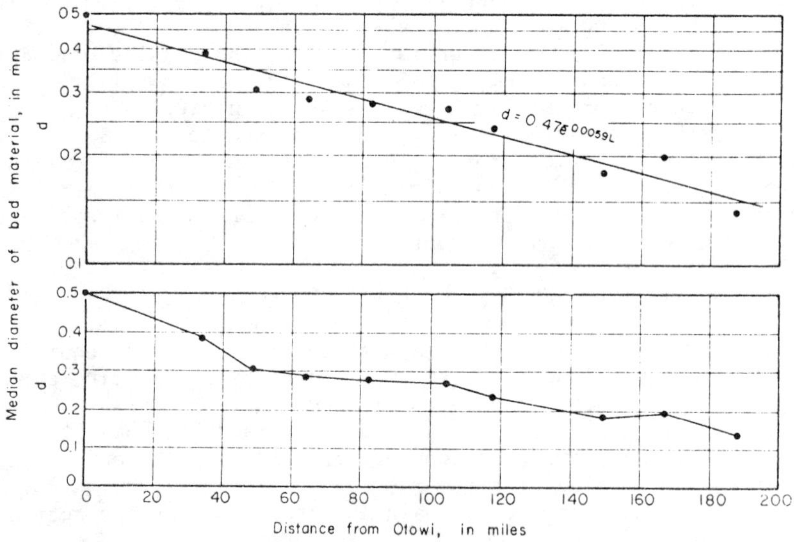

Fig. 5-38. Particle-Size Reduction Along Rio Grande, after Rafay (1964).

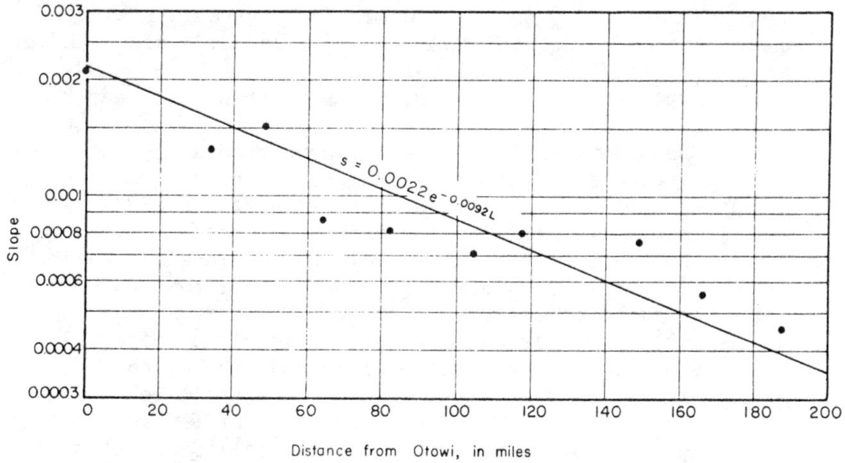

Fig. 5-39. Slope Variation Along Rio Grande, after Rafay (1964).

from equilibrium than that which originally prevailed. Good engineering design must invariably seek to enhance the natural tendency of the stream toward poised conditions. To do so, an understanding of the direction and magnitude of change in channel characteristics caused by the actions of man and nature is required. This understanding can be obtained by: (1) studying the river in a natural condition; (2) having knowledge of the sediment and water discharge; (3) being able to predict the effects and magnitude of man's future activities; and (4) applying to these a knowledge of geology, soils, hydrology, and hydraulics of alluvial rivers.

Predicting the response to channel development is a very complex task. There are a large number of variables involved in the analysis that are interrelated and can respond to changes in a river system and in the continual evolution of river form. The channel geometry, bars, and forms of bed roughness all change with changing water and sediment discharges. Because such a prediction is necessary, useful methods have been developed to predict both qualitative and quantitative response of channel systems to change.

Qualitative Prediction of General River Response to Changes

Quantitative prediction of response can be made if all of the required data are known with sufficient accuracy. Usually, however, the data are not sufficient for quantitative estimates, and only qualitative estimates are possible.

In more general terms, Lane (1955) studied the changes in river morphology in response to varying water and sediment discharge. Similarly, Leopold and Maddock (1953), Schumm

(1971), and Santos-Cayudo and Simons (1972) have investigated channel response to natural and imposed changes. These studies support the following general relationships:

(1) Depth of flow y is directly proportional to water discharge Q.
(2) Channel width W is directly proportional to both water discharge Q and sediment discharge Q_s.
(3) Channel shape, expressed as width to depth W/y ratio is directly related to sediment discharge Q_s.
(4) Channel slope S is inversely proportional to water discharge Q and directly proportional to both sediment discharge Q_s and grain size D_{50}.
(5) Sinuosity s is directly proportional to valley slope and inversely proportional to sediment discharge Q_s.
(6) Transport of bed material Q_s is directly related to stream power $\tau_o V$ and concentration of fine material C_F, and inversely related to the fall diameter of the bed material D_{50}.

A very useful relation for predicting system response was developed by Simons et al. (1975) establishing a proportionality between bed-material transport and several related parameters.

$$Q_s \sim \frac{(\tau_o V) W C_F}{D_{50}} \tag{5-57}$$

where τ_o = bed shear; V = cross sectional average velocity; C_F = concentration of fine material load. Utilizing the relation for boundary shear stress $\tau = \gamma DS$ and the continuity equation

$$Q = AV = WDV \tag{5-58}$$

one obtains

$$Q_s \sim \frac{(\gamma DS) WV}{D_{50}/C_F} = \frac{\gamma QS}{D_{50}/C_F} \tag{5-59}$$

If specific weight γ is assumed constant and the concentration of fine material C_F is incorporated in the fall diameter, this relation can be expressed simply as

$$QS \sim Q_s D_{50} \tag{5-60}$$

Equation (5-60) is essentially the relation proposed by Lane (1955), except fall diameter, which includes the effect of temperature on transport, has been substituted for the physical median diameter.

Equations (5-60) and (5-61) are useful for qualitative prediction of channel response to natural or imposed changes in canal and river systems. To use a classic example, consider the downstream response of a river to the construction of a dam

(Fig. 5-40). Aggradation in the reservoir upstream of the dam will result in relatively clear water being released downstream of the dam, that is, Q_s will be reduced to Q_s^- downstream.

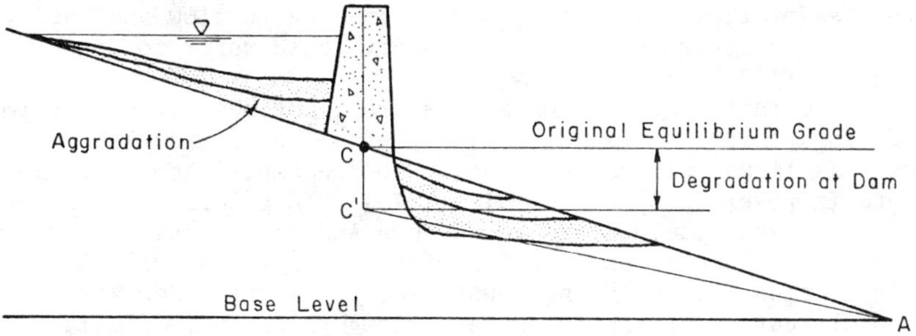

Fig. 5-40. Downstream Response of a River to the Construction of a Dam (after Simons and Sënturk, 1976).

Assuming fall diameter and water discharge remain constant, slope must decrease downstream of the dam to balance the proportionality of Eq. (5-60)

$$Q_s^- \; D_{50}^{\circ} \sim Q^{\circ} \; S^- \qquad (5\text{-}61)$$

In Fig. 5-40 the original channel gradient between the dam and a downstream geologic control (line CA) will be reduced to a new gradient (line C'A) through gradual degradation below the dam. With time, of course, the pool behind the dam will fill and sediment would again be available to the downstream reach. Then, except for local scour, the gradient C'A would increase to the original gradient CA to transport the increase in sediment load. Upstream, the gradient would eventually parallel the original gradient, offset by the height of the dam. Thus, dams with small storage capacity may induce scour and then deposition over a relatively short time period.

The engineer is also interested in quantitative results in addition to qualitative indications of trends. The geomorphic relation $QS \sim Q_s D_{50}$ is only an initial step in analyzing long-term channel response problems. However, this initial step is useful, because it warns of possible future difficulties in designing channel improvement and flood protection works and provides a good first-order estimate of response to all types of river development.

Qualitative Prediction of River Response to Specific Changes

Consider several relatively simple situations commonly encountered by engineers, scientists, and geologists in the river environment. Each case is introduced by a sketch which shows the physical situation prior to a selected natural or man-induced change. Below the sketch, some of the major local effects, upstream effects, and downstream effects resulting from natural processes or development activity are given. It is necessary to emphasize that only the gross local, upstream and downstream effects are identified. For more detail on regimes of flow, bed forms, and resistance to flow and their effects on river response, the reader could refer to Simons and Sentürk (1976).

The initial river conditions are sometimes given in terms of storage dams, water diversions, etc. These examples are used as illustrations relating to common experience. For more details refer to Simons et al. (1975). In general, the effect of a storage reservoir is to cause a sudden increase of base level for the upstream section of the river. The result is aggradation of the channel upstream, degradation downstream and a modification of the downstream flow hydrograph. Similar changes in the channel result if the base level is raised by some other mechanism, say a tectonic uplift. The effect of diversions from rivers is to decrease the river discharge downstream of the diversion with or without an overall reduction of the sediment transport. Similarly, changes in water and sediment input to a river reach often occur due to river development projects upstream from the reach under consideration as a result of natural causes.

Figure 5-41 illustrates the confluence of a tributary stream with the main stem river. The average water surface elevation in the main channel acts as the base level for the tributary. It is assumed here that the base level in the main channel has been lowered by a natural change in the river environment or by man-induced change such as the lowering of a reservoir level on the main stem. Applying Eq. (5-60), $QS \sim Q_s D_{50}$, to the tributary stream it can be seen that the increase in the slope S^+ must be balanced by an increase in sediment transport Q_s^+. Thus, under the new imposed condition, the local gradient of the tributary stream is significantly increased. This increased energy gradient induces headcutting and causes a significant increase in water velocities in the tributary stream. The result is bank instability, possible major changes in the geomorphic characteristics of the tributary stream and increased local scour.

Response to the converse situation, raising the base level, can be illustrated by considering river response to construction of a dam (Fig. 5-42). Whenever the base level of a channel is raised a pool is created extending a considerable distance upstream depending on the amount of change. This "backwater" effect results in the M_1 curve gradually varied flow. As the water and sediment being transported by the river encounters this pool, most of the sediments drop out forming a delta-like formation at the head of the pool which slowly advances downstream. The deposition of sediment at the entrance to the pool induces aggradation in the channel upstream. This aggradation may extend many miles upstream after a long period of time, producing significant changes in river geometry, and

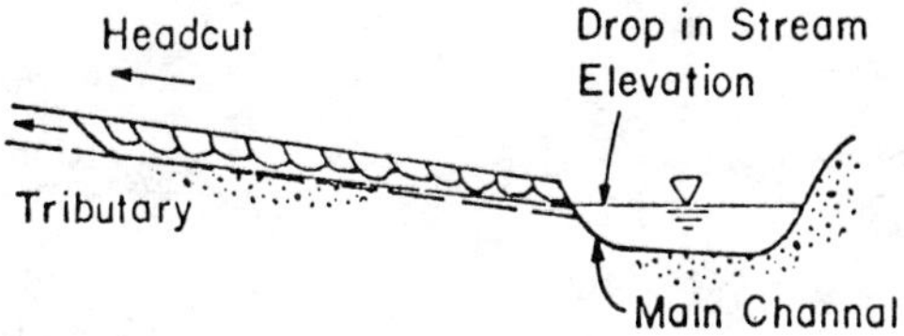

Local Effects	Upstream Effects	Downstream Effects
1. Headcutting	1. Increased velocity	1. Increased transport to main channel
2. General scour	2. Increased bed material transport	2. Aggradation
3. Local scour	3. Unstable channel	3. Increased flood stage
4. Bank instability	4. Possible change of form of river	4. Possible change of form of river
5. High velocities		

Fig. 5-41. Lowering of Base Level for Tributary Stream.

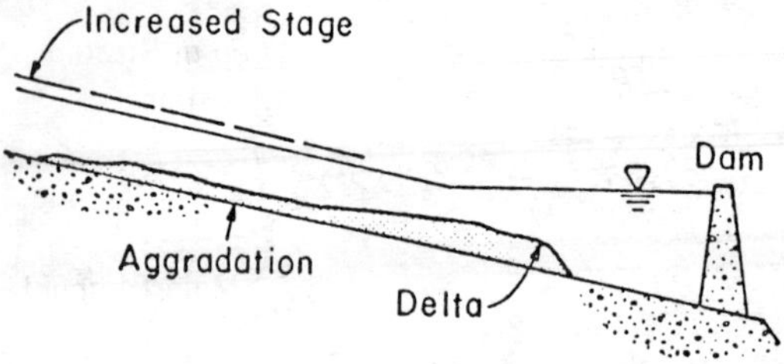

Local Effects	Upstream Effects	Downstream Effects
1. Aggradation of bed	1. See local effects	1. See downstream effects, Fig. 14
2. Loss of waterway capacity	2. Change in base level for tributaries	
3. Change in river geometry	3. Deposition in tributaries near confluences	
4. Increased flood stage	4. Aggradation causing a perched river channel to develop or changing the alignment of the main channel	

Fig. 5-42. Raising Base Level in Main Channel.

increasing flood stages. Again, Eq. (5-60) provides an indication of the response. The decrease in slope S^- must be accompanied by a decrease in transport capacity Q_s or $Q_s^- \sim Q_s^- D_{50}^o$. In the extreme it is possible that the river may become sufficiently perched that at some high flow it could abandon the old channel and adopt a new one. As noted in Fig. 5-42, the effects of raising the base level of the main channel include an increase in base level for any tributaries entering the pool formed by the main stem dam. The impact of this change on the tributaries is shown in Fig. 5-43.

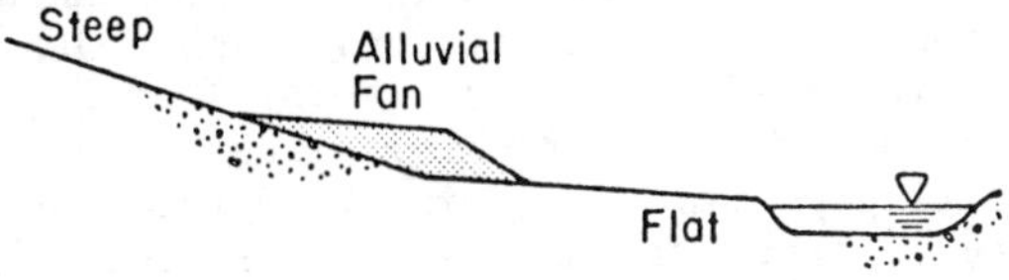

Local Effects	Upstream Effects	Downstream Effects
1. Alluvial fan reduces waterway	1. Erosion of banks	1. Aggradation
2. Channel location is uncertain	2. Unstable channel	2. Flooding
	3. Large transport rate	3. Development of tributary bar in the main channel

Fig. 5-43. Raising Base Level for Tributary Stream.

The change in gradient of the tributary stream in most cases causes significant deposition. This can be seen from Eq. (5-60) where a decrease in slope is accompanied by a decrease

$$Q^o S^- \sim Q_s^- D_{50}^o$$

assuming constant conditions of water discharge and size of bed material. In this case illustrated, an alluvial fan develops which in time can divert the river or reduce the waterway. In general, streams on alluvial fans shift laterally so that the future location of the channel is uncertain. A similar situation occurs naturally where a steep tributary stream draining an upland region reaches the flatter floodplain of the parent stream.

The impact of the construction of a dam on the reach upstream of the dam is outlined in Fig. 5-42. Construction of an upstream storage dam provides a desilting basin for the water flowing in the system (Fig. 5-42). In most instances all of the bed-material load coming into a reservoir drops out within the reservoir. Water released from the reservoir is quite clear. The existing river channel is the result of its interaction with normal water-sediment flows over a long period of time. With the sediment-free flows the channel below the dam is too steep and sediments are entrained from the bed and the banks bringing about significant degradation. The channel banks may become unstable due to degradation and there is a possibility that the river, as its profile flattens, may change its plan form. Figure 5-44 illustrates the possible impact of a significant decrease in slope on channel pattern. Assuming that prior to dam construction the reach below the dam plotted as an intermediate stream (Point 1 of Fig. 5-43b), the decrease in slope at constant water discharge could move the stream's plotting position to Point 2 in the meandering region of the chart. In the extreme case, it is possible that the degradation may cause failure of the dam and the release of a flood wave.

Figure 5-45 illustrates a situation where artificial cutoffs have straightened the channel below a given reach. It is obvious that straightening the channel downstream of Reach A significantly increases the channel slope. In general, this causes higher velocities, increased bed material transport degradation and possible headcutting through Reach A. This can result in unstable river banks and a braided stream form as shown on Fig. 5-45b. Here, the original plotting position (Point 1) is moved to Point 2 in the braided region by the increase in channel slope. In addition, the straightening of the main channel brings about a drop in base level and any tributary streams flowing into the affected reach of the main channel are subjected to conditions outlined in Fig. 5-41.

On the other hand, if the straightened section is designed to transport the sediment loads that the river is capable of carrying both upstream and downstream of the straightened reach, bank stability may not be endangered. Such a channel should not undergo significant change over either short or long periods of time. It is possible to build modified reaches of main channels that do not introduce major adverse responses due to local steepening of the main channel. In order to design a straightened channel so that it behaves essentially as the natural channel in terms of velocities and magnitude of bed-material transport, it is necessary, in general, to build a wider shallower section.

The development of an alluvial island below the confluence of a steep tributary is a common feature of the river environment (Fig. 5-46). The tributary introduces relatively large quantities of bed material into the main channel. As a result of island formation in the main channel, divided flow exists. In an attempt to maintain navigation depths in the main channel it is common practice to close the chute channel by construction of a dike across the subchannel to the island or bar formed by

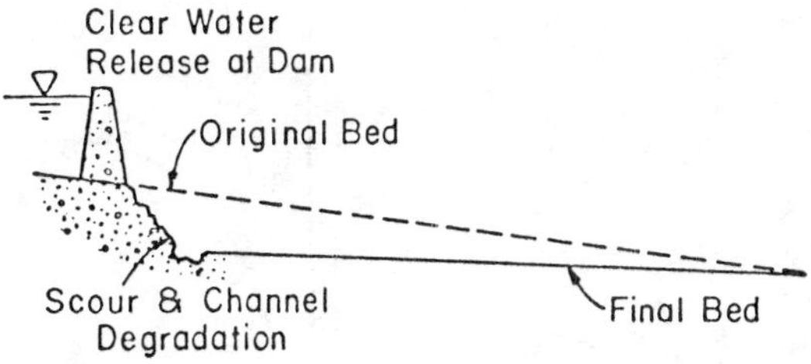

(a)

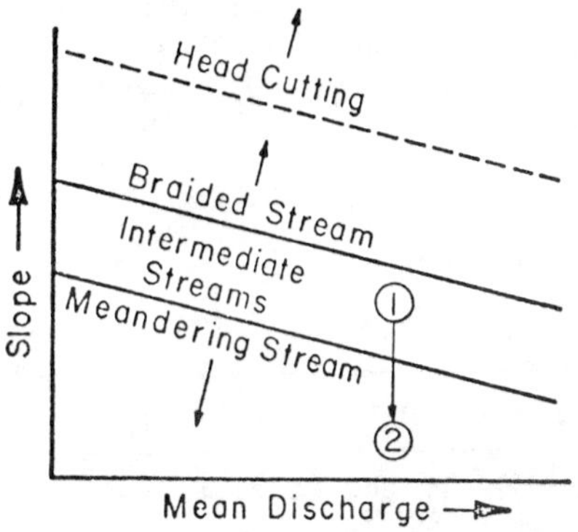

(b)

Local Effects	Upstream Effects	Downstream Effects
1. Channel degradation	1. See upstream effects, Fig. 5-42	1. Degradation
2. Possible change in river form		2. Reduced flood stage
3. Local scour		3. Reduced base level for tributaries, increased velocity and reduced channel stability causing increased sediment transport to main channel
4. Possible bank instability		
5. Possible dam failure		

Fig. 5-44. Clear Water Release Below a Dam.

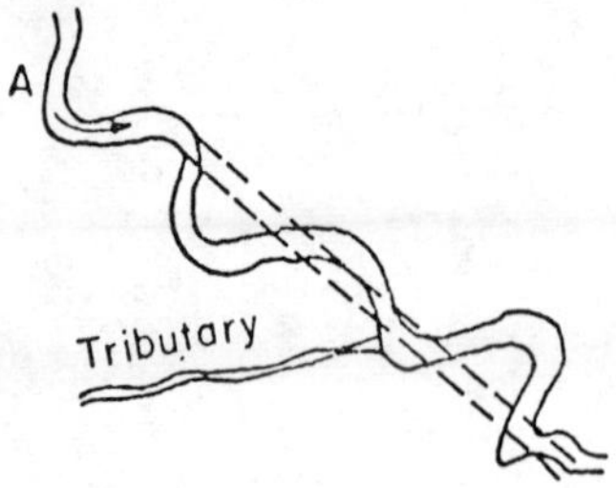

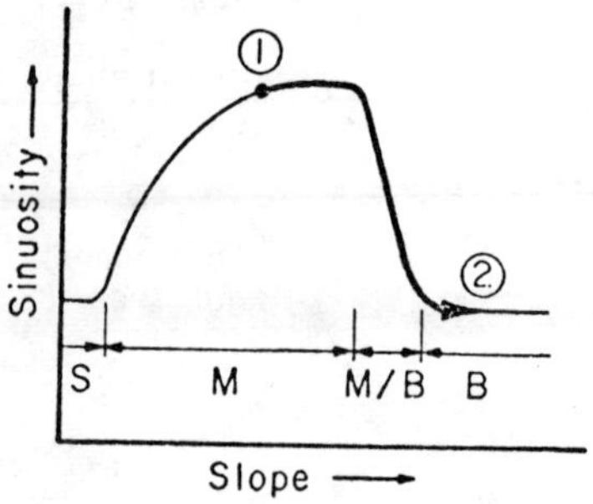

Local Effects	Upstream Effects	Downstream Effects
1. Steeper slope	1. See local effects	1. Deposition downstream of straightened channel
2. Higher velocity		2. Increased flood stage
3. Increased transport		3. Loss of channel capacity
4. Degradation and possible headcutting		
5. Banks unstable		
6. River may braid		
7. Degradation in tributary		

Fig. 5-45. Straightening of a Reach by Cutoffs.

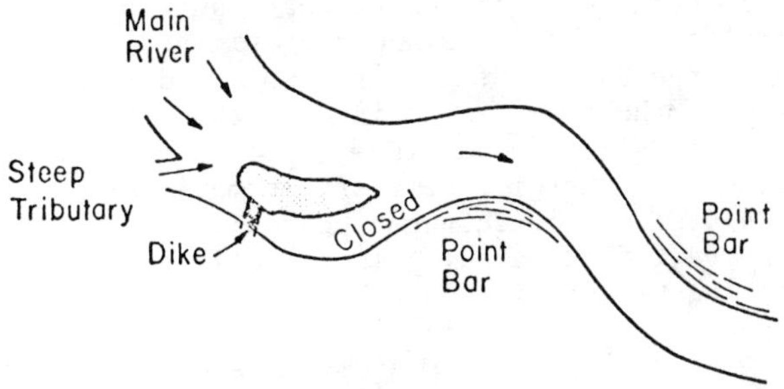

Local Effects	Upstream Effects	Downstream Effects
1. Contraction of the river	1. Aggradation	1. Deposition of excess sediment downstream of the closure
2. Increased velocity	2. Backwater at flood stage	2. More severe attack at first bend downstream
3. General and local scour	3. Changed response response of the tributary	3. Possible development of a chute channel across the next point bar downstream
4. Bank instability		

Fig. 5-46. Closure of a Chute Canal.

deposition. Such a procedure forces all of the water and sediment to pass through a reduced width. This contraction of the river in general increases the local velocity, increases general and local scour, and may significantly increase bank instability.

In addition, the contraction can change the alignment of the flow in the reach and thus would affect the downstream main channel for a considerable distance. A chute channel can develop across the next point bar downstream and its effect may extend several meander loops downstream. Upstream of the reach there is aggradation and its amount depends on the magnitude of water and sediment being introduced from the tributary. Also, there is significant increase in the backwater upstream of the reach at high flows which in turn affects other tributaries farther upstream. With this analysis the continuity equation, the relationship between discharge and bed-material transport, concepts of rapidly varied flow and the backwater curves all provide indicators of the response to be anticipated.

Among the natural phenomena that can impact the river environment are earthquakes and tectonic activity. Large portions of the United States are subjected to at least infrequent earthquakes. Associated with earthquake activity are severe landslides, mud flows, uplifts in the terrain, and liquefaction of otherwise semi-stable materials all of which can have a profound effect upon channels and structures located within the earthquake area. Historically, several rivers have completely changed their course as a consequence of earthquakes. For example, the Brahmaputra River in Bangladesh and India shifted its course laterally a distance of some 200 miles as a result of earthquakes that occurred approximately 200 years ago. Although it may not be possible to design for earthquake effects, knowledge of the probability of its occurrence is important so that certain aspects of the induced effects from earthquakes can be taken into consideration when evaluating river response.

Figure 5-47 illustrates a more complicated set of circumstances. In this case a reach of river is affected by Dam A constructed upstream as well as Dam B constructed downstream. As documented in Fig. 5-44, Dam A causes significant degradation in the main channel. Dam B causes aggradation in the main channel (Fig. 5-42). The final condition in Reach C is estimated by summing the effects of both dams on the main channel and the tributary flows. The scour below Dam A would make some sedimentary material available for deposition in the reservoir above Dam B, further complicating the situation. Normally, this analysis requires water and sediment routing techniques studying both long- and short-term effects of the construction of these dams.

River response to upstream and downstream storage reservoirs on the same stream as analyzed in Fig. 5-47 can be quite complex. Another situation that is amenable to a basic qualitative analysis involves the response of reaches on two major tributaries a considerable distance upstream of their confluence (Fig. 5-48). Upstream of Reach A a diversion structure is built to divert essentially clear water by canal to the adjacent tributary on which Reach B is located. Upstream of Reach B the clear water diverted from the other channel plus water from the tributary is released through a hydropower plant. Ultimately, it is anticipated that a larger storage reservoir may be constructed downstream of the tributary confluence on the main stem at C. These changes in normal river flows give rise to several complex responses in Reaches A and B on the tributary systems as well as on the main stem. Reach A may aggrade due to the excess of sediment left in that tributary when clear water is diverted:

$$Q^- S^+ \sim Q_s^\circ D_{50}^\circ$$

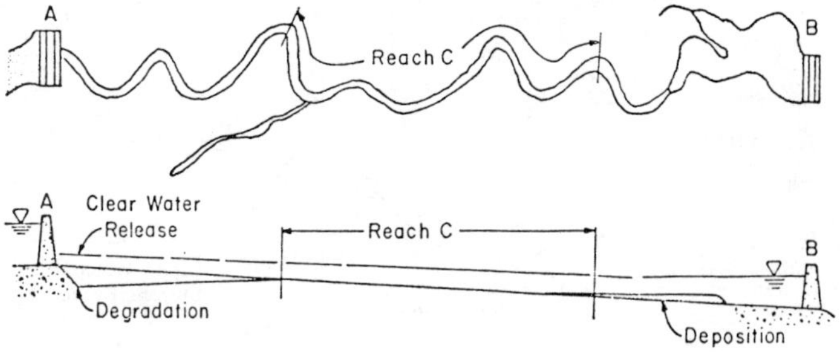

Local Effects	Upstream Effects	Downstream Effects
1. Dam A causes degradation	1. Channel could aggrade or degrade with effects similar to Figs. 5-42 and 5-44.	1. See upstream effects
2. Dam B causes aggradation		
3. Final condition in Reach C is the combined effect of (1) and (2). Situation is complex and combined interaction of dams, main channel and tributaries must be analyzed using water and sediment routing techniques and geomorphic factors.		

Fig. 5-47. Combined Increase of Base Level and Reduction of Upstream Sediment Load.

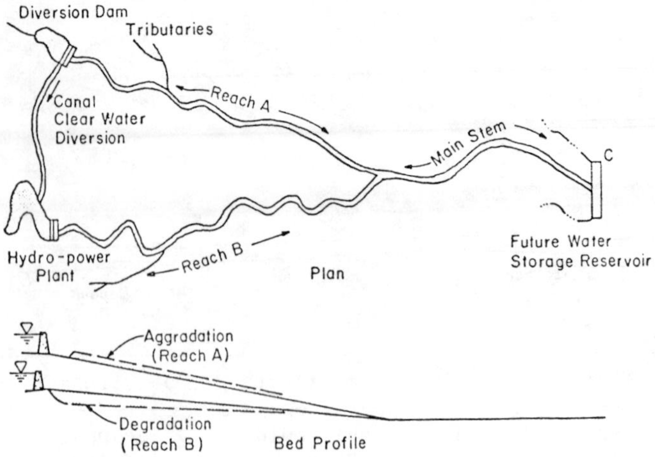

Local Effects	Upstream Effects	Downstream Effects
1. Reach A may be be subjected to aggradation due to excess sediment left in the channel by diversion of clear water and degradation in tributaries caused by lowering of their base level	1. Upstream of Reach A--aggradation and possible change of river form	1. See upstream effects
2. Reach B may be subjected to degradation due to increased discharge in the channel	2. Upstream of Reach B--degradation and change of river form	2. Construction of reservoir C could induce aggradation in the main channel and in the tributaries
3. If a storage reservoir was constructed at C it could induce aggradation in both tributaries	3. Channel instabilities	
	4. Significant effects on flood stage	

Fig. 5-48. Clear Water Diversion and Release Combined with Downstream Storage.

However, initially there may be a lowering of the channel bed in the vicinity of the diversion structure because of the deposition upstream of the diversion dam and the release of essentially clear water for the relatively short period of time until the sediment storage capacity of the reservoir is satisfied. Reach B is subjected to degradation due to the increased discharge and an essentially clear water release:

$$Q^+ S^- \sim Q_s^- \, D_{50}^{\circ}$$

However, the degradation of the channel could induce degradation in the tributaries causing them to provide additional sediment to the main channel. This response would to some degree counteract the degrading situation in this reach of river. Such changes in river systems are not uncommon and introduce complex responses throughout the system. Any complete analysis must consider the individual effects and sum them over time to determine a response in the reaches of concern.

5.16 MODELING OF RIVERS

There are many problems in hydraulic engineering for which the basic equations are known but which are geometrically so complicated that the direct application of the equations is difficult. Many such problems can be solved by the use of models which duplicate this complicated geometry and in which the resulting flow patterns can be observed directly. The models may be physical models; that is small-scale physical replications. They may also be mathematical consisting of mathematical abstractions of the phenomena. Models are used to test the performance of a design or to study the details of a phenomenon. The performance tests of proposed structures can be made at moderate costs and small risks on small-scale (physical) models. Similarly, the interaction of a structure and the river environment can be studied in detail.

The natural phenomena are governed by appropriate sets of governing equations. If these equations can be integrated, the prediction of a given phenomenon in time and space domains can be made mathematically. In many cases related to river engineering, all the governing equations are not known. Also, the known equations cannot be directly treated mathematically for the geometries involved. In such cases, models are used to physically integrate the governing equations. Similitude between a prototype and a model implies two conditions:

1. To each point, time and process in the prototype, a uniquely coordinated point, time and process exists in the model.
2. The ratios of corresponding physical magnitudes between prototype and model are constant for each type of physical quantity.

Mobile Bed Models

In modeling response to development works in the river environment, three-dimensional mobile models are often used. These models have the bed and sides molded of materials that can be moved by the model flows. Similitude in mobile bed models implies that the model reproduces such fluvial processes as bed scour, bed deposition, lateral channel migration, and varying boundary roughness. It has not been considered possible to faithfully simulate all of these processes simultaneously on scale models. Distortions of various parameters are often made in such models.

Mobile bed models are more difficult to design and their theory is extremely complicated as compared to clear water rigid bed models. However, many successful examples of their use are available. In general, all important river training and control works are studied on physical models. The interpretation of results from a mobile bed model requires a basic understanding of the fluvial processes and some experience with such models. In many cases, where it is possible to obtain only qualitative information from mobile bed models, this information is of great help in comparing the performance of different designs.

Mathematical Models

A physical scale model is a means for extracting information from some source other than from the prototype. With a distorted physical model the geometrical analogy is weakened considerably, but still, under most conditions, the analogy of the overall behavior is strong.

Once one gets used to the idea of looking at a model as an analog computer the next logical step would be to model the process under study on a digital computer in numerical form. This of course requires that a "complete" set of governing equations (some of them differential equations) is available. Such equations would include basic flow equations, the differential equation of nonuniform and unsteady flow, the sediment transport equation, the differential equation formulating continuity of sediment transport, and criterion to predict the bed deformations, just to mention the most obvious equations involved. It is clear that the interaction between these equations is complex. This, after all, is the reason for attempting to model these processes physically. But with the availability of high speed digital computers it becomes entirely feasible to study some of the characteristics of a river system numerically. Of course the results cannot be better than the basic equations used in the analysis, and most equations available are for one- or two-dimensional flow fields only. But when an overall river system is considered, a river can be viewed as a highly two-dimensional system, and with certain simplifying assumptions a river can be modeled as a one-dimensional system. It is only when one starts looking into the details that three-dimensional processes become important.

Calibration of a mathematical model involves evaluation and modification of the supplementary relations to the basic equations from field data and/or theories such that the mathematical model will reproduce the historical response of the modeled river system. This is similar to calibration of a physical model. To perform the mathematical model calibration, the following information is required: (a) hydrographic maps of the modeled river reach; (b) hydrographs of stage, flow and sediment discharge, and (c) geological and physical properties of the bed and bed material.

From (a), one can evaluate the geometric properties of the river reach. The relations for S_f, Q_s, q_ℓ and V_ℓ can then be evaluated from (b) and (c). If part of data is not available, relations based on experimental, empirical, or theoretical approaches can be used. However, calculated results are only as good as the calibration relations. More specifically, the resistance function for S_f and the sediment transport function for Q_s must be tested and modified to accomplish the model calibration, that is, until the historical data along the river can be reproduced by the mathematical model.

The application of mathematical modeling is not limited to the main channels of a river system. The management of watersheds and river basins requires, in general, a complete knowledge of the interrelations between ecology and environment. The watershed response to developments, either natural or man-induced, must be anticipated correctly if progress is to be made towards wise use of our nation's natural resources.

The physical quantities which describe the major watershed response are the water hydrograph and yield, the sediment hydrograph and yield, and the resultant watershed stream morphology. Because the physical processes governing watershed behavior are very complicated many past studies have utilized a statistical interpretation of observed watershed response data. The Unit Hydrograph Method for water routing and the Universal Soil-loss Equation for estimating soil erosion are examples of these types of statistical studies. However, it is difficult to predict the response of a watershed to various land developments or treatments using these methods because they are based on the assumption of homogeneity in both time and space. Numerical modeling using equations describing the physical processes provides a viable method of estimating time-dependent watershed response. In recognition of the necessity of such models, Colorado State University has developed a number of numerical computer programs employing the formulation of the basic physical processes to determine water and sediment hydrographs and yields from small watersheds as described by Dr. R. M. Li and Dr. Y. H. Chen in their presentations.

5.17 REFERENCES

Albertson, M. L., Simons, D. B., and Richardson, E. V., 1958, "Discussion of mechanics of sediment-ripples formation, by H. K. Liv," Jour. of the Hydraulics Div., ASCE, Vol. 84, No. HY1, February, 1558 p.

Ashida, K., and Tanaka, Y., 1967, "A statistical study of sand waves," Proc. 12th Congress of IAHR, Vol. 2.

Assifi, A. T., 1966, "Hydraulics and geometry of rivers," in partial fulfillment of the requirements for the Degree of Master of Science in Civil Engineering, Colorado State University, Fort Collins, Colorado.

Athaullah, M., 1968, "Prediction of bed forms in erodible channels," Ph.D. dissertation, Dept. of Civil Eng., Colorado State University, Fort Collins, Colorado.

Beckman, E. W., and Furness, L. W., 1962, "Flow characteristics of Elkhorn River near Waterloo, Nebraska," U.S. Geological Survey Water-Supply Paper 1498-B, 34 p.

Bogardi, J. L., 1958, "Some recent advances in the theory of sediment movement," Hydrological Jour., Vol. 38, No. 4, 241-252 pp.

Colby, B. C., 1964, "Discharge of sands and mean-velocity relationships in sand-bed streams," U.S. Geological Survey Professional Paper 462-A, 47 p.

Colby, B. C., and Christensen, R. P., 1956, "Visual accumulation tube for size analysis of sand," Am. Soc. Civil Engineers Jour., V. 82, No. HY-3, 17 p.

Corey, A. T., 1949, "Influence of shape on the fall velocity of sand grains," M.S. Thesis, Colorado A and M College, 102 p.

Culbertson, D. M., Young, L. E., and Brice, J. C., 1967, "Scour and fill in alluvial channels," U.S. Geological Survey Open-file Report, 58 p.

Culbertson, J. K., and Dawdy, D. R., 1964, "A study of fluvial characteristics and hydraulic variables, Middle Rio Grande, New Mexico," U.S. Geological Survey Water-Supply Paper 1498-F, 74 p.

Einstein, H. A., 1950, "The bed load function for sediment transportation in open channel flows," U.S. Dept. Agri. Tech. Bull. 1026, 70 p.

Friedkin, J. F., 1945, "A laboratory study of the meandering of alluvial rivers," United States Waterways Experiment station, Corps of Engineers, Vicksburg, Mississippi, May.

Garde, R. J., and Albertson, M. L., 1959, "Sand waves and regimes of flow in alluvial channels," Proc., IAHR, Vol. 4, 28-511, 7 p., Montreal.

Garde, R. J., and Raju, K. G. R., 1963, "Regime criteria for alluvial channels," Jour. of the Hydraulics Div., ASCE, Vol. 89, November.

Guy, H. P., Simons, D. B., and Richardson, E. V., 1966, "Summary of alluvial channel data from flume experiments, 1956-1961, " U.S. Geological Survey Professional Paper 462-I, 96 p.

Hubbell, D. W., and Matejka, D. Q., 1959, "Investigation of sediment transport, Middle Loup River at Dunning, Nebraska," U.S. Geological Survey Water-Supply Paper 1476.

Ippen, A. T., and Drinker, P. A., 1962, "Boundary shear stresses in curved trapezoidal channels," Proc. ASCE, Vol. 88, No. HY5, 143 p.

Jopling, A. V., 1964, "Laboratory study of sorting processes related to flow separation," Jour. of Geophysical Research, 69:3403-3418.

Jordan, P. R., 1965, "Fluvial sediment of the Mississippi River at St. Louis, Missouri," U.S. Geological Survey Water-Supply Paper 1802.

Kalinski, A. A., 1942, "Role of turbulence in river hydraulics," Hydraulic Conference, Iowa City, 2:226-79, (Iowa University, studies in Engineering, Bull. 27).

Lane, E. W., 1955, "The importance of fluvial morphology in hydraulic engineering," Proc. ASCE, Vol. 81, No. 745, 1-17 pp.

Lane, E. W., 1957, "A study of the shape of channels formed by natural streams flowing in erodible material," Missouri River Division Sediments Series No. 9, U.S. Army Engineer Division, Missouri River, Corps of Engineers, Omaha, Nebraska.

Leopold, L. B. and Langbein, W. B., 1962, "The concept of entropy in landscape evolution," U.S. Geological Survey Professional Paper 500-A, 20 p.

Leopold, L. B. and Maddock T., Jr., 1953, "The hydraulic geometry of stream channels and some physiographic implication," U.S. Geological Survey Professional Paper 252, 57 p.

Liu, H. K., 1957, "Mechanics of sediment-ripple formation," Jour. of the Hydraulics Div., ASCE, Vol. 183, HY2, April.

Nemenyi, P. F., 1946, "Discussion on paper by Vanoni, Transportation of suspended sediment by water," Paper 2267, ASCE Trans., 116-124 pp.

Nordin, C. F., Jr., 1964, "Aspects of flow resistance and sediment transport: Rio Grande near Bernalillo, New Mexico," U.S. Geological Survey Water-Supply Paper 1498-11.

Nordin, C. F., Jr., 1968, "Statistical properties of dune profiles," Ph.D. dissertation, Dept. of Civil Eng., Colorado State University, Fort Collins, Colorado.

Nordin, C. F., Jr. and Algert, J. H., 1966, "Spectral analysis of sand waves," Jour. of the Hydraulics Div., ASCE, Vol. 92, No. HY5, 95-114 pp.

Rafay, T., 1964, "Analysis of change in size of bed material along alluvial channels," M.S. thesis, Colorado State University.

Rafay, T., and Simons, D. B., 1965, "Particle-size reduction along streams," presented at the Gological Society of America, Inc., Rocky Mountain Section Eighteenth Annual Meeting, Colorado State University, Fort Collins, Colorado.

Rozovskii, I. L., 1957, "Flow of water in bends of open channels," Academy of Sciences of the Ukrainian SSR, Kiev, Translation published by the Clearinghouse for Federal Scientific and Technical Information of the U.S. Department of Commerce.

Santos-Cayudo, J. and Simons, D. B., 1972, "River response," Environmental Impacts on Rivers, H. W. Shen, ed., Chap. 1, Water Resources Publications, Fort Collins, Colorado.

Schumm, S. A., 1960, "River adjustment to altered hydrologic regimen-Murrumbidgee River and Paleochannels," Australia, Geological Survey Professional Paper 598.

Schumm, S. A., 1971, "Fluvial geomorphology--the historical perspective," River Mechanics, H. W. Shen, ed., Vol. 1, Chap. 4, Water Resources Publications, Fort Collins, Colorado.

Sentürk, F., 1973, "A new category of bed configurations: anti-ripples," Transaction, IAHR, VOl. 5, Istanbul Congress.

Shukry, A., 1950, "Flow around bends in open flume," Transaction, ASCE, Vol. 115, 615 p.

Shulits, Samual, 1936, "Fluvial morphology in terms of slope, abrasion, and bed load," Am. Geophysical Union, Transactions, 17:440-444.

Simons, D. B., 1957, "Theory and design of stable channels in alluvial material," Ph.D. dissertation, Dept. of Civil Eng., Colorado State University, Fort Collins, Colorado.

Simons, D. B. and Albertson, M. L., 1960, "Uniform water conveyance channels in alluvial material," Jour. of the Hydraulics Div., ASCE, Vol. 86, No. HY5, May.

Simons, D. B. and Albertson, M. L., 1963, "Uniform water convergence channels in alluvial material," Am. Soc. of Civil Engineers Trans., V. 128, 65-167 pp.

Simons, D. B., Li, R. M., and Stevens, M. A., 1975, "Development of models for predicting water and sediment routing and yield from storms on small watersheds," prepared for U.S. Dept. Agri. Forest Service, Rocky Mountain Forest and Range Experiment Station, Flagstaff, Arizona.

Simons, D. B. and Richardson, E. V., 1962, "Resistance to flow in alluvial channels," Am. Soc. Civil Engineers Trans., V. 127, 927-1006 pp.

Simons, D. B. and Richardson, E. V., 1966, "Resistance to flow in alluvial channels," U. S. Geological Survey Professional Paper 422J.

Simons, D. B. and Sënturk, F., 1976, "Sediment transport technology," Water Resources Publications, Fort Collins, Colorado.

Straub, L. G., 1935, "Some observations of sorting of river sediments," Am. Geophysical Union, Transactions: 16:463-467.

U.S. Waterways Experiment Station, 1935, "Studies of river bed materials and their movement, with special reference to the Lower Mississippi River," USWES, Vicksburg, Paper 17.

U.S. Waterways Experiment Station, 1965, "Geological Influences on Bank Erosion along Meanders of the Lower Mississippi River," by E. L. Krinitzsky, USWES, Vicksburg, Potamology Investigation Report 12-15, September.

Vanoni, V. A., 1946, "Transport of suspended sediment by water," Am. Soc. Civil Engineers Trans., V. 111, 67-102 pp.

Woodward, S. M., 1920, "Hydraulics of Miami flood control project," Technical reports, Miami Conservancy District, Dayton, Ohio, pt. VII, 264 p.

Chapter 6

PRINCIPLES OF PHYSICAL MODELING

by

H. W. Shen, Professor of Civil Engineering, Colorado State University, Fort Collins, Colorado

Chapter 6

PRINCIPLES OF PHYSICAL MODELING

6.1 INTRODUCTION

Hydraulic modeling has contributed significantly to design of hydraulic structures, training of rivers, and even basic research. It is a common practice to conduct hydraulic model tests of spillways, stilling basins, gates, pump pits, etc., to verify or modify the design of prototype structures. Hydraulic model tests are particularly useful in the study of complex flow phenomena for which no completely satisfactory analysis is available.

Some hydraulic tests are rather routine; many others are complex. For simple situations, hydraulic model tests provide accurate information that can be applied directly to prototype situations. However, for complex situations, hydraulic modeling is still more of an art than a science. One must thoroughly understand the basic principles of modeling as well as the fundamental knowledge governing the phenomena so that model results can be adjusted to provide reliable information on the prototype.

Hydraulic modeling is usually divided into four categories: rigid boundary with undistorted scales, rigid boundary with distorted scales, movable bed with undistorted scales, and movable bed with distorted scales.

This chapter describes the basic principles of modeling and the techniques for adjusting model results to provide useful information for prototype situations; the four model categories will be discussed as well as methods of dealing with dissimilar models.

Historical Development

Leonardo da Vinci (1452-1519) is most widely known for his artistic achievements, but he showed equal genius in many other fields--including hydraulic engineering. Although he may or may not have been the first to use hydraulic modeling, he stressed the necessity of observation:

> "I will treat of such a subject. But first of all I shall make a few experiments and then demonstrate why bodies are forced to act in this manner.
>
> "This is the method that one has to pursue in the investigation of phenomena of nature. It is true that nature begins by reasoning and ends by experience; but, nevertheless, we must take the opposite route: as I have said, we must begin with experiment and try through it to discover the reason."[1]

[1] from Rouse and Ince (1963)

Da Vinci conducted model tests to study the rate of flow over weirs, projectories of liquid jets, eddy formations, and open channel flows. He remarked that

> "A river in each part of its length in an equal time gives passage to an equal quantity of water, within the width, the depth, the slope, the roughness, the tortuosity..."[2]

Dubuat (1816) and Weisbach (1855) were among the first to author fundamental treatises on hydraulics, including both theoretical and experimental analysis. The first large-scale river model was constructed by H. Engels at the River Hydraulics Laboratory in Dresden, Germany, in 1913. Freeman (1929) edited a most comprehensive treatise on hydraulic laboratory practice, including descriptions of various European and American laboratories and notes on other theory of experiments with models. Although this book was published 50 years ago, much knowledge can still be gained from it. As stated by Engels in Freeman's book,

> "The proper interpretation of experimental results from models and their transference to the full-size structure is conditioned not only on the model being made exactly similar, but on the flow of water in the model being exactly similar to that in the full-size structure; also, on the understanding, on our part, of the limitations under which this analogy exists. To be sure, a scientifically exact similarity between model and full-size structure need never be looked for, but in most practical examples one or another motive force plays such a striking role as to enable us to speak of analogy."

Now there are probably hundreds, if not thousands, of hydraulic models tests being conducted each year by various governmental agencies, universities, research institutes, and private laboratories all over the world. The International Association for Hydraulic Research, Delft, Netherlands, has published a reasonably complete list of existing hydraulic laboratories with brief descriptions of their facilities.

<u>List of References</u>

Principles of hydraulic modeling have been discussed in many fluid mechanics books. Numerous books are addressed only to modeling. Some of the latter are listed in the appendix of this chapter.

---

[2]ibid

6.2 DIMENSIONAL ANALYSIS

Basic Dimensions

In general, dimensional analysis is a mathematical operation of units or dimensions used to obtain a functional relationship between the dependent variables and the independent variables for analysis of physical phenomena. The four primary units or dimensions involved in hydraulic modeling are length, time, temperature, and either weight or mass. Engineers usually deal with weight or force and scientists prefer mass. The relationship between mass and weight is

$$F = W = Mg \tag{6-1}$$

where F = force, W = weight, M = mass, and g = the gravitational acceleration.

The international standard of the kilogram is a piece of platinum-iridium, carefully preserved at the International Bureau of Weights and Measures near Paris. In 1954, the Tenth Conference on Weights and Measures established a new Kelvin Scale of temperature as the international standard, with 273.16 degrees Kelvin as the arbitrarily designated temperature of the triple point of water. The international standard meter was originally a platinum-iridium bar with two scratch marks at 1 meter apart. The Eleventh Conference on Weights and measures specified the unit of 1 meter as 1,650,763.73 wave lengths of the orange-red line of krypton-86. At the Twentieth General Conference on Weights and Measures in 1964, the primary standard of frequency was tentatively defined by the frequency of radiation that is emitted by cesium-133 in a quantum transition between two of its fundamental hyperfine energy levels. The standard emitted frequency is 9,192,631,770 hertz, where 1 cycle per second equals 1 hertz. Time signals are broadcast periodically by the Naval Observatory and the National Bureau of Standards.

Normally in hydraulic problems, the effects of temperature are secondary and therefore can be neglected. However, a recent study by the writer conducted with personnel from the U.S. Army Corps of Engineers indicates that temperature significantly affects stage-discharge relationships in the Missouri River near Omaha.

Formulation of Dimensionless Parameters

Let L, T, and M be the units for length, time and mass. A given quantity (or variable) a may be expressed as

$$[a] = \phi(L,T,M) \tag{6-2}$$

or dimensions of

$$[a] = L^{\alpha}T^{\beta}M^{\gamma} \tag{6-3}$$

If [a] is a length, $\alpha = 1$, $\beta = \gamma = 0$. If [a] is a velocity, $\alpha = 1$, $\beta = -1$ and $\gamma = 0$. If [a] is a force, $\alpha = 1$, $\beta = -2$, $\gamma = 1$, etc. A variable a is said to be

(i) a dimensionless quantity, if $\alpha = \beta = \gamma = 0$

(ii) a geometric quantity, if $\alpha \neq 0$; $\beta = \gamma = 0$

(iii) a kinematic quantity, if $\alpha \neq 0$, $\beta \neq 0$, $\gamma = 0$

(iv) a dynamic quantity, if $\alpha \neq 0$, $\beta \neq 0$; $\gamma \neq 0$

A group of dimensional variables may form a dimensionless parameter if certain conditions are satisfied. This will be illustrated in the following example:

Example 6-1: To form a dimensionless number from the product of these four variables: i) a_1, Velocity (V); ii) a_2, length (L); iii) a_3, mass density (ρ); and iv) a_4, fluid dynamic viscosity (μ).

Following Eq. 6.3, the dimensions of

$$[a_i] = L^{\alpha_i} T^{\beta_i} M^{\gamma_i} \tag{6-4}$$

In our cases,

$$\alpha_1 = 1, \ \beta_1 = -1, \ \gamma_1 = 0 \tag{6-5}$$

$$\alpha_2 = 1, \ \beta_2 = 0, \ \gamma_2 = 0 \tag{6-6}$$

$$\alpha_3 = -3, \ \beta_3 = 0, \ \gamma_3 = 1 \tag{6-7}$$

$$\alpha_4 = -1, \ \beta_4 = -1, \ \gamma_4 = 1 \tag{6-8}$$

One is attempting to form a dimensionless parameter

$$\begin{aligned} a_5 &= (a_1)^{k_1} (a_2)^{k_2} (a_3)^{k_3} (a_4)^{k_4} \\ &= V^{k_1} L^{k_2} \rho^{k_3} \mu^{k_4} \end{aligned} \tag{6-9}$$

in such a way that

$$\alpha_5 = \beta_5 = \gamma_5 = 0 \tag{6-9a}$$

Essentially there are three equations to solve, i.e.,

$$\alpha_1 k_1 + \alpha_2 k_2 + \alpha_3 k_3 + \alpha_4 k_4 = \alpha_5 = 0 \tag{6-10}$$

$$\beta_1 k_1 + \beta_2 k_2 + \beta_3 k_3 + \beta_4 k_4 = \beta_5 = 0 \tag{6-11}$$

and

$$\gamma_1 k_1 + \gamma_2 k_2 + \gamma_3 k_3 + \gamma_4 k_4 = \gamma_5 = 0 \tag{6-12}$$

Substituting the values of α, β, γ in Eqs. (6-5) to (6-8) into Eqs. (6-10), (6-11), and (6-12) one obtains, respectively,

$$k_1 + k_2 - 3k_3 - k_4 = 0 \qquad (6\text{-}13)$$

$$- k_1 - k_4 = 0 \qquad (6\text{-}14)$$

$$k_3 + k_4 = 0 \qquad (6\text{-}15)$$

At this point, one should note that there are always three equations—one for each dimension. The number of unknowns should always be the same as the number of variables. In our example, there are four variables (V, L, μ and ρ) and three dimensions (L, T and M). Since one is only interested in the relative ratios of the k's, one can always assign any k value as one. Thus, the number of solutions would be equal to the difference between n, total number of variables, and r, total number of dimensions. This is the famous pi-theorem which was introduced by Edgar Buckingham (1914). In our example, if one assigns $k_1 = 1$ in Eqs. (6-13), (6-14), and (6-15), the solution would be

$$k_1 = 1,\ k_2 = 1,\ k_3 = 1 \text{ and } k_4 = -1 \qquad (6\text{-}16)$$

In other words, the dimensionless number would be $\frac{\rho VL}{\mu}$.

If there are five variables (n=5) and three dimensions (r=3), two sets of solutions can be obtained and in that case two independent dimensionless parameters can be formed.

Applications of Dimensionless Analysis

The following example illustrates how dimensionless analysis is used in experimental studies:

Example 6-2: To investigate the average shear stress in a river channel or closed conduit.

This is an important problem because shear stress is directly related to friction force and energy drop. From experience, one may speculate that the pertinent variables and their dimensions are as shown in Table 6-1.

Table 6-1. List of Common Variables

Variables	Name	Dimensions
τ	Boundary shear stress	$ML^{-1}T^{-2}$
R	Hydraulic radius	L
k_s	Roughness	L
ρ	Density of fluid	ML^{-3}
μ	Dynamic viscosity of fluid	$ML^{-1}T^{-1}$
V	Flow velocity	

In order to study the effect of each independent variable on τ, the dependent variable, one should allow τ and one other independent variable a to vary and keep all other independent variables constant in order to know the effect of this independent variable a on τ under a given condition. Conceivably, if one keeps all other independent variables at other sets of constant values, the effect of a on τ could very well be different and it would be a tremendous task to complete this experimental analysis. However, with the help of dimensional analysis and the pi-theorem, a total of three dimensionless parameters can be formulated from six variables and three dimensions. According to the procedure described in Example 6-1 (page 6-4), the dimensionless parameters can be $\tau/\rho V^2$, k_s/R, and $\rho VR/\mu$. Since τ is the variable under investigation, $\tau/\rho V^2$ should be the dependent variable. This can be expressed as follows:

$$\frac{\tau}{\rho V^2} = \phi\left(\frac{k_s}{R}, \frac{\rho VR}{\mu}\right) \tag{6-17}$$

The boundary shear stress can also be expressed as

$$\tau = \frac{1}{c^2}\rho V^2 \tag{6-18}$$

Combining Eq. (6-17) and (6-18) one obtains

$$\frac{1}{c^2} = \left(\frac{k_s}{R}, \frac{\rho VR}{\mu}\right) \tag{6-19}$$

Experimentally, one can hold the value of k_s/R constant to investigate the variation of either $\tau/\rho V^2$ or c^2 with $\rho VR/\mu$. Gradually, the varieties of either $\tau/\rho V^2$ or c^2 with $\rho VR/\mu$ for different constant values of k_s/R can be determined and the entire problem is solved. This analysis can lead to the extremely useful Darcy-Weisbach friction coefficient and the famous Moody diagram, discussed in virtually every basic fluid mechanics book. This clearly demonstrates the usefulness of dimensional analysis in experimental investigation.

6.3 SIMILITUDE

General Discussion

Similitude is the study of similarity. The main function of a hydraulic model is to produce a flow situation in the model similar to that in the prototype. Therefore, observations of the flow situation may be used to predict the situation in the prototype.

In order for the model to reproduce the results of the prototype, the model must be "similar" to the prototype. There are normally three types of similarities: i) geometric similarity; ii) kinematic similarity; and iii) dynamic similarity.

Geometric similarity exists if all aspects of the model are geometrically similar to the corresponding dimensions of the prototype. The subscripts m, p, and λ indicate the corresponding quantities in the model; prototype; and ratio between model and prototype, respectively. In other words, L_m and L_p are the corresponding lengths in model and prototype, respectively.

$$L_r = L_m/L_p \tag{6-20}$$

Geometrical similarity requires that all corresponding length ratios L_r's be the same.

Kinematic similarity exists if all kinematic quantities in the model, such as velocity and acceleration, are similar to the corresponding quantities in the prototype.

Dynamic similarity exists if all dynamic quantities in the model, such as force and mass, are similar to the corresponding quantities in the prototype.

As stated earlier, in order for a model to reproduce results in prototype, the model must be "similar" to the prototype. This similar condition requires that one or more dimensionless parameters governing flow phenomena must be the same in the model as in the prototype. These criteria can be derived from analysis.

Derivation and Selection of Similarity Criteria

Two methods for derivation and selection of similarity criteria will be presented here. The first method is based on physical reasoning; the second is by inspectional analysis from a flow governing equation.

Physical Reasoning: If one wishes to study the variation of sediment bed load transport rate with flow condition, the first question he asks himself is, "What is the dominant flow parameter governing sediment bed load transport rate?" After some consideration, he probably will conclude that the sediment bed load transport rate should depend on the force acting on the particle by the flow and the resistance of that particle to motion. The flow could either exert a drag or lift force, or both on the particle. In both cases the force can be represented by the following equation:

$$F_1 = \frac{1}{2} c_1 \rho A V_1^2 = \frac{1}{2} c_1 \rho c_2 d^2 V_1^2 \tag{6-21}$$

where F_1 is either the drag or lift force, c_1 is a coefficient, ρ is the density of the fluid, A is the projector area of the sediment particle, V_1 is the flow velocity near the sediment particle, d is the average sediment particle size and c_2 is a shape factor such that $c_2 d^2 = A$.

The resistance force of the particle to motion can be represented by

$$F_2 = k(\gamma_s - \gamma)c_3 d^3 \tag{6-22}$$

where F_2 is the resistance force by the sediment particle to motion, k is a coefficient of friction, γ_s and γ_f are the specific weights of the sediment particle and fluid respectively, d is the average diameter of the particle, and c_3 is a slope factor such that the product of c_3 and d^3 is the volume of the sediment particle.

The sediment bed load transport rate could depend on the ratio of F_1/F_2 or

$$\frac{F_1}{F_2} = \frac{\frac{1}{2} c_1 \rho c_2 d^2 V_1^2}{k(\gamma_s - \gamma_f) c_3 d^3} = c \frac{\gamma_f}{\gamma_s - \gamma_f} \frac{V_1^2}{gd} \tag{6-23}$$

when

$$c = \frac{\frac{1}{2} c_1 c_2}{k c_3} \tag{6-24}$$

and

$$\gamma_f = \rho g \tag{6-25}$$

V_1 could be taken as an average shear velocity U_* at the boundary and also in steady uniform flow.

$$U_* = \sqrt{gRS} \tag{6-26}$$

Equation (6-23) becomes

$$\frac{F_1}{F_2} = c \frac{\gamma_f}{\gamma_s - \gamma_f} \frac{V_*^2}{gd} = c \frac{\gamma_f}{\gamma_s - \gamma_f} \frac{RS}{d} \tag{6-27}$$

Equation (6-27) is the famous Shields parameter and also the inverse of Einstein's (1950) ψ "intensity of shear in particle" or "flow intensity number". According to their analysis, F_1/F_2 should be a governing factor to describe sediment bed load transport rate.

If the above analysis is correct, the value of F_1/F_2 in the model must be the same as that in the prototype for the study of sediment bed load transport rate. Thus F_1/F_2, as expressed by Equation 6-27, should be used as a similarity criterion for modeling sediment bed load transport rate. F_1/F_2 can also be called the ratio between an inertial force and a gravitational force. It is commonly referred to as a Froude number.

Inspectional Analysis

The x-component of the incompressible Navier-Stokes equation for motion is

$$\frac{\partial u}{\partial t} + u\frac{\partial u}{\partial x} + v\frac{\partial u}{\partial y} + w\frac{\partial u}{\partial z} = -g\frac{\partial h}{\partial x}$$
$$-\frac{1}{\rho}\frac{\partial p}{dx} + \frac{\mu}{\rho}\left(\frac{\partial^2 u}{\partial x^2} + \frac{\partial^2 u}{\partial y^2} + \frac{\partial^2 u}{\partial z^2}\right) \qquad (6\text{-}28)$$

where u, v, and w are the flow velocities in directions x, y, and z, respectively; h is the flow depth in the vertical direction; p is the pressure; ρ is the fluid density; and μ is the fluid dynamic viscosity.

One may define a set of dimensionless quantities as follows:

$$x' = \frac{x}{L},\ y' = \frac{y}{L},\ z' = \frac{z}{L},\ h' = \frac{h}{L} \qquad (6\text{-}29)$$

$$u' = \frac{u}{V_o},\ v' = \frac{v}{V_o},\ w' = \frac{w}{V_o} \qquad (6\text{-}30)$$

$$t' = \frac{t}{t_o} = \frac{t}{L/V_o},\ p' = \frac{p}{\rho V_o^2} \qquad (6\text{-}31)$$

ρ, μ, g are constants.

Substitute Equations (6-29), (6-30) and (6-31) into Equation (6-28).

$$\left(\frac{V_o^2}{L}\right)\frac{\partial u'}{\partial t'} + \left(\frac{V_o^2}{L}\right)u'\frac{\partial u'}{\partial x'} + \left(\frac{V_o^2}{L}\right)v'\frac{\partial u'}{\partial y'} + \left(\frac{V_o^2}{L}\right)w'\frac{\partial u'}{\partial z'}$$
$$= -g\frac{\partial h'}{\partial x'} - \left(\frac{V_o^2}{L}\right)\frac{\partial p'}{\partial x'} + \frac{\mu V_o}{\rho L^2}\frac{\partial^2 u'}{\partial (x')^2} + \left(\frac{\partial^2 u'}{\partial (y')^2} + \frac{\partial^2 u'}{\partial (z')^2}\right) \qquad (6\text{-}32)$$

Dividing the entire Equation (6-32) by $\frac{V_o^2}{L}$, one obtains

$$\frac{\partial u'}{\partial t'} + u'\frac{\partial u'}{\partial x'} + v'\frac{\partial u'}{\partial y'} + w'\frac{\partial u'}{\partial z'}$$
$$= -\left(\frac{gL}{V_o^2}\right)\frac{\partial h'}{\partial x'} - \frac{\partial p'}{\partial x'} + \left(\frac{\mu}{\rho V_o L}\right)\nabla^2 (u') \qquad (6\text{-}33)$$

where $$\nabla^2 = \frac{\partial^2}{\partial (x')^2} + \frac{\partial^2}{\partial (y')^2} + \frac{\partial^2}{\partial (z')^2} \qquad (6\text{-}34)$$

Now there are two flow systems. One is for the prototype, with subscript p. The flow system in prototype must obey the equation of motion; thus

$$\frac{Du'_p}{Dt'_p} = -\left(\frac{gL_p}{V^2_{op}} \frac{\partial h'_p}{\partial x'_p}\right) - \frac{\partial p'_p}{\partial x'_p} + \left(\frac{\mu_p}{\rho_p V_{op} L_p}\right) \nabla^2 (u'_p) \qquad (6\text{-}35)$$

where D is the total derivation. Similarly, the flow system in the model must obey the same set of equations.

$$\frac{Du'_m}{Dt'_m} = -\left(\frac{gL_m}{V^2_{om}}\right) \frac{\partial h'_m}{\partial x'_m} - \frac{\partial p'_m}{\partial x'_m} + \left(\frac{\mu_m}{\rho_m V_{om} L_m}\right) \nabla^2 (u'_m) \qquad (6\text{-}36)$$

In order to have dynamic similarity between the model and the prototype, all corresponding values of x', y', z', h', u', v', w', t' and p' between model and prototype should be the same. In other words, Equation (6-35) should be the same as Equation (6-36). The conditions are

$$\frac{gL_p}{V^2_{op}} = \frac{gL_m}{V^2_{om}} \qquad (6\text{-}37)$$

and

$$\frac{\mu_p}{\rho_p V_{op} L_p} = \frac{\mu_m}{\rho_m V_{om} V_m} \qquad (6\text{-}38)$$

Therefore, in this case, the similarity criteria are expressed by the above two equations.

Equation (6-37) expresses the ratio between a gravitational force and an inertial force. As stated previously, this ratio is commonly known as a Froude number. Froude number is a significant factor in studying open channel flow where the gravitational force and the inertial force are both important.

Equation (6-38) expresses the ratio between an inertial force and a viscous force. This ratio is commonly known as a Reynolds number. Reynolds number is a significant factor in the study of fluid flow resistance. Strictly speaking, Reynolds number represents the ratio between inertial force and viscous force for flows under geometrically similar conditions.

Selection of Similarity Criteria

Selection of a similarity criterion or criteria is frequently the most difficult part of planning a model test because usually there is more than one criterion involved. It is difficult to plan a model test with more than one set of similarity criteria (the technique of dealing with this situation will be discussed later in this chapter). Experience, theoretical knowledge, and even intuition are all useful in the selection of similarity criteria.

Some Commonly Used Similarity Criteria in Hydraulics

$$\frac{\text{Inertial Force}}{\text{Gravitational Force}} = \frac{V}{\sqrt{L\gamma/\rho}} \quad \text{(Froude Number)} \qquad (6\text{-}39)$$

$$\frac{\text{Inertial Force}}{\text{Viscous Force}} = \frac{\rho VL}{\mu} \quad \text{(Reynolds Number)} \qquad (6\text{-}40)$$

$$\frac{\text{Inertial Force}}{\text{Surface Energy Force}} = \frac{\rho V^2 L}{\sigma} \quad \text{(Weber Number)} \qquad (6\text{-}41)$$

where σ is the surface energy.

$$\frac{\text{Inertial Force}}{\text{Elastic Force}} = \frac{V}{\sqrt{E/\rho}} \quad \text{(Mach Number)} \qquad (6\text{-}42)$$

$$\frac{\text{Local Absolute Pressure - Vapor Pressure}}{\text{Stagnation Pressure}} = \frac{p-p_v}{\rho V^2/2}$$

(Cavitation Number) (6-43)

6.4 MODEL SCALES

After the proper model similarity criteria have been selected, the next step is to investigate the model-to-prototype scale ratios of length, time, velocity, discharge, force, etc.

Derivation of Model Scale Ratios for Flow Characteristics

Since Froude number is important in rivers, the following derivations will be based on a Froude number criterion. The main principle is to derive the prototype-to-model ratios for flow characteristics as a function of L_r , μ_r , ρ_r. Then, an attempt is made to select the proper fluid for the model study according to the density and viscosity requirements. In many cases $\mu_r = 1 = \rho$; in other words, water is used in the model.

Example 6-3: To derive model scale ratios according to a Froude Criterion:

$$\left(\frac{V}{\sqrt{\frac{\gamma}{\rho} L}}\right)_p = \left(\frac{V}{\sqrt{\frac{\gamma}{\rho} L}}\right)_m$$

or

$$\frac{V_r}{\sqrt{\frac{\gamma_r}{\rho_r} L_r}} = 1 \qquad (6\text{-}44)$$

Velocity Scale: (L/T)

$$V_r = \sqrt{\frac{\gamma_r}{\rho_r} L_r} \tag{6-45}$$

Time Scale: (T)

$$T_r = \frac{L_r}{V_r} = \sqrt{\frac{\rho_r L_r}{\gamma_r}} \tag{6-46}$$

Discharge Scale: (L^3/T)

$$Q_r = L_r^3 \sqrt{\frac{\gamma_r}{\rho_r L_r}} = L_r^{5/2} \left(\frac{\gamma_r}{\rho_r}\right)^{1/2} \tag{6-47}$$

Mass Scale: (M)

$$\rho_r = \rho_r \tag{6-48}$$

Force Scale: (ML/T^2)

$$F_r = \rho_r \; L_r/T_r^2 = L_r^3 \; \gamma_r \tag{6-49}$$

Momentum: (ML/T)

$$M_r = \rho_r \; L_r/T_r^2 = L_r^{7/2} \; (\rho_r \gamma_r)^{1/2} \tag{6-50}$$

Energy at Work: (ML^2/T^2)

$$E_r = \rho_r \; L_r^2/T_r^2 = L_r^4 \; \gamma_r \tag{6-51}$$

Power: (ML^2/T^3)

$$P_r = \rho_r \; L_r^2/T_r^3 = \frac{L_r^{7/2} \; \gamma_r^{3/2}}{\rho_r^{1/2}} \tag{6-52}$$

Summary of Model Scale Ratios

Table 6-2 summarizes scale model ratios according to the criteria of Reynolds number, Froude number and Weber number. In this table the subscript r is omitted.

Application of Model Scale Ratios

Example 6-4: A dam is modeled with a model scale of 1/60. The prototype is an agee spillway designed to carry a flood of 3200 cubic meters per second. What should be the required model discharge in cubic feet per second for the design flood? What time in the model represents one day in the prototype? Water is used in the model.

Table 6-2. Scale Ratios for Models

Parameter	Dimension	Scale Ratios: Reynolds	Froude	Weber
Volume	L^3	L^3	L^3	L^3
Area	L^2	L^2	L^2	L^2
Length	L	L	L	L
Discharge	L^3/T	$L\mu/\rho$	$L^{5/2}(\frac{\gamma}{\rho})^{1/2}$	$L^{3/2}(\sigma/\rho)^{1/2}$
Acceleration	L/T^2	μ^2/ρ^2L^3	γ/ρ	$\sigma/L^2\rho$
Velocity	L/T	$\mu/L\rho$	$(L\gamma/\rho)^{1/2}$	$(\sigma/L\rho)^{1/2}$
Time	T	$\rho L^2/\mu$	$(L\rho/\gamma)^{1/2}$	$(L^3\rho/\sigma)^{1/2}$
Power	ML^2/T^3 = FL/T	$\mu^3/L\rho^2$	$\frac{L^{7/2}\gamma^{3/2}}{\rho^{1/2}}$	$\sigma^{3/2}(L/\rho)^{1/2}$
Energy and work	ML^2T^2 = FL	$L\mu^2/\rho$	$L^4\gamma$	$L^2\sigma$
Impulse and momentum	ML/T	$L^2\mu$	$L^{7/2}(\rho\gamma)^{1/2}$	$L^{5/2}(\rho\sigma)^{1/2}$
Pressure	M/LT^2 = F/L^2	$\mu^2/L^2\rho$	$L\gamma$	σ/L
Specific weight	M/L^2T^2 = F/L^3	$\mu^2/L^3\rho$	γ	σ/L^2
Density	M/L^3	ρ	ρ	ρ
Force	ML/T^2 = F	μ^2/ρ	$L^3\gamma$	$L\sigma$
Mass	M	$L^3\rho$	$L^3\rho$	$L^3\rho$

According to Table 6-2 for Froude a criterion,

$$Q_m = L_r^{5/2}\ Q_p = 4.05 \text{ cfs}$$

$$T_r = \frac{L_r^3}{Q_r} = L_r^{1/2} = 0.129$$

$$T_m = 0.129 \times 24 = 3.12 \text{ hours}$$

Example 6-5: If for some reason a Reynolds criterion is used instead of a Froude criterion in the preceding example, what would be the discharge and time ratios?

According to Table 6-2 for a Reynolds criterion,

$$Q_m = L_r Q_p = 1883 \text{ cfs}$$

$$T_r = L_r^2 = 0.000278$$

$$T_m = 0.000278 \times 24 = 0.007 \text{ hours}$$

Possibility of Satisfying More Than One Similarity Criteria

As indicated by Examples 6-4 and 6-5, significant dilemmas can result if difficult similarity critieria are used. The next question is: "What is the possibility of satisfying more than one similarity criterion?" This request certainly will occur if both similarity criteria are equally important in a problem, as indicated by Equation (6-33).

If the Froude criterion is satisfied,

$$\frac{V_r}{\sqrt{\frac{\gamma_r}{\rho_r} L_r}} = 1 \tag{6-53}$$

and if the Reynolds criterion is satisfied,

$$\frac{\rho_r V_r L_r}{\mu_r} = 1 \tag{6-54}$$

For one case, both Equation (6-53) and (6-54) must be satisfied

$$\frac{V_r}{\sqrt{\frac{\gamma_r}{\rho_r} L_r}} = \frac{\rho_r V_r L_r}{\mu_r} \tag{6-55}$$

Thus

$$\frac{\rho_r^{1/2} \gamma_r^{1/2} L_r^{3/2}}{\mu_r} = 1 \tag{6-56}$$

One must select ρ, the fluid, to be used in the model according to the above equation.

Model With Distorted Scales

Quite frequently the model is much smaller than the prototype. The vertical dimension scale (involving flow depth) cannot follow the horizontal dimension scale because the flow depth would be much too small to take measurements. A model with different vertical dimension scale from the horizontal dimension scale will be used. This is a distorted model. In this case the vertical scale ratio h_r must be considered independently from the horizontal scale ratio L_r. The discharge model scale ratio would be $h_r L_r^2/T_r$ and the longitudinal velocity scale ratio remains as L_r/T_r.

Since h_r is usually much larger than L_r, the slope of the model is much steeper than that of the prototype. Additional roughness is required in a distorted model in order to reproduce flow profiles similar to that of the prototype. When additional roughness is introduced, usually by metal strips, concrete blocks, etc., it is almost impossible to study vertical velocity distribution in a distorted model. It is difficult to correctly predict the roughness of a metal strip; thus, verification of a distorted model is particularly important. Normally at least three prototype flow profiles (high, medium and low flows) should be duplicated correctly by a model.

The next section is devoted to a general analysis of distorted river models with movable beds.

6.5 SIMILARITY OF DISTORTED RIVER MODELS WITH MOVABLE BEDS

It is rather unfortunate that an outstanding scholarly contribution on this subject by Einstein and Chien (1956) has not received proper attention. There must have been hundreds of distorted model tests conducted in the past twenty years; it would be a great contribution to our knowledge if someone would test the applicability of Einstein and Chien's principles on a distorted model. Actually, the principles introduced in their paper are reasonably simple and flexible. Perhaps one difficulty of applying their principles is the lack of information on the derivation of their Table 2.

Discussion of the Einstein/Chien Principles

Much of the following information is directly from Einstein and Chien.

In their study several distortions were contemplated:

1. If the ratio of horizontal lengths L_r is independent of h_r, the model is vertically distorted.

2. If the grain-size ratio d_r is different than both L_r and h_r, a third length scale is introduced and with it a second distortion.

3. If the slope ratio S_r is chosen independently of L_r and h_r, the model is assumed to be tilted in addition to other distortions.

4. If the ratio of effective densities of the sediment, $(\rho_s-\rho_f)_r$, is assumed to be different than the ratio of the fluid densities, ρ_{fr}, which is unity, there is a fourth distortion.

5. A fifth distortion is introduced if t_{1r}, the hydraulic-time ratio for the time values involved in the determination of velocities and sediment rates, is chosen differently from t_{2r}, the sedimentation-time ratio of durations for individual flow conditions, indicating the speed at which flow-duration curves are duplicated.

6. A sixth distortion is a result of the impossibility of obtaining suspended-load rates in a model in the same scale at which the bed-load rates are reproduced; that is, the bed-load-rate ratio q_{Br} is different from the total-load-rate ratio q_{Tr}.

7. A seventh and last distortion permits the ratio of settling velocities of corresponding grains, V_{sr}, to be different than the ratio of corresponding flow velocities.

Relationships Describing Flows in Alluvial Rivers

The following formulas can be transformed into relationships between the various ratios, and thus represent conditions which must be satisfied when the various distortions are chosen. The choice of the equations in the exact form presented herein is not a strong restriction to the generality of the method. Even if one prefers to substitute different formulas for the ones offered herein, one will find that it is not possible to change the number of equations which must be satisfied in a particular problem. It is thus apparent that one must have the same number of conditions and, therefore, the same number of degrees of freedom in choosing the ratios. It will be found that the substitute equations will contain the same variables and that the entire difference will be a slightly changed set of exponents.

Friction Criterion. The flow in an alluvial channel cannot generally be described by one formula with universal constants, but must be interpreted as a composite effect, the various parts of which follow different and independent laws. Einstein and Chien proposed

$$V = \frac{C\sqrt{g}}{d^m} S^{1/2}\, h^{(1/2+m)} \qquad (6\text{-}57)$$

which is a generalized Manning equation. In Equation (6-57), V represents flow velocity, C is a constant, g denotes gravitational acceleration, h is the hydraulic radius and is characteristic of a depth, and the superscript m is an unknown. Equation (6-57) becomes identical to the Manning equation when $m = 1/6$ and if one uses the relationship $n \sim d^{1/6}$, in which n is a friction factor. If it is assumed that the exponent m can be used to describe both the prototype and the model relationships, at least for the most important range of discharges, the equation between ratios can be written as

$$V_r^2 S_r^{-1} h_r^{-1-2m} d_r^{2m} C_r^{-2} = \Delta_V \tag{6-58}$$

The value of Δ_V is unity if the similarity between prototype and model is exactly satisfied, but may indicate a small deviation from exact similarity if such a deviation is necessary for any practical reason.

Whereas V_r, S_r, h_r, and d_r are ratios which will recur in other equations, m and C_r are the exponent and the ratio of the constants in the generalized Equation (6-58)

$$\frac{V}{\sqrt{R_T S_g}} = C \left(\frac{R_T}{K_s} \right)^m \tag{6-59}$$

in which R_T is the hydraulic radius of the total section with the bottom width as the wetted perimeter, and K_s denotes the grain size of the bed which is representative of its grain roughness. If one can assume that the grain mixture equals that of the grain sizes d, the values of C must be determined individually for model and prototype, and the ratio must be found for a representative average channel section. The hydraulic radius R_T assumes the value

$$R_T = \frac{A_T}{P_b} = \frac{A_b' + A_b'' + A_W}{P_b} = \frac{R_b' P_b + R_b'' P_b + R_W P_W}{P_b}$$

$$= R_b' + R_b'' + \frac{R_W P_W}{P_b} \tag{6-60}$$

In Equation (6-60), A_T is the total cross-sectional area, P_b denotes the wetted perimeter of the bed, A_b' is the cross-sectional area pertaining to the grains, A_b'' represents the cross-sectional area pertaining to the irregularities, A_W is that part of the cross section pertaining to the banks, R_b' denotes the hydraulic radius with respect to the grains, R_b'' is the hydraulic radius for the channel irregularities, R_W represents the hydraulic radius with respect to the bank, and P_W is the wetted perimeter of the banks. The model values of C and m can only be determined by a trial-and-error method, as they depend on the choice of the remaining scale ratios.

Froude Criterion. The Froude law may be written in ratios as

$$V_r \, h_r^{-1/2} = \Delta_F \tag{6-61}$$

Again $\Delta_F = 1$ indicates exact similarity, and a deviation from unity signifies a possible necessary deviation from the exact solution.

Sediment-Transport Criterion. In order to have similar sediment-transport conditions near the bed it is, in general, necessary that both the intensity of transport, Φ_*, and the intensity of shear, Ψ_*, for the individual grain sizes be equal in model and prototype, as these two quantities are not connected by a power-type equation. Only if the transport rates are restricted to a very narrow range of values is it possible to combine the two conditions. The equality of the values,

$$\Phi_* = \frac{i_B}{i_b} \frac{q_B}{g(\rho_s - \rho_f)} \left(\frac{\rho_f}{\rho_s - \rho_f}\right)^{1/2} \left(\frac{1}{gd^3}\right)^{1/2} \qquad (6\text{-}62)$$

is possible for all fractions of a mixture only if the two mixtures are similar so that the ratios of the i-values become equal to unity. In Equation (6-62), i_B is that fraction of bed load in a given grain-size range and i_b denotes that fraction of bed material in a given grain-size range. With ρ_f equal in model and prototype, the equation of equal Φ_*-values can be written as

$$q_{Br} \, (\rho_s - \rho_f)_r^{-3/2} \, d_r^{-3/2} = 1 \qquad (6\text{-}63)$$

No deviation from Equation (6-63) is usually desired.

Zero Sediment-Load Criterion. The parameter Ψ_* can be used as a criterion for both model and prototype to describe the beginning of sediment motion. This parameter can be defined as

$$\Psi_* = \frac{\rho_s - \rho_f}{\rho_f} \frac{d}{RS} \qquad (6\text{-}64)$$

Thus, in ratios, equality of corresponding Ψ_*-values is expressed by

$$(\rho_s - \rho_f)_r \, d_r \, \eta^{-1} \, (R_{br})^{-1} \, S_r^{-1} = 1 \qquad (6\text{-}65)$$

in which η is the ratio of the hydraulic radius, R_b', to the surface drag of the entire radius, R_T.

Laminar-Sublayer Criterion. For similarity, this can be written in ratios as

$$d_r \, \eta^{1/2} \, (R'_{br})^{1/2} \, S_r^{1/2} = \Delta\delta \qquad (6\text{-}66)$$

Again, a slight deviation of the value Δ_δ from unity can be permitted, especially in cases in which the bulk of the bed is considerably coarser than δ, the sublayer thickness, and therefore not directly affected by its value. In the derivation of Equation (6-66), it should be noted that δ is dependent on the surface drag only and thus must contain

$$u_* = \sqrt{R_b' Sg} \tag{6-67}$$

which has the ratio, $\eta_r^{1/2}$, $h_r^{1/2}$, $S_r^{1/2}$. In Equation (6-67), u_* is the shear velocity with respect to the grain. The viscosity ν is assumed to be equal in model and prototype, but can be introduced differently if necessary.

Bed Load to Total Load Criterion. By computing some characteristic flows in model and prototype, the ratio of q_T to q_B, the average ratio of these ratios, q_{Tr}/q_{Br}, can be determined; it may be termed B. This value can be used to give a general relationship between the two load ratios

$$q_{Br}\, q^{-1}_{Tr}\, B = 1 \tag{6-68}$$

Hydraulic Time. Hydralic time t_1 may be defined as the time required by a water particle to move with velocity V through a distance L:

$$V_r\, t_{1r}\, L^{-1}_r = 1 \tag{6-69}$$

Sediment Transport Rate Time. The time t_2 may indicate the duration of individual flows. Ratios of this time must be such that corresponding time intervals are required by corresponding sediment rates, q_T, to fill corresponding volumes. Expressed in ratios, this equation can be written for the unit width as

$$q_{Tr} t_{2r} L^{-1}_r h^{-1}_r (\rho_s - \rho_f)^{-1}_r = 1 \tag{6-70}$$

assuming the porosity of the deposits to be equal in model and prototype; the sediment rates q_T are measured in weight under water. The time t_{2r} is the time scale at which the prototype hydrographs must be repeated in the model.

Tilting of Model. Independent ratios have been introduced for length, height, and slope, implying that the model is not only vertically distorted but also tilted. Because the tilt is applied to the model during construction and because it is assumed to be proportional to the prototype slope, it can be applied only to flows which have, at all points, water-surface slopes and energy-line slopes which are constant with time. This condition is not fulfilled when the flow reverses direction, such as under the influence of the tide or in most overbank flows. In all such cases no additional tilt can be permitted.

$$S_r L_r h_r^{-1} = \Delta_N \qquad (6\text{-}71)$$

in which Δ_N is equal to unity when there is zero tilt. A small tilt is represented by a small deviation of Δ_N from unity.

Solution of the Similarity Equations

Table 6-3 gives the nine independent equations which have been introduced herein as products of powers of the various ratios equal to unity; the values in Table 6-3 are the exponents of the various ratios. There appear thirteen ratios (of which ten are free), the exponent m, and four Δ-values. The exponent m and the three last ratios must be determined by auxiliary computations (as shown in the subsequent problem), and the Δ-values can be chosen to suit the conditions.

Table 6-3. Exponents for Model Laws for River Models With Sediment Motion [a]

Eq.	L_r	h_r	V_r	S_r	d_r	$(\rho_s-\rho_f)r$	q_{Br}	q_{Tr}	t_{1r}	t_{2r}	B	C_r	η_r	Δv	ΔF	Δδ	ΔN	
6-58		-1-2m	2	-1	2m							-2		-1				=1
6-61		-1	2												-2			=1
6-63					-3	-3	2											=1
6-65		-1		-1	1	1							-1					=1
6-66		1		1	2								1			-2		=1
6-68							1	-1			1							=1
6-69	-1		1						1									=1
6-70	-1	-1				-1		1		1								=1
6-71	-1	-1		1													-1	=1

[a]The values of B, C_r, η_r and m are determined from auxiliary computations. Values of Δ may be chosen to suit the conditions.

Between the ten free ratios there are nine equations to be satisfied, so that only one of the ten can be chosen freely if the Δ-values are held rigidly to unity or to some other fixed value. If some of the Δ-values are permitted to deviate from unity by a certain margin, it is possible to choose two or more of the free ratios within certain limits.

The nine equations are solved in Table 6-3 in three different ways: first, with the vertical scale h_r chosen freely; then with the horizontal scale L_r chosen; and finally, with the density of the model sediment chosen. One of the solutions will usually satisfy the particular need in designing the model. The exact solution is obtained with all Δ-values equal to unity.

The ten variables in these nine equations are L, h, V, S, d, $(\rho_s-\rho_f)$, q_B, q_T, t_1 and t_2.

If the inertial scale h_r is chosen freely, one can write

$$L_r = h_r^{x_1} \tag{6-72}$$

$$V_r = h_r^{x_2} \tag{6-73}$$

$$S_r = h_r^{x_3} \tag{6-74}$$

$$d_r = h_r^{x_4} \tag{6-75}$$

$$(\rho_s - \rho_f)r = h_r^{x_5} \tag{6-76}$$

$$q_{Br} = h_r^{x_6} \tag{6-77}$$

$$q_{Tr} = h_r^{x_7} \tag{6-78}$$

$$t_{1r} = h_r^{x_8} \tag{6-79}$$

$$t_{2r} = h_r^{x_9} \tag{6-80}$$

Substitute into Eqs. (6-72) through (6-80) the values $B = C_r = \eta_r = \Delta_r = \Delta F = \Delta\delta = \Delta N = 1$. One can solve Equations (6-60), (6-61), (6-63), (6-65), (6-66), (6-68), (6-69), (6-70) and (6-71) by equating the exponents of h_r to be zero. The final solutions should be

$$x_1 = \frac{4m+1}{m+1} ; \quad x_2 = \frac{1}{2} ; \quad x_3 = \frac{-3m}{m+1} ;$$

$$x_4 = \frac{2m-1}{2(m+1)} ; \quad x_5 = \frac{3(1-2m)}{2(m+1)} ; \quad x_6 = \frac{3(1-2m)}{2(m+1)} ;$$

$$x_7 = \frac{3(1-2m)}{2(m+1)} ; \quad x_8 = \frac{7m+1}{2(m+1)} ; \quad x_9 = \frac{5m+2}{m+1} .$$

These values are also given in column 1 of Table 6-4. Similarly, one can obtain another set of values if the length ratio L_r is chosen freely. Results of this solution are given in column 2 of Table 6-4. Column 3 of Table 6-4 gives the results of the sediment density ratio $(\rho_s - \rho_f)r$ that is chosen freely. In all three columns, the solutions given in Table 6-4 are exact solutions with all Δ-values equal to unity. Remember, if Manning's equation is used, m = 1/6. Here one has the freedom to use other values for m, if a frictional relationship better than Manning's equation can be determined by the field data. The investigator has the freedom also to replace any equation in the group of equations listed in Table 6-3; the emphasis is on the basic principle and the procedure. Each investigator can choose the variables and equations for the type of investigation involved.

Now suppose for some reason the model results do not exactly duplicate the corresponding values in the prototype. Einstein and Chien's (1956) method also provides many possible relaxations to adjust model results to prototype applications. The most commonly used method of model verification is by verifying water surface profiles (in stages) in the model against those of the prototype. Let us again use Einstein and Chien's example, analyzing data collected from Big Sand Creek in Mississippi. According to them, the prototype data may show (for detailed analysis see Einstein and Chien, 1956)

$$\frac{V^2}{R_T S g} = 21.05 \left(\frac{R_T}{K_s} \right)^{0.372} \qquad (6\text{-}81)$$

and the model flow resistance may be

$$\frac{V^2}{R_T S g} = 30.76 \left(\frac{R_T}{K_s} \right)^{0.372} \qquad (6\text{-}82)$$

Comparing these two equations with Equation (6-58), one obtains

$$C_p = 4.595, \quad C_m = 5.55, \quad C_r = 0.827 \quad \text{and} \quad m = 0.186$$

How does one use the values given in the first column of Table 6-4 if some of these values--C_r, η_r, and Δ's--are not one? As an example, let us try to adjust column 2 (with L_r chosen freely) of Table 6-4 with $C_r \neq 1$ and $\eta_r \neq 1$. Instead of using Equation (6-72) through (6-80), one now assumes the following

Table 6-4. Model Ratios for Open-Channel Flows with Sediment Motion (Taken from Einstein and Chien, 1956)

Symbol	h_r	L_r	$(\rho_s - \rho_f)_r$	C_r	η_r	B	Δ_F	Δ_δ	Δ_N	Δ_v
(a) DEPTH RATIO, h_r, CHOSEN										
L_r	$\frac{4m+1}{m+1}$	...	...	$\frac{2}{m+1}$	$\frac{m}{m+1}$	...	$\frac{-2}{m+1}$	$\frac{-2m}{m+1}$	1	$\frac{1}{m+1}$
V_r	$\frac{1}{2}$	...	...	...	...	...	1	...	...	...
S_r	$\frac{-3m}{m+1}$	...	...	$\frac{-2}{m+1}$	$\frac{-m}{m+1}$	...	$\frac{2}{m+1}$	$\frac{2m}{m+1}$	...	$\frac{-1}{m+1}$
d_r	$\frac{2m-1}{2(m+1)}$	...	...	$\frac{1}{m+1}$	$\frac{-1}{2(m+1)}$	...	$\frac{-1}{m+1}$	$\frac{1}{m+1}$	...	$\frac{1}{2(m+1)}$
$(\rho_s - \rho_f)_r$	$\frac{3(1-2m)}{2(m+1)}$	...	...	$\frac{-3}{m+1}$	$\frac{3}{2(m+1)}$	...	$\frac{3}{m+1}$	$\frac{2m-1}{m+1}$	...	$\frac{-3}{2(m+1)}$
q_{Br}	$\frac{3(1-2m)}{2(m+1)}$	...	...	$\frac{-3}{m+1}$	$\frac{3}{2(m+1)}$	...	$\frac{3}{m+1}$	$\frac{3m}{m+1}$	...	$\frac{-3}{2(m+1)}$
q_{Tr}	$\frac{3(1-2m)}{2(m+1)}$	...	...	$\frac{-3}{m+1}$	$\frac{3}{2(m+1)}$	1	$\frac{3}{m+1}$	$\frac{3m}{m+1}$	...	$\frac{-3}{2(m+1)}$
t_{1r}	$\frac{7m+1}{2(m+1)}$	...	...	$\frac{2}{m+1}$	$\frac{m}{m+1}$	...	$\frac{-(m+3)}{m+1}$	$\frac{-2m}{m+1}$	1	$\frac{1}{m+1}$
t_{2r}	$\frac{5m+2}{m+1}$	...	...	$\frac{2}{m+1}$	$\frac{m}{m+1}$	−1	$\frac{-2}{m+1}$	$\frac{-(3m+1)}{m+1}$	1	$\frac{1}{m+1}$
(b) LENGTH RATIO, L_r, CHOSEN										
h_r	...	$\frac{m+1}{4m+1}$	...	$\frac{-2}{4m+1}$	$\frac{-m}{4m+1}$	...	$\frac{2}{4m+1}$	$\frac{2m}{4m+1}$	$\frac{-(m+1)}{4m+1}$	$\frac{-1}{4m+1}$
V_r	...	$\frac{m+1}{2(4m+1)}$	...	$\frac{-1}{4m+1}$	$\frac{-m}{2(4m+1)}$	...	$\frac{2(2m+1)}{4m+1}$	$\frac{m}{4m+1}$	$\frac{-(m+1)}{2(4m+1)}$	$\frac{-1}{2(4m+1)}$
S_r	...	$\frac{-3m}{4m+1}$	...	$\frac{-2}{4m+1}$	$\frac{-m}{4m+1}$	...	$\frac{2}{4m+1}$	$\frac{2m}{4m+1}$	$\frac{3m}{4m+1}$	$\frac{-1}{4m+1}$

Table 6-4. (continued)

D_r	...	$\frac{2m-1}{2(4m+1)}$	...	$\frac{2}{4m+1}$	$\frac{-(2m+1)}{2(4m+1)}$	...	$\frac{-2}{4m+1}$	$\frac{2m+1}{4m+1}$	$\frac{1-2m}{2(4m+1)}$	$\frac{1}{4m+1}$
$(\rho_s-\rho_f)_r$	...	$\frac{3(1-2m)}{2(4m+1)}$	...	$\frac{-6}{4m+1}$	$\frac{3(2m+1)}{2(4m+1)}$	...	$\frac{6}{4m+1}$	$\frac{2m-1}{4m+1}$	$\frac{3(2m-1)}{2(4m+1)}$	$\frac{-3}{4m+1}$
q_{Br}	...	$\frac{3(1-2m)}{2(4m+1)}$	...	$\frac{-6}{4m+1}$	$\frac{3(2m+1)}{2(4m+1)}$	...	$\frac{6}{4m+1}$	$\frac{6m}{4m+1}$	$\frac{3(2m-1)}{2(4m+1)}$	$\frac{-3}{4m+1}$
q_{Tr}	...	$\frac{3(1-2m)}{2(4m+1)}$	...	$\frac{-6}{4m+1}$	$\frac{3(2m+1)}{2(4m+1)}$	1	$\frac{6}{4m+1}$	$\frac{6m}{4m+1}$	$\frac{3(2m-1)}{2(4m+1)}$	$\frac{-3}{4m+1}$
t_{1r}	...	$\frac{7m+1}{2(4m+1)}$	...	$\frac{1}{4m+1}$	$\frac{m}{2(4m+1)}$	...	$\frac{-2(2m+1)}{4m+1}$	$\frac{-m}{4m+1}$	$\frac{m+1}{2(4m+1)}$	$\frac{1}{2(4m+1)}$
t_{2r}	...	$\frac{5m+2}{4m+1}$	...	$\frac{-2}{4m+1}$	$\frac{-m}{4m+1}$	−1	$\frac{2}{4m+1}$	$\frac{-(2m+1)}{4m+1}$	$\frac{-(m+1)}{4m+1}$	$\frac{-1}{4m+1}$
(*c*) SEDIMENT-DENSITY RATIO, $(\rho_s-\rho_f)_r$, CHOSEN										
L_r	...	...	$\frac{-2(4m+1)}{3(2m-1)}$	$\frac{-4}{2m-1}$	$\frac{2m+1}{2m-1}$	...	$\frac{4}{2m-1}$	$\frac{2}{3}$	1	$\frac{-2}{2m-1}$
h_r	...	...	$\frac{-2(m+1)}{3(2m-1)}$	$\frac{-2}{2m-1}$	$\frac{1}{2m-1}$	...	$\frac{2}{2m-1}$	$\frac{2}{3}$	...	$\frac{-1}{2m-1}$
V_r	...	...	$\frac{-(m+1)}{3(2m-1)}$	$\frac{-1}{2m-1}$	$\frac{1}{2(2m-1)}$	...	$\frac{2m}{2m-1}$	$\frac{1}{3}$	...	$\frac{-1}{2(2m-1)}$
S_r	...	...	$\frac{2m}{2m-1}$	$\frac{2}{2m-1}$	$\frac{-2m}{2m-1}$	...	$\frac{-2}{2m-1}$	...	...	$\frac{1}{2m-1}$
D_r	...	...	$-\frac{1}{3}$	...	...	...	...	$\frac{2}{3}$	...	...
q_{Br}	...	...	1	...	...	...	...	1	...	...
q_{Tr}	...	...	1	...	...	1	...	1	...	...
t_{1r}	...	...	$\frac{-(7m+1)}{3(2m-1)}$	$\frac{-3}{2m-1}$	$\frac{4m+1}{2(2m-1)}$	...	$\frac{2(2-m)}{2m-1}$	$\frac{1}{3}$	1	$\frac{-3}{2(2m-1)}$
t_{2r}	...	...	$\frac{-2(5m+2)}{3(2m-1)}$	$\frac{-6}{2m-1}$	$\frac{2(m+1)}{2m-1}$	−1	$\frac{6}{2m-1}$	$\frac{1}{3}$	1	$\frac{-3}{2m-1}$

[a] Solution with known effects of omitting Eqs. 2, 6, 11, and 16.

$$h_r = L_r^{x_1} C_r^{x_2} \eta_r^{x_3} \qquad (6\text{-}83)$$

$$V_r = L_r^{x_4} C_r^{x_5} \eta_r^{x_6} \qquad (6\text{-}84)$$

$$S_r = L_r^{x_7} C_r^{x_8} \eta_r^{x_9} \qquad (6\text{-}85)$$

$$d_r = L_r^{x_{10}} C_r^{x_{11}} \eta_r^{x_{12}} \qquad (6\text{-}86)$$

$$(\rho_s-\rho_f)r = L_r^{x_{13}} C_r^{x_{14}} \eta_r^{x_{15}} \qquad (6\text{-}87)$$

$$q_{Br} = L_r^{x_{16}} C_r^{x_{17}} \eta_r^{x_{18}} \qquad (6\text{-}88)$$

$$q_{Tr} = L_r^{x_{19}} C_r^{x_{20}} \eta_r^{x_{21}} \qquad (6\text{-}89)$$

$$t_{1r} = L_r^{x_{22}} C_r^{x_{23}} \eta_r^{x_{24}} \qquad (6\text{-}90)$$

$$t_{2r} = L_r^{x_{25}} C_r^{x_{26}} \eta_r^{x_{27}} \qquad (6\text{-}91)$$

Substitute the above nine equations with $B = \Delta_r = \Delta_F = \Delta\delta = \Delta_r = 1$, into Equations (6-60), (6-61), (6-63), (6-65), (6-66), (6-68), (6-69), (6-70) and (6-61), and equate the exponents of L_r, C_r, η_r to be zero. One obtains the results as shown in columns 2, 4 and 6 in section (b) of Table 6-4.

Now let us go back to Einstein and Chien's (1956) example from data in Big Sand Creek. Assuming $\eta_r = 1$,

$$h_r = L_r^{(m+1)/(4m+1)} C_r^{-2/(4m+1)} \eta_r^{-m(4m+1)}$$

$$= 150^{0.68} \times 0.827^{-1.147} = 37.3$$

$$S_r = L_r^{(-3m)/(4m+1)} C_r^{-2/(4m+1)} \eta_r^{-m/(4m+1)}$$

$$= 150^{-0.32} \times 0.827^{-1.147} = 0.25$$

$$d_r = L_r^{(2m+1)/[2/4m+1)]} C_r^{2/(4m+1)} \eta_r^{-(2m+1)/[2(4m+1)]}$$

$$= 150^{-0.18} \times 0.827^{1.147} = 0.326$$

and $$(\rho_s - \rho_f)r = d_r^{-3} = 28.6 \quad .$$

From these ratios, the model values obtained are

$$S_m = \frac{0.00105(Sp)}{0.25} = 0.0042$$

$$\rho_{sm} = 1.059 \text{ g per cubic centimeter}$$

$$d = 0.00288 \text{ feet}$$

Einstein and Chien (1956) also discussed the problem of how to adjust to side wall friction and how to separate grain roughness and form roughness.

An important point is that Einstein and Chien have provided a powerful procedure to adjust model-prototype relationships. Which equations are used is not critical. One has the flexibility of using a particular set of equations and adjusting the model-prototype relationship as proposed by the investigators.

6.6 COMMENTS

1. A physical model is very useful in the study of characteristics of complex flow phenomena involving significant flow variations in all three dimensions where no theoretical analysis is available.

2. Dimensional analysis, physical reasoning and inspectional analysis are essential approaches to the selection of the governing similarity criteria. If more than one similarity criterion are needed, extensive knowledge of the basic process under investigation is necessary to deal with the situation.

3. If the prototype is large, a distorted model may be necessary. In a distorted model the vertical model scale is usually smaller than the horizontal scale, in which case the model roughness should be much greater than that of the prototype in order to dissipate a relatively greater amount of energy in a relatively shorter reach. The increase of model slope should, therefore, be a function of distortion.

4. A movable bed model may be necessary if a significant three-dimensional variation of sediment movement occurs in the prototype. Since movable bed model results are difficult to interpret, it can be advantageous to first investigate general flow variations in a fixed bed model. The movable bed model

can be constructed later to study particular details of a smaller section or reach.

5. According to the Froude criterion, if flow is slower in a model than in the prototype sediment would move at a much slower rate (if at all) in the model, and light-weight sediment material may be needed.

6. A popular similarity criterion used in the selection of the model sediment material when the sediment transport rate is important is the Shields number of Einstein's flow intensity as given in Eq. (6-2). If this rule is followed, the correct sediment to use in a movable bed model under normal conditions would be that with a density of 1.05 (such as plastic pellets). One model-day would be equivalent to about one year of prototype time.

7. If bed form roughness is important in the model study, selection of the sediment bed material should consider the factor Wd/ν, where W is the fall velocity of the sediment particle, d is the average sediment particle size, and ν is the kinematic viscosity of the fluid. According to Shen (1962), sediments with the same Wd/ν will produce approximately the same bed form roughness.

8. The distortion model scale of a movable bed model should be kept at a minimum (normally below three).

9. Both the entrance and exit sections of a model should be relatively stable reaches and their respective flow distributions should be similar to those of the prototype.

10. The verification of model results is absolutely necessary. Model results are usually verified with at least three flow conditions: high, medium and low flow. It is rather difficult to determine a dominant river discharge for all purposes, and even if a dominant discharge can be found, there are serious doubts as to the applicability of the model results to the prototype when only the dominant discharge is verified.

11. In order to design a river model study correctly, one must decide the purpose of the model tests, know the principles of modeling thoroughly, and also have a thorough technical knowledge of the subject matter.

6.7 REFERENCES

Some of these references were not described in the chapter, but are listed here for the reader's information.

American Society of Civil Engineers, 1942. Hydraulic Models. Manuals of Engineering Practice, No. 25.

Baker, W.E., P.S. Westine, and F.T. Didge, 1973. Similarity methods in engineering dynamics. Spartan Books, Rochelle Park, New Jersey.

Bogardi, J., 1974. Hydraulic similarity in sediment transport. Chapter 2.5, Sediment Transport in Alluvial Streams (Transl. by Z. Szilvassy), Akademiai Kiado, Budapest.

Bridgman, P.W., 1922. Dimensional Analysis. Yale University Press, New Haven. (Second printing of revised edition, 1937).

Buckingham, Edgar, 1914. On physically similar systems. Physics Review, Vol. 4, p. 345.

Buckingham, Edgar, 1915. The principle of similitude. Nature, Vol. 95, 2406, pp. 396-397 (December).

DuBuat, M., 1816. Principes d'Hydraulique et de Pyrodynamique," Vol. I, p. 103, Didot, Paris.

Duncan, W.J., 1953. Physical similarity and dimensional analysis. Edwald Arnold, London.

Einstein, H.A., 1950. The bed-load function for sediment transportation in open channel flows, Technical Bulletin 1026, Soil Conservation Service, United States Department of Agriculture.

Einstein, H.A., and Chien, N., 1956. Similarity of distorted river models with movable beds, Transactions of the American Society of Civil Engineers, Vol. 121, pp. 440-457.

Focken, C.M., 1953. Dimensional methods and their applications. Edwald Arnold, London.

Freeman, J.R., ed., 1929. Hydraulic Laboratory Practice, The American Society of Mechanical Engineers, New York.

Gukhman, A.A., 1965. Introduction to the theory of similarity. Academic Press, New York.

Holt, M., 1961. Dimensional analysis. Chapter 15, Handbook of Fluid Dynamics, (ed. V.L. Streeter), McGraw-Hill, New York.

Huntley, H.E., 1951. Dimensional analysis. Rinehart, New York.

Ipsen, D.C., 1960. Units, dimensions, and dimensionless numbers. McGraw-Hill, New York.

Jupp, E.W., 1962. An introduction to dimensional method. Cleaver-Hume Press, London.

Keulegan, G.H., 1966. Model laws for coastal and estuarine models. Chapter 17, Estuary and Coastline Hydrodynamics, (ed. A.T. Ippen), McGraw-Hill, New York.

King, C.A.M., 1966. Experiment and theory--models. Chapter 4, Techniques in Geomorphology, St. Martin's Press, New York.

Kline, S.J., 1965. Similitude and approximation theory. McGraw-Hill, New York.

Lanchester, F.W., 1936. Theory of dimensions and its application for engineers. Crossby, Lockwood and Sons, London.

Langhaar, H., 1951. Dimensional analysis and theory of models. Wiley, New York (eighth printing, 1967).

LeCorbeiller, P., 1966. Dimensional analysis. Appleton-Century-Crofts, New York.

Murphy, G., 1950. Similitude in engineering. Ronald Press, New York.

Palacios, J., 1964. Dimensional analysis. MacMillan, London.

Pankhurst, R.C., 1964. Dimensional analysis and scale factors. Reinhold, New York.

Porter, A.W., 1946. The method of dimensions. Methuen, London.

Rayleigh, J.W.S., 1914. The principle of similitude. Nature, Vol. 95, no. 2368, pp. 66-68 (March).

Rouse, H., 1951. Model techniques in meteorological research. Compendium of Meteorology, American Meteorological Society, Boston, Massachusetts.

Schuring, D.J., 1977. Scale models in engineering. Pergamon Press, Elmsford, New York.

Shen, H. W., 1962. Development of bed roughness in alluvial channels. Journal of Hydraulics Division, ASCE, Proc. Paper 3113, Vol. 88, No. HY-3, May.

Simmons, H.B., 1966. Tidal and salinity model practice. Chapter 18, Estuary and Coastline Hydrodynamics (ed. A.T. Ippen), McGraw-Hill, New York.

Warnock, J.E., 1950. Hydraulic similitude. Chapter 11, Engineering Hydraulics, Proc. Fourth Hydraulics Conference, Wiley, New York.

Weber, E., 1961. Dimensional analysis. Chapter 3, Handbook of Engineering Fundamentals (ed. O.W. Esback), Wiley, New York.

Weisbach, J., 1855. Die Experimental-Hydraulik, Freiburg.

Yalin, M.S., 1971. Theory of hydraulic models. MacMillan, New York.

Young, D.F., 1971. Basic principles and concepts of model analysis. Experimental Mechanics, Vol. 11, no. 7, pp. 325-336 (July).

Zierep, J., 1971. Similarity laws and modeling. Gasdynamics Series, Vol. 2, Marcel Dekker, New York.

Chapter 7

SEDIMENT SOURCES AND IMPACTS IN THE FLUVIAL SYSTEM

by

Daryl B. Simons, Associate Dean of Engineering Research and Professor of Civil Engineering, Colorado State University, Fort Collins, Colorado

Timothy J. Ward, Assistant Professor of Civil Engineering, Colorado State University, Fort Collins, Colorado

and

Ruh-Ming Li, Associate Professor of Civil Engineering, Colorado State University, Fort Collins, Colorado

Chapter 7

SEDIMENT SOURCES AND IMPACTS IN THE FLUVIAL SYSTEM

7.1 INTRODUCTION

The management of river and irrigation systems is an important task in proper use of water resources. In order to optimize utilization of water resources, it is essential to understand the basic physical processes that govern watershed and river behavior and their responses to natural changes and changes imposed by man's development. The short-term and long-term behavior of river systems and their subcomponents is greatly affected by sediment load eroded, carried, and deposited in the systems. It is essential to consider the sources of sediment that are derived from various components of the watershed systems, how these sediments are transported to the river systems, and how developments associated with the river are impacted by sediment. This paper attempts to identify some of the major sources of sediment that impact on watersheds, rivers, canals, industrial developments, and other natural and developed aspects of the fluvial system.

Natural and man influenced environments are a non-linear combination of various physical processes. These processes can be categorized as hydrologic-hydraulic, atmospheric-meteorologic, geologic-geomorphic, biologic and man's influence. The interaction between processes dictates the type of environment, either terrestrial, aquatic, or atmospheric, that will be prevalent at a given time or location. The fluvial system is one important part of the larger aquatic environment, and is itself composed of various segments. These segments can be classified by size as one proceeds from the headwaters of a river basin to the ocean. Segments that often receive attention in erosion and sedimentation studies are upland watersheds on one end of the spectrum and large rivers on the other. In this paper sediment sources will be discussed as they relate to these two connected but different segments of the fluvial systems. The division of the fluvial system into upland watersheds and rivers is made to bring out effects of scale when studying sediment sources. These effects of scale dictate the importance of each sediment source. For example, a poorly designed roadway in a small forested watershed may have severe sediment impact on the stream draining that watershed, but unmeasurable effect on the river draining many undisturbed streams.

Interrelated factors work to dictate the source and impact of sediment in the fluvial system. These factors include the type of sediment, precipitation, runoff characteristics of the streams and rivers, vegetation and man's alteration of the natural system. As these and other factors vary from area to area and with time so do the sources and impacts of sediment. A conceptualization of the relationship of environmental processes and other factors is shown in Fig. 7-1.

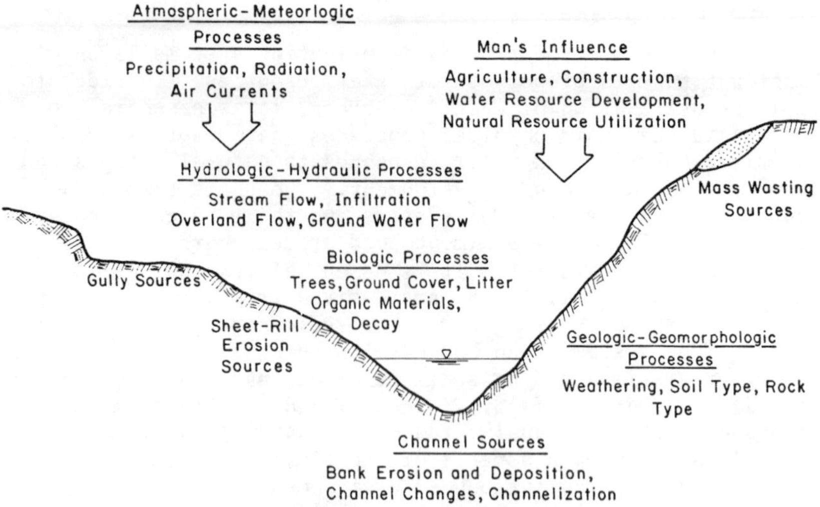

Fig. 7-1. Relationship of Environmental Processes Affecting the Fluvial System.

Sediment sources can be categorized into two broad groups based on the segment of the fluvial system under consideration. These groups are upland watersheds sources and river channel sources. Upland watersheds are those drainage areas roughly ranging in size from a few acres up to about 50 square miles. Rivers may be described as draining areas larger than 100 square miles. Although this is a arbitrary size classification, it does serve to provide a basis for discussion.

7.2 UPLAND WATERSHED SOURCES

Upland watershed sediment sources can be classified as natural and man-caused. Natural sources include sheet-rill erosion, gullying, mass wasting, and channel erosion, while man caused sources include the natural types as influenced by man plus road construction, logging operations, urban development, and energy resource acquisition. Because the processes controlling channel sediment sources are similar for both upland watersheds and large river channels, upland channel sources will not be covered here but are discussed in a following section.

Sources and Related Processes

Complex variables influence the erosion and transport of sediment overland through rills and subchannels to streams and rivers. A flow chart developed in connection with water sediment routing models clearly identifies the significant variables that affect the delivery of sediment from the watershed to the channel system and subsystems (Fig. 7-2). The amount of sediment derived from the watershed is influenced by many factors. These factors include watershed characteristics and storm characteristics. Characteristics of the watershed include the geometry of the watershed, the soil types, the vegetation and ground cover, and flow resistance as water and sediments traverse the watershed to the collection system. Another important factor is the characteristics of storms that occur over the watershed. Some of the principal variables in this case include rainfall intensity, rainfall areal distribution, snowmelt rates, and evaporation rates. In a similar manner, antecedent conditions can have a very significant impact on the delivery of sediments from the watershed. It is important to consider interception storage, antecedent soil moisture, initial loose soil storage and related factors.

In general, the data required to evaluate the quantity of sediment delivered from a watershed are delineated in the foregoing flow chart. The flow chart clearly identifies the basic data needed in order to evaluate the amount of sediment derived from the watershed.

Some of the factors that influence the amount of sediment derived from a watershed must be considered next. The climate of the area plays a significant role as one would need to consider whether or not the area was humid, somewhat humid, arid or perhaps some transitional state. Similarly, one can identify many different variables that influence the amount of sediment derived from watersheds. Different variables have various degrees of importance, depending upon such factors as the type of climatic conditions, soil types, and vegetative cover found in a particular watershed. In order to determine which of these variables must be included in the analysis of a particular system, it may be necessary to evaluate the importance of each variable by conducting a sensitivity analysis. By this means it can be determined which variables are sufficiently important that they must be carefully evaluated in order to accurately determine watershed sediment yield. In some instances it may be necessary to collect actual field data to insure a more accurate evaluation of watershed response considering the sediment delivery under different storm and flow conditions. For certain types of analysis, data required to make estimates of the sediment delivery and the sediment yield may be simulated with adequate accuracy to meet the needs

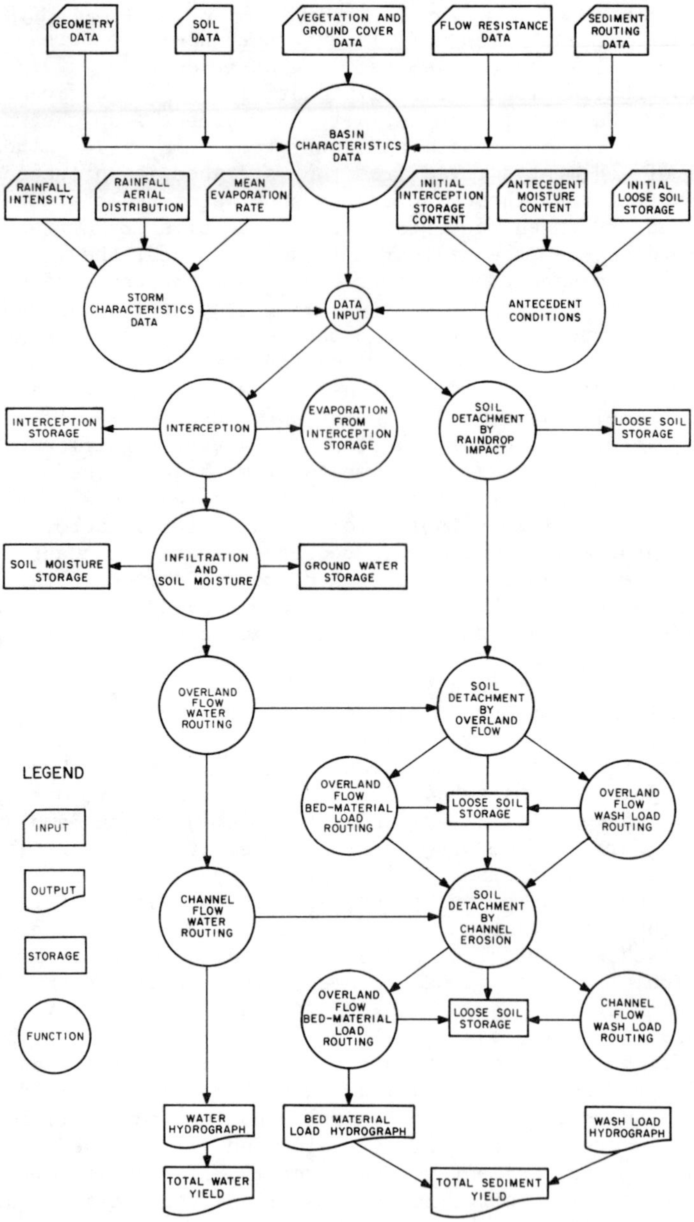

Fig. 7-2. Flow Chart for the Water and Sediment Routin Model.

of the particular application. Another alternative, when there is a shortage of information necessary to evaluate a system, is to resort to physical modeling of portions of the system. This provides the needed data to determine the quantities of sediment that will be eroded and transported from the watershed for a particular event or for a series of events.

Of all the sources of sediment in an upland watershed, sheet flow-rill erosion is probably the most widely recognized. Sheet flow occurs when the rate at which water is being applied to a surface exceeds the rate of infiltration. Applied water may be provided from either rainfall or snowmelt. Runoff usually flows only a short distance as sheet flow before it concentrates into small channels or rills. Concentration of flow can result from previous erosion, vegetation effects, microrelief or alteration of drainage patterns by man. It is often advantageous to separate the sheet flow and rill flow components of this combined sediment source (Foster and Meyer, 1975), however, both are considered as a source here because they are related as overland flow types. Meyer et al. (1975) have shown that as down slope distance increases the relative importance of rill flow as a sediment source increases while that of sheet flow decreases. This is in part related to the decrease in flow volume in the sheet flow portion with a corresponding increase in the rill flow component.

Because applied water rate must exceed infiltration rate before sheet-rill flow occurs, this type of flow is not observed in all regions of the country. Areas where sheet-rill flow does occur and acts as a sediment source includes the semi-arid West and humid Southeast. In the Western United States storms occur that quickly saturate the thin soils thus producing surface runoff. In the Southeast high intensity convective storms associated with passage of frontal systems often surpass the ability of the deeper soils to infiltrate the rainfall. In heavily forested areas, overland flow may be occurring, but not be visible, as a type of interflow between the soil surface and the overlaying organic debris. Therefore, this type of sediment source and process is probably the most prevalent type in most upland watersheds whether recognized or not.

Another sediment source is gully erosion. Gullies result from one portion of a channel reaching base level much more rapidly than the channel upstream of the gully. Causes leading to the development of gullies in watersheds are complex (Leopold et al., 1964) but are again controlled by basic physical processes common to all environments. Predominant processes that have been recognized in gully development include mass wasting of gully scarps and surface runoff transport of the resulting materials (Bradford et al., 1973; Piest et al., 1975; Bradford and Piest, 1977). Sediment yields from gullies may vary from about 0.5 to over 6 inches per acre per year depending on gully conditions (Miller et al., 1962; McDowell et al.,

1967). Although such rates have tremendous significance in headwater areas, Miller et al., (1962) show that only 30% of total gross erosion was attributable to gullies when the drainage area grew to about 10 square miles. However, they further noted that in some cases, sediment production from one gully could fill a channel that was five feet deep, 30 feet wide and 760 feet long in five years. As more gullies provide sediment for a channel, the significance of gully source areas becomes evident.

A third sediment source in upland watersheds is provided by mass wasting, or landslides, from falls, slides, and flows (Varnes, 1958). All three types are important in upland watersheds with slides and flows being prominent sources. Common slide types are termed rotational or planar depending on the overall shape of the failure plane. Flow types that occur in upland watersheds include dry raveling, mudflows, and debris avalanches. All of these types of landslides provide direct or indirect sediment supply to the stream channels. A landslide may disturb the ground surface, destroy protective vegetation and alter prevailing drainage patterns thus producing indirect sites for sheet-rill on gully erosion. A landslide mass can directly enter a stream channel. If the channel is dry, the next flow event may transport some or all of the material away. If the channel has flowing water in it the mass may substantially increase the sediment load downstream. Either directly or indirectly, the sediment source created by landslides can be extremely high. Dyrness (1967) estimated that over 340,000 cubic yards of material were disturbed by landsliding during one winter in the H. J. Andrews Experimental Watershed in Oregon, all of which could become a future sediment source.

Watershed Classification Based on Sediment Source

The three general sediment sources of sheet-rill erosion, gullying, and landsliding are not totally independent and do interact with one another. However, in upland watersheds one of these types will probably be the primary sediment source. Therefore, upland watersheds can be classified on a sediment basis as sheet-rill erosion, gully-head cut, or landslide dominated watersheds. Combinations of these three sources could lead to a complex classification of a watershed. Examples of watersheds in each type are numerous There are three United States Department of Agriculture Forest Service watersheds which may be typical of each of those types.

Beaver Creek Experimental Watershed Near Flagstaff, Arizona: This group of watersheds ranging in size up to about 15,000 acres typifies sheet-rill watersheds. Surface soils are stony with low infiltration rates thus producing flashy runoff during high intensity rainfall events. Typical channels in this area are naturally lined with cobbles and small gravel. Sediment yields from these areas are composed of materials eroded from the land surface by sheet-rill erosion.

Pine Watersheds Near Coffeeville, Mississippi, part of Southern Forest Hydrology Laboratory, Oxford, Mississippi: These watersheds are excellent examples of gully-headcut sediment sources. Although small in size, up to about seven acres, these watersheds are characterized by gullies and headcuts ranging in height from a few inches to about twenty feet. Heavy ground cover and permeable sandy soils in the area prohibit sheet-rill erosion from developing but provide essentially clear water to contribute to the gullying processes.

H. J. Andrews Experimental Watershed Near Blue River, Oregon: This 15,000 acre watershed encompasses smaller basins on the order of a few hundred acres. Dense vegetation and permeable soils preclude development of observable sheet-rill erosion. Gully-headcut erosion is minor or non existant. Landsliding, however, is a dominant source of sediment in this area. Numerous episodic events have occurred over the past 25 years, partly as natural occurrences, but predominantly as man-influenced occurrences (Swanston and Swanson, 1976). Whatever the cause, the occurrence of landslides can rapidly increase the sediment production in a watershed (Swanson and Dyrness, 1975).

Watersheds could also be classified as to whether the sediment yield was primarily coming from the channel or the contributing area. However, only those sources which may contribute to the channel system are considered and not the channel itself. The channel system is considered under river channel sources.

These three examples illustrate the types of naturally occurring sediment sources in upland watersheds. Other areas may be characterized as a combination of one or more sources, but most can be classified by one of these source types.

Man Caused Sources in Upland Watersheds

Man's activities in upland watersheds provide increased rates of sediment yield from the basic sediment sources and sources specific to man's activities. Guy (1974) suggests seven activities that influence sediment sources. These are: 1) agricultural tillage, 2) domestic animal grazing, 3) highway construction and maintenance, 4) timbering, 5) mining, 6) urbanization, and 7) recreational land development. Each of these activities influences the basic sediment sources to some extent. For example, Dickerson (1968, 1975) in Mississippi studied sheet-rill erosion; Swanston and Swanson (1976), in Northern California and Oregon, and Megahan and Kidd (1972) have documented the effects of timber harvesting; Weber and

Reed (1976) have recently examined sediment yields from highway construction; and Meyer (1974) has discussed sedimentation processes and urban erosion.

Man's activities that increase sediment yield often do so by altering the existing landscape. These landscape alterations include changing slope gradients, realigning hydrologic systems, removal of vegetation or disturbing the surface soil. Such activities are associated with road building and preparation of construction and agricultural sites. By making these alterations in the natural landscape and associated physical processes, man can, if care is not taken, accelerate sediment yield from the different sediment sources or create new sediment sources.

One of the factors mentioned above that can significantly affect the supply of sediment is overgrazing. An example of this is on the Indus River watershed in Pakistan, where overgrazing has resulted in high rates of erosion from the watersheds and significant sediment problems downstream. There are many other land use problems that can be identified, such as the construction of trails and people using the area for recreation that can impact unfavorably on the fluvial system. More specifically, other land uses that can significantly affect the supply of sediment are related to farming and methods of farming, timber harvests, and perhaps even the use of vegetative barriers to minimize the transport of sediments overland into the stream systems.

Mining activities play a significant role on the supply of sediment to river systems. Tailings from mines discharged into channels are a significant man caused source of sediments. It is necessary to look at these wastes and tailings and at the overall mining activity to determine the importance of mining as a source of sediment to our channel systems. If waste materials resulting from mining operations are relatively coarse they may simply extend into the river and divert it. Conversely, where the sizes are relatively small, fairly large quantities may be transported downstream to cause other impacts. In any event, this is an issue that is currently being considered by many of the Federal agencies, particularly in terms of energy resource development.

Several aspects of irrigation may contribute to sediment problems. In many instances excessive runoff from irrigated land carries water and sediments back to the channel system where they are deposited. In the past, wastewater channels from irrigated areas were uncontrolled. Permission to transport these waters back to the channel system was obtained by easement across properties. In many instances, procurement of these easements provided an easy way to get wastewaters back to the river system, but because of the steepness of these channels, erosion problems developed and large areas of irrigated land have been lost.

7.3 CHANNEL SOURCES

Sources and Geology

One other source that is found in watersheds, but was not discussed above, is channels. As the size of the drainage areas increases, the importance of this source also increases as the effects of other upland watershed sources diminishes. Channel sources become extremely important on larger rivers where the other, smaller sources such as sheet-rill erosion are relatively negligible when compared to the total sediment load being transported by the river. Stream channel sources are the stream bed and the stream banks. The relative importance of each of these sources is dependent on the erosive power of the flowing water and the types of materials composing the bed and banks. In upland watersheds, the stream bed and banks may be composed of large cobbles or a hard erosion resistant material. In this case, relatively little sediment is derived from these sources. Alluvial rivers, however, flow in materials that, except for local variations, are not erosion resistant. This is particularly true for the sand bed streams. When both the bed and banks are composed of erodible materials, the flowing water can readily change its channel by meandering or braiding depending on other conditions. Where the banks are more resistant than the bed, entrenchment of the channel may occur. If local geology dictates an erosion-resistant bed and weak erodible banks, lateral migration may occur. The effects of geology on the character of the stream can be seen in the following example.

In the upland areas of northern Mississippi there are numerous streams that have clay beds and sand banks. These banks often cave into the stream during changes in flow stage and thus provide both clay wash load, which is quickly transported through the system, and sand which forms a layer over the clay bed. However, beneath these clay layers is another sequence of sands. When and if the downcutting of the stream channels exposes the sand strata the channel is rapidly degraded until dynamic equilibrium or a new, less erosive layer is reached. Where these "break throughs" occur, potholes may develop which are many times deeper than the depth of flow and provide an almost unlimited source of sand. As the flow recedes, however, these potholes refill with sands from upstream and are not apparent in the dry channels. Although this is only one effect that local geology has on the relative importance of each stream channel source, it establishes the need to carefully consider all influencing variables when analyzing sediment sources in channels.

Related Processes and Impacts

There are numerous processes and activities that affect sediment from channel sediment sources. These include bank erosion, channel changes, hydraulic structures, channelization, dredging, dam breaches and removal and gravel mining.

Bank Erosion: Channel bank erosion can be a source of significant quantities of sediment reaching downstream areas. Quantities of sediment from this process are difficult to estimate. However, several methods are available to help evaluate the magnitude of sediment delivered as a consequence of the erosion of channels and subchannel systems.

It is possible to evaluate the stability of the channel systems to determine if they will be subjected to significant erosion. There are a number of methods that can be utilized. One of the most widely used methods utilizes the Shields diagram or Shields type analysis (Simons and Sĕnturk, 1977). This method provides a means of evaluating the conditions under which the bed materials will just begin to move. It is possible to include in the Shields criteria the additional forces that influence the grains on the banks of the channel. By this technique bank stability can be evaluated. The principal addition is the gravitational force which acts on the particles resting on the bank slope. This type of analysis is presented and illustrated by Simons and Sĕnturk (1977) not only for Shields' criteria, but other methods of analyzing bank stability that are widely recognized and utilized. This concept can be extended so that a suitable method to stabilize and hold the alignment of a particular channel, such as utilizing riprap, can be developed. Other factors that affect the erodibility of the banks are the cohesion of the bank materials, the type and extent of vegetation, seepage forces, wave action, and local geology. Root systems developed in the banks by vegetation generally increase the stability of small rills, subchannels and rivers. They may, however, have adverse effects in large systems where the river currents attack the banks below the root zones. In this latter case, the added surcharge weight of the vegetation may contribute to failure of banks. Once trees slide into the river, debris dams may cause deflection of currents that further concentrate the flows in such a manner as to attack the banks, causing increased erosion. The accumulation of trees and debris in the channels can significantly affect channel deposition. The amount of sediment contributed by bank erosion can presently be most accurately evaluated by the analysis of aerial photographs of the system collected over a period of several years. Analyses of these photos reveal the extent of bank erosion and the degree to which the channel system has shifted. In addition, analyses of photographs will provide information regarding major changes in channel geometry, such as widening, narrowing, entrenchment, or braiding. Bank erosion, however, does not imply continuous widening of the channel system. River channels generally develop certain width, depth and slope relations. In most eroding channels, material eroded from the banks is usually transported only a short distance down the stream before it is redeposited in the form of bars, new bankline or bed material accumulations. As an example, refer to Fig. 7-3. This figure illustrates the generalization and exception to the rule of an established river width. At grid mile 2 note that the channel width has remained constant over a long time span which is the generalized case. The river width at grid mile 9 has, however, changed significantly in the same time span, the exception to the rule.

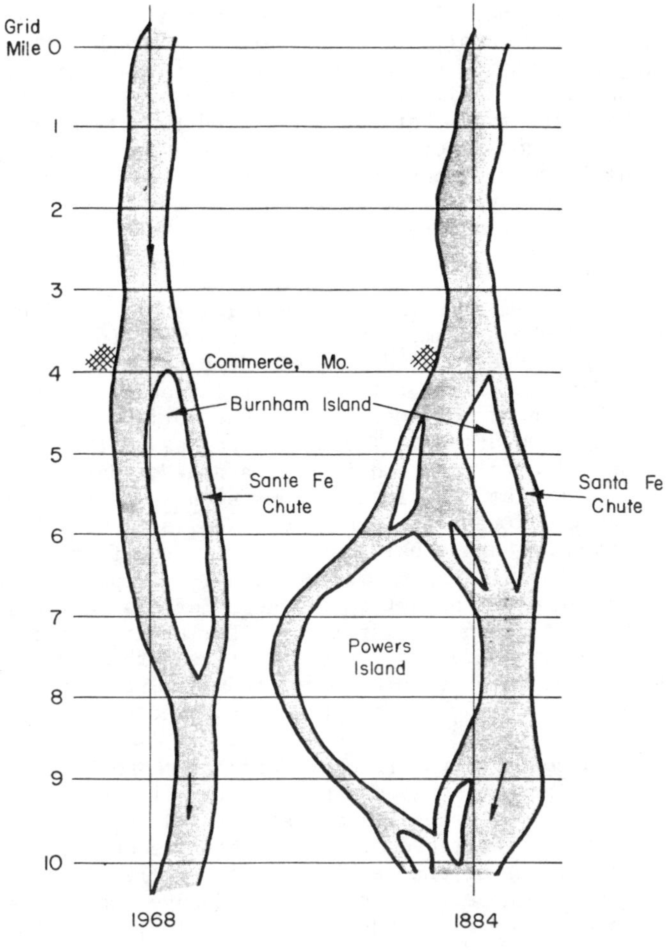

Fig. 7-3. Loss of Chute Channels Near Powers Island

Many different forces can erode stream banks. One of the major forces causing bank erosion is the attack by the stream currents. The direction of these currents can vary with the alignment of the stream, the stage of the water discharge and local disturbances. In order to evaluate the erosive effects of the currents it is essential to consider the channel geometry, the alignment of the flow and how the alignment of the flow can change with time. Another significant force that influences bank erosion is waves generated by wind or ships at the air-water interface. As these waves impact upon the banks a whole new set of forces are brought to bear upon the banks and rapid erosion can occur.

The magnitudes of waves depends upon the fetch, distance, and depths of flow. Generated waves must also be examined from the viewpoint of amplitude and duration. Work has been done relating the erosive effects of wind and ship-generated waves to the supply of sediments. Currently several problems exist along the Ohio River that are attributed to raising the pool levels by the construction of locks and dams, increasing the ship traffic recreational uses, and wind fetch which all contribute to the development of waves that can cause additional erosion and sedimentation problems.

Other types of bank erosion include sloping and landslides. The sloping of the channel banks may be a surface phenomena with small thin layers working their way from the surface into the stream. In other instances, relatively large elements may actually slide out of the bank into the channel system. Bank landslide types of bank instability can be analyzed by use of soil mechanics. One form of bank erosion by landsliding is mudflows. The actual banks, as they begin to slide, may liquify and then flow as a non-Newtonian fluid into the channel system, causing problems such as aggradation, deflection of currents, blockage of flow, or mudfloods. Another form of bank erosion can occur in the form of creep.

Bank materials may be moving at a very slow rate into the current because of the nature of the materials themselves and the forces acting on the materials. Different factors are involved when considering the stability and the erosion of the banks and, in particular, magnitudes of sediment that must be planned for downstream of these areas where erosion occurs.

Channel Changes: There are many forms of channel changes that can occur. These changes may have a significant effect on the amounts of sediment supplied downstream and how the downstream portions of the watershed or river system react. Some of the most common forms of channel changes are accretion and avulsions. Accretions are those types of channel changes where one side of the bank system may erode relatively slowly while an equivalent amount of material may be depositing on the opposite bank. Most meandering river systems, during periods when they are changing slowly (this excludes cutoffs), would generally fall in this category of channel change. It is also common to have channel changes that can be classified as avulsions. This implies a more rapid, violent change in the channel system such as may be associated with the development of a natural cutoff in a meandering system. Other forms of avulsions could be caused by landslides that cause blockages and diversions of the flow from the original channels to new channels.

Both natural cutoffs and man-made cutoffs should also be considered. In many instances, man has taken the liberty of speeding up the natural processes by making several cutoffs which have significantly increased the channel gradient, increased flow gradients, increased flow velocities, and consequently increased the channel's ability to transport relatively large quantities of sediment and water. This method of modifying channel systems to achieve a lowering of flood stage is quite effective in those reaches of the channel where the slope of the system is actually increased. However, one must remember that neither the total channel system is straightened nor is the channel gradient increased overall. Hence, in the region where natural or man-made cutoffs occur there is a large increase in the slope of energy gradient and an accompanying increase in the flow velocity of the water and sediment. Because sediment transport is proportional to about the fourth power of the velocity, larger quantities of sediment can be transported in this section of the river. When the water and sediment being transported from this particular steepened section of the river reaches a portion of the river further downstream where the slope is lower and more consistent with the natural river gradient, aggradation may occur. This aggradation can cause increased stages through channel filling. The aggradation may also cause lateral diversions so that the banks are attacked or that the channel is forced into a new alignment. Therefore, the role of cutoffs is very significant in terms of the overall behavior of river systems and the supply of sediments that must be considered throughout the system.

Hydraulic Structures and Water Diversion: There are a wide variety of hydraulic structures that are utilized to improve and develop river systems. One of the most common types of structures utilized in connection with river systems is dams and their reservoirs. Reservoirs are sites for storage of sediment being carried into them from upstream watersheds. It is essential in designing dams and reservoirs to consider the rate of sediment inflow in order to determine the impact on the life and usefulness of the reservoir. Such factors as storage of water for hydropower, reducing downstream flood stages, improving the system for environmental purposes and effects on recreation must be considered. For example, Lazenby and Nelson (1976) report that sediment inflow into Lake Powell, Arizona-Utah, has a volume rate of about 85,400 acre-feet per year. The desilting of the water in the reservoir and the subsequent release of clear water downstream of the dam can produce general degradation below structures. Essentially clear water releases from the reservoirs to the downstream channel allows this water to pick up a natural sediment load by degrading the bed and attacking the banks until perhaps inhibited by armoring. Downcutting of the bed can have a very important impact on the local tributaries. As the main channel tends to degrade the slope of the tributary systems are increased and the increased gradients in these tributaries cause them to discharge larger quantities of sediment. If the increase in gradient is sufficient, the tributary form may change from meandering to braided. In an extreme case, if the increase in gradient is sufficiently large, head cutting can occur back through the tributary watershed, bringing large quantities of sediment down to the main channel. The main channel may then aggrade again causing backwater pools and decreasing the tributary gradients.

There are several interesting, related phenomena such as the response of canal systems downstream of the storage reservoirs. Canals that formerly diverted water and sediment from the river system may contain only clear water after construction of a reservoir. Removal of the fine sediments from the water flowing in the canal, into its laterals and on into the irrigation ditches, creates potential problems. First, the clearer water can cause significant erosion of the banks and bed of the canal system. With this erosion there is the exposure of more permeable materials that allow a significant increase in seepage to occur. Second, as the clearer water is delivered to the farm units, the total irrigation system may require redesign because the clear water will only run a relatively short distance before totally infiltrating into the soil surface. Hence, in order to maintain some significant delivery of irrigation water, an increase in fine sediment load is needed.

Another form of hydraulic structure that can impact on the sediment load in the river system is the type that can be classified as transbasin diversion. In this type, the discharge of water in a particular basin is increased or decreased by the transfer of water between adjacent river basins. This redistribution of water affects the ability of both systems that are involved to transport sediments. In the case of the watershed that is subjected to a depletion of flows by the transbasin diversion, the resultant channel gradient is too flat to transport the sediments entering from the watersheds and there is a tendency for aggradation throughout the system. Conversely, in the river basin that receives the diverted water the channel system is generally too steep to accommodate the larger flow of water and, under these circumstances, there can be significant degradation of the bed and increase in bank erosion until limited by armoring or other controls. A smaller steep gradient system that is subjected to increased flow may be totally altered in river form, perhaps from meandering to braided. There are many development plans, however, that require these types of diversions. It may be worthwhile to transfer water from one basin to another to improve the supply for irrigation, for municipalities, for industries, or for hydropower production. An example of this interbasin diversion problem is the James Bay Hydropower Scheme in Canada. In this case, diversions from two adjacent basins supply the LaGrande River which discharges into James Bay, a part of Hudson Bay. The amount of water diverted, coupled with the leveling effects of storage on the LaGrande River will significantly increase the base flow from on the order of 30,000 to 40,000 cfs to as much as 140,000 cfs. Such increases from transbasin diversions can have significant impacts on the stability of systems and consequently on the sediment derived from these systems. Another example of transbasin diversion is the North Fork of South Boulder Creek in Colorado. In this case the flow in the base stream was approximately tripled by diversions from the Western Slope to the Eastern Slope. This transfer of water to the smaller basin was done without making any adjustments in the particular basin. When these larger flows were released into the system, significant degradation and bank erosion resulted and a considerable volume of sediment reached storage reservoirs, seriously depleting their capacity.

Channelization: Another form of hydraulic structure is that associated with improving river systems for navigation and flood control. The classic example is the development of the middle Mississippi River. In this instance, the channel has been significantly reduced in width by using combinations of revetments, dike fields, and other measures. As one narrows the flow width, the depth increases. This generally reduces the relative roughness and increases the velocity and increases the capacity of the system to transport sediment. The usual consequence of channelization projects of this nature is to bring about some degradation of the bed. Degradation of the bed can cause the low flows in the channel system to take place at a significantly lower stage than in the natural river system. For example, Simons et al. (1974) have shown the effects of encroachment in the form of dikes on the Middle Mississippi River (Fig. 7-4). Narrowing of channel cross sections has led to degradation of the channel bottom and significant changes in high and low river stages. Under these circumstances channelization can significantly alter the gradients of tributary streams and affect the amounts of sediments that are derived from these systems as a consequence of such a change. It is common to use locks and dams in connection with channelization. These locks and dams are usually designed so that they can increase the depths of flow to improve navigation conditions during the low flow period while, during high flows the gates on the dams can be opened and the river system will run essentially as an open river. The utilization of locks and dams can have a significant effect on the flow of sediments through the system. During periods of low flow, as sediments

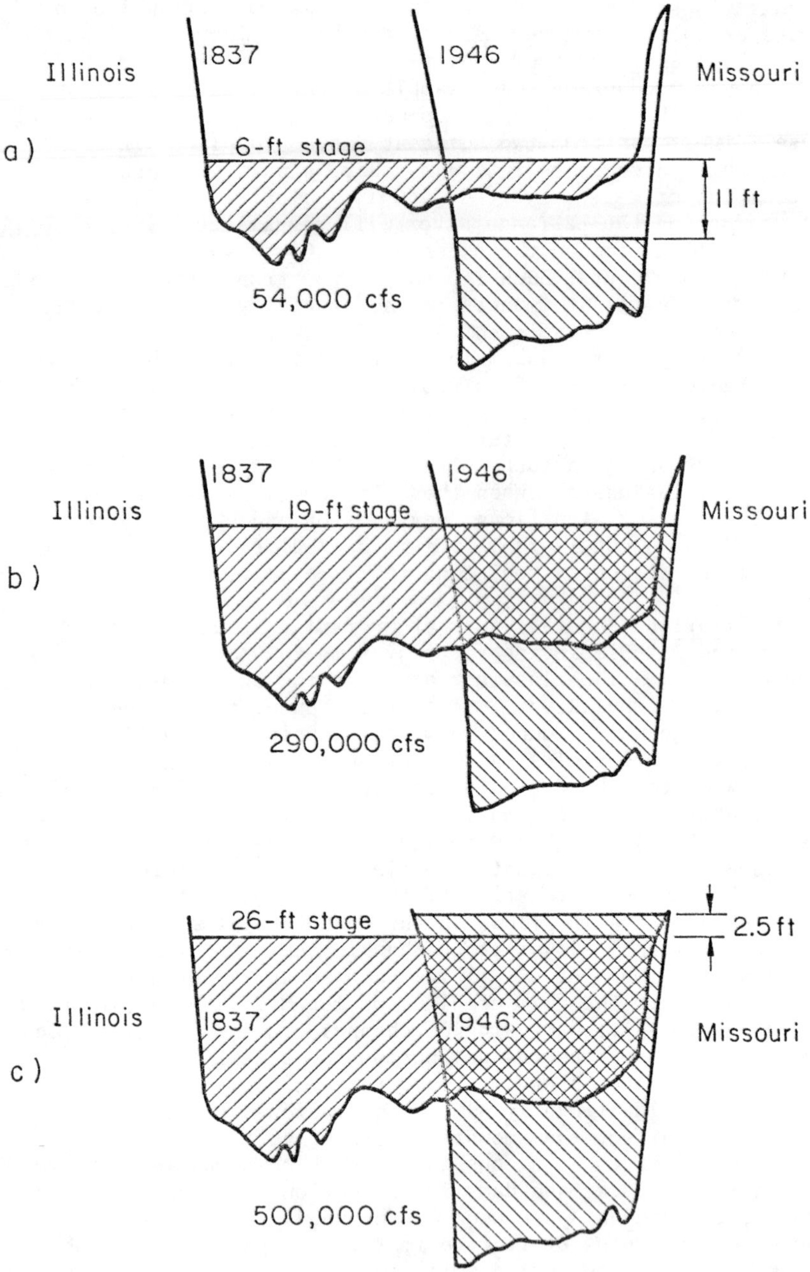

Fig. 7-4. Flow Areas at St. Louis

are transported down the main channel or into the pools from the tributaries, there is usually a tendency for aggradation as these pools act as sediment sinks. As a certain level of flow is exceeded and the gates on the dams are opened, the river returns to an open river operation and the sediments which were stored are once again available for transport downstream. To balance out inabilities that the river system may have to transport sediments, even with locks and dams, a certain amount of dredging may be necessary. The dredging problem is a totally different, but complimentary, problem to channel improvement for navigation and flood control.

Dredging: Dredging is a common practice utilized to improve channels for navigation and flood control. Lagasse (1975) reports that 9.8 million cubic yards of sediment are dredged from about 50 miles of the Columbia River annually. Large quantities of sediment can be removed from the channel system during dredging operations. This may deplete the normal sediment supply and create a reservoir that serves as a sink to store sediment. As such quantities of sediment are removed from the system, downstream portions of the system generally have less sediment to carry and consequently some degradation and bank erosion may result. In any dredging type activity, it is important to look at the benefits and the disadvantages. One of the significant concerns of environmentalists and ecologists in terms of dredging is the wasting of the dredged materials. If these dredged materials are placed on the bank along the channel system, they can be transported into the wet land and marsh areas and the floodplain during subsequent high flow periods. Significant quantities may also be washed back into the main channel, reducing the effectiveness of dredging. If the dredged materials are improperly placed, they may adversely impact on biologically active areas that are important to fisheries and wildlife. There is a growing concern over this particular method of solving river channel problems, and considerably more attention will be paid to the impact of the removal of these sediments by dredging, the wasting of them, and the ultimate fate of these dredged materials as they are subjected to higher flows in subsequent years.

Dam Breaches and Removal: An activity associated with river engineering is the construction of dams. There is always the opportunity for failure or the ultimate need to remove the dam because it no longer fulfills original purposes. Considering the dam breach problem, it is important to know that for a short period of time after a dam breaches there is a huge increase in the water discharge down through the river system and a great increase in the ability of that system to transport sediment. Therefore this particular problem should be analyzed in order to deduce the amounts of sediment that would be delivered downstream, the impact of these sediments moving with the water flood, and other subsequent effects. It is not only essential to understand that the breaching of dams can provide large quantities of sediment, but it must also be recognized that the sediments moving with the water may cause higher stages than would otherwise occur. In this regard it is important to reanalyze the dam breach problem so that we can route simultaneously the water and sediments to get a more realistic picture of the impacts of such activities.

Another typical type of problem is when dams are removed by man after they have served their useful purpose. Many cases have been documented where small dams have been constructed on river systems to serve a particular purpose such as hydropower, municipal supply, or irrigation. Over time these storage facilities have filled with sediment and no longer meet the needs for which they were constructed. In some instances, attempts have been made to remove these structures in order to regain the use of the river for such uses as the spawning of fish. One significant problem that must be considered is how to handle the sediment stored in the reservoir. If the dam is simply removed allowing the water to interact with the sediments that are stored within the reservoir large quantities of these sediments are transported downstream where they can aggrade the bed, thus causing increases in flood stage and perhaps doing damage to fisheries and other habitats. This other type of dam problem is one that can provide large quantities of sediment which can greatly affect the hydrology and hydraulic performance of a river system. In many instances it has been noted that when dams have been removed releasing these waves of sediment, sewage intakes and water outfalls have been buried by the resultant wave of sediment moving downstream. In advance of the removal of such structures it is possible to assess the factors that would affect the movement of the sediments deposited in the reservoir system enabling one to carry out a water-sediment routing scheme to determine the downstream impacts and what may be done to avoid these impacts.

Another dam breach problem is the breaching of "tributary" dams. These "tributary" dams can be developed across a main channel as a consequence of large quantities of sediment and water coming out of a tributary. When these flows encounter the flatter gradient, a tributary bar is developed that may significantly shift the main channel away from the mouth of the tributary or, in extreme cases, sufficient sediments may be delivered from the tributary to totally dam the mainstem for a short period of time. This type of phenomena is quite common on many of our river systems. For example, this is a common phenomena on the Rio Grande in New Mexico and Texas. Relatively large quantities of coarse material are carried from the tributary and deposited in the mainstem, causing the tributary dam. Then large quantities of water are stored in the main channel until the tributary dam is breached. With the breaching of the dam, a wave of water moves downstream creating the aforementioned impacts.

Mining of Gravels from Channels: It is a practice throughout the U.S. and the rest of the world to mine gravels from the streambeds of rivers in order to provide materials for construction activities. These gravels are an integral part of the sediment transport in the river systems, and the removal of large quantities of gravels may reduce the stability of the channel system. Instabilities would show up through increased degradation in the channel and by the increased erodibility of bars and islands within the channel system. In many instances, bars and islands are held in position by a relatively thin layer of coarse materials on the upstream end. These materials tend to armor the upstream ends and prevent erosion up to some critical discharge level. By removal of these coarse materials, bar formations and islands may be subject to attack. Segments of the streambed are, in many instances, also armored with gravel. These gravel bars and armoring act as semi-controls that help contain or control the profile of the stream. If excessive mining of these gravels is allowed, the controls can be destroyed causing the whole channel system to be susceptible to significant geomorphic adjustment.

7.4 NATURAL DISASTERS

Apart from everyday or man-caused processes that affect sediment yield are those which occur at random intervals. Oftentimes these processes create conditions that are significantly different from the everyday environment that encompasses the watershed and river system. These processes, when they affect man, are termed natural disasters.

There are several types of natural disasters that can cause huge increases in the supply of sediments in watersheds and river systems. Storms and resulting floods are a principal cause of increased sediment yield from watersheds. With large storms, there is an excess delivery of water and sediments from the watersheds to the river systems. With these large flows there is a further increase in the ability of the river system to transport sediments. Another factor that is important in terms of major floods is the flood duration. If the discharge is continuously changing, the impact is usually less than if a flood occurs that is of relatively long duration at a constant discharge. When the flow line remains essentially stable and high for a significant period of time, relatively large amounts of bank erosion often result.

Another natural diaster is earthquakes. Earthquakes can cause slides, mudflows, loss of structures such as dams and levies and even complete tilting of a segment of river system. It can be historically documented that earthquakes have caused major shifts in river systems. A classic example is the Bhamaputra River in Bangladesh. About 200 years ago an earthquake caused large landslides that filled the Bhamaputra channel, diverting it westward. That channel runs approximately 200 miles west of where it formerly operated prior to the time of this major earthquake. In some instances, earthquakes can cause the formation of lakes. Quake Lake in the Yellowstone area of Wyoming is a classic example of a landslide formed lake. Another is the large landslide that occurred in Peru (Lee and Duncan, 1975). In this instance a huge block of earth filled the Mantaro River canyon and caused large quantities of water to be stored. If these landslide dams cannot be stabilized so that they will not fail, there is the ultimate potential for yet another disaster. When landslide dams are breached, large quantities of water and sediment can be released that do serious flood damage downstream as was the case in the Peruvian landslide.

Another disaster that can affect the supply of sediment is fire, both natural and man caused. As watersheds are denuded by fire, there can be a change in the soil characteristics and vegetative cover. Soils subjected to extreme heat generally react chemically in such a way that they can be more easily transported or have decreased infiltration rates. The sheltering effect of the canopy cover and the stabilizing effect of the root systems is lost. Therefore, fire can be a significant factor in creating favorable conditions for erosion of upland soils.

Another disaster that affects the supply of sediment is drought. In the western United States, it has been noted that during wet periods of time when the vegetation is more abundant, the sediment supply is generally less. During periods of drought in such arid and semi-arid areas, a large portion of the vegetation dies back. When rains do occur, large quantities of sediment are carried from these watersheds down into the channel systems, creating further impacts. An example is the floods that occurred out of the Wasatch Range in Utah during the drought period of the 1930's.

7.5 QUALITATIVE IMPACTS OF SEDIMENT

Sediments not only interfere with the quantitative aspects of the fluvial system such as reservoir storage and channel maintenance, they also affect many of the qualitative aspects. One example of a change in the quality of the fluvial system is caused by sediment impacts on fisheries. O'Connor and Sherk (1976) list the sediment characteristics that affect organisms as 1) concentration of sediment, 2) composition of sediment, 3) sorbed materials on sediment, and 4) tolerance limits of the organism. Phillips (1971) lists specific effects on fish as 1) reducing food chain production by blocking light; 2) damaging gill membranes; 3) filling gravel beds, thus interfering with oxygen flow into and out of egg beds; and 4) blocking fry emergence from gravel beds. Hansen and Alexander (1976) noted distinct changes in fish habitat in the form of the loss of spawning areas and changes in thermal character of a stream artificially loaded with sediment. It is apparent then that sediment can and does have severe impacts on our fishery resources.

One final impact that must be mentioned is the effect of sediment on water quality. The removal of sediment from water for domestic use costs millions of dollars every year. There is a more disturbing aspect of sediment impact on water quality that is becoming more important in this pollution age; the relationship between sediment and attached deleterious substances . Recent studies have shown that uptake of hazardous substances such as mercury (Kudo and Hart, 1974; Townsend et al., 1974) and DDT (Oloffs and Albright, 1974; Hargrave and Phillips, 1974; Gillot et al., 1975) are inversely related to particle size. That is, smaller particles (clay sizes) have higher uptake rates. These findings have serious consequences as fine particles travel rapidly with the stream flow and are difficult to remove from domestic water.

Associated with the transport of deleterious substances in the fluvial system is the transport of these same substances in the atmospheric system. In atmospheric processes there is the potential transport of mass through the wind aerosolization of dusts--with increased susceptibility from such man-related activities as road building, grazing, recreation, mechanical site preparation and timber harvesting. There also exists the real possibility of pollution from potentially deleterious materials such as aerially applied pesticides or herbicides that fail to deposit effectively on foliage because of inappropriate size distribution produced from spraying systems or unfavorable meteorological conditions during spraying. Natural occurrences such as scavenging processes that may wash out airborne gaseous or particulate matter from the atmosphere should be noted including nucleation processes that occur within clouds causing pollutants to be associated with droplets having non-negligible settling velocities and forest fires which create a tri-modal particle-size distribution which indicates significant drift potential.

The characteristic bi-modal volume distribution (also mass unit density) of atmospheric particles is one with modal peaks at approximately .4 μm and 50 μm. It would be reasonable to anticipate similar size ranges of particles to occur in watersheds and streams. It has been reasonably well established that particles having mean diameters of 20-40 μm characterize wind aerosolized dusts under moderate wind conditions although considerable work is needed in this area especially in actual forest ecosystems.

It is entirely feasible to develop models for predicting the mass input to river ecological systems considering air as a transport medium. Models and methodologies are intimately related to the physical principles which govern the mass transport processes. Existing and future models would be applicable to a wide variety of physical environments including forest ecosystems, rangelands and urban watersheds. Model parameters such as velocity profiles and wind shear effects, fluid and particle diffusivities, mass concentrations and particle size distribution, terrain roughness elements, and local meteorology could be modified to adapt predictive capabilities to different environments and geographical locations. This is possibly due to the universal principles of governing physical processes. The use of physical process simulations has proved useful in such applications and has begun to replace the use of statistical models that are mathematically and physically unrealistic. With a limited field data base which could be relatively easily increased with a well managed field sampling effort in both air and water, the use of physical principles and processes in model simulation could be quite accurately applied without the usual assumptions made in statistically based models such as homogeneity in temporal and spatial considerations as well as broad based assumptions of local meteorology and momentum parameters.

Another example of a prediction on mass transport through air, much of which could be available for deposition on trees and ground for subsequent runoff or direct deposition into waterways, is to consider the cloud concentration directly. Using a conservative figure of about 1 $\mu g/m^3$ for cloud concentration and assuming an open swath of land, as would arise from harvesting or selective cuttings in a forest, 100 m wide with a mixing depth of 30 m, 260 grams of particulate matter would pass any given point in a day's time or ~100 kilograms in a year at a mean wind speed of 1 m/sec, a fairly substantial figure.

Aerially delivered pesticides are another concern. A good deal of the airborne spray droplets either miss the canopy foliage or drift to undesirable locations due to inappropriate size distributions produced by spray atomizers. A recent wind tunnel modeling effort for a Douglas Fir canopy has identified a critical Stokes number above which increased efficiency of deposition ceases and waste occurs.

Air transport of pollutants can be modeled both qualitatively and quantitatively. Transport rate, transport volumes, eventual deposition sites, concentrations and absorption uptake rates by organic and inorganic particles are important characteristics of airborne pollutants. Models describing the controlling processes that govern these characteristics can be used in conjunction with other physical process models that describe the movement of water, sediment, and organic material in the watersheds and river systems.

Linkage of the physical process models of atmospheric and terrestrial transport will enable the development of a more complete nonpoint source pollution analysis methodology. An example of how such models may be linked is as follows: Misapplied sprays are air transported from the target area and settled in a downwind area. The concentration of sprays and location of this area can be predicted with air transport models. Uptake rate of the pollutant sprays as they are adsorbed to sediment and organic matter may then be utilized as well as decay rates for the pollutant. If a hydrologic event occurs, the pollutant can be routed by overland surface flow, subsurface flow and even ground water flow techniques into the channel system. The pollutant may reach downstream either dissolved in the water flow or attached to sediment particles.

By linking air transport and terrestrial transport process models, a more complete simulation of pollutant movement from nonpoint sources may be realized. Only by utilizing combinations of component process models can realistic assessment of environmental impacts associated with sediment be developed.

7.6 SUMMARY AND CONCLUSIONS

The interactions of the fluvial system are complex and not completely understood. This is particularly true in erosion and sedimentation. However, history and experience provide insight into the impacts of sediment in the fluvial system. Some of the impacts associated with man's various activities are listed in the following table. Although the table lists selected activities, it does create an appreciation of the numerous factors that must be considered in assessing the impact of sediment.

Table 7-1. Impacts from Man's Activities on Sediment Sources and Yields

Activities Associated with Man	Impact
Upland Watersheds	
Agriculture	Removes native vegetation cover Disturbs soil surface Alters drainage patterns
Grazing and Timbering	Removes vegetation cover Disturbs soil surface Alters drainage patterns
Road Construction and Recreational Development	Disturbs soil surface Changes slope gradients Alters drainage patterns
Mining	Disturbs soil surface Generates loose material sources Creates soil instability
Urbanization	Alters drainage patterns Disturbs land surfaces Increases sediment yield

Table 7-1 (continued)

Activities Associated with Man	Impact
Channels	
Bank Erosion	Increased sediment load to channel Loss of property near the channel
Channel Changes	Degradation upstream of straightened section Aggradation downstream of section Change in flow depths at same discharge
Hydraulic Structures and Water Diversions	Create artificial sediment sinks Produce clear water degradation below dams Create aggradation and degradation problems in tributaries Changes in discharge and sediment transport by interbasin diversion
Channelization	Alters cross section of channel Changes stage discharge relation Produces changes in sediment transport capability of the channel
Dredging	Creates a sediment sink Induces degradation and bank erosion downstream Adversely impacts on biologically active areas
Dam Breaches	Movement of large sediment waves down channel changing channel geometry and flood flow level
Mining of Gravels from Channels	Changes size distribution of sediments Removes armoring protection from banks and bars

All components and processes of a fluvial system interact to deliver water and sediment to the ocean or inland basins. Sediment sources can be delineated and the processes controlling sediment yield from these sources studied. Research on upland sources of sediment associated with river processes is lagging. It must progress to meet the increased growth pressures that will eventually affect these sediment sources. Many of the impacts associated with sediment in the fluvial system revolve around reservoir and channel filling, the effect on aquatic fauna and flora, and the necessity of removal of sediment from drinking water. In this chemical age, however, sediment interacts with man-made pollution to provide a transport medium that both carries and stores the pollution. The topics raised here are broad in scope and contain many fine details. There is much left to understand about sediment sources and impacts in the fluvial system.

7.7 REFERENCES

Bradford, R. G., Farrell, D. A., and Larson, W. E., 1973. Factors effecting gully stability. Soil Science Society of America Proceedings, Vol. 37.

Bradford, J. M., and Piest, R. F., 1977. Gully wall stability in Loess-derived alluvium. Soil Science Society of America Journal, Vol. 41.

Dickerson, B. P., 1968. Logging disturbance on erosive sites in North Mississippi. USDA Forest Service, Research Note SO-72.

Dickerson, B. P., 1975. Stormflows and erosion after tree-length skidding on coastal plain soils. Transactions American Society of Agricultural Engineers, Vol. 18, No. 5.

Dyrness, C. J., 1967. Mass soil movements in H. J. Andrews experiment forest. USDA Forest Service Research Paper, PNW-42.

Foster, G. R., and Meyer, L. D., 1975. Mathematical simulation of upland erosion by fundamental erosion mechanics. in Present and Prospective Technology for Predicting Sediment Yield and Sources, Agricultural Research Service, ARS-S-40.

Gillot, M. A., Floyd, G. L., and Ward, D. V., 1975. The role of sediment as a modifying factor in pesticide-algae interactions. Environmental Entomology, Vol. 4, No. 4.

Guy, H. P., 1974. An overview of urban sedimentology. Proceedings National Symposium on Urban Rainfall and Runoff and Sediment Control, University of Kentucky.

Hansen, E. A., and Alexander, G. R., 1976. Effect on an artificially increased sand bedload on stream morphology and its implications on stream habitat. Proceedings of Third Federal Interagency Sedimentation Conferences, Denver, Colorado, March 22-25, Water Resources Research Council, Washington, D.C.

Hargrave, B. T., and Phillips, G. A., 1974. Adsorption of C14-DDT to particle surfaces. Proceedings of International Conference of Transport of Persistant Chemicals in Aquatic Ecosystems, Ottawa, Canada.

Kudo, A., and Hart, J. S., 1974. Uptake of mercury by bed sediments. Journal of Environmental Quality, Vol. 3, No. 3.

Lagasse, P. F., 1975. Interactions of river hydraulics and morphology with riverine dredging operations. Dissertation, Colorado State University, Fort Collins, Colorado.

Lazenby, J. F., and Nelson, L., 1976. Lake Powell sediment surveys. Proceedings of Third Federal Interagency Sedimentation Conferences, Denver Colorado, March 22-25, Water Resources Research Council, Washington, D.C.

Lee, K. L., and Duncan, J. M., 1975. Landslide of April 25, 1974 on the Mantaro River, Peru. Committee on Natural Disasters, National Academy of Sciences, Washington, D.C.

Leopold, L. B., Wolman, M. G., and Miller, J. P., 1964. Fluvial processes in geomorphology. W. H. Freeman and Company, San Francisco.

McDowell, L. L., Bolton, G. C., and Ryan, M. E., 1967. Sediment production from a Lafayette County, Mississippi Gully. Paper presented at Second Mississippi Water Resources Conference, Jackson, Mississippi.

Megahan, W. F., and Kidd, W. J., 1972. Effect of logging roads on sediment production in the Idaho Batholith. USDA Forest Service, Research Paper INT 123.

Meyer, L. D., 1974. An overview of the urban erosion and sedimentation processes. Proceedings National Symposium on Urban Rainfall and Runoff and Sediment Control, University of Kentucky.

Meyer, L. D., Foster, G. R., and Romkens, M. J. M., 1975. Source of soil eroded by water from upland slopes. <u>in</u> Present and Prospective Technology for Predicting Sediment Yield and Sources, Agricultural Research Service, ARS-S-40.

Miller, C. R., Woodburn, R., and Turner, H. R., 1962. Upland gully sediment production. Symposium of Bari, Commission of Land Erosion, International Association of Scientific Hydrology, Publication 59.

O'Connor, J. M., and Sherk, J. A., 1976. Effects of sedimentation on coastal game organisms. Proceedings of Third Federal Interagency Sedimentation Conferences, Denver, Colorado, March 22-25, Water Resources Research Council, Washington, D.C.

Oloffs, P. C., and L. J. Albright, 1974. Transport of some organochlorines in B.C. waters. Proceedings of International Conference of Transport of Persistant Chemicals in Aquatic Ecosystems, Ottawa, Canada.

Phillips. R. W., 1971. The effects of sediment on the gravel environment and fish production. Proceedings of a Symposium on Forest Land Use and Stream Environment, October 19-21, 1970, Oregon State University, Corvallis, Oregon.

Piest, R. F., Bradford, J. M., and Spomer, R. G., 1975. Mechanisms of erosion and sediment movement. in Present and Prospective Technology for Predicting Sediment Yield and Sources, Agricultural Research Service, ARS-S-40.

Simons, D. B., Schumm, S. A., and Stevens, M. A., 1974. Geomorphology of the Middle Mississippi River. Civil Engineering Report, Colorado State University, Fort Collins, Colorado. Prepared for U.S. Army Corps of Engineers Waterways Experiment Station, Vicksburg, Mississippi.

Simons, D. B., and Sënturk, F., 1977. Sediment transport technology. Water Resources Publications, Fort Collins, Colorado.

Swanson, F. J., and Dyrness, C. J., 1975. Impact of clear-cutting and road construction on soil erosion. by Landslides in the Western Cascade Range, Oregon, Geology, July.

Swanston, D. N., and Swanson, F. J., 1976. Timber harvesting, mass erosion and steepland forest geomorphology in the Pacific Northwest. in Gemorphology and Engineering, D. R. Coates, ed., Dowden, Hutchinson and Ross, Inc., Stroudsburg, Pennsylvania.

Townsend, D. R., Kudo, A., Sayeed, H., and Miller, D. R., 1974. Mercury transport by bed sediment movement. Proceedings of International Conference of Transport of Persistent Chemicals in Aquatic Ecosystems, Ottawa, Canada.

Varnes, D. J., 1958. Landslide types and processes. Landslides and Engineering Practice, Highway Research Board, Special Report 29, Publication No. 544.

Weber, W. G., Jr., and Reed, L. L., 1976. Sediment runoff during highway construction. Civil Engineering, Vol. 46, No. 3.

Chapter 8

COMPUTER MODELING OF RIVERS: HEC 1-6

by

William A. Thomas, Hydraulic Engineer, Waterways Experiment Station, Corps of Engineers, Vicksburg, Mississippi

Chapter 8

COMPUTER MODELING OF RIVERS: HEC 1-6

8.1 INTRODUCTION

In the late 1950's and early 60's the developers of computer programs had as their objective "let's use the computer to make calculations which are being done by hand." Many such programs were developed and the cost was staggering. However, that cost was envisioned as a one-time investment which thereafter would return rich dividends because of the speed and accuracy of the electronic computer.

As it turned out the hand calculations which were programmed often utilized approximations and shortcuts which were appropriate only for the particular study for which the program code was developed. Consequently, code changes were required for each new application. Often the author of the program was no longer around and neither were documents explaining what he had done. It was easier for a person to develop a completely new program than attempt to reconstruct someone else's logic from listings of assembly language code. Visions of rich dividents faded away and managers were able to mount convincing arguments against the use of computers in engineering studies.

As if these types of problems were not enough, the computer hardware manufacturers were advancing at such a rapid pace that successful program developers were continually modifying codes to take advantage of the "bigger and faster" machines. This process of program evolution was as expensive as program development in terms of manpower. That is, engineers who could have worked productively on projects were committed to a capital investment for developing a better tool in anticipation that others in their organization would be much more productive because of that capital investment. Engineers could sense the power of the computer, but they could not demonstrate economical advantages from using it.

Some felt strongly that general purpose codes were the essential missing link in making the computer economically attractive. In 1964 the Hydrologic Engineering Center was established and assigned the mission of upgrading and systemizing hydrologic engineering methods. Although not specifically established as a computer program software development center, the concept of "generalized computer programs" fits the requirements of systematic methods like a glove fits a hand. The director of HEC, Mr. Leo R. Beard, established the basic philosophy and structure of the center as

- methods systemization would utilize the electronic computer
- research and development would provide the latest tools for the methods systemization program
- cooperative assistance to Corps of Engineers field offices would reveal the type of tools needed as well as strengths

and weaknesses in newly developed computer programs before they were released to other offices for use.

training courses of one or two week durations would facilitate the transfer of newly developed technology to the field office personnel

Mr. Beard believed that general purpose computer codes could be developed, that a single code could be applied to a wide variety of problems and that codes could be developed in such a manner as to be easily transported from one computer to another. However, the codes would have to be carefully designed, programmed in standard Fortran and documented. The name given to this concept was "Generalized Computer Programs."

A group of such programs were developed for the computers of that day--the IBM 1620 and GE-225 class machines. In most cases these programs performed only one step of a study because the small computers would allow only that much code. However these programs demonstrated portability--from one computer to another--and useability because any competent engineer could use the programs with the aid of the documents and an occasional telephone call to HEC.

Although used by the HEC staff, these programs were developed for all Corps of Engineers offices. Standard 8-column field formats were adopted for coding data which reduced the number of "job aborts" caused by coding errors, simplified keypunching and simplified documentation. Each card was identified with an alphabetical letter in the description of data. Programs were designed for batch processing.

Method systemization was not construed to mean centralization or elimination. Consequently, several different methods were programmed for making the same calculation because all field offices did not utilize the same method. For example, some offices used the Modified Puls Routing Method while others used Tatum or Straddle-Stagger. Several small computer programs were soon available for routing floods and for many other such calculations.

A second characteristic of these early programs was the manual intervention required while performing a study. Rainfall/runoff analysis required three separate runs: unit hydrograph analysis, runoff hydrograph calculations and hydrograph routing and combining. This approach required too much study time because of the slow turnaround of jobs. Generalization would not permit program segmentation and overlays because of their system dependence.

It was obvious that bigger computers were required to successfully perform engineering studies. All of the small codes required or desired for analyzing rainfall/runoff from an entire basin, subdivided into many subbasins, could be combined and the logic for linking all computations together could be included. This concept became known as "package" programs and in 1967 HEC introduced the "Flood Hydrograph Package" (HEC-1). This was followed by "Water Surface Profiles" (HEC-2) a short time later. Both of these were second generation programs. Not only could an entire study analysis be conducted by the single submission to the computer, but also several options were available for many of the calculations.

By 1968 the HEC staff had grown to 21. Since the main effort was methods systemization for the hydrologic engineering studies most of the HEC staff were engineers. They were well versed in Fortran, had considerable insight into hydrologic engineering problems and field office constraints and were optimistic about the power of the computer. Today, June 1977, there are 32 full-time employees on the HEC staff. Under the direction of Mr. Bill S. Eichert the computer continues to be the hub of all activities. The 1976 annual report lists 28 Fortran computer programs in the active library (see Appendix). Over 2500 source decks have been distributed from that program library. These are being used in government, academic and private sectors of this country and abroad. These programs are maintained and supported by the HEC staff.

The six "Package Programs" on the list have been identified with special numbers: HEC-1 through HEC-6. These were developed by the HEC staff. The center is commissioned to provide systematic methods, however, and not to develop computer programs. Therefore, hydrology-related programs developed by others are always being considered as potentially useful methods. Several such programs are included in the HEC library and they meet all the problem solving requirements for "package programs"; however, they are not given the "HEC" label. Instead, appropriate credit is given to the program developers in the user's manuals.

8.2 SCOPE OF A COMPUTER PROGRAM

HEC operates under the guidance and policies established by the Office, Chief of Engineers, U.S. Army Corps of Engineers, Washington, D.C. The scope of hydrologic engineering associated with the Corps mission embraces

- water yield and reservoir capacity requirements
- performance of systems of reservoirs to meet multiple objective usage requirements
- rainfall to runoff volumes and flow rates
- water surface profiles
- the analysis of and adjustments to streamflow records
- discharge capacity of a stream
- hypothetical floods
- statistics of streamflow records
- groundwater movement
- water quality
- sedimentation
- snowmelt/runoff
- flood damage calculations
- synthetic streamflow generation
- non-structural, alternative evaluation

The level of detail required for hydrologic studies ranges from the early feasibility requirements through design requirements and project operation requirements. The cost of computer program development, maintenance and support makes it necessary to address as broad a scope as possible in terms of both processes involved and level of detail. On the other hand programs must be as small as possible to be manageable. These conflicting requirements are satisfied by making trade-offs between level of detail required to model the physical processes involved and the extent to which specific processes impact on the decision-making process. For example, the essential physical processes in rainfall/runoff are the rainfall pattern, amount and rate; losses; rainfall excess/runoff transformation; and phasing of runoff from different portions of the basin as flows combine to produce the runoff hydrograph(s) at the point(s) of interest. These same physical processes are pertinent to water yield studies, but not to the decision making process of determining water yield. Therefore, computer programs which deal with water yield (or conservation studies) do not need detailed rainfall/runoff calculations--only the volume sequence of available water. Matching level of detail of calculations with the type of pertinent results required for the decision making process is essential in establishing the scope of a computer program.

In summary, the scope of a computer program should be study oriented--not process oriented. Most studies involve several processes. For example, open channel flow is primarily the physical process of energy dissipation due to friction along natural channels, but this process is interrupted by flow at bridges, culverts and other hydraulic structures. A study requiring water surface elevations involves all of the above processes. Therefore, they should all be included in the computer program designed for water surface profile studies.

A second major consideration in establishing the scope of a computer program is to identify weak links connecting the physical processes. For example, inland rivers flow in one direction. Water surface profiles can be calculated for steady flow discharges even when those discharges occur at the peak of a discharge hydrograph. Near the coast, tides and wind cause flow to change directions. Steady flow approximations are frequently inadequate in such problems because the flow of information is not in one direction. A computer program must have a strong feedback loop to match that flow of information, and unsteady flow equations provide that feedback. (Of course the unsteady flow equations may be used to calculate steady flow profiles, but not as economically as the steady flow equations.) Although programs are study oriented the developer must understand the physical processes involved so important feedback loops will be included.

To the maximum extent possible the results of hydrologic engineering studies must be expressed in dollars. Cost and damage functions are an important consideration in establishing the scope of a computer program. These are included in several of HEC's programs.

8.3 SELECTION OF COMPUTATION TECHNIQUES

Having established general classes of studies and thereby the types and scopes of computer programs to accomplish those studies, attention may be given to the selection of analytical techniques. The objective of a hydrologic engineering study differs considerably from that of scientific hydrology. Hydrologic engineering requires combining knowledge of scientific hydrology with that of engineering principles in such a way that alternative plans for utilizing water resources may be properly ranked. Consequently, the goal for hydrologic engineering studies is to evaluate the performance of a physical system--not to simulate the physical processes involved. Analytical techniques which emulate the response of the physical system are more cost effective and require less detailed (spatially distributed) data than those techniques which attempt to simulate the processes.

The form, availability and cost of basic data are key points of consideration when selecting computational techniques. It is also important to consider how much of a physical process is understood and how much cannot even be modeled because of state-of-the-art limitations. In all hydrologic engineering studies it is the use of empirical coefficients which, in the end, bridges the gap between theory and practical applications. This gap is often associated with the spatial and temporal variability of physical conditions over a watershed and cannot be bridged with definitive data sets within present funding levels and instrumentation. Therefore, it is sound engineering practice to bridge that gap after having let nature integrate as many of the processes as possible to reduce variability.

Technology transfer is another important point of consideration when selecting analytical techniques. The program library at HEC is being developed for Corps-wide usage (i.e. more than 50 different offices). It is the analytical technique which establishes the minimum level of training required to equip personnel in these offices to use the programs; that is, basic theory and the adaptation of this theory for the computer must be understood by program users.

The empirical coefficients which are a part of any hydrologic engineering study must be "calibratable," which is a major effort in most cases. Automatic optimization techniques are often utilized to expedite this step. Analytical techniques are more easily accommodated than are numerical integration techniques.

Dependability is an important consideration. Some computation techniques tend to be more sensitive than others with the trade-off being speed for stability. For example, water surface profile calculations in HEC-6 require successive approximations to solve the non-linear energy equation. The most stable solution scheme converges asymptotically toward the final result; however, it is slow, often requiring more than four iterations. Studies which require the calculation of one or two thousand water surface profiles over several hundred miles of river cannot afford the computer time for such slow convergence. Therefore, an over-relaxation factor was introduced to speed up convergence. The objective was to converge in one or two iterations in most cases and only rarely require more than three. This resulted in an oscillation type of convergence which was fast when it worked. However, it frequently failed to converge at all. Consequently, a hierarchy of convergence techniques was developed whereby speed, stability and finally brute force work together to insure dependability.

Computer program development and maintenance are the final points of consideration. Again, analytical techniques are more attractive than numerical integration techniques because the latter require extensive testing to isolate and correct potential stability problems. Program maintenance requires that strong consideration be given to methods which are most likely to have widespread usage and, thereby, familiarity. Although vital, maintaining a computer program is less challenging than developing one. Consequently, this tends to be a rotating assignment, and new people are continually being trained to perform this function.

Program maintenance is an expensive activity. Consequently, "limited purpose" calculation techniques are avoided if they mean an extra computer program to maintain and support. For example, unsteady flow computations by solving the full Saint Venant equations are somewhat complicated because of the non-linear equations, and some researchers leave out the non-linear terms. The resulting computer program will not accommodate the full scope of problems in unsteady flow studies. Therefore, two unsteady flow programs are required: one having the complete equations for special cases and the other having abbreviated forms. Any gain in program development or execution cost is surely lost to maintenance and support cost in such a situation.

HEC is interested not only in computer programs developed by others but also in computation techniques developed by others because the research program of the center is designed to provide--not develop--methods. The Engineering Manuals and Technical Reports published for the Corps of Engineers are always considered when selecting computational techniques. Recent advances in the academic community, research organizations and government agencies are also considered. Whatever technique is selected, the center's mission to develop systematic methods for use by others requires considerations for technology transfer, maintenance and support.

8.4 SELECTION OF COMPUTER FACILITIES AND PROCESSING MODE

General purpose "package programs" tend to be large and to require large volumes of input. Likewise they produce large volumes of output. To develop such programs efficiently HEC turned to large, high speed computers. Otherwise, exorbitant amounts of time would have been required just to "pack" the programs into small machines.

The center buys computer time from the government-owned computers at Lawrence Berkeley Laboratory (LBL), Berkeley, California. These CDC 7600/6600 mainframes support almost all of HEC's needs (about 100 jobs a day). In addition, one commercial firm provides backup to the LBL facility and another commercial firm provides special time sharing and graphics support. Both use UNIVAC 1108 hardware. The HEC programs are also run on IBM and Honeywell equipment by Corps offices.

All programs are designed to run in batch mode. (On the average the HEC staff transmits 30,000 cards per day from their office in Davis, California to the LBL facility 75 miles away. At the same time on-site files at LBL store most input data). This mode of operation takes advantage of the cheaper rates at facilities which do not have time-sharing overheads plus the greater portability of codes from one computer system to another. Therefore, HEC has been reluctant to modify programs for time sharing.

On the average the HEC staff receives 120,000 lines of printout a day. On occasion novices in computer applications question that volume, "Who reads all of that printout?" The answer, "Hopefully, no one." The simple explanation: "That is the most cost-effective way to develop and use a computer program at this time." A box of paper, costing ten to fifteen dollars, will provide 180,000 lines of output. The alternative is additional expense for program development, inefficient use of manpower, and non-portable codes costing many times more than the present procedure and printer paper.

8.5 UTILITY

The type and amount of required basic data, rules which one must follow to change an existing data deck, mode of computer processing, special optimization features for coefficient calibration, error diagnostics, result display, program size and execution costs all influence the utility of a computer program. Documentation has long been recognized as an essential ingredient in utility. There are two other ingredients, however, which are essential and yet often do not receive sufficient consideration. These are:

- maintenance - the periodic modification of an existing computer code to correct errors, add alternate computation options, increase scope, add flexibility, enhance the display of results, etc.

- support - on-call service for program users having difficulty with coding data or using a program

Both of these areas have been recognized by the computer hardware industry. They are even more essential for computer software because it is largely hand-made and extremely expensive.

HEC is currently producing two levels of documentation for the package programs--"Users Manuals" and "Programmers Manuals." The Users Manual gives engineers the necessary information on theory and data requirements. The Programmers Manual gives computer system requirements and other details needed to modify a code or make one operational on a new computer system. Only a few programs have Programmers Manuals at this time (see references in Section 8.12).

The center continually seeks feedback from program users to identify strengths and weaknesses in codes and design concepts. They provide on-call support to all Corps of Engineers users.

The rest of this chapter is devoted to the six package programs in the HEC library. These are

- Flood Hydrograph Package, HEC-1
- Water Surface Profiles, HEC-2
- Reservoir System Analysis (for conservation), HEC-3
- Monthly Streamflow Generation, HEC-4
- Simulation of Flood Control & Conservation Systems, HEC-5C
- Scour and Deposition in Rivers and Reservoirs, HEC-6

In addition to these package programs, the HEC program library contains a water quality program for rivers or reservoirs, a storm water treatment, overflow and runoff model, a peak discharge frequency program, a program for groundwater flow, one for analyzing dam break flood waves and several other programs. Information may be obtained by writing the center: The Hydrologic Engineering Center, 609 Second Street, Davis, California 95616. Pricing information is available for non-government requests.

8.6 FLOOD HYDROGRAPH PACKAGE, HEC-1

Purpose

This program calculates the runoff hydrograph from a watershed when given a specified precipitation pattern and precipitation rates for a single storm event. In addition, hydrographs at any number of intermediate points in the watershed may be produced. Precipitation may be in the form of rainfall or snowfall.

This program is discussed along with its hydrologic methods in Bowers (1972). The "Flood Hydrograph Package" Users Manual (1973) and Programmers Manual (1973) present the theory, methods of usage and source deck description for the program.

Basic Theory

Rainfall is distributed and rainfall excess is calculated; a unit hydrograph is used to transform rainfall excess into runoff; hydrologic methods of flood routing are used to route and combine runoff hydrographs from various portions of the watershed.

Precipitation Rates: The first major analytical requirement involves the precipitation rate. If measured data are available, precipitation can be supplied by entering the observed amount during each computation interval. In most cases, measured precipitation is not available for every basin and it is necessary to utilize nearby gaging stations. A time pattern may be specified as well as the total quantity of precipitation from nearby gages and rates for the watershed in question will be calculated using a weighted average. Basin average precipitation may be adjusted using data gage and basin values of normal annual precipitation factors. Another technique for developing precipitation data is the hypothetical storm precipitation criteria (U.S. Army Corps of Engineers, 1952).

Loss Rates: The second major analytical requirement is the calculation of losses. A simple relationship expressing an initial plus uniform loss rate is available. A Holtan type of loss rate function is available in which the initial loss rate is related to available storage in the soil column. A soil moisture decay function reduces the loss rate to an ultimate uniform value. Finally, a loss rate function which relates loss rates to rainfall intensity and accumulated soil moisture is available.

Unit Hydrograph: The third major analytical requirement is the rainfall/runoff calculation. Rainfall excess is converted to runoff using the Clark unit hydrograph method. (The unit graph characteristics may be prescribed using Snider unit graph coefficients, but the program converts these to parameters for the Clark method.) The Clark method offers two advantages: (1) the computations provide a direct solution rather than requiring trial and error adjustments of the computed unit hydrograph and (2) the heart of the method is a time area curve which may be adjusted for changes in drainage characteristics and pattern without requiring "postage stamp-sized" subareas.

The time area relationship is used to calculate a runoff rate from a subarea by the following equation

$$I_i = K\, a_i/\Delta t \qquad (8\text{-}1)$$

a_i = volume of rainfall excess over an increment of drainage area which is equal in size to $(\Delta t/t_c)$ times the total drainage area and all points of which require the same travel time to reach the outflow point. (Units are inch · square mile or mm · square kilometer).

I_i = the time-area runoff volume, a_i, expressed as a runoff rate at the end of the ith computation period. (Units are cfs or m^3/s).

i = number of current computation period

K = conversion factor for units (645 for English units or 0.278 for metric units)

Δt = the computation interval in hours

t_c = the time of concentration, or that time required for water particles to travel from the most upstream point in the basin to the outflow point.

The time-area runoff volumes from all incremental subareas are then routed to the basin outflow point with a Muskingum type of routing equation in which X, the attenuation coefficient, is set equal to zero. The routing equation is

$$O_i = CI_i + (1-C)O_{i-1} \qquad (8\text{-}2)$$

O_i = basin outflow at the end of period i (cfs or m^3/s)

I_i = increment runoff rate from each subarea at the end of period i (cfs or m^3/s)

C = routing coefficient (dimensionless)

The routing coefficient is calculated as

$$C = 2 \cdot \Delta t/(2 \cdot R + \Delta t) \qquad (8\text{-}3)$$

R = attenuation constant having the dimensions of time

The magnitude of R can be evaluated at the point of inflection of the recession limb of the runoff hydrograph as illustrated in Figure 8-1.

Routing and Combining: The fourth major analytical requirement is the routing of flow hydrographs and the combining of hydrographs from different portions of the drainage basin. Five routing techniques are available: Muskingum, Modified-Puls, Working R & D, Straddle-Stagger and Tatum. The first three are storage-routing methods and the last two are decay-type routing methods. The storage-routing techniques all solve the continuity of mass equation.

The Muskingum equation relates storage to inflow and outflow as follows:

$$S = K(X \cdot I + (1-X) \cdot O \qquad (8\text{-}4)$$

I = inflow to the reach

K = flood wave travel time through the reach

O = outflow from the reach

S = storage in the reach

X = attenuation coefficient

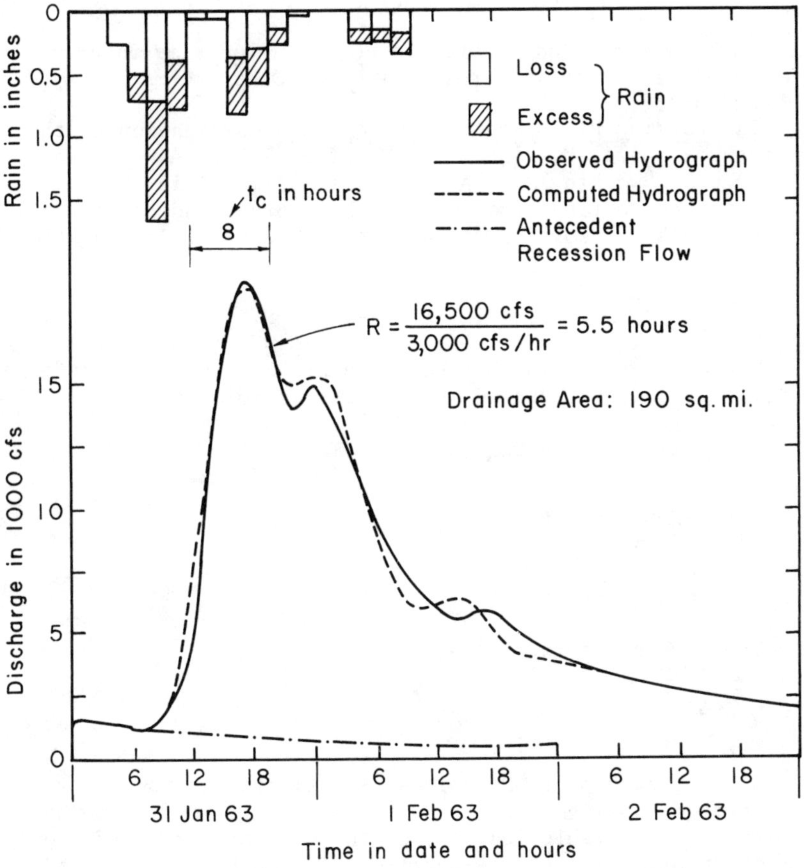

Fig. 8-1. Determination of Clark Coefficients and Flood Reconstitution (used with permission of HEC)

The Modified-Puls routing is a direct solution of a storage-indication transformation $(\frac{S}{\Delta t} + \frac{O}{2})$ of the continuity equation

$$(I-O) = \Delta S/\Delta t \qquad (8\text{-}5)$$

The storage-indication transformation is

$$.5(I_1 + I_2) + (\frac{S}{\Delta t} + \frac{O}{2})_1 - O_1 = (\frac{S}{\Delta t} + \frac{O}{2})_2 \qquad (8\text{-}6)$$

where subscripts 1 and 2 refer to beginning of time interval and end of time interval, respectively. Other terms are as defined earlier.

The Working R & D Routing Method is a combination of the two methods above--Muskingum and Modified-Puls. A weighted discharge or "working discharge", D, is utilized in the place of O in the routing equation and the storage-indication term is replaced by a "working storage," R. The resulting equation is

$$.5(I_1 + I_2) + R_1 - D_1 = R_2 \qquad (8\text{-}7)$$

The other terms in this equation are as previously defined. The relationship between O and D is as follows

$$O_2 = D_2 - \frac{X}{(1-X)} \cdot (I_2 - D_2) \qquad (8\text{-}8)$$

D = working discharge
I = inflow
O = outflow
X = Muskingum coefficient for attenuation

The working storage is

$$R = (1-X) \cdot (\frac{S}{\Delta t} + \frac{O}{2}) \qquad (8\text{-}9)$$

A relationship of D vs. R is utilized in the analysis. The range of Muskingum X is 0 to 0.5. This routing method combines the desirable features of the Muskingum X-coefficient with the nonlinear storage--outflow relationship in Modified Puls.

Runoff hydrographs from any number of subbasins may be routed and combined to produce the outflow hydrograph for the basin.

Hydrograph Balancing: A specified hydrograph can be converted to another hydrograph having a specified volume within each of several durations. The process requires successive approximations whereby the newest adjusted shape becomes the pattern for the next derived hydrograph at the start of each new cycle of computations.

Stream System Computations: This is a rather neat approach for analyzing an entire drainage basin, by subbasins, without the great proliferation of computations usually required to preserve consistent depth-area-duration information, storm centerings and total precipitation-excess volumes.

Up to five index runoff hydrographs are computed for each subbasin and the runoff hydrograph consistent with the drainage area is calculated by interpolation. Moreover, the index hydrographs may be routed and combined to produce families from which hydrographs, which are consistent with respect to the drainage basin depth-area relationship, may be interpolated.

Snowmelt: Either the degree-day or the energy-budget method may be used for snowmelt computations. Up to 10 elevation zones may be specified. Both the rain and rain-free periods are simulated, and the amount of forest cover is considered.

Computational Methods

The unit hydrograph, loss rate, precipitation distribution and flood routing calculations utilize analytical expressions. Consequently, solutions are stable and do not exhibit the oscillation tendencies associated with numerical techniques. Most computations do not require successive approximations.

The hydrograph balancing algorithm does require successive approximations and ten iterations are permitted.

Automatic optimization is available to determine loss rate coefficients, unit hydrograph coefficients or Muskingum routing coefficients. The objective function weights errors toward peak flows by the equation $(Q + \overline{Q})/2 \cdot \overline{Q}$ where $\overline{Q}$ is the average flow. The objective function minimizes the weighted root-mean-square error between computed and observed flows. A steepest ascent method is used. However, as in all automatic optimization schemes the user is cautioned to appraise the calculated results very carefully before accepting the values.

Automatic optimization is a very attractive capability and is used extensively.

Program Design

This program was written for flood hydrograph analysis as its name states. It is a single event model, but permits several separate rainfall events to be processed in a single job. The program is particularly useful in planning and design studies, but its single event logic makes soil moisture accounting and initial loss volumes somewhat difficult to establish in day-to-day real time operation or streamflow forecasting studies.

The program does not restrict the number of subbasins in a watershed. Up to 150 hydrograph ordinates are permitted, and the computation interval may be in minutes or hours. The user shapes the program to fit specific problem situations by input data. There is no program/user interaction during the computation.

Data is coded in standard 8-column fields. Tables of data are used rather than curves or other functional relationships.

Because the program can handle a sizable piece of the watershed in a single subbasin and because detail geometry is not required for flood routing, the amount of input data required for HEC-1 is not excessive.

The program requires less than 40,000 words of storage and four tape or disk units. The seven tests in the test deck require 141 seconds to execute on UNIVAC 1108 equipment and about 10 percent of that on CDC 7600 equipment. Compilation times are 30 seconds on UNIVAC 1108 and 3 seconds on CDC 7600 machines. These times depend, strongly, on the operating system of the host computer.

8.7 WATER SURFACE PROFILES, HEC-2

Purpose

This program calculates the water surface profile for a steady flow discharge. Pertinent hydraulic parameters are determined in the process: depth of flow, water surface width, elevation of the total energy line, friction slope, flow velocity, critical depth and volume of water beneath the computed profile.

A secondary purpose of HEC-2 is to calculate the permissible encroachment width for a specified rise in the water surface elevation. Details are contained in the HEC-2 (Water Surface Profiles) Users Manual and Programmers Manual.

Basic Theory

This is a steady flow solution of the one-dimensional energy equation.

$$\frac{\partial H}{\partial X} + \frac{\partial\left(\frac{\alpha V^2}{\partial g}\right)}{\partial X} = S_\ell \tag{8-10}$$

α = velocity distribution coefficient, horizontal
g = acceleration of gravity
H = water surface elevation
S_ℓ = slope of energy line
V = velocity
X = distance in direction of flow

Friction loss is calculated with the Manning equation

$$V = \frac{C}{n} \cdot R^{2/3} \cdot S_f^{1/2} \tag{8-11}$$

C = 1.486 for English units; 1.0 for metric
n = Manning n-value
R = hydraulic radius
S_f = friction slope
V = velocity of flow

Computations of hydraulic radius utilize wetted perimeter rather than a top width approximation. This insures the program will perform satisfactorily on small as well as large streams.

The steady flow continuity equation links velocity and discharge.

$$Q = V \cdot A \tag{8-12}$$

A = cross-sectional area
Q = discharge
V = velocity

Since cross-sectional geometry is not rectangular it is necessary to calculate critical depth by solving the specific energy equation rather than utilizing one of the analytical expressions. The basic equation is

$$E = y + \frac{\alpha V^2}{2g} \tag{8-13}$$

α = velocity distribution coefficient, horizontal
g = acceleration of gravity
E = specific energy
y = depth of flow
V = velocity of flow

The value of y which produces the minimum value of E is selected as critical depth. Details are presented in Eichert, "Critical Water Surface by Minimum Specific Energy Using the Parabolic Method."

Minor losses due to contraction and expansion are related to the change in velocity head as

$$h_o = C \cdot \left| \frac{\alpha V_u^2}{2g} - \frac{\alpha V_d^2}{2g} \right| \tag{8-14}$$

The coefficient C is an expansion coefficient when the upstream velocity is greater than the downstream velocity. Otherwise, the coefficient is a contraction coefficient.

Energy loss at bridges is calculated with one of several equations depending on the type of flow which exists. If subcritical flow exists, Yarnell's equation is used.

$$\Delta h = 2 \cdot K \cdot (K+10\cdot\omega - 0.6)(m+15m^4)\frac{V_3^2}{2g} \tag{8-15}$$

g = acceleration of gravity
Δh = drop in water surface between the upstream and downstream sides of the bridge
K = pier shape coefficient (See HEC-2)
m = contracted opening coefficient = $\frac{\text{obstructed area}}{\text{total unobstructed area}}$
V_3 = velocity downstream from the bridge
ω = ratio of velocity head to depth of flow downstream from bridge

If critical or supercritical flow exists a momentum equation is used.

$$m_1 - m_{p1} + \frac{Q^2}{gA_1^2}(A_1 - A_{p1}) = m_2 + \frac{Q^2}{gA_2^2} = m_3 - m_{p3} + \frac{Q^2}{gA_3^2} \quad (8\text{-}16)$$

A = cross section area, unobstructed
g = accleration
m = $A \cdot \bar{y}$
Q = water discharge
$\bar{y}$ = vertical distance from water surface to the centroid of A

Subscripts

1,2,3 = upstream section, throat of bridge, downstream section, respectively.
p = pier(s) obstructing flow

The weir equation is included for flow over a roadway

$$Q = CLH^{3/2} \quad (8\text{-}17)$$

C = discharge coefficient
L = weir crest length
H = total energy line of approaching flow minus the weir crest elevation
Q = discharge

If the water spills over the top of the bridge, flow is treated as pressure flow under the bridge and weir flow over it. The equation for pressure flow through an orifice is

$$Q = A\sqrt{\frac{2gH}{K}} \quad (8\text{-}18)$$

A = net area of bridge opening
g = acceleration of gravity
H = total energy line of approaching flow minus the downstream water surface elevation
K = discharge coefficient

Flow at culverts is handled with the same equations as flow at bridges. The culvert roof is input by coordinates, as is the lower half of the culvert section; therefore, any cross section shape can be modeled.

Computational Methods

The energy equation is solved by the numerical integration scheme called the standard step method. Cross sections are considered to define the ends of reaches (i.e., computation points in the numerical integration scheme). The finite difference form of Eq. (8-10) is

$$H_2 - H_1 + \left(\frac{\alpha_2 V_2^2}{2g} - \frac{\alpha_1 V_1^2}{2g}\right) = \overline{S}_\ell \cdot L \tag{8-19}$$

Subscripts 1 and 2 stand for downstream and upstream sections, respectively, and $\overline{S}_\ell$ is the representative slope in the reach between Sections 1 and 2 and includes friction plus minor losses.

Values at Section 1 are boundary values and are specified. Combining Eqs. (8-11) and (8-12), the friction slope can be written in terms of Q as follows

$$S_f = \frac{Q^2}{\left(\frac{C}{n} AR^{2/3}\right)^2} \tag{8-20}$$

let

$$K = \frac{C}{n} \cdot A \cdot R^{2/3} \tag{8-21}$$

then

$$S_f = \left(\frac{Q_1 + Q_2}{K_1 + K_2}\right)^2 \tag{8-22}$$

$$\overline{S}_\ell = \overline{S}_f + \frac{h_o}{L} \tag{8-23}$$

Equations (8-10) and (8-23) provide the necessary relationship to solve for H_2 once Q and H_1 have been specified. Computations move upstream, one reach at a time, by repetitively solving these nonlinear equations.

The solution technique employs successive approximations. Results at each cross section are checked to insure they are on the correct side of critical depth. That is, if the boundary value was subcritical flow, the correct side is above critical depth. If the prescribed boundary value was supercritical, the correct side will be less than or equal to critical. These details are handled automatically by the computer.

The convergence equations estimate the next trial water surface based on the present and the preceding values of trial and computed water surface elevation. Solutions prove to be stable, and they converge to the .01 foot tolerance rapidly in most cases. Occasionally, flows near bank full will display an oscillation between in-banks and out-of-banks elevations rather than converging. After 20 such trials the program leaves the trial and error loop with the water surface elevation equal to top bank elevation. A note is printed.

The solution of the critical depth equation requires successive approximations, also. Since large amounts of computer time can be consumed in calculating critical depth, an approximation is used to first test the need for critical depth. If calculations are near critical, the full solution is made.

Minor losses may be neglected if zero loss coefficients are prescribed. In any case they are simply added to the friction loss to get total loss.

Flow at bridges is a complex hydraulic process which is approximated by one-dimensional equations and discharge coefficients. Usually, more than one type of flow exists (i.e. flow beneath the bridge plus flow over the roadway approaches or flow over the entire bridge structure plus beneath it and over the roadway). Two routines are available for analyzing bridge problems. The program can handle very complex situations, and interested readers are referred to HEC-2, "Water Surface Profiles"; HEC Training Document No. 6, "Computation of Water Surface Profiles through Bridges Using HEC-2"; and HEC Technical Paper #20, "Computer Determination of Flow Through Bridges."

Program Design

This program is adaptable to a wide variety of water surface profile problems with no changes to the Fortran. It is the most widely used of all the HEC programs, and useful for survey, design and operational levels of detail.

Geometric Data: The three-dimensional digital representation of the river channel and floodplains is developed by the computer program from cross sections and the distances between cross sections. Sections are coded by coordinate points (station versus elevation). Natural levees may be identified. Skewed cross sections may be easily corrected to normal with a single coefficient.

Starting Water Surface Elevation: The first water surface elevation may be entered, it may be calculated by slope-area analysis or it may be calculated as critical depth.

Water Discharge: A constant value of discharge may be prescribed for the entire study reach, or it may be changed at any cross section to account for a tributary inflow or a diversion.

Hydraulic Roughness: Manning n-values may be prescribed at the first cross section and held constant throughout the study area or they may be changed at any or all cross sections. Each cross section may be divided into 3 subsections and a different n-value prescribed for each. The overbank subsections may be further partitioned for greater detail in modeling overbank roughness. This option permits up to 20 subintervals for prescribing n-values. Another option permits n-values in the channel subsection to vary in the vertical. An additional option permits changing all n-values by a single ratio. If the value of that ratio is coded as a negative number, all channel n-values will be changed but overbank n-values will not.

Solving for n-Values: The n-values may be calculated directly by entering a known water discharge and at each cross section the observed water surface elevation for that discharge. Since distributed n-values are used in the computations, e.g., overbank and channel n-values are specified separately rather than assigning one "composite" n-value to the entire cross section, it is necessary to specify the ratio between channel and overbank n-values. This is established for the first cross section and is retained until changed at some subsequent section(s). Because of the sensitivity of calculated results to slight errors in observed water surface elevations, a weighted n-value is also calculated at each cross section. It is the average value from all previous cross sections. Once the n-value calculation is complete only minor restructuring of the data deck is required before one can proceed with the water surface profile study. If the trial and error method for determining n-values is preferred, observed high water marks may still be entered at sections where they are available. They do not enter into the computations but are available for easy reference when evaluating the selected n-values by comparing calculated with observed water surfaces.

Reach Lengths: In discussing the geometric model, the distance between cross sections was not mentioned in detail. This distance is commonly referred to as "reach length" whether it is along a straight portion of the river or not. Three values may be entered for each reach: left overbank, main channel, and right overbank.

Data Structure

Water surface profile computations require large amounts of data. The structure of HEC-2 was developed to accomodate as much or as little data as desired. In addition, the cards are identified with alphanumeric characters in columns 1 and 2. The computer reads these letters to determine what type of data is on the card. This eliminates the need for control variables and reduces the number of job aborts caused by changing a data table and neglecting to change a control or counter variable on another card. Standard 8-column field formats are used.

Computer Requirements

This program was written for CDC 6600/7600 computer equipment having five or more tape drives and requires 32,000 words of memory. Versions for UNIVAC 1108, IBM 360 50+ and Honeywell 635 equipment may be easily converted from the library version.

8.8 RESERVOIR SYSTEM ANALYSIS, HEC-3

Purpose

The purpose of this program is to perform a conservation routing of flows through a system of multipurpose reservoirs. Inflows, water demands and reservoir operating rules are prescribed. System performance is measured in terms of storage requirements and in terms of dollars with shortages declared during any period when a demand is not met. Details are contained in HEC-3, "Reservoir System Analysis."

The approach is the simulation of system requirements during long periods of time and the computation interval is usually monthly. Automatic optimization is not available for adjusting system configuration, size and operating rules; these all have to be input values.

In CDC 6600 class hardware up to 30 reservoirs may be analyzed for serving 40 control points. Up to 8 levels of storage may be specified in each reservoir. Inflows may be prescribed at each control point and up to 25 diversions and 20 power plants may be included.

This is considered to be a planning level program with unique capabilities for including system power demands as an energy demand in addition to specific project power. It is an extremely powerful program and has been used on numerous studies.

Basic Theory

This is a simulation model; consequently, the heart of the model is the logic which controls the automatic selection of operational criteria by the computer program during the simulation of the system operation.

Power generation is computed with the following equation:

$$P = C \cdot E \cdot H \cdot Q \tag{8-24}$$

where

C = 0.08464 for English units, 9.817 for metric units
P = Power in kilowatts
E = Power plant efficienty ratio
H = Effective head on turbine in feet or meters
Q = Flow through turbines in fps or m^3/s

The energy head, H, is automatically corrected for reservoir level and changes in tailwater level.

A "control point" is assigned to each potential project location. Reservoirs, power plants or diversion points are assigned the identification number of their project control point. Those projects which are to be operated specifically for a control point are prescribed as well as the diversions from the unregulated area above each control point. More than one river basin may be included in a single analysis, which is essential when designing a study to consider system power demands.

Minimum flow requirements and maximum flow limits are assigned to each project. Data describing these and other physical characteristics of the system may be entered in any sequence of control points, but computations are performed by sweeping the system in the downstream direction. Since end-of-period erservoir levels are required before the water demand for power and evaporation can be computed, at least two iterations are made for each time period. When system power demands are specified, three iterations are required for each time period.

To start the system routing, preliminary flow requirements are estimated for each control point in accordance with power and other demands. The maximum flow constraint is satisfied. Diversions are made with releases from the respective projects. Minimum flow constraints are satisfied. A system search is then made to establish releases which, when excess volume is combined with uncontrolled runoff, will satisfy the preliminary flow requirements. A storage index level is used to keep projects in balance.

Economic benefits are computed from tables of dollars versus flow rate for streamflow, versus reservoir storage for flood control and versus kilowatt-hours for power generation. Values are scaled by 1000 to prevent numbers from overflowing field size. Benefits are summarized by location and type utilizing five separate tables: (1) project benefits at control points, (2) project benefits allocated to projects, (3) project plus preproject benefits, (4) total potential benefits and (5) remaining potential benefits not obtained in the system operation. Monthly allocated benefits are printed for each function and location to permit a detailed examination.

Flood routing is not included in this program; therefore, the time translation necessary for large systems is accomplished by lagging inflows and labeling each with a different time base. System computations are made as though all inflows were simultaneous; therefore, flows at upstream locations should be those occurring earlier in time than inflows at downstream locations. Time of travel must be estimated from other sources.

Computational Methods

Except for hydropower and benefit calculations all equations are mass balance equations and are solved analytically.

Only one calculation, the firm yield calculation, is subject to a convergence criterion. In making this calculation a location is prescribed and the program begins an iteration procedure using shortage or surplus ratios to adjust firm yield after each successive iteration. When the magnitude of the adjustment becomes less than 1 percent, the iteration process stops and the calculated value is accepted as the maximum firm yield. Consequently, stability, oscillation and convergence are not problem areas with this program.

Program Design

This program was developed to study potential systems of reservoirs during the planning phase of project development. Actually, it has been used for both planning and operational studies but the operational studies involved analyzing existing systems with historical flows and water demands as opposed to real time reservoir regulation.

Both small and large systems have been analyzed. Special logic permits hydrologic output to be rearranged or summarized in a variety of ways to highlight the following: unregulated flows, river flows, diversion flows, diversion shortages, shortage in desired flows, shortages in meeting the minimum flow demand, reservoir storage (end of period) changes in reservoir storage, reservoir elevation, reservoir data in general (storage, elevation, inflow, outflow, evaporation and generated power). The summary may be by control point for all years, by year for all control points, or both.

This is a batch program and requires large amounts of input and output. Standard input utilizes 8-column fields, 10 per card, with the card type designated by alphanumeric characters. Inflows are an exception to this rule as they are coded in 6-column fields.

Physical characteristics of projects are described with tables of data. These include relationships of reservoir storage, surface area and outlet capacity to elevation. To minimize the number of input data cards those relationships which may or may not be constants have optional cards for coding tables and standard input by a single constant value. Power plant efficiency is an example of this latter type of data structure.

Special coefficients are not required; however, evaporation is a key physical quantity and one which is difficult to measure. Special care is required to insure that evaporation rates are "representative."

8.9 MONTHLY STREAMFLOW SIMULATION, HEC-4

Purpose

The purpose of this program is to generate a sequence of monthly flow volumes of any desired length. Statistical characteristics used in the generation are calculated from observed monthly streamflows or from regionalized statistics. Missing data are calculated based on concurrent flows at other stations. Maximum and minimum monthly flows are flagged in the observed, reconstituted and generated flow records.

Basic Theory

The mean, standard deviation and skew coefficients are calculated for the log transform of monthly flows using the following equations.

$$X_{i,m} = \log(Q_{i,m} + q_i) \tag{8-25}$$

$$\overline{X}_i = \sum_{m=1}^{N} X_{i,m}/N \tag{8-26}$$

$$S_i = \sqrt{\sum_{m=1}^{N} (X_{i,m} - \overline{X}_i)^2/(N - 1)} \tag{8-27}$$

$$g_i = N \sum_{m=1}^{N} (X_{i,m} - \overline{X}_i)^3/(N - 1)\cdot(N - 2)\cdot S_i^3 \tag{8-28}$$

g = unbiased estimate of population skew coefficient
i = month number
m = year number
N = total years of record
Q = observed monthly streamflow
q = small increment of flow to insure a positive value for the log transform
S = unbiased estimate of population standard deviation
X = logarithm of incremental streamflow
$\overline{X}$ = mean logarithm of monthly, incremental streamflow

Filling in Missing Record: Stations with incomplete records are compared to each long-record station to find which will contribute the most toward the reliability of the statistics of the short-record stations. Short-record mean and standard deviations are then adjusted.

Each monthly flow is converted to a normalized standard variate using the following approximation of the Pearson Type III distribution:

$$t_{i,m} = (X_{i,m} - \overline{X}_i)/S_i \qquad (8\text{-}29)$$

$$K_{i,m} = 6/g_i[((g_i t_{i,m}/2) + 1)^{1/3} - 1] + g_i/6 \qquad (8\text{-}30)$$

t = Pearson Type III standard deviate
K = Normal standard deviate

Simple correlation coefficients between all pairs of stations for each current and preceding calendar month are computed for the normalized flows by using the following equation

$$R_i = \{1-[1-(\sum_{m=1}^{N} x_{i,m}\cdot x_{i-1,m})^2/(\sum_{m=1}^{N}(x_{i,m}^2)\cdot\sum_{m=1}(x_{i-1,m}^2))]\cdot\frac{(N-1)}{(N-2)}\}^{1/2} \qquad (8\text{-}31)$$

where $x = X - \overline{X}$

If there are insufficient simultaneous observations of any pair of variables to compute a required correlation coefficient that coefficient must be estimated. Utilizing pairs of values in the current and preceding month designated by subscripts i, j and k, the missing value is estimated for the month in question by averaging the two limits given by the following:

$$R_{ij} = R_{ki}R_{kj} \pm \sqrt{(1 - R_{ki}^2)\cdot(1 - R_{kj}^2)} \qquad (8\text{-}32)$$

Missing monthly flows are estimated by stepping through the record, month by month, and analyzing all stations before proceeding to the next month. A missing value is reconstituted by solving the regression equation composed of coefficients from the completed correlation matrix for the most current month from which data is available by the Crout method. The missing value is modified by a random component.

Since all correlation coefficients must agree with the data which is substituted into the equations and since they must be mutually consistent, the program logic corrects the matrix before beginning the generation step. The least significant independent variable is dropped until the determination coefficient becomes less than 1. To protect against inconsistency, all affected correlation coefficients are recomputed after each element of the missing record is calculated.

Monthly Streamflow Generation: Hypothetical monthly streamflow volumes are generated computing a regression equation, by the Crout Method, for each station and month and then computing streamflows for each station for one month at a time using the following equation

$$K'_{i,j} = \beta_i \cdot K'_{i,1} + \beta_2 \cdot K'_{i,2} + \ldots \beta_{j-1} \cdot K'_{i,j-1} + \beta_j \cdot K'_{i-1,j} + \ldots$$

$$+ \beta_n \cdot K'_{i-1,n} + Z_{i,j} \sqrt{1-R^2_{i,j}} \qquad (8\text{-}33)$$

K' = logarithm of generated monthly flow expressed as standard deviate
β = coefficient from correlation matrix
i = month number
j = station number
n = number of interrelated stations
R = multiple correlation coefficient
Z = random number from normal, standard population

A common problem in regionalized streamflow generation studies is unequal length of records. This problem is handled automatically by the program logic.

Computational Methods

The regression equations are linear and coefficients form a matrix which is symmetrical about the diagonal. The solution technique developed by P. D. Crout (1941) solves such a matrix very efficiently by a double sweep method. Precision is important in this mathematical technique and 60-bit word lengths should be utilized. In some computers this means double precision.

Because of the linear nature of the equations the solution technique is stable. Program logic handles searches and successive approximations in the coefficient consistency test without convergence problems.

Program Design

This program was designed for and has been used on studies to fill in missing records and to generate long sequences of monthly streamflows. It operates in batch mode. Flags are printed to draw attention when automatic correction procedures were selected by the program.

Data Structure

Data is read from cards and any number of jobs may be "stacked" in a single urn. Each job is treated separately. Data is coded in 6 column fields so an entire year can be punched on 1 card.

No coefficients have to be supplied by the user, but they may be if they are available. Given the observed monthly values the program calculates all required coefficients.

Computer Requirements

This program is designed for CDC 6000-class equipment and requires 64000 words of memory. In addition a card reader, line printer, punch and two "scratch" tapes are required.

Any number of stations may be utilized but a maximum of 10 are permitted on any one run.

8.10 SIMULATION OF FLOOD CONTROL & CONSERVATION SYSTEMS, HEC-5C

Purpose

The purpose of this program is to determine the releases from each reservoir project in a system of reservoirs. Both conservation and flood control purposes may be included. The system will be operated automatically according to prescribed criteria. Inflows may be deterministic, as in planning studies, or they may be forecasted with an assigned contingency factor for error. Details are contained in the HEC-5C, "Simulation of Flood Control and Conservation Systems," Users Manual.

Basic Theory

Since this is a simulation model the heart of the model is the logic which controls the automatic selection of operational criteria by the computer program during the simulation of the system operation. The following details were copied from the HEC-5C Users Manual with permission of HEC.

> Reservoir Operational Criteria: "Reservoirs are operated to satisfy constraints at individual reservoirs, to maintain specified flows at downstream control points, and to keep the system in balance. Constraints at individual reservoirs are as follows:
>
> "When the level of a reservoir is between the top of conservation pool and the top of flood pool, releases are made to attempt to draw the reservoir to the top of conservation pool without exceeding the designated channel capacity at the reservoir or at downstream control points for which the reservoir is being operated.
>
> "Releases are made equal to or greater than the minimum desired flows when the reservoir storage is greater than the top of buffer storage, and equal to the required flow if between level one and the top of buffer pool. No releases are made when the reservoir is below level one (top of inactive pool). Releases calculated for hydropower requirements will override minimum flows if they are greater than the controlling desired or required flows.
>
> "Releases are made equal to or less than the designated channel capacity at the reservoir until the top of flood pool is exceeded, then all excess flood water is dumped

if sufficient outlet capacity is available. If insufficient capacity exists, a surcharge routing is made. Input options permit channel capacity releases (or greater) to be made prior to the time that the reservoir level reaches the top of the flood pool if forecasted inflows are excessive.

"The reservoir release is never greater (or less) than the previous period release plus (or minus) a specified percentage of the channel capacity at the dam site, unless the reservoir is in surcharge operation.

"Operational criteria for specified downstream control points are as follows:

"Releases are not made (as long as flood storage remains) which would contribute to flooding at one or more specified downstream locations during a predetermined number of future periods. The number of future periods considered is the lesser of the number of reservoir release routing coefficients or the number of local flow forecast periods.

"Releases are made, where possible, to exactly maintain downstream flows at channel capacity (for flood operation) or for minimum desired or required flows (for conservation operation). In making a release determination, local (intervening area) flows can be multiplied by a contingency allowance (greater than 1 for flood control and less than 1 for conservation) to account for uncertainty in forecasting these flows.

"Operational criteria for keeping a reservoir system in balance are as follows:

"Where two or more reservoirs are in parallel operation for a common control point, the reservoir that is at highest index level (see next paragraph), assuming no releases for the current time period, will be operated first to try to increase the flows in the downstream channel to the target flow. Then the remaining reservoirs will be operated in a priority established by index levels to attempt to fill any remaining space in the downstream channel without causing flooding during any of a specified number of future periods.

"If one of two parallel reservoirs has one or more reservoirs upstream whose storage should be considered in determining the priority of releases from the two parallel reservoirs, then an equivalent index level is determined for the tandem reservoirs based on the combined storage in the tandem reservoirs.

"If two reservoirs are in tandem, the upstream reservoir can be operated for control points between the two reservoirs. In addition, when the upstream reservoir is being operated for the downstream reservoir, an attempt is made to bring the upper reservoir to the same index level as the lower reservoir based on index levels at the end of the previous time period.

"Reservoir operational priority for different purposes is shown in Table 8-1.

Index Levels: "As indicated in previous paragraphs, index levels assigned by the program user to each reservoir are used to determine priority of releases among reservoirs. The program operates to meet specified constraints throughout the system and then to keep all reservoirs in the system in balance if possible. A system is 'in balance' when all reservoirs are at the same index level. In establishing the reservoir index level at a given point in time during a system operation, the program interpolates linearly on an input table of index level versus storage. In balancing levels among reservoirs, priority for releases is governed by index levels such that reservoirs at the highest levels at the end of the current time period (assuming no releases) are given first priority for the current time period...index levels can be used to specify operating rules for a system."

Flood Routing Methods: Both short interval releases (flood routing theory) and long interval releases (volume conservation) may be utilized in a single simulation. Establishing releases during floods requires successive approximations whereby a linear routing equation of the following form is used to make the first approximation:

$$Q_n = C_i R_n + C_2 \cdot R_{n-1} + C_3 \cdot R_{n-3} + \dots \qquad (8\text{-}34)$$

C = routing coefficient calculated from the prescribed computation interval, travel time and attenuation coefficient
Q = routed reservoir release at downstream control point
R = reservoir release

Subsequent approximations are made with the particular routing method adopted for the reach in question. Muskingum Modified-Puls, Working R & D, Straddle-Stagger, and Tatum are available. These were discussed in Section 8.6 in connection with HEC-1.

Flow Diversions: Several options are available for establishing the quantity of diversions:

$$D = f(Q)$$

or

$$D = f(S)$$

or

Note: Used by permission of HEC.

Table 8-1. Reservoir Operational Priority

Condition:	Normal Priority:	Optional Priority:
During flooding at downstream location	No release for power requirements	Release for primary power
If primary power releases can be made based on downstream flooding	Release down to top of buffer pool	Release down to top of inactive pool (level 1)
During flooding at downstream location	No release for minimum flow	Release minimum desired flow
If minimum desired flows can be made based on downstream flooding	Release from top of conservation to top of buffer pool	Same as Normal
If minimum required flows can be made based on downstream flooding	Release from top of conservation to top of inactive pool	Same as Normal
Diversions from reservoirs (except when diversion is a function of storage)	Divert down to top of buffer pool	Divert down to top of inactive pool (level 1)

Note: Used by permission of HEC.

$$D = \text{Const}$$

or

$$D = \text{specified by time period}$$

Hydroelectric Power Generation: Hydroelectric power production is based on specified, installed nameplate capacity plus 15 percent overload and is calculated for individual projects--not power systems. Plant efficiency is assumed to be 86 percent. Discharge requirement considers both tailwater elevation (as for run-of-river plants) and headwater elevation (as for high storage plant fluctuations) in computing the available head. Both the overload factor and the plant efficiency factor may be changed with input data if the default values are not correct for the control point being evaluated.

Evaporation Rates: As in HEC-3, evaporation rates are input. Monthly evaporation for the entire basin may be utilized or if more detailed information is available the rates for specific reservoirs may be utilized.

Nonstructural Alternative Evaluation: By assigning proposed channel discharge capacities (or non-damaging discharge values) the resulting impact of nonstructural flood control alternatives can be evaluated. The results are presented in terms of either average annual flood damages or single event damages.

Flood Damage Calculations: Expected annual flood damages are calculated at all control points of interest. Damage is assumed to be a single valued function of peak discharge and each event is assumed to have a fixed exceedance frequency which is a unique function of the peak discharge. If a base condition is available for an existing or previously calculated project the expected annual damage value can be input and all subsequent alternative operations will be compared to that base condition.

Flood damages can be computed for each flood or the expected annual damages can be computed for these conditions: natural, regulated and uncontrolled local runoff.

Computational Methods

The basic solution technique in HEC-5C is analytical rather than numerical. Consequently stability and oscillation are not problems. Utilizing linear functions as the first estimate of reservoir release rates helps insure convergence of the successive approximations which follow. Utilizing the hydrologic methods of flood routing permits any length of computation interval to be used.

An automatic optimization technique is available to routing criteria from observed hydrographs. Both Muskingum and Modified-Puls criteria may be optimized.

Program Design

This program was written for planning, design and operational studies and it has been used for all three. In general these studies all require the same basic capability; i.e. to take inflows, to operate a system of reservoirs to achieve specified objectives and to display the results of the computations. In particular, however, accomplishing all three types of studies with a single computer program requires a great deal of flexibility in designing the input data. A program written for planning level studies will usually not accept design level information and a program written for design studies will usually require such detailed information that it is not useful for planning level studies. On the other hand, a program written for operational studies requires minute detail for reconstituting historical events and a "crystal ball" for forecasting the future.

These conflicting requirements were resolved by carving out that portion of the overall problem which is common to all studies and which may be neatly packaged and interfaced with the different, specific requirements of the different levels of study. Consequently, HEC-5C is the system operation package. It looks to other programs for providing input data on reservoir inflows and for handling data management and display requirements.

Not only the source, amount and dependability of data but also the very nature of problems is different between planning/ design type studies and real time operational studies. In the former type of studies, the size and location of projects is the unknown. This program is designed so projects may be included in the system or deleted from it without restructuring the input data deck. Both cost and damage calculations can be made.

Flood control and conservation operations are two competitive requirements in terms of storage usage and also in terms of programming. Long periods of time are required for analyzing conservation requirements whereas short computation intervals are required for flood control analysis. Consequently, a multiflood option is available to operate a system for a continuous period of record using a mixture of computation time intervals. Any number of hydrograph values may be coded.

Systems are described with tables of data using either English or metric units. Calculations are made in batch mode but preprocessing of input data and postprocessing of calculated results can be done with conversational time sharing programs. Output from this program can be selected for display on any width of paper and at any speed. Files may be plotted using CRT graphics (cathode ray tube) or using the printer plot option in the program.

Data Structure

Input/output files are structured so several alternative futures can be quickly analyzed by a person seated at a CRT graphical input/output terminal.

The cards are identified by a two character alphanumeric code and the type and amount of data is determined from this code by the computer. Control character, end-of-data flags, etc. are not required. This simplifies data deck preparation and modification. The standard 8-column field format is utilized.

There are no special coefficients. In addition to HEC-3 type coefficients, however, this program requires flood routing coefficients (essentially storage versus outflow relationships) and constraints on rate of change of releases from projects.

Computer Requirements

This is the largest program in the HEC library. It requires 65,000 words of central memory and 500,000 words of extended core on CDC 7600 hardware. The UNIVAC 1108 version requires 65,000 words of memory and can accomodate 15 control points, 10 reservoirs, 11 diversions, and 9 power plants.

8.11 SCOUR AND DEPOSITION IN RIVERS AND RESERVOIRS, HEC-6

Purpose

This program calculates the water surface profile and the streambed profile. The water velocity, water depth, energy slope, sediment load, gradation of the sediment load and gradation of the bed surface are computed also. Because a large number of water surface profiles are calculated, a completely different algorithm to HEC-2 is utilized.

Basic Theory

The computational procedure is presented in Exhibit 3 of HEC-6, "Scour and Deposition in Rivers and Reservoirs," and summarized below.

Water surface profile and sediment movement calculations are fully coupled using an explicit computation scheme. First the conservation of energy equation is solved to determine the water surface profile and pertinent hydraulic parameters (velocity, depth, width and slope) at each cross section along the study reach. The basic hydraulic equation is identical to Eq. (8-10) in Section 8.7 on the HEC-2 program.

$$\frac{\partial H}{\partial X} + \frac{\partial (\alpha \frac{V^2}{2g})}{\partial X} = S_\ell \qquad (8\text{-}36)$$

g = acceleration of gravity
H = water surface elevation
S_ℓ = friction and other losses
V = average, flow velocity
X = direction of flow
α = coefficient for the horizontal distribution of velocity

In addition, the continuity of sediment material is expressed by

$$\frac{\partial G}{\partial X} + B \cdot \frac{\partial y_s}{\partial t} = q_s \qquad (8\text{-}37)$$

B = width of movable bed, ft
G = rate of sediment movement in ft^3/day
q_s = lateral inflow of sediment ft^3/ft/day

t = time in days
x = distance in direction of flow
y_s = change in bed surface elevation

The third equation relates the rate of sediment movement to hydraulic parameters as follows:

$$G = f\ (V,D,W,S,T,f,\frac{df}{dt},d) \tag{8-38}$$

V = velocity
D = depth
W = width
S = slope
T = water temperature
f = fraction of bed material in a size class
t = time
d = particle size

The numerical technique used to solve Eq. (8-36) is commonly called the standard step method. Equation (8-37) has both time and space domains. An explicit form of a 6-point finite difference scheme is utilized. Several equations of the form of Eq. (8-38) are available. These transport capacity equations are empirical and "G" is determined analytically.

Equation (8-37) is the only explicit equation, but it controls the entire analysis by imposing stability constraints. Several different computation schemes were tested and the 6-point scheme proved the most stable. No stability criteria have been developed for this scheme. The rule of thumb is to observe the amount of bed change during a single computation interval and reduce the computation time until that bed change is tolerable.

Oscillation in the bed elevation is the key factor in selecting a suitable computation interval. The computation time interval may be made short enough to eliminate oscillation. On the other hand, computer time increases as the computation interval decreases. The proper value to use is determined by successive approximations.

Several supporting equations are required in transforming the field data for the computer analysis.

Friction and Other Losses: The Manning Equation is used to evaluate friction loss. Average geometric properties are combined, using an average end area approach, into an average conveyance for the reach using the following equation:

$$K_T = \sum K_i = \sum_{I=1}^{NSS} \frac{1.486}{n_i} \frac{(A_u + A_d)_i}{2} \cdot \left(\frac{(R_u + R_d)}{2}\right)^{2/3} / \sqrt{L_i} \tag{8-39}$$

A = cross-sectional area

K_i = conveyance of the ith strip

K_T = total conveyance of the section

L = reach length of the ith strip

n_i = Manning n-value of the ith strip

R_i = Hydraulic radius of the ith strip

d = downstream end of the reach

u = upstream end of the reach

I = subsection (left overbank or channel or right overbank, etc.)

Contraction and expansion losses are calculated as "other" losses by multiplying a coefficient times the change in velocity head. All geometric properties are calculated from cross section coordinates.

Critical Depth: The specific energy curve is traced to its minimum value to locate critical depth. As in HEC-2, critical depth is always monitored to insure computations remain on the correct side.

Program Design

This program is adaptable to both stream and reservoir problems with no changes to Fortran Code. The preparation of input data is straightforward and documented. In most cases sediment data is not as plentiful as hydraulic or hydrology data and this program is designed to use as much or as little as is available.

Units: English units are used entirely.

Subcritical Flow: Only subcritical flow may be analyzed; however zones of critical or supercritical flow may occur within the study reach. The program treats these zones as "critical" and utilizes subcritical-flow, sediment transport functions to move sediment through them.

Geometric Model: The primary data set is the HEC-2 data set. In addition, the movable bed portion of each cross section must be specified. There is an alternate geometry which permits subdividing the channel or overbanks into several subsections if the standard 3 are inadequate.

Spacing of cross sections is somewhat more critical for this program than it is with HEC-2. Long reach lengths are desirable because reach length and computation interval are related. The computation interval, and consequently the number of discharges which have to be analyzed during a study, is controlled by the shortest reach length in the data set. Pressure flow is not accommodated.

Hydraulic Roughness: Manning n-values are entered at the first cross section and may be changed at any cross section. The standard HEC-2 format permits assigning 3 subsections to each cross section: left overbank, channel and right overbank. The n-value in each of these may be a constant or it may vary in the vertical as a function of elevation or discharge.

Automatic calibration is not available. The reconstitution of high water marks is recommended for determining n-values, but the trial and error process requires manual intervention to code and analyze results from each successive trial.

Reach Length: As in HEC-2 "reach length" refers to the distance between cross sections. The standard HEC-2 data structure permits three values at each cross section. If more detail is needed an alternate data structure for HEC-6 permits up to seven reach lengths for each cross section.

Starting Elevation: The water surface elevation at the downstream end may be interpolated from a rating curve or it may be entered from a rule curve. If neither of these options is suitable, a critical depth start may be used. Always prescribe a "hard bottom channel" at the starting cross section when utilizing a critical depth start.

Water Discharge: The river flow may be prescribed as a constant value for the entire study area or it may be changed up to 20 times to account for tributary inflow or diversions. The flow rate may change from virtually zero to 9999999 cfs. The maximum value is limited by the size of fields in printout statements.

A sequence of water discharges may be utilized so the response of the streambed profile through the study area may be simulated over a long period of time. During a 50 year period 2,000 to 4,000 water discharges are frequently analyzed.

Flow Through Bridges: No special provisions are available to calculate head loss at bridges. The contracted opening may be modeled such that scour and deposition are simulated during the passing of a flood event but calculated results must be interpreted with the aid of a great deal of engineering judgment and sensitivity analysis.

Sediment Material: Four different sediment properties are required: (1) the total concentration of suspended and bed loads, (2) grain size distribution for the total concentration, (3) grain size distribution for sediment in the streambed and (4) the unit weight of deposits. A wide range of sediment material may be accomodated in the transport calculations

(.004 mm to 64 mm). The maximum size of material may exceed 64 mm so far as armoring is concerned. However, the basic assumption that the supply of large material just equals the amount transported away during each flood permits the elimination of these sizes from the transport function.

Water Temperature: The temperature of water is an important parameter in sediment transport and, consequently, may be prescribed with each water discharge in the hydrograph. Flexibility of input permits a value to be entered as needed to change from a previous entry.

Coefficients Required: The usefulness of a calculation technique depends a great deal upon the coefficients which must be supplied. As in HEC-2, Manning n-values, contraction coefficients and expansion coefficients must be provided to accomplish the water surface profile calculations. Several other coefficients are required for sediment calculations as follows:

1. The specific gravity and shape of sediment particles must be specified.
2. The bed shear stress at which silt or clay particles begin to move is a required coefficient.
3. The unit weight of silt, clay and sand deposits is somewhat like a coefficient because of the difficulty in measuring it. Also the density changes with time.

All of the sediment-related coefficients have default values because sediment data seems to be much more scarce than hydraulic data. There are fewer sources for generalized coefficients. All of the default values should be replaced by field data where possible and the input data is structured for such a process.

Data Structure

Data is coded in standard 8-column fields. Default values are provided for sediment data wherever possible.

Computer Requirements

This program is written for batch processing and requires about 40,000 words of core. Double precision is programmed in some calculations for execution on IBM and UNIVAC hardware (i.e., word lengths = 32-36 bits). All input is by cards and output is on the line printer; however, three scratch tapes are utilized in the computations. Execution time ranges from a few seconds for a single pass through the computations to a thousand or so seconds when simulating a 50 year hydrograph of flows in a 150 cross section model.

8.12 REFERENCES

Bowers, C. E., A. F. Pabst, and S. P. Larson, 1972. Computer programs in hydrology. Bulletin 44, Water Resources Research Center, University of Minnesota, Minneapolis, Minnesota.

Clark, C. O., 1945. Storage and the unit hydrograph. Transactions of the American Society of Civil Engineers, Vol. 110, pp. 1419-1488.

Eichert, B. S. Critical water surface by minimum specific energy using the parabolic method. Hydrologic Engineering Center, U.S. Army Corps of Engineers, Davis, California.

Eichert, B. S. and J. Peters, 1970. Computer determination of flow through bridges. HEC Technical Paper #20, Journal of the Hydraulic Division, ASCE, Vol. 96, No. HY7, July.

HEC-1. Flood Hydrograph Package, Generalized Computer Program, Users Manual, Hydrologic Engineering Center, U.S. Army Corps of Engineers, Davis, California, Jan. 1973.

HEC-1. Programmers Manual.

HEC-2. Water Surface Profiles. Users Manual, Oct. 1973.

HEC-2. Programmers Manual.

HEC-3. Reservoir System Analysis, Users Manual, Feb. 1971.

HEC-4. Monthly Streamflow Simulation, Users Manual, Feb. 1971.

HEC-5C. Simulation of Flood Control and Conservation Systems, Users Manual, March 1976.

HEC-6. Scour and Deposition in Rivers and Reservoirs, Users Manual, March 1977.

HEC Training Document No. 6. 1974. "Computation of Water Surface Profiles Through Bridges Using HEC-2," Draft, June 1974.

U.S. Army Corps of Engineers. EM 1110-2-1405. Flood Hydrograph Analysis, 31 August 1959.

U.S. Army Corps of Engineers, EM 1110-2-1411. Standard Project Flood Determinations. 26 March 1952.

8.13 APPENDIX

While the Government is not responsible for the results obtained when using these programs, assistance in resolving any malfunctioning of the programs is furnished by the Center to the extent that time and funds are available. The following is a list of programs that are currently available.

HEC-1 Flood Hydrograph Package
Basin Rainfall and Snowmelt Computation
Unit Graph and Hydrograph Computation
Unit Graph and Loss Rate Optimization
Streamflow Routing Optimization
Hydrograph Combining and Routing
Balanced Hydrograph

HEC-2 Water Surface Profiles
Gradually Varied Unsteady Flow Profiles
Spillway Rating and Flood Routing
Spillway Rating - Partial Tainter Gate Openings
Spillway Gate Regulation Curve

HEC-3 Reservoir System Analysis for Conservation
HEC-5C Simulation of Flood Control and Conservation Systems
Reservoir Yield
Reservoir Area-Capacity Tables by Conic Method

HEC-4 Monthly Streamflow Simulation
Daily Streamflow Simulation
Regional Frequency Computation
Average Annual Damage Computation

Storage, Treatment, Overflow, Runoff Model (STORM)
Water Quality for River-Reservoir Systems (WQRRS)
Reservoir Temperature Stratification

HEC-6 Scour and Deposition in Rivers and Reservoirs
Suspended Sediment Yield
Reservoir Delta Sedimentation
Deposit of Suspended Sediment

Finite Element Solution of Steady State Potential Flow Problems

---

Note: Used by permission of HEC (FY 1976 Annual Report)

Chapter 9

WATER AND SEDIMENT ROUTING FROM WATERSHEDS

by

Ruh-Ming Li, Associate Professor of Civil Engineering,
Colorado State University, Fort Collins, Colorado

Chapter 9

WATER AND SEDIMENT ROUTING FROM WATERSHEDS

9.1 INTRODUCTION

As population increases, development of land and water resources continues to grow. Coinciding with the increased activities in the natural resources acquisition, there is an ever growing concern for preserving the quality of our natural environment.

The management of watersheds and river basins for the optimum benefit of the people in general requires a complete knowledge of the interrelations between ecology and environment. The watershed response to developments, either natural or man-induced, must be anticipated correctly if progress is to be made towards wise use of our natural resources. The increasing interests in predicting watershed response has accelerated the progress in the mathematical modeling of water and sediment yields. Many water yield and routing models are available, however, very few models have been developed for simulating sediment routing and yield from watersheds. The watershed sediment yield is a direct indication of the land erosion rates and is also the primary source of fine sediment being carried in most rivers. With the growing concern for protecting the natural environment, the watershed sediment yield study is receiving increased attention. Degradation, aggradation and movement of sediment and other pollutants in watersheds are closely related to water movement. No sediment yield can be predicted without the knowledge of water routing and yield. This chapter describes several methodologies for water and sediment routing from watersheds being developed at the Engineering Research Center, Colorado State University, under the general direction of Daryl B. Simons and Ruh-Ming Li. Chapter 7 reported various sources and impacts of sediment in the upland watershed. This chapter predominantly deals with surface erosion (sheet-sill erosion). This type of soil erosion is perhaps the most important and fundamental type of soil loss in forest, agricultural, and urban watersheds. The management activities such as mechanical site preparation for timber planting and/or harvesting, fire, road and trail construction, grazing, agricultural cropping, urban constructions, facility and recreational site development all contribute directly or indirectly to surface erosion to a large degree.

The physical processes governing watershed response are very complicated. Many past studies have utilized a statistical interpretation of observed response data. The unit hydrograph method for water routing, the universal soil loss equation for soil erosion, and the hydraulic geometry equations for stream morphology are examples of these types of studies. It is often difficult to predict the response of a watershed to various watershed developments or treatments using these methods, because they are based on the assumption of homogeneity in time and space. However, numerical modeling using the governing physical process is a viable way to estimate the time-dependent response of watersheds to precipitation with varying vegetative covers and land use. This paper describes several methodologies for water and sediment routing from watersheds.

9.2 GEOMETRIC REPRESENTATION OF A WATERSHED

Watershed Segmentation

The first problem encountered in numerical modeling of watershed response is to determine representative response units for mathematical computations. Simons and Li (1975) have approached this problem by developing a watershed segmentation program based on a grid system. The grid size is chosen so that the watershed boundary and channel segments can be approximated by grid lines (Fig. 9-1). The overland flow units are the grid units inside the watershed boundary and the channel units are segments of channel between grid intersection points.

From the contour lines, the elevations of the land surface at the grid points are determined (Fig. 9-2). These elevations, along with the locations and bed elevation of the stream channel, are input to the developed computer program. In addition to elevation data, vegetation and soil code numbers can be input for each grid point (Fig. 9-2). The computer program then performs the following functions:

1. The slope and the slope azimuth of each overland flow unit are computed.

2. It is assumed that the water flows in the opposite direction of the slope gradient to the next overland flow unit or to the adjacent channel. Thus, water cascades from overland unit to overland unit and then into the channel system. The program identifies the cascade sequence (arrows in Fig. 9-1).

3. The computational sequence for the flow is established by the program. The method employed is simply to follow the logics of gravity flow and flow continuity.

4. If data on the vegetation type, soil type, canopy cover density, and ground cover density are available, the variations of these factors inside a watershed can be established in the program. This is executed by decoding the vegetation and soil codes and assigning previously input parameters to each type code. These parameters may include oil porosity, soil depth, and selected vegetation measures.

In order to save computer storage capacity and processing time in the water and sediment routing computations, an additional computer program to combine small grid units into larger response units is developed. With this treatment, the

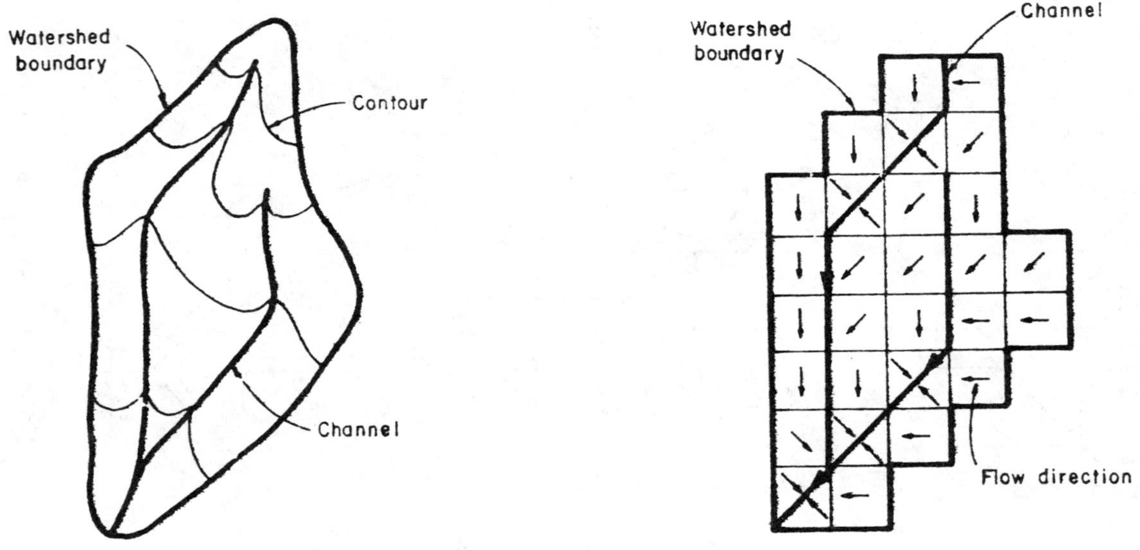

a. Topographic features

b. Segmentated watershed

Fig. 9-1. Example of Watershed Segmentation

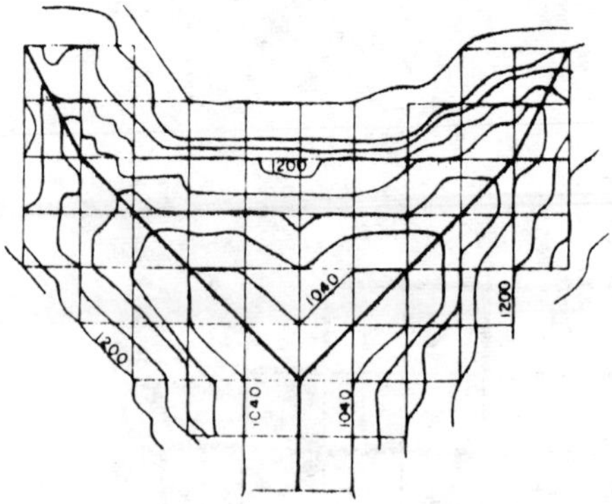

(a) Topographic Map Contour Interval = 40'

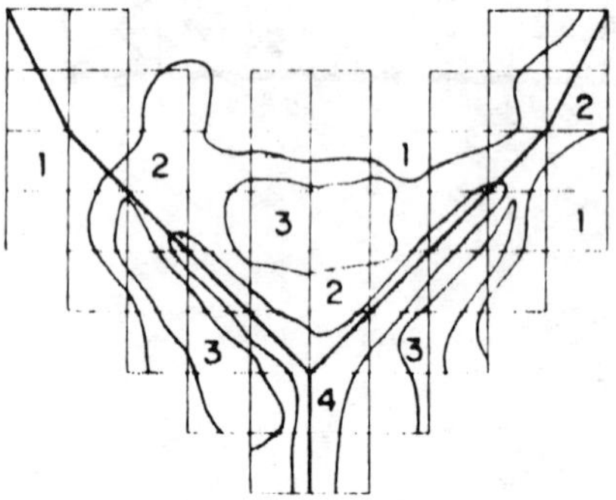

(b) Soils Map Numbers are soil type codes

(c) Vegetation Map Numbers are vegetation type codes

Fig. 9-2. Input Data for Hypothetical Watershed

flow is conceptually routed from overland flow units to channel units and to the selected watershed outlet.

This segmentation method is essential not only in water and sediment routing, but also for introducing the information from snowmelt computations, landslide hazard mapping, forest fire hazard mapping, forest inventory studies, and snow avalanche hazard identification into the routing model. Moreover, if such factors as soil properties, vegetation cover, type of management treatment, or rainfall vary within the watershed, these variations can be handled easily and with the least manual input by segmenting the watershed with a grid system. The developed segmentation method provides input data on watershed geometry and computational sequence required for the simulation model to predict water and sediment routing and yield from small watersheds.

Often the manual determination of the response unit is preferable for non-computer oriented personnel. Such a manual determination should follow the similar logic used in the computer segmentation method. The flow path can be drawn perpendicular to contour lines. The manual determination of hydrologic response unit is subject to the individual's perception and is time consuming for applications to complicated watersheds. It is recommended that the manual determination be limited in application to small and simple watersheds.

Channel Geometry

Another characteristic of the channel system that must be numerically defined is the wetted perimeter-flow area relationship for each channel segment. This relationship is needed for use in water and sediment routing in the channel. The relationship is most often expressed in the power form as

$$p = a_1 A^{b_1} \tag{9-1}$$

where p is the wetted perimeter, A is the flow area, a_1 and b_1 are statistically determined values. Data needed for development of Eq. (9-1) are measurements on the channel cross-section. These measurements are the horizontal distance from a datum mark or a bank to a point in the channel and the elevation change between the point and the mark.

9.3 MODEL FORMULATION

Once the watershed has been numerically defined by the above segmentation procedure, overland flow units and channel flow units in the watershed can be determined. Simons and Li (1975) developed a watershed sediment model which is primarily applicable for surface erosion simulation. It simulates the land surface hydrologic cycle, sediment production, and water and sediment movement on small watersheds. Conceptually, the watershed is divided into an overland flow part and a channel system part. Different physical processes are important for the two different environments. In the overland flow loop, processes of interception, evaporation, infiltration, raindrop impact detachment of soil, erosion by overland flow, and overland flow water and sediment routing to the nearest channel are simulated. In a channel system loop, water and sediment contributed by overland flow are routed and the amount of channel erosion or sediment deposition through the channel system is determined. A flow chart presenting the interrelationship of these processes is shown in Fig. 9-3. A brief summary of the components is given below.

Overland Flow Loop

There are four components in the overland flow loop: interception, infiltration, overland surface water, and overland flow sediment routing.

Interception Component: In this component the interception amounts due to the crown and forest floor are computed and the net rainfall is determined from the rainfall input. The interception loss includes the constant interception storage and the continuous evaporation from the interception surfaces. The evaporation is usually negligible during the storm. The interception storage is formulated to be a function of canopy cover density, ground cover density, and vegetation type.

Infiltration Component: This component of the model simulates the process of infiltration. The infiltration rate is computed by an approximation of Darcy's Law assuming that a distinct wetting front exists and is formulated to be a function of saturated hydraulic conductivity, average capillary suction pressure, soil porosity, antecedent moisture content, and moisture content in the wetted zone. Therefore, the rate of rainfall excess can be determined from the net rainfall and infiltration rates.

Overland Surface Water Routing Component: With this component the overland surface water runoff resulting from the mean rainfall excess is routed to the nearest channel. The routing procedure is based on the continuity of water, a momentum equation of kinematic wave approximation, and a set of resistance functions for different hydraulic conditions. The total resistance to flow is assumed to be a sum of the drag resistance due to ground cover and the shear stress acting on the soil bed. The computation is carried out by a nonlinear finite difference scheme developed by Li et al. (1975a) and the computation results include the mean flow depth, bed shear stress and flow discharge at computation points as a function of time and space point.

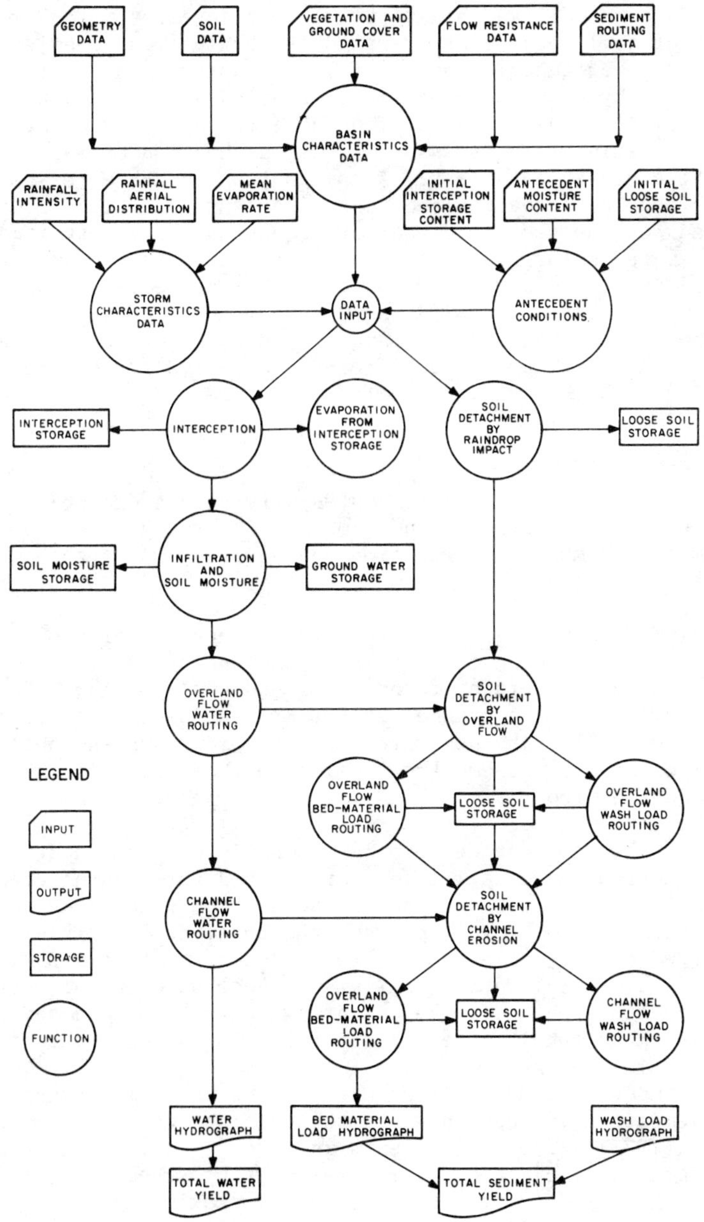

Fig. 9-3. Flow Chart for the Watershed Sediment and Routing Model

Overland Flow Sediment Component: This component of the model computes the amount of soil detachment by raindrop splash and by overland flow, the amount of wash load pickup and transport by surface runoff, and bed-material load movement. The amount of soil detachment by raindrop splash is assumed to be a simple power function of rainfall intensity. The soil detachment by surface runoff is considered as the result of bed-material load movement. The local transporting capacity of bed-material load is assumed to be a function of local effective bed shear stress. A combination of the Meyer-Peter, Müller bed load equation and Einstein's suspended load procedure is used for the sediment transport equation. The wash load pickup rate is formulated to be a function of bed shear stress and the available amount of loose soil. The sediment routing procedure is based primarily on the continuity equation for sediment (wash load and bed-material load), and the computation is carried out by a finite difference numerical procedure coupling with the overland surface water routing.

Channel System Loop

The channel system loop routes water and sediment contributed from all overland flow areas through the channel system, and computes degradation and aggradation in the channels.

Channel Water Routing Component: This component of the model routes the water down the creeks in the channel system and computes the hydrograph at the watershed outlet. The lateral water inflows to the channel system are the overland surface water flows. The channel water routing procedure and the finite difference scheme are similar to those used in the overland flow loop.

Channel Sediment Routing Component: With this component, the wash load and bed-material load are routed through the channel system. The computation results include the wash load and bed-material load hydrographs and the total sediment yield at the watershed outlet. The procedures of routing wash load and bed-material load are similar to those used in the overland flow sediment routing. The amount of degradation and aggradation in the channel system is determined by using the continuity equation for sediment.

Although the model structure remains the same, significant improvements have been added to the original work by Simons et al. (1975). The improvements include: (1) to develop a four-point implicit nonlinear water routing scheme, (2) to extend the validity of kinematic wave approximation by considering the full momentum equation, (3) to add overland subsurface flow routing, (4) to route sediment by size fractions to account for different pollutant absorption rates for different sizes and the development of an armor layer, and (5) to develop various simplified methods for practical applications. The following sections will describe the main feature of the current efforts on routing water and sediment from watersheds.

9.4 ESTIMATION OF EXCESS RAINFALL

Water is one of the major factors in erosion and sediment movement. Kinetic energy associated with falling and flowing water produces the sediment supply that is transported by the flowing water. It becomes necessary to first determine the rainfall excess or the amount of water available for the processes of erosion and sediment yield. Rainfall excess is that amount of rainfall available after all losses have been subtracted. Commonly considered losses are interception, depression storage, infiltration, and evapotranspiration. In the storm watershed simulation, losses due to evapotranspiration and evaporation are considered small and assumed negligible during the storm. Losses due to depression storage can be combined into ground cover interception. The two remaining losses are then interception and infiltration.

Interception

The amount of interception loss depends on the percentage of the ground that is covered by canopy and ground cover, their respective water holding capacities, as well as their initial moisture levels. The total intercepted volume can be written as:

$$V_i = C_c V_c + C_g V_g \tag{9-2}$$

in which V_i is the total intercepted volume in depth, C_c is the canopy cover density ($0 \leq C_c \leq 1$), V_c is the potential volume of canopy cover interception in depth, C_g is the ground cover density ($0 \leq C_g \leq 1$) and, V_g is the potential volume of ground cover interception in depth. The value of V_i is dependent on the type of vegetation, height of vegetation (related to leaf areas), and the initial moisture level. The value of V_g is a function of forest litter, grass and rock mulch, and the initial moisture level.

As the computations proceed through time, the rainfall is all intercepted until the potential volumes for ground and canopy cover are filled. This would constitute a series of net rainfall rates. Although interception losses are continuous over the storm period, it is assumed that the losses occur during the beginning period of the storm. A more detailed simulation of the interception process is given by Simons et al. (1975).

Infiltration

The Green-Ampt infiltration model is a simple, physically based, two parameter infiltration equation first proposed in 1911. It may be derived by direct application of Darcy's Law under the following assumptions:

1. A distinct piston wetting front exists.
2. The hysteresis effects in the soil properties are negligible.
3. The pressure effect of ponded water is negligible.

A Green-Ampt type equation may be written as

$$\frac{F}{\delta} - \ell n(1 + \frac{F}{\delta}) = \frac{Kt}{\delta} \tag{9-3}$$

in which F is the infiltrated volume, K is the hydraulic conductivity of the soil in the wetted zone, t is the time and δ is the potential head parameter and defined as

$$\delta = (\theta_w - \theta_i)\psi_{ave} \tag{9-4}$$

in which θ_w is the moisture content of the soil after wetting, θ_i is the antecedent moisture of content, and ψ_{ave} is the average suction head across the wetting front.

If at any time, t, the infiltrated volume is F(t), then at some later time $t + \Delta t$

$$F(t + \Delta t) = F(t) + \Delta F \tag{9-5}$$

in which ΔF is the change in infiltrated volume which occurred during the time increment, Δt. An expression for ΔF is obtained from Eq. (9-5)

$$\Delta F = F(t + \Delta t) - F(t) \tag{9-6}$$

Li et al. (1976a) developed the following method of solving for the infiltration rate. Their derivation yields

$$\frac{\Delta F}{\delta} - \ell n \left[\frac{\delta + F(t) + \Delta F}{\delta + F(t)}\right] = \frac{K}{\delta} \Delta t \tag{9-7}$$

Eq. (9-7) is implicit with respect to ΔF. However, the equation is simplified by expanding the logarithmic term in a power series (see Li et al., 1976b).

$$\ell n(1 + \frac{\Delta F}{\delta + F}) = \ell n\ 1 + \frac{\frac{2\Delta F}{\delta + F}}{2 + \frac{\Delta F}{\delta + F}} + \ldots \tag{9-8}$$

Truncating Eq. (9-8) after the second term and substituting into Eq. (9-7) one obtains that

$$\frac{\Delta F}{\delta} - 2\left[\frac{\dfrac{\Delta F}{\delta + F}}{2 + \dfrac{\Delta F}{\delta + F}}\right] = \frac{K}{\delta}\Delta t \qquad (9\text{-}9)$$

Eq. (9-9) is simplified into a quadratic whose solution is (see Li et al., 1976a):

$$\Delta F = -\frac{(2F-K\Delta t)}{2} + \frac{[(2F-K\Delta t)^2 + 8K\Delta t(\delta + F)]^{1/2}}{2} \qquad (9\text{-}10)$$

Only the positive root of Eq. (9-10) has any physical significance. The average infiltration rate, $\bar{f}$, is obtained by dividing ΔF by Δt, or

$$\bar{f} = \frac{\Delta F}{\Delta t} \qquad (9\text{-}11)$$

After accounting for the interception and infiltration losses, the rainfall excess, i_e, can be determined. A rainfall event is commonly reported as a hyetograph, i.e., a series of net rainfall intensities, I, each lasting for a time increment, Δt. Thus, if the net rainfall intensity is greater than the infiltration rate; the infiltration rate is subtracted from the net rainfall intensity to give the excess rainfall. If the net rainfall intensity is less than the corresponding infiltration rate, the infiltration rate equals the net rainfall intensity, and there is no excess.

9.5 WATER ROUTING

Governing Equation

Water runoff can be described by the equation of continuity, the equation of motion, and equations describing resistance to flow.

Continuity Equation: The equation of continuity for water is

$$\frac{\partial Q}{\partial x} + \frac{\partial A}{\partial t} = q_\ell \qquad (9\text{-}12)$$

in which Q is the discharge, x is the downslope distance, A is the cross-sectional area of flow, t is the time, and q_ℓ is the lateral inflow or outflow rate per unit length of channel. For overland flow, q_ℓ is the rainfall excess, A is the depth of flow, Q is the discharge per unit width of channel.

Momentum Equation: The momentum equation for a prismatic channel can be expressed as (Henderson, 1966)

$$S_f = S_o - \frac{\partial y}{\partial x} - \frac{1}{gA}\frac{\partial Q}{\partial t} - \frac{1}{gA}\frac{\partial}{\partial x}\left(\frac{Q^2}{A}\right) \qquad (9\text{-}13)$$

in which S_f is the friction slope, S_o is the channel bed slope, y is the depth of flow, and g is the gravitational acceleration.

The assumption of the kinematic wave approximation is that the friction slope is equal to the channel bed slope. That is, the gradients due to local and convective accelerations are assumed to be negligible and the water surface slope is assumed to be equal to the bed slope. Then the simplified momentum equation is

$$S_o \approx S_f = f \frac{Q^2}{8gRA^2} \qquad (9\text{-}14)$$

in which f is the Darcy-Weisbach friction factor, and R is the hydraulic radius. By definition

$$R = \frac{A}{P} \qquad (9\text{-}15)$$

in which P is the wetted perimeter. Usually the wetted perimeter can be expressed as a power function of flow area, i.e., (see Eq. (9-1)),

$$P = a_1 A^{b_1} \qquad (9\text{-}16)$$

where a_1 and b_1 are constants.

If Manning's equation is used, the simplified momentum equation is

$$S_o \approx S_f = \frac{n^2 Q^2}{2.21\ R^{4/3} A^2} \qquad (9\text{-}17)$$

in which n is Manning's roughness coefficient.

Resistance Equations: In a natural watershed, the form resistance due to the ground cover is a very important component of the resistance to flow. The dependence of flow resistance on the ground cover becomes further complicated depending on whether the ground cover is submerged or not. Rarely is the ground cover submerged in overland flow units. Therefore, in overland flow units, one only needs to consider the resistance as that caused by flow through ground cover. In channel flow units, the probability of submerging the ground cover is apparently large. The resistance is then considered as the resistance caused by flow through ground cover and flow over ground cover simultaneously.

Resistance to flow for the overland flow response units is generally expressed as a function of surfacing material, vegetation type and density of vegetation. Palmer (1946), Ree (1949), and Ree and Palmer (1949) conducted a series of experiments in channels with various types of grasses. More recently, Kouwen and Unny (1969), Phelps (1970), Wenzel (1970), Li and Shen (1973), and Chen (1976) carried the experimental studies further. Results indicate a functional relationship between the overall Darcy-Weisbach friction factor and flow characteristics as

$$f = \frac{K_t}{N_r} \tag{9-18}$$

in which f is the overall Darcy-Weisbach friction coefficient, N_r is the flow Reynolds number, K_t is a constant describing the overall resistance. According to Chen (1976), this type of relationship can be used for a Reynolds number up to 10^5. This would practically cover all of the possible overland flow conditions on the natural surfaces.

The flow Reynolds number is defined

$$N_r = \frac{QR}{\nu A} \tag{9-19}$$

in which ν is the kinematic viscosity of water.

Assuming that the factors describing resistance to flow are independent, the overall resistance can be expressed as

$$f = f_g + f_d \tag{9-20}$$

in which f_g is the Darcy-Weisbach friction factor for grain resistance only which is a function of grain size, flow Reynolds number and rainfall intensity, and f_d is the Darcy-Weisbach friction factor due to form drag resistance which is a function of ground cover density, size of ground cover, drag coefficient, depth of flow and flow Reynolds number. In overland flow cases, the value of f_g can be further assumed as

$$f_g = \frac{K_g}{N_r} \tag{9-21}$$

in which K_g is the parameter describing Darcy-Weisbach friction factor for grain resistance only. Laboratory experiments show that K_g is between 30-60 (Chow, 1959 and Woolhiser, 1975). It is assumed that K_g is equal to 50 for practical purposes.

The review of literature suggests that the overall flow resistance of overland flow can be assumed as

$$K_t = K_\ell + (K_h - K_\ell) C_g^2 \tag{9-22}$$

in which K_ℓ is the parameter describing the minimum resistance ($C_g = 0.0$) and K_h is the parameter describing the maximum resistance ($C_g = 1.0$).

In the channel flow, Darcy-Weisbach friction factor is often considered as a constant. Sometimes, the Chezy function equation is used in evaluating the overall resistance for overland flow and flow in rivers. By definition

$$C = \sqrt{\frac{8g}{f}} \tag{9-23}$$

in which C is the Chezy friction factor. Eq. (9-23) shows that a constant value of f will give a constant value of C.

Manning's equation is frequently used by hydraulic engineers to describe flow in open channels. The Manning roughness coefficient is usually determined by measurement and can be expressed as a power function of flow discharge, i.e.,

$$n = a_2 Q^{b_2} \tag{9-24}$$

in which a_2 and b_2 are constant.

Discharge and Flow Area Relation: In general, the cross-sectional area of the flow can be expressed as a power function of discharge

$$A = \alpha Q^\beta \tag{9-25}$$

in which α and β are coefficients whose values depend on the shape of the channel, the friction slope, and the roughness of the wetted perimeter.

If the Darcy-Weisbach friction factor is used, the values of α and β can be determined by substituting Eqs. (9-14), (9-15), (9-16), (9-18), and (9-19) into Eq. (9-25). The solutions are

$$\alpha = \left(\frac{K_t \nu a_1^2}{8gS_f}\right)^{1/(3-2b_1)} \tag{9-26}$$

and

$$\beta = \frac{1}{3-2b_1} \tag{9-27}$$

For overland flows or flows in very wide channels the wetted perimeter is constant so that $b_1 = 0$ and $\beta = 1/3$.

If Manning's equation is applied, the corresponding α and β values are determined from Eqs. (9-15), (9-16), (9-17), (9-24), and (9-25). The results are:

$$\alpha = \left(\frac{a_1^{4/3} a_2^2}{2.21 S_f}\right)^{3/(10-4b_1)} \tag{9-28}$$

and

$$\beta = \frac{3-3b_2}{5-2b_1} \tag{9-29}$$

If the Chezy friction equation is used, the corresponding α and β values are determined by using Eqs. (9-14), (9-15), (9-16), (9-23), and (9-25). The values are:

$$\alpha = \left(\frac{a_1}{C^2 S_f}\right)^{1/(3-b_1)} \tag{9-30}$$

$$\beta = \frac{2}{3-b_1} \tag{9-31}$$

For the kinematic wave approximation, the friction slope S_f in Eqs. (9-26), (9-28), and (9-30) is equal to the bed slope S_o.

The problem of water routing is now a matter of solving Eqs. (9-12) and (9-25). These two equations can be solved either by an analytical method or a numerical method. The solution techniques follow.

Analytical Solutions

The analytical solutions of Eqs. (9-12) and (9-25) are available for some special cases. (Eagleson, 1970, Kibler and Woolhiser, 1970, Harley et al., 1970, and Li et al., 1975b). For analytical solutions, it is common to express Eq. (9-25) in the following form:

$$Q = \alpha' A^{\beta'} \tag{9-32}$$

where

$$\alpha' = \left(\frac{1}{\alpha}\right)^{1/\beta} \tag{9-33}$$

and

$$\beta' = \frac{1}{\beta} \tag{9-34}$$

From Eqs. (9-12) and (9-32), one obtains

$$\alpha'\beta' A^{\beta'-1} \frac{\partial A}{\partial x} + \frac{\partial A}{\partial x} = q_\ell \tag{9-35}$$

This partial differential equation may be solved by the method of characteristics. The total differential of $A(x,t)$ is given by

$$dA = \frac{\partial A}{\partial t} dt + \frac{\partial A}{\partial x} dx \tag{9-36}$$

Eqs. (9-35) and (9-36) form a system of two equations in two unknowns, and may be written in matrix form as,

$$\begin{bmatrix} \alpha'\beta' A^{\beta'-1} & 1 \\ dx & dt \end{bmatrix} \begin{bmatrix} \frac{\partial A}{\partial x} \\ \frac{\partial A}{\partial t} \end{bmatrix} = \begin{bmatrix} q_\ell \\ dA \end{bmatrix} \tag{9-37}$$

The characteristic paths along which the solution is valid are found by determining the loci of indeterminancy of the solution. These loci are obtained by equating the determinant of the coefficient matrix with zero. The resulting characteristic equation is

$$\frac{dx}{dt} = \alpha'\beta' A^{\beta'-1} \tag{9-38}$$

Integrating Eq. (9-38) with respect to time yields

$$x - x_o = \alpha'\beta' \int_{t_o}^{t} A^{\beta'-1} \, dt' \tag{9-39}$$

in which x_o is the initial distance and t_o is the initial time.

The invariants of this solution are found by substituting the right-hand side of Eq. (9-37) for each column of the coefficient matrix and equating the determinant of the resulting matrix equal to zero. The invariants are

$$\frac{dA}{dt} = \frac{dQ}{dx} = q_\ell \tag{9-40}$$

or by integrating Eq. (9-40)

$$A - A_o = \int_{t_o}^{t} q_\ell \, dt' \tag{9-41}$$

or

$$Q - Q_o = \int_{x_o}^{x} q_\ell \, dx' \tag{9-42}$$

in which A_o and Q_o are initial values of area and discharge respectively. By substituting Eq. (9-41) into Eq. (9-39) the following expression for the characteristic is obtained,

$$x = x_o + \alpha'\beta' \int_{t_o}^{t} [A_o + \int_{t_o}^{\eta} q_\ell(\sigma)\, d\sigma]^{\beta'-1} d\eta \tag{9-43}$$

Theoretically, if the functional form of $q_\ell(\sigma)$ is known, then Eq. (9-43) will give the characteristics in the x-t plane. Harley et al. (1970) noted that when $q_\ell(\sigma)$ is a histogram, then with reference, Eq. (9-41) and (9-43) may be evaluated in pieces.

The procedure as outlined by Harley et al. (1970) is as follows (see Fig. 9-4). From any point (x_i, t_i) on characteristic C, the value of x_{i+1} on C corresponding to time t_{i+1} may be calculated by Eq. (9-43).

$$x_{i+1} = x_i + \alpha'\beta' \int_{t_i}^{t_{i+1}} [A_i + q_{\ell i+1}(\eta - t_i)]^{\beta'-1} d\eta \tag{9-44}$$

or

$$x_{i+1} = x_i + \frac{\alpha'}{q_{\ell i+1}} \{[A_i + q_{\ell i+1}(t_{i+1} - t_i)]^{\beta'} - A_i^{\beta'}\} \tag{9-45}$$

for $q_{\ell i+1} \neq 0$, and

$$x_{i+1} = x_i + \alpha'\beta' A_i^{\beta'-1} (t_{i+1} - t_i) \tag{9-46}$$

for $q_{\ell i+1} = 0$.

From Eq. (9-41) one obtains

$$A_{i+1} = A_i + q_{\ell i+1} (t_{i+1} - t_i) \tag{9-47}$$

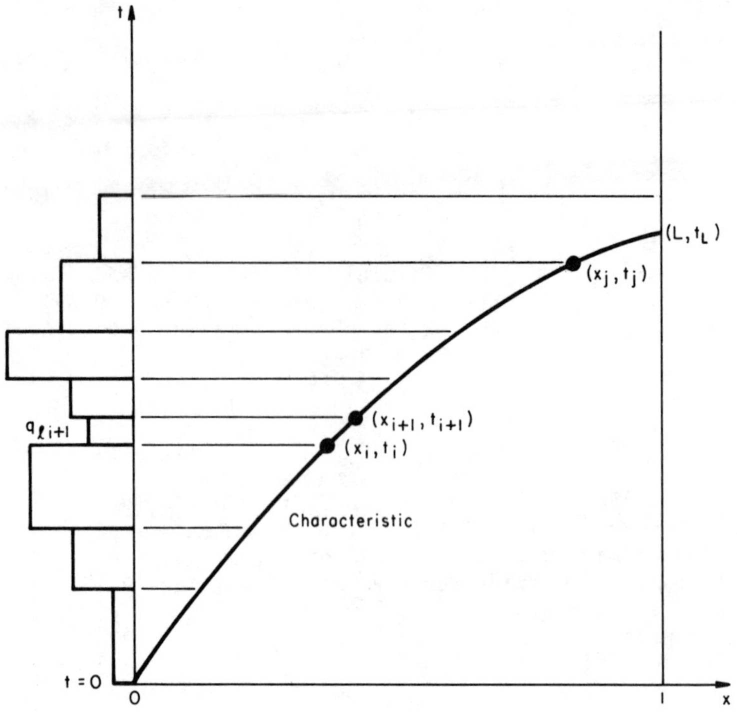

Fig. 9-4. Schematic Diagram of Characteristic Path

For a given plane of length L, values of x_{i+1} are calculated until $x_{i+1} \geq L$. The time of arrival of the characteristic at the downstream boundary, t_L, is found by solving Eq. (9-45) or Eq. (9-46) for t_L in terms of t_j and x_j.

$$t_L = t_j + \frac{1}{q_{\ell i+1}} \left\{ \left[\frac{q_{\ell i+1}(L-x_j)}{\alpha'} + A_j^{\beta'} \right]^{1/\beta'} - A_j \right\} \tag{9-48}$$

for $q_{\ell i+1} \neq 0$, and

$$t_L = \frac{L-x_j}{\alpha'\beta' A_j^{\beta'-1}} + t_j \tag{9-49}$$

Then, the discharge at the end of channel or overland flow reach is

$$Q_L = \alpha' A_L^{\beta'} \tag{9-50}$$

Repeating the above solution, it is not difficult to route the water in terms of time and space. However, the timing of the outflow at the downstream boundary, t_L, cannot be predetermined. Li et al. (1975b) suggested to trace the characteristic path backward from the downstream boundary for a desired t_L. Their method is simply to solve Eq. (9-43) in terms of t_o. Or

$$\frac{L}{\alpha'\beta'} = \int_{t_o}^{t_L} \left[\int_{t_o}^{\eta} q_\ell(\sigma)\, d\sigma \right]^{\beta'-1} d\eta \tag{9-51}$$

Since $q_\ell(\sigma)$ generally has the form of a histogram, Eq. (9-51) can be expressed in terms of a polynomial equation. Applying the second order Newton method, the value of t_o can be obtained (see Li et al., 1977a). Then, the cross-sectional area and the discharge can be determined utilizing Eqs. (9-41) and (9-50).

The above analytical solution is certainly desirable since there is no problem encountered in computational stability and convergence. Unfortunately, this analytical solution is often restricted for practical applications because of the formulation of kinematic "shock-wave." The "shock" is represented by the intersections of characteristics in the x-t plane. This produces an abrupt increase in flow depth (Kibler and Woolhiser, 1970). An example of the kinematic wave shock is given in Fig. 9-5. Li et al. (1976c) gives an in-depth discussion of this limitation. However, for a small watershed with simple geometry having two planes and one channel the above analytical procedure can be used. Simons et al. (1977a) reported a simple watershed model using small programmable calculations based on the two-planes and one-channel representation of watersheds.

Numerical Solution

Li et al. (1977b) and Simons (1977) developed a nonlinear scheme with an iterative procedure to solve a four-point implicit formulation of Eqs. (9-12) and (9-25). A linear scheme is used to obtain the initial estimate of the unknown discharge for the nonlinear scheme. The linear scheme may be used as is, with no iterations to solve for the unknown discharge providing accuracy is satisfactory.

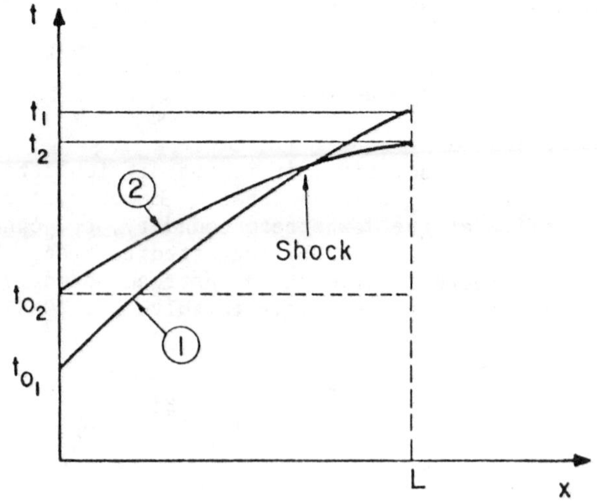

Fig. 9-5. Schematic Diagram of Kinematic-Wave Shock Formation

Nonlinear Scheme: The finite difference form of Eq. (9-12) using quantities at all four points (see Fig. 9-6) of the box scheme can be written as

$$\left[\frac{Q_{j+1}^{n+1}-Q_j^{n+1}}{\Delta x}(1-a) + \frac{Q_{j+1}^{n}-Q_j^{n}}{\Delta x}(a)\right] + \left[\frac{A_{j+1}^{n+1}-A_{j+1}^{n}}{\Delta t}(1-b) + \frac{A_j^{n+1}-A_j^{n}}{\Delta t}(b)\right]$$

$$= \frac{(1-b)q_{\ell_{j+1}}^{\ n+1} + bq_{\ell_j}^{\ n+1} + (1-b)q_{\ell_{j+1}}^{\ n} + bq_{\ell_j}^{\ n}}{2} \qquad (9\text{-}52)$$

in which Q_j^n is the quantity Q at grid point $x = j\Delta x$, $t = n\Delta t$, where Δx is the space increment and Δt is the time increment. Also, a is the time weighting factor and b is the space weighting factor where $0 \leq a \leq 1$ and $0 \leq b \leq 1$.

The unknowns in Eq. (9-52) are Q_{j+1}^{n+1}, and A_{j+1}^{n+1} but they are related according to Eq. (9-25). With the two equations, the two unknowns may be calculated. The boundary conditions required for this particular formulation are as follows: $Q(0,t)$ and $Q(x,0)$ or in other words, the discharge must be known or assumed along both the space axis $[Q(x,0)]$ and the time axis $[Q(0,t)]$. No downstream boundary conditions are required. However, if there is a strong downstream control, the kinematic wave approximation no longer applies.

Either Q or A can be selected as the independent variable in the numerical procedure. As customary in backwater computations the depth of flow (equivalent to A above) is

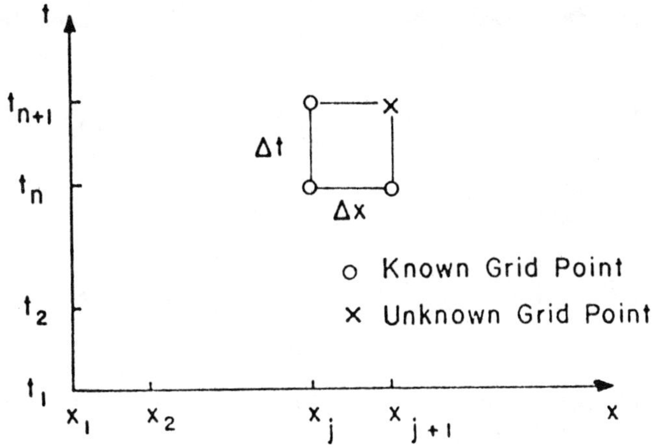

Fig. 9-6. Rectangular Network in x-t Plane

chosen as the independent variable (see Henderson, 1966). However, Q is a better choice for the following reason. From Eq. (9-25) one can derive,

$$\frac{dA}{A} = \beta \frac{dQ}{d} \tag{9-53}$$

Since the value of β is generally less than 1.0, if the discharge is computed incorrectly, the relative error in the flow area is smaller than the relative error in the discharge. On the other hand, the error in the discharge estimation is magnified if the numerical computations are performed on the flow area. From the physical viewpoint it is more appropriate to consider routing unit volumes of water rather than areas of flow.

From Eq. (9-25)

$$A_{j+1}^{n+1} = \alpha \, (Q_{j+1}^{n+1})^{\beta} \tag{9-54}$$

$$A_{j}^{n+1} = \alpha \, (Q_{j}^{n+1})^{\beta} \tag{9-55}$$

$$A_{j+1}^{n} = \alpha \, (Q_{j+1}^{n})^{\beta} \tag{9-56}$$

$$A_{j}^{n} = \alpha \, (Q_{j}^{n})^{\beta} \tag{9-57}$$

Equations (9-54) through (9-57) are substituted into Eq. (9-52) and rearranged to yield,

$$\frac{\Delta t}{\Delta x} Q_{j+1}^{n+1}(1-a) + \alpha(Q_{j+1}^{n+1})^{\beta}(1-b) = \frac{\Delta t}{\Delta x}\left\{Q_j^{n+1}(1-a)-[Q_{j+1}^{n}-Q_j^{n}](a)\right\}$$

$$+ \alpha(Q_{j+1}^{n})^{\beta}(1-b) - [\alpha(Q_j^{n+1})^{\beta} - \alpha(Q_j^{n})^{\beta}](b)$$

$$+ \frac{\Delta t}{2}\left\{(1-b)q_{\ell_{j+1}}^{n+1} + bq_{\ell_j}^{n+1} + (1-b)q_{\ell_{j+1}}^{n} + bq_{\ell_j}^{n}\right\} \tag{9-58}$$

The right-hand side of Eq. (9-58) consists of known quantities and is represented by Ω, i.e.,

$$\Omega = \frac{\Delta t}{\Delta x}\left\{Q_j^{n+1}(1-a) - [Q_{j+1}^{n}-Q_j^{n}](a)\right\} + \alpha(Q_{j+1}^{n})^{\beta}(1-b)$$

$$- [\alpha(Q_j^{n+1})^{\beta} - \alpha(Q_j^{n})^{\beta}](b)$$

$$+ \frac{\Delta t}{2}\left\{(1-b)q_{\ell_{j+1}}^{n+1} + bq_{\ell_j}^{n+1} + (1-b)q_{\ell_{j+1}}^{n} + bq_{\ell_j}^{n}\right\} \tag{9-59}$$

Let $r = Q_{j+1}^{n+1}$ and $\theta = \frac{\Delta t}{\Delta x}$ so that the left-hand side of Eq. (9-58) can be written as

$$f(r) = \theta(1-a)r + \alpha(1-b)r^{\beta} \tag{9-60}$$

The solution to Eq. (9-58) is the solution r^* which satisfies the condition,

$$f(r^*) = \theta(1-a)r^* + \alpha(1-b)r^{*\beta} = \Omega \tag{9-61}$$

Equation (9-61) is nonlinear in r^*. An approximate solution to this nonlinear equation is obtained by the following iterative technique.

Let r^k be the value of r at the kth iteration. The Taylor series expansion of the function $f(r)$ around r^k is

$$f(r) = f(r^k) + (r-r^k)f'(r^k) + 1/2(r-r^k)^2 f''(r^k)$$

$$+ 1/6(r-r^k)^3 f'''(r^k) + ----. \tag{9-62}$$

in which $f'(r^k)$, $f''(r^k)$, and $f'''(r^k)$ are the first, second and third derivatives of the function at r^k.

Dropping terms higher than third order one obtains,

$$f(r) \approx f(r^k) + (r-r^k)f(r^k) + 1/2(r-r^k)^2 f''(f^k) \qquad (9\text{-}63)$$

Iteration forces $f(r^{k+1})$ to approach the value of Ω

$$\Omega = f(r^k) + (r^{k+1}-r^k)f'(r^k) + 1/2(r^{k+1}-r^k)^2 f''(r^k) \qquad (9\text{-}64)$$

The solution of Eq. (9-64) is

$$r^{k+1} = r^k - \frac{f'(r^k)}{f''(r^k)} \pm \left\{ \left[\frac{f'(r^k)}{f''(r^k)}\right]^2 - \frac{2[f(r^k)-\Omega]}{f''(r^k)} \right\}^{1/2} \qquad (9\text{-}65)$$

in which

$$f(r^k) = \theta(1-a)r^k + \alpha(1-b)(r^k)^\beta \qquad (9\text{-}66)$$

$$f'(r^k) = \theta(1-a) + \alpha\beta(1-b)(r^k)^{\beta-1} \qquad (9\text{-}67)$$

$$f''(r^k) = \alpha\beta(\beta-1)(1-b)(r^k)^{\beta-2} \qquad (9\text{-}68)$$

There are two solutions to Eq. (9-65). It is advisable to choose the solution which gives the smaller value of $|f(r^{k+1})-\Omega|$. The iterations are continued until the absolute error $|f(r^{k+1})-\Omega|$ is less than some preassigned tolerance ε

$$|f(r^{k+1})-\Omega| \leq \varepsilon \qquad (9\text{-}69)$$

An appropriate value for ε is 0.01 Ω.

The initial guess r^o is the key to the speed of convergence to the correct numerical solution. The best way to determine r^o is to use a linear scheme. The linear scheme may be used as the solution instead of going through the iterative, nonlinear scheme when accuracy is of less importance and when computer time is to be conserved.

Linear Scheme: The term $\frac{\partial A}{\partial t}$ in Eq. (9-12) can be expressed as

$$\frac{\partial A}{\partial t} = \frac{\partial A}{\partial Q}\frac{\partial Q}{\partial t} \qquad (9\text{-}70)$$

and from Eq. (9-25)

$$\frac{\partial A}{\partial Q} = \alpha\beta Q^{\beta-1} \qquad (9\text{-}71)$$

Substitution of Eqs. (9-70) and (9-71) into (9-12) yields

$$\frac{\partial Q}{\partial x} + \alpha\beta Q^{\beta-1} \frac{\partial Q}{\partial t} = q_\ell \tag{9-72}$$

The finite difference form of Eq. (9-72) can be expressed as

$$\frac{Q_{j+1}^{n+1} - Q_j^{n+1}}{\Delta x}(1-a) + \frac{Q_{j+1}^{n} - Q_j^{n}}{\Delta x}(a)$$

$$+ \alpha\beta\left[\frac{Q_j^{n+1} + Q_{j+1}^{n}}{2}\right]^{\beta-1}\left[\frac{Q_{j+1}^{n+1} - Q_{j+1}^{n}}{\Delta t}(1-b) + \frac{Q_j^{n+1} - Q_j^{n}}{\Delta t}(b)\right]$$

$$= \frac{1}{2}\left[(1-b)\, q_{\ell_{j+1}}^{\ n+1} + b q_{\ell_j}^{\ n+1} + (1-b) q_{\ell_{j+1}}^{\ n} + b q_{\ell_j}^{\ n}\right] \tag{9-73}$$

Solving for $r^o = Q_{j+1}^{n+1}$,

$$Q_{j+1}^{n+1} = \left[\frac{1-a}{\Delta x} + \alpha\beta\left(\frac{Q_j^{n+1} + Q_{j+1}^{n}}{2}\right)^{\beta-1}\frac{(1-b)}{\Delta t}\right]^{-1}\left\{\frac{Q_j^{n+1}}{\Delta x}(1-a) - \frac{Q_{j+1}^{n} - Q_j^{n}}{\Delta x}(a)\right.$$

$$- \alpha\beta\left(\frac{Q_j^{n+1} + Q_{j+1}^{n}}{2}\right)^{\beta-1}\left[-\frac{Q_{j+1}^{n}}{\Delta t}(1-b) + \frac{Q_j^{n+1} - Q_j^{n}}{\Delta t}(b)\right]$$

$$\left. + \frac{1}{2}\left[(1-b) q_{\ell_{j+1}}^{\ n+1} + b q_{\ell_j}^{\ n+1} + (1-b) q_{\ell_{j+1}}^{\ n} + b q_{\ell_j}^{\ n}\right]\right\} \tag{9-74}$$

Eq. (9-74) provides the best estimate r^o for the nonlinear scheme. However, Eq. (9-74) is not applicable if both Q_{j+1}^{n} and Q_j^{n+1} are zero. In this case, use $\beta = 1$ in Eq. (9-61)

$$r^o = \frac{\Omega}{(1-a)\theta + (1-b)\alpha} \tag{9-75}$$

Li et al. (1977b) proved that the weighting factor a and b must be less than or equal to 1/2 in order to insure stability. When a and b are both zero, this is a fully implicit scheme. This scheme was developed by Li et al. (1975a). When a and b equal to 1/2, it is a central implicit scheme. The central implicit formulation has been tested using Iwagaki's experiment data (Iwagaki, 1955). Fig. 9-7 shows good comparison between experimental and computed results for the three events during different time duration.

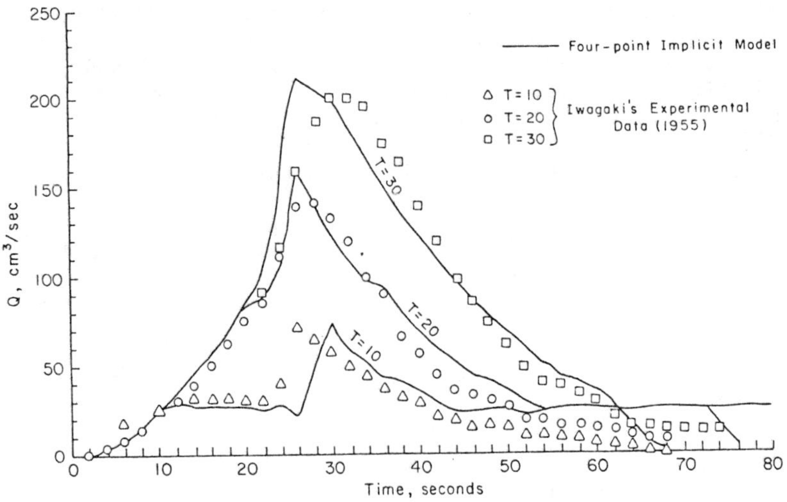

Fig. 9-7. Comparison of the Discharge Hydrographs Given by the Four-Point, Central Implicit Model to Iwagaki's Experimental Data

Li et al. (1975a) successfully applied the fully implicit scheme to a variety of flow conditions.

Other available numerical methods for kinematic wave approximation are given by Houghton and Kasahara (1968), Brakensiek (1967), Kibler and Woolhiser (1970), and Singh (1975). However, these methods are much more complicated and sometimes unstable in solution.

Modification of the Kinematic Wave Assumption

The kinematic wave approximation states that the friction slope equals the bed slope. The pressure gradient, local and convective acceleration terms in the full dynamic momentum equation are assumed negligible. It has been reported that the kinematic wave approximation may be applied when the channel gradients range from moderate to steep (see Simons, 1977). The rate of rise and fall of a flood wave is another determining factor when considering dynamic effects (described by the pressure gradient and acceleration terms in the full dynamic momentum equation) are significant (see Fread, 1973). In order to account for dynamic effects due to the unsteady, nonuniform flow encountered in natural channels, the friction slope can be solved for in terms of the channel geometry, resistance to flow and input hydrograph. The friction slope changes each time and space step in order to model the variable friction slope due to the passage of a flood wave. This modification is particularly necessary if studying a large watershed. As pointed out in Chapter 11, most watersheds subject to legal requirements for water resources or water quality management are quite large and often encompass complete river basins or sub-basins. In these large watersheds, the water routing should consider the pressure gradient and the gradients due to local and convective accelerations in the momentum equation. Li et al. (1977b) developed the following modified scheme.

From Eqs. (9-16) and (9-17) one obtains

$$Q = \left[\frac{1.486}{na_1^{2/3}}\right] A^{(5-2b_1)/3} S_f^{1/2} \tag{9-76}$$

or

$$A = \left[\frac{na_1^{2/3}}{1.486}\right]^{3/(5-2b_1)} (S_f^{-1/2})^{3/(5-2b_1)} Q^{3/(5-2b_1)} \tag{9-77}$$

The depth of flow can be expressed as a power function of the flow area.

$$y = a_3 A^{b_3} \tag{9-78}$$

in which a_3 and b_3 are constants.

By taking the partial derivative of Eq. (9-78) with respect to x one obtains

$$\frac{\partial y}{\partial x} = a_3 b_3 A^{b_3 - 1} \frac{\partial A}{\partial x} \tag{9-79}$$

An expression for $\partial A/\partial x$ is found by taking the partial derivative of Eq. (9-77) with respect to x.

$$\frac{\partial A}{\partial x} = \left\{\left(\frac{na_1^{2/3}}{1.486}\right)^{3/(5-2b_1)} (S_f^{-1/2})^{3/(5-2b_1)} \left(\frac{3}{5-2b_1}\right) Q^{(-2+2b_1)/(5-2b_1)}\right\} \frac{\partial Q}{\partial x} \tag{9-80}$$

Next, the term $\partial Q/\partial x$ is expressed in terms of the input discharge hydrograph through the continuity equation for the case of no lateral inflow or outflow (see Chow, 1959).

$$\frac{\partial Q}{\partial x} + \frac{\partial A}{\partial t} = 0 \tag{9-81}$$

For convenience and simplicity in mathematics, Eq. (9-81) is used to relate the space and time domains for determination of friction slope, but is not used in the water routing. The omission of lateral inflows in this development is not critical because the amount of lateral inflows is small compared to the total discharge for most large rivers having significant dynamic effects.

The partial derivative of the flow area with respect to time is determined using Eq. (9-77)

$$\frac{\partial Q}{\partial x} = -\frac{\partial A}{\partial t} =$$

$$-\left\{\left[\frac{na_1^{2/3}}{1.486}\right]^{3/(5-2b_1)} (S_f^{-1/2})^{3/(5-2b_1)} \left(\frac{3}{5-2b_1}\right) Q^{(-2+2b_1)/(5-2b_1)}\right\} \frac{\partial Q}{\partial t} \tag{9-82}$$

This expression is now substituted into Eq. (9-80) which is then related to the pressure gradient using Eq. (9-79). The resultant expression for the water surface slope is,

$$\frac{\partial y}{\partial x} = a_3 b_3 A^{b_3-1} \left(\frac{1.486}{na_1^{2/3}}\right)^{-6/(5-2b_1)} \left(\frac{3}{5-2b_1}\right)^2 Q^{(-4+4b_1)/(5-2b_1)} S_f^{-3/(5-2b_1)} \frac{\partial Q}{\partial t} \tag{9-83}$$

The next term in the full momentum equation to be found is the convective acceleration term, $\frac{1}{gA}\frac{\partial Q^2/A}{\partial x}$. Using Eq. (9-76) an expression for $\frac{\partial Q^2/A}{\partial x}$ is derived.

$$\frac{\partial Q^2/A}{\partial x} = \left(\frac{1.486}{na_1^{2/3}}\right)^2 \frac{7-4b_1}{3} S_f A^{(4-4b_1)/3} \frac{\partial A}{\partial x} \tag{9-84}$$

The term $\frac{\partial A}{\partial x}$ is given in Eq. (9-80). From Eq. (9-81) one obtains

$$\frac{\partial Q}{\partial x} = -\frac{\partial A}{\partial t} \tag{9-85}$$

The term $\partial A/\partial t$ is found by taking the partial derivative of Eq. (9-77) with respect to time.

$$\frac{\partial A}{\partial t} = \left\{\left[\frac{na_1^{2/3}}{1.486}\right]^{3/(5-2b_1)} (S_f^{-1/2})^{3/(5-2b_1)} \left(\frac{3}{5-2b_1}\right) Q^{(-2+2b_1)/(5-2b_1)}\right\} \frac{\partial Q}{\partial t} \tag{9-86}$$

Substitution of Eq. (9-86) into Eq. (9-85) and then substitution of the resulting expression into Eq. (9-80) and finally substitution of this result into Eq. (9-84) gives the desired equation.

$$\frac{\partial Q^2/A}{\partial x} = -\left(\frac{7-4b_1}{3}\right)\left(\frac{3}{5-2b_1}\right)^2 \frac{\partial Q}{\partial t} \tag{9-87}$$

The final term of the full dynamic momentum equation is the local acceleration, $\partial Q/\partial t$. This term is known since the input hydrograph $Q(0,t)$ is given.

$$\frac{\partial Q}{\partial t} \approx \frac{\Delta Q}{\Delta t} \tag{9-88}$$

Expressions for all of the terms in the full dynamic momentum equation have been derived and are substituted into Eq. (9-12). The result is

$$S_f = S_o + a_3 b_3 A^{b_3 - 1}\left(\frac{1.486}{na_1^{2/3}}\right)^{-6/(5-2b_1)}\left(\frac{3}{5-2b_1}\right)^2 Q^{(-4+4b_1)/(5-2b_1)} S_f^{-3/(5-2b_1)} \frac{\partial Q}{\partial t}$$
$$+ \frac{1}{gA}\left(\frac{7-4b_1}{3}\right)\left(\frac{3}{5-2b_1}\right)^2 \frac{\partial Q}{\partial t} - \frac{1}{gA}\frac{\partial Q}{\partial t} \tag{9-89}$$

Before Eq. (9-89) can be solved for S_f, the area term in the factor $1/gA$ and in the pressure gradient term are converted to discharge through Eq. (9-76). Then, the resultant equation is

$$S_f - S_o - a_3 b_3 \left(\frac{1.486}{na_1^{2/3}}\right)^{(-3-3b_3)/(5-2b_1)}\left(\frac{3}{5-2b_1}\right)^2 Q^{(-7+4b_1+3b_3)/(5-2b_1)} S_f^{(-3-3b_3)/(10-4b_1)} \frac{\partial Q}{\partial t}$$
$$+ \frac{1}{g}\frac{\partial Q}{\partial t}\left[1 - \left(\frac{7-4b_1}{3}\right)\left(\frac{3}{5-2b_1}\right)^2\right]\left\{\left(\frac{na_1^{2/3}}{1.486}\right)^{-3/(5-2b_1)} Q^{-3/(5-2b_1)} S_f^{1/2(3/(5-2b_1))}\right\} = 0 \tag{9-90}$$

This above equation can be solved by a similar second order Newton method as presented in the last section.

Once the friction slope has been determined for that particular point in the space-time plane, it is used instead of the bed slope in relating the discharge to the area in Manning's equation. Therefore, the dynamic effects are accounted for, and the friction slope is no longer assumed equal to the bed slope under the kinematic wave approximation. The friction slope replaces the bed slope in Eq. (9-17). The numerical scheme is solved using the same technique as the kinematic wave, but with incorporation of the full dynamic momentum equation. In this way the water-routing scheme does not require a downstream boundary condition.

This modified method is tested on a large river of the Yazoo Basin, Mississippi. A flood hydrograph is routed on the Tallahatchie River from Swan Lake to Money in Mississippi. The two stations are 26.2 miles apart. The geometry is described by a cross-section survey taken in 1970. Using observed stage discharge curves at Swan Lake and Money, as well as the bed slope, Manning's equation is computed as a power function of discharge. The input discharge hydrograph (at Swan Lake) is shown in Fig. 9-8. Observed and computed stages at Money

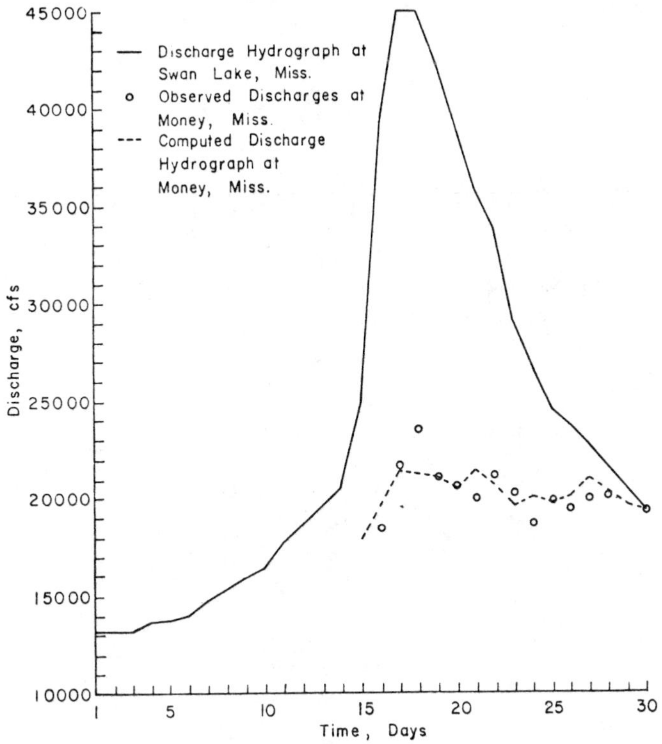

Fig. 9-8. Discharge Hydrographs on the Tallahatchie River

are compared in Fig. 9-9. The error is within acceptable limits. The observed and computed discharges at Money are compared in Fig. 9-8 for the second half of the flood when discharges were measured. The computed and observed stage-discharge relation is plotted in Fig. 9-10 showing the significance of the dynamic effects due to the unsteady flow accelerations. The procedure for converting discharge to

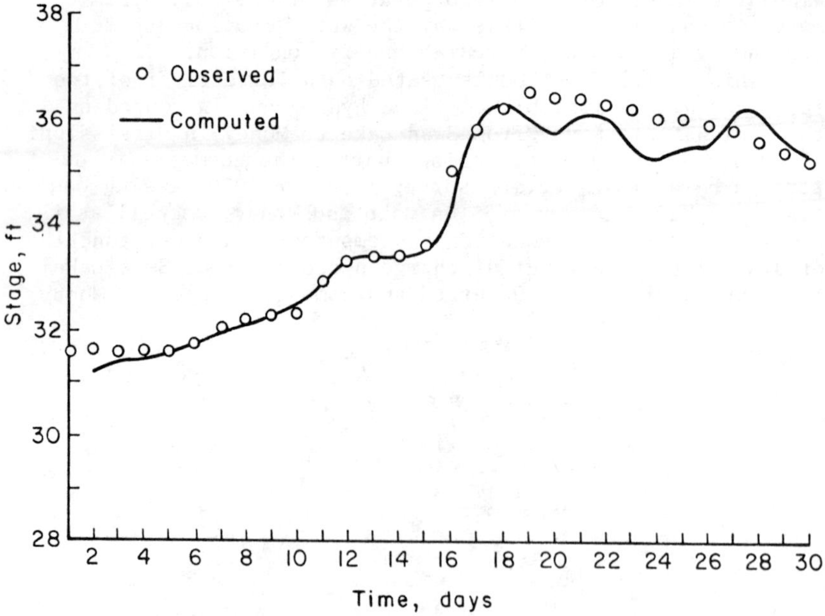

Fig. 9-9. Stage Hydrograph on the Tallahatchie River at Money, Mississippi

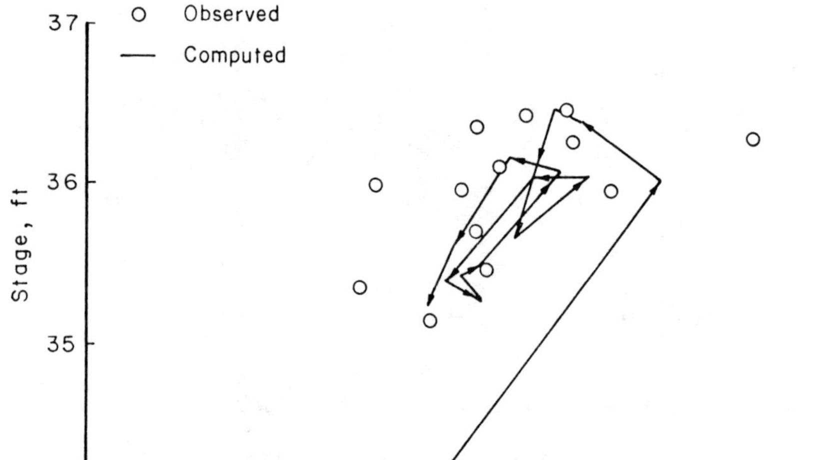

Fig. 9-10. Stage-Discharge Relationship on the Tallahatchie River at Money, Mississippi

stage can be used separately without water routing. The example of the comparison of the simulated and observed stage-discharge relation is shown in Fig. 9-11 (Mississippi River, Tarbert Landing, Louisiana). For the rising limb of a floodwave the friction slope is larger than the bed slope, and hence the corresponding stages are smaller. During the falling limb of a floodwave, the friction slope is smaller than the bed slope, and hence the corresponding stages are larger (Simons et al., 1977).

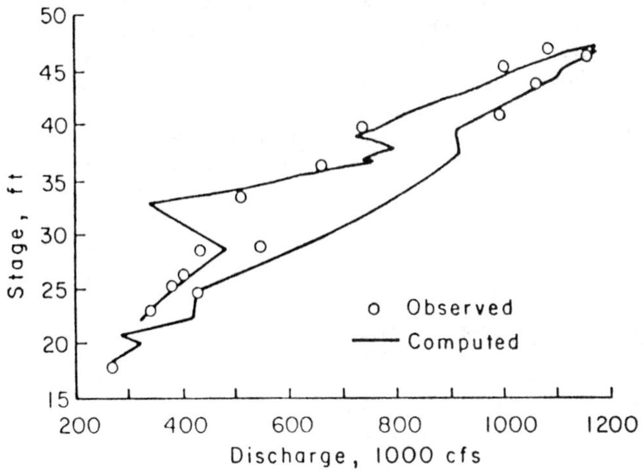

Fig. 9-11. Stage-Discharge Relation for Mississippi River, Tarbert Landing, Louisiana

Limitations

The applicability of kinematic wave method is limited by the assumption that the friction slope is equal to the channel bed slope. Furthermore, the analytical solution is very restricted in practical applications because of kinematic "shock" formation. In the analytical solution, water is conceptually routed along a particular characteristic path which is independent of the other characteristic paths. This additional assumption can cause the formation of "shock." Numerical solutions have not been developed under this additional assumption. Besides, numerical solutions usually provide artificial "damping" and "smoothing" effects. Thus, there is usually no kinematic wave "shock" problem when a numerical solution is used. The modification to consider dynamic effects in the kinematic wave routing greatly extends the applicability of such a simple routing scheme. However, if there exists strong downstream effects, this revised procedure is also not valid. This results because the river response is governed by not only the continuity equation and momentum equation but also the downstream control equation, such as weir or bridge, etc., which is not considered in the formulation.

9.6 SEDIMENT ROUTING

Governing Equations for Sediment Routing

The sediment routing portion of the current model is similar to the formulation used by Simons et al., (1975). However, this model is further developed to consider the routing of sediment by sizes. The attention to routing sediment by sizes is increasing because different sizes of sediment have different uptake rates of other pollutants.

The movement of sediment from watersheds is governed by the equation of continuity for sediment and sediment transport equations. The amount of soil that could be transported is described by equations of sediment supply which are the amount of soil detachment from raindrop impact and surface runoff. The equations used in the model are described below.

Continuity for Sediment: The equation of continuity for sediment can be expressed as

$$\frac{\partial G_s}{\partial x} + \frac{\partial CA}{\partial t} + (1-\lambda)\frac{\partial PZ}{\partial t} = g_s \qquad (9\text{-}91)$$

in which

$$C = \frac{G_s}{Q} \qquad (9\text{-}92)$$

and G_s is the total sediment transport rate by volume per unit time, C is the sediment concentration by volume, Z is the depth of loose soil, λ is the soil porosity, and g_s is the lateral sediment inflow.

Sediment load can be further subdivided into different sizes. Then the continuity equation can be divided into N parts.

$$\frac{\partial G_{si}}{\partial x} + \frac{\partial C_i A}{\partial t} + (1-\lambda)\frac{\partial PZ_i}{\partial t} = g_{si} \qquad (9\text{-}93)$$

$$i = 1, \ldots, N$$

in which i is the size fraction index and N is the number of size fractions

$$C_i = \frac{G_{si}}{Q} \qquad (9\text{-}94)$$

$$C = \sum_{i=1}^{N} C_i \tag{9-95}$$

and

$$Z = \sum_{i=1}^{N} Z_i \tag{9-96}$$

Sediment Transport Equations: The sediment transport equation is used to determine the sediment transport capacity of a specific flow condition. Different transport capacities can be expected for different sediment sizes. For each sediment size, the transport rate includes the bed-load transport rate and the suspended load transport rate. The following equations are adopted to determine the sediment transport capacity for each size.

The Meyer-Peter, Müller equation is a simple and commonly used bed-load transport equation (see USBR, 1960). It is

$$q_b = \frac{12.85}{\sqrt{\rho}\,\gamma_s} (\tau_o - \tau_c)^{1.5} \tag{9-97}$$

in which

$$\tau_c = \delta_s(\gamma_s - \gamma)d_s \tag{9-98}$$

Here q_b is the bed-load transport rate in volume per unit width for a specific size of sediment, τ_c is the boundary shear stress acting on the grain, τ_c is the critical tractive force, ρ is the density of water, γ_s is the specific weight of sediment, γ is the specific weight of water, d_s is the size of sediment and δ_s is a constant dependent on flow conditions. Gessler (1965) shows that δ_s should be 0.047 for most flow conditions. If rilling develops on the overland flow surface, the value of δ_s should be lower.

It is not necessary to use the Meyer-Peter, Müller equation but an equation should be chosen which is applicable to the field condition. Other bed-load transport equations follow a similar form as Eq. (9-97). According to Shen (1971), the general bed-load function can be expressed as the following function form:

$$q_b = a_4(\tau_o - \tau_c)^{b_4} \tag{9-99}$$

in which a_4 and b_4 are constants.

The flow discharge Q and flow area A are determined in time and space by the water routing procedure described in Section 9.5. The corresponding value of τ_o is computed as follows. The mean flow velocity is

$$V = Q/A \tag{9-100}$$

Then, the boundary shear stress acting on the grain using Eq. (9-21) is

$$\tau_o = \frac{1}{8} \rho f_g V^2 = \frac{1}{8} \rho \frac{K_g}{N_r} V^2 \tag{9-101}$$

The sediment concentration profile which relates the sediment concentration with depth above the bed (see Einstein, 1950) can be written as

$$\frac{C_\xi}{C_a} = \left(\frac{R-\xi}{\xi} \frac{a'}{R-a'}\right)^w \tag{9-102}$$

in which C_ξ is the sediment concentration at the distance ξ from the bed, C_a is the known concentration at a distance a' above the bed, and w is a parameter defined as

$$w = \frac{V_s}{\kappa U_*} \tag{9-103}$$

Here V_s is the settling velocity of the sediment particles, κ is the Karman constant (assumed 0.4), and U_* is the shear velocity of the flow defined as

$$U_* = \left(\frac{\tau_*}{\rho}\right)^{1/2} \tag{9-104}$$

Note that

$$\tau_* = \frac{1}{8} f \rho V^2 \tag{9-105}$$

f is the overall Darcy-Weisbach resistance factor which is given in Section 9.5.

A logarithmic velocity profile is commonly adopted to describe the velocity distribution in turbulent flows. For simplicity, a logarithmic velocity profile is assumed in this study. The equation is

$$\frac{u_\xi}{U_*} = B + 2.5 \,\ell n\left(\frac{\xi}{\eta_s}\right) \tag{9-106}$$

in which u_ξ is the point mean velocity at the distance ξ from the bed, B is a constant dependent on roughness, and η_s is the roughness height.

The integral of suspended load above the a' level in the flow is obtained by combining Eqs. (9-102) and (9-106) or

$$q_s = \int_{a'}^{R} u_\xi C_\xi d\xi = C_a U_* \int_{a'}^{R} [B + 2.5 \ln(\frac{\xi}{\eta_s})](\frac{R-\xi}{\xi}\frac{a'}{R-a'})^w d\xi \tag{9-107}$$

Let

$$\sigma = \frac{\xi}{R} \tag{9-108}$$

and

$$G = \frac{a'}{R} \tag{9-109}$$

The equation becomes

$$q_s = C_a U_* a' \frac{G^{w-1}}{(1-G)^w} \{[B + 2.5 \ln(\frac{R}{\eta_s})] \int_G^1 (\frac{1-\sigma}{\sigma})^w d\sigma + 2.5 \int_G^1 \ln\sigma (\frac{1-\sigma}{\sigma})^w d\sigma\} \tag{9-110}$$

According to Einstein (1950), the concentration near the "bed layer" C_a is related to the bed-load transport rate q_b by the expression

$$q_b = 11.6\, C_a U_* a' \tag{9-111}$$

in which a' is now defined as the thickness of the bed layer which is twice the size of sediment.

The average flow velocity V is defined by the equation

$$V = \frac{\int_0^R u_\xi d\xi}{\int_0^R d\xi} \tag{9-112}$$

Using Eq. (9-106)

$$\frac{V}{U_*} = B + 2.5 \ln(\frac{R}{\eta_s}) - 2.5 \tag{9-113}$$

Einstein (1950) defined the two integrals in Eq. (9-110) as

$$J_1 = \int_G^1 (\frac{1-\sigma}{\sigma})^w d\sigma \tag{9-114}$$

and

$$J_2 = \int_G^1 (\frac{1-\sigma}{\sigma})^w \ln\sigma \, d\sigma \tag{9-115}$$

The integrals J_1 and J_2 cannot be integrated in closed form for most values of w so a numerical integration is necessary. An efficient numerical method of determining J_1 and J_2 was developed by Li (1974) and is adopted in this study.

The substitution of Eqs. (9-111), (9-113), (9-114), and (9-115) into Eq. (9-110) yields

$$q_s = \frac{q_b}{11.6} \frac{G^{w-1}}{(1-G)^w} [(\frac{V}{U_*} + 2.5)J_1 + 2.5 J_2] \tag{9-116}$$

The total load per unit width of channel is

$$q_t = q_b + q_s \tag{9-117}$$

Eq. (9-117) gives the total load per unit width of channel for a uniform size of sediment in the bed. When considering the transport by different sizes and the entire width of channel, the sediment transport capacity of the ith size fraction of a sediment mixture would be

$$G_{ci} = P F_{ai} q_t \tag{9-118}$$

in which G_{ci} is the sediment transport capacity for the ith size, and F_{ai} is the adjusted fraction of the ith size sediment. This will be determined later in the next section.

Equations for Sediment Supply

The sediment supply rate is another determining factor in the actual sediment transport rate. Simons et al. (1975) states that the sediment supply depends on the initial depth of loose soil left from previous storms, the amount of soil detachment by raindrop impact, and the amount of soil detachment by flow.

Soil Detachment by Raindrop Impact: Raindrop impact is a primary source of kinetic energy for detaching soil from any unprotected land surface. Ellison (1944) made a comprehensive study about the raindrop splash. Law and Parson (1943), Young and Wiersma (1973), and Mutchler (1967) greatly contributed to the study on detachment by raindrop impact. The conclusion drawn from these studies is that soil detachment is a function of erosivity of rainfall and the erodibility of the soil particles. The erosivity is directly related to the energy produced by raindrop impact and is generally formulated as a power function of rainfall intensity, size of droplet, cover condition, and terminal velocity of the drop. More recently, Carter et al., (1974) reported energy produced by raindrop impact is also a function of air temperature, season of the year, and storm duration. In this study the potential rate of soil detachment by raindrop impact is assumed as a power function of rainfall intensity as given by Meyer (1971).

$$D_r = a_5 I^{b_5} (1 - \frac{Z_w}{Z_m})(1-C_g)(1-C_c) \qquad (9\text{-}119)$$

in which I is the rainfall intensity, a_5 is a parameter depending on soil characteristics and b_5 is a constant (b_5 = 2.0, see Meyer, 1971). The term Z_w is the depth of water plus the loose soil depth. The term Z_m is the maximum penetration depth of raindrop splash. According to Mutchler and Young (1975), Z_m can be equal to three times the median raindrop size. The median raindrop size is often expressed as a power function of rainfall intensity. Therefore, Z_m can be written as:

$$Z_m = 3(2.23I^{0.182}) \qquad (9\text{-}120)$$

in which Z_m is in millimeters.

Eq. (9-119) is valid when Z_w is less than Z_m. When the depth of loose soils plus the water depth is greater than Z_m, D_r is zero, i.e.,

$$D_r = 0 \ , \quad \text{if} \quad Z_w > Z_m \qquad (9\text{-}121)$$

If the tree is very tall, the drops from leaves can regain the terminal velocity and have the erosive potential. However, it is common that the ground cover density under trees is usually very high which would protect the surface. The canopy cover density, C_g , in Eq. (9-119) should be the portion that can protect the surface effectively from raindrop impact.

The potential rate of loose soil detachment, D_r, is expressed in units of depth per unit time. Thus, the new amount of loose soil available for transport is:

$$Z_i = Z_i + F_i \ D_r \ \Delta t \qquad (9\text{-}122)$$

where Z_i is respectively the amount of loose soil available and F_i is the original percentage of sediment size in a given size fraction designated by i.

The percentage in each size fraction on the surface changes over time due to armoring. Water transports the smaller sediment sizes more easily and leaves the larger size fractions behind. Thus, the percentages of surface material need adjustment each time step. If the total loose soil depth is greater than D_a (the thickness of the armor layer), the adjusted percentages, F_{ai} can be written as:

$$F_{ai} = \frac{Z_i}{Z} \qquad (9\text{-}123)$$

If the total loose soil depth, Z, is less than D_a , then the adjusted percentages must account for the layer of undisturbed soil which is distributed according to the original percentages plus the loose soil which covers it.

$$F_{ai} = \frac{1}{D_a} [Z_i + F_i (D_a - Z)] \qquad (9\text{-}124)$$

The thickness of the armor layer can be determined as the maximum size of particles in motion. It is usually assumed that D_{84} (the size of sediment for which 84 percent of the sample is finer by weight) is the representative size of the armor layer.

<u>Soil Detachment by Surface Runoffs</u>

The amount of soil detachment by surface runoff is determined by comparing the total sediment transport capacity to the total available amount of loose soil. The total sediment transport capacity is

$$G_c = \sum_{i=1}^{N} G_{ci} \qquad (9\text{-}125)$$

By substituting the total sediment transport capacity G_c into the transport rate in Eq. (9-91), the total potential change in loose soil can be determined.

$$\Delta Z^p = \frac{\partial Z}{\partial t} \Delta t \qquad (9\text{-}126)$$

If $\Delta Z^p \geq -Z$, the loose soil storage is enough for transport and no detachment of soil by surface runoff is expected. Soil is detached if $\Delta Z^p \leq -Z$ and the amount of detachment is

$$D = -D_f[\Delta Z^p + Z] \qquad (9\text{-}127)$$

in which D is the total amount of detached soil and D_f is defined as a detachment coefficient with values ranging from 0.0 to 1.0 depending on soil erodibility. As an example, if the flow were over a nonerodible surface, the value for D_f

would be zero. If the surface is completely paved with gravel, D_f is equal to zero. If one were considering flow in a river where the riverbed is always loose, the value for D_f would be unity.

The new amount of loose soil should be further modified as follows:

$$Z_i = Z_i + D F_i \tag{9-128}$$

in which Z_i is calculated for each size fraction of sediment.

Numerical Procedure for Sediment Routing

The following numerical procedure for sediment routing is used to couple the equations governing sediment motion with the water routing procedure described in Section 9.5.

The sediment transport capacity is determined by Eq. (9-118) for a given sediment size for the flow conditions obtained by water routing. The potential sediment load concentration for a given sediment size i is then

$$C_i^p = \frac{G_{ci}}{Q} \tag{9-129}$$

Using the same finite difference approximation as that in the water routing procedure (see Fig. 9-6) the potential change in loose soil storage for a given sediment size is determined by utilizing Eqs. (9-93) and (9-129). That is

$$\Delta z_i^p = \frac{1}{(1-\lambda)p} \{\theta[G_{si_j}^{\ n+1} - C_i^p Q_{j+1}^{n+1})(1-a) + (G_{s_i}^{\ n} - G_{si_j}^{\ n})a]$$

$$- [(C_i^p A_{j+1}^{n+1} - C_{i_{j+1}}^{\ n} A_{j+1}^n)(1-b) + (C_{i_j}^{\ n+1} A_j^{n+1} - C_{i_j}^{\ n} A_j^n)b]$$

$$+ \frac{\Delta t}{2} [(1-b)g_{si_{j+1}}^{\ n+1} + bg_{si_j}^{\ n+1} + (1-b)g_{si_{j+1}}^{\ n} + bg_{si_j}^{\ n}]\} \tag{9-130}$$

The sediment trnasport rate is dependent on both the available supply of sediment and the transporting capacity of the flow. If $\Delta Z_i^p \geq -Z_i$, the availability is greater than the transporting capacity. Consequently, the bed-material load transport rate is equal to its transporting capacity or

$$C_{i_{j+1}}^{\ n+1} = C_i^p \tag{9-131}$$

and the actual change in Z_i is

$$\Delta Z_i = \Delta Z_i^p \qquad (9\text{-}132)$$

If $\Delta Z_i^p < -Z_i$, the availability of material is less than the transporting capacity. The transport rate is limited by the availability of loose soil and the bed material is, therefore,

$$C_{i_{j+1}}^{n+1} = \{(1-\lambda)pZ_i + \frac{\Delta t}{2}[(1-b)g_{si_{j+1}}^{n+1} + bg_{si_j}^{n+1}$$

$$+ (1-b)g_{si_{j+1}}^{n} + bg_{si_j}^{n}] + \theta[Gsi_j^{n+1}(1-a)-(G_{si_{j+1}}^{n} - G_{si_j}^{n})a]$$

$$+ C_{i_{j+1}}^{n} A_{j+1}^{n}(1-b)-(C_{i_j}^{n+1}A_j^{n+1} - C_{i_j}^{n} A_j^{n})b\} \;/\; [A_{j+1}^{n+1}(1-b)+\theta Q_{j+1}^{n+1}(1-a)]$$

(9-133)

and

$$\Delta Z_i = -Z_i \qquad (9\text{-}134)$$

The sediment transport rate G_{si} is determined by

$$G_{si_{j+1}}^{n+1} = C_{i_{j+1}}^{n+1} Q_{j+1}^{n+1} \qquad (9\text{-}135)$$

and the amount of loose soil available at the next time increment is:

$$Z_i = Z_i + \Delta Z_i \qquad (9\text{-}136)$$

Again, if a and b are zero the above sediment routing scheme becomes identical to that reported by Simons et al. (1975).

9.7 EXAMPLES OF APPLICATION

Examples of successful applications of the above water and sediment routing methods are: (1) a mathematical simulation of water and sediment routing and yields from small watersheds in Beaver Creek Watershed, Arizona (see Simons et al., 1975), (2) a mathematical simulation of water and sediment yields from roads (see Simons et al. 1977b), (3) mathematical models for predicting on-site soil erosion (see Li et al., 1977c and Simons et al., 1977c), (4) a mathematical simulation of infiltration and subsurface flow routing (see

Li et al., 1976a), and (5) a simple watershed model using a small programmable calculator (see Simons et al., 1977a).

Water Routing

Some examples of application of the water routing schemes were demonstrated in Section 9.5. For practical applications, Simons et al. (1977a) developed a simple watershed model using a small programmable calculator (Hewlett Packard HP-97) to simulate storm runoff from a two-plane and one-channel representation of watersheds. The infiltration response is calculated using an explicit formulation of Green-Ampt (see Section 9.4) and the routing is accomplished by an analytical solution of the kinematic wave approximation (see Section 9.5). The definition sketch of this simple watershed geometry is shown in Fig. 9-12. An example of the comparison between

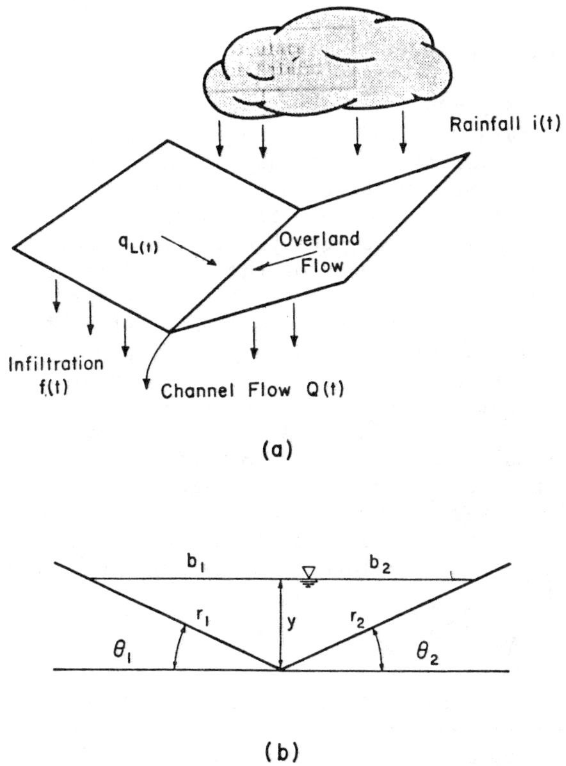

Fig. 9-12. Definition Sketch of Simple Watershed Geometry

the measured and calculated runoff of watershed 2H, an agricultural watershed in Nebraska, is shown in Fig. 9-13. The agreement appears to be quite good.

On-Site Soil Erosion

Li et al. (1977c) applied the water and sediment routing scheme to evaluate on-site soil erosion. The soil erosion data was collected by Van Lieu (1976) on mine spoil areas on Edna Mine in the Yampa Coal Field of Routt County, Colorado. Fig. 9-14 shows good agreement of measured and computed water and sediment yields. The water and sediment hydrographs run number 1 are plotted in Figs. 9-15 and 9-16. The results are acceptable. Fig. 9-17 gives the original sediment size distribution curve for the mine spoil plot. It is apparent that the size distribution curve of the transported material is shifted to the left showing the armoring effect. The larger size fractions are not transported and thus the size distribution curve is expected to shift. The actual versus simulated size distribution curves agree very well.

The second set of data tested was collected by Kilinc (1972). The data was from a soil plot of adjustable slope in an artificial rainfall facility at Colorado State University. Fig. 9-18 shows the simulated and observed water yields to agree closely. Fig. 9-19 shows the comparison of the observed and simulated sediment yield. The simulated results agree reasonably well except for those runs with the lowest rainfall intensity (1.25 in./hr). The errors at the lowest rainfall intensity may be due to a consistantly high percentage of the infiltrated volume to the application rate (21 percent to 40 percent of rainfall) in these runs. The relatively small errors in the infiltration prediction will result in a significant error in sediment estimation. Another source of error may be due to uneven slope profiles. Fig. 9-20 shows the comparison of measured and simulated sediment concentrations over the 60 minute rainfall duration. Rainfall intensity was 3.65 in./hr, and all slopes were used for comparison. The simulated results agree well with the measured data.

Road Sediment Yield

Following the same procedures of routing water and sediment as outlined earlier, a road sediment yield model has been developed for estimating the amount of water and sediment at any road location at any time (see Simons et al., 1977b).

To demonstrate the utility of the road sediment model, some hypothetical examples are used. The following three types of road prism which are commonly found in the forest

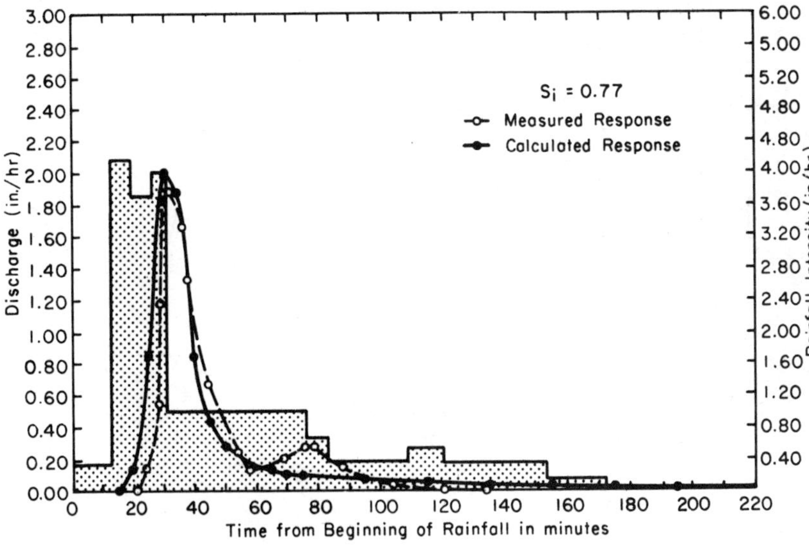

Fig. 9-13. Rainfall-Runoff for Watershed 2H, Hastings, Nebraska, Using the Two Planes-One Channel Model

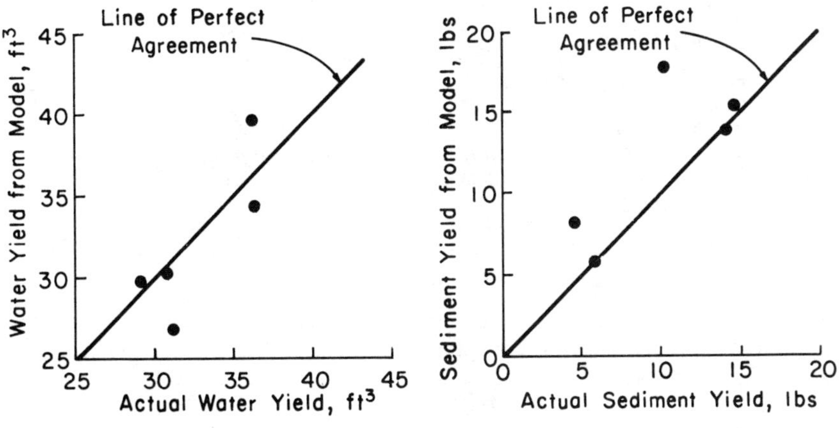

Fig. 9-14. Comparison of Measured and Simulated Water and Sediment Yields of Edna Mine Soil Plot

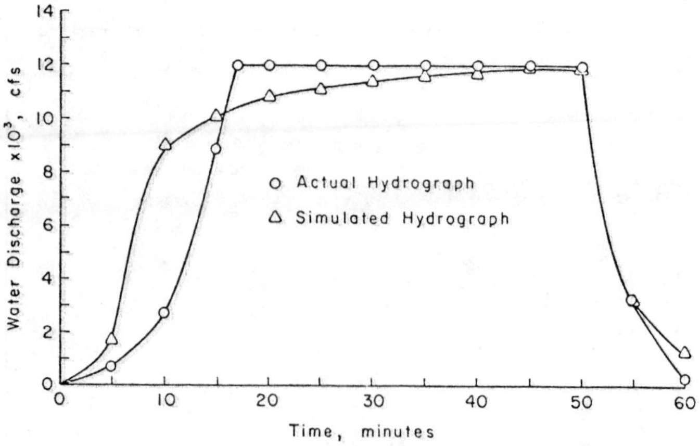

Fig. 9-15. Comparison of Measured and Simulated Water Hydrographs for Run Number 1 of Edna Mine Soil Plot

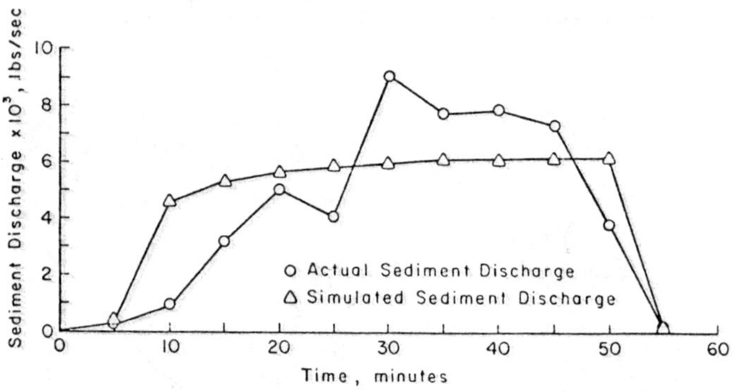

Fig. 9-16. Comparison of Measured and Simulated Sediment Hydrographs for Run Number 1 of Edna Mine Soil Plot

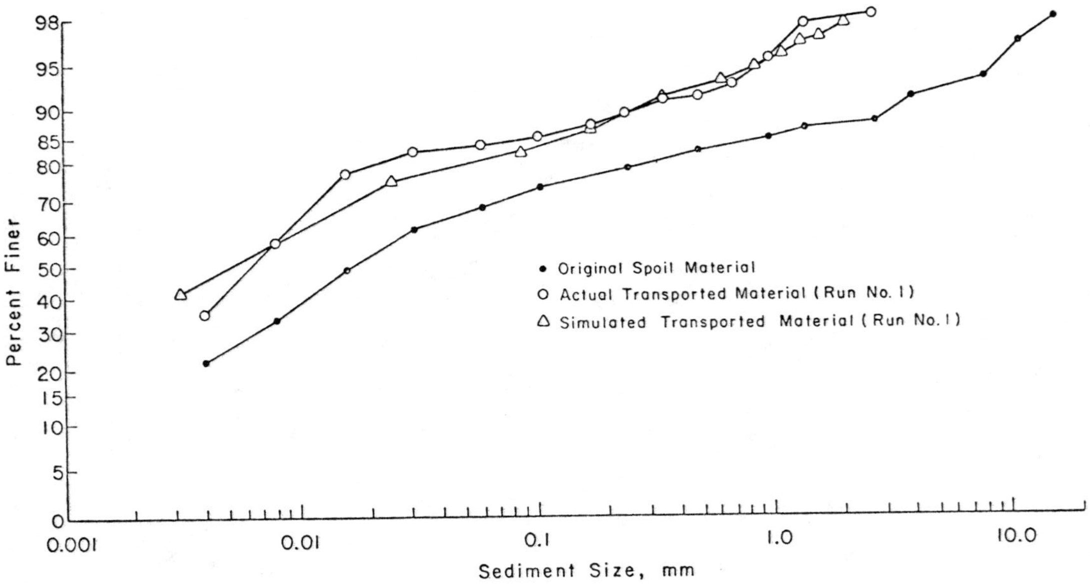

Fig. 9-17. Comparison of Sediment Size Distribution Curves for Run Number 1 of Edna Mine Soil Plot

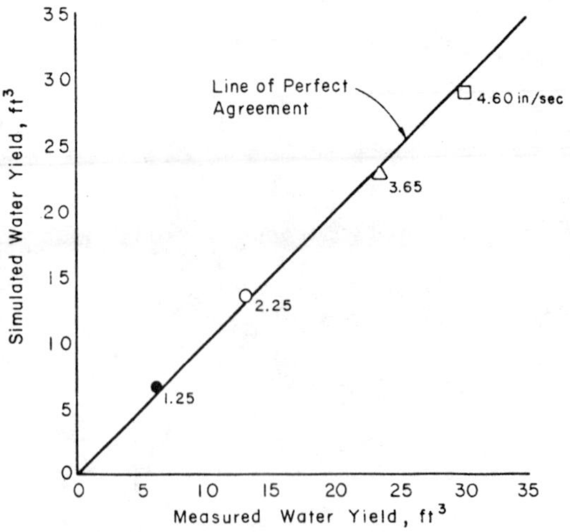

Fig. 9-18. Comparison of Measured and Simulated Water Yields (Kilinc's Data)

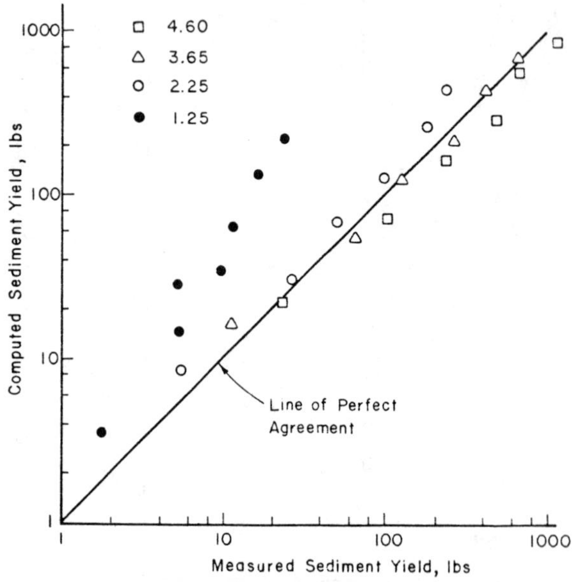

Fig. 9-19. Comparison of Observed and Simulated Sediment Yields Using Kilinc's Data

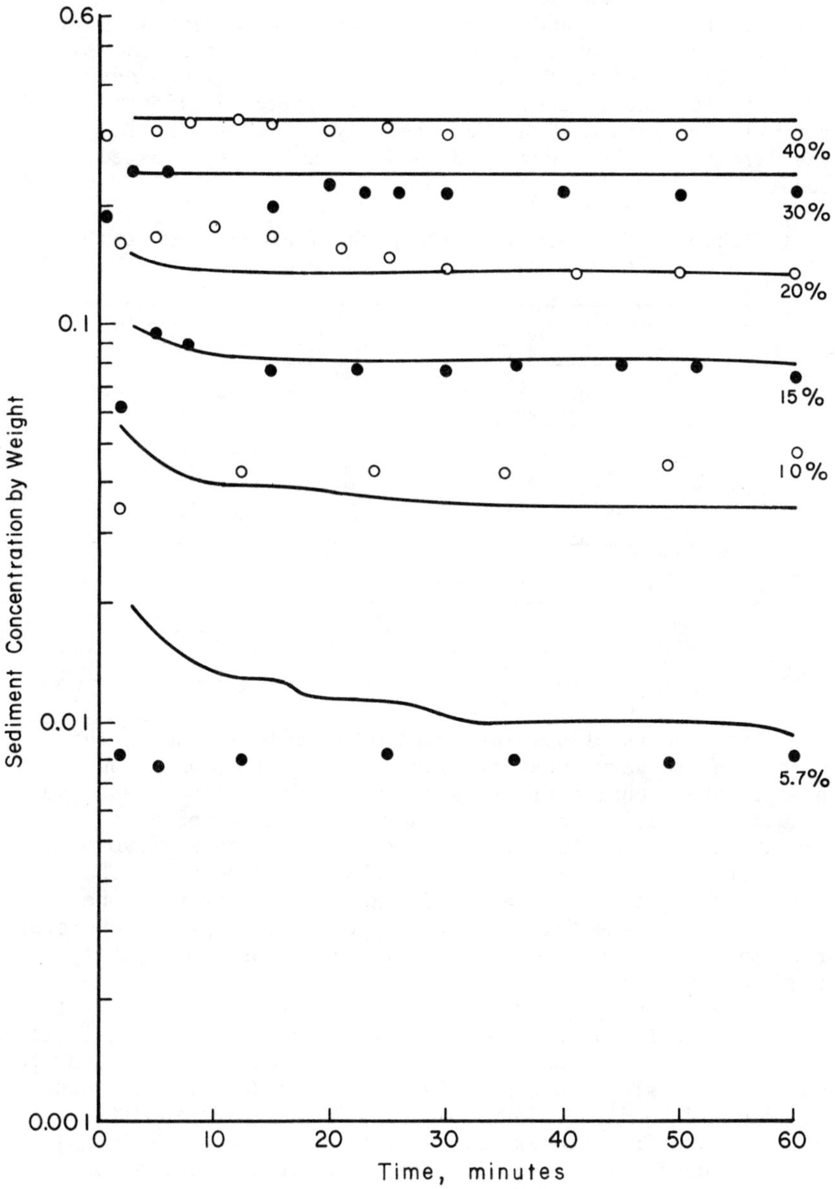

Fig. 9-20. Comparison of Measured and Simulated Sediment Concentrations (Kilinc's Data, Rainfall Intensity = 3.65 in./hr)

logging roads were selected for demonstration: Types A, B, and C in Fig. 9-21. Type A road prism has only overland flow units. Type B road prism has overland flow, ditch flow and culvert flow units, and Type C road prism is essentially the same as Type B layout, except the flow direction of the road surface is different. Table 9-1 gives a summary of the geometry for the road prism. For simplicity, a uniform intensity rainstorm with the intensity of 3.0 in./hr and the duration of 55 minutes is used for all of the tested conditions.

Table 9-1. Dimension of Hypothetical Road Section

	Type A	Type B	Type C
Length	30.5 m	30.5 m	30.5 m
Width	30.5 m	30.5 m	30.5 m
Slope	0.316 (3:1 horizontal to vertical)	0.316 (3:1 horizontal to vertical)	0.316 (3:1 horizontal to vertical
Length	30.5 m	30.5 m	30.5 m
Width	12.2 m	12.2 m	12.2 m
Slope	0.10	0.10	0.10
Side Slope	-	3.0	3.0
Longitudinal Slope	-	0.15	0.15
Length	-	30.5 m	30.5 m
Diameter	-	0.9 m	0.9 m
Slope	-	0.25	0.25
Length	-	42.7 m	42.7 m

The amounts of simulated sediment yield at the outlet were found to vary substantially with the flow conditions, physiographic characteristics of cut or fill slopes, and the ground cover conditions. In order to demonstrate the variations of sediment yield due to these factors, five hypothetical cases are studied. Table 9-2 summarizes the simulated results. The water discharge hydrograph at the outlet is the same for all three types of road prism. However, the amount of simulated sediment yield varies substantially from case to case.

Case 1 has no cover for all flow units. For Case 2, the road surface is paved with gravel while both cut and fill slopes have no cover. In Case 3, there is 30 percent ground cover for cut and fill slopes and no cover for the road surface. Case 4 has a paved road surface with gravel. The cover conditions on both cut and fill slopes are the same as in Case 3. In Case 5, the ground cover condition for both cut and fill slopes is increased to about 100 percent, and the road surface is paved with gravel. It is worthwhile to note that the simulation results seem to agree well with physical reality. For all cases, Type A produces the largest amount of sediment as

Type A:

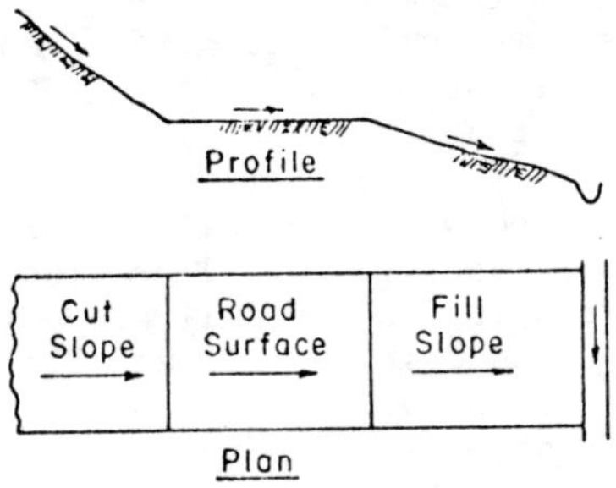

Type B:

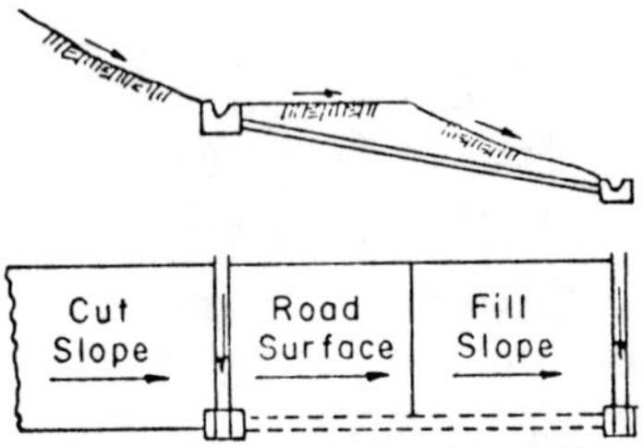

Type C:

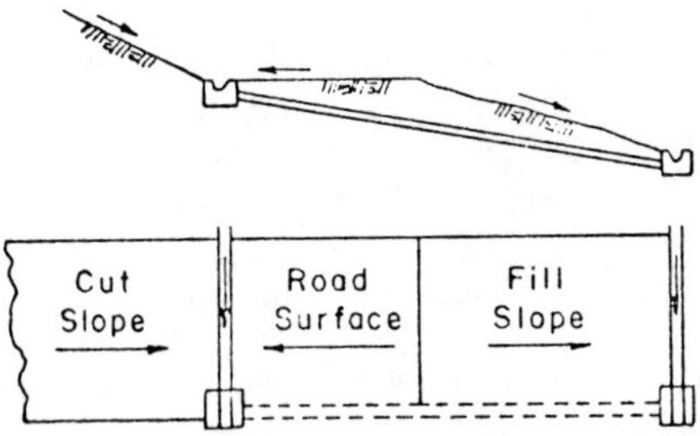

Fig. 9-21. Typical Road Prism

expected. This is because the outslope design of Type A prism will provide the highest degree of runoff concentration, this usually causes the severe soil erosion if no control measure such as grass or gravel paving is implemented. Other interesting points are that Type B will yield more sediment than Type C if the fill slope is not protected by ground cover, and the trend is reversed when the fill slope is protected by grass. This is because Type B will provide a larger runoff rate and cause more erosion on the fill slope if the surface is not protected as assumed in Cases 1 and 2. If the fill slope is protected by grass as in Cases 3, 4, and 5 the flow on the slope is substantially retarded. In these cases, part of the sediment eroded in the road surface is trapped in the fill slope because of the low transport capacity. Thus, the Type B prism will produce a smaller amount of sediment if the fill slope is protected.

Table 9-2. Simulated Sediment Load (in kilograms)

Case No.	Type A	Type B	Type C
1	7390	5275	4368
2	6550	5058	4150
3	1265	599	665
4	689	445	449
5	207	109	199

Note: Case 1 = No ground cover.
Case 2 = No ground cover on the cut and fill slopes. Road surface is paved with gravel.
Case 3 = 30% ground cover on cut and fill slope. No ground cover on road surface.
Case 4 = 30% ground cover on the cut and fill slopes and the road surface is paved with gravel.
Case 5 = Close to 100% ground cover on the cut and fill slopes. The road surface is paved with gravel.

To test the effect of ground cover on the sediment yield, different percents of ground cover are used for the Type C road prism. Fig. 9-22 gives the variation of sediment yield versus ground cover density in terms of percentage. The result indicates that the reduction in sediment yield tends to reach an equilibrium value when ground cover reaches about 60 percent.

The spacing of cross drains is an important factor in the design of a road system. Fig. 9-23 shows a power function relationship between the spacing of cross drains and the sediment yield. Note that Fig. 9-23 gives the ratio of sediment yield to that of 100 ft cross drain spacing. In general, the total sediment yield at an outlet will increase

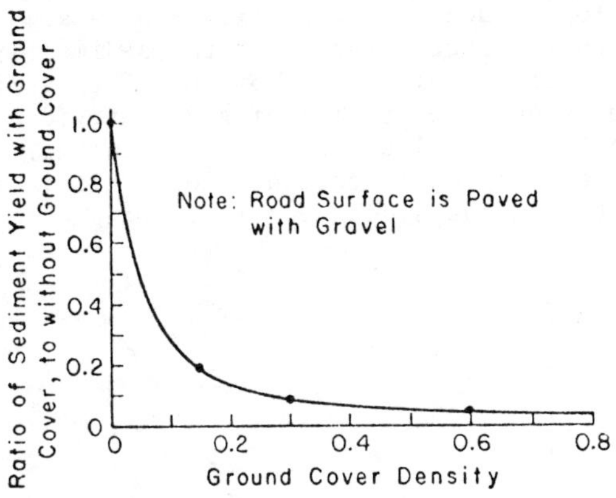

Fig. 9-22. Variation of Sediment Yield Versus Ground Cover Density

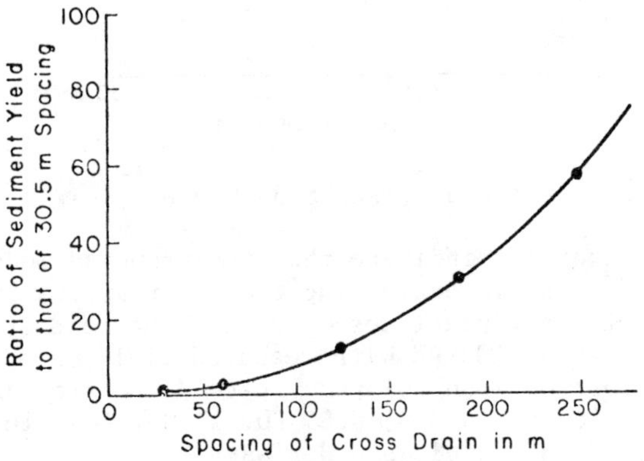

Fig. 9-23. Spacing of Cross Drain in Feet

as cross drain spacing increases due to enlargement of the total area.

The side slopes such as cut or fill slopes are also important factors in determining the total sediment yield. Let the sediment yield from a 5:1 (horizontal to vertical) slope be a base value. The ratios of sediment yields from other slopes to that of 5:1 slope are given in Fig. 9-24. For this example case, the total sediment yield will increase about 25 percent when the side slope is increased to 2:1.

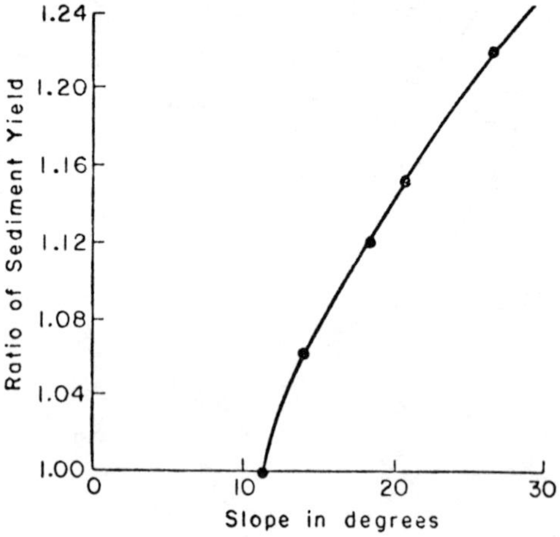

Fig. 9-24. Sediment Yield Versus Side Slope

The above examples indicate that the proposed model has the potential for use in evaluating the alternatives of road designs or maintenance practices. As yet, the road sediment model has not been validated with measured field data. A cooperative effort between Colorado State University and Coweeta Hydrologic Laboratory, U.S. Forest Service, in testing and refinement of the model is under way.

Watershed Water and Sediment Yield

Simons et al. (1975) tested the applicability of their model on the Beaver Creek Watershed, Arizona. Five runoff events in Watershed 1 and one runoff event in Watershed 17 were used.

Watershed 1 is a small drainage catchment with an area of 313.6 acres which has been clear cut. The five storm events in Watershed 1 used in this study area occurred on November 22, 1965, November 24, 1965, November 25, 1965, September 6, 1967, and September 5, 1970. The latter is known as the "Labor Day" storm.

Watershed 17 has an area of 287.4 acres. The only storm available for testing is the "Labor Day" storm of September 5, 1970.

An example of the comparison between the simulated and the measured water hydrographs is shown in Fig. 9-25. The agreement between the measured and the simulated water hydrographs is, in the most part, satisfactory. Because there is no measured sediment hydrograph, no comparison can be made. Other comparisons on water yield, peak water flow, time to peak water flow, and sediment yield are given respectively in Figs. 9-26, 9-27, 9-28, and 9-29. These figures show that the water and sediment routing model simulated the shape, volume, peak flow, and time to peak flow of the six water hydrographs, and the sediment yield from Watershed 1 and Watershed 17 within approximately 30 percent. Satisfactory results were obtained for different size storms and in different watersheds by using only one set of model parameters. This verifies that the model could be used to synthesize missing data and to predict the response of watersheds to various types of watershed management practices. Also, it has been demonstrated that the model could be used to estimate flood flows from ungaged watersheds.

The transferability of the model is one of the main advantages of this physical process simulation model over the conventional methods such as the Unit Hydrograph and Universal Soil loss equation. For example, the Labor Day (September 5, 1970) storm produced approximately 2.2 times the surface runoff in Watershed 17 than in Watershed 1, but only about 0.4 times the yield of sediment. There was more surface runoff and less sediment yield in Watershed 17 because it is longer and narrower and its average slope is less than in Watershed 1.

The computed sediment hydrographs for a large Labor Day storm(September 5, 1970) and for a small storm (November 22, 1965) on Watershed 1 are given in Figs. 9-30 and 9-31, respectively. The shapes of sediment and water hydrographs are similar for the large storm but different for the small storm. For the small storm tested, the peak of the sediment hydrograph occurs earlier than that of the water hydrograph because the sediment supply rate is greater in the early stage of surface runoff. This effect is not significant for a larger storm.

An important aspect in watershed management is to predict watershed treatment effects. The vegetation treatment effects on water and sediment yields are estimated by changing the canopy cover density and the ground cover density in overland flow units. Based on the storms of September 5, 1970 and September 6, 1967, the effects of vegetation treatment on the water and sediment yields from Watershed 1 have been evaluated. As shown in Figs. 9-32 and 9-33, for a constant and undisturbed ground cover of 65 percent, water yield, sediment yield and the peak flow rates from these two storms are increased as the canopy cover density is decreased.

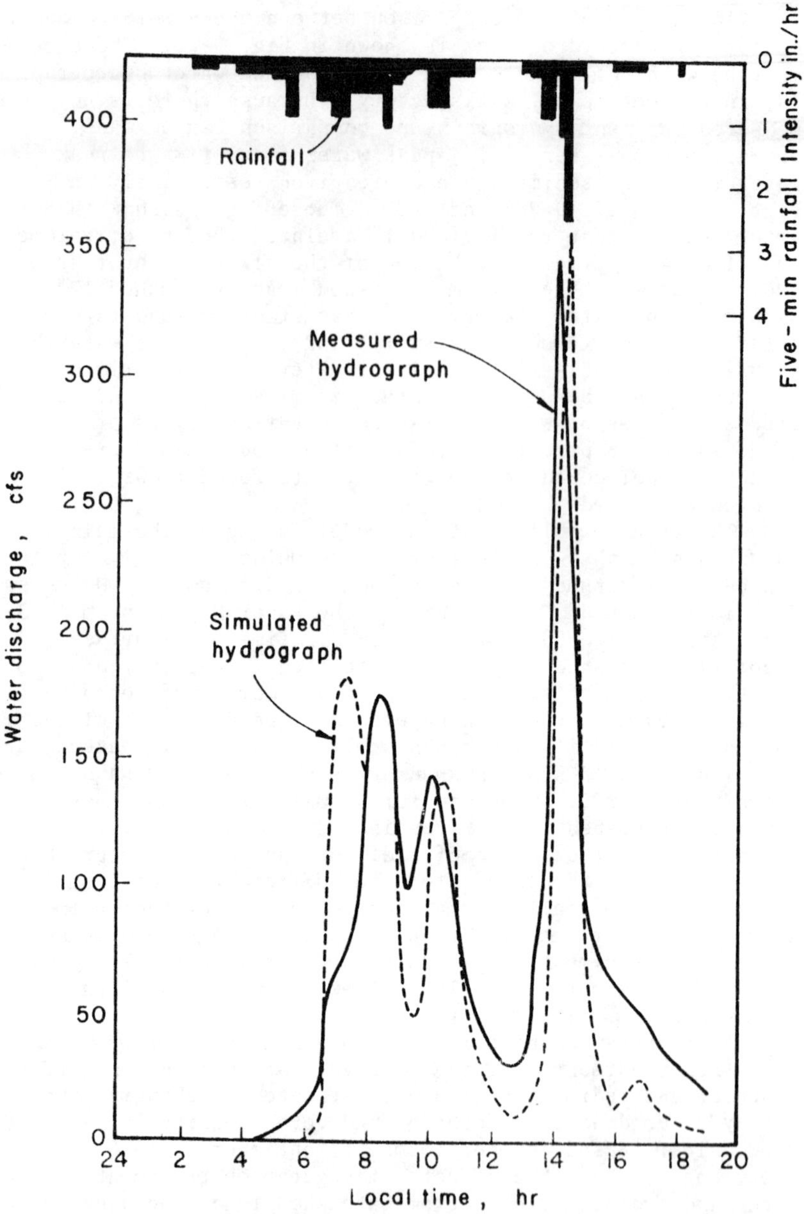

Fig. 9-25. Water Hydrograph from Watershed 17 for the September 5, 1970 Storm (after Simons et al, 1975)

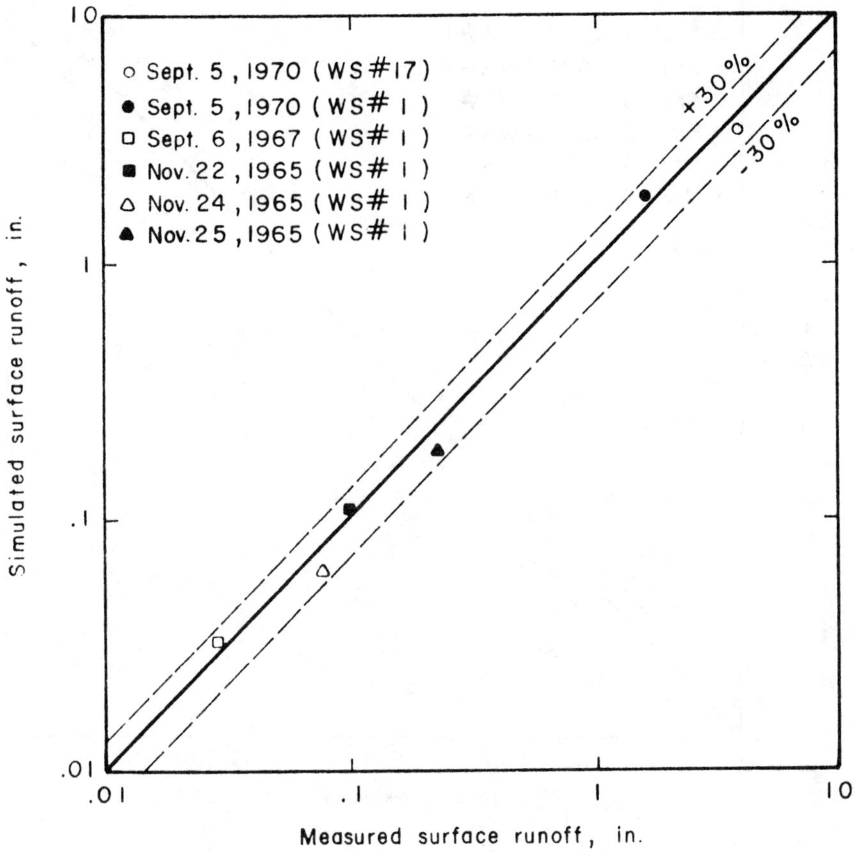

Fig. 9-26. Comparison of Measured and Simulated Water Yield (after Simons et al., 1975)

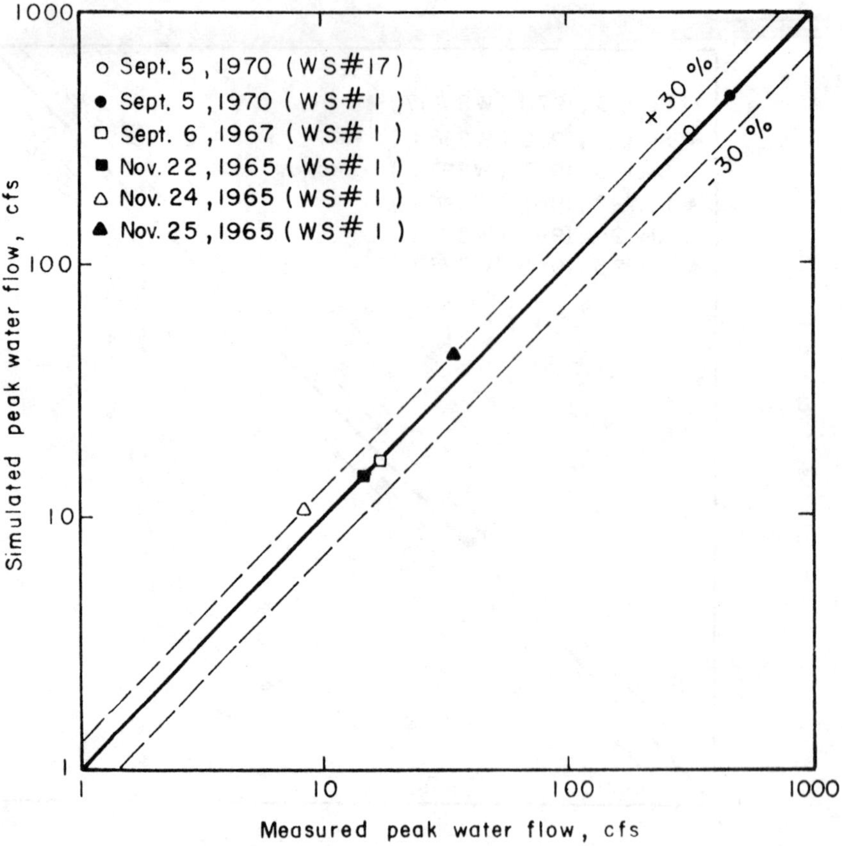

Fig. 9-27. Comparison of Measured and Simulated Peak Water Flows (after Simons et al., 1975)

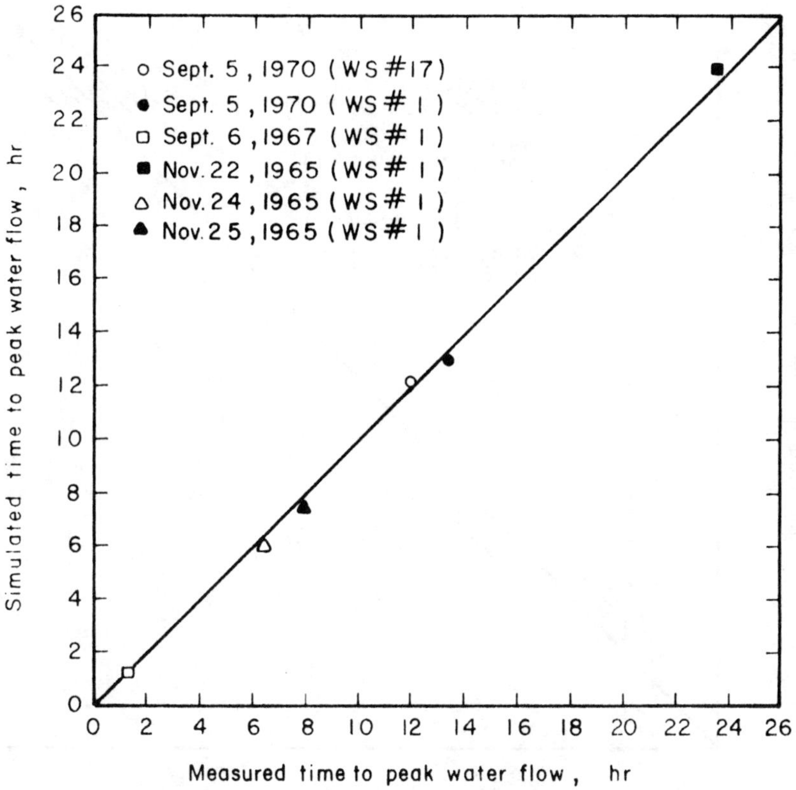

Fig. 9-28. Comparison of Measured and Simulated Time to Peak Water Flow (after Simons et al., 1975)

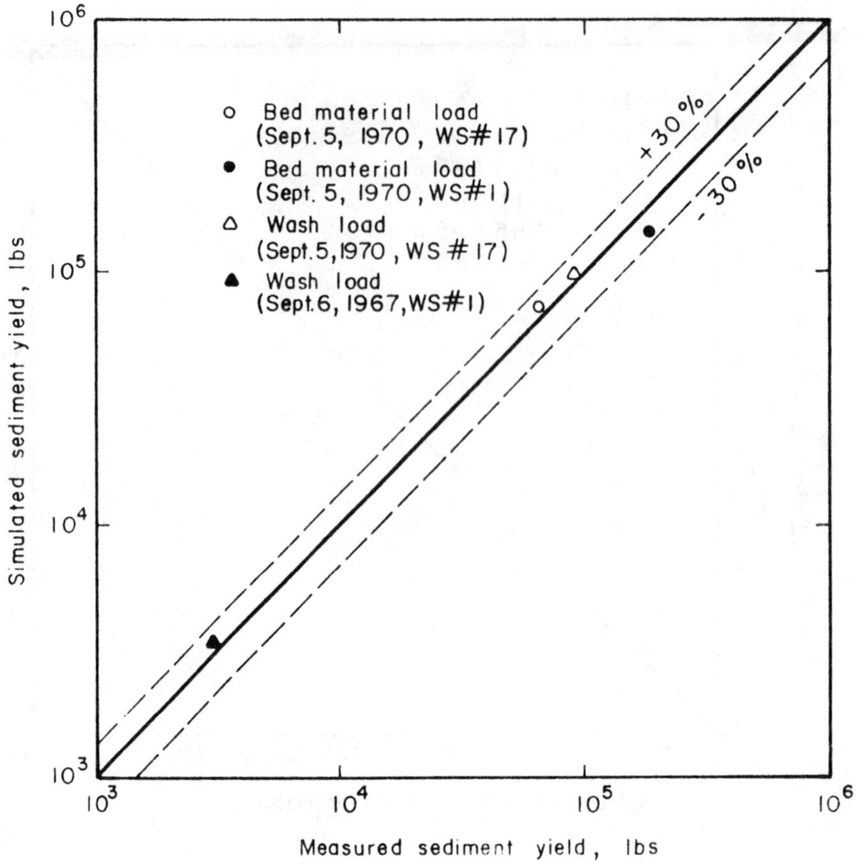

Fig. 9-29. Comparison of Measured and Simulated Sediment Yields (after Simons et al., 1975)

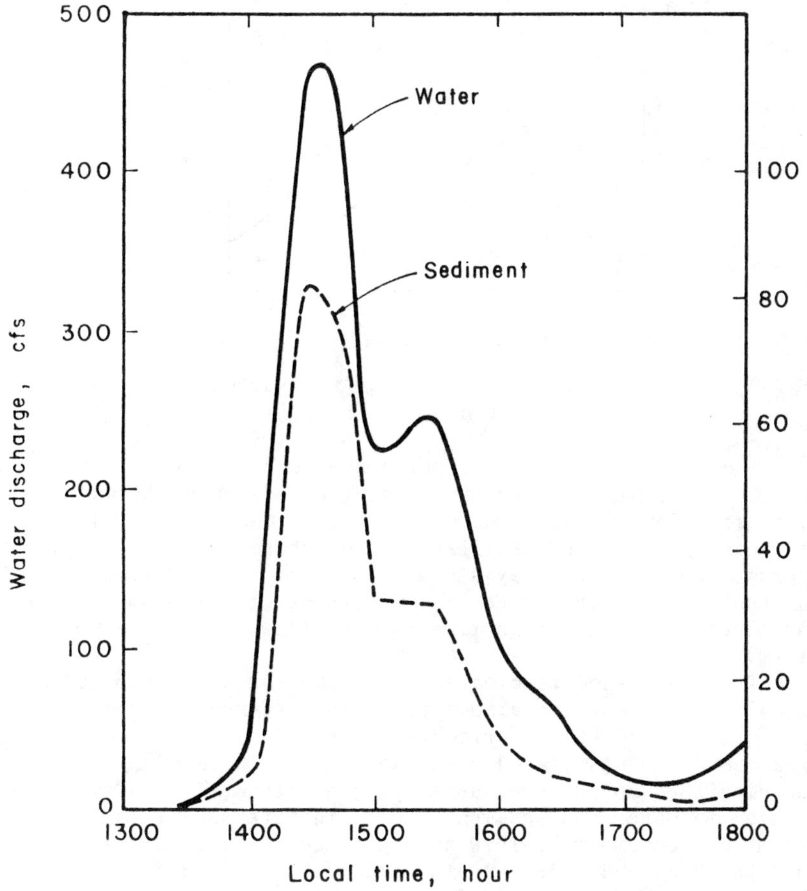

Fig. 9-30. Computed Hydrographs for the September 5, 1970 Storm on Watershed 1 (after Simons et al., 1975)

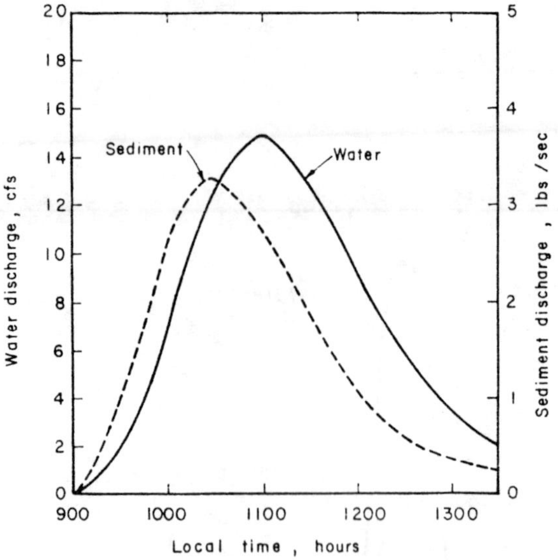

Fig. 9-31. Computed Hydrographs for the November 22, 1965 Storm on Watershed 1 (after Simons et al., 1975)

The reduction in interception caused by removing the vegetation results in the increase of excess rainfall and loose soil detachment. These effects are much more pronounced in Watershed 1 for the smaller size of storm than for large storms like the Labor Day storm. The time to peak flow is shortened as the canopy cover is decreased for the small storm, but there is no change in time to peak flow for the large storm.

If a watershed is clear-cut and the forest litter, tree mulch, rocks, etc. are also removed in different degrees, or if the ground cover is seriously destroyed by a burning treatment, the associated response can be estimated by changing the ground cover density in overland flow units. The changes in water and sediment yields in Watershed 1 for the storms of September 5, 1970 and September 6, 1967 are shown in Figs. 9-34 and 9-35. As the ground cover is decreased, the total surface runoff and peak water flow are increased moderately, the sediment yield and peak sediment flow are increased greatly and the time to peak flow shortened slightly. The effect on water yield is more pronounced for a smaller storm but nearly the same effects on sediment yield are obtained for both small and large storms.

As explained, the ground cover density is an important factor in controlling sediment yield. If a watershed is under the clear-cut treatment but there is a proper management of ground cover, the sediment yield may not be increased significantly.

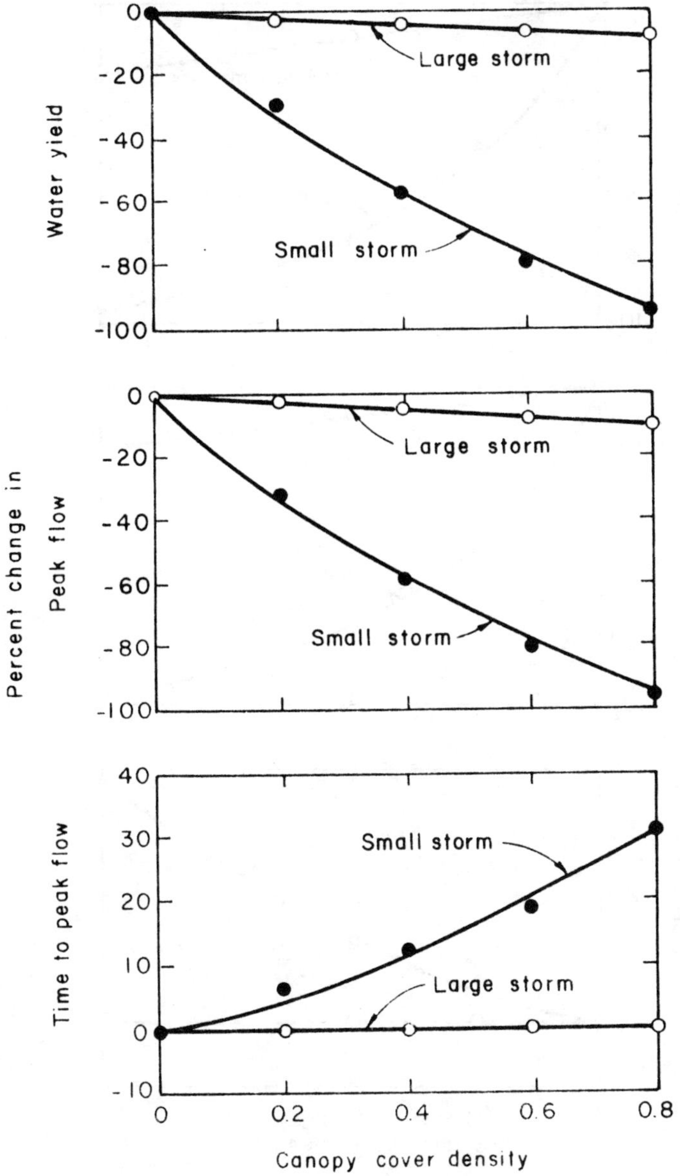

Fig. 9-32. Effect of Canopy Cover Density on the Water Hydrograph from Watershed 1 (after Simons et al., 1975)

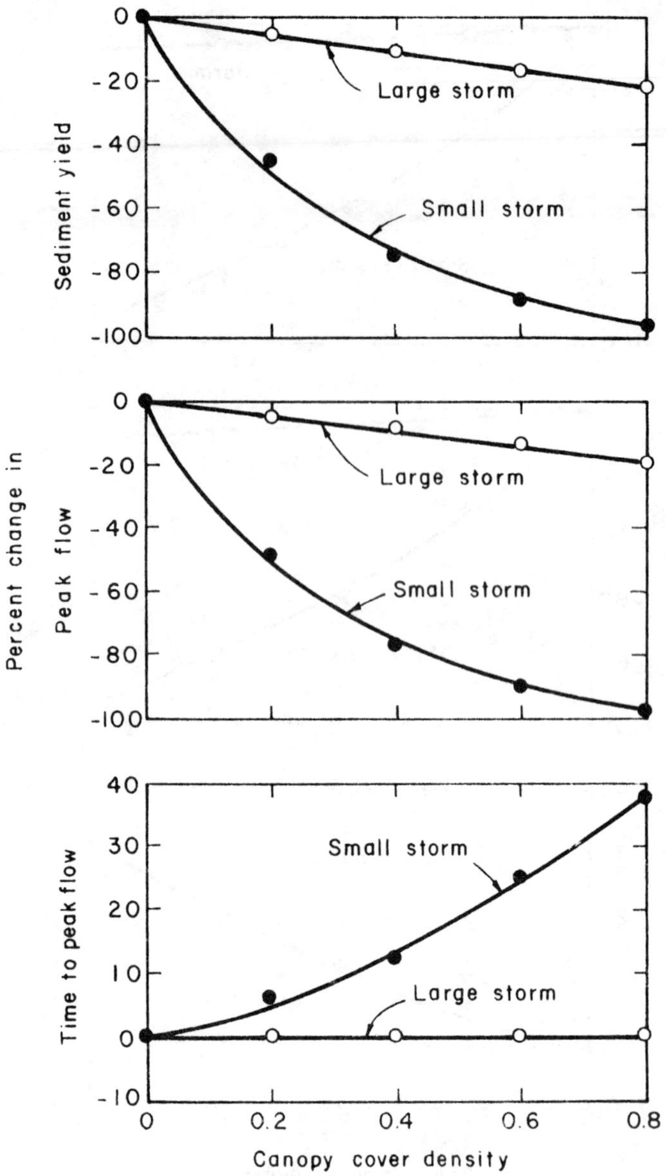

Fig. 9-33. Effect of Canopy Cover Density on the Sediment Hydrograph from Watershed 1 (after Simons et al., 1975)

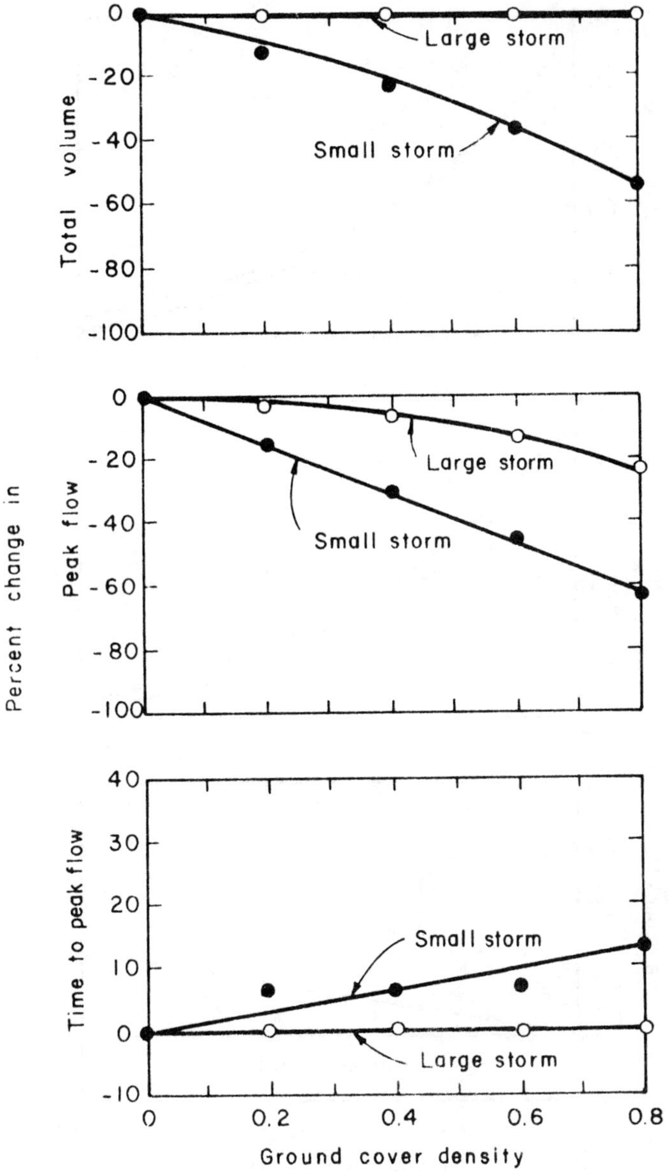

Fig. 9-34. Effect of Ground Cover Density on the Water Hydrograph from Watershed 1 (after Simons et al., 1975)

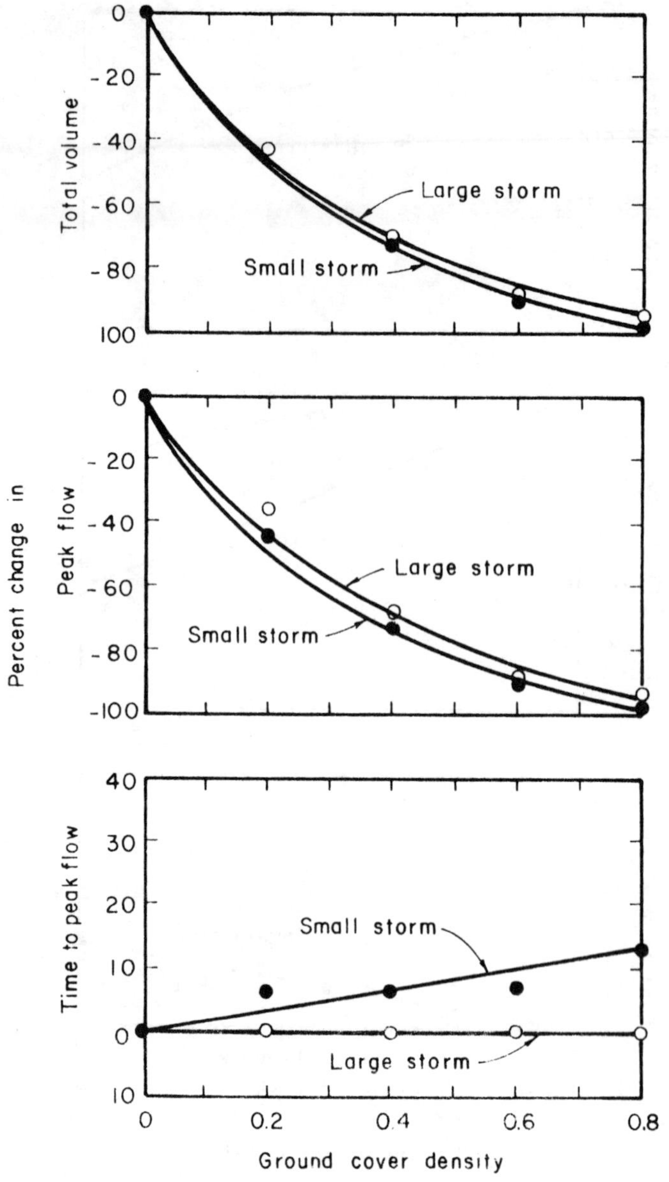

Fig. 9-35. Effect of Ground Cover Density on the Sediment Yield from Watershed 1 (after Simons et al., 1975)

9.8 SIMPLE METHOD FOR ESTIMATING ON-SITE SOIL EROSION

Basic Concepts and Equations

Water runoff, erosion, and sediment yield are common hydraulic and hydrologic phenomena that are governed by complex physical processes. Mathematical models can be constructed that accurately simulate these complex processes in time and space. These are described earlier. However, such models are often as complex and difficult to understand and use as the process system they intend to simulate. Model simplification can reduce this complexity, and if such simplification leaves the basic physical processes intact, the model may not necessarily lose its accuracy. Simplification while leaving the process components intact is the basis for the simplified on-site water and sediment yield model. The processes in simplified and complex models include interception losses, infiltration, water runoff, sediment runoff, erosion by raindrop splash and erosion by overland flow (see Simons et al., 1977c).

Assumptions

In order to develop this simplified procedure the following assumptions are made: (1) the design storms can be represented by a constant intensity and duration, (2) the flow reaches maximum discharge instantaneously, (3) the sediment yield can be approximated by examining the overall sediment availability during the storm and the total sediment transport capacity for the whole runoff period, and (4) the armoring effect of water layer and loose soil is negligible. In general, these assumptions result in a conservative oversestimation of sediment and water yields.

Effective Rainfall Duration and Time of Ponding

The total volume of potential interception V_i can be determined by using Eq. (9-2). The length of time of effective rainfall is computed as

$$D_e = D_t - V_i/I \tag{9-137}$$

in which D_e is the effective rainfall duration and D_t is the total storm duration and I is the rainfall intensity. Although interception losses are continuous over the storm period, it is assumed that the losses occur at storm initiation (see Fig. 9-36).

Under a constant rainfall with steady soil parameters the time of ponding can be estimated by (Mein and Larson, 1973).

$$t_p = \frac{\delta}{I(\frac{I}{K} - 1)} \quad (9\text{-}138)$$

in which t_p is the time to ponding since the beginning of effective rainfall, where δ and K are defined in Eq. (9-3).

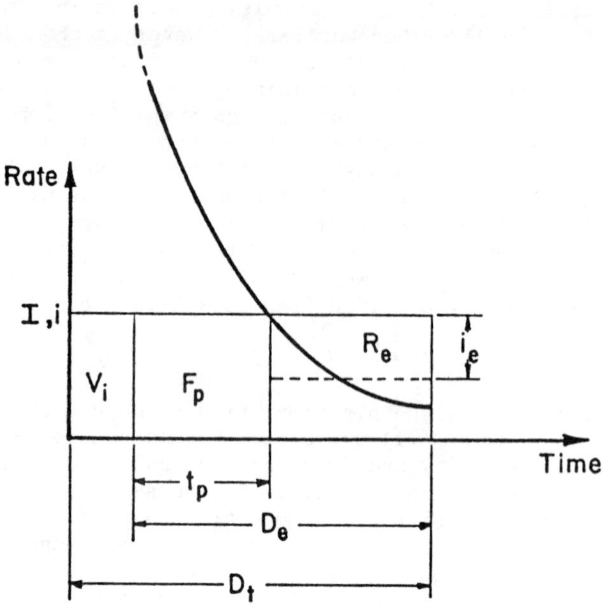

Fig. 9-36. Definition Sketch for Hydrologic Processes

Runoff Determination

Once the infiltration volume is determined, the average rainfall excess rate and the total runoff volume can be computed.

Rainfall Excess Rate: The average rainfall excess rate can be determined by

$$I_e = I - \frac{R_e}{D_e - t_p} \quad (9\text{-}139)$$

where I_e is the rainfall excess rate and R_e is the infiltrated volume. R_e can be determined by using the method recommended by Mein and Larson (1973).

Runoff Rate: The runoff rate, q, at the end of overland flow plot is

$$q = \int_0^L I_e dx = I_e L \tag{9-140}$$

where L is the length of plot.

Water Yield: The total runoff volume or water yield is

$$Y_w = W \int_0^{D_e - t_p} q\, dt = Wq(D_e - t_p) \tag{9-141}$$

where Y_w is the water yield in volume, and W is the width of overland plot.

Sediment Transport Capacity

The sediment transport rate for a level of discharge q (discharge per unit width) can be determined by using the procedure reported in Section 9.6. Considering the transport rate by size fractions, one obtains

$$q_t = \sum_{i=1}^{N} q_{ti} F_i \tag{9-142}$$

in which q_{ti} is determined by using Eq. (9-117).

The potential transport capacity can then be found as

$$\forall_t = \frac{W}{\gamma_s} \int_0^{D_e - t_p} q_t\, dt$$

or

$$\forall_t = \frac{q_t (D_e - t_p) W}{\gamma_s} \tag{9-143}$$

where $\forall_t$ is the nonporous volume of potential transport.

Determination of Sediment Supply

The potential sediment transport represents the capacity of the system. The supply of sediment comes from two mechanisms, detachment by raindrop splash and detachment by overland flow. The raindrop splash detachment can be formulated as

$$\mathcal{V}_r = D_r\ LW(1 - \lambda) \qquad (9\text{-}144)$$

where $\mathcal{V}_r$ is the nonporous volume of detached material by raindrop splash. The value of D_r was determined in Eq. (9-119).

Sediment supply by overland flow detachment is determined by

$$\mathcal{V}_f = D_f(\mathcal{V}_t - \mathcal{V}_r) \qquad (9\text{-}145)$$

where $\mathcal{V}_f$ is detachment by overland flow, and D_f is the flow detachment coefficient which must be calibrated. The value of D_f is generally larger for this simplified model than that of the complicated simulation.

If $\mathcal{V}_t < \mathcal{V}_r$ then there is no overland flow detachment because the transport rate is limited by the transporting capacity. The total available sediment supply, $\mathcal{V}_a$, is then

$$\mathcal{V}_a = \mathcal{V}_r + \mathcal{V}_f \qquad (9\text{-}146)$$

Determination of Controlling Process

Sediment yield is controlled by either supply or transport capacity. If supply is greater than capacity, capacity controls and vice versa. As particle size changes, so does the capacity and supply. Therefore, supply and capacity must be compared for each particle size. The individual capacity or potential yield is given as

$$\mathcal{V}_{ti} = \frac{q_{ti}\ F_i(D_t - t_p)W}{\gamma_s} \qquad (9\text{-}147)$$

where $\mathcal{V}_{ti}$ is the individual demand for the particle size.

The available supply is

$$\mathcal{V}_{ai} = F_i\ \mathcal{V}_a \qquad (9\text{-}148)$$

where $\mathcal{V}_{ai}$ is the available supply for the ith particle size. Values of $\mathcal{V}_{ti}$ and $\mathcal{V}_{ai}$ can be compared. If $\mathcal{V}_{ti}$ is greater than $\mathcal{V}_{ai}$ then supply controls, if $\mathcal{V}_{ai}$ is greater than $\mathcal{V}_{ti}$ then capacity controls or

$$\mathcal{V}_{yi} = \mathcal{V}_{ai} \quad \text{if} \quad \mathcal{V}_{ti} > V_{ai} \qquad (9\text{-}149)$$

and

$$\mathcal{V}_{yi} = \mathcal{V}_{ti} \quad \text{if} \quad \mathcal{V}_{ti} < \mathcal{V}_{ai} \qquad (9\text{-}150)$$

where $\mathcal{V}_{yi}$ is the volume yield for the particle size fraction. The total yield will then be

$$Y_s = \gamma_s \sum_{i=1}^{N} \forall_{yi} \tag{9-151}$$

where Y_s is the sediment yield by weight.

Comparison of Simplified and Complex Model Simulations

Simulations of water and sediment yields for the five runs were generated by a complex water and sediment routing model. This model was developed by Simons et al. (1975) and recently upgraded by Li et al. (1977c). Results of these simulations are presented in Fig. 9-37.

The comparison of the simplified and complex models is quite good. This is expected as the simplified model closely simulated the physical processes. Similar close comparisons were obtained by Simons et al. (1976) from simplified and complex model simulations for several hypothetical cases. These results further indicate the applicability of the simplified model.

The computer time required to simulate using the complex model is on the order of 10-15 times more than the simplified model. This time savings alone negates any improvement in accuracy obtained by the complex model for these simple cases. For larger or more complicated planes or watersheds and routing application, however, the complex simulation model is superior to the simplified model because of its time-space routing structure.

9.9 ROAD SEDIMENT PROCEDURAL GUIDE

Outgrowths from development of the road sediment model have been the generation of a qualitative procedural guide consisting of a series of graphs. These graphs were generated utilizing the road sediment model. The graph relates such variables as infiltration rate, soil detachment rate, and rainfall intensity, road gradient, cut and fill slope, sediment discharge and water discharge. These ground cover conditions generated graphs that can be used by the forest planner or engineer to quickly estimate water and sediment yield from roadways of different designs. Because both the road sediment model and the procedural guide have not been validated, the present procedural guide can only be used to assess the relative amounts of sediment production considering different design alternatives. A set of these generated graphs is presented as Figs. 9-38, 9-39, 9-40, and 9-41. Using these graphs sediment yield from a roadway can be rapidly estimated as indicated by the following example (see Simons et al., 1976).

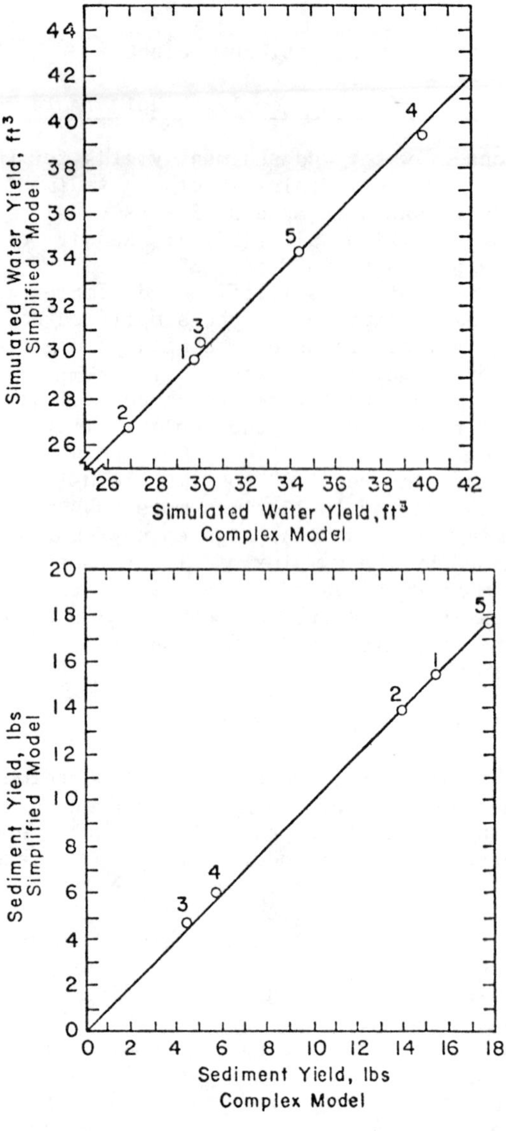

Fig. 9-37. Simplified Versus Complex Model Simulations for On-Site Soil Erosion

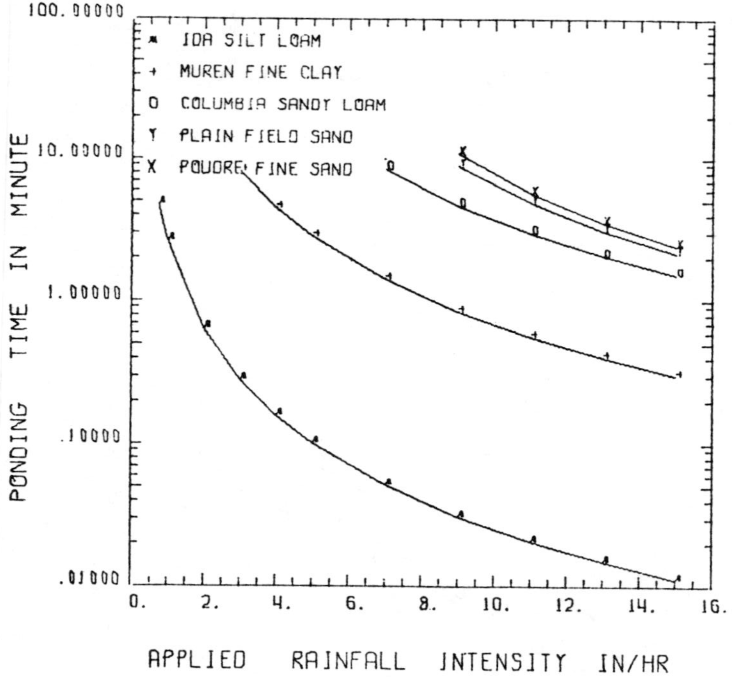

Fig. 9-38. Ponding Time for Different Rainfall Intensity

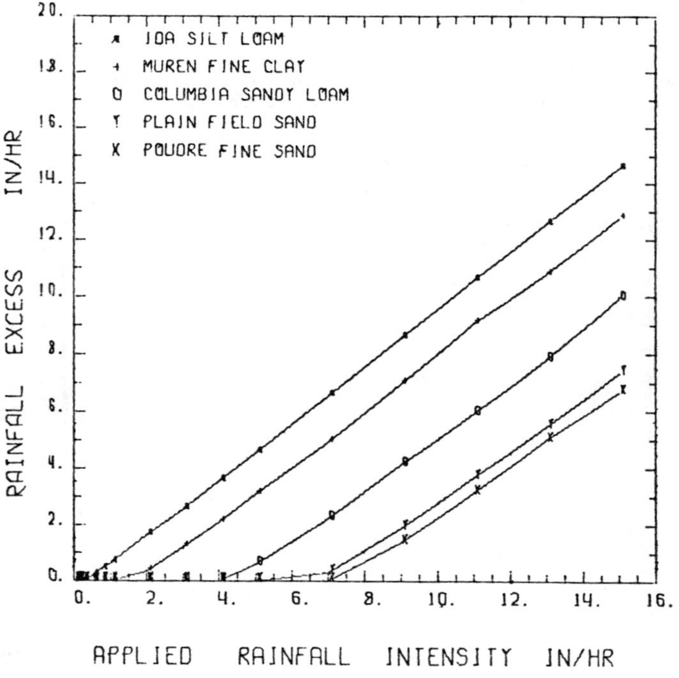

Fig. 9-39. Rainfall Excess Rate Versus Rainfall Intensity for Storm Duration of 30 Minutes

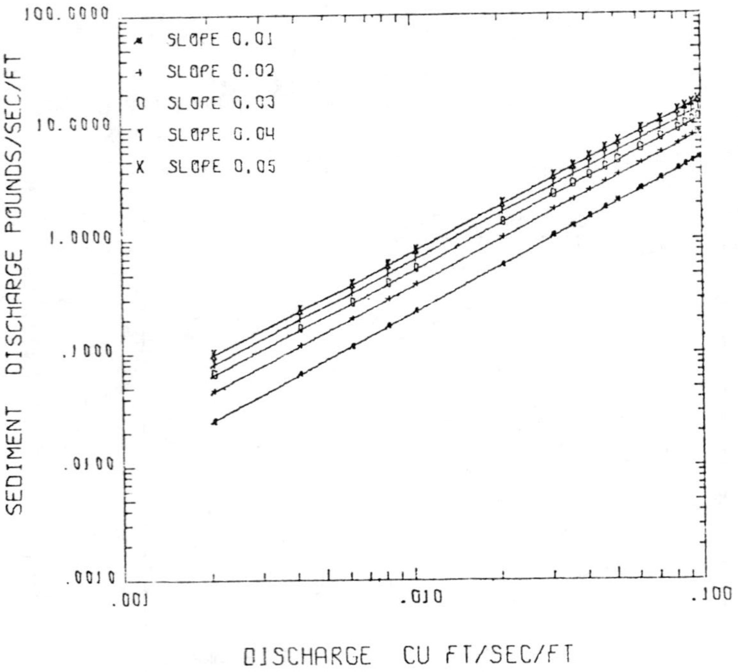

Fig. 9-40. Sediment Discharge Versus Water Discharge for Bare-Soil Road Bed with Slopes from 0.01 to 0.05 and for Sediment Size of 0.02 mm

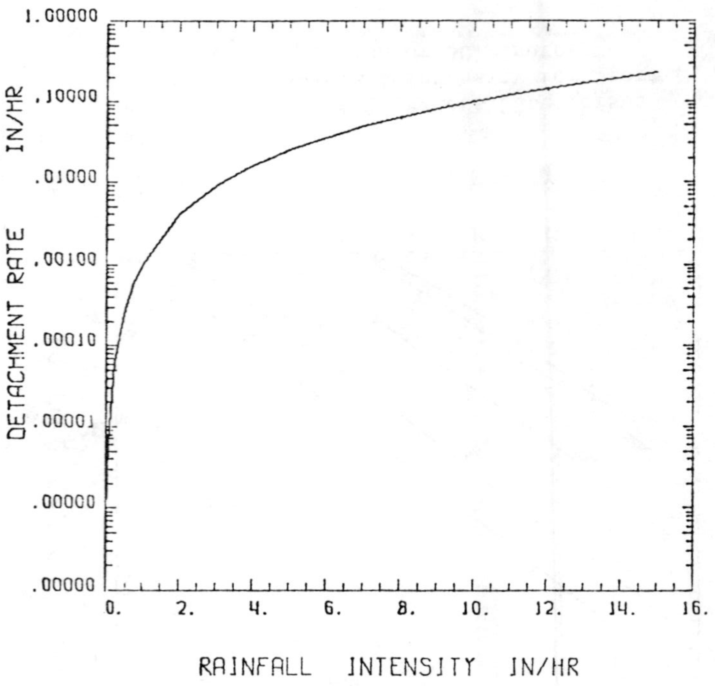

Fig. 9-41. Assumed Soil Detachment by Raindrop Splash

Example I. Water Yield

Given:
Road surface longitudinal gradient: 3 percent
Length: 500 ft
Width: 10 ft
Soil: fine clay, bare soil surface, sediment size is 0.02 mm and porosity is 0.5 (Muren fine clay)
Design Storm: intensity 3 in./hr; duration 30 min

What is the total water yield of the storm?

The procedure follows:

Step 1: From Fig. 9-38 with rainfall intensity 3 in./hr and Muren fine clay one can estimate the ponding time:

$$t_p = 8 \text{ min}$$

The ponding time is less than the duration of storm, and surface runoff occurs. Then the duration of excess rainfall is

$$D_e = D_t - t_p = 30 - 8$$
$$= 22 \text{ min}$$

Where t_p is the ponding time from which runoff begins, D_e is the duration of excess rainfall, and D_t is the duration of storm.

Step 2: From Fig. 9-39 with rainfall duration of 30 min and Muren fine clay the excess rainfall rate:

$$I_e = 1.25 \text{ in./hr}$$

Where I_e is the excess rainfall rate.

Step 3: The total water yield is

$$Y_w = I_e D_e$$
$$= 1.25 \times 22/60$$
$$= 0.46 \text{ in.}$$

Where Y_w is the water yield in depth of water.

Example II. Sediment Yield from Bare-Soil Road Surface

What would be the sediment yield of the storm using the data given in Example I?

Step 1: Determine the duration of excess rainfall and excess rainfall rate as those in Example I.

Step 2: The maximum discharge per unit width of road is

$$q = I_e L = 1.25 \times 500/43{,}200$$
$$= 0.0144 \ ft^3/sec/ft$$

Where q is the discharge per unit width of road, L is the length of road, or cross drain spacing, and 43,200 is a conversion factor from inches per hour to feet per second.

Step 3: With slope $S_o = 0.03$, $q = 0.0144 \ ft^3/sec/ft$ and sediment size $d_s = 0.02$ mm, Fig. 9-40 shows that,

$$q_t = 0.88 \ lb/sec/ft$$

Where q_t is the sediment transport rate per unit width of the road and S_o is the road gradient.

Step 4: The total transport capacity for the entire width of road surface G_c is

$$G_c = q_t \quad W = 0.88 \times 10$$
$$= 8.8 \ lb/sec$$

Where W is the width of road surface.

Step 5: The total potential transport capacity for the storm expressed as volume is,

$$\forall_t = G_c \, D_e/\gamma_s$$

$$= 8.8 \times 22 \times 60/165$$

$$= 70.4 \ ft^3$$

Where γ_s is the specific weight of sediment, it is assumed to be 165 lb/ft^3 in this report.

Step 6: From Fig. 9-41 the volume of loose soil available from raindrop impact detachment during the storm can be estimated by

$$\forall_r = D_r \, D_t \, A_s(1-\lambda)$$

$$= 0.009 \times 30/60 \times 1/12 \times 500 \times 10 \times 0.5$$

$$= 0.94 \ ft^3$$

Where $\not{V}_r$ is the available loose soil by raindrop splash in volume, D_r is the raindrop-splash soil detachment rate, A_s is the area, and n is the porosity of soil.

<u>Step 7</u>: Determine the volume of loose soil available from runoff detachment by comparing $\not{V}_t$ and $\not{V}_r$.

Because $\not{V}_r < \not{V}_t$, the transport capacity is greater than the availability, soil detachment by runoff occurs with the amount

$$\not{V}_f = D_f(\not{V}_t - \not{V}_r)$$

$$= 0.06 \times (70.4 - 0.94)$$

$$= 4.17 \text{ ft}^3$$

Where $\not{V}_f$ is the available loose soil by runoff detachment, D_f is the overall runoff detachment coefficient. For clay and silt, D_f is assumed to be 0.06.

<u>Step 8</u>: Determine the total volume of loose soil available for transport during the storm by

$$\not{V}_a = \not{V}_r + \not{V}_f$$

$$= 0.94 + 4.17$$

$$= 5.11 \text{ ft}^3$$

Where $\not{V}_r$ is the total available loose soil in volume.

<u>Step 9</u>: Determine the amount of soil erosion or sediment yield from road surface by comparing $\not{V}_t$ and $\not{V}_a$.

Because $\not{V}_a < \not{V}_t$ the availability of loose soil determines the yield. The sediment yield is

$$Y_s = \not{V}_a = 5.11 \text{ ft}^3$$

$$= 843 \text{ lb}$$

This simple example illustrates the type of useful product that can be generated using a computer based physical simulation model.

In addition to the example shown here, the model can also estimate sediment yield from cut and fill road slopes with or without vegetative erosion control treatment. The use of computer based models to develop field user guides, such as those presented above, is an attractive tool for transferring the information from the research level to the practicing level.

9.10 DATA NEEDS FOR MODELING

The data needed for mathematical modeling of watershed response are dependent on the type of watershed, the objective of modeling and the type of model utilized. The data requirements summarized below are primarily for studying the major watershed response (water yield, sediment yield, and stream morphology) from small watersheds. The emphasis is made on the input data requirements of the watershed models presented in this chapter. However, additional data for other potential studies related to the watershed response are also identified.

There are two distinct types of data involved in the analysis. They are the model input data, which are specifically required for the model input, and the basic data (raw data), which are directly taken from field measurements. Generally, the model input data are obtained from retrieving and or analyzing the raw data. The required model input data and basic data, and the supplementary data that are important to watershed study are summarized as follows.

Model Input Data

The input data required for the simulation are different for different models. Described below are those for general purposes.

Geometry Data: These data include slope length, bed slope, wetted perimeter versus flow area relations, and computation sequence.

Soil Data: These data include the saturated hydraulic conductivity, average capillary suction pressure, soil porosity, degree of saturation in the wetted zone and particle size distribution.

Vegetation and Ground Cover Data: These data include canopy cover density, ground cover density, the typical interception storage capacity of canopy cover, and ground cover.

Flow Resistance Parameters: These parameters are constants describing grain resistance and overall resistance. The overall resistance parameters for overland flow units should include parameters for both minimum and maximum resistance.

Sediment Routing Parameters: These parameters include constants describing the sediment transporting capacity, parameters describing the potential soil detachment rate, the detachment coefficient of soil by surface runoff (soil erodibility), and the thickness of armor layer.

Storm Characteristics: These data include rainfall intensity and areal distribution of rainfall intensity if studying a large watershed.

Antecedent Conditions: These data include antecedent soil moisture content, and initial loose soil storage.

Basic Data

In order to have the model input data given in the previous section, the following basic data (raw data) are needed.

Topographic Data: The topographic data include the topography survey map, aerial photographs, and channel geometry survey results. With these data, all the geometry data required for the model input can be determined for both overland flow areas and channels.

Soil Data: These data include soil survey map indicating the aerial distribution of soil types, and saturated hydraulic conductivity, average capillary suction pressure, soil porosity, degree of saturation in the wetted zone, and particle size distribution for different types of soil.

Forest Inventory Data: The forest inventory data include total vegetation volume, total litter volume (or litter index), canopy cover density, ground cover density, areal distribution of vegetation type (can be estimated from aerial or color infrared photos), size and aerial distribution of rock mulch, and interception storage capacity of a typical tree and an unit area of forest floor. From these forest inventory data, the vegetation and ground cover data for the model input can be easily determined.

Hydraulic Data: The hydraulic data include flow discharge, depth, velocity, sediment load for various flow conditions at some locations. These data are used to estimate the flow resistance parameters and sediment routing parameters and to evaluate the performance of the mathematical model.

Climatologic Data: The climatologic data include the rainfall record in either strip charts or hyetographs. If the inter-storm water balance simulation is incorporated, the following additional climatologic data are needed: wind, temperature, humidity, evaporation, and incoming short-wave radiation. The antecedent conditions can only be estimated if the interstorm water balance simulation is available.

Hydrographic Data: The hydrographic data are needed for model calibration and simulation evaluation. These data include the water and sediment hydrographs at the watershed outlet, and the resultant channel profiles.

Watershed Management Data: The watershed management data are very important for evaluating the watershed treatment effects. These data include the vegetation treatment data (clear-cut, patch-cut, or burning treatment) and the mechanical treatment data (check dam, logging road, windrow and skid trail, etc.).

Supplementary Data

The following supplementary data are needed when some other studies such as water quality management are taken into consideration.

Historical Data: Whenever a long-term water resources planning or designing of hydraulic structures is necessary, the prediction of the response should be based on the stochastic input. The historical records of precipitation, temperature, wind flow, radiation, and floods are needed to perform a time-series analysis or frequency analysis.

Supplementary Water Quality Data: When monitoring water quality and controlling water pollution are essential, the supplementary data regarding biological oxygen demand, dissolved oxygen, turbidity, salt level, nutrient level and other chemical and biological water pollutants are also necessary in routing these pollutants by a mathematical model.

Wildlife Management Data: If the interactions between the wildlife and the environment are being studied, the information regarding game, fish, and other wildlife habitat is certainly important.

Model Parameter Calibration

In the mathematical modeling of system response, the identification of model parameters has often relied on an optimization scheme. The dependency on the optimization technique may be reduced if the model is formulated according to the physical significance. For the water routing problem, the parameters describing flow resistance are usually unknown, but their ranges are known from measured data. However, the optimum values of the parameters which reproduce correct model response are usually not available. Hence, the model calibration is inevitable. The simplest calibration technique is the trial and error method. Except for some models which contain parameters with very narrow searching ranges, the trial and error procedure is inefficient for most problems. An efficient procedure is apparently needed for the model calibration.

There are many optimization techniques available for the purpose of model calibration. However, the usefulness of a particular optimization technique is very much dependent on the formulation of the model being calibrated. Oftentimes Rosenbrock's (1960) optimization technique is recommended for use to find the optimum set of parameters because it is by far the most promising and efficient method for fitting a hydrologic model. However, in order to have an efficient method of calibration, one should make a sequence of calibration runs considering the sequence of physical events which govern the water and sediment yield. The three sequential calibration runs are: (1) rainfall excess to calibrate the total runoff volume, (2) water routing to calibrate the peak time, the time to peak, and the shape of the hydrograph, and (3) sediment supply and routing to calibrate the sediment yield. Of course, there are some low levels of interactions among the above three processes. Neglecting the interactions, these sequential runs can only provide a very good estimate of the "optimum" parameters. Oftentimes those calibrated parameters already give good simulation results. The sequential calibration runs would make the calibration job much easier because fewer parameters are involved in each calibration effort.

9.11 SUMMARY

Presented in this chapter are some of the efforts at Colorado State University in the mathematical modeling of water and sediment routing from watersheds.

Models being developed, used, and improved are based on physical processes and therefore reflect changes in the natural system much more meaningfully than models based on regression analysis. This allows these process models to be widely used and easily transferable if the limitations of each model are recognized.

Computer based process models are a very useful tool in predicting and planning. However, more research is needed to perfect and simplify these models before they are widely accepted.

9.12 LIST OF SYMBOLS

A = Flow Area

A_o = Initial Value of Area

a = Time Weighting Factor

a_1, a_2, a_3, a_4, a_5 = Parameters or Constants Describing Coefficient of Power Functions

a' = Thickness of the Bed Layer

b = Space Weighting Factor

b_1, b_2, b_3, b_4, b_5 = Parameters or Constants Describing Exponent of Power Functions

C = Sediment Concentration by Volume

C_a = Known Concentration at a Distance a' Above Bed

C_c = Canopy Cover Density

C_g = Ground Cover Density

C_i^p = Transporting Capacity in Concentration by Volume

D = Total Amount of Detached Soil

D_a = Thickness of Armor Layer

D_e = Effective Rainfall Duration

D_f = Flow Detachment Coefficient

D_r = Soil Detachment Rate

D_t = Total Storm Duration

d_s = Size of Sediment

F = Infiltrated Volume

ΔF = Change in Infiltrated Volume

F_{ai} = Adjusted Fraction of the ith Size

F_i = Original Percentage of Sediment Size in a Given Size Fraction Designated by i

f = Overall Darcy-Weisbach Friction Coefficient

$\overline{f}$ = Average Infiltration Rate

f_d = Darcy-Weisbach Friction Factor Due to Form Drag Resistance

f_g = Darcy-Weisbach Friction Factor for Grain Resistance

G = a'/R

G_{ci} = Sediment Transport Rate for ith Size

G_s = Total Sediment Transport Rate by Volume per Unit Time

g = Gravitational Acceleration

g_s = The Lateral Sediment Inflow

I = Rainfall Intensity

I_e = Rainfall Excess

J_1, J_2 = Integrals

K = Hydraulic Conductivity of Soil in the Wetted Zone

K_g = Darcy-Weisbach Friction Factor for Grain Resistance

K_h = Parameter Describing Maximum Resistance

K_ℓ = Parameter Describing Minimum Resistance

K_t = Constant Describing Overall Resistance

L = Length of a Given Plan

N = Number of Size Fractions

N_r = Flow Reynolds Number

n = Manning's Roughness Coefficient

P = Wetted Perimeter

Q_L = Discharge at the End of Channel of Overland Flow Reach

Q_o = Initial Value of Discharge

q = Runoff Rate Per Unit Width of Channel

q_b = Bed Load Transport Rate in Volume Per Unit Width

q_ℓ = Lateral Inflow or Outflow Rate Per Unit Length of Channel

q_s = Suspended Load in Volume Per Unit Width

q_t = Total Sediment Load Per Unit Width of Channel

R = Hydraulic Radius

r^k = Value of r at the kth Iteration

r^o = Initial Guess of γ

S_o = Channel Bed Slope

S_f = Friction Slope

t = Time

Δt = Increment of Time

t_o = Initial Time

t_L = Timing of the Outflow at the Downstream Boundary

t_p = Time of Ponding Since the Beginning of Effective Rainfall

U_* = Shear Velocity

V = Mean Flow Velocity

V_i = Total Volume of Potential Interception

V_c = Potential Volume of Canopy Cover Interception in Depth

V_g = Potential Volume of Ground Cover Interception in Depth

V_s = Settling Velocity of Sediment Particles

W = Width of Overland Plot

ℓ = Parameters Defined as $\frac{V_s}{\kappa U_*}$

x = Downslope Distance

Δx = Increment of Distance

x_o = Initial Distance

Y_s = Sediment Yield by Weight

Y_w = Water Yield in Volume

y = Depth of Flow

Z = The Depth of Loose Soil

Z_i = Amount of Loose Soil Available of ith Sediment Size

Z_m = Maximum Penetration Depth of Rainfall Splash

Z_w = Depth of Water Plus the Loose Soil Depth

α = Coefficient Whose Value Depends on Shape of the Channel, Resistance to Flow and Channel Gradient

β = Coefficient Whose Value Depends on Shape of the Channel, Resistance to Flow and Channel Gradient

γ = Specific Weight of Water

γ_s = Specific Weight of Sediment

ε = Tolerance

δ = The Potential Head Parameter

δ_s = Constant Depending on Flow Conditions

η_s = Roughness Height

θ = $\frac{\Delta t}{\Delta x}$

θ_i = Antecedent Moisture Content

θ_w = Moisture Content of the Soil After Wetting

κ = Karman Constant

γ = Kinematic Viscosity of Water

ξ = Distance from Bed

c_ξ = Sediment Concentration at the Distance ξ from the Bed

μ_ξ = Mean Velocity at the Distance ξ from the Bed

ρ = Density of Water

τ_c = Critical Tractive Force

τ_o = Boundary Shear Stress Acting on the Grain

τ_* = Total Shear Stress

ψ_{ave} = Average Suction Head Across the Wetting Front

$\forall_a$ = Total Available Sediment Supply in Volume

$\forall_f$ = Soil Detachment by Overland Flow in Volume

$\forall_r$ = Non-Porous Volume of Detached Material by Raindrop Splash in Volume

$\forall_t$ = Non-Porous Volume of Potential Transport in Volume

$\forall_{yi}$ = Sediment Yield for ith Particle Size Fraction in Volume

9.13 REFERENCES

Brakensiek, D.L., 1967. Kinematic Flood Routing, ASAE, Vol. 10, No. 3, pp. 340-343.

Carter, C.E., Greer, J.D., Brand, H.J., and Floyd, J.M., 1974. Raindrop characteristics in South Central United States, Trans. ASAE, pp. 1033-1037.

Chen, C.L., 1976. Flow resistance in broad shallow greased channels. Journal of the Hydraulics Division, ASCE, Vol. 102, No. HY3, pp. 307-322, March.

Chow, V.T., 1959. Open channel hydraulics. McGraw-Hill Book Company, New York.

Eagleson, P.S., 1970. Dynamic hydrology. McGraw-Hill, New York.

Einstein, H.A., 1950. The bed load function for sediment transportation in open channel flows. USDA, Technical Bulletin, No. 1026.

Ellison, W.D., 1944. Studies of raindrop erosion. Agricultural Engineering, April, pp. 131-136; May pp. 181-182.

Fread, D.L., 1973. A dynamic model of stage-discharge relations affected by changing discharge. NOAA Technical Memorandum NWS Hydro 16, November.

Gessler, J., 1965. The beginning of bedload movement of mixtures investigated as natural armoring in channels. (Translated by Prych, E.A., W.M. Keck Laboratory of Hydraulics and Water Resources, CIT).

Green, W.H., and G.A. Ampt, 1911. Studies on soil physics part 1: the flow of air and water through soils. Journal of Agriculture Science, May.

Harley, B.M., Perkins, F.E., and Eagleson, P.S., 1970. A modular distributed model of catchment dynamics, MIT, Department of Civil Engineering, Hydrodynamics Laboratory Report No. 133, June.

Henderson, F.M., 1966. Open channel flow. Macmillan, New York.

Houghton, D.P., and Kasahara, A., 1968. Nonlinear shallow fluid flow over an isolated ridge. Communication on Pure and Applied Mathematics, Vol. XXI.

Iwagaki, Y., 1955. Fundamental studies on runoff analysis by characteristics. Disaster Prevention Research Institute Bulletin No. 10, Kyoto University, Kyoto, Japan.

Kibler, D.F., and Woolhiser, D.A., 1970. The kinematic cascade as a hydrologic model. Hydrology Paper No. 39, Colorado State University, Fort Collins, Colorado.

Kilinc, M.Y., 1972. Mechanics of soil erosion from overland flow generated by simulated rainfall. Ph.D. dissertation presented to Colorado State University, Fort Collins, Colorado.

Kouwen, N., and Unny, T.E., 1969. Flow retardance in vegetated channels. Journal of irrigation and drainage division, ASCE, Vol. 95, No. IR2, June, pp. 329-342.

Law, O.J., and Parson, D.A., 1943. The relation of raindrop size to intensity. Trans. AGU.

Li, R.M., and Shen, H.W., 1973. Effect of tall vegetation on flow and sediment. Journal of Hydraulics Division, ASCE, Vol. 99, No. HY5, May, pp. 793-814.

Li, R.M., 1974. Mathematical modeling of response from small watershed. Ph.D. dissertation, Department of Civil Engineering, Colorado State University, Fort Collins, Colorado.

Li, R.M., Simons, D.B., and Stevens, M.A., 1975a. Nonlinear kinematic wave approximation for water routing. Water Resources Research, Vol. 11, No. 2.

Li, R.M., Simons, D.B., and Stevens, M.A., 1975b. On overland flow water routing. Proceedings of the National Symposium on Urban Hydrology and Sediment Control, Lexington, Kentucky, July.

Li, R.M., Simons, D.B., and Eggert, K.G., 1976a. Process model of infiltration and subsurface flow routing. Colorado State University Report CER76-77RML-DBS-KGE20. Prepared for USDA Forest Service, Rocky Mountain Forest and Range Experiment Station, Flagstaff, Arizona, December.

Li, R.M., Stevens, M.A., and Simons, D.B., 1976b. Solution to Green-Ampt infiltration equation. Journal of Irrigation and Drainage Division, ASCE, Vol. 102, No. IR2, June, pp. 239-248.

Li, R.M., Simons, D.B., Shiao, L.S., and Chen, Y.H., 1976c. On kinematic wave approximation for flood routing. RIVERS 76, Vol. 1, 3rd Annual Symposium on Inland Waterways for Navigation, Flood Control and Water Diversion, Colorado State University, Fort Collins, Colorado, pp. 377-398, August.

Li, R.M., Simons, D.B., and Eggert, K.G., 1977a. ANAWAT--A simple watershed model based on analytical solutions. Colorado State University Report (in preparation).

Li, R.M., Simons, R.K., and Simons, D.B., 1977b. A generalized kinematic wave approximation for flood routing. Paper submitted to Journal of Hydraulic Division, ASCE for possible publication.

Li, R.M., Simons, R.K., and Shiao, L.Y., 1977c. Mathematical modeling of on-site soil erosion. Paper to be presented at International Symposium on Urban Hydrology, Hydraulics and Sediment Control, University of Kentucky, Lexington, Kentucky, July 25-28.

Mein, R.G., and Larson, C.L., 1973. Modeling infiltration during a steady rain. Water Resources Research, Vol. 9, No. 2, April.

Meyer, L.D., 1971. Soil erosion by water on upland areas. In River Mechanics, Vol. II, edited by Hsieh Wen Shen, Fort Collins, Colorado.

Mutchler, C.K., 1967. Parameters for describing raindrop splash. Journal of Soil Conservation, May-June, pp. 91-94.

Mutchler, C.K., and Young, R.A., 1975. Soil detachment by raindrops. Proceedings of the Sediment-Yields Workshops, Oxford, Mississippi, USDA, ARS-S-40, June, pp. 113-117.

Palmer, V.J., 1946. Retardance coefficients for low flow in channels lined with vegetation. Trans. AGU, Vol. 27, pp. 187-197.

Phelps, H.O., 1970. The friction coefficient for shallow flows over a simulated turf surface. Water Resources Research, Vol. 6, No. 4, August, pp. 1220-1226.

Ree, W.O., 1949. Hydraulic characteristics of vegetation for vegetated waterways. Agricultural Engineers, Vol. 30, No. 4, pp. 184-189.

Ree, W.O., and Palmer, V.J., 1949. Flow of water in channels protected by vegetative linings. U.S. Soil Conservation Technical Bulletin No. 967.

Rosenbrock, H.H., 1960. An automatic method for finding the greatest or least value of a function. Computer Journal, Vol. 4, pp. 175-184.

Shen, H.W., 1971. Wash load and bed load. Chapter 11, River Mechanics, edited and published by H.W. Shen, Colorado State University, Fort Collins, Colorado.

Simons, D.B., and Li, R.M., 1975. Watershed segmentation by a digital computer for mathematical modeling of watershed response, Colorado State University Report, CER75-76DBS-RML9, Prepared for USDA Forest Service, Rocky Mountain Forest and Range Experiment Station, Flagstaff, Arizona.

Simons, D.B., Li, R.M., and Stevens, M.A., 1975. Development of models for predicting water and sediment routing and yield from storms on small watershed. USDA Forest Service, Rocky Mountain Forest and Range Experiment Station, August.

Simons, D.B., Li, R.M., and Shiao, L.Y., 1976. Preliminary procedural guide for estimating water and sediment yield from roads in forests. Draft prepared for USDA Forest Service, Rocky Mountain Forest and Range Experiment Station, Flagstaff, Arizona.

Simons, D.B., Li, R.M., and Eggert, K.G., 1977a. Simple water routing and yield model using a small programmable calculator. Colorado State University Report for USDA Forest Service, Rocky Mountain Forest and Range Experiment Station, Flagstaff, Arizona, CER76-77DBS-RML-KGE 52, April.

Simons, D.B., Li, R.M., and Shiao, L.Y., 1977b. Formulation of road sediment model. Colorado State University Report for USDA Forest Service, Rocky Mountain Forest and Range Experiment Station, Flagstaff, Arizona, CER76-77DBS-RML-LYS50, March.

Simons, D.B., Li, R.M., and Ward, T.J., 1977c. A simple procedure for estimating on-site erosion. Paper presented at the International Symposium on Urban Hydrology, Hydraulics, and Sediment Control, Lexington, Kentucky, July 25-28.

Simons, R.K., 1977. A modified kinematic approximation for water and sediment routing. M.S. Thesis, Colorado State University, Fort Collins, Colorado.

Simons, R.K., Li, R.M., and Simons, D.B., 1977. On stage-discharge relations of rivers. Paper presented at 17th Congress of International Association of Hydraulic Research, Baden-Baden, Germany, August 15-19.

Singh, V.P., 1975. A distributed approach to kinematic wave modeling of watershed surface runoff. Proceedings of National Symposium on Urban Hydrology and Sediment Control, July 28-31, University of Kentucky, pp. 227-236.

USBR, 1960. Investigation of Meyer-Peter, Muller bedload formulas. Sedimentation Section, Hydrology Branch, Division of Project Investigations, U.S. Department of the Interior, Bureau of Reclamation, June.

Van Liew, M.W., 1976. Sediment production from short test plots in mine spoils. M.S. Thesis, Colorado State University, Fort Collins, Colorado.

Wenzel, H.G., 1970. The effects of raindrops impact and surface roughness on sheet flows. WRC Research Report No. 34, University of Illinois, Water Resources Center, Urbana, Ill.

Woolhiser, D.A., 1975. Simulation of unsteady overland flow. In Unsteady Flow in Open Channels, edited by K. Mahmood and V. Yevjevich, Water Resources Publications, Fort Collins, Colorado.

Young, R.A., and Wiersma, J.L., 1973. The role of rainfall impact on soil detachment and transport. Water Resources Research, Vol. 9, pp. 1629-1636.

Chapter 10

WATER AND SEDIMENT ROUTING IN RIVERS

by

Yung Hai Chen, Assistant Professor of Civil Engineering, Colorado State University, Fort Collins, Colorado

Chapter 10

WATER AND SEDIMENT ROUTING IN RIVERS

10.1 INTRODUCTION

Many problems in water resource systems, river mechanics and hydraulic structures are directly related to the transport of material by flow. In general, the flows in natural and man-made systems are time-variant. Therefore the related transport phenomena are time-variant. However, even when the flows are steady or nearly steady, the transport phenomena may be unsteady. Examples are river systems that have not achieved equilibrium conditions after an initial disturbance.

In solving the unsteady transport problems analytical solutions are usually not available, and numerical methods using iteration procedures and digital computers are used. A number of numerical solution techniques are available, e.g., references (Amein and Fang, 1970; Chang and Richards, 1971; Contractor and Wiggert, 1972; Chen, 1973), among others. This chapter describes one-dimensional water and sediment routing models applied to study responses of rivers to development.

The principal limitation of the one-dimensional model is its assumption of one-dimensional flow. In a relatively non-uniform river, only the general pattern of the river geomorphology can be considered. To perform a more detailed study, either a modification of the one-dimensional model can be made by using a multiple stream or compound approach, or a two-dimensional model can be applied. These three types of models are also presented in this chapter. The advantages and disadvantages of various types of models for water and sediment routing in rivers are also discussed.

10.2 ONE-DIMENSIONAL WATER AND SEDIMENT ROUTING MODEL

Introduction

In solving the unsteady flow problems in canals and rivers, the unsteady flow of sediment-laden water is formulated with one-dimensional partial differential equations representing the continuity of mass for sediment, the continuity of mass for sediment-laden water and the equation of motion for sediment-laden water. The set of equations solved by numerical methods using a digital computer constitutes a mathematical model for studying various types of unsteady flow problems in canals and rivers.

Basic Equations

The one-dimensional differential equations of gradually varied unsteady flow in natural alluvial channels can be derived based on the following assumptions:

1. The channel is sufficiently straight and uniform in the reach so that the flow characteristics may be physically represented by a one-dimensional model.

2. The velocity is uniformly distributed over the cross section.

3. Hydrostatic pressure prevails at every point in the channel.

4. The water surface slope is small.

5. The density of the sediment-laden water is constant over the cross section.

6. The resistance coefficient is assumed to be the same as that for steady flow in alluvial channels and can be approximated from resistance equations applicable to alluvial channels or from field survey.

The three basic equations derived (Chen, 1973) are

$$\frac{\partial Q_s}{\partial x} + p\,\frac{\partial A_d}{\partial t} + \frac{\partial AC_s}{\partial t} - q_s = 0 \qquad (10\text{-}1)$$

the flow continuity equation

$$\frac{\partial Q}{\partial x} + \frac{\partial A}{\partial t} + \frac{\partial A_d}{\partial t} - q_\ell = 0 \qquad (10\text{-}2)$$

and the flow momentum equation

$$\frac{\partial \rho Q}{\partial t} + \frac{\partial \beta\rho QV}{\partial x} + gA\,\frac{\partial \rho y}{\partial x} = \rho gA(S_o - S_f + D_\ell) \qquad (10\text{-}3)$$

or

$$\frac{\partial \rho Q}{\partial t} + V\,\frac{\partial \beta\rho Q}{\partial x} + \beta\rho V\,\frac{\partial Q}{\partial x} - \beta\rho V^2 T\,\frac{\partial y}{\partial x} + gA\,\frac{\partial \rho y}{\partial x}$$

$$= \rho gA\,(S_o - S_f + D_\ell) + \beta\rho V^2 A_x^y \qquad (10\text{-}4)$$

where (see Fig. 10-1)

x = horizontal distance along the channel

t = time

Q_s = sediment discharge

p = volume of sediment in a unit volume of bed layer given by ρ_b/ρ_s

ρ_b = bulk density of sediment forming the bed

ρ_s = density of sediment

A_d = volume of sediment deposited on channel bed per unit length of channel, the value of which is negative when bed erosion occurs

A = water cross-sectional area

C_s = mean sediment concentration on a volume basis given by Q_s/Q

Q = flow discharge

q_s = lateral sediment flow per unit length of channel. A positive quantity indicates inflow and a negative value denotes outflow.

q_w = lateral water flow per unit length of channel. A positive quantity indicates inflow and a negative value denotes outflow.

q_ℓ = lateral flow per unit length of channel, given by $q_s + q_w$

ρ = density of sediment-laden water given by $\rho_w + C_s(\rho_s - \rho_w)$

ρ_w = density of water

β = momentum coefficient

V = mean flow velocity

T = $\partial A/\partial y$

y = flow depth

g = acceleration of gravity

S_o = bed slope

S_f = friction slope

D_ℓ = dynamic contribution of lateral discharge given by $q_\ell V_\ell / Ag$

V_ℓ = velocity component of lateral inflow in the main flow direction

A_x^y = departure from a prismatic channel given by $(\partial A/\partial x)_y$

h = water surface elevation

z = riverbed elevation

Δz = change in riverbed elevation

B = top width

The effect of the variation of sediment-laden water density ρ on the water and sediment movement is usually negligible. Then the variable ρ in Eq. (10-3) can be factored out. Also, to adequately account for the effect of reversed flow acceleration for momentum, the term $\partial\beta\rho QV/\partial x$ should be replaced by $\partial\beta\rho Q|V|/\partial x$.

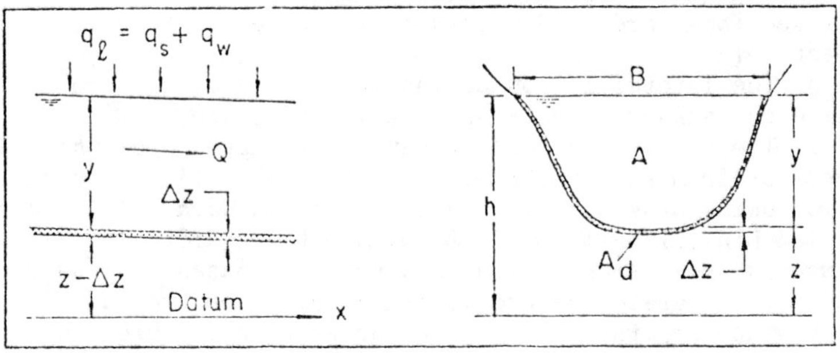

Fig. 10-1. Definition Sketch of an Alluvial Channel

Supplemental Equations

The three equations contain three basic unknowns: Q, y, and A_d. The other variables in the equations must be expressed as a function of the three unknowns in order to obtain a solution. These supplementary equations describe the physical properties of the prototype.

1. The geometric properties of cross sections are expressed as a function of stage from the unknown channel geometry.

2. The mean bed slope

$$S_o = -\partial z/\partial x \tag{10-5}$$

in which the initial bed elevation z is known.

3. The friction slope S_f is a function of flow and channel characteristics. The resistance functions such as Manning's or Chezy's equations can be employed to relate to S_f with the basic unknowns.

That is

$$S_f = \frac{Q|Q|}{K_i^2} \tag{10-6}$$

where K_i = total conveyance for section i at a given stage h_i. Considering a channel with nonuniform cross-sectional shape and resistance, the conveyance for a subsection k (e.g., see Fig. 10-2) is defined by

$$K_{ki} = \frac{1.486}{n_{ki}} A_{ki}^{5/3} P_{ki}^{-2/3} \tag{10-7}$$

where A_{ki} and P_{ki} are cross-sectional area and wetted perimeter at subsection k, respectively. The total conveyance is then

$$K_i = \sum_k K_{ki} \tag{10-8}$$

A simplified relation for K_i is

$$K_i = \frac{1.486}{n_i} A_i^{5/3} P_i^{-2/3} \tag{10-9}$$

where n_i, A_i and P_i are overall Manning's roughness coefficient, cross-sectional area and wetted perimeter at section i, respectively.

4. The lateral inflow discharges q_ℓ and q_s are considered as known functions of time and space.

5. The sediment discharge can be estimated from the field surveys or from the available theories. Shen (1971) suggested the following procedure for analyzing field data: (a) Use the modified Einstein method (Colby and Hembree, 1955) to estimate the unmeasured suspended load and bed load based on measured data. (b) Separate the bed-material load from the wash load and analyze them separately. (c) Decide which available sediment transport equations best agree with the measured data and use it to estimate the transport load for the given flow conditions. When no measured data are available, Shen suggested: (a) Use

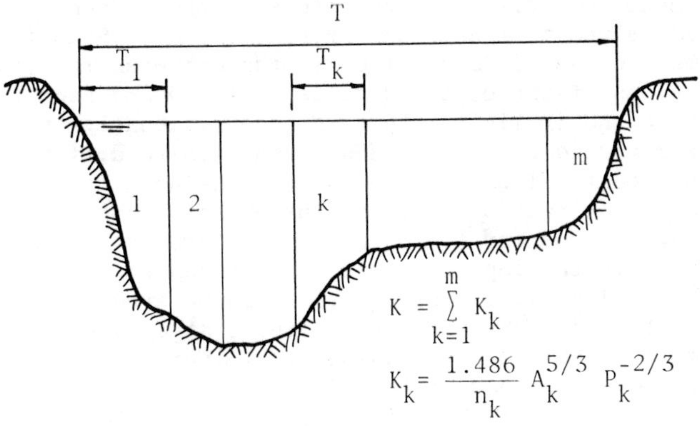

Fig. 10-2. Channel Elements to Calculate the Section Conveyance of a Composite Cross Section

Einstein's (1950) procedure if bed load is a significant portion of the total bed-material load. (b) Use Colby's (1964) method for rivers with flow depth less than or about 10 feet. (c) Use Toffaleti's (1969) method for large rivers. For mathematical modeling, it may be more efficient to relate the sediment discharge to the water velocity and depth as a power function.

Numerical Solution of Basic Equations

The numerical solution of Eqs. (10-1), (10-2), and (10-3) can proceed in two directions. Either an attempt can be made to convert the original system of partial differential equations into an equivalent system of ordinary differential equations by using the method of characteristics (Chang and Richards, 1971; De Vries, 1965), or one can replace the partial derivatives in the original system with quotients of finite-differences by using the explicit method or the implicit method (Amein and Fang, 1970; Contractor and Wiggert, 1972). The various solution techniques frequently used for flood routing purposes (Chen, 1973; Price, 1974) are:

1. Characteristic equations in explicit or implicit finite-difference form, solved in a curvilinear network of time and distance;

2. Characteristic equations in explicit or implicit finite-difference form, solved in a fixed rectangular time-distance network;

3. The original equations in implicit finite-difference form, solved in a fixed rectangular time-distance grid; and

4. The original equations in explicit finite-difference form, solved in a fixed rectangular time-distance network.

The choice of a numerical technique for a particular case should be made with regard to its accuracy, stability and convenience in use. Accuracy refers to the closeness of the computed values to the actual or true values. Stability problems are usually identified in the computed results as oscillations of the computed values. Amein and Fang (1970) found that the implicit method is unconditionally stable and is faster and more accurate than other finite-difference methods when applied to flood routing in open channels. Price (1974) also came up with a similar conclusion after comparing four numerical methods for flood routing. Therefore, in this chapter, only the implicit method is presented in detail.

The finite-difference equations formulated from Amein and Fang's method are a system of nonlinear algebraic equations, which must be solved simultaneously by iteration methods. This reduces the efficiency of their method in terms of computing time. Preissmann (1960) and Richtmyer and Morton (1967) presented a widely accepted implicit finite-difference scheme to approximate a function f(x,t) and its derivatives of $\partial f/\partial x$ and $\partial f/\partial t$ at Point B in Fig. 10-3 as

$$f_B(x,t) \simeq \lambda \frac{f_i^{j+1} + f_{i+1}^{j+1}}{2} + (1-\lambda) \frac{f_i^j + f_{i+1}^j}{2}$$

$$\left.\frac{\partial f}{\partial x}\right|_B \simeq \lambda \frac{f_{i+1}^{j+1} - f_i^{j+1}}{\Delta x} + (1-\lambda) \frac{f_{i+1}^j - f_i^j}{\Delta x} \qquad (10\text{-}10)$$

$$\left.\frac{\partial f}{\partial t}\right|_B \simeq \frac{1}{2} \left\{\frac{f_i^{j+1} - f_i^j}{\Delta t} + \frac{f_{i+1}^{j+1} - f_{i+1}^j}{\Delta t}\right\}$$

where $f_i^j = f(x_i,t^j)$, $f_{i+1}^j = f(x_{i+1},t^j)$, etc., and $\lambda = \delta t/\Delta t$, δt is the distance of the point B on the time axis from the old time line, $\Delta x = x_{i+1} - x_i$ and $\Delta t = t^{j+1} - t^j$. Such an approach of finite-differencing permits the use of varying values of t and x which may be required to cover the specified points in the x-t network of Fig. 10-3. For $\lambda = 1/2$, the scheme is center-implicit; for $\lambda = 1$, the scheme is fully-implicit; and for $\lambda = 0$, the scheme is fully-explicit. Chen (1973) modified Preissmann's and Amein and Fang's methods to derive an unconditionally stable and convenient linear implicit method. His method applied to solve the unsteady flow Eqs. (10-1), (10-2), and (10-3) is described below. Stability and convergency of numerical schemes will be discussed in a later section.

The finite-difference approximations employed by Chen (1973) to express the values and the partial derivatives of a function f within a four-point grid formed by the intersections of the space lines x_i and x_{i+1} with the time lines t^j and t^{j+1} are given by (see Fig. 10.3)

$$f \approx \frac{1}{2}(f_i^j + f_{i+1}^j)$$

$$\frac{\partial f}{\partial x} \approx \frac{1}{\Delta x}(f_{i+1}^{j+1} - f_i^{j+1}) \qquad (10\text{-}11)$$

and

$$\frac{\partial f}{\partial t} \approx \frac{1}{2\Delta t}[(f_i^{j+1} - f_i^j) + (f_{i+1}^{j+1} - f_{i+1}^j)]$$

in which f represents Q, A, T, etc. All the variables are known at all nodes of the networks on the time line t^j. The unknown values of the variables on time line t^{j+1} can be found by solving the system of linear algebraic equations formulated from substitution of the finite-difference approximations (10-10) or (10-11) into Eqs. (10-1), (10-2), and (10-3).

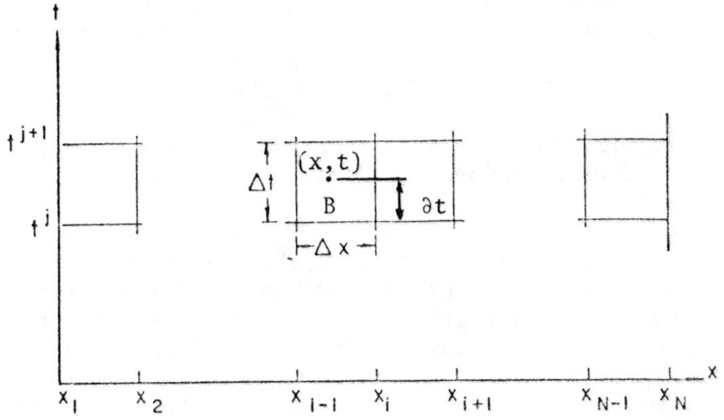

Fig. 10-3. Network for the Implicit Method

Four different methods of solution are examined in the following sections. They are (1) the complete solution, (2) the uncoupled unsteady solution, (3) the known discharge solution, and (4) the uncoupled steady solution. The latter three methods yield simplified numerical solutions to Eqs. (10-1), (10-2), and (10-3).

Complete Solution: As the expressions of the finite-difference approximations are substituted into the basic Eqs. (10-1),

(10-2), and (10-3) by assuming $\partial A_d/\partial z = \partial A/\partial y = T$ and $Q_s^{j+1} \approx Q_s^j + (\partial Q_s/\partial Q)^j (Q^{j+1}-Q^j) + (\partial Q_s/\partial y)^j (y^{j+1}-y^j) + (\partial Q_s/\partial z)^j (z^{j+1}-z^j)$, three linear algebraic equations are formed. They can be written as

$$a_{1i}Q_i + b_{1i}y_i + c_{1i}z_i + d_{1i}Q_{i+1} + e_{1i}y_{i+1} + f_{1i}z_{i+1} = r_{1i} \tag{10-12}$$

$$a_{2i}Q_i + b_{2i}y_i + c_{2i}z_i + d_{2i}Q_{i+1} + e_{2i}y_{i+1} + f_{2i}z_{i+1} = r_{2i} \tag{10-13}$$

$$a_{3i}Q_i + b_{3i}y_i + c_{3i}z_i + d_{3i}Q_{i+1} + e_{3i}y_{i+1} + f_{3i}z_{i+1} = r_{3i} \tag{10-14}$$

in which the coefficients a, b, c, d, e, f and r are functions of variables evaluated at the time step t^j and therefore are known. The expressions for these coefficients are given in Appendix 10A. To assure the stability of the scheme, the friction slope, S_f, is taken on the time line t^{j+1} (Streloff, 1970). The complete solution method was developed and used by Chen and Simons (1974) and Simons, et al. (1975).

There are six unknowns in Eqs. (10-12), (10-13), and (10-14) at the time step t^{j+1}, which cause the system to be indeterminate. However, three unknowns are common for any two neighboring rectangular grids. Consequently, (N-1) sets of equations contain 3N unknowns. Three additional equations supplied by the boundary conditions make this system of equations mathematically determinable. The boundary conditions may be any of the following relations: (1) the stage hydrograph, (2) the flow discharge hydrograph, (3) the sediment discharge hydrograph, (4) the depth-discharge rating curve, or (5) the known bed elevations at the controlled boundary section.

If two upstream boundary conditions and one downstream boundary condition are given, they can be linearized in the forms

$$d_{20}Q_1 + e_{20}y_1 + f_{20}z_1 = r_{20} \tag{10-15}$$

$$d_{30}Q_1 + e_{30}y_1 + f_{30}z_1 = r_{30} \tag{10-16}$$

and

$$a_{1N}Q_N + b_{1N}y_N + c_{1N}z_N = r_{1N} \tag{10-17}$$

The boundary conditions are discussed in more detail in the section titled "Boundary Conditions" forthcoming. Equations (10-12) through (10-17) constitute a system of 3N linear algebraic equations in 3N unknowns. Any of the standard methods, such as the Gaussian elimination method or the matrix inversion method can be used for its solution. A double-sweep method is applied here for solving this system of linear equations (see Appendix 10B). After the flow condition at each node section on the time line t^{j+1} is computed by solving Eqs. (10-1), (10-2), and (10-3) numerically, the computation is moved to the next time step.

The change of sediment area over a time step from t^j to t^{j+1} is given by

$$\Delta A_d = T(z^{j+1} - z^j) \tag{10-18}$$

in which ΔA_d is assumed to be uniformly distributed over the unit width when formulating the finite-difference Eqs. (10-12) through (10-17). However, a certain distribution of ΔA_d can be assumed from theoretical or empirical information. A discussion of the method for distributing ΔA_d over a cross section is given in the section titled "Sediment Distribution Function" forthcoming.

Uncoupled Unsteady Solution: If the change in the sediment area A_d within a short period of time is much smaller than in the cross-sectional area A, the solution obtained by solving Eqs. (10-1), (10-2), and (10-3) simultaneously can be approximated by solving the flow continuity Eq. (10-2) and the flow momentum Eq. (10-3) first. The solution is then refined by using the sediment continuity Eq. (10-1).

Let it be assumed that the effects of changes in A_d and S_o on the propagation of a flood wave within a time step are negligible; then Eqs. (10-2) and (10-3) can be written in finite-difference form applying Eqs. (10-10) or (10-11) as

$$a_{1i}\Delta Q_i + b_{1i}\Delta y_i + c_{1i}\Delta Q_{i+1} + d_{1i}\Delta y_{i+1} = r_{1i} \tag{10-19}$$

and

$$a_{2i}\Delta Q_i + b_{2i}\Delta y_i + c_{2i}\Delta Q_{i+1} + d_{2i}\Delta y_{i+1} = r_{2i} \tag{10-20}$$

in which Δ denotes the change in values from t^j to t^{j+1} and the values of coefficients in the equations are known. The expressions for these coefficients are given in Appendix 10C. The (N-1) pairs of Eqs. (10-19) and (10-20) plus two boundary conditions constitute a system of 2N equations in 2N unknowns. This system of equations can be solved by using the double sweep method presented in Appendix 10B. The two boundary conditions may be: (1) the stage hydrograph, (2) the flow discharge hydrograph, or (3) the depth-discharge rating curve.

After the flow conditions at each node section on the time line t^{j+1} are computed by solving Eqs. (10-2) and (10-3) numerically, their values are used for computing the sediment load Q_s^{j+1} from the sediment transport function. These computed sediment loads are then used to estimate the change in the channel bed or cross-sectional area by applying the finite-difference approximation of Eq. (10-1). Again, the method discussed in the section titled "Sediment Distribution Function" can be used to determine the changes in cross-sectional geometry caused by bed movement. After some adjustments of flow characteristics due to sediment motion, the computation is moved to the next time step. This method was developed by Chen (1973) to study various river problems.

Known Discharge Solution: For many practical cases the celerities of disturbance at the water level are much larger than those at the bed. It can be assumed that the flow is quasi-steady for the geomorphic computations, and the flow discharge in a channel reach is known. The number of basic unknowns can then be reduced to two, namely, A_d and y. These unknowns can be evaluated by solving the sediment continuity Eq. (10-1) and the flow momentum Eq. (10-3), where in Eq. (10-3) $\partial\rho Q/\partial t$ is known. Eqs. (10-1) and (10-3) written in finite-difference form are:

$$A_i \Delta y_i + B_i \Delta y_{i+1} + C_i \Delta z_i + D_i \Delta z_{i+1} = Q_i \qquad (10\text{-}21)$$

and

$$E_i \Delta y_i + F_i \Delta y_{i+1} + G_i \Delta z_i + H_i \Delta z_{i+1} = R_i \qquad (10\text{-}22)$$

in which the values of coefficients are known. The system of linear equations obtained by assembling Eqs. (10-21), (10-22) and the boundary conditions is similar to that for the uncoupled solution and can be solved by using the same double sweep method. The two boundary conditions may be: (1) the sediment discharge hydrograph, (2) the known stage, or (3) the known bed elevation at the controlled boundary section. This method was used by Cunge and Perdreau (1973), Chen and Simons (1974), and Mahmood and Ponce (1976) to study sediment movement in open channels by approximating the actual discharge hydrograph using a step function (e.g. A-B-C-D in Fig. 10-4). Owens (1977) modified the method of approximating the hydrograph with a view to better simulate the unsteady flow effect (e.g. E-f in Fig. 10-4).

Uncoupled Steady Solution: Let it be assumed that the effects of changes in A_d and S_o on the flow conditions within a time step are negligible. Also, let it be assumed that the flow is quasi-steady for the geomorphic computations and the flow discharge in a channel reach is known. Then the solution of Eqs. (10-1), (10-2), and (10-3) can be approximated by solving Eq. (10-3) for a given known discharge to compute the

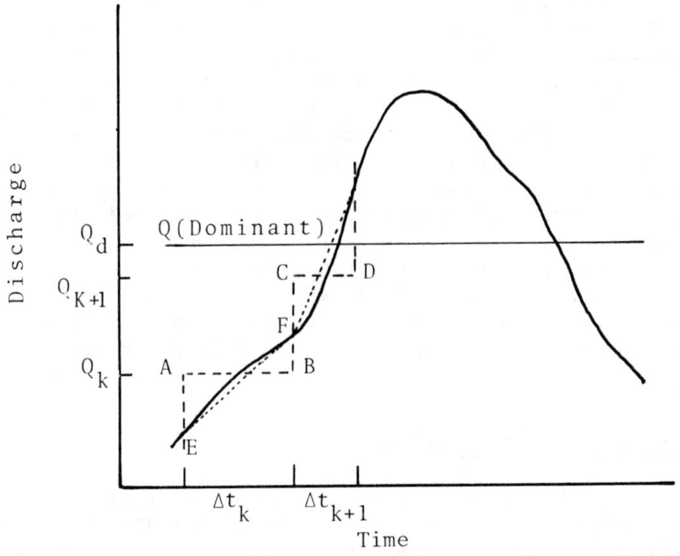

Fig. 10-4. Simulation of Unsteady Flow Hydrograph

water surface profile, and then solve Eq. (10-1) uncoupled to determine sediment movement. This approximation has been used by Chang and Hill (1976), the Army Corps of Engineers (1976), and recently by Li (1977).

Discussion of the Four Types of Numerical Solution Methods

The four types of methods can be categorized into two groups: coupled method (complete solution and known discharge solution) and uncoupled method (unsteady solution and steady solution). The advantage of the coupled method is its better consideration of the continuous interaction of the hydraulic and sediment transport phases than the uncoupled method. This clearly makes better physical sense. Also, the coupled method can generally be used with a longer time increment. However, the formulation of the complete solution method is the most complex one among the four methods. The known-discharge solution is developed solely for sediment routing.

The uncoupled method is simpler to formulate than the coupled method. However, in the uncoupled solutions, the length of a time increment is restricted in that the bed elevation changes over one time increment must be so small that they do not significantly affect the solution of the hydraulic variables that have been determined in the first step of the uncoupled procedure. Also, the sediment routing scheme is an explicit scheme which has a stability limitation of time increment. The stability criteria will be discussed in a forthcoming section "Stability and Accuracy."

Since the numerical testing and comparison of coupled and uncoupled solutions are not well known, it is suggested that for a stable channel where the channel changes are very mild, the uncoupled method can be applied. In this case, the uncoupled-unsteady method can be used for water and sediment routing. For sediment routing only, the uncoupled-steady method can be used.

If the channel is quite active, the coupled method is preferable. For both water and sediment routing, the complete solution method is desirable, while the known discharge solution can be utilized when only the sediment routing is interested.

Sediment Distribution Function

The erosion and/or deposition on a channel cross section is a two-dimensional process. The one-dimensional mathematical model can only compute cross-sectional area changes but not specifically the changes in cross-sectional shapes. It is necessary to properly distribute the computed area changes over each cross section in order to obtain realistic changes in bed elevations. Several researchers (e.g. Chen, 1973; the Corps of Engineers, 1976) assumed a uniform distribution of bed changes. In some cases this assumption is not adequate. To improve the accuracy of the sediment routing model some nonuniform distributions of scour or deposition on the bed have been developed.

Chang and Hill (1976) distributed the area changes according to the effective shear stress $\tau - \tau_c$, where τ is the actual bed shear and τ_c is the shear stress at incipient motion of the bed material. If over a time increment their uncoupled-steady solution indicates scour, the scour depth in each subsection is calculated using the following formula:

$$\Delta z_{ki} = \frac{(\tau - \tau_c)_{ki}}{\sum_{k=1}^{M} (\tau - \tau_c)_{ki} T_{ki}} \Delta A_{d_i} \tag{10-23}$$

where ΔA_{d_i} is the scoured area, M is the number of subsections in Section i, and Δz_{ki}, $(\tau-\tau_c)_{ki}$, and T_{ki} are the bed elevation change, effective shear stress and top width of subsection k at Section i, respectively (see Fig. 10-2). If the computed results indicate deposition, the sediment is assumed to deposit on horizontal layers in the lower part(s) of the cross section. It should be emphasized that this scheme for distributing the bed elevation change is not based on an empirical or theoretical law, but instead on the author's experience and knowledge. The model was able to reproduce fairly well the scour and deposition measured in a flooded prototype drainage canal.

To further simplify the sediment distribution process, it appears to be feasible to use a single relation to distribute both the erosion and deposition. Owens (1977) suggested the use of the following simplified relation:

$$\frac{\Delta A_{d_{ki}}}{\Delta A_{d_i}} = \left(\frac{K_{ki}}{\Sigma K_{ki}}\right)^M \tag{10-24}$$

where M is a constant exponent and K_{ki} is the conveyance of subsection k at Section i defined by Eq. (10-6). The validity of this simplified method was checked using the compound stream model developed by Dass (1975), and was also checked preliminarily by field data from Greenwood Bend in the Yazoo River. It is found that this method is valid in a river reach where the channel alignment is reasonably smooth (see Figs. 10-5 and 10-6). If the river reach encounters an abrupt change in channel alignment such as the presence of dikes, then both scour and deposition could occur simultaneously in different subsections at a given cross section. In this case, the suggested simplified method is not valid. A compound stream model similar to Dass's model or a two-dimensional model is required to study this type of river problem. Compound stream models and a two-dimensional river model are presented in Sections 10.3 and 10.4, respectively.

Boundary Conditions

There are two kinds of boundary conditions for one-dimensional water and sediment routing models: (1) exterior boundary conditions, at the ends of the model, and (2) interior boundary conditions.

Exterior Boundary Conditions: The exterior boundary conditions vary with the character of the flow. If the flow is tranquil (subcritical), the water routing solution is possible only when one condition is specified at the downstream end of the reach. If the flow is supercritical the condition at the downstream boundary is redundant. For this type of flow the conditions cannot influence the upstream situation and the

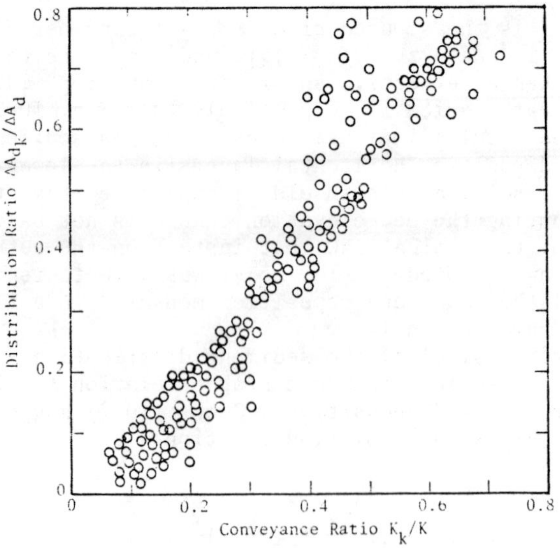

Fig. 10-5. Verification of the Simplified Sediment Distribution Function Using the Compound Stream Model

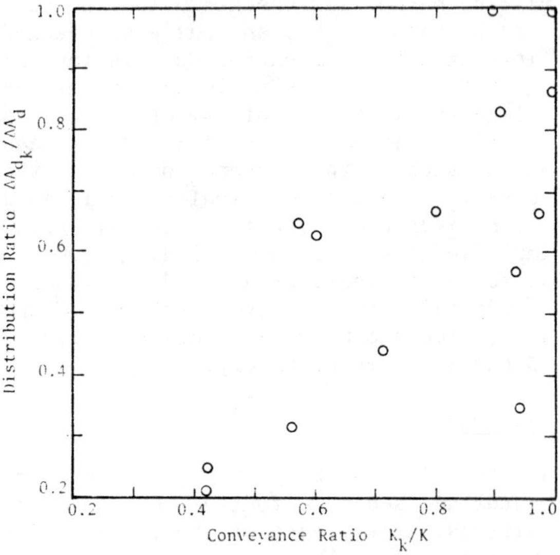

Fig. 10-6. Verification of the Simplified Sediment Distribution Function Using the Field Data in Greenwood Bend

boundary conditions should be specified at the upstream boundary. The exterior boundary conditions are usually of five kinds: (1) the stage hydrograph; (2) the water discharge hydrograph; (3) the sediment discharge hydrograph; (4) the depth-discharge rating curve; and (5) the known bed elevations at a controlled boundary section. Conditions (1), (2), and (4) are for water routing, while conditions (3) and (5) are for sediment routing. In general, two boundary conditions are required for water routing and one condition is required for sediment routing.

These conditions can be linearized or formulated in a linear relation as given in Eqs. (10-15), (10-16), etc. For example, if the upstream boundary condition is a stage hydrograph, h = f(t), then the coefficient in Eq. (10-15) becomes

$$d_{20} = 0., \ e_{20} = 1., \ f_{20} = 1. \text{ and } r_{20} = f(t)$$

For a sediment discharge hydrograph, $Q_S = f(t)$, a linearization process is imposed to linearize the known sediment discharge inflow to a form of Eq. (10-16) as follows:

$$Q_s^{j+1} = Q_s^j + \left(\frac{\partial Q_s}{\partial Q}\right)^j (Q^{j+1} - Q^j) +$$

$$\left(\frac{\partial Q_s}{\partial y}\right)^j (y^{j+1} - y^j) + \left(\frac{\partial Q_s}{\partial z}\right)^j (z^{j+1} - z^j) \qquad (10\text{-}25)$$

$$= f(t^{j+1})$$

Eq. (10-25) can be rearranged as

$$\left(\frac{\partial Q_s}{\partial Q}\right)^j Q^{j+1} + \left(\frac{\partial Q_s}{\partial y}\right)^j y^{j+1} + \left(\frac{\partial Q_s}{\partial z}\right)^j z^{j+1}$$

$$= f(t^{j+1}) - Q_s^j + \left(\frac{\partial Q_s}{\partial Q} Q\right)^j + \left(\frac{\partial Q_s}{\partial y} y\right)^j \qquad (10\text{-}26)$$

$$+ \left(\frac{\partial Q_s}{\partial z} z^j\right)$$

where the superscripts j and j+1 indicate that the variables are evaluated at the time t^j and t^{j+1}, respectively. Those variables evaluated at time t^j are all known. Comparing Eq. (10-16) with Eq. (10-26), the coefficients in Eq. (10-16) can be formulated.

Equation (10-26) is used in the coupled method (i.e., complete solution and known discharge solution). For the uncoupled method, the original sediment discharge hydrograph is used for sediment routing. In summary, for a subcritical flow, the complete solution method requires two upstream boundary conditions (on water and sediment, respectively) and one downstream boundary condition (on water). For uncoupled-unsteady method, one upstream and one downstream boundary condition is required for water routing. A boundary condition related to sediment is then used to route sediment. For known-discharge method, the inflow water discharge hydrograph is used to determine the water discharge in the study reach. Then an upstream boundary condition (on sediment) and a downstream boundary condition (on water) are used for the coupled sediment routing. Finally, for the uncoupled-steady method, again the inflow water discharge hydrograph is used to determine the water discharge in the study reach. Then a downstream boundary condition (on water) is used to determine the water surface profile for a given discharge. Thereafter, a boundary condition related to sediment is used for sediment routing.

In general, there is no necessity for the initial conditions to satisfy the flow equations, since the influence of the initial state on the solution diminishes with time and the boundary conditions become prevailing. However, if there is no link between the stages and discharges in the boundary conditions, such as imposing a condition of the type Q(t) at the two boundaries of the model, the initial volume will always be contained in the channel and influence the final results.

Significant error may be introduced when the relationship Q = f(h) is imposed at the downstream boundary in subcritical flow. This relation is a single-valued function corresponding to the steady flow. The computed results of the unsteady flow in the reach upstream of this condition are biased within the range of the backwater influence. In this case one can try to prolong the reach fictitiously so as to displace the unsatisfactory condition as far downstream as possible.

Interior Boundary Conditions: The interior boundary conditions are all similar. There are always continuity conditions for discharges and another condition usually concerning the water stage or energy balance for water routing. Some examples are given below:

(1) Sudden change of the cross section between two neighboring sections (Fig. 10-7). The interior conditions are:

$$\text{(a)} \quad Q_1 = Q_2 \qquad (10\text{-}27)$$

$$\text{(b)} \quad h_1 + \frac{V_1^2}{2g} = h_2 + \frac{V_2^2}{2g} + h_f \qquad (10\text{-}28)$$

If the energy loss and kinetic energy head can be neglected, the second condition becomes

$$h_1 = h_2 \qquad (10\text{-}29)$$

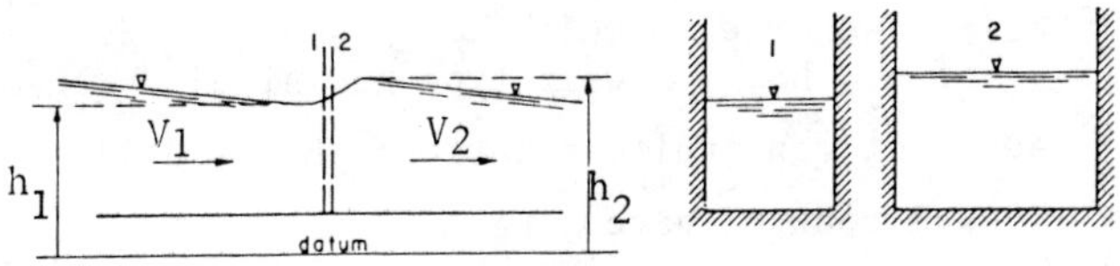

Fig. 10-7. Cross Section Variation Between Two Reaches

(c) The sediment continuity equation (Eq. 10-1) is generally valid in the interior boundary

(2) Junction of two rivers (Fig. 10-8):

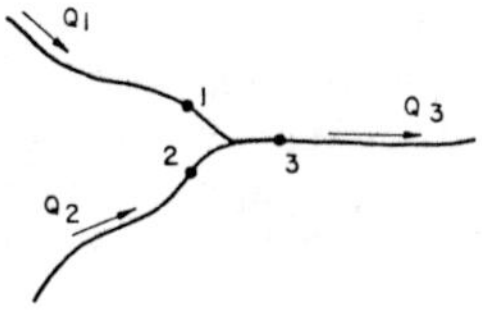

Fig. 10-8. Junction of Two Rivers

(a) $Q_3 = Q_1 + Q_2$ (10-30)

(b) $h_1 + \frac{V_1^2}{2g} = h_3 + \frac{V_3^2}{2g} + h_{f_{1-3}}$, or $h_1 = h_3$ (10-31)

(c) $h_2 + \frac{V_2^2}{2g} = h_3 + \frac{V_3^2}{2g} + h_{f_{2-3}}$, or $h_2 = h_3$ (10-32)

(d) sediment continuity equation: Eq. (10-1).

(3) Flow over weirs (Fig. 10-9):

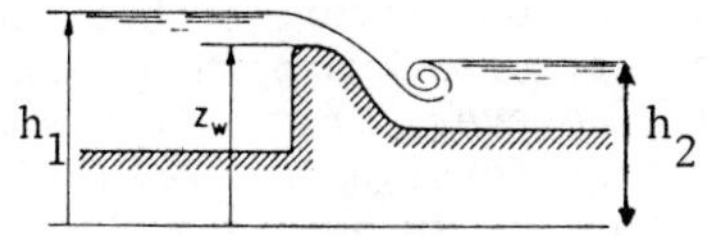

Fig. 10-9. Flow Over Weirs

(a) $Q_1 = Q_2$ (10-33)

(b) Weir discharge formula

$Q = f(h_1, h_2, z_w, \text{weir type and size})$ (10-34)

(c) sediment continuity equation: Eq. (10-1)

(4) Flow through control gates

(a) $Q_1 = Q_2$ (10-35)

(b) Gate discharge equation

$Q = f(h_1, h_2, z_w, \text{gate type and size, gate opening})$ (10-36)

(c) Sediment continuity equation: Eq. (10-1)

Stability and Accuracy

A finite difference procedure for calculating time-dependent phenomena is considered stable when small numerical errors of truncation and round-off inevitably introduced at time t_o are not amplified during successive applications of the procedure, and at subsequent time t have not grown so large as to obscure the valid part of the solution. In practical computational experience, instability of a process for calculating unsteady open-channel flows shows up as wildly oscillating values of the dependent variables with respect to t and x. In practice, this means that limitations in the maximum size of the space and time intervals must be imposed to the numerical scheme in order to avoid stability problems.

Implicit methods of finite differences are generally regarded as unconditionally stable, the maximum size of the time step being limited only by accuracy requirements (Fread, 1974; Liggett and Cunge, 1975; Ponce et al., 1978). On the other hand, explicit methods are limited in the maximum size of the time step by a strict stability criterion called the Courant condition,

$$\Delta t \leq \frac{\Delta x}{c} \quad (10\text{-}37)$$

where c is the celerity of the small gravity wave given by $c = \sqrt{gy} + V$, V is the mean flow velocity, and Δx and Δt are the space and time increments.

Accuracy is the degree of difference between observed real-life data (such as measured hydrographs) and computed results. Unfortunately, however, true values are not available since the equations in consideration are analytically unsolvable and the field or laboratory measurements may be subject to uncertainties just as great as those in the mathematical model solutions. Liggett and Cunge (1975) point out that there might be several reasons for discrepancy between a mathematical model and the prototype such as:

1) Inaccurate simplifications and approximations in the basic equations fail to simulate the complexity of the prototype.

2) Insufficiently accurate measuring techniques, e.g., surveying errors, badly located gages, etc.

3) Insufficient data, e.g., unknown tributary discharges, exchange of discharge with underground water, etc.

4) Phenomena which are not taken into account, e.g., infiltration into the dried-up soil at the beginning of a flood, variation of roughness coefficients with vegetation, etc.

5) Poor schematization of topographic features.

An estimate of accuracy, often used in numerical schemes, is introduced by the concept of convergence. Convergence refers to the ability of a numerical scheme to reproduce the terms of the differential equation without introducing extraneous terms that impair the accuracy of the solution. In theory, the numerical solution converges to the analytical solution as Δt and Δx tend to zero. In practice, however, Δx and Δt equal to zero is an impossibility. Therefore, when using mathematical models it is necessary to introduce the concept of degree of convergence. This concept refers to the approximation of the numerical solution to the analytical solution.

Since the flow equations (10-1), (10-2) and (10-3) are analytically unsolvable, the best measure of convergence of a particular numerical scheme is to check against a simplified problem which has an analytical solution. Price (1974) used this approach to compare results from four of the more important numerical methods for flood routing with exact analytical solutions for the monoclinal wave. He found that the Preissmann type implicit scheme is the most efficient method for flood routing problems, and the optimum accuracy is obtained when the finite difference time step is chosen approximately equal to the space step divided by the kinematic wave speed

$$\Delta t = \frac{\Delta x}{c_k} \qquad (10\text{-}38)$$

in which the kinematic wave speed is given by $c_k = 1.5\ V$.

The importance of the Courant criterion in numerical modeling is apparent when the similarity of equations (10-37) and (10-38) is recognized. In explicit schemes, the Courant number cannot exceed one for stability reasons, i.e., $c\ \Delta t/\Delta x \leq 1$. In implicit schemes, no limitations of stability are imposed and the Courant number can take values greater than one depending on the spatial resolution and the accuracy requirements. In addition, there is a definite computational advantage of the implicit over the explicit scheme because the explicit Courant condition is based on the celerity of small disturbances while the implicit Courant "condition" (Eq. 10-38) is based on the flood wave celerity (Ponce et al., 1978). This means that for a comparable degree of accuracy, the implicit time step can be many times larger than the explicit time step. Liggett and Cunge (1975) also obtained the same conclusion. They applied the Preissmann implicit scheme of finite differences to a highly simplified set of linear differential equations with a known Fourier series analytical solution, and showed that each component of the series obtained by the numerical method is characterized by its damping factor and its celerity, and that the convergence criterion is a function of the Fourier series component wavelengths and also of the ratio $\Delta t/\Delta x$. They reported better accuracy for the implicit scheme than for other well-known explicit schemes of finite differences.

Based on the above considerations, the Preissman type implicit scheme of finite differences, as far as stability is concerned, allows for the use of any time step in the solution domain. In such a case, however, the time step will be governed by the other requirements, such as accuracy and convenience in use, for instance. However, for an uncoupled method of solution to the three basic differential equations using an implicit scheme for the flow routing and an explicit scheme for the sediment routing, the selection of the time step will be governed by the stability of the latter one, rather than the former one.

In order to obtain a stability criterion for the adopted explicit scheme of finite differences Eq.(10-11) ($Q = 0$), Lopez (1978) provided the following analysis. First, an expression for the celerity of the sand wave is developed and further adjusted by numerical experiments. Second, the migration of a bed wave specified as initial condition on the channel bed is studied and a critical value of $\Delta x/\Delta t$ for which the scheme becomes unstable is obtained.

Consider a simplified form of the sediment continuity equation:

$$\frac{\partial Q_s}{\partial x} + \frac{\partial A_d}{\partial t} = 0 \qquad (10\text{-}39)$$

in which the lateral flow and the variation of suspended sediment concentration with time have been neglected.

Assuming that A_d is a unique function of Q_s we can write:

$$\frac{\partial A_d}{\partial t} = \frac{dA_d}{dQ_s}\frac{\partial Q_s}{\partial t} \qquad (10\text{-}40)$$

Substituting Eq. (10-40) into Eq. (10-39):

$$\frac{dQ_s}{dA_d}\frac{\partial Q_s}{\partial x} + \frac{\partial Q_s}{\partial t} = 0 \qquad (10\text{-}41)$$

which is a nonlinear equation since $\frac{dQ_s}{dA_d}$ is itself a function of Q_s.

The celerity of bed waves can be represented by

$$\frac{dx}{dt} = \frac{dQ_s}{dA_d} \qquad (10\text{-}42)$$

which describes a positive characteristic direction in the x-t plane indicating that any perturbation on the bed can only be propagated downstream.

An expression for the celerity can be obtained by assuming that the sediment discharge is a power function of the velocity

$$Q_s = mV^n\ T \simeq m\ \frac{Q^n}{(h-z)^n}\ T^{1-n} \qquad (10\text{-}43)$$

in which T is the cross-section top width, D is the hydraulic depth, and m and n are constants which can be obtained by calibration or from the available equilibrium data. Its derivative with respect to z is

$$\frac{\partial Q_s}{\partial z} \simeq mV^n\ T\ \frac{n}{y} = \frac{nQ_s}{y} \qquad (10\text{-}44)$$

where it was assumed that $\partial Q/\partial z = \partial h/\partial z = \partial T/\partial z = 0$. Since

$$\frac{\partial Q_s}{\partial A_d} = \frac{1}{T}\ \frac{\partial Q_s}{\partial z} \qquad (10\text{-}45)$$

we have

$$\frac{\partial Q_s}{\partial A_d} = \frac{nQ_s}{Ty} \qquad (10\text{-}46)$$

Equation (10-46) gives the celerity of the small bed waves indicating that it is proportional to the sediment discharge Q_s and inversely proportional to the flow depth y.

Equation (10-46) is adjusted by carrying out some test runs with the numerical model. A bed wave of trapezoidal shape was imposed as initial condition on the channel bed and its velocity of propagation was determined for different initial flow conditions. It was found that the model celerity c_m and the theoretical celerity c_c computed from Eq. (10-46) have a relation: $c_m = 2.33\ c_c^{0.96}$. Equation (10-46) then takes the form:

$$c = 2.33\ \left(\frac{nQ_s}{Ty}\right)^{0.96} \qquad (10\text{-}47)$$

To insure the stability of the scheme, the value of $\Delta x/\Delta t$ should be smaller than the bed wave celerity computed from Eq. (10-47).

Application of Mathematical Models

Introduction: The procedure of applying the one-dimensional water and sediment routing model to study river problems is summarized. More detailed information is given by Chen (1973) and by Cunge (1975).

1. Planning and selection of models

There are a large number of different mathematical models available. A model which is efficient and is able to accomplish the objectives of the study should be adopted. Therefore, the advantages, disadvantages and limitations of various models should be realized.

2. Construction of the mathematical model

   a. Select a series of computational points (channel cross sections) to simulate the water course in such a way that the points represent all important topographical and hydraulical features of the study reach. As a rule of thumb, the computational points may be chosen at all irregular sections and/or interesting locations. These computational points are a function of the model and its study objectives. For example, when computing river bed changes in the Mississippi River the distance between the points was varied from 0.5 to 4 miles. The model itself was 200 miles long. To study a local problem such as dredging the space increment was reduced to as small as 0.07 miles by introducing additional sections in the modeled river reach.

   b. Evaluate the physical features at each section (computational point) by defining the supplemental relations as described in the section titled "Supplemental Equations."

Some guidelines are given in that section and also below:

(i) The area and top width of the cross section is computed for different water levels in the section and tabulated starting with the zero depth. Thus, a tabulated or fitted function which represents the cross-sectional geometry is built. When there are cross-sectional geometries measured between the computational points, the representative channel geometry at the computation points may be obtained by averaging these cross-sectional geometries. This case is especially true when simplified routing methods are used or computational points are far apart from each other. A method developed by Simons, et al. (1971), can be used to determine the geometry of the representative section. The method is physically correct and easy to apply.

The volume of water within the reach can be approximated as

$$\forall = \frac{1}{2} \sum_{i=1}^{m-1} (A_i + A_{i+1})(L_{i+1} - L_i) \qquad (10\text{-}48)$$

where A_i is the cross-sectional area of the water passage at a distance L_i downstream along the watercourse from some arbitrary reference point and m is the number of cross sections describing the reach.

The length of the river reach is

$$L = L_m - L_1 \qquad (10\text{-}49)$$

The average cross-sectional area is the volume divided by the length of the reach or

$$A = \frac{\sum_{i=1}^{m-1} (A_i + A_{i+1})(L_{i+1} - L_i)}{2(L_m - L_1)} \quad (10\text{-}50)$$

Similarly, the average wetted perimeter of the reach is

$$P = \frac{\sum_{i=1}^{m-1} (P_i + P_{i+1})(L_{i+1} - L_i)}{2(L_m - L_1)} \quad (10\text{-}51)$$

and the average top width is

$$T = \frac{\sum_{i=1}^{m-1} (T_i + T_{i+1})(L_{i+1} - L_i)}{2(L_m - L_1)} \quad (10\text{-}52)$$

(ii) The conveyance factor, using Manning's equation is easily defined as a function of water level for the given section as shown in Fig. 10-2:

$$K = \sum_{k=1}^{M} K_k \quad \text{with} \quad K_k = \frac{1.486}{n_k} A_k R_k^{2/3}$$

Thus, it is possible to construct the function representative of the hydraulic characteristics of a given section $K = K(y)$. Again, sometimes the analytic, most often a polynomial or power function is used instead of the tabulated one. On many occasions it may be difficult to obtain a good approximation of the function $K(y)$ by any fitting procedure.

For sediment routing, it may be advantageous to input the original cross-sectional geometries of the computation points in the model. In this way, the cross-sectional geometries and conveyances can be determined directly from the cross-sectional properties. Then the computed cross-sectional changes can be distributed on the cross-sectional points using a sediment distribution function similar to Eq. (10-24). However, if the number of the computational points is large, this method becomes quite tedious and time consuming. A combination of this method for computing Δz and the above method for computing A, T and K may be a good alternative.

(iii) The lateral inflow discharges can be evaluated from field data, empirical relations and/or theories. For example, water yield and sediment yield inflowing to rivers from watersheds can be determined from methods described in Chapter 9.

3. Calibration of the mathematical model

It is desired to reproduce the historical flow characteristics and geomorphic changes in the study reach to insure existence of similarity between the mathematical model and the modeled river reach such that the model can be used to predict future changes. The important features to be simulated include:

a. The water discharges and water surface profiles, where the hydrographs need to be discretized according to variation rate of the hydrographs and computation time steps. Usually, for water routing a time step varying from minutes to hours is used. For sediment routing a longer time step in days is employed.
b. The cross-sectional area and bed elevation changes.
c. The sediment transport rates.

Efforts should be made to modify the resistance function and the sediment transport function at each section until the known historical changes are reproduced.

4. Operation of the mathematical model

The calibrated mathematical model can be operated to predict the river responses to changes. It is necessary to check the calculated results with the principles of river mechanics. If the results make sense, confidence in the model is reinforced.

5. Evaluation of development plans

By comparing the river responses to various development plans obtained from operation of the mathematical model, the feasible plans can be identified and their corresponding effects can be determined.

6. Improving the mathematical model.

The mathematical model should be updated using the new measured data and/or improved theories. This process is designed to:

a. Verify the applicability of the mathematical model.
b. Refine the model for predicting future responses of the river basin to development.
c. Finalize a methodology for assessing watershed and river response to water resources control and development.

7. Documentation of the mathematical model and the model results

A flow chart is given in Fig. 10-10 to show the principal programming steps. Some application examples are presented in the following sections.

Sand Wave Propagation in a Laboratory Flume (Chen and Simons, 1974): An experiment was conducted in a 1.961-ft wide and 60-ft long flume to study the mechanics of degradation and aggredation. The experiment was started from the condition of incipient motion with a constant water discharge of 0.5 cfs. Sediment

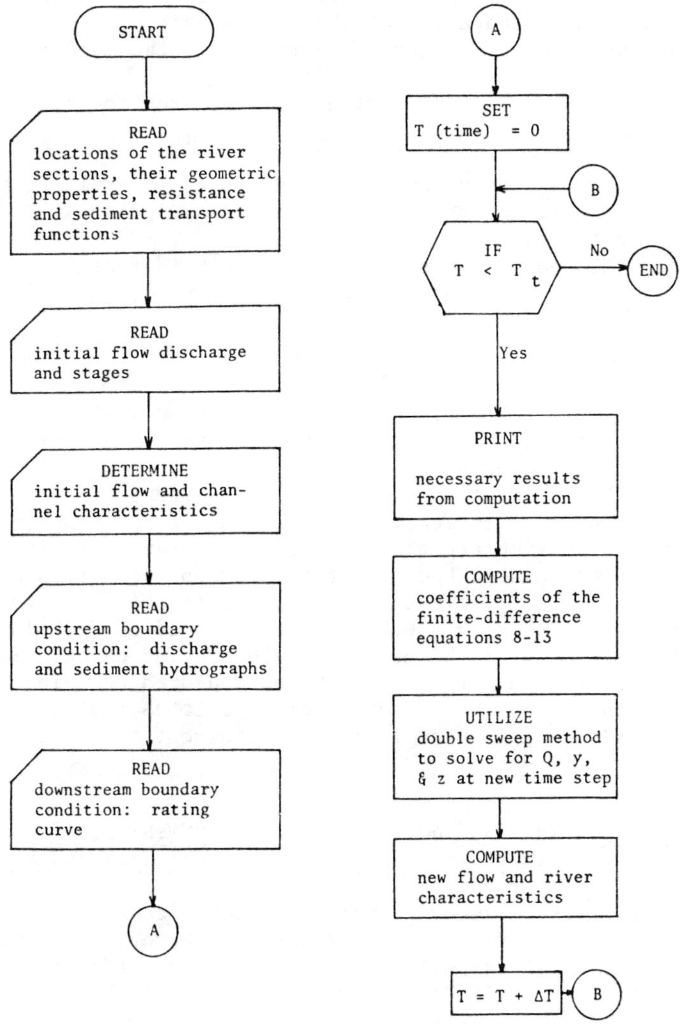

Fig. 10-10. Flow Chart of the Mathematical Model

was supplied at the upstream end of the flume at a constant rate of 0.0337 lb/sec to cause aggradation. A mathematical model was constructed using the flow and flume characteristics to simulate the propagation of the sand wave.

The supplementary relations derived for the mathematical model were:

1. The geometric properties of cross section are

   $T = \partial A/\partial y = B = 1.961$ ft and $A = Ty$.

2. The Manning's equation is employed to evaluate the friction slope with the Manning's roughness coefficient equal to 0.0185.

3. The sediment discharge is calculated from

   $$c_s = KV^m y^n \tag{10-53}$$

   and

   $$Q_s = C_s\, Q \tag{10-54}$$

   in which K = 0.0000302, m = 3, and n = -1 are so chosen that the calculated results closely fit the measured data.

4. The sediment volume ratio p is found to be 0.59.

The complete solution method and the known discharge model were applied to simulate the propagation of sand wave in the flume. The calculated bed elevations for upstream boundary conditions of Q = 0.5 cfs and Q_s = 0.0337 lb/sec and downstream boundary condition of h = 0.939 ft are plotted on Fig. 10-11 and are compared with the measured values. The calculated results are in good agreement with the measured bed elevations especially near the wave front. This indicates that the mathematical model is useful to study sediment transport phenomena.

Study of Pools 24, 25, and 26 in the Upper Mississippi and Lower Illinois Rivers (Simons, et al., 1975b; Chen and Simons, 1975):

Description of study reach: A mathematical model was constructed to cover the Upper Mississippi River from above Lock and Dam 22 (River mile 304.8) to St. Louis (River mile 179.6), the Lower Illinois River from River mile 80 to the mouth of the Illinois River, and the Lower Missouri River below Hermann (River mile 98) (see Fig. 10-12). The effects of locks and dams and the interactions between the Mississippi River and its main tributaries on the geomorphology of rivers and adjacent lands were considered. The model can be used to study the impacts of different operation schemes for the locks and dams, the effects of the pools on the behavior and form of the tributary rivers, the impact of change in the delivery of sediment and water to the study reach on the morphology of the river and adjacent lands, and the impacts of dredging and dredged material disposal on the hydraulic response and sedimentation patterns in the main channel.

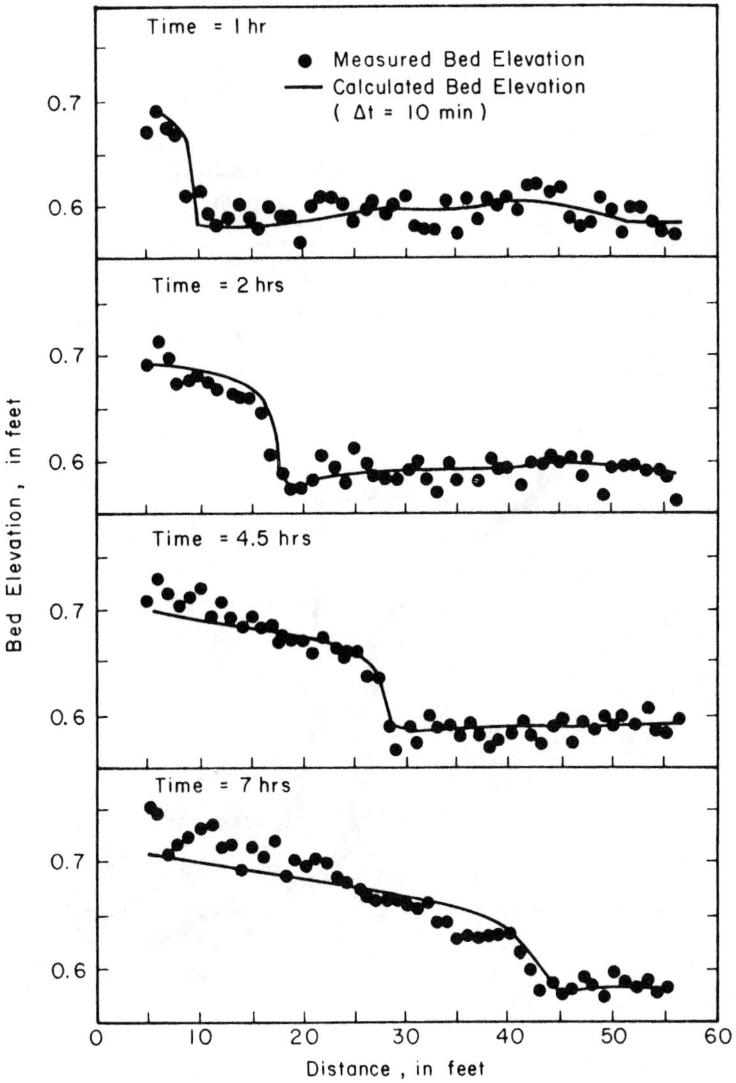

Fig. 10-11. Propagation of Sand Wave in a Rectangular Flume

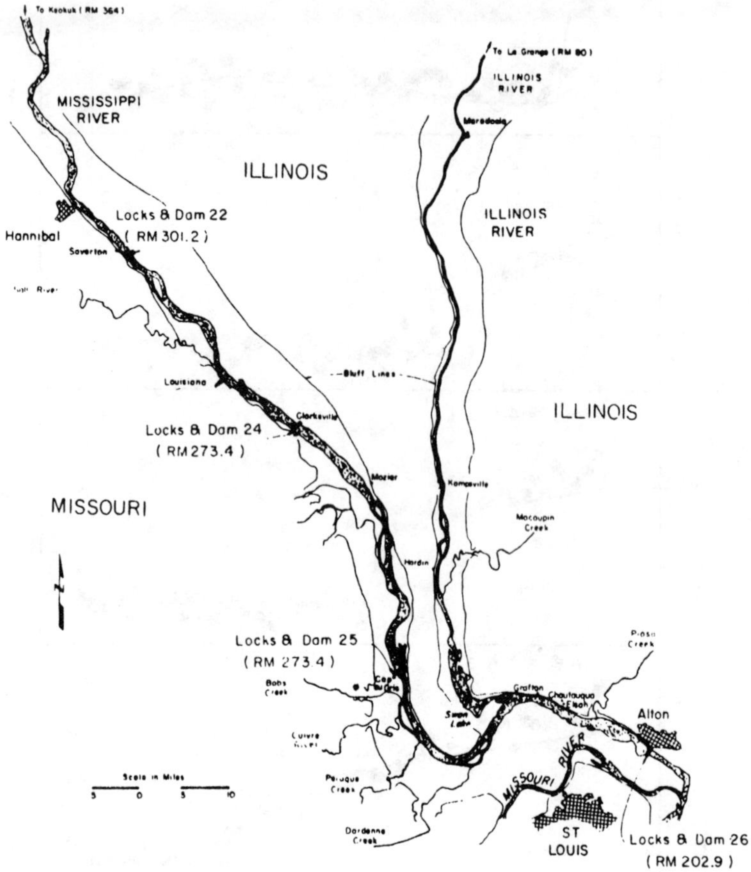

Fig. 10-12. Index Map of Pools 24, 25 and 26 in the Upper Mississippi and Lower Illinois River

Objectives of the study: The model permits an analysis of future geomorphic change. What will Pools 24, 25, and 26 in the Upper Mississippi and Lower Illinois Rivers look like 50 years from now? Will the side channels fill with sediment? Will the riverbed aggrade making the maintenance of the 9-foot channel project more expensive? Are there any viable alternatives to the present-day operations that would enhance the environmental aspects of the pools and at the same time maintain the navigation channel? The mathematical model was applied to answer these and other questions relative to the Pool 24, 25, and 26 reach.

Construction and calibration of the model: The mathematical model was constructed and calibrated using the following information: 1. hydrographic maps of the river reach, 2. hydrographs of stage, flow and sediment discharge, and 3. geological and physical properties of the bed and bed materials. The model was formulated using the basic equations (10-1), (10-2), and (10-3), coupled with the interior conditions simulating the interaction between the Upper Mississippi River and its tributaries (Eqs. (10-30) to (10-32)). Also, Eqs. (10-35) and (10-36) were utilized to simulate sediment-laden water flowing through the locks and dams. These equations were solved by the complete solution method (Appendices 10A and 10B) after evaluating the supplemental relations as follows. From the hydrographic maps, one can evaluate the geometric properties of the river reach by tabulating cross-sectional areas and top widths versus stage at each selected computation point. The relations for q_ℓ, S_f, and Q_s can then be evaluated from 2 and 3. In this study, q_ℓ was assumed negligible. Initial Manning's roughness coefficients were obtained from solving the steady momentum equation for a given water discharge and stage profile. Q_s is assessed from Toffaleti's method (1969). The S_f and Q_s relations were modified until the mathematical model reproduced faithfully the observed changes in water surface and bed profiles between 1939 and 1971 when routing the 1939 to 1971 discharge hydrographs through the modeled river reach.

The calculated and verified 1965 and 1967 water surface profiles are compared with the measured stages in Figs. 10-13 through 10-16. These figures show an agreement between the measured and calculated values. In Tables 10-1 and 10-2 the calculated changes in bankfull cross-sectional areas and in average river bed elevation (average of the river bed elevations in the deepest 1000-ft width of river channel) between 1939 and 1971 are compared with the measured changes. Data in Table 10-1 shows an agreement between the measured and calculated changes in bankfull cross-sectional areas except in the lower part of Pool 25. The results in Table 10-2 are less satisfactory than the results in Table 10-1. The difference is due primarily to

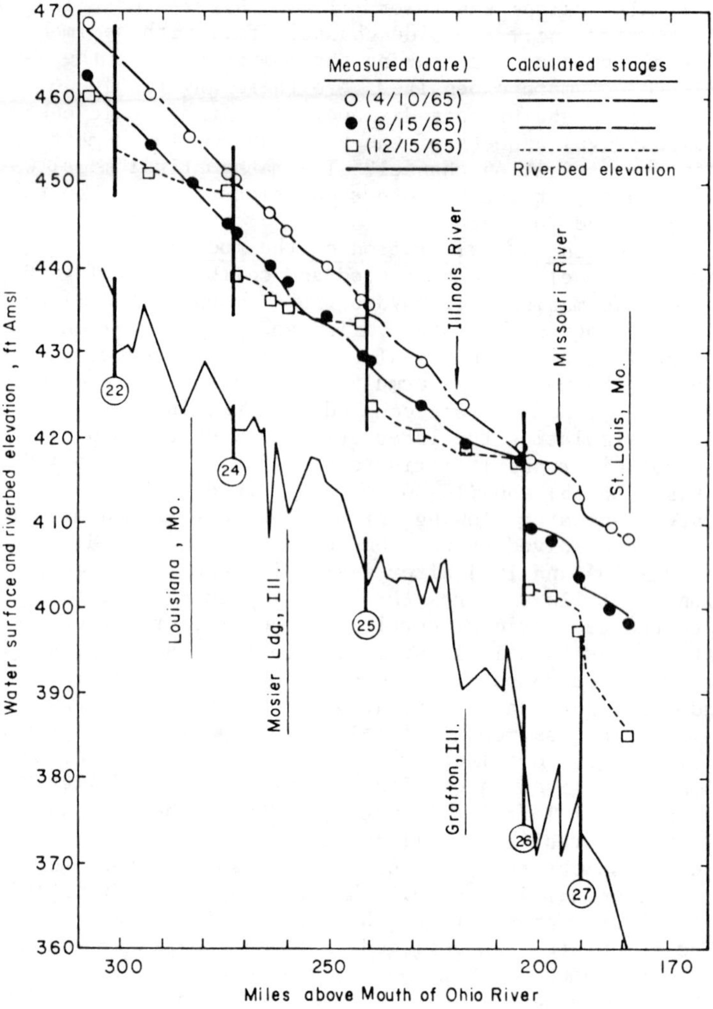

Fig. 10-13. Mathematical Model Reproduction of 1965 Water Surface Profile in the Upper Mississippi River

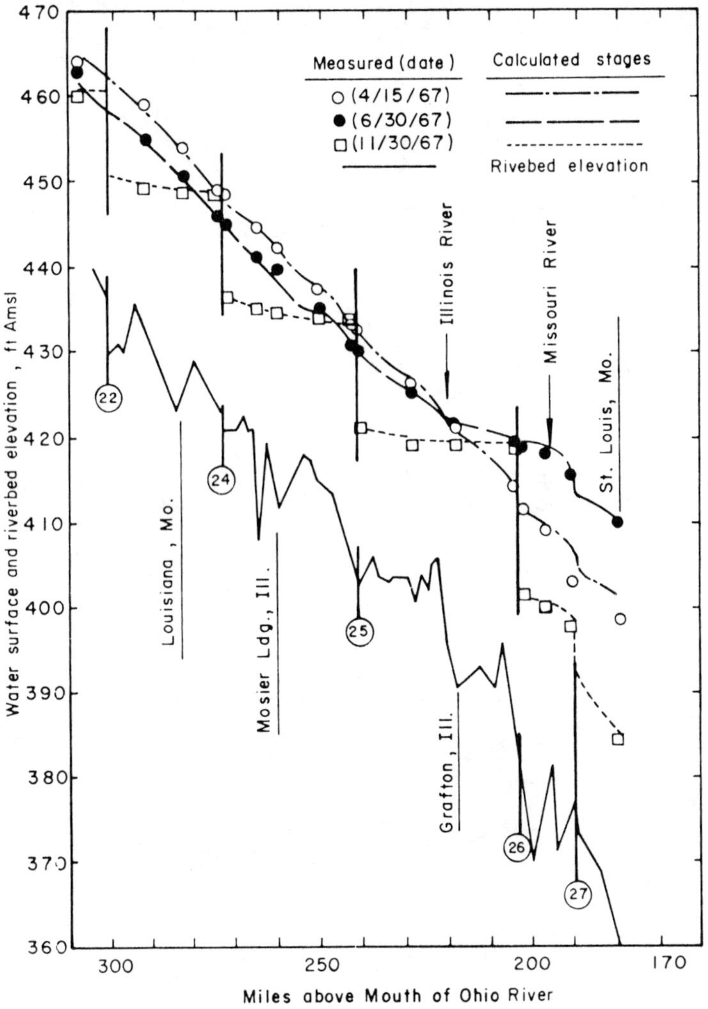

Fig. 10-14. Mathematical Model Reproduction of 1967 Water Surface Profile in the Upper Mississippi River

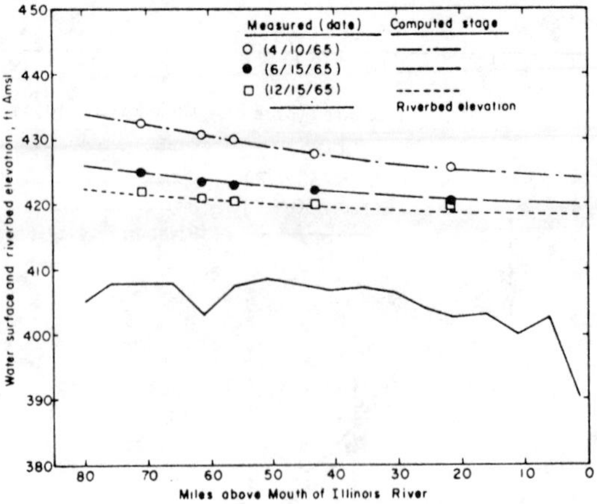

Fig. 10-15. Mathematical Model Reproduction of 1965 Water Surface Profile in the Lower Illinois River

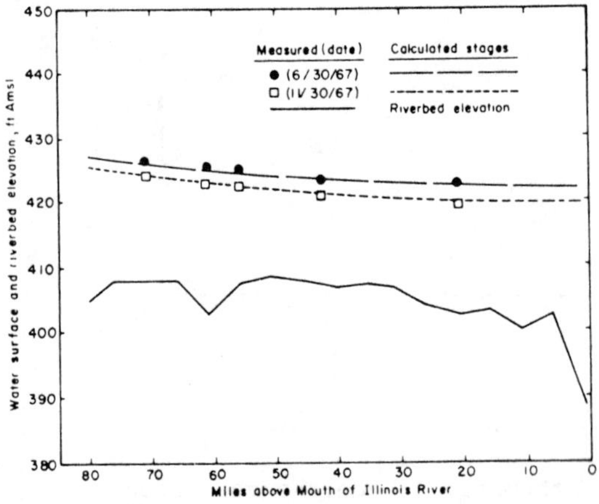

Fig. 10-16. Mathematical Model Reproduction of 1967 Water Surface Profile in the Lower Illinois River

Table 10-1. Change in Bankfull Cross-Sectional Area in the Mississippi River from 1939 to 1971

Location	Bankfull Cross-Sectional Area,* 100 ft^2		
	1939	Measured Change	Calculated Change
Pool 26:			
Below Illinois River	783	-44	-50
Middle third	546	-45	-25
Next eighth	521	-36	-33
Upper eighth	459	+34	+60
Pool 25:			
Lower quarter	500	-12	+11
Lower middle quarter	494	- 4	+20
Upper middle quarter	459	+20	+40
Next eighth	565	-68	-71
Upper eighth	470	+40	+31
Pool 24:			
Lower quarter	528	-85	-88
Lower middle quarter	425	-32	-30
Upper middle quarter	399	-25	-19
Upper quarter	396	-19	- 6

*Cross-sectional area at bankfull stage.

Table 10-2. Change in Average Riverbed Elevations in the Mississippi River from 1939 to 1971

Location	Average Riverbed Elevation,* ft Amsl		
	1939	Measured Change	Calculated Change
Pool 26:			
Below Illinois River	390.2	-0.4	-0.9
Middle third	400.7	+0.1	+0.7
Upper quarter	405.7	-3.3	-0.9
Pool 25:			
Lower quarter	410.9	-2.7	-1.7
Middle half	414.5	+0.9	-1.0
Upper quarter	421.5	-2.5	-0.2
Pool 24:			
Lower quarter	424.7	+2.7	+2.9
Middle half	427.2	+2.0	+0.6
Upper quarter	430.8	+1.9	-0.4

*Average of the riverbed elevations in the deepest 1000-ft width of river channel. Amsl = above 1929 adjusted mean sea level.

the assumption of a uniform distribution of sediment over the channel width. A better agreement could be obtained if a proper sediment distribution function was used. Since the calculated trend agrees with the measured value and since the amount of sediment deposition on the floodplain from 1939 to 1971 appears reasonable, it was concluded that the mathematical model as calibrated was as good as the available field information, and could be employed to study the river's response to future development.

Study cases: Considering possible future changes of the amounts of water and sediment delivered to the reach, the time sequence in which these amounts are delivered and man's activities within the reach, five major 50-year simulations were conducted to assess future geomorphic changes in the river reach. They are: 1. present scheme of operation to maintain the 9-foot channel, 2. holding the pool level 1 ft above the normal pool, 3. holding the pool level 1 ft below the normal pool, 4. zero sediment inflow into Pool 24, and 5. maximum sediment inflow into Pool 24. Simulations 1 through 3 were conducted to assess the effects of different operational schemes for the locks and dams on the geomorphology of the study reach in the next 50 years. The 50-year hydrographs used in the models were synthesized from the 1932 to 1973 peak discharge and flow volume frequency curves. The effects of other alternative operational schemes can be determined in a similar way. Simulations 4 and 5 were performed to estimate the effects caused by the upper and lower limits of changes in the delivery of sediment to the study reach. From these two simulations and Simulation 1, the impact of changes in the delivery of sediment to the study reach on the morphology of rivers and adjacent lands in next 50 years were assessed. The effects of the pools on the behavior and form of the Illinois River were also estimated during the simulations.

Simulation 1 was performed by simply routing the 50-year series of flow through the model. To conduct the latter four simulations, some minor modifications of the control statements of the mathematical model were made. For Simulations 2 and 3, the control limits of the pool levels were raised one foot and lowered one foot, respectively. For Simulation 4, the sediment discharge entering Pool 24 was set to be zero. For Simulation 5, the gates at Dam 22 were raised entirely out of water at all times.

A longer prediction period of the future geomorphic changes in the study reach may be assessed from the operation of the mathematical model, provided that the input flow discharge hydrographs are adequately synthesized.

On the basis of study of past geomorphic changes in the river reach and with the mathematical simulation of future river response, it is concluded that the river scene in the study reach 50 years into the future will be essentially as it is today. The present-day manner of operation does not have serious detrimental effects on the geomorphology or hydraulics of the river system in the study reach. Holding the normal pool level one foot higher or lower does not create any sedimentation problems. The impact of changes in the delivery of sediment to the study reach will not go beyond Lock and Dam 24, at least for 50 years.

This model was also applied to study the feasibility of thalweg disposal of dredged material. In many occasions, disposal of dredged material on floodplain, islands or dike fields along a river involves serious environmental impacts. One significant portion of the riverine environment that offers potential disposal locations is the main channel or thalweg region of the river itself. In a meander stream the bed profile appears as an irregular series of high points and low points. The high points of the profile generally correlate with the crossings and the low points with the deep bendway pools. It is these high points at crossing areas that require repetitive dredging. The concept of thalweg disposal, then, involves dredging a cross area and disposing of the dredged material in a downstream pool area.

A dredged cut 3 feet deep and 950 feet long (from RM 268.91 to RM 268.72) was made in the crossing area over the channel width in Pool 25. This resulted in approximately 200,000 cubic yards of dredged material which was disposed of in the downstream pool area (RM 268.46 to RM 268.28). The cut was made at the beginning of the low-water season and riverbed level changes in the modeled reach were computed during the next year for the 2-year annual hydrograph. Results are shown in Fig. 10-17.

The initial reference bed level with dredged cut and disposal site is shown in Fig. 10-17a (dotted line). The difference between the dotted and the solid line indicates the change in bed elevation. The dark vertical bars represent the upstream and downstream locks and dams. The discharge hydrograph, number of days after dredging, and discharge volume in thousands of cfs are shown in each figure. After a 132 day low-flow period, the bed level showed very small changes (Fig. 10-17b). With the flood entering the reach (Fig. 10-17c), the dredged cut was filled in rapidly and the bar at the disposal site was moved to a downstream crossing as shown in Fig. 10-17c. After one year, both crossing and pool returned to nearly the natural state.

To test the impact of the quantity of material disposed of in the thalweg of the downstream pool, model runs were made under the same conditions as described above, but with, first, all the material disposed of on the adjacent floodplain, and, second, half disposed of on the floodplain and half in the downstream pool. Figure 10-18 shows the general location of the dredging and disposal operation and the impacts of each alternative. With all the dredged material disposed of on the floodplain

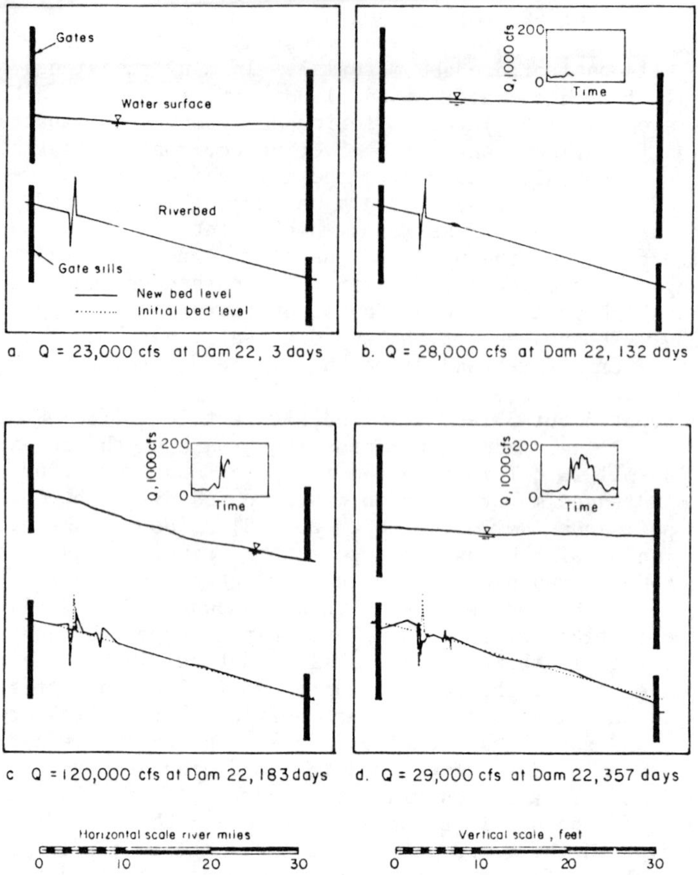

Fig. 10-17. Riverbed Level Changes During the Year after Dredging and Disposing

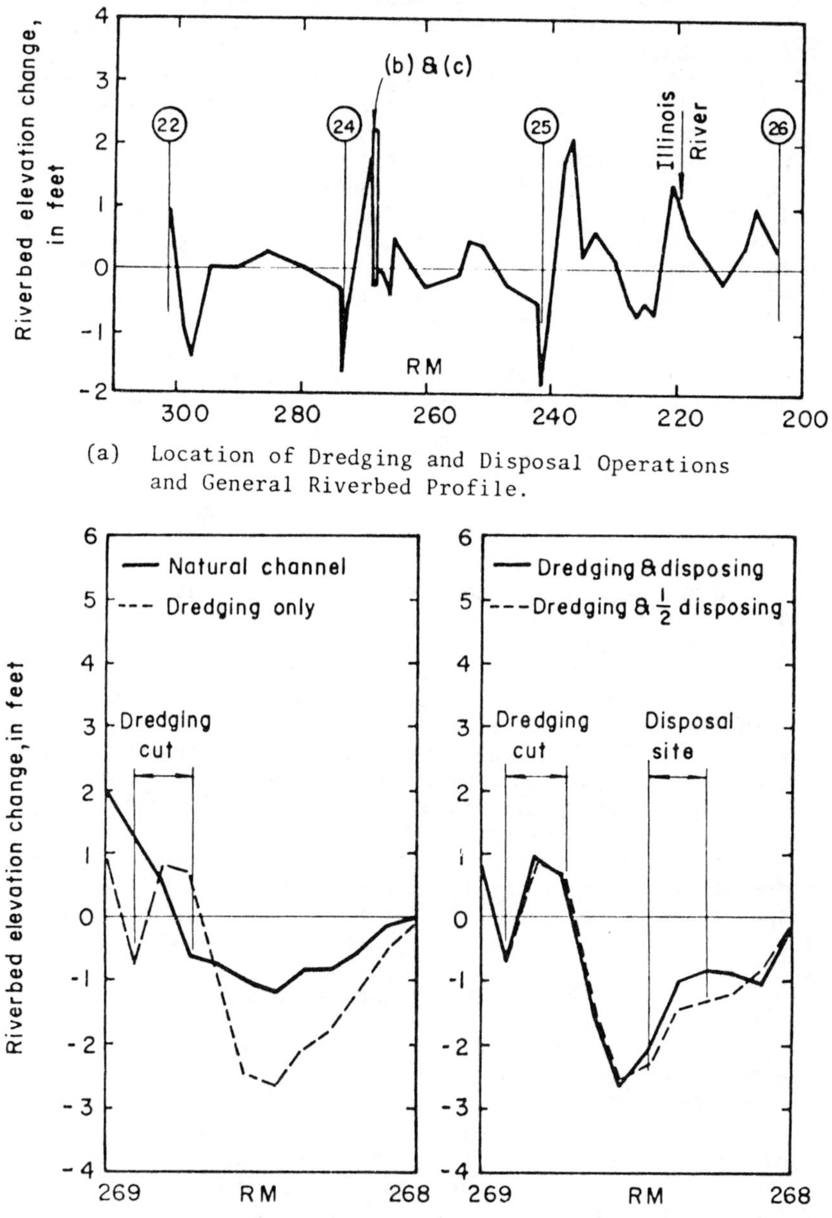

(a) Location of Dredging and Disposal Operations and General Riverbed Profile.

(b) Dispose All on Floodplain.

(c) Dispose 1/2 on Floodplain and 1/2 in Pool.

Fig. 10-18. Average Riverbed Elevation Changes One Year after Dredging with Alternate Disposal Options (2-Year Annual Hydrograph)

(Fig. 10-18b), the downstream pool scoured much more during the year following dredging than if no dredging had been done. The sediment derived from this scour may be deposited, in part, on a downstream crossing. Disposing of half the dredged material in the downstream pool (Fig. 10-18c) reduced the amount of scour in the pool area. The riverbed profile one year after disposal of half the dredged material in the downstream pool is quite similar to the profile resulting from disposal of all the dredged material in the downstream pool (Fig. 10-18c). It should be noted that very little accretion occurs on the downstream crossing (RM 268.28 to RM 268.0) for the average conditions represented by the 2-year annual hydrograph.

To investigate the impact of thalweg disposal under a different flow condition, the study area was subjected to a one-year annual hydrograph. The results of subjecting the model to this annual hydrograph with small flow volume and reduced peak are shown in Fig. 10-19. With all the dredged material deposited on the floodplain (Fig. 10-19b), the dredged cut remains essentially unchanged during the year following dredging, and the profile through the downstream pool and crossing coincides closely with that expected under natural conditions. However, with disposal of all the dredged material in the downstream pool under conditions of a small annual hydrograph, the next crossing downstream is strongly impacted. Figure 10-19c shows accretion of almost 2.5 feet over natural river conditions at the downstream crossing. With disposal of half the dredged material in the pool, accretion on the downstream crossing is reduced by about 1.0 feet (Fig. 10-19c). There is a risk, then, that if the annual hydrograph which follows dredging and thalweg disposal is small, the material disposed of in the pool area will be scoured and will deposit on the next crossing downstream. The result could be inadequate navigation depths and consequent dredging requirements on that crossing. This is particularly true if the crossing below the disposal pool is already experiencing dredging problems, and could also be true if a divided reach or unstable straight reach is located below the disposal pool.

In summary, this study indicates that dredging from a crossing and disposal of the dredged material in a downstream pool can constitute a feasible alternative to the disposal problem. The process involves a degree of risk of impacting the integrity of the navigation channel downstream from the pool, particularly if dredging is followed by a small discharge hydrograph. However, the risks incurred would be outweighed by the potential environmental benefits at many locations. Based on data currently available, the direct biologic impacts of disposal in the main channel as well as possible secondary impacts from turbidity generated by the disposal process appear minimal. In addition, the serious ecological problems associated with open water disposal on marshlands and near

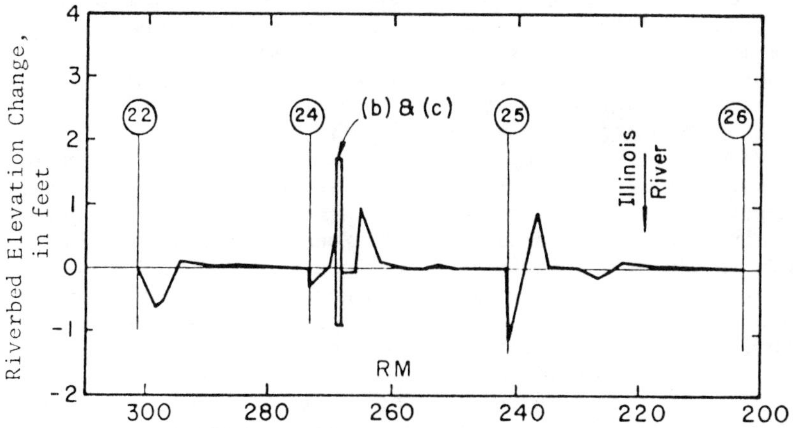

(a) Location of Dredging and Disposal Operations and General Riverbed Profile.

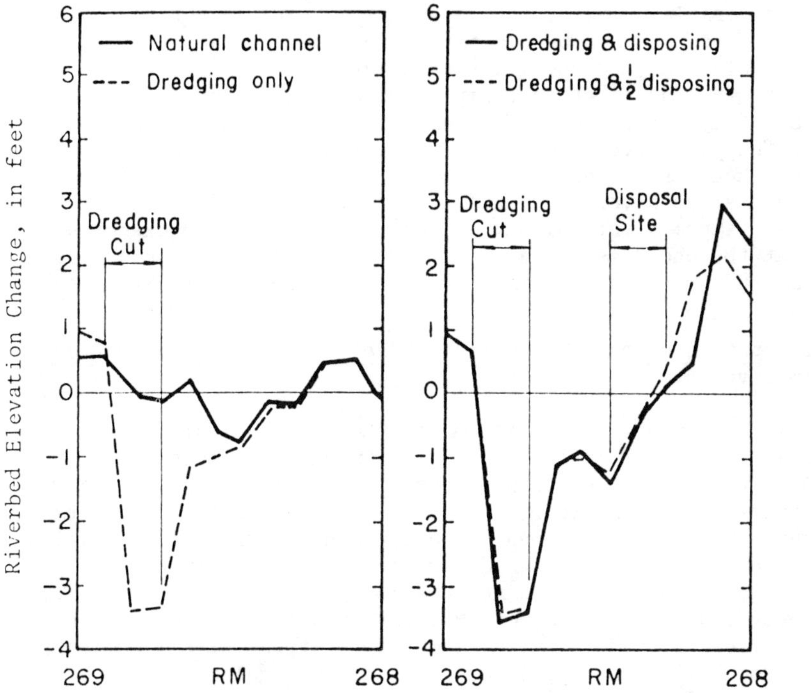

(b) Dispose All on Floodplain.

(c) Dispose 1/2 on Floodplain and 1/2 in Pool.

Fig. 10-19. Average Riverbed Elevation Changes One Year after Dredging with Alternate Disposal Options and a Small Annual Hydrograph chute channels, sloughs, and backwater areas are avoided by the process of thalweg disposal. Although conditions downstream of a proposed disposal site may preclude thalweg disposal at certain locations, in many cases depositing only a portion of the dredged material along the thalweg would still result in reduced environmental impacts.

The concept of thalweg disposal may be modified to reduce the existing detrimental impacts of dredged material disposal. The process involves redredging and placing the material, previously dredged and deposited on the undesirable areas, back in the main channel during high flow (Goodell,* 1975), in hope that the placed material may be flushed out of the problem reach without causing unacceptable impacts. This process again involves a degree of risk and should be carefully evaluated before its application.

The concept of thalweg disposal appears to offer a viable alternative to both long-term and emergency disposal requirements. As an emergency procedure where unexpected shoaling threatens the integrity of the navigation channel and immediate corrective dredging is required, thalweg disposal holds considerable promise. Long-term implementation of the practice should be preceded by a detailed analysis of the hydraulics and morphology of the specific reach involved.

Other application examples of one-dimensional water and sediment routing models are an investigation of the effects of Chippewa River erosion and silt reduction measures (Simons and Chen, 1977) and assessment of hydraulic impacts of the Atchafalaya River Flood Control Program on the Lower Mississippi River (Simons and Chen, 1978). These examples show that the one-dimensional mathematical model is a useful tool to study water and sediment related problems in river systems.

Limitation of the Mathematical Model and Recommendation for Improvement

The principal limitation of the mathematical model constructed herein is its assumption of one-dimensional flow. Only the general pattern of the river geomorphology can be considered in a nonuniform channel. For a river reach with fairly uniform channel alignment, the sediment distribution relation presented can be used to determine the changes in the cross-sectional shape at each computation point. However, to obtain more detailed information, either a modification of the present model can be made by using a multiple stream or compound stream approach (Section 10.3), or a two-dimensional model can be applied (Section 10.4).

*Goodell, H., 1975, personal communication.

10.3 ONE-DIMENSIONAL MATHEMATICAL MODEL OF NONUNIFORM CHANNELS

Introduction

In natural rivers, the shape of a given cross section is irregular and varies from section to section. Also, the bed material, bed form and resistance are not uniform along a cross section. If these variations in a river reach are relatively small, then the one-dimensional mathematical model presented in Section 10.2 is adequate for study of the water and sediment related problems. However, if the channel is highly irregular or if its alignment changes abruptly such as in the presence of a dike field, the one-dimensional model presented can only provide the general patterns of river hydraulics and geomorphology. The sediment distribution function developed is no longer valid to determine the changes in cross-sectional geometries. To obtain more detailed information related to water and sediment movement adjacent to the highly irregular river reaches without using a sophisticated two-dimensional model, Dass (1975) introduced two models based on the one-dimensional governing equations that calculate nonuniform cross-sectional changes. These two models are the multiple stream flow model and the compound stream flow model. The models were subsequently modified by Owens (1977) and Lopez (1978) to study various water and sediment problems in nonuniform channels.

Multiple Stream Flow Model

The multiple stream flow model divides each cross section along a reach into a constant number of subsections (Dass, 1975). The subsections form a set of continuous streams in the reach (see Fig. 10-20). The governing equations (Eqs. 10-1, 10-2 and 10-3) are applied to each stream separately under the assumption that the stage is the same for all subsections at a given cross section. Lateral flows between streams are allowed, so the boundaries between adjacent streams need not follow streamlines.

Assuming that a river reach is divided into streams (Fig. 10-20), the water and sediment transfer terms between any two adjoining streams, say streams k and (k+1), are denoted by q_k and q_{sk}, respectively and are assumed to be positive in the direction marked by the arrows in the figure. For any assumed stream, the equation of continuity for sediment and the equation of continuity for sediment-laden water can be rewritten for the kth stream as

$$\frac{\partial Qs_k}{\partial x_k} + p_k \frac{\partial Ad_k}{\partial t} + \frac{\partial As_k}{\partial t} - Sa_k = 0, \text{ and} \qquad (10\text{-}55)$$

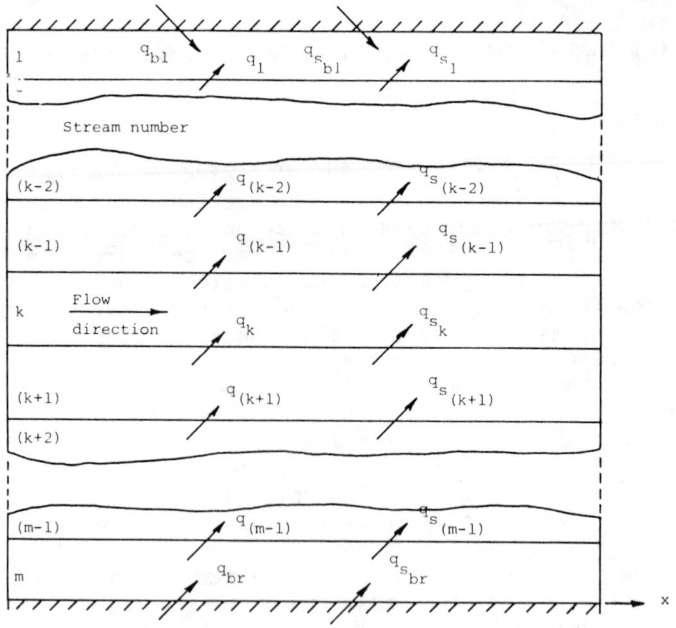

Fig. 10-20. Flow Interaction Terms for Water and Sediment Movement in a Multi-Stream Flow Approach

$$\frac{\partial Q_k}{\partial x_k} + \frac{\partial A_k}{\partial t} - Fa_k = 0 \tag{10-56}$$

$$\frac{\partial Q_k}{\partial t} + \frac{\partial}{\partial x_k}(\beta QV)_k + gA_k \frac{\partial h}{\partial x_k} + gA_k \, S_{f_k} - Mc_k = 0 \tag{10-57}$$

The sediment discharge per unit length added to the kth stream can be expressed by:

$$Sa_k = q_{s_k} - q_{s_{(k-1)}} \tag{10-58}$$

The sediment-laden water discharge per unit length added to the kth stream can be expressed by

$$Fa_k = q_{b_k} - q_{(k-1)} + q_k \, , \tag{10-59}$$

where:

q_{b_k} = seepage inflow into the kth stream,

$q_{(k-1)}$ = bulk inflow into the (k-1) stream from k^{th} stream, and q_k = bulk inflow into the kth stream from $(k+1)^{th}$ stream.

The momentum correction term Mc_k due to lateral flow term Fa_k can be expressed by

$$M_{ck} = \psi_k q_{bk} V_k - q_{(k-1)} V_{(k-\frac{1}{2} \pm \frac{1}{2})} + q_k V_{(k+\frac{1}{2} \pm \frac{1}{2})} \tag{10-60}$$

where ψ_k = 0 if lateral flow comes into the stream

ψ_k = 0.5 if lateral flow goes out of the stream.

The expression $V_{(k-\frac{1}{2} \pm \frac{1}{2})}$ in Eq. (10-60) uses "+" when $q_{(k-1)} \geq 0$ and vice-versa. Similarly, the expression $V_{(k+\frac{1}{2} \pm \frac{1}{2})}$ uses "+" when $q_k \geq 0$ and vice versa. The unknown quantities in the expressions of Eqs. (10-55), (10-56) and (10-57) for any one segment of the reach are

Q_k; k = 1, 2, 3, ..., m

z_k; k = 1, 2, 3, ..., m

q_k; k = 1, 2, 3, ..., (m-1), and

y.

Since the total number of unknown quantities is 3m and the total number of equations available (Eqs. 10-55, 10-56 and 10-57, where k = 1, 2, 3, ..., m) is also 3m, the system of equations is uniquely solvable.

The requirement of using a constant number of subsections at each computation point by the multiple stream flow model is a major limitation to its field application. Dass (1975) realized this problem and he developed the compound stream flow model.

Compound Stream Flow Model

The compound stream flow model developed by Dass (1975) calculates the water routing solution using the usual single approach, but the sediment routing is performed in subsections. The computation is based on a uncoupled-unsteady solution method to solve the basic governing equations (Eqs. 10-1, 10-2, and 10-3). The following steps are utilized to route the sediment after the water routing solution has been calculated: (1) the water discharge in each subsection is calculated by

$$Q_{ki} = Q_i \frac{K_{ki}}{K_i} \tag{10-61}$$

where Q_{ki} is the subsection discharge, Q_i is the total discharge at section i known from the water routing solution, and K_{ki} and K_i are given by Eqs. (10-6) and (10-7), respectively.
(2) The water continuity equation (Eq. 10-2) is applied to each stream in the following form:

$$\frac{\partial Q_k}{\partial x} + \frac{\partial A_k}{\partial t} = q_{\ell_k} \qquad (10\text{-}62)$$

where the subscript k refers to the kth stream. From the water routing solution and step 1, the left side of Eq. (10-62) is known. It is solved for the net lateral inflow to stream k, $q_{\ell k}$.

(3) Using a sediment transport function, the sediment load in each subsection Q_{s_k} and the lateral sediment flow associated with the lateral water flow, q_{sk}, are calculated.

(4) The sediment continuity equation (Eq. 10-1) is applied to each stream in a similar form to Eq. (10-62):

$$\frac{\partial Q_{s_k}}{\partial x} + p\frac{\partial A_{d_k}}{\partial t} + \frac{\partial}{\partial t}(A_k C_{s_k}) = q_{s_{k+1}} - q_{s_k} \qquad (10\text{-}63)$$

From the previous steps, all the terms are known except the second term. Eq. (10-63) can then be solved for A_{dk}, the scour or deposition in subsection k.

Dass (1975) tested the multiple stream flow model and the compound stream flow model. He found that the compound stream model is easier to formulate and more economical to use, while it gives practically the same results as the multiple stream model. In addition, the compound stream model, unlike the multiple stream model, can have a variable number of subsections at different cross sections. This provides the necessary flexibility for the practical application of the compound stream model.

To study river water and sediment entering a reservoir and to predict the sediment distribution pattern in reservoirs, Lopez (1978) incorporated a submerged jet theory developed by Albertson and others, (1950) to Dass's compound stream model. He also considered routing the nonuniformly distributed sediment in fraction to study the sorting process and its effect on the delta formation.

Recently, in order to study sediment routing in rivers in a more efficient way, Owens (1977) modified Dass's model to develop a known discharge coupled solution. The basic principles of his model are similar to those presented in the section titled "Known Discharge Solution."

Assuming that the flow is quasi-steady for the sediment routing and the water discharge in a channel reach is known, the water depth and the cross-sectional area change at each cross section in the modeled river reach can be computed from the sediment continuity equation (Eq. 10-1) and the flow momentum equation (Eqs. 10-3). Then by applying Dass's compound stream approach for sediment routing with a constraint that the sum of cross-sectional area changes at each subsection of a cross section should equal the total cross-sectional area changes determined from the known discharge solution, the bed elevation change at subsections of each cross section or the changes in the cross-sectional geometries can be determined.

Applications of the Compound Stream Flow Model

Introduction: To test the feasibility of the original and modified compound stream flow model, Dass (1975), Lopez (1978) and Owens (1977) all presented some test problems. Dass used a hypothetical river reach to simulate the effect of dikes and dredging on channel deformation and checked the computed results with the principles of river mechanics. Lopez compared the computed results of a mathematical model with experimental data obtained in a laboratory flume having a sudden expansion. Owens applied his modified compound stream model to study river responses to development in Pool 4 in the Upper Mississippi River. Lopez's and Owens' examples are briefly discussed in the following sections.

Modeling of Sediment Deposition in a Laboratory Flume with a Sudden Expansion (Lopez, 1978): Prediction of the sediment distribution pattern in reservoirs is a complex task because of numerous interrelated factors involved in the process. To simulate the deposition of sediment in reservoirs using a mathematical model, it is necessary to develop a model that simulates both the velocity and sediment fields.

Lopez (1978) considered the reservoir as a set of multiple channels and the inflow river as a single channel. He used a single one-dimensional model to model the river and then connected the river model to a compound stream flow simulating the reservoirs, together with a two-dimensional jet theory (Albertson, et al., 1950). He then tested the feasibility of his model using experimental data.

These experimental data were collected from a flume 2 ft wide and 60 ft long. A small channel 8 in. wide and 35 ft long was built in the sand bed flume to create a sudden expansion in the channel as shown in Fig. 10-21. Eight sections were taken along the channel bed in the expansion reach. The space increment was selected to be 2 ft and a time increment of 1 min was selected to insure the stability of the numerical scheme.

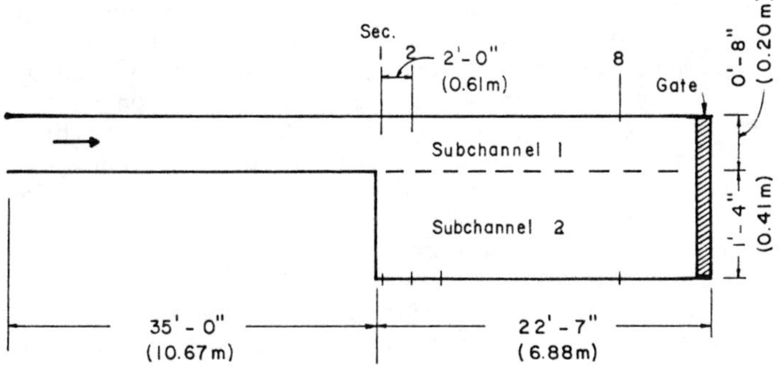

Fig. 10-21. Sudden Expansion in a Laboratory Flume

The mathematical model was established for this flume system by assuming uniform grain size distribution with no correction for the specific weight of the sediment deposits. The variations in bed elevations at all sections were measured for each subchannel. The computed bed profiles at different time intervals together with the observed bed elevations are indicated in Fig. 10-22 and Fig. 10-23. The computed and measured results show fairly good agreement. It was found that the delta moved downstream and a submerged levee formed along subchannel 2. The levee was formed by the buildup of sediment along channel 2 as shown in Fig. 10-24. In nature, the formation of natural levees at the channel mouth has been reported by many observers when rivers discharge into reservoirs or oceanic basins. Similar phenomena were produced by the mathematical model. Lopez also applied his compound stream model to study sedimentations in reservoirs in the Colorado River system.

<u>A Compound Stream Model of Pool 4 in the Upper Mississippi and Lower Chippewa Rivers</u>: Owens (1977) modified Dass's compound stream flow model using a known-discharge solution. The principles of his model have been described in previous sections. His model was applied to assess the response of a riverbed near Reading Landing and above Crats Island to a modification of dike fields in the study reach, with a hope of reducing the dredging requirements in Pool 4 in the Upper Mississippi River.

Seventy-one computation sections (cross sections) were selected to cover the study river reaches. The space increment varied from 0.08 to 1 mile. The computation sections in the Mississippi River were subdivided into three subsections, while the Chippewa River was considered as a single channel reach. In general, the three subsections cover fractions of a cross section within dike field, within a dredged cut and beyond the dredged cut, respectively, as shown in Fig. 10-25. Using the hydrographic maps, the cross-sectional geometries were evaluated.

The first phase of the calibration is the estimation of parameters of the resistance and sediment transport functions at cross sections that were modeled using only the single stream sediment routing. A constant Manning's n and a sediment transport function for each cross section were selected based on Simons and Chen's (1976) model calibration results. These values were used as initial estimates of the unknown parameters in subsections where the compound stream sediment routing is calibrated.

An optimization technique based on a linearized gradient method developed by Becker and Yeh (1973) was then used to calibrate the model. This calibration proceeded from the most upstream cross section downstream to each adjacent cross section. The general procedure was to vary only the resistance coefficient at each subsection at a given cross section until

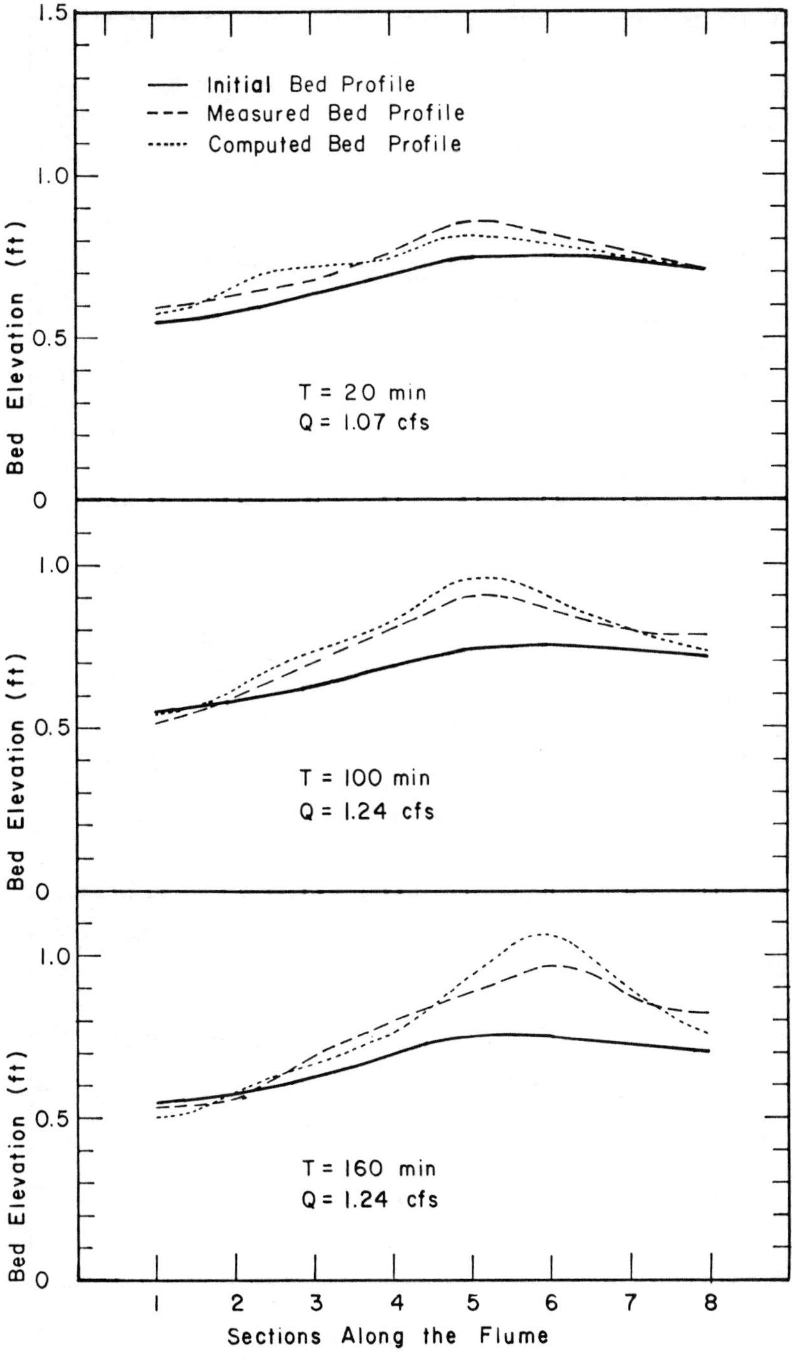

Fig. 10-22. Bed Elevation Changes on Subchannel 1

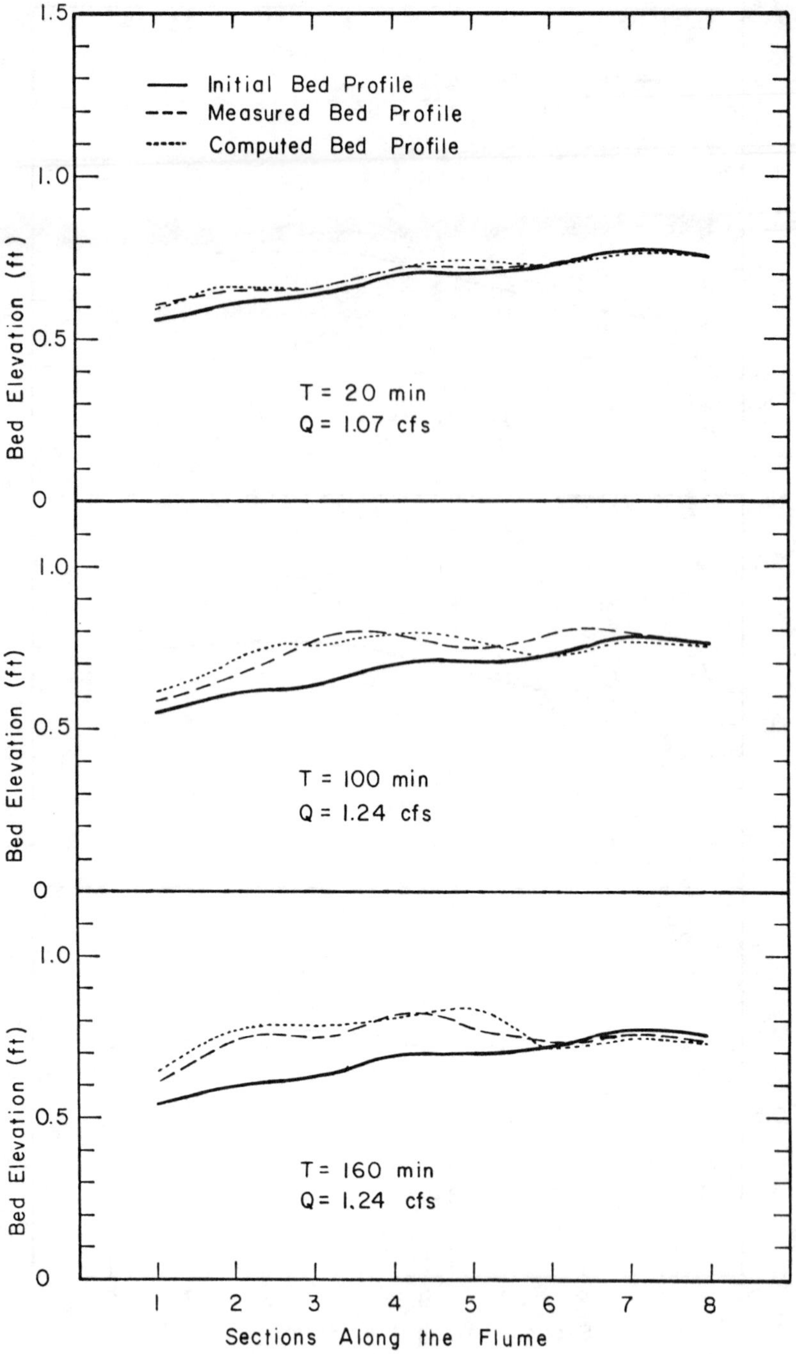

Fig. 10-23. Bed Elevation Changes on Subchannel 2

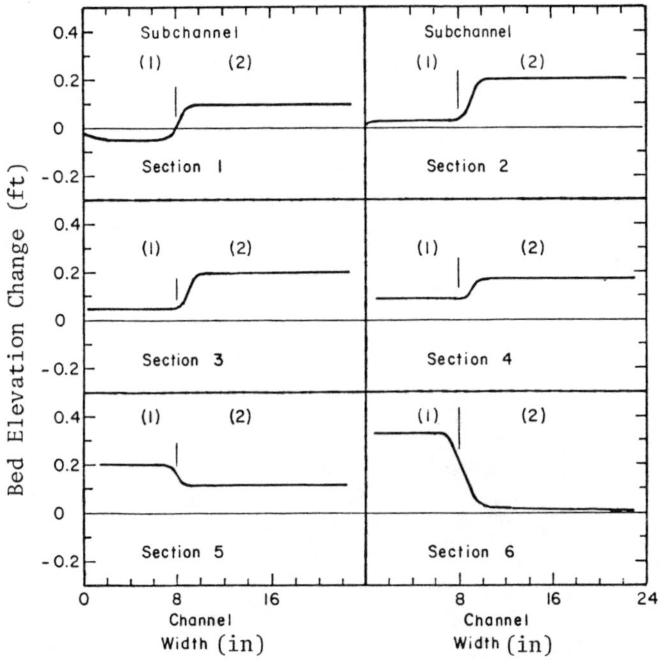

Fig. 10-24. Vertical Cross-Sectional Geometries Showing the Computed Bed Elevation Changes at Time 160 Minutes for Different Sections Along the Channel

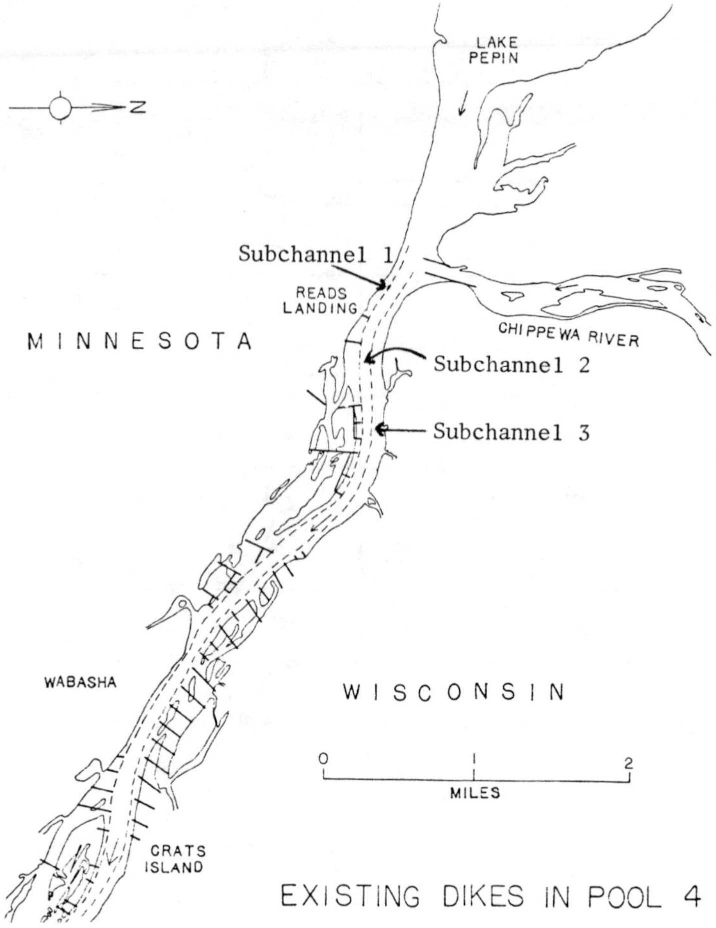

Fig. 10-25. A Portion of Pool 4 Showing Subsections of the Compound Stream Flow Model

the calculated results approached the measured data, and then vary only the transport coefficient until the calculated bed elevation changes were within 0.5 ft of the measured data. Some calculated and measured bed elevation changes in the subsections at Reads Landing are shown in Fig. 10-26. To check the estimations of the resistance and sediment transport parameters in Pool 4, the calibration was verified using measured field data over the period of about one year following the one year period that was used in the calibration as shown in Fig. 10-27. The calculated and measured values agree fairly well. It is therefore concluded that the compound stream model so calibrated can be used to predict the future response of the study reach.

The calibrated compound stream model was used to study the effects of constructing new dikes or raising or extending dikes at Reads Landing and above Crats Island on the riverbed elevation and thereby to answer some of the following questions: how many dikes should be constructed, where should they be placed, how much will the dredging requirement be reduced, and is there a better overall solution that may be used to maintain the navigation channel?

Some study results from two alternatives are shown in Fig. 10-28. Proposal 1 raised the two existing dikes (RM 762.9 and RM 762.82) on the Minnesota bank to elevation 665 ft. Proposal 2 constructed a new set of five dikes (RM 762.9 - 762.56) on the Wisconsin bank. The results in Subsection 2 (dredge part of the cross section and the navigation channel) show that raising the two existing dikes or constructing new dikes would reduce the deposition near the construction sites but would increase deposition in the dike fields and downstream of the construction sites. These case studies were made to give an example of how the compound stream model can be used to investigate various alternatives, and to compare several alternatives using dikes. The compound stream model is particularly useful when studying river responses to training activities using unsymmetrical structures such as dike fields.

Limitation of the Compound Stream Model

The compound stream flow model shows great promise for field use wherever nonuniform cross sections are encountered; e.g., it can be used in the simulation of water and sediment routing in channels with floodplains by accounting for the sediment transfer between the main channel and the floodplain channels, or in the evaluation of erosion and deposition in a channel due to the effect of dike construction in the channel.

However, basically the compound stream model is still a one-dimensional model. It cannot be expected that the compound stream model be able to reproduce all nonuniform distributions of scour and deposition that occur in natural alluvial channels. Many such distributions are caused by two- or three-dimensional

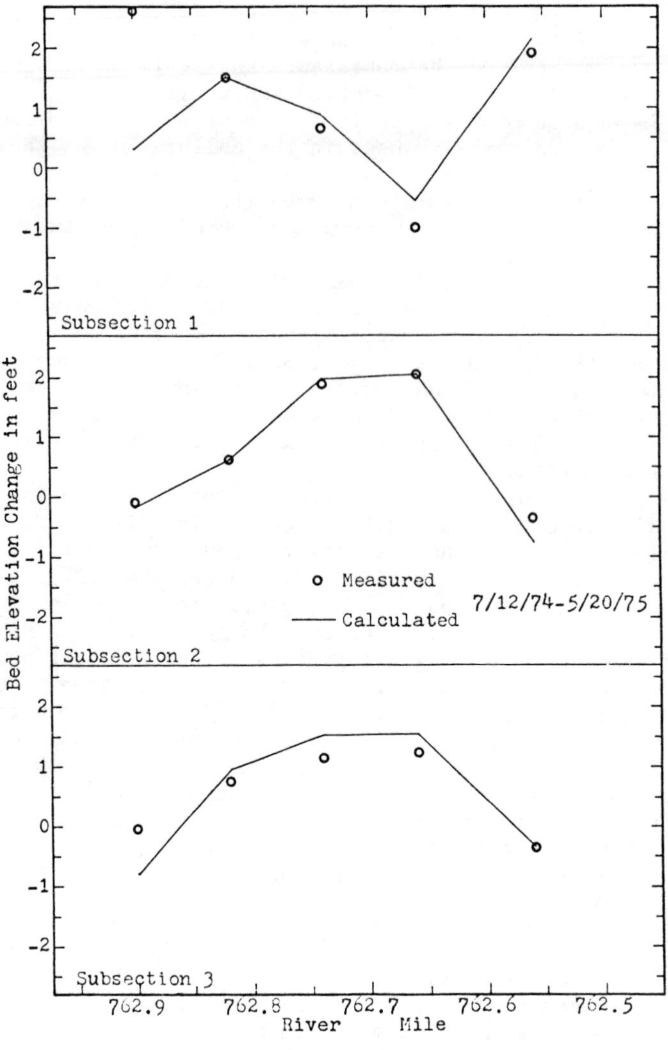

Fig. 10-26. Compound Stream Calibration at Reads Landing

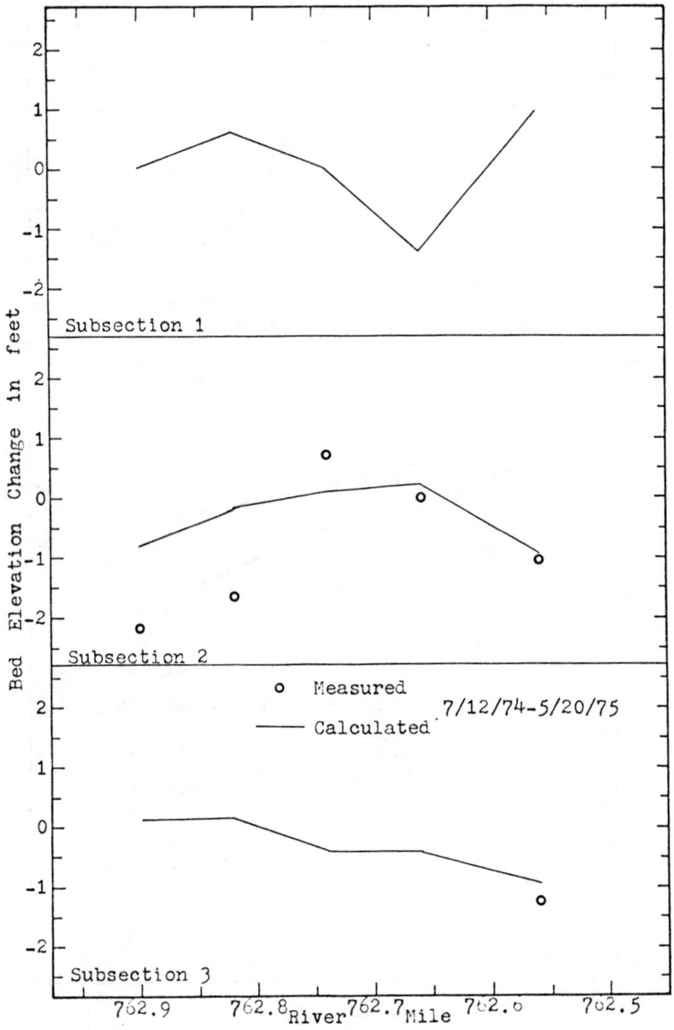

Fig. 10-27. Verification at Reads Landing

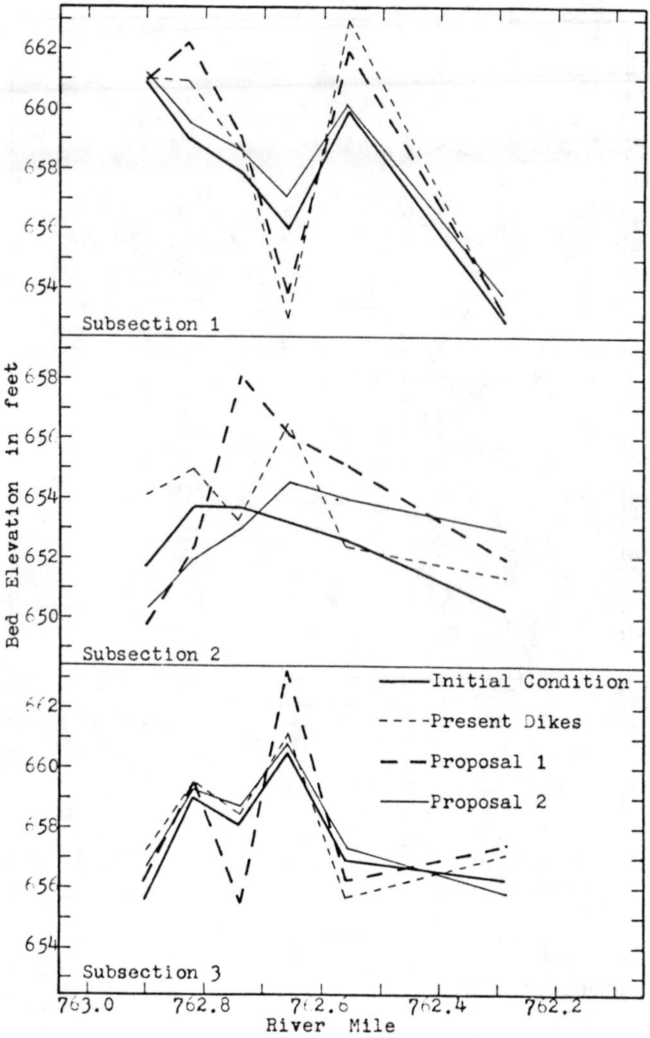

Fig. 10-28. Case Study at Reads Landing

flow patterns, and hence cannot be accurately determined by a one-dimensional model. An example is the scour that occurs on the outer concave bank of an alluvial channel bend and the coincident deposition and point bar formation on the inner bank. Only general trends in this distribution that can be related to nonuniformity in channel geometry, resistance and sediment transport through the assumption of one-dimensional flow can be indicated by this model. Another major problem in applying the model to a natural river is that a large number of unknown parameters of the resistance and sediment transport functions must be estimated when calibrating the model. This requires extensive effort.

10.4 TWO-DIMENSIONAL RIVER MODEL

In many instances, such as flow in estuaries, in highly irregular channels, on floodplains, etc., the flow is predominantly two-dimensional. That is, the flow velocity in both the longitudinal and transverse directions are comparable. As stated above, the one-dimensional model can only determine general trends in nonuniform channels. If more detailed information is to be obtained, a two-dimensional mathematical model can be utilized. Colorado State University is developing a two-dimensional water and sediment routing model, specifically designed for applications to river routing. This model is briefly presented in the following sections. The model is subjected to further modifications and refinement.

Basic Equations

Flow in an alluvial river is governed by the conservation laws, which state the conservation of mass and momentum quantities. In two-phase flow the conservation of mass can be expressed analytically by two equations: 1) the water continuity equation, and 2) the sediment continuity equation. Momentum, a vector quantity, has to be conserved in any direction. In cartesian coordinates, the momentum balance is expressed along three orthogonal space directions.

In open channel flow, an expedient approximation to the three-dimensional flow problem is the assumption of bidimensionality along two perpendicular horizontal directions, averaging flow properties in the vertical direction. The pressure distribution along a vertical is assumed hydrostatic, and only time-averaged turbulent flow motions are taken into account in the main flow. However, the simulation of secondary flow patterns, such as large eddies, requires that the effective stresses in vertical planes be taken into account in the two-dimensional analog of the three-dimensional flow problem. No conclusive theory has yet been formulated to evaluate the effective stresses, although simplified approximations have shown to be promising from a practical standpoint.

The equation of water continuity in two horizontal space coordinates is:

$$\frac{\partial q_x}{\partial x} + \frac{\partial q_y}{\partial y} + \frac{\partial d}{\partial t} = 0 \qquad (10\text{-}64)$$

where

q_x = the unit-width water discharge in the x-direction
q_y = the unit-width water discharge in the y-direction
d = the depth of flow
x and y = horizontal space variables
and t = the time variable.

The equation of sediment continuity in two horizontal space coordinates is

$$\frac{\partial g_x}{\partial x} + \frac{\partial g_y}{\partial y} + \gamma_s \frac{\partial z}{\partial t} = 0 \qquad (10\text{-}65)$$

where

g_x = the unit-width bed material transport rate in the x-direction
g_y = the unit-width bed material transport rate in the y-direction
z = the elevation of the channel bed, with reference to an arbitrary datum
and γ_s = the unit weight of the material deposited in the channel bed.

The bed material transport rate is expressed in units of weight per unit of time, per unit of width. The unit weight of the bed deposit is expressed as:

$$\gamma_s = (1 - \lambda)\, \rho_s g \qquad (10\text{-}66)$$

where λ is the bed porosity
ρ_s is the mass density of sediment
and g is the acceleration of gravity.

The momentum equations for two-dimensional horizontal flow can be obtained by vertical integration of Reynolds' equations for turbulent flow in three dimensions. Taking into account the boundary conditions at the bottom and at the surface, they yield

$$\frac{\partial}{\partial t}(q_x) + \frac{\partial}{\partial x}\left(\frac{q_x|q_x|}{d}\right) + \frac{\partial}{\partial y}\left(\frac{q_x|q_y|}{d}\right)$$

$$-f_g q_y + \frac{g}{2}\frac{\partial}{\partial x}(d^2) + gd\frac{\partial z}{\partial x} - \frac{1}{\rho}(\tau_{wx} - \tau_{bx})$$

$$-\frac{1}{\rho}\frac{\partial}{\partial x}(d\ \tau_{xx}) - \frac{1}{\rho}\frac{\partial}{\partial y}(d\ \tau_{xy}) = 0 \qquad (10\text{-}67)$$

$$\frac{\partial}{\partial t}(q_y) + \frac{\partial}{\partial x}\left(\frac{|q_x|q_y}{d}\right) + \frac{\partial}{\partial y}\left(\frac{q_y|q_y|}{d}\right)$$

$$+ f_g q_x + \frac{g}{2}\frac{\partial}{\partial y}(d^2) + gd\frac{\partial z}{\partial y} - \frac{1}{\rho}(\tau_{wy} - \tau_{by})$$

$$-\frac{1}{\rho}\frac{\partial}{\partial x}(d\ \tau_{xy}) - \frac{1}{\rho}\frac{\partial}{\partial y}(d\ \tau_{yy}) = 0 \qquad (10\text{-}68)$$

where

$q_x = ud$
$q_y = vd$

and u is the depth-integrated velocity in the x-direction
v is the depth-integrated velocity in the y-direction
f_g is the geostrophic parameter
ρ is the mass density of the water
τ_{wx} and τ_{wy} are the components of surface shear stress due to wind
τ_{bx} and τ_{by} are the components of bottom shear stress
τ_{xx}, τ_{xy} and τ_{yx} are the effective stresses in vertical planes, and all other terms are as defined previously.

Equations (10-64), (10-65), (10-67) and (10-68) are the governing equations of the two-dimensional water and sediment flow problem. In river flow, the influence of the geostrophic term is usually small, and f_g is made equal to zero. Likewise, the surface shear stresses are small as compared to the bottom shear stresses, and are usually neglected.

Supplemental Equations

The four basic equations contain four basic unknowns: q_x, q_y, d and z. The other variables in the equations must be expressed as a function of the four unknowns in order to obtain a solution. Supplementing equations for the bed material transport rate, the bottom shear stress and the effective stresses in vertical planes are needed in order to make the system of governing equations amenable to a mathematical solution.

The bed material transport is a vector quantity. Therefore, a model for its prediction should be capable of being resolved into two perpendicular horizontal directions. For this reason, a simple model based on an empirical power function of mean velocity is preferred over the more sophisticated phenomenological models. Furthermore, its simplicity allows for a considerable speed in the transport aspect of the computations, a significant asset of a two-dimensional numerical model where the coefficient and inversion algorithms may take large amounts of computational time.

The selected bed material transport function is of the form

$$g_w = C_g \rho w^m \tag{10-69}$$

where g_w is the bed material transport rate per unit width in the direction of the velocity w (averaged in the vertical), and C_g and m are an empirical coefficient and exponent, respectively. Equation (10-69) is resolved along two perpendicular directions according to

$$g_x = C_g \rho u w^{m-1} \tag{10-70}$$

$$g_y = C_g \rho v w^{m-1} \tag{10-71}$$

where u and v are the velocities (averaged in the vertical) in the x and y directions, respectively.

The usual assumption regarding bottom stress τ_b is that its magnitude is the same as that corresponding to steady uniform flow, and that its direction is the same as that of the depth-averaged velocity. For any direction

$$\tau_{bw} = f \rho w^2 \tag{10-72}$$

where τ_{bw} is the bottom stress in the direction of the depth-averaged velocity w, and f is a dimensionless friction factor defined as

$$f = \frac{g}{C^2} \quad \text{or} \quad f = \frac{g n^2}{(1.486)^2 d^{1/3}} \tag{10-73}$$

where C is the Chezy coefficient, and n is Manning's roughness coefficient.

The depth-averaged velocity w is resolved along two perpendicular horizontal coordinates such that

$$w = (u^2 + v^2)^{1/2} \tag{10-74}$$

In two horizontal space dimensions, the resistance equations are expressed as

$$\tau_{bx} = f \rho uw \qquad (10\text{-}75)$$

$$\tau_{by} = f \rho vw \qquad (10\text{-}76)$$

where τ_{bx} and τ_{by} are the bottom shear stresses along the x and y directions, respectively.

In the two-dimensional flow problem, computational convenience dictates that the eddy viscosity term be substituted for the terms accounting for the effective stresses in vertical planes of Eqs. (10-67) and (10-68). The eddy viscosity term for the x-momentum equation is

$$- \varepsilon d \left(\frac{\partial^2 u}{\partial x^2} + \frac{\partial^2 v}{\partial y^2}\right) \qquad (10\text{-}77)$$

and for the y-momentum

$$- \varepsilon d \left(\frac{\partial^2 v}{\partial x^2} + \frac{\partial^2 v}{\partial y^2}\right) \qquad (10\text{-}78)$$

where ε is the coefficient of eddy viscosity.

As a first approximation, the coefficient of eddy viscosity can be related to the mean flow properties by (Vreugdenhil, 1973)

$$\varepsilon = \varepsilon_q (q_x^2 + q_y^2)^{1/2} \qquad (10\text{-}79)$$

where ε_q is a dimensionless eddy viscosity coefficient, approximately equal to $6\sqrt{f}$.

The eddy viscosity terms, Eqs. (10-77) and (10-78), although approximations of the effective stress terms, have far-reaching consequences in that they enable the two-dimensional treatment of the predominantly three-dimensional features of the flow phenomena, with the computational advantages associated with it.

Numerical Solution Method

Equations (10-64), (10-65), (10-67), and (10-68) are the conservation equations governing the flow in an alluvial river. Their solution can be carried out by use of a finite-difference scheme that approximates the functions and their partial derivations in terms of their values in a tridimensional orthogonal time-space grid system, or by use of a finite-element scheme to approximate the solution by discretizing a solution domain with a finite number of points at which values of variables are obtained.

Finite-difference solutions to problems involving various forms of the shallow water equations have been obtained by Leendertse (1967, 1970); Leendertse and Gritton (1971); Abbott, Danesgaard and Rodenhuis (1973); Apelt, Gont and Szewezyki (1974) and many other researchers. Finite-element solutions for similar problems have also been presented by several authors including Taylor and Davis (1973); Connor and Wang (1974); Brebbia and Partridge (1976); Partridge and Brebbia (1976); and Brebbia and Connor (1976). Most of their solutions have dealt with bays or estuaries where the discretization in space can be fairly coarse. Generally speaking the larger the space grid dimension the larger the time step that can be used. Even so the problem of limited length of time step has continued to restrict the time span of solution because of the computational costs. This aspect has been the subject of much recent research for finite-difference and finite-element methods. The length of time step in the transient integration of the equations is of particular significance in the study of river problems since changes in the channel bed, which are of major concern, tend to occur much more slowly than changes in the water surface. In addition, flood waves in rivers often have periods of several days and in some cases weeks.

In order to compute the effects of such waves and still retain feasible computer costs, the length of time step must be large. Alternatively, a series of steady-state solutions (uncoupled steady solution) could be obtained with averaging of transport processes between successive steady states. Two methods of solution are thus proposed: (1) unsteady water routing uncoupled with the sediment continuity equation (uncoupled unsteady solution), and (2) steady water-flow computation uncoupled with the sediment continuity equation (uncoupled steady solution). For the unsteady solution, attention has been directed toward implicit solution methods with a hope to use a large time step. Detailed description of the numerical formulation of the two-dimensional mathematical models and their applications is given by Simons et al. (1977, 1978). Only conceptual designs of numerical methods are described below.

Finite-Difference Scheme of the Uncoupled-Unsteady Solution Method: The finite-difference scheme of the uncoupled-unsteady solution method is formulated in a multioperational mode. A review of the available literature (Leendertse, 1967; Abbott and Cunge, 1975; Abbott, Danesgaard and Rodenhuis, 1973) indicated that an alternating direction-implicit finite-difference formulation could be used as a basis for the model. The principle of the solution method is to use several operations to advance the values of the dependent variables to a new time step. In the first operation, the water continuity equation and the momentum equation in the x-direction are solved simultaneously, resulted in the values of water discharge q_x and depth d being advanced to time level (n + 1/2). Variables q_y and z are taken at the known time level n. This operation is performed one row at a time, for as many rows as required by the boundary configuration.

In the second step, the water continuity equation and the momentum equation in the y-direction are solved simultaneously, resulting in the values of water discharge q_y and depth d being advanced to time level (n + 1). Variables q_x and z are taken at the known time levels (n + 1/2) and n, respectively. This operation is performed one column at a time, for as many columns as required by the boundary configuration.

Then in the third step, the sediment continuity equation is used to advance the values of z, explicitly, from time level n to (n + 1). This operation completes the first time step. Figure 10-29 shows a schematic illustration of the locations where the various dependent variables are calculated in the time domain.

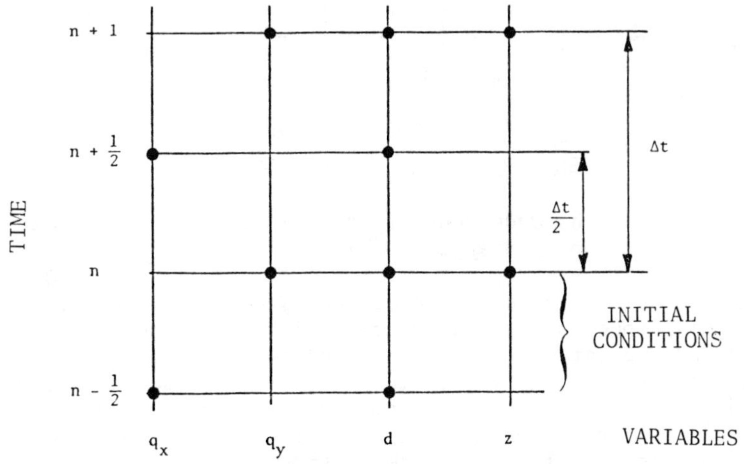

Fig. 10-29. Variable Location in the Time Domain

These dependent variables q_x, q_y, d and z are defined on a space-staggered grid system to reduce the number of unknowns to be solved. That is, they are defined only at their respective subgrids as shown in Fig. 10-30. This grid system is designed so that the multioperation solution method can be effectively applied to compute the unknown variables. This figure also

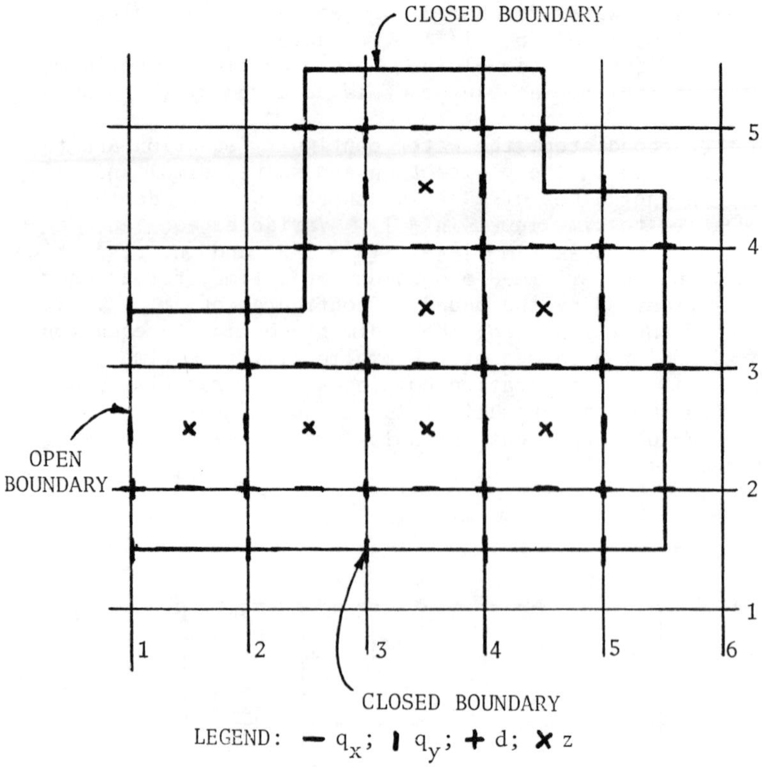

Fig. 10-30. The Open Boundary and Closed Boundary in the Space-Staggered Grid

shows two types of boundary conditions specified in the two-dimensional space-staggered system: (1) open boundary and (2) closed boundary. In an open boundary, the water discharge and depth are allowed to vary in time. In a closed boundary, the water discharge component perpendicular to the boundary is set equal to zero. A river configuration can be described by this two-dimensional grid system as shown in Fig. 10-31. Some approximations need to be made along the closed boundaries. It

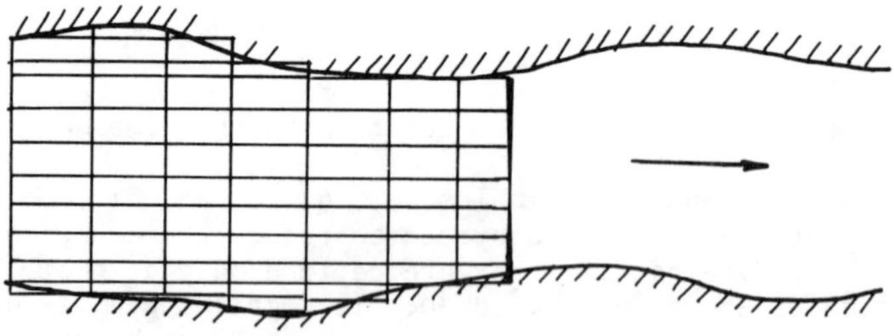

Fig. 10-31. Finite-Difference Grids Describing a River Reach

is possible to describe the river configuration to a better degree using an orthogonal curvilinear grid system, and then transform this system into a cartesian coordinate. The equations are then solved on this transformed coordinate (Wanstrath et al., 1976). However, the formulation of the computation processes is quite complex. Also, within-river boundaries such as islands cannot be transformed. These limitations restrict the applicabilities of the coordinate transformation method.

The uncoupled solution of the sediment continuity equation requires that the bed level change Δz within a time step Δt be small as compared with the changes in discharge and depth. In subcritical flow, where the bed level changes in time at a much slower rate than the water depth, this requirement is readily satisfied.

Finite-Element Scheme: In the finite-element method the differential equation to be solved is expressed in terms of a variation or extremum principle which is valid over the region for solution. A quantity E is obtained by integrating the unknowns over the region and since E is a function of unknowns, it is termed a functional. Minimization of E with respect to the variable value at each nodal point leads to the set of simultaneous equations to be solved.

In some differential equations it is possible to use the calculus of variations to obtain a mathematical exact integral whose minimization is equivalent to solving the original differential equation. Consider the general variational problem:

$$E(\phi) = \iint_A G(\phi, \frac{\partial \phi}{\partial x}, \frac{\partial \phi}{\partial y}, x, y)\, dx\, dy \qquad (10\text{-}80)$$

The Euler equation of variational calculus for minimization of E in region A is

$$\frac{\partial G}{\partial \phi} - \frac{\partial}{\partial x}\left(\frac{\partial G}{\partial (\frac{\partial \phi}{\partial x})}\right) - \frac{\partial}{\partial y}\left(\frac{\partial G}{\partial (\frac{\partial \phi}{\partial y})}\right) = 0 \qquad (10\text{-}81)$$

When such a functional can be vigorously established, normally the resulting simultaneous equations have a symmetric coefficient matrix which allows a saving to be made in computer time and storage. However, for some differential equations such as Eqs. (10-64), (10-65), (10-67) and (10-68) it may not be possible to derive the functional. In this case the method of weighted residuals can be involved.

Consider the general differential equation

$$D(\phi) = 0 \qquad (10\text{-}82)$$

Then in general for trial values ϕ_a of ϕ the equation may not be satisfied so that

$$D(\phi_a) = R \neq 0 \tag{10-83}$$

in which R is termed the residual. If this residual could be reduced to zero by changing ϕ_a, suitably the equation would be solved.

The method of weighted residuals applies weighting coefficients to the residual evaluated at each node of the discretized domain and then minimize these weighted residuals. Obviously if R were identically zero everywhere the integral of the weighted residuals would be zero. Thus the solution is approached by requiring

$$\int_A W_i R dA = 0 \tag{10-84}$$

in which W_i is the weighting function at node i. If there are m nodes, then m linearly independent functions W_i are chosen so that m simultaneous equations can be written since Eq. (10-84) is equivalent to

$$\int_A W_i D(\phi) dA = 0 \tag{10-85}$$

In the Galerkin weighted residual approach the weighting functions W_i are set equal to the shape function N_i which defines the distribution of the variable. Thus over the area of one single element, the equation written for node I is

$$\int_A N_I D\left(\sum_{j=1}^{n} N_j \phi_j\right) dA = 0 \tag{10-86}$$

in which ϕ_j is the trial value of ϕ at node j and the summation is for all nodes n in that particular element. Summing all such contributions from all elements yields the required number of simultaneous equations in ϕ.

By using the principles of the finite-element formulation, Eqs. (10-64), (10-65), (10-67) and (10-68) can be solved. Linear triangle elements are used here to describe the solution procedure. Other shapes of elements such as quadratic-triangle elements can also be applied.

Consider first the momentum equation for the x-coordinate direction and its discretization for one element IJK shown in Fig. 10-32. Let u be x velocity component at any interior point. Using the finite-element printing, u can be represented as

$$u = [N]\{u\}^e \tag{10-87}$$

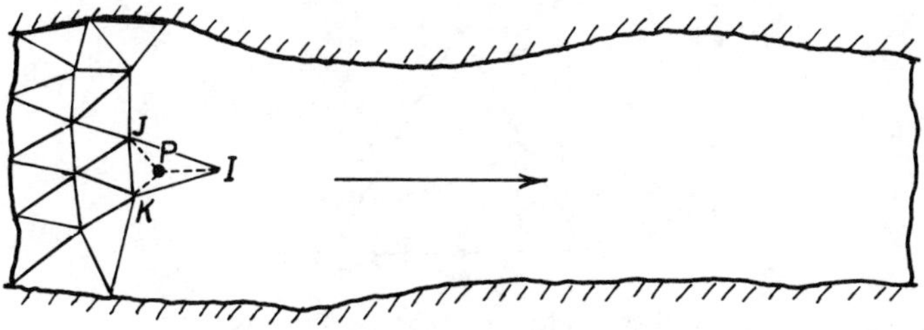

Fig. 10-32. Triangle Elements Describing a River Reach

in which [N] is the row vector of shape functions and $\{u\}^e$ is the column vector of variable values at nodes on the element. Thus

$$[N] = [N_I, N_J, N_K] \tag{10-88}$$

and

$$\{u\}^e = \begin{Bmatrix} u_I \\ u_J \\ u_K \end{Bmatrix} \tag{10-89}$$

For a linear triangle element in Fig. 10-32 it can be shown that convenient expressions for the function N_i are:

$$N_I = L_I; \; N_J = L_J; \; N_K = L_K \tag{10-90}$$

where L_I, L_J, and L_K are area coordinates of the point at which ϕ is being evaluated. Thus for example at point P (Fig. 10-32)

$$L_I = \frac{\text{Area JPK}}{\text{Area IJK}}$$

$$L_J = \frac{\text{Area IPK}}{\text{Area IJK}} \tag{10-91}$$

$$L_K = \frac{\text{Area IPJ}}{\text{Area IJK}}$$

Higher order triangular elements can be formulated and other element shapes can also be used (e.g., Fig. 10-33).

Similar to Eq. (10-87), the variables v, d and z can be represented as

$$v = [N]\{v\}^e \tag{10-92}$$

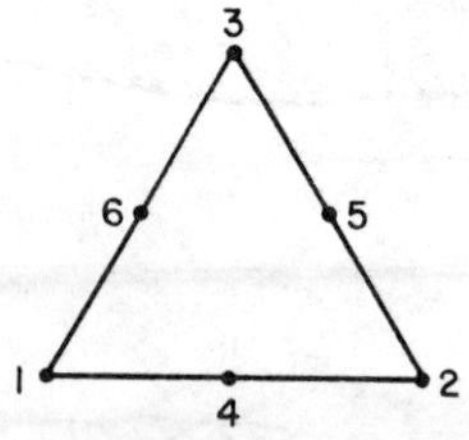

(a) Six Node Quadratic Triangle

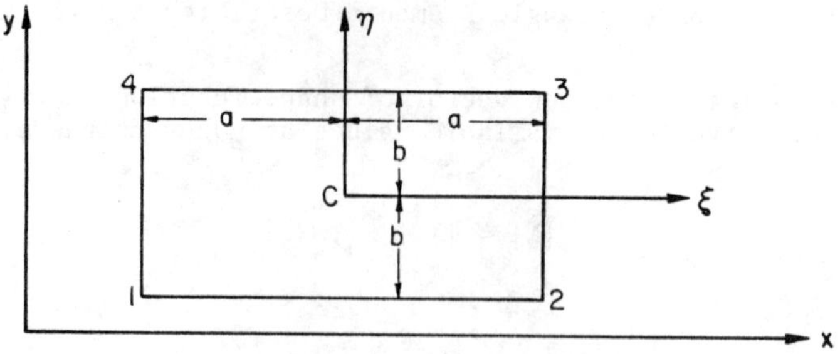

(b) Four Node Element

Fig. 10-33. Various Shapes of Finite Elements

$$d = [N]\{d\}^e \tag{10-93}$$

$$z = [N]\{z\}^e \tag{10-94}$$

Applying the Galerkin formulation (Eq. 10-86) to the x-direction momentum equation in a simplified form

$$\frac{\partial u}{\partial t} + u\frac{\partial u}{\partial x} + v\frac{\partial u}{\partial y} + g\frac{\partial h}{\partial x} + \frac{g}{C^2}\frac{wu}{(h-z)} = 0 \tag{10-95}$$

for Node I yields:

$$\int_A N_I\left[\frac{[N]\{u^{t_1}\}^e - [N]\{u^{t_o}\}^e}{\Delta t} + [N]\{u\}^e\frac{\partial}{\partial x}[N]\{u\}^e\right.$$

$$+ [N]\{v\}^e\frac{\partial}{\partial y}[N]\{u\}^e + g\frac{\partial}{\partial x}[N]\{h\}^e$$

$$\left. + \frac{g[N]\{u\}^e w}{C^2([N]\{h\}^e - [N]\{z\}^e)}\right]dA = 0 \qquad (10\text{-}96$$

in which

$$w = [([N]\{u\}^e)^2 + ([N]\{v\}^e)^2]^{1/2} \tag{10-97}$$

Now if the u^{t_o} terms are transposed the coefficients of each of the variables can be obtained. For a three node linear triangle element the result is:

$$\int_A \frac{N_I}{\Delta t}[N_I N_J N_K] + N_I(N_I u_I + N_J v_J + N_K u_K)[\frac{\partial N_I}{\partial x}\ \frac{\partial N_J}{\partial x}\ \frac{\partial N_K}{\partial x}]$$

$$+ N_I(N_I v_I + N_J v_J + N_K v_K)[\frac{\partial N_I}{\partial y}\ \frac{\partial N_J}{\partial y}\ \frac{\partial N_K}{\partial y}]$$

$$+ N_I \frac{g\ w\ [N_I\ N_J\ N_K]}{C^2(N_I h_I + N_J h_J + N_K h_K - N_I z_I - N_J z_J - N_K z_K} \begin{matrix} u_I \\ u_j \\ u_K \end{matrix}\ dA$$

$$+ \int N_I g[\frac{\partial N_I}{\partial x}\ \frac{\partial N_J}{\partial x}\ \frac{\partial N_K}{\partial x}] \begin{matrix} h_I \\ h_J \\ h_K \end{matrix}\ dA = \int_A \frac{N_I}{\Delta t}(N_I u_I^{t_o} + N_J u_J^{t_o} + N_K u_K^{t_o})dA \tag{10-98}$$

in which w in expanded form is given by:

$$w = [(N_I u_I + N_J u_J + N_K u_K)^2 + (N_I v_I + N_J v_J + N_K v_K)^2]^{1/2} \tag{10-99}$$

Equation (10-93) can be regrouped as

$$[K1]\{q_x\}^e + [K2]\{q_y\}^e + [K3]\{d\}^e = Q1 \tag{10-100}$$

where [K1], [K2] and [K3] are the matrix of coefficients and Q1 is a function of trial values of variables. Similarly, the y-direction momentum equation, the water continuity equation and the sediment continuity equation can be expressed as:

$$[K1]\{q_y\}^e + [K2]\{q_y\}^e + [K4]\{d\}^e = Q2 \tag{10-101}$$

$$[K1]\{d\}^e - [K5]\{q_x\}^e - [K6]\{q_y\}^e = Q3 \tag{10-102}$$

and

$$\frac{\gamma_s}{C_s \rho}[K1]\{z\}^e + [K7]\{q_x\}^e + [K8]\{q_y\}^e + [K9]\{q_x\}^e + [K10]\{q_y\}^e = Q4 \tag{10-103}$$

Thus Eqs. (10-100), (10-101), (10-102) and (10-103) are the finite-element formulations of the four governing partial differential equations for node I, in one particular element. Similar equations are written for all nodes in the finite-element grid unless a prescribed boundary condition overrules any of the four equations on the boundary. The final result is the same number of simultaneous equations as there are unknown values of q_x, q_y, d and z. The element coefficient matrices written in the preceding equations are in integral form and require numerical integrations.

The total number of simultaneous equations to be solved is equal to the number of nodes times the number of basic unknown variables. The complete matrix formulation results in a square coefficient matrix of dimensions. For a realistic number of nodes to adequately define the solution domain and inversion, such a matrix would produce prohibitive computer costs in storage and processor time. However, the storage and time can be greatly reduced by taking advantage of the fact that the matrix is banded about the diagonal. Since the matrix coefficients involve nonlinear terms, an iteration method similar to the Newton-Raphson procedure can be used to solve the finite-element equations.

Concluding Remarks

Two-dimensional water and sediment routing models are applied when detailed knowledge of water and sediment movement and bed elevation changes in a river reach due to passage of flood waves is required, and when the boundary geometry is complex. For example, with a midstream island or with irregular banklines the one-dimensional approach cannot reflect the detailed flow patterns occurring.

Again, four types of methods can be applied to develop the two-dimensional water and sediment routing models. They are the complete solution, the uncoupled-unsteady solution, the known discharge solution and the uncoupled-steady solution.

The numerical solutions on which these models are based can be categorized into finite-difference and finite-element methods as presented in preceding sections.

As is readily seen in Figs. 10-31 and 10-32, the finite-element method can better represent the irregular configuration of a water course. Therefore, its computed results are generally more reliable near the boundaries than the finite-difference method. However, the computer time required by the finite-element method currently used to determine the unknown variables at one time step is much larger than that required by the finite-difference method. Recent research related to the finite-element method has been concentrated on increasing the time step length and improving the efficiency of matrix solution. Until these problems can be solved, the application of finite-element methods to study two-dimensional unsteady river problems is not practical.

Colorado State University is developing various types of two-dimensional river models and testing their efficiencies and applicabilities. At present, an unsteady-uncoupled finite-difference model and a complete finite-element model are under testing. The former method has been applied to simulate the flow patterns in a physical model. The testing result is promising. After further testing, this and other models will be applied to study water and sediment related problems in Pool 4 in the Upper Mississippi River. The model can be easily modified to study other river systems.

10.5 SUMMARY AND CONCLUSIONS

Mathematical models for water and sediment routing in rivers have been described in this chapter. The developed models have been successfully applied to study various types of water and sediment related problems in rivers. It is not the intention to present all the models available in the literature. Instead, the chapter presents some useful models that have been tested and verified.

The models presented include one-dimensional models, multiple stream and compound stream models, and two-dimensional models. These models are based on the conservation laws: sediment continuity, water continuity and momentum equations. Considering types of governing equations used for model formulations, the models can be subcategorized into four types of solution models. They are: (1) complete solution: solve all the governing equations simultaneously, (2) uncoupled-unsteady solution: solve the water continuity and momentum equations for routing water first and then solve the sediment continuity equation explicitly for sediment routing, (3) known-discharge solution: assuming the water discharge in the river reach is known, solve the momentum and sediment continuity equations simultaneously for routing sediment, (4) uncoupled-steady solution: assuming the water discharge is steady and is known in the river reach, solve the momentum equation to determine the water surface profile and corresponding flow conditions and then solve the sediment equation uncoupled to determine bed elevation changes. As discussed in the section titled, "Discussion of the Four Types of Numerical Solution Methods," each solution method has its advantage and disadvantage, and should be used properly considering its applicability and limitations.

In general, a one-dimensional water and sediment routing model is adequate to study water and sediment related problems and to predict river responses to development in a relatively uniform river. Changes in the cross-sectional geometries at the computational sections can be determined from properly distributing the computed changes in cross-sectional area to subsections at a cross section. These computational sections are generally located at irregular points in a study reach and at particularly interesting points.

For a highly irregular river reach, the one-dimensional model can only determine general river patterns and changes to development. To obtain more detailed information without getting into a sophisticated two-dimensional model, a multiple stream model or a compound stream model is recommended. Applicability of the multiple stream model is limited by its requirement that each computational section in a river reach must be divided into a constant number of subsections to form continuous subchannels within the river banks. A compound stream model is thus developed to relax this limitation. However, it should be realized that the compound stream flow model is still based on the one-dimensional flow equations. If a river reach has significant flow in both longitudinal and transverse directions, detailed knowledge of water and sediment movement can only be obtained from a two-dimensional or even only from a three-dimensional model.

It is apparent that the model complexities and computational efficiencies depend on the number of governing equations being solved in the dominative direction(s). A relatively simple model that can provide adequate information is always desired. This requires some engineering judgment and experience with mathematical modeling. Also, on some occasions unreasonable results may be obtained due to wrong data input and/or limitation of models. The computed results from a model need to be carefully interpreted using basic principles of river mechanics and sediment transportation.

10.6 NOTATIONS

A	water cross-sectional area
A_d	volume of sediment deposited on channel bed per unit of length of channel, the value of which is negative when bed erosion occurs
As_k	volume of sediment suspended in the water over the cross section in the kth stream of the multiple stream model
A_x^y	departure from a prismatic channel given by $(\partial A/\partial x)y$
B	top width
C	Chezy coefficient
Cg	empirical coefficient in the sediment transport equation
C_s	mean sediment concentration on a volume basis given by Q_s/Q
c	wave celerity
D_ℓ	dynamic contribution of lateral discharge given by $q_\ell V_\ell/Ag$
d	water depth
Fa_k	sediment-laden water discharge per unit length added to the kth stream of the multiple stream model
f	dimensionless friction factor; a function

f_g geostrophic parameter
g acceleration of gravity
g_w, g_x, g_y unit-width bed material transport rate in the directions of w, x, and y respectively
h water surface elevation
h_f energy loss
K conveyance
[K1]-[Kq] matrix coefficient formulated in the finite-element method
Mc_k momentum correction term due to lateral flow term Fa_k
m total number of subsections at a composite or compound cross section; exponent in the sediment transport equation
N shape function defined in the finite-element method
n Manning's roughness coefficient
P wetted perimeter
p volume of sediment in a unit volume of bed layer given by ρ_b/ρ_s
Q flow discharge
Q_s sediment discharge
q_ℓ lateral flow per unit length of channel, given by $q_s + q_w$
q_s lateral sediment flow per unit length of channel, a positive quantity indicates inflow and a negative value denotes outflow
q_w lateral water flow per unit length of channel, a positive quantity indicates inflow and a negative value denotes outflow
q_x and q_y unit-width water discharge in the x and y directions, respectively
Sa_k sediment discharge per unit length added to the kth stream of the multiple stream model
S_o bed slope
S_f friction slope
T $\partial A/\partial y$
u, v and w depth-averaged velocity in the x, y, and resultant directions, respectively
V mean flow velocity
V_ℓ velocity component of lateral inflow in the main flow direction
x horizontal distance along the channel
y flow depth; horizontal space variable for the two-dimensional model
z riverbed elevation
β momentum coefficient
Δ change in a variable from old time to new time, e.g., $f = f^{j+1} - f^j$
ε coefficient of eddy viscosity
ε_g dimensionless eddy viscosity coefficient
λ bed porosity

ρ	density of sediment-laden water given by $\rho_w + C_s(\rho_s - \rho_w)$
ρ_b	bulk density of sediment forming the bed
ρ_s	density of sediment
ρ_w	density of water
r_s	unit weight of the material deposited on the channel bed
τ	bed shear stress
τ_{bw}, τ_{bx}, τ_{by}	bottom shear stresses in the directions of w, x, and y, respectively
τ_c	shear stress at incipient motion of the bed material
τ_{wx} and τ_{wy}	components of surface shear stress due to wind
τ_{xx}, τ_{xy}, τ_{yx}	effective stresses in vertical planes

10.7 REFERENCES

Abbot, M.B., and Cunge, J.A., 1975. Two-dimensional modeling of tidal deltas and estuaries. Unsteady Flow in Open Channels, K. Mahmood and V. Yevjevich, ed., Vol. 2, Water Resources Publications, Fort Collins, Colorado, 1975.

Abbott, M.B., Damsgaards, A., and Rodenhuis, G.S., 1973. System 21 "Jupiter", a design system for two-dimensional nearly horizontal flows. Journal of Hydraulic Research, Vol. 11, No. 1, pp. 1-28.

Abertson, M.L., Day, Y.B., Jensen, R.A., and Rouse, H., 1950. Diffusion of submerged jets. Transactions, ASCE, Vol. 115, pp. 639-664.

Amein, M., and Fang, C.S., 1970. Implicit flood routing in natural channels. Journal of the Hydraulics Division, ASCE, Vol. 96, No. HY12, Proc. Paper 7773, December pp. 2481-2500.

Apelt, C.J., and Gout, J.J., 1974. Numerical modeling of tidal phenomena in bays and estuaries with inter-tidal flats. Proceeding 5th Australas Conference on Hydraulics and Fluid Mechanics, Christchurch, New Zealand.

Becker, L. and Yeh, W.W., 1973. Identification of multiple reach channel parameters. Water Resources Research, American Geophysical Union, Vol. 9, No. 2.

Brebbia, C.A., and Connor, J.J., 1976. Introduction to finite element techniques for fluid flow. Butterworths, London, England.

Brebbia, C.A., and Partridge, P.W., 1976. Finite element simulation of water circulation in the North Sea. Appl. Math Modeling, Vol. 1, No. 2, September, pp. 101-107.

Chang, F.F.M., and Richards, D.L., 1971. Deposition of sediment in transient flow. Journal of the Hydraulics Division, ASCE, Vol. 97, No. HY6, Proc. Paper 8191, June, pp. 837-849.

Chang, H.H., and Hill, J.C., 1976. Computer modeling of erodible flood channels and deltas. Journal of the Hydraulics Division, ASCE, Vol. 102, No. HY10, Proc. Paper 12485, October, pp. 1461-1478.

Chen, Y.H., 1973. Mathematical modeling of water and sediment routing in natural channels. Dissertation presented to the Department of Civil Engineering, Colorado State University, Fort Collins, Colorado, in partial fulfillment of the requirements for the degree of Doctor of Philosophy.

Chen, Y.H., and Simons, D.B., 1974. Routing of water and sediment in canals and rivers. Proceedings, Contribution of Irrigation and Drainage to the World Food Supply Specialty Conference, ASCE, August 14-16, pp. 121-146.

Chen, Y.H., and Simons, D.B., 1975. Mathematical modeling of alluvial channels. Symposium on Modeling Techniques, ASCE, Vol. 1, September 3-5, pp. 466-483.

Colby, B.R., 1964. Discharge of sands and mean-velocity relationships in sand-bed streams. U.S. Geological Survey Professional Paper, 462-A.

Colby, B.R., and Hembree, C.H., 1955. Computation of total sediment discharge Niobrara River near Cody, Nebraska. U.S. Geological Survey Water-Supply Paper, 1357.

Connor, J.J., and Wang, J., 1974. Finite element modeling of hydrodynamic circulation. Numerical Methods in Fluid Dynamics, C.A. Brebbia and J.J. Connor, eds., Peutech Press, London, England, pp. 355-387.

Contractor, D.N., and Wiggert, J.M., 1972. Numerical studies of unsteady flow in the James River. Bulletin 51, Water Resources Research Center, Virginia Polytechnic Institute and State University, Blacksburg, Virginia.

Cunge, J.A., 1975. Applied mathematical modeling of open channel flow. Unsteady Flow in Open Channels, Edited by K. Mahmood and V. Yevjevich, Vol. 1, Chapter 10, Water Resources Publications, Fort Collins, Colorado.

Cunge, J.A., and Perdreau, N., 1973. Mobile bed fluvial mathematical models. La Houille Blanche, No. 7.

Dass, P., 1975. Water and sediment routing in nonuniform channels. Dissertation presented to the Department of Civil Engineering, Colorado State University, Fort Collins, Colorado, in partial fulfillment of the requirements for the degree of Doctor of Philosophy.

DeVries, M., 1965. Considerations about non-steady bed-load-transport in open channels. Proceedings of the Eleventh Congress of the International Association for Hydraulic Research, Vol. 3, No. 3.8, Leningrad, U.S.S.R.

Einstein, H.A., 1950. The bed load function for seaiment transportation in open channel flows. U.S. Department of Agriculture, Technical Bulletin, No. 1026.

Fread, D.L., 1974. Numerical properties of implicit four-point finite difference equations of unsteady flow. NOAA Technical Memorandum NWS HYDRO-18.

Leendertse, J.J., 1967. Aspects of a computational model for long period water-wave propagation. Memorandum RM-5294-PR, The Rand Corporation, Santa Monica, California, May.

Leendertse, J.J., 1970. A water quality simulation model for well-mixed estuaries and coastal seas: Principles of computation. Memorandum RM-6230-RC, Vol. 1, The Rand Corporation, Santa Monica, California, February.

Leendertse, J.J., and Gritton, E.C., 1971. A water quality simulation model for well-mixed estuaries and coastal seas: Computational procedures. Memorandum R-708-NYC, Vol. 2, The New York City Rand Institute, New York, New York, July.

Li, R.M., 1977. An uncoupled-steady water and sediment routing model. Unpublished report, Colorado State University, Engineering Research Center, Fort Collins, Colorado.

Liggett, J.A. and Cunge, J.A., 1975. Numerical methods of solution of the unsteady flow equations. Chapter 4, Vol. 1, Unsteady Flow in Open Channel, Water Resources Publications, Fort Collins, Colorado.

Lopez, S.J.L., 1978. Mathematical modeling of sediment deposition in reservoirs. Ph.D. Dissertation, Department of Civil Engineering, Colorado State University, Fort Collins, Colorado.

Mahmood, K., and Ponce, V.M., 1976. Mathematical modeling of sedimentation transients in sand-bed channels. Report No. CER75-76KM-VMP28, Colorado State University, Engineering Research Center, Fort Collins, Colorado.

Owens, E.M., 1977. One-dimensional sediment routing in rivers. M.S. Thesis, Department of Civil Engineering, Colorado State University, Fort Collins, Colorado.

Partridge, P.W., and Brebbia, C.A., 1976. Quadratic finite-elements in shallow water problems. Journal of Hydraulic Division, ASCE, Vol. 102, No. HY9, September, pp. 1299-1313

Ponce, V.M., Indlekofer, H., and Simons, D.B., 1978. The convergence of implicit bed transient models. Approved for publication in the Journal of Hydraulics Division, ASCE.

Preissmann, A., 1960. Propagationdes intumescences dans les cannaux et rivieres, 1st Congres de l'Assoc. Francaise de Calcul, Grenoble, printed 1961, p. 433-442.

Price, R.K., 1974. Comparison of four numerical methods for flood routing. Journal of the Hydraulic Division, ASCE, Vol. 100, No. HY7, pp. 879-899.

Richtmyer, R.D. and Morton, K.W., 1967. Solution of the implicit equations. Difference Methods for Initial Value Problems, Interscience Publishers, New York.

Shen, H.W., 1971. Total sediment load. River Mechanics, Edited by H.W. Shen, Vol. 1, Chap. 13, Water Resources Publications, Fort Collins, Colorado.

Simons, D.B., and Chen, Y.H., 1976. A mathematical model study of pool 4 in the Upper Mississippi and Lower Chippewa Rivers. Colorado State University, Report No. CER76-77-DBS-YHC8, prepared for the U.S. Fish and Wildlife Service, Twin Cities, Minnesota.

Simons, D.B., and Chen, Y.H., 1977. Investigation of the effects of Chippewa River erosion and silt reduction measures. Colorado State University, Report No. CER76-77-DBS-YHC54, prepared for the U.S. Army Engineer District, St. Paul, Minnesota.

Simons, D.B. and Chen, Y.H., 1978. Assessment of hydraulic impacts of the Atchafalaya River flood control program on the lower Mississippi River. Colorado State University, Report No. CER77-78DBS-YHC38, prepared for the U.S. Army Engineer District, New Orleans, Louisiana.

Simons, D.B., Lagasse, P.F., Chen, Y.H., and Schumm, S.A., 1975. The river environment. Colorado State University, Report No. CER75-76-DBS-PFL-YHC-SAS-14, prepared for the U.S. Fish and Wildlife Service, Twin Cities, Minnesota.

Simons, D.B., Ponce, V.M. and Y.H. Chen, 1978. A two-dimensional water and sediment routing model in river basins. Progress Report, Prepared for the U.S. Fish and Wildlife Service, Twin Cities, Minnesota.

Simons, D.B., Richardson, E.V., Stevens, M.A., Duke, J.H., and Duke, V.S., 1971. Geometric and hydraulic properties of the rivers. Venezuelan International Meteorological and Hydrological Experiment (VIMHEX), Hydrology Report, Vol. III, Colorado State University Report CER70-71DBS-EVR-MAS-JHD-VCD36, Fort Collins, Colorado.

Simons, D.B., Schumm, S.A., Stevens, M.A., Chen, Y.H., and Lagasse, P.F., 1975. A geomorphic study of pools 24, 25, and 26 in Upper Mississippi and Lower Illinois Rivers. Colorado State University, Report No. CER75-76-DBS-SAS-MAS-YHC-PFL8, prepared for the Waterways Experiment Station, Vicksburg, Mississippi.

Simons, D.B., Volker, R.E. and Chen, Y.H., 1977. A two-dimensional river flow model based on a finite-element method. Progress Report, Engineering Research Center, Colorado State University, Fort Collins, Colorado.

Streloff, T., 1970. Numerical solution of Saint-Venant equations. Journal of the Hydraulics Division, ASCE, Vol. 96, No. HY1, Proceedings Paper 7043, January, pp. 223-252.

U.S. Army Corps of Engineers, 1976. HEC-6, Scour and deposition in rivers and reservoirs, users manual. Hydrologic Engineering Center, Computer Program 723-G2-L2470, Davis, California.

Taylor, C., and Davis, J.M., 1973. Finite element numerical modeling of flow and dispersion in estuaries. Proceedings, International Symposium on River Mechanics, Asian Institute of Technology, Bangkok, Thailand, Vol. 3, January, pp. 465-476.

Toffaleti, F.B., 1969. Definitive computation of sand discharge in rivers. Journal of the Hydraulics Division, ASCE, Vol. 95, No. HY1, Proc. Paper 6350, January, pp. 225-248.

Vreugdenhil, C.B., 1973. Secondary flow computations. Publication No. 1973, Delft Hydraulics Laboratory, Delft, The Netherlands, November.

Wanstrath, J.J., and Whitaker, R.E., Reid, R.O., and Vastano, A.C., 1976. Storm surge simulation in transformed coordinates. Technical Report No. 76-3, U.S. Army, Corps of Engineers, Coastal Engineering Research Center, Fort Belvoir, Virginia.

Appendix 10A

FINITE-DIFFERENCE EQUATIONS OF THE COMPLETE SOLUTION METHOD

Introduction

When the finite-difference approximations (Eqs. 10-10 or 10-11) are substituted into the basic unsteady flow equations (Eqs. 10-1, 10-2 and 10-3), three linear algebraic equations (Eqs. 10-12, 10-13 and 10-14) are formed. The coefficients of these three linear algebraic equations are presented in the appendix.

Sediment Continuity Equation

Substituting Eq. (10-11) into Eq. (10-1), the sediment continuity equation is approximated:

$$\frac{1}{\Delta x}\Big[Q_{s_{i+1}}^{j} + \Big(\frac{\partial Q_s}{\partial Q}\Big)_{i+1}^{j} \Delta Q_{i+1} + \Big(\frac{\partial Q_s}{\partial y}\Big)_{i+1}^{j} \Delta y_{i+1} + \Big(\frac{\partial Q_s}{\partial z}\Big)_{i+1}^{j} \Delta z_{i+1} - Q_{s_i}^{j}$$

$$- \Big(\frac{\partial Q_s}{\partial Q}\Big)_{i}^{j} \Delta Q_{i} - \Big(\frac{\partial Q_s}{\partial y}\Big)_{i}^{j} \Delta y_{i} - \Big(\frac{\partial Q_s}{\partial z}\Big)_{i}^{j} \Delta z_{i}\Big] + \frac{p}{4\Delta t}(T_i^j + T_{i+1}^j)(\Delta z_i$$

$$+ \Delta z_{i+1}) + \frac{1}{4\Delta t}\Big[\Big(\frac{1}{V}\Big)_i^j + \Big(\frac{1}{V}\Big)_{i+1}^j\Big](\Delta Q_{s_i} + \Delta Q_{s_{i+1}}) - \frac{1}{2}\Big[\Big(\frac{Q_s}{AV^2}\Big)_i^j$$

$$+ \Big(\frac{Q_s}{AV^2}\Big)_{i+1}^j\Big]\Big\{\frac{1}{2\Delta t}(\Delta Q_i + \Delta Q_{i+1}) - \frac{1}{4\Delta t}[(VT)_i^j + (VT)_{i+1}^j](\Delta y_i + \Delta y_{i+1})\Big\}$$

$$= \frac{1}{2}\Big(q^{j}_{s_{i+(\frac{1}{2})}} + q^{j+1}_{s_{i+(\frac{1}{2})}}\Big) \qquad (10A\text{-}1)$$

where Δx is the distance between sections i and $(i+1)$, Δt is the time increment from t^j to t^{j+1}, and Δ indicates the difference between the values at t^{j+1} and at t^j. Rearranging Eq. 10A-1, the following equation is obtained

$$a_{1i}Q_i^{j+1} + b_{1i}y_i^{j+1} + c_{1i}z_i^{j+1} + d_{1i}Q_{i+1}^{j+1} + e_{1i}y_{i+1}^{j+1} + f_{1i}z_{i+1}^{j+1} = \gamma_{1i} \qquad (10A\text{-}2)$$

This equation is the same as Eq. (10-12). The coefficients in Eq. (10A-2) are expressed as:

$$a_{1i} = \left(\frac{\partial Q_s}{\partial Q}\right)_i^j k_2 - k_3 \quad (10A\text{-}3)$$

$$b_{1i} = \left(\frac{\partial Q_s}{\partial y}\right)_i^j k_2 + k_4 \quad (10A\text{-}4)$$

$$c_{1i} = \left(\frac{\partial Q_s}{\partial z}\right)_i^j k_2 + k_5 \quad (10A\text{-}5)$$

$$d_{1i} = \left(\frac{\partial Q_s}{\partial Q}\right)_{i+1}^j k_6 - k_3 \quad (10A\text{-}6)$$

$$e_{1i} = \left(\frac{\partial Q_s}{\partial y}\right)_{i+1}^j k_6 + k_4 \quad (10A\text{-}7)$$

$$f_{1i} = \left(\frac{\partial Q_s}{\partial z}\right)_{i+1}^j k_6 + k_5 \quad (10A\text{-}8)$$

$$\gamma_{1i} = 2\Delta t[q^j_{s_{i+(\frac{1}{2})}} + q^{j+1}_{s_{i+(\frac{1}{2})}}] - 4\theta(Q^j_{s_{i+1}} - Q^j_{s_i}) + a_{1i}Q^j_i + b_{1i}y^j_i$$

$$+ c_{1i}z^j_i + d_{1i}Q^j_{i+1} + e_{1i}y^j_{i+1} + f_{1i}z^j_{i+1} \quad (10A\text{-}9)$$

where $$\theta = \frac{\Delta t}{\Delta x}$$

$$k_1 = \left(\frac{1}{V}\right)_i^j + \left(\frac{1}{V}\right)_{i+1}^j$$

$$k_2 = -4\theta + K_1$$

$$k_3 = \left(\frac{C_s}{V}\right)_i^j + \left(\frac{C_s}{V}\right)_{i+1}^j$$

$$k_4 = (C_sT)_i^j + (C_sT)_{i+1}^j$$

$$k_5 = p(T^j_i + T^j_{i+1})$$

$$k_6 = 4\theta + K_1$$

and $q_{s_{i+(\frac{1}{2})}}$ is the lateral sediment inflow per unit length of channel between Sections i and $(i+1)$

Flow Continuity Equation

Applying the finite-difference approximations, Eq. (10-2) becomes:

$$\frac{1}{\Delta x}(Q_{i+1}^{j+1} - Q_i^{j+1}) + \frac{1}{4\Delta t}(T_i^j + T_{i+1}^j)(y_i^{j+1} - y_i^j + y_{i+1}^{j+1} - y_{i+1}^j)$$

$$+ \frac{1}{4\Delta t}(T_i^j + T_{i+1}^j)(z_i^{j+1} - z_i^j + z_{i+1}^{j+1} - z_{i+1}^j) = \frac{1}{2}[q_{\ell_{i+(\frac{1}{2})}}^j + q_{\ell_{i+(\frac{1}{2})}}^{j+1}]$$

(10A-10)

Let $k_7 = T_i^j + T_{i+1}^j$

$$k_8 = y_i^j + y_{i+1}^j + z_i^j + z_{i+1}^j$$

$$a_{2i} = -4\theta$$

$$b_{2i} = k_7$$

$$c_{2i} = k_7$$

$$d_{2i} = 4\theta$$

$$e_{2i} = k_7$$

$$f_{2i} = k_7$$

and $r_{2i} = 2\Delta t[q_{\ell_{i+(\frac{1}{2})}}^j + q_{\ell_{i+(\frac{1}{2})}}^{j+1}] + k_7 k_8$

Eq. (10A-10) becomes

$$a_{2i}Q_i^{j+1} + b_{2i}y_i^{j+1} + c_{2i}z_i^{j+1} + d_{2i}Q_{i+1}^{j+1} + e_{2i}y_{i+1}^{j+1} + f_{2i}z_{i+1}^{j+1} = r_{2i}$$

(10A-11)

This equation is the same as Eq. (10-13).

Momentum Equation

Applying the finite-difference approximations, the terms in Eq. (10-3) can be approximated by using the finite-difference equations (Eq. 10-11):

$$\frac{\partial \rho Q}{\partial t} \simeq \frac{1}{2\Delta t}\{\rho_i^j Q_i^{j+1} - \rho_i^j Q_i^j + \rho_{i+1}^j Q_{i+1}^{j+1} - \rho_{i+1}^j Q_{i+1}^j\} \quad (10A\text{-}12)$$

$$\frac{\partial \beta\rho QV}{\partial x} \simeq \frac{1}{\Delta x}\{(\beta\rho V)_{i+1}^j Q_{i+1}^{j+1} - (\beta\rho V)_i^j Q_i^{j+1}\} \quad (10A\text{-}13)$$

$$gA\frac{\partial \rho y}{\partial x} \simeq g(A)^j_{i+(\frac{1}{2})}\frac{1}{\Delta x}\{\rho_{i+1}^j y_{i+1}^{j+1} - \rho_i^j y_i^{j+1}\} \quad (10A\text{-}14)$$

$$\rho gAS_o \simeq g(\rho A)^j_{i+(\frac{1}{2})}\frac{1}{\Delta x}\{z_i^{j+1} - z_{i+1}^{j+1}\} \quad (10A\text{-}15)$$

$$\rho gAS_f \simeq g\{\frac{1}{2}[2\rho A\frac{S_f}{Q}(Q^{j+1} - Q^j) - 2\rho A\frac{S_f}{K}\frac{\partial K}{\partial y}(y^{j+1} - y^j)$$

$$- 2\rho A\frac{S_f}{K}\frac{\partial K}{\partial z}(z^{j+1} - z^j) + \rho S_f T(y^{j+1} - y^j)_i$$

$$+ \frac{1}{2}[2\rho A\frac{S_f}{Q}(Q^{j+1} - Q^j) - 2\rho A\frac{S_f}{K}\frac{\partial K}{\partial y}(y^{j+1} - y^j)$$

$$- 2\rho A\frac{S_f}{K}\frac{\partial K}{\partial z}(z^{j+1} - z^j) + \rho S_f T(y^{j+1} - y^j)]_{i+1}$$

$$+ \frac{1}{2}[(\rho AS_f)^j_i + (\rho AS_f)^j_{i+1}]\} \quad (10A\text{-}16)$$

$$\rho q_\ell V_\ell \simeq \frac{1}{2}[(\rho q_\ell V_\ell)^j_{i+(\frac{1}{2})} + (\rho q_\ell V_\ell)^{j+1}_{i+(\frac{1}{2})}] \quad (10A\text{-}17)$$

Substituting Eqs. (10A-12) through (10A-17) to Eq. (10-3) and rearranging the resulting equation, the following equation is obtained:

$$a_{3i}Q_i^{j+1} + b_{3i}y_i^{j+1} + c_{3i}z_i^{j+1} + d_{3i}Q_{i+1}^{j+1} + e_{3i}y_{i+1}^{j+1} + f_{3i}z_{i+1}^{j+1} = r_{3i} \quad (10A\text{-}18)$$

The coefficients in Eq. (10A-18) are formulated as:

$$a_{3i} = \rho_i^j(1 + 2k_{17}) - k_{10} \quad (10A\text{-}19)$$

$$b_{3i} = -\rho_i^j(k_{13} + k_{14}(\frac{\partial K}{\partial y})_i^j - k_{19}] \qquad (10A\text{-}20)$$

$$c_{3i} = -k_{16} - \rho_i^j k_{14}(\frac{\partial K}{\partial z})_i^j \qquad (10A\text{-}21)$$

$$d_{3i} = \rho_{i+1}^j(1+2k_{18}) - k_{11} \qquad (10A\text{-}22)$$

$$e_{3i} = \rho_{i+1}^j[k_{13} - k_{15}(\frac{\partial K}{\partial y})_{i+1}^j + k_{20} \qquad (10A\text{-}23)$$

$$f_{3i} = k_{16} - \rho_{i+1}^j k_{15}(\frac{\partial K}{\partial z})_{i+1}^j \qquad (10A\text{-}24)$$

$$r_{3i} = (\rho Q)_i^j\ (1 + k_{17}) + (\rho Q)_{i+1}^j\ (1 + k_{18})$$

$$+ \rho_i^j\{-k_{14}[(\frac{\partial K}{\partial y} y + \frac{\partial K}{\partial z} z)_i^j] + k_{19} y_i^j\}$$

$$+ \rho_{i+1}^j\{-k_{15}[(\frac{\partial K}{\partial y} y + \frac{\partial K}{\partial z} z)_{i+1}^j] + k_{20} y_{i+1}^j\}$$

$$+ \frac{\Delta t}{2}(\rho_i^j + \rho_{i+1}^j)\ [(q_\ell V_\ell)_{i+(\frac{1}{2})}^j + (q_\ell V_\ell)_{i+(\frac{1}{2})}^{j+1}]$$

where

$$\theta = \Delta t/\Delta x$$

$$k_9 = g\Delta t$$

$$k_{10} = 2\theta(\beta\rho V)_i^j$$

$$k_{11} = 2\theta(\beta\rho V)_{i+1}^j$$

$$k_{12} = \theta[(\beta\rho V^2 T)_i^j + (\beta\rho V^2 T)_{i+1}^j]$$

$$k_{13} = \theta g[A_i^j + A_{i+1}^j]$$

$$k_{14} = 2k_9[AS_f/K]_i^j$$

$$k_{15} = 2k_9[AS_f/K]^j_{i+1}$$

$$k_{16} = \theta g[(\rho A)^j_i + (\rho A)^j_{i+1}]$$

$$k_{17} = k_9(S_f/V)^j_i$$

$$k_{18} = k_9(S_f/V)^j_{i+1}$$

$$k_{19} = k_9(S_fT)^j_i$$

$$k_{20} = k_9(S_fT)^j_{i+1}$$

Appendix 10B

DOUBLE-SWEEP METHOD

Introduction

The system of linear algebraic equations formulated from the finite-difference approximations of the basic partial differential equations can be solved by a double-sweep method. The principles and formulations of the double-sweep methods for the complete solution method and the known-discharge solution method are described in this appendix. The latter method can also be applied to solve the system of equations developed from the uncoupled-unsteady solution method.

Double-Sweep Method for the Complete Solution

The principles of the double-sweep method can be explained by the following example. Consider a river reach being covered by three computation points and the linear equations derived are:

$$d_{2,0}Q_1 + e_{2,0}y_1 + f_{2,0}z_1 = r_{2,0} \tag{10B-1}$$

$$d_{3,0}Q_1 + e_{3,0}y_1 + f_{3,0}z_1 = r_{3,0} \tag{10B-2}$$

$$a_{1,1}Q_1 + b_{1,1}y_1 + c_{1,1}z_1 + d_{1,1}Q_2 + e_{1,1}y_2 + f_{1,1}z_2 = r_{1,1} \tag{10B-3}$$

$$a_{2,1}Q_1 + b_{2,1}y_1 + c_{2,1}z_1 + d_{2,1}Q_2 + e_{2,1}y_2 + f_{2,1}z_2 = r_{2,1} \qquad (10B\text{-}4)$$

$$a_{3,1}Q_1 + b_{3,1}y_1 + c_{3,1}z_1 + d_{3,1}Q_2 + e_{3,1}y_2 + f_{3,1}z_2 = r_{3,1} \qquad (10B\text{-}5)$$

$$a_{1,2}Q_2 + b_{1,2}y_2 + c_{1,2}z_2 + d_{1,2}Q_3 + e_{1,2}y_3 + f_{1,2}z_3 = r_{1,2} \qquad (10B\text{-}6)$$

$$a_{2,2}Q_2 + b_{2,2}y_2 + c_{2,2}z_2 + d_{2,2}Q_3 + e_{2,2}y_3 + f_{2,2}z_3 = r_{2,2} \qquad (10B\text{-}7)$$

$$a_{3,2}Q_2 + b_{3,2}y_2 + c_{3,2}z_2 + d_{3,2}Q_3 + e_{3,2}y_3 + f_{3,2}z_3 = r_{3,2} \qquad (10B\text{-}8)$$

$$a_{1,3}Q_3 + b_{1,3}y_3 + c_{1,3}z_3 = r_{1,3} \qquad (10B\text{-}9)$$

Where Eqs. (10B-1) to (10B-2), Eqs. (10B-3) to (10B-8), and Eq. (10B-9) have the form of the upstream boundary Eqs. (10-15) and (10-16), of the interior Eqs. (10-12) to (10-14), and of the downstream boundary Eq. (10-17), respectively.

Equations (10B-1) and (10B-2) with three unknowns can be reduced to

$$Q_1 = m_1 + n_1 z_1 \qquad (10B\text{-}10)$$

and

$$y_1 = p_1 + q_1 z_1 \qquad (10B\text{-}11)$$

in which the coefficients m, n, p and q are functions of d, e, f and r. Substituting Eqs. (10B-10) and (10B-11) into the first set of three interior Eqs. (10B-3) to (10B-5) yields

$$s_1 z_1 + d_{1,1}Q_2 + e_{1,1}y_2 + f_{1,1}z_2 = t_1 \qquad (10B\text{-}12)$$

$$Q_2 = m_2 + n_2 z_2 \qquad (10B\text{-}13)$$

and

$$y_2 = p_2 + q_2 z_2 \qquad (10B\text{-}14)$$

Thus, Eqs. (10B-1) to (10B-5) are reduced to Eqs. (10B-10) to (10B-14). The same procedure can be repeated to reduce the next set of three interior Eqs. (10B-6) to (10B-8) by substituting Eqs. (10B-13) and (10B-14) into them yielding

$$s_2 z_2 + d_{1,2} Q_3 + e_{1,2} y_3 + f_{1,2} z_3 = t_2 \qquad (10B-15)$$

$$Q_3 = m_3 + n_3 z_3 \qquad (10B-16)$$

and

$$y_3 = p_3 + q_3 z_3 \qquad (10B-17)$$

All the coefficients m, n, p, q, s and t in Eqs. (10B-10) to (10B-17) can be computed by recurrence equations and therefore can be easily programmed. The procedure of using the recurrence equations to compute the values of the coefficients in Eqs. (10B-10) to (10B-17) is called "Forward Sweep."

Equations (10B-16) and (10B-17) derived from the forward sweep can be combined with Eq. (10B-9) to form a set of three equations in 3 unknowns. The values of Q_3, y_3 and z_3 can be easily determined. Thereafter, the values of z_2, Q_2, y_2, z_1, Q_1, y_1 can be determined backward from Eqs. (10B-15) to (10B-10). Again, the recurrence equations can be easily formulated for programming. This procedure of using the recurrence equations to compute the values of unknowns is called "Backward Sweep." The whole procedure is thus designated as the "Double-Sweep" method. The method can be extended to solve a set of linear equations formulated for any number of sections in a channel reach.

The recurrence equations to compute the values of the coefficients during "Forward Sweep" are:

1. Define initial values of

(a) $g_{1,1} = d_{2,0}$, $h_{1,1} = e_{2,0}$, $k_{1,1} = f_{2,0}$, and $\ell_{1,1} = r_{2,0}$

(b) $g_{2,1} = d_{3,0}$, $h_{2,1} = e_{3,0}$, $k_{2,1} = f_{3,0}$, and $\ell_{2,1} = r_{3,0}$

2. Compute

$$m_i = \frac{\ell_{1,i} g_{2,i} - \ell_{2,i} g_{1,i}}{h_{1,i} g_{2,i} - h_{2,i} g_{1,i}} \qquad (10B-18)$$

$$n_i = -\frac{k_{1,i} g_{2,i} - k_{2,i} g_{1,i}}{h_{1,i} g_{2,i} - h_{2,i} g_{1,i}} \qquad (10B-19)$$

$$p_i = \frac{\ell_{1,i} h_{2,i} - \ell_{2,i} h_{1,i}}{g_{1,i} h_{2,i} - g_{2,i} h_{1,i}} \qquad (10B-20)$$

$$q_i = -\frac{k_{1,i} h_{2,i} - k_{2,i} h_{1,i}}{g_{1,i} h_{2,i} - g_{2,i} h_{1,i}} \qquad (10B-21)$$

$$s_i = k_{1,i} = a_{1,i}q_i + b_{1,i}n_i + c_{1,i} \tag{10B-22}$$

$$K_2 = a_{2,i}q_i + b_{2,i}n_i + c_{2,i} \tag{10B-23}$$

$$K_3 = a_{3,i}q_i + b_{3,i}n_i + c_{3,i} \tag{10B-24}$$

$$t_i = k_{4,i} = r_{1,i} - a_{1,i}p_i - b_{1,i}m_i \tag{10B-25}$$

$$g_{1,i+1} = d_{1,i}K2 - d_{2,i}k_{1,i} \tag{10B-26}$$

$$h_{1,i+1} = e_{1,i}K2 - e_{2,i}k_{1,i} \tag{10B-27}$$

$$k_{1,i+1} = f_{1,i}K2 - f_{2,i}k_{1,i} \tag{10B-28}$$

$$\ell_{1,i+1} = k_{4,i}K_2 - (r_{2,i} - a_{2,i}p_i - b_{2,i}m_i)k_{1,i} \tag{10B-29}$$

$$g_{2,i+1} = d_{2,i}K_3 - d_{3,i}K_2 \tag{10B-30}$$

$$h_{2,i+1} = e_{2,i}K_3 - e_{3,i}K_2 \tag{10B-31}$$

$$k_{2,i+1} = f_{2,i}K_3 - f_{3,i}K_2 \tag{10B-32}$$

$$\ell_{2,i+1} = (r_{2,i} - a_{2,i}p_i - b_{2,i}m_i)K_3 - (r_{3,i} - a_{3,i}p_i - b_{3,i}m_i)K_2 \tag{10B-33}$$

where i increases from 1 to N and N is the total number of the computational sections. At $i = N$, the computation must stop at Eq. (10B-21) to complete the "Forward Sweep."

The recurrence equations to compute the values of the unknown variables during "Backward Sweep" are:

1. At the last computational section,

$$z_N = \frac{r_{1,N} - a_{1,N}p_N - b_{1,N}m_N}{c_{1,N} + a_{1,N}q_N + b_{1,N}n_N} \tag{10B-34}$$

2.

$$Q_i = m_i + n_iz_i \tag{10B-35}$$

$$y_i = p_i + q_iz_i \tag{10B-36}$$

$$z_{i-1} = \frac{1}{k_{1,i-1}}(k_{4,i-1} - d_{1,i-1}Q_i - e_{1,i-1}y_i - f_{1,i-1}z_i) \tag{10B-37}$$

3. Repeat Step 2 from $i = N$ to 1, but neglect Eq. (10B-37) when $i = 1$.

Double Sweep Method for the Known-Discharge Solution

The finite-difference equations obtained from the known-discharge solution are of the form:

$$A_i \Delta y_i + B_i \Delta y_{i+1} + C_i \Delta z_i + D_i \Delta z_{i+1} = Q_i \tag{10B-38}$$

$$E_i \Delta y_i + F_i \Delta y_{i+1} + G_i \Delta z_i + H_i \Delta z_{i+1} = R_i \tag{10B-39}$$

for $1 \leq i \leq N - 1$. The coefficients A_i, B_i, C_i, D_i, E_i, F_i, G_i, H_i, Q_i, and R_i are known for each cross section, and Δy_i and Δz_i are unknown.

There are N - 1 equations of the form of both Eqs. (10B-38) and (10B-39), yielding 2(N-1) equations. The remaining two equations are supplied by the boundary conditions. The upstream boundary condition can be written in the general form:

$$\Delta y_1 = S_1 \Delta z_1 + T_1 \tag{10B-40}$$

The coefficients of Eq. (10B-40) can be specified depending on whether ΔQ_{s_1} or Δz_1 is used as the boundary condition. If ΔQ_{s_1} is used, it can be expressed by the following truncated Taylor's series expansion:

$$\Delta Q_{s_1} = \left(\frac{\partial Q_s}{\partial Q}\right)_1^j \Delta Q_1 + \left(\frac{\partial Q_s}{\partial y}\right)_1^j \Delta y_1 + \left(\frac{\partial Q_s}{\partial z}\right)_1^j \Delta z_1 \tag{10B-41}$$

Eqs. (10B-40) and (10B-41) are the same with the coefficients of Eq. (10B-40) defined by

$$S_1 = -\left(\frac{\partial Q_s}{\partial z}\right)_1^j / \left(\frac{\partial Q_s}{\partial y}\right)_1^j \tag{10B-42}$$

$$T_1 = \left(\Delta Q_{s_1} - \left(\frac{\partial Q_s}{\partial Q}\right)_1^j \Delta Q_1\right) / \left(\frac{\partial Q_s}{\partial y}\right)_1^j \tag{10B-43}$$

where Δ shows the change in a variable between time t^{j+1} and time t^j. If Δz_i is known, the coefficients of Eq. (10B-40) can be defined as

$$S_1 = \text{some large number } (10^{10}) \tag{10B-42}$$

$$T_1 = -S_1 \Delta z_1 \tag{10B-43}$$

The downstream boundary condition will specify the change in stage Δh_N over the time increment Δt. By definition,

$$\Delta y_N + \Delta z_N = \Delta h_N \tag{10B-44}$$

Eqs. (10B-40) and (10B-44) are the two additional boundary equations.

The coefficient matrix of this linear system of equations is a sparse diagonal matrix. This type of system can be solved efficiently by the double-sweep algorithm. The following linear relationship can be shown to exist between the two unknowns at a given cross section:

$$\Delta y_i = S_i \Delta z_i + T_i \tag{10B-45}$$

If Eq. (10B-45) is substituted into Eqs. (10B-38) and (10B-39), the following equations result when written in terms of Δz_i:

$$\Delta z_i = (R_i - A_i T_i - B_i \Delta y_{i+1} - D_i \Delta z_{i+1})/(C_i + A_i S_i) \tag{10B-46}$$

$$\Delta z_i = (Q_i - E_i T_i - F_i \Delta y_{i+1} - H_i \Delta z_{i+1})/(G_i + E_i S_i) \tag{10B-47}$$

The right-hand sides of Eqs. (10B-46) and (10B-47) can be equated, which after rearranging yields the following:

$$\Delta y_{i+1} = \frac{D_i U_i - H_i}{V_i} \Delta z_{i+1} + \frac{R_i - E_i T_i - U_i (Q_i - A_i T_i)}{V_i} \tag{10B-48}$$

where

$$U_i = (G_i + E_i S_i)/(C_i + A_i S_i) \tag{10B-49}$$

$$V_i = F_i - U_i B_i \tag{10B-50}$$

Equation (10B-48) is in the same form as Eq. (10B-45), and the coefficients of Eq. (10B-45) are therefore defined by:

$$S_{i+1} = \frac{D_i U_i - H_i}{V_i} \tag{10B-51}$$

$$T_{i+1} = \frac{R_i - E_i T_i - U_i(Q_i - A_i T_i)}{V_i} \tag{10B-52}$$

Since the upstream boundary condition supplies S_1 and T_1, Eqs. (10B-51) and (10B-52) can be used as recursive formulas to calculate S_i and T_i for $i = 2,\ldots,N$. This would complete the forward sweep.

At the downstream boundary, the following equation exists

$$\Delta y_N = S_N \Delta z_N + T_N \tag{10B-53}$$

where S_N and T_N are known from the forward sweep. Equations (10B-44) and (10B-53) can be combined to solve for the unknowns at the downstream boundary. These solutions are:

$$\Delta z_N = \frac{\Delta h_N - T_N}{S_N + 1} \tag{10B-54}$$

$$\Delta y_N = \Delta h_N - \Delta z_N \tag{10B-55}$$

Now the backward sweep begins with the calculation of the unknowns at the first cross section upstream of the downstream boundary. First, either Eq. (10B-46) or (10B-47) with $i = N - 1$ is used to calculate Δz_{N-1}. Then Eq. (10B-45) is used to calculate Δy_{N-1}. This two-step procedure is continued for each cross section upstream to the upstream boundary. The backward sweep and the solution are now complete.

This double-sweep method can be applied to solve the linear algebraic equations developed from the uncoupled-unsteady solution method.

Appendix 10C

FINITE-DIFFERENCE EQUATIONS OF THE UNCOUPLED-UNSTEADY SOLUTION METHOD

Introduction

The uncoupled-unsteady solution method has been utilized to formulate a water and sediment routing model. The expressions for the coefficients in finite-difference equations are presented in this appendix.

Numerical Analysis of Water Routing

Substituting Eq. (10-10) into Eqs. (10-2) and (10-3) and assuming ρ is constant yields:

$$\frac{\lambda}{\Delta x}(\Delta Q_{i+1} - \Delta Q_i) + \frac{1}{\Delta x}(Q^j_{i+1} - Q^j_i) + \frac{T_{i+1}}{2\Delta t}\Delta y_{i+1} + \frac{T_i}{2\Delta t}\Delta y_i$$

$$- \frac{1}{2}(q^j_{\ell\, i+(\frac{1}{2})} + q_{\ell\, i+(\frac{1}{2})}^{j+1}) = 0 \qquad (10C\text{-}1)$$

and

$$\frac{1}{2\Delta t}(\Delta Q_i + \Delta Q_{i+1}) + \frac{\lambda}{\Delta x}[2(\beta V)^j_{i+1}\Delta Q_{i+1} - (\beta V^2 T)^j_{i+1}\Delta y_{i+1}$$

$$-2(\beta V)^j_i \Delta Q_i + (\beta V^2 T)^j_i \Delta y_i]$$

$$+ \frac{1}{\Delta x}[(\beta QV)^j_{i+1} - (\beta QV)^j_i] + \frac{g\lambda}{2\Delta x}[(A^j_{i+1} + A^j_i) + (y^j_{i+1} - y^j_i)T^j_{i+1}]\Delta y_{i+1}$$

$$+ \frac{g\lambda}{2\Delta x}[(y^j_{i+1} - y^j_i)T^j_i - (A^j_{i+1} + A^j_i)]\Delta y_i$$

$$+ \frac{g}{2\Delta x}(A^j_{i+1} + A^j_i)(y^j_{i+1} - y^j_i) = \frac{g}{2\Delta x}(A^j_{i+1} + A^j_i)(z^j_i - z^j_{i+1})$$

$$- \frac{g\lambda}{4}[(S_{f_{i+1}}^{\ j} + S_{f_i}^{\ j})T^j_{i+1} + (A^j_{i+1} + A^j_i)(\frac{\partial S_f}{\partial y})^j_{i+1}]\Delta y_{i+1}$$

$$- \frac{g\lambda}{4}[(S_{f_{i+1}}^{\ j} + S_{f_i}^{\ j})T^j_i + (A^j_{i+1} + A^j_i)(\frac{\partial S_f}{\partial y})^j_i]\Delta y_i$$

$$- \frac{g\lambda}{4}(A^j_{i+1} + A^j_i)(\frac{\partial S_f}{\partial Q})^j_{i+1}\Delta Q_{i+1}$$

$$- \frac{g\lambda}{4}(A^j_{i+1} + A^j_i)(\frac{\partial S_f}{\partial Q})^j_i \Delta Q_i$$

$$- \frac{g}{4}(A^j_{i+1} + A^j_i)(S_{f_{i+1}}^{\ j} + S_{f_i}^{\ j}) + \frac{1}{2}[(q_\ell V_\ell)^j_{i+(\frac{1}{2})} + (q_\ell V_\ell)^{j+1}_{i+(\frac{1}{2})}]$$

(10C-2)

in which λ is the weighting factor, and

$$\frac{\partial S_f}{\partial Q} \simeq 2\, S_f\,(\frac{1}{Q} + \frac{1}{n}\frac{\partial n}{\partial Q}) \qquad (10C\text{-}3)$$

and

$$\frac{\partial S_f}{\partial y} \simeq -2\, S_f \left[\frac{1}{A}\left(\frac{5}{3}\, T - \frac{2}{3}\, R\, \frac{dP}{dy}\right) - \frac{1}{n}\frac{\partial n}{\partial y}\right] \tag{10C-4}$$

Rearranging Eqs. (10C-1) and (10C-2) yields:

$$a_{1i}\, \Delta Q_i + b_{1i}\, \Delta y_i + C_{1i}\, \Delta Q_{i+1} + d_{1i}\, \Delta y_{i+1} = \gamma_{1i} \tag{10C-5}$$

and

$$a_{2i}\, \Delta Q_i + b_{2i}\, \Delta y_i + C_{2i}\, \Delta Q_{i+1} + d_{2i}\, \Delta y_{i+1} = \gamma_{2i} \tag{10C-6}$$

in which the coefficients a, b, c, and d are functions of variables evaluated at the time step t^j and therefore are known. The expression for the coefficients are given below:

$$a_{1i} = -2\theta\lambda \tag{10C-7}$$

$$b_{1i} = T_i \tag{10C-8}$$

$$C_{1i} = -a_{1i} \tag{10C-9}$$

$$d_{1i} = T_{i+1} \tag{10C-10}$$

$$\gamma_{1i} = -2\theta\, (Q^j_{i+1} - Q^j_i) + \Delta t\, (q_{\ell}{}^{j}_{i+(\frac{1}{2})} + q_{\ell}{}^{j+1}_{i+(\frac{1}{2})}) \tag{10C-11}$$

$$a_{2i} = 1. - 4\theta\lambda\, (\beta V)^j_i + \frac{1}{2}\, K_1 K_2 \lambda \left(\frac{\partial S_f}{\partial Q}\right)^j_i \tag{10C-12}$$

$$b_{2i} = 2\theta\lambda\, (\beta V^2 T)^j_i + g\theta\lambda\, (K_3 T^j_i - K_2)$$

$$+ \frac{1}{2}\, K_1\, \lambda \left[K_4 T^j_i + K_2 \left(\frac{\partial S_f}{\partial y}\right)^j_i\right] \tag{10C-13}$$

$$C_{2i} = 1 + 4\theta\lambda\, (\beta V)^j_{i+1} + \frac{1}{2}\, K_1 K_2 \lambda \left(\frac{\partial S_f}{\partial Q}\right)^j_{i+1} \tag{10C-14}$$

$$d_{2i} = -2\theta\lambda\, (\beta V^2 T)^j_{i+1} + g\theta\lambda\, (K_2 + K_3\, T^j_{i+1})\, \Delta y_{i+1}$$

$$+ \frac{1}{2}\, K_1\, \lambda \left[K_4 T^j_{i+1} + K_2 \left(\frac{\partial S_f}{\partial y}\right)^j_{i+1}\right] \tag{10C-15}$$

$$\gamma_{2i} = -2\theta\,[(\beta QV)^j_{i+1} - (\beta QV)^j_i] - g\theta K_2 K_3$$

$$+ g\theta K_2\,(Z^j_i - Z^j_{i+1}) - \frac{1}{2}K_1 K_2 K_4 + \Delta t\,[(q_\ell V_\ell)^j_{i+(\frac{1}{2})} + (q_\ell V_\ell)^{j+1}_{i+(\frac{1}{2})}] \tag{10C-16}$$

in which

$$\theta = \frac{\Delta t}{\Delta x}$$

$$K_1 = g\Delta t$$

$$K_2 = A^j_i + A^j_{i+1}$$

$$K_3 = y^j_{i+1} - y^j_i$$

$$K_4 = S_{f_i}^{\ j} + S_{f_{i+1}}^{\ j}$$

At the upstream boundary, it is assumed that the gauging station flood hydrograph provides the flow depth or discharge as a function of time. The relation can be written as

$$c_{2_0}\Delta Q_1 + d_{2_0}\Delta y_1 = r_{2_0} \tag{10C-17}$$

in which c_{2_0} or d_{2_0} is usually zero and the other is unity, and r_{2_0} equals $\Delta Q_1(t)$ or $\Delta y_1(t)$.

At the downstream boundary, a depth hydrograph, a discharge hydrograph, or a depth-discharge relationship can be used as a downstream boundary condition if the flow is subcritical.

This relationship can be expressed in segment form as

$$a_{1_N}\Delta Q_N + b_{1_N}\Delta y_N = r_{1_N} \tag{10C-18}$$

Equations (10C-5), (10C-6), (10C-17) and (10C-18) constitute a system of 2N linear algebraic equations in 2N unknowns. Any of the standard methods, such as the Gaussian elimination method or the matrix inversion method, can be used for its solution. A double-sweep method similar to the one presented in Appendix 10B can be applied for solving this system of linear equations.

Numerical Analysis of Sediment Routing

In natural alluvial channels, the phenomena of sediment motion are of a three-dimensional nature. If the channel is relatively uniform so that the flow characteristics may be physically represented by a one-dimensional model and the net deposition or erosion of sediment at a section can be distributed according to a known function (for example, Eq. 10-24), Eq. (10-1) can be applied to route sediment in natural channels.

Applying a finite-difference approximation to Eq. (10-1) yields

$$\frac{1}{2\Delta x}[(Q_{s_i}^{j} - Q_{s_{i-1}}^{j}) + (Q_{s_i}^{j+1} - Q_{s_{i-1}}^{j+1})] + \frac{p}{\Delta t}\Delta A_{d_{i-(\frac{1}{2})}} + \frac{1}{2\Delta t}[(AC_s)_{i-1}^{j+1} - (AC_s)_{i-1}^{j} + (AC_s)_{i}^{j+1} - (AC_s)_{i}^{j}] - \frac{1}{2}[q_{s_{i-(\frac{1}{2})}}^{j} + q_{s_{i-(\frac{1}{2})}}^{j+1}] = 0 \tag{10C-19}$$

in which $\Delta A_{d_{i-(\frac{1}{2})}}$ is the net deposited volume of sediment in Δt per unit length of channel bed between sections (i-1) and i; $q_{s_{i-(\frac{1}{2})}}$ is the lateral sediment inflow between sections (i-1) and i.

The expression for $\Delta A_{d_{i-(\frac{1}{2})}}$ is obtained by rearranging Eq. (10C-19). Similarly, a relation for $\Delta A_{d_{i+(\frac{1}{2})}}$ can be obtained. Then, the values of ΔA_{d_i} at the interior sections can be estimated from

$$\Delta A_{d_i} = \frac{1}{2}[\Delta A_{d_{i-(\frac{1}{2})}} + \Delta A_{d_{i+(\frac{1}{2})}}]$$

$$= \frac{1}{4p}\{\Delta t[q_{s_{i-(\frac{1}{2})}}^{j} + q_{s_{i-(\frac{1}{2})}}^{j+1} + q_{s_{i+(\frac{1}{2})}}^{j} + q_{s_{i+(\frac{1}{2})}}^{j+1}]$$

$$- [2(AC_s)_{i}^{j+1} - 2(AC_s)_{i}^{j} + (AC_s)_{i-1}^{j+1} - (AC_s)_{i-1}^{j}$$

$$+ (AC_s)^{j+1}_{i+1} - (AC_s)^{j}_{i+1}] - \theta(Q^j_{s_{i+1}} - Q^j_{s_{i-1}} + Q^{j+1}_{s_{i+1}} - Q^{j+1}_{s_{i-1}})\} \qquad (10C\text{-}20)$$

in which $\theta = \Delta t/\Delta x$.

At the upstream boundary, let $Q^j_{s_o}$ denote the inflow sediment discharge upstream of the reach at time step t^j. This value can be estimated from the inflow sediment discharge hydrograph. If this hydrograph is not available, $Q^j_{s_o}$ is assumed equal to $Q^j_{s_1}$. Assuming $\Delta A_{d_1} \approx A_{d_{1+(\frac{1}{2})}}$ at the upstream boundary then

$$\Delta A_{d_1} = \frac{1}{2p}\{\Delta t[q^j_{s_{1+(\frac{1}{2})}} + q^{j+1}_{s_{1+(\frac{1}{2})}}]$$

$$- [(AC_s)^{j+1}_1 - (AC_s)^j_1 + (AC_s)^{j+1}_2 - (AC_s)^j_2$$

$$- \theta[(Q^j_{s_2} - Q^j_{s_o}) + (Q^{j+1}_{s_2} - Q^{j+1}_{s_o}]\} . \qquad (10C\text{-}21)$$

At the downstream boundary, it is assumed that $\Delta A_{d_N} \approx \Delta A_{d_{N-(\frac{1}{2})}}$. Then ΔA_{d_N} can be estimated from Eq. (10C-19).

Chang and Hill (1976) determined the value of ΔA_{d_i} by assuming

$$\Delta A_{d_i} \approx \Delta A_{d_{i-(\frac{1}{2})}} \qquad (10C\text{-}22)$$

This assumption simplifies the formulation and calibration of the model. Nevertheless, the conservation of sediment mass is not well satisfied as Eq. (10C-20).

After ΔA_{d_i} is obtained for each computational section, the change in bed elevation can be determined from a sediment distribution function (for example, Eq. 10-24).

Appendix 10D

FINITE-DIFFERENCE EQUATIONS OF THE KNOWN DISCHARGE SOLUTION

Introduction

The known-discharge solution has been utilized to formulate a sediment routing model. The expressions for the coefficients in the finite-difference equations are presented in this appendix.

Determination of the Water Discharge

Assuming that variation in the magnitude of the discharge along a reach is due only to the flow being spatially varied, and not due to the unsteadiness, the water continuity equation (Eq. 10-2) reduced to the form for a steady spatially varied flow is:

$$\frac{\partial Q}{\partial x} = q_\ell \qquad (10D\text{-}1)$$

The water discharge at each computational section can be determined from the finite-difference equation of Eq. (10D-1) as follows

$$Q_{i+1} = Q_i + (q_\ell \Delta x)_{i+(\frac{1}{2})} \qquad (10D\text{-}2)$$

where $q_{\ell_{i+(\frac{1}{2})}}$ is the lateral inflow per unit length along the channel between Sections i and $(i+1)$. If the inflow discharge at the upstream boundary is known, then the discharge at the other sections can be determined from Eq. (10D-2).

Formulation of Finite-Difference Equations

The sediment continuity equation (Eq. 10-1) and the momentum equation (Eq. 10-3) can be transformed to the following forms:

$$\frac{\partial Q_s}{\partial x} + pT\,\frac{\partial z}{\partial t} + C_s T\,\frac{\partial y}{\partial t} + \frac{1}{V}\,\frac{\partial Q_s}{\partial t} - \frac{C_s}{V}\,\frac{\partial Q}{\partial t} = q_s \qquad (10D\text{-}3)$$

and

$$\frac{\partial Q}{\partial t} + 2V\,\frac{\partial Q}{\partial x} - V^2\left(T\,\frac{\partial y}{\partial x} + A_x^y\right) + gA\left(\frac{\partial h}{\partial x} + S_f\right) = 0 \qquad (10D\text{-}4)$$

where the density of sediment-laden water is assumed to be a constant. Substituting Eq. (10-11) into Eqs. (10D-3) and (10D-4) yields

$$\frac{1}{\Delta x_i}(Q_{s_{i+1}}^{j} - Q_{s_i}^{j} + \Delta Q_{s_{i+1}} - \Delta Q_{s_i}) + \frac{p}{\Delta t} T_{i+1}^{j} \Delta z_{i+1}$$

$$+ \frac{1}{4\Delta t}(C_{s_i}^{j} T_i^{j} + C_{s_{i+1}}^{j} T_{i+1}^{j}) \cdot (\Delta y_i + \Delta y_{i+1}) + \frac{1}{4\Delta t}\left(\frac{1}{V_i^{j}} + \frac{1}{V_{i+1}^{j}}\right)$$

$$\cdot (\Delta Q_{s_i} + \Delta Q_{s_{i+1}}) - \frac{1}{4\Delta t}\left(\frac{C_{s_i}^{j}}{V_i^{j}} + \frac{C_{s_{i+1}}^{j}}{V_{i+1}^{j}}\right)(\Delta Q_i + \Delta Q_{i+1}) = \frac{1}{2}(q_{s_i}^{j} + q_{s_i}^{j+1}) \qquad \text{(10D-5)}$$

$$\frac{1}{2\Delta t}(\Delta Q_i + \Delta Q_{i+1}) + (V_i^{j} + V_{i+1}^{j}) \frac{1}{2\Delta x_i}(- Q_i^{j+1} - Q_i^{j} + Q_{i+1}^{j+1} + Q_{i+1}^{j})$$

$$- \frac{1}{2\Delta x_i}(V_i^{j\,2} T_i^{j} + V_{i+1}^{j\,2} T_{i+1}^{j})(y_{i+1}^{j} - y_i^{j} + \Delta y_{i+1} - \Delta y_i)$$

$$- \frac{1}{2}(V_i^{j\,2} A_{x_i}^{y^j} + V_{i+1}^{j\,2} A_{x_{i+1}}^{y^j}) + \frac{g}{2\Delta x_i}(A_i^{j} + A_{i+1}^{j})(h_{i+1}^{j} - h_i^{j} + \Delta h_{i+1} - \Delta h_i)$$

$$+ \frac{g}{2}(A_i^{j} S_{f_i}^{j+1} + A_{i+1}^{j} S_{f_{i+1}}^{j+1}) = 0 \qquad \text{(10D-6)}$$

In Eqs. (10D-5) and (10D-6), ΔQ_i and ΔQ_{i+1} are known from the known-discharge assumption. The momentum equation Eq. (10D-6) can be written as

$$A_i \Delta y_i + B_i \Delta y_{i+1} + C_i \Delta z_i + D_i \Delta z_{i+1} = Q_i \qquad \text{(10D-7)}$$

The coefficients in Eq. (10D-7) are expressed as

$$A_i = C_1 - C_2 + C_4 \frac{\partial S_f^{j}}{\partial y}_i$$

$$B_i = C_2 - C_1 + C_5 \frac{\partial S_f^{j}}{\partial y}_{i+1}$$

$$C_i = -C_2$$

$$D_i = C_2$$

$$Q_i = C_1(y^j_{i+1} - y^j_i) - C_2(h^j_{i+1} - h^j_i) + \Delta x_i ({V^j_i}^2 {A_{x_i}^{y}}^j + {V^j_{i+1}}^2 {A_{x_{i+1}}^{y}}^j)$$

$$- 2C_3(Q^j_{i+1} - Q^j_i) - C_4 {S_{f_i}}^j - C_5 {S_{f_{i+1}}}^j$$

where

$$C_1 = {V^j_i}^2 T^j_i + {V^j_{i+1}}^2 T^j_{i+1}$$

$$C_2 = g(A^j_i + A^j_{i+1})$$

$$C_3 = V^j_i + V^j_{i+1}$$

$$C_4 = g\Delta x_i A^j_i$$

$$C_5 = g\Delta x_i A^j_{i+1}$$

$$\theta = \frac{\Delta x_i}{\Delta t}$$

Similarly the sediment continuity equation (Eq. 10D-5) can be written

$$E_i \Delta y_i + F_i \Delta y_{i+1} + G_i \Delta z_i + H_i \Delta z_{i+1} = R_i \qquad (10D\text{-}8)$$

where

$$E_i = C_1 + C_4 \left.\frac{\partial Q_s}{\partial y}\right|^j_i$$

$$F_i = C_1 + C_5 \left.\frac{\partial Q_s}{\partial y}\right|^j_{i+1}$$

$$G_i = C_4 \left.\frac{\partial Q_s}{\partial z}\right|^j_i$$

$$H_i = C_5 \left.\frac{\partial Q_s}{\partial z}\right|^j_{i+1} + p \frac{\Delta x_i}{\Delta t} T^j_{i+1}$$

$$R_i = Q_{s_i}^{\,j} - Q_{s_{i+1}}^{\,j} + C_3(\Delta Q_i + \Delta Q_{i+1}) - C_4 \frac{\partial Q_s^{\,j}}{\partial Q_i}\Delta Q_i - C_5 \frac{\partial Q_s^{\,j}}{\partial Q_{i+1}}\Delta Q_{i+1}$$

$$+ \frac{\Delta x_i}{2}(q_{s_i}^{\,j} + q_{s_{i+1}}^{\,j})$$

$$C_1 = \theta(C_{s_i}^{\,j} T_i^j + C_{s_{i+1}}^{\,j} T_{i+1}^j)$$

$$C_2 = \theta\left(\frac{1}{V_i^j} + \frac{1}{V_{i+1}^j}\right)$$

$$C_3 = \theta(C_{s_i}^{\,j}/V_i^j + C_{s_{i+1}}^{\,j}/V_{i+1}^j)$$

$$C_4 = C_2 - 1$$

$$C_5 = C_2 + 1$$

$$\theta = \frac{\Delta x_i}{4\Delta t}$$

When a given reach is to be modeled, it is described by some number of cross sections N. At each section, Δy and Δz are unknown, so the total number of unknowns is $2N$. For each adjacent pair of cross sections, two equations of the form of Eqs. (10D-7) and (10D-8) exist, yielding a total of $2(N-1)$ equations. The remaining equations are supplied by the boundary conditions. These conditions may be in the form of

$$\Delta y_1 = S_1 \Delta z_1 + T_1 \tag{10D-9}$$

$$\Delta y_N + \Delta z_N = \Delta h_N \tag{10D-10}$$

where $i = 1$ indicates the upstream boundary and $i = N$, the downstream boundary. The system of equations is now complete and can be solved uniquely. The coefficient matrix of this linear system is a sparse diagonal matrix, which can be solved efficiently by the double sweep algorithm as described in Appendix 10B.

Chapter 11

ANALYSIS OF WATERSHEDS AND RIVER SYSTEMS

by

Daryl B. Simons, Associate Dean for Engineering Research and Professor of Civil Engineering, Colorado State University, Fort Collins, Colorado

and

Ruh-Ming Li, Associate Professor of Civil Engineering, Colorado State University, Fort Collins, Colorado

Chapter 11

ANALYSIS OF WATERSHEDS AND RIVER SYSTEMS

11.1 INTRODUCTION

The increasing interest in land and water resource planning has stimulated the development of specific and general watershed and river system response models. The models, whether physical or conceptual, are formulated to estimate physical quantities that describe the major ecosystem responses to precipitation such as water yield, sediment yield, yields of other water pollutants and stream morphology. Methods to estimate these watershed and river responses are needed for analyzing the economic feasibility and trade-offs of any proposed water resources or land use development and for predicting possible adverse environmental impacts associated with the projects.

The physical processes governing watershed and river responses are very complicated. Many past studies have utilized a statistical interpretation of observed response data. The unit hydrograph method for water routing, and the Universal Soil Loss Equation for soil erosion are examples of these types of studies. It is often difficult to predict the response of a watershed to various watershed developments or treatments using those methods, because they are based on the assumption of homogeneity in time and space. Mathematical models simulating the governing physical process is a viable way to estimate the time-dependent response of watershed and river systems to precipitation with varying land use and water resources development. By analyzing basic ecosystem processes and the impacts of management activities on specific processes governing production of water, non-point source pollution and stream morphology, it is possible to predict the cause-effect relationships between management activities and ecosystem responses. With the aid of systems analysis techniques, a desirable mix of management activities can then be selected taking into consideration both the environmental and resource goals.

Chapter 7 identified some of the major sources of sediment that impact on watersheds, rivers, canals and other natural and man-affected aspects of the fluvial system. Chapter 9 described mathematical models for routing water and sediment from watersheds and Chapter 10 reported mathematical models of water and sediment routing in rivers. This chapter presents a brief assessment of the current methods utilized to analyze watershed and river systems, an overview of some case studies and problems, criteria related to the development of mathematical models for various applications, procedural steps involved in model development, and examples of applications of models to watershed, river and basin systems.

11.2 STATUS OF KNOWLEDGE ASSESSMENT

Mathematical models used to simulate the effects of management activities on ecosystem responses can be classified as being one of three types: regression, "black box" simulation, or physical process simulation. In cases where models are actually hybrids of components that fall into two or all three of these categories, they would be classified according to the dominant traits of the model as a whole.

A general weakness of regression models available for use in water and land resources management, such as the Musgrave approach to soil erosion (U.S. Forest Service, 1976), is that the variables representing water and land uses and conditions are not specific enough to reflect the effects of many individual management activities. In addition, the regression models usually require sufficient observed data to develop meaningful relations. This is often the most serious drawback for this type of approach. Furthermore, it is very difficult to predict time and space dependent processes using regression equations.

The lumped parameter, "black box" or "simulation programming" type of model interprets input-output relations using oversimplified forms which may or may not have physical significance. All processes related to movement of the water and sediment through the watershed are "lumped" together into several coefficients. The classic example of a lumped parameter model is the rational formula for estimating peak discharge, i.e., $Q = CIA$ where Q is the peak discharge, I is the rainfall input, A is the drainage area, and C is the runoff coefficient that represents the major hydrologic processes. Such a model is easy to use, but has limited physical meaning and is often inaccurate. In most cases, their ability to predict the effects of alternative mixes and sequences of management activities is very limited. Further, simulation programming models can only be made to represent a particular watershed by calibrating numerous internal model coefficients using water and sediment flow data derived under existing or past vegetation and soil conditions. In the absence of calibration data for significantly different future vegetation and soil conditions, the predicted future water and sediment flow using simulation programming models must be suspect.

Physical process simulation models, however, avoid the "lumping" of physical significant variables by decomposing the overall hydrologic and hydraulic phenomena into their respective components such as infiltration and sediment detachment from raindrop splash. By simulating the selected phenomena into separate components, each individual process can be analyzed and refined or altered to meet the needs of the user. Consequently, as each process component is upgraded, the model becomes more representative of the physical system. Use of component process models also allows input of variables that have physical significance to the user and the field situation. All of the above characteristics of component process models allow for greater flexibility than other types of models. Advantages of physical process component models over other types are numerous. In general, physical process simulation models are superior to regression type models or "black box" type mathematical models. This is because they require less data to develop and calibrate. The input variables to process models are physically significant. They indicate system response to management activities by changing one or more physically significant variables. The physical process simulation models are "dynamic simulation systems." They are not assumed stationary in either time or space and therefore with very little calibration they can be used for predicting the future response of the system in time and space. Furthermore, because these models are formulated according to physical significance, they would be applicable to any area where the governing physical processes are basically the same as those considered in the model.

There are numerous mathematical models available for predicting the response of watershed and river systems. A comprehensive review of non-point water quality modeling in wildland management was conducted by U.S. Forest Service (1976). A brief review of available component models for predicting system response from watershed and river systems is presented.

Degradation, aggradation and transport of pollutants in watershed and river systems are closely related to water movement. The streamflow predictive and water routing models have received the most intensive study of all non-point predictive techniques. The ability of the majority of available streamflow and water routing models to relate wildland activities to their unique environments and to account for spatial diversity is not well demonstrated. Efforts to further develop a promising model which will be useful in predicting effects of management activities and also represent spatial and temporal variability of both activities and processes is necessary.

Existing sediment models for watershed systems mainly consider surface erosion. No process models exist for unstable channel erosion, nor are any models available for predicting both mass wasting and its interaction with channels. Almost all existing surface erosion models are based on either the Musgrave approach or the Universal Soil Loss Equation (U.S. Forest Service, 1976). These models are difficult to use because they are insensitive to both the spatial and temporal variability of management activities. Recently, Simons et al. (1975a) developed a numerical model to simulate the physical processes governing sediment movement on small watersheds. Their models can predict the effects of management activities on sediment yield in both time and space. However, the applicability of this model is limited to surface erosion on fairly stable land and channels, small watersheds, and a single storm. This model is being modified to accommodate unstable channel erosion and deposition, a more complex watershed system, and long-term simulation of response considering inter-storm periods. In addition, for coupling the nitrogen, phosphorus, etc., movement with water and sediment, it is necessary to route the sediment considering sediment size. This is necessary because different sizes of sediment have different uptake rates for different chemicals and substances (see Chapter 9).

It is important to recognize that a river is a dynamic part of a dynamic system. An alluvial river is generally changing its position and shape continuously as a consequence of hydraulic forces acting on its bed and banks and as a result of the interaction of these forces with the biological processes of the river environment. These changes may be slow or rapid and may result from natural events or from man's activities. When a river channel is modified locally, the change frequently causes modification of channel characteristics both up and downstream. These changes can be propagated for long distances. Many available river routing models either neglect the dynamic response due to sediment movement or are insensitive to man's activities. In a study recently completed at Colorado State University by Simons et al. (1975b), a one-dimensional water and sediment routing model was used to simulate the dynamic response of river-bed elevation resulting from operations of locks and dams. Their model can be used to study the short and long term impacts of different operational schemes for the locks and dams; the effects of the pools on the behavior and form of the tributary rivers; the impacts of changes in water and sediment inflows on the river morphology and adjacent lands; and the impacts of dredging and dredged material disposal on the hydraulic response and sedimentation in the channel (see Chapter 10).

11.3 OVERVIEW OF SOME CASE STUDIES

General

The following describes several case histories that illustrate the complexity of water and river systems, and the associated development and utilization of mathematical modeling techniques to achieve improved planning, analysis and design.

Variation of a Canal Cross Section with Time

A canal was constructed to carry a discharge of approximately 2,000 cu ft per second. The canal cross section was approximately 80 ft wide and had an average depth of approximately 10 ft. The side slopes were originally constructed at 1½:1 (see Fig. 11-1). The segment of this particular canal was constructed through a sand hill area. The bank material is very erodible. When a canal is constructed with banks consisting of such material, large width to depth ratios result as the channel adjusts. This can be illustrated by looking at wetted perimeter versus discharge

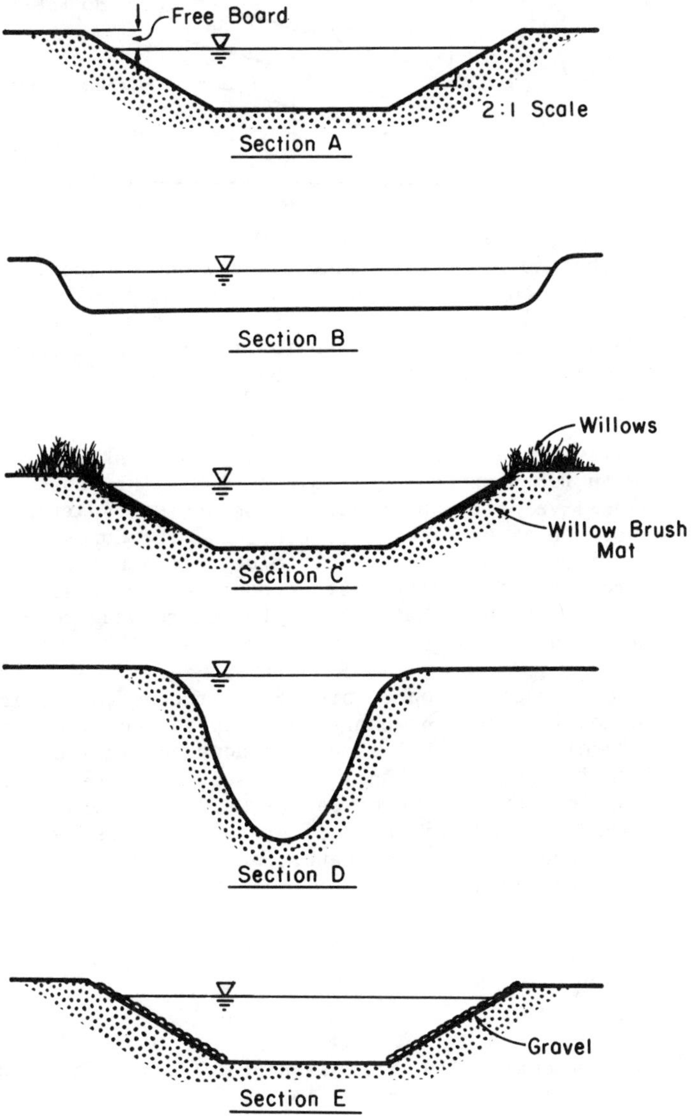

Fig. 11-1. Variation of a Channel Cross Section with Time

A. Original channel constructed in sand hills area

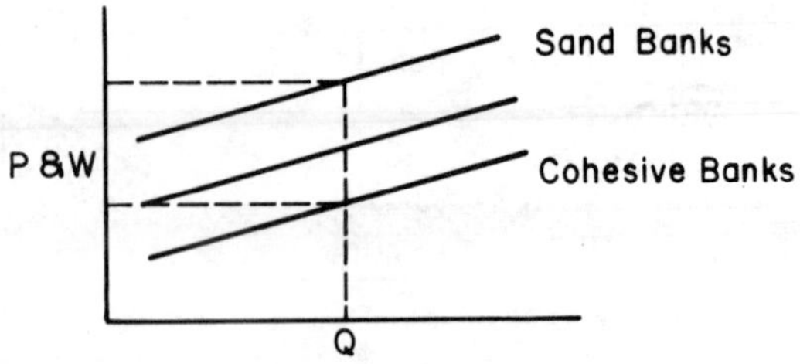

B. With sandy erodible banks the channel widened, and encroached on the right-of-way.

C. Channel was reconstructed to its original dimensions. A willow brush mat was placed on the bank and willows were planted.

D. Willow growth developed rapidly, the banks were reinforced with roots extending into the flow. The roots slowed the local velocity and trapped fine sediment particles that were being carried in suspension. The deposition of fine sediment caused the canal to develop a narrow and deep cross section which restricted the flow causing a backwater upstream that reduced freeboard and caused local flooding.

E. The channel was once more reconstructed to its original dimensions and the banks were rip-rapped with local river gravel. The gravel armor has done an excellent job of stabilizing the channel cross section. The Lane, Simons, Shields, etc., methods can be used to determine the size, gradation and thickness of gravel blankets to stabilize the channel and river banks, and reduce the supply of sediment downstream.

Fig. 11-1 (Continued). Variation of a Channel Cross Section with Time relationships for channels with different types of bank materials ranging from highly erodible sands that will yield wide channels to cohesive materials that tend to resist erosion and may be quite narrow. When flow was introduced into this sand bank channel, rapid erosion of the banks occurred. This caused the channel to widen from its original width of 80 ft. to a width on the order of 150-200 ft. There was a corresponding reduction in depth. The widening of the channel encroached on the right-of-way and introduced numerous other problems. Hence, it was decided to reconstruct the canal cross section to its original dimensions and stabilize the banks into that position, using willow mattresses. At the same time the growth of willows in the banks was encouraged. Willow growth developed rapidly. Roots from the willows extended into the flow. The roots from the willows slowed the local velocity and trapped the fine sediments consisting of silts and clays that were transported in the flow. The result was deposition of fine sediments on the banks. As deposition occurred, the root system continued to extend itself therefrom into the channel. The net result was the development of a very narrow, deep channel that restricted the flow to such an extent that backwater effects caused serious encroachment on the free board and the canal upstream. In some instances, the backwater caused overflow and bank failure. Because of these problems it was decided to reconstruct the canal once more to its original dimensions. This was done and the banks were riprapped with local river gravel to provide protection from erosion of the flowing water. The gravel armor has done an excellent job of stabilizing the canal cross section. Furthermore, the water losses introduced by extensive growths of vegetation along the bank line have been eliminated. This type of problem can be easily handled in terms of designing the size of riprap required. Reference can be made to Simons and Senturk (1977) that outlines the Lane method and other methods, including those recently developed at Colorado State University, to determine the size, gradation and thickness of gravel bank cuts required to stabilize such systems.

The importance of the effects of silts and clays, or the so-called washload, on river and canal geometry cannot be overstressed. Back in the early 1900s when irrigation systems were being developed in the Imperial Valley of California the water diverted from the Colorado River to these systems carried large quantities of washload. The channels were sized according to current engineering knowledge. The concentrations of silts and clays in the flowing water diverted to the canals was such that berming occurred in the canals. The berms consisted of materials that held water well, were fairly fertile and encouraged rapid growth of vegetation and grasses. As in the foregoing case, the channel narrowed dramatically with time as a consequence of the growth of vegetation at the water line, the extension of roots into the water and the trapping of fine sediments. In some instances these channels became almost pipe-like. The berms would grow out over the flowing water partly covering the channel. This change in channel geometry constricted the water cross section and increased the resistance to flow. Hence, it was necessary to regularly restore these channels to their original dimensions in order for them to convey the required quantities of water to the irrigation system.

In subsequent years there has been a rapid development of large storage facilities on the Colorado River. More and more of the sediments that normally flowed downstream are now trapped in upstream reservoirs. As a consequence, the irrigation system alluded to that originally received large quantities of silts and clays with the water diverted to it no longer does so. In fact, the upstream reservoirs trap the majority of the sediments and most of these canal systems now operate with almost a clear water environment. That is, the water diverted from the river system and its reservoirs is largely free of sediment. This change in the quality of the water brought about by upstream storage and control caused the reverse of the situation to occur in these canals. With the removal of silts and clays the water entering the canals had an opportunity to attack the banks and bed, eroding them. Where the materials were sufficiently erodible, the channels widened, depths reduced, and another set of adverse conditions developed. In this case it was possible to reconstruct these canals to desirable dimensions and armor them with properly sized gravel or riprap. This type of protection has very effectively held the channels in the desired location with the desired cross-sectional dimensions. Again, design of stabilization works for such systems can be handled as was cited.

An illustration of the effects of silts and clays on channel geometry is provided by the following example. A river system in the western United States carried large quantities of silts and clays during periods of high flow until the early 1900's. During these high flow events the channel was actively changing form. Other times the channel was relatively stable because of vigorous vegetative growth on bars and banks. In the early 1900's a major earthquake in the area caused a large landslide to form a dam of considerable size across the river valley. This resulted in a lake. The lake is of sufficient size that it has essentially removed the silts and clays that normally flow in the river at high stages. In the absence of silts and clays downstream, during high flow events the banks and bars were attacked and new bars were formed, but these new bars and new bank lines had little silts and clays in them as a consequence of the desilting upstream and they were unable to hold moisture for a sufficiently long period to allow the growth of vegetation. Consequently, in recent decades the bank lines have changed, new bank lines have formed, but without silts and clays in the newly deposited material, no new vegetation has been able to grow. The end result has been the elimination of a significant amount of riparian vegetation along the stream line, greatly reducing the beauty of the stream system and in many ways deteriorating it from the viewpoint of recreational use, fisheries habitat, etc. Once again, the potential for change exists when the washload concentrations flowing in either canals or rivers are altered.

As a final example of the importance of washloads on river and canal morphology, a study was conducted by Colorado State University in Venezuela that extended over a period of years. This study dealt primarily with the geomorphology and the hydraulics of rivers and the relationship of these river systems to the watersheds. It was found that vegetation developed and grew along certain segments of the bank line and floodplain. In other sections or locations it could be identified that vegetation sometimes started growing but died before a root system could be developed due to an inadequate supply of water and nutrients. Conducting extensive soil sampling along the banks and in the floodplains of these rivers showed that in every case where a substantial growth of vegetation was occurring, there was a significant percentage of silts and clays. Hence, where a river system experiences natural cutoffs that subsequently fill with finer sediments, an environment is developed that can support rapid and vigorous growth of vegetation. In those segements of the floodplain that consist largely of gravels and sands, the area drains so rapidly that significant vegetation usually does not develop.

Impact of Construction of a Large Dam on a River System Upstream of a Diversion to an Irrigated Area

A particular river system and its watershed was of such a nature that during the majority of the year the water flowing in the river channel carried significant percentages of fine sediments comprised of silts and clays. An irrigation system was developed that diverted water from this river long before there was construction of dams upstream of the diversion structure (see Fig. 11-2). The main canal diverts water from the river system to the irrigated area at a diversion structure. Thereafter, it flows around the side hill comprised of rather permeable materials including sands and gravels, passes through a city area where there is relatively dense development along both sides of the channel and then continues downstream to serve the distant irrigated area.

Recently, a large storage dam was constructed on a river immediately upstream of the diversion structure that serves the irrigation system. The reservoir is in effect a desilting basin and water released from there is essentially clear. This clear water released from the dam, flows down the short reach of river channel and is then diverted by means of the

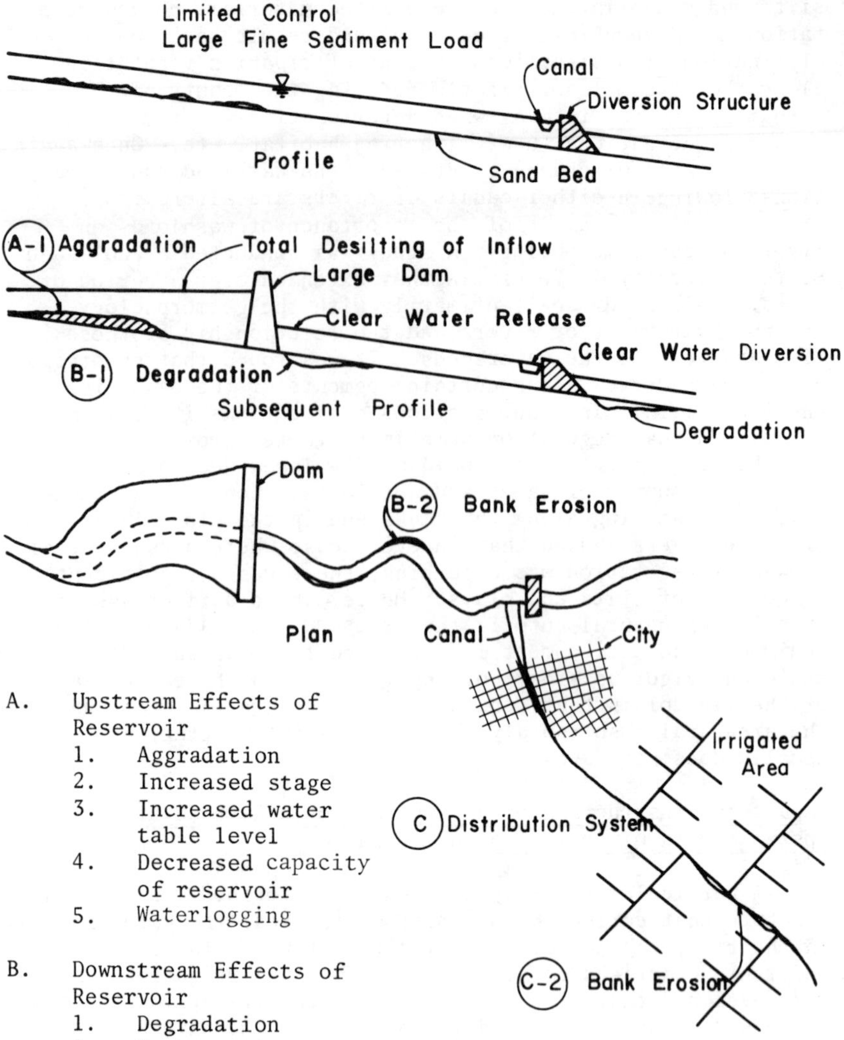

A. Upstream Effects of Reservoir
   1. Aggradation
   2. Increased stage
   3. Increased water table level
   4. Decreased capacity of reservoir
   5. Waterlogging

B. Downstream Effects of Reservoir
   1. Degradation
   2. Bank erosion

C. Effects on Distribution System
   1. Degradation
   2. Bank erosion
   3. Increase seepage losses
   4. Growth of aquatic plants
   5. Reduce efficiency of irrigation system requiring redesign

Fig. 11-2. Impact of Construction of a Large Dam on a River System Upstream of a Diversion to an Irrigated Area diversion dam to the irrigation canal. This release of clear water may cause some degradation immediately downstream of the dam as one would anticipate. Then the essentially clear water enters the canal. There has been some degradation in the main channel and bank erosion that has exposed more permeable bed and bank materials. This has increased seepage losses. There are several major detrimental factors as a consequence of the increased seepage from the canal.

1. A significant decrease in the amount of water that can be delivered to the farm units for irrigation;

2. The seepage from the canal system has a potential to cause a rise in the water table. The increase in the level of the water table may cause extensive damage to adjacent buildings by filling the basement levels with water.

There are, of course, other potentially detrimental factors that can adversely affect the irrigation system. For example, with clear water in the canal system it will be an ideal environment for the growth of aquatic plants. With these plants choking the channels, there will be added water losses and it will be difficult to convey the water required to the irrigation systems. Therefore, some type of periodic treatment will be required to keep the channels free of vegetation. There is another important issue on the farm units. Where land has been irrigated by diversions from head ditches into furrows, the head ditches were spaced so that a fairly efficient application of water could be made with water containing silts and clays. With silt and clay in the water, relatively long runs were possible. When one attempts to use these same lengths of runs with clear water there are excess seepage losses and, in fact, before the water reaches the ends of the furrows, all of it is infiltrated thus building up the groundwater level. This means that the distribution systems on the farm units must be dramatically changed. Length of runs may have to be cut in half. This will require more head ditches and more non-productive space. The numerous ditches may also add to problems of harvesting.

As referred to previously, the clear water tends to scour the banks and degrade the bed of the canal system. If the attack is sufficiently severe, channel stabilization of some nature may be required or, as another possibility, some type of lining may be required to minimize the losses of water and to provide stability to the system.

In summary, it is important to stress that canals and river systems develop in accordance with the forces to which they are subjected. When conditions are altered, a variety of responses are initiated in rivers, canals and watersheds that supply water to the river systems. It is necessary, then, in carrying out these developments to utilize methods of analysis such as mathematical models to investigate the potential for change and to help identify both the favorable and adverse factors that may result from the development and subsequent changes to the system. In addition, it should be stressed that systems are continually being subjected to changes and once one has an operational model for the system one can periodically evaluate long-term response resulting from changes in upstream and downstream developments in the basin. The necessity for overall basin planning cannot be over-emphasized. To carry out a meaningful investigation for basins and sub-basins the mathematical approaches utilizing process models, and management models are essential.

The Response of a River System and Intervening Reaches of Channel to the Construction of Major Reservoirs

In the analysis of river systems it is common to find water resources development projects that involve the constuction and use of major reservoirs. Such a case is outlined in Fig. 11-3. The reservoir, identified as "A", has been constructed on a major river. Reservoir "B", further downstream, is planned for subsequent development. At location "C", there is a city that will be affected by the backwater from reservoir "B." With this general background, consider the responses of the system to the development. To begin with, reservoir "A" is large. It stores on the order of 10,000,000 acre feet of water. However, even though it is large, it only stores a small percent of the flow that is discharged from its watershed annually. This is illustrated by the small inserted hydrograph. This particular hydrograph points out that approximately five times the volume of the reservoir runs down the river system each year. Literally all of the sediment carried into the reservoir is trapped and the quantity of water that is desilted is approximately equal to five times the volume of the reservoir, about 50,000,000 acre feet per year. The sediment load carried by the river is average to large. This means that desilting of such a large volume of water causes a large volume of sediment to be deposited within reservoir "A" each year. This deposition can significantly decrease the storage capacity of the reservoir for hydropower, irrigation, and flood control in a relatively short time. Furthermore, essentially clear water is released downstream of the dam. The channel is comprised of materials ranging from fine sand, silts and clays, to coarse gravel. The fact that there is coarser materials in the bed of the stream will tend to limit the degradation induced downstream of the dam by the release of clear water. Nevertheless, it is essential not only to evaluate the rate at which the loss of live storage will occur in the reservoir, but also to document the rate and magnitude of ultimate degradation so that the structure's safety can be adequately considered.

Next consider reservoir "B." This particular reservoir has approximately the same storage capacity as reservoir "A." The waters released into this particular reservoir will be essentially clear insofar as the water coming through reservoir "A" is concerned. On the other hand, in the vicinity of city "C" and on the upper side of Fig. 11-3 two other river systems

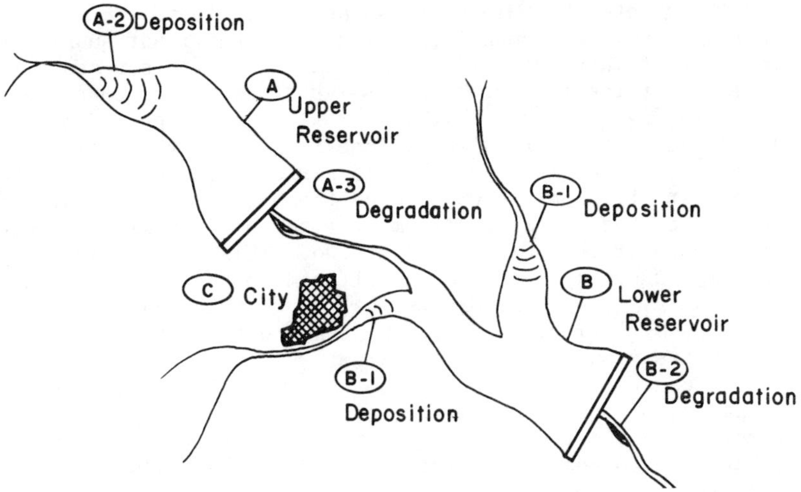

A. Upper Reservoir
   1. Large volume of runoff
   2. Desilting of 5-6 times the storage capacity of the reservoir, hence rapid filling with sediment
   3. Downstream degradation limited by armoring of bed, lateral movement and bank erosion

B. Lower Reservoir
   1. Deposition in upper reaches, including below reservoir "A" after it has filled with sediment
   2. Degradation downstream of the structure

C. Effects on City
   1. Backwater of reservoir
   2. Deposition of sediment
   3. Increase in river stage
   4. Flooding
   5. Increase in ground water level
   6. Flood protection required. Designed to consider changing stage with time

D. Conclusion

   Analysis should consider routing sediment through the second major reservoir by releasing first part of runoff through reservoir and storing last part

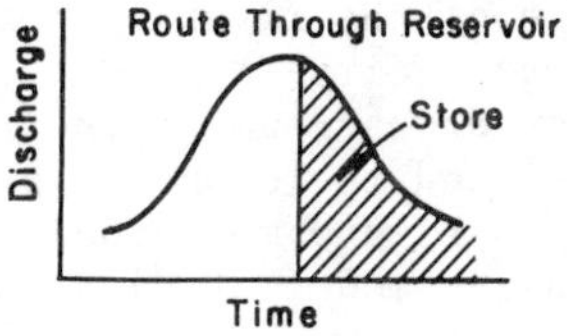

Fig. 11-3. The Response of a River System and Intervening Reaches of Channel to the Construction of Major Reservoirs

that flow into the reservoir can be seen. These are not desilted and, consequently, will carry significant quantities of sediment into the reservoir. Once again it is necessary to consider the rate of loss of live storage as a consequence of inflowing sediments. It should also be recognized that ultimately reservoir "A" will lose its capacity to store sediment and sediments may be discharged from reservoir "A" to reservoir "B." On the other hand there is the possibility that other reservoirs upstream of reservoir "A" will be constructed over the next decade or two. This type of activity can significantly increase the lifetime of reservoirs such as "A" and "B." This is very typical of what has happened to downstream reservoirs due to the construction of upstream reservoirs along the Colorado River.

Because of the importance of sediment in terms of the economics of operation of the reservoirs over long periods of time, an interesting problem with respect to reservoir "B" is to attempt to construct it in such a way that relatively large quantities of sediment can be discharged annually from it instead of being stored within it. This would involve building a large gated structure, put together in such a way that reservoir "B" could essentially run as an open river during those periods of high flow when water could be passed through the system. Then, toward the end of the runoff period, it would be necessary to close the gates in sufficient time to allow filling of the reservoir. With this type of operation it would be necessary to empty the reservoir annually. Otherwise, it would not be possible to pass excess sediments through the reservoir system. Hence, it can be seen that there are many tradeoffs that have to be considered when determining the economics of these systems as well as the response of the rivers and impacts that may result.

Next, refer to location "C" on Fig. 11-3. Here a city is located on a river discharging into an arm of the reservoir. The backwater from the reservoir may require construction of flood protection works for the city. In addition, just determining water levels in the vicinity of the city is not adequate. It is necessary to route the water and sediment down the river system into the reservoir to determine the rates of deposition, the locations of deposition, and how these deposits of sediment may increase river stage, aggravate the flooding situation, increase the groundwater level, and perhaps cause other adverse responses. Once again it is stressed that the response of these systems is complicated and the best possible way of evaluating the response of the river systems involving rates of deposition in the reservoirs, aggradation upstream of the reservoirs, degradation downstream of the reservoirs, the lifetime of the reservoirs, and the economic production of water for irrigation, hydropower, and municipal use, can best be evaluated by comprehensive model studies such as those identified earlier.

In order to carry out these comprehensive model studies involving reservoirs and the responses of river systems and watersheds it is essential that there be an adequate data base. In most instances the data base is probably inadequate. There are various alternatives that can be utilized to help alleviate this problem. One is to involve the use of physical models with the mathematical models where detailed information on the performance of portions of the structure or short river reaches is desired. It is, in fact, entirely feasible to predict how the system will respond using physical models. However, certain field data are necessary in order to validate the physical model. Consequently, whether it is a physical model, a mathematical model, or a combination of both, it is necessary to carefully overview the system, determine those data that exist, and conduct a study to determine how important actual data are to obtain accurate results. With this knowledge, a data base can be formulated. It is recognized that to obtain an adequate data base of all the hydrologic and climatological variables may require years. Hence, in most instances river development proceeds without all of the desired data. This makes the physical process model more applicable to the analysis. With properly constituted physical process models, one can minimize the amount of data required and can arrive at logical conclusions regarding evaluation of river development. Note that it is possible to gather the existing data, synthesize missing data, then formulate process models and analyze the system to gain a better answer in terms of the design, long range planning, and spatial and temporal response of the system, using physical process models, than by any other methodology presently available. Furthermore, it should be stressed that these models, once properly constituted, can be easily coupled with economic, social and related type management models. Furthermore, these models, since they are based upon physical processes, can be used continuously throughout the construction of the project, and after implementation of it, to predict and adjust methodologies related to operation of the system to obtain optimum economic results.

Conceptual Distribution of the Ganges River Water and Associated Problems

Record low flows occurred on the Ganges River during the dry seasons of 1975, 1976 and 1977 causing grave concern that upstream development and diversions could seriously alter an important natural resource and reduce an important part of the downstream water supply which could in turn adversely affect the economic development of the nation. Again, this is a type of problem (see Fig. 11-4) that is amenable to mathematical analysis coupled perhaps with physical models of segments where complexities of the system are such that detailed information is required. The problem of the diversion of the Ganges River flows encompasses many facets, including environmental and ecological considerations, groundwater development, irrigation, water for municipal supplies, saltwater intrusions and the actual fate of spill rivers. In many instances only limited data are available with the exception that relatively long flow records in terms of river stage have been gathered. However, the accuracy of these records needs to be verified. In connection with the analysis of the Ganges River problem sediment related problems should be evaluated. A quick survey of the river system and the use of the river's waters verifies that sediment is a major problem affecting diversions structures, spill channels and navigation. The Ganges River is subjected to a wide range of flow conditions and it reacts accordingly. The river channel has characteristics that are transitional between that of a meandering and braided river. Low flow gradients and low discharges fall in a range where meandering occurs and the characteristics of a meandering river develop. The low flow meandering channel is destroyed and replaced by a transitional condition at intermediate and higher flow stages. The velocity and sediment transport in the channel are large and the system is quite unstable as indicated by rapid channel changes, bed erosion, and formation of new bars and fresh bank erosion.

The bed material consists of a small percent of clay, some silt, and predominantly fine sand with a median diameter ranging from 0.15 to 0.30 mm depending upon sampling techniques. This material is highly susceptible to erosion and transport at intermediate and high flows. The bank material consists of lenses of sands and mixtures of clay, silt and sand. This material is easily eroded at all stages of flow but in particular during intermediate and high stages. Segments of banks often slide into the channel during periods of falling stage.

The river can rapidly shift laterally, cutting out old sand bar deposits, forming new ones, changing the flow alignment and channel geometry. There are few, if any, segments of channel banks that are relatively stable unless protected by riprap or training works. The bed profile of the Ganges River is quite flat and on the order of 0.2 ft per mile. This slope is approximately equal to the slope of the delta toward the bay. The flatness of the delta makes it difficult to design distribution channels that are sufficiently steep to transport the sediment carried into them by the river water. This is mainly a problem that arises during periods of intermediate and high flow. The smaller sediment loads in the river at low flow can usually be handled by the distribution channels provided by the river. This is an important factor seriously affecting the design, construction and operation of supply channels for irrigation and for redistribution of other flows. More specifically, the flow of water is accompanied by the transport of sediment. When the water discharge is small and

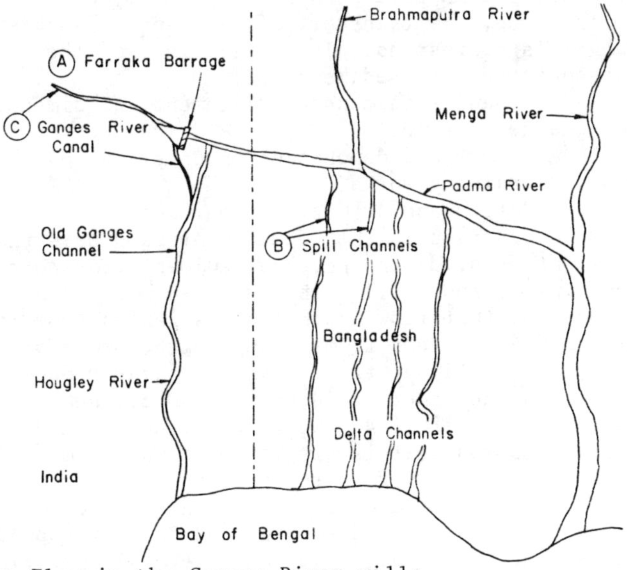

A. Low Flow in the Ganges River will:
   1. Reduce river stage
   2. Reduce depth
   3. Adversely affect:
      a. Navigation
      b. Irrigation diversions
      c. Spill channels

B. Farraka Barrage
   1. Canal diverts water to the Hougly River channel
   2. Slope of canal is about 0.2 ft/mi
   3. Low flow diversion is 15,000-20,000 cfs
   4. High flow diversion is 40,000 cfs
   5. Possibility of sediment deposition in the canal

C. Spill Channels
   1. Flow varies with conditions on the Ganges
   2. Flow affected by alignment of the Ganges
   3. Diversions from the Ganges must consider:
      a. Headworks
      b. Canal design
      c. Barrage, etc.

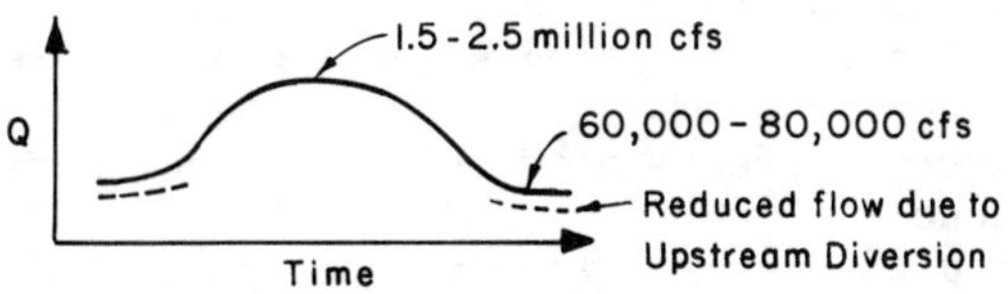

Fig. 11-4. Distribution of the Ganges River Water and Associated Problems

the corresponding transport is also very small, when the water discharge is large, the transport of sediment in such systems as the Ganges is correspondingly large. Some of the major problems associated with sediment are:

1. Major changes in channel alignment and dimensions with varying discharge in time,

2. Local scour and deposition endangering control structures and training works,

3. Aggradation and/or degradation, and

4. Major changes in the location of sand bars that can affect flow patterns, discharge, and sediment transport into irrigation, inlets and spill channels.

Records show that flow in the Ganges ranges from minimums on the order of 60,000 cfs to maximums on the order of 2.6 million cfs. Presently upstream diversions may reduce the minimum flows on the order of 20,000-25,000 cfs and may reduce peak flows on the order of 40,000 cfs. If upstream diversion channels can accommodate this discharge, downstream users must consider the possibility of increased future diversions that will result in further reduction in available water.

The reduction in water discharge and the corresponding changes in sediment discharge can cause several specific responses that are tentatively identified. Such responses must be considered in future water resources planning and development. The most significant possible responses follow:

1. Lower stages during the low flow season and possibly minor reductions in stage at high flow if upstream diversions are made during the high flow season,

2. A significant reduction in capacity to transport sediment during the low flow season. However, this effect would be partly offset by storage of part of the low flow sediment discharge in the pools formed by upstream diversion works,

3. Increased potential for bank erosion during the low flow period due to shallow depths, higher banks, and increased seepage forces where adjacent floodplain lands drain into the river channel,

4. An increased lateral shifting of the main channel of the river during the low flow season.

5. Possible releases of sediment in waves from upstream barrage pools when control gates are opened to pass flood flows. This may affect both river stage and channel stability,

6. Increased flow of sediment from minor tributaries of the main channel due to the reduction in river stage in the main channel and the resulting steeper gradient in the tributary channels,

7. Adverse effects of water level and entrance conditions at spill channels,

8. Reduced depth of flow in both pools and crossings that will adversely affect navigation during the low flow season,

9. Possibly increased intrusion of brackish and saline water in the delta channels and adjacent areas,

10. Require construction of distribution channels to improve availability of water for irrigation, industry, etc.

Other possible adverse effects related to agriculture, fisheries, forests, navigation and ecology could be alluded to here, but because of limited time and space, must be eliminated.

The foregoing paragraphs simply attempt to identify some of the possible responses within this river system and some of the impacts. Once again, this type of problem can be best approached using physical process type mathematical models.

Other Cases

There are many other cases of interest that can be identified and discussed. Figure 11-5 introduces a common problem associated with removal of dams from river systems either by man or by other causes. Discussion of the case is limited to the figure.

11.4 PHYSICAL PROCESSES IMPORTANT TO THE ANALYSIS OF WATERSHEDS AND RIVERS

The overview of case studies given in the last section clearly identifies the complexity of the ecological system in the watersheds and rivers, as well as the significance of man's activities. The source and impact of sediment in fluvial systems was discussed in Chapter 7. In order to demonstrate the complexity and the interaction of the governing processes, a conceptualization of the primary inputs and the important physical processes is given in Fig. 11-6. The two primary inputs to the ecosystem are climatic inputs and man's activities. The climatic inputs include precipitation (rain or snow), solar radiation, air currents and moisture. Man's activities important to watershed and river analysis include weather modification, urban development, watershed management, fire, energy resource acquisition, water resources development, land use zoning, mining activities, water supply and irrigation, recreational development, the development and operation of transportation systems, man-made pollutants, locks and dam construction and operation, river channelization, river training, transbasin water diversions, timber planting and harvesting, and grazing and browsing, etc. The processes which govern the responses from watersheds and rivers are numerous. The key physical processes are atmospheric, soil, and groundwater processes. The atmospheric processes deal with aerosolization, transport phenomena, diffusion, dispersion, convection, and particle dynamics. The vegetative processes include interception, evaporation, transpiration, nitrogen fixation, litter production, ground cover functions, plant succession, and thermal balance modulation. The ground surface processes account for infiltration, depression storage, evaporation, thermal balance, litter accumulation and decay, and raindrop splash soil erosion. The overland flow processes

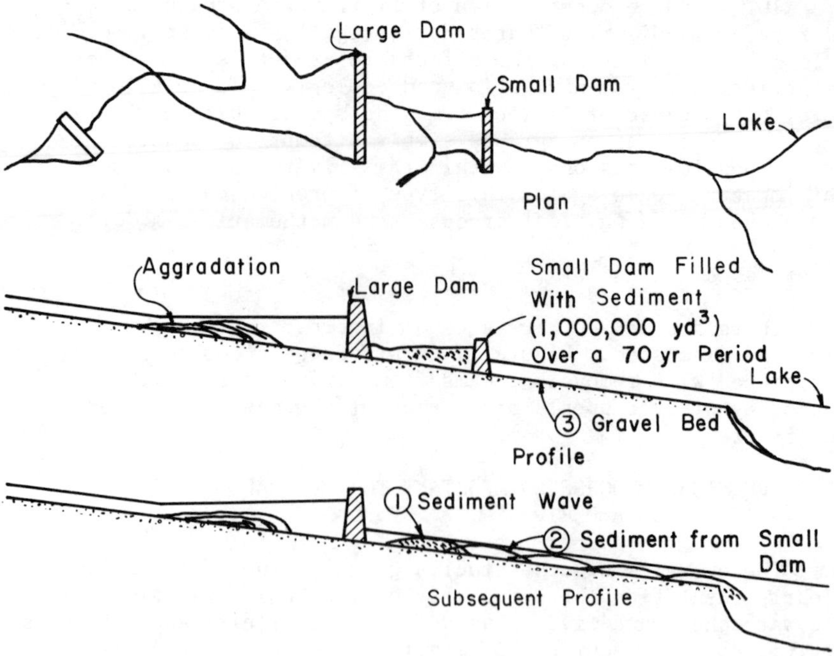

1. Released wave of sediment (1,000,000 yd^3)

2. Sediment wave moved downstream

3. Converted gravel bed stream to sand bed stream

4. The sand wave will form sand bars, dunes, etc. that may increase or decrease resistance to flow, velocity, transport, stage, flooding, etc.

5. It is necessary to eliminate the rate at which the wave of sediment will move downstream and what will be the short and long-term effects?

Fig. 11-5. Response of River to Removal of Small Dam

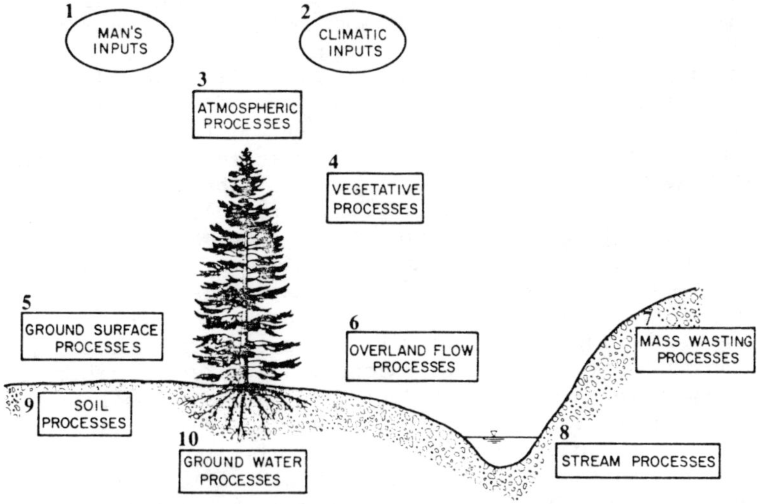

1. Weather modification, urban development, watershed management, fire, energy resource acquisition, water resources development, land use zoning, mining activities, recreational development, transportation systems, man-made pollutants, locks and dams, river channelization, transbasin diversion, timber planting and harvesting, water supply and irrigation, river training structures, grazing and browsing.

2. Precipitation, temperature, radiation, air currents, moisture.

3. Transport phenomena, diffusion, dispersion, convection, particle dynamics, aerosolization.

4. Interception, evaporation, transpiration, nitrogen fixation, litter production, groundcover functions, plant succession, thermal balance modulation.

5. Infiltration, depression storage, evaporation, thermal balance, litter accumulation and decay, raindrop splash soil erosion.

6. Water flow, sediment detachment and transport, pollutant transport, pollutant adsorption, rills development, thermal diffusion.

7. Mud flows, slides, debris avalanches, streambank failure, soil creep, dry gravel, soil piping.

8. Water flow, sediment transport, gully development, degradation, aggradation, geomorphology, bank erosion, deltas and fan formation, pollutant uptake and storage, pollutant transport, dissolved oxygen balance, dissolved solids, routing of forest litter.

9. Weathering, pollutant storage, nitrate reactions, pollutant adsorption, thermal balance, water percolation.

10. Flow movement, pollutant transport, diffusion, dispersion, recharge, heat exchange, nitrogen and phosphorous transport, dissolved oxygen.

Fig. 11-6. Primary Input and Processes Involved in the Analysis of Watersheds and Rivers involve water flow routing (both surface and subsurface), sediment detachment and transport, pollutant adsorption and transport, rill development and thermal diffusion. The mass wasting processes are only important to the watershed with relatively unstable soil. They include mud flows, slides, debris, avalanches, stream bank failure, soil creep, dry ravel, and soil piping. The stream processes are of the greatest concern to both engineers and environmentalists. The river response is often referred to as the indicator for the impact of watershed and river basin development. The stream processes include water routing, sediment transport, gully development, degradation, aggradation, geomorphology, bank erosion, deltas and fan formation, pollutant uptake and storage, pollutant transport, dissolved oxygen balance, dissolved solids, and routing of forest litter and debris. The soil processes involve weathering, pollutant storage, nitrate reactions, pollutant adsorption, thermal balance, and water percolation. The ground water processes include flow movement, pollutant transport, diffusion, dispersion, recharge, heat exchange, nitrogen and phosphorous transport, and dissolved oxygen.

It is important to recognize that different variables and physical processes have different degrees of importance, depending on the type of climatic conditions, soil types, geology, vegetative cover, land use, man's activities, size of drainage area, water and sediment inflows, and the purpose of analysis, on a particular watershed and river ecosystem. In order to determine which of these variables and/or processes must be included in the analysis, it may be necessary to evaluate the importance of each variable and/or process by a sensitivity analysis.

As previously discussed, degradation, aggradation, and movement of pollutants are closely related to water and sediment movement. Therefore, the most important physical processes for the analysis of watersheds and rivers are probably those related to water and sediment routing. Chapters 9 and 10 discussed physical processes, governing equations, and solution techniques for water and sediment routing to a great extent, no in-depth discussion regarding those techniques will be made in this chapter. In spite of the complexity of problems involved in water and sediment routing in watershed and river ecosystems, the governing equations are the same. They are the continuity equation of water, the flow momentum equation, the continuity equation of sediment, the flow energy equation, and some supplementary equations such as flow resistance relations, channel geometry equations, and sediment supply equations.

11.5 CRITERIA OF USEFUL MATHEMATICAL MODELS

For all practical purposes, a useful mathematical model for predicting system response from watershed and river systems should meet the following criteria: (1) The temporal resolution should be both short and long term. Management practices usually have short- and long-term effects on the environment. Thus, the temporal resolution should be oriented on both short- and long-term basis. (2) The spatial resolution should be flexible. Specific management plans and activities are very often limited to small watersheds. Most watersheds subject to legal requirements for environmental quality and water resources management are quite large and often encompass complete river basins or sub-basins. As a consequence, the spatial resolution should accommodate both small watersheds and river systems as a whole. (3) The method should be widely applicable. That is, although the model parameters may be area or regionally specific, the model itself should not be area regionally specific. The only feasible way to develop such a model is to consider the physically based model can provide an appropriate cause effect relationship between management practice and ecosystem response in a general way. (4) The model should be sensitive to the desired management activities. That is, it must be possible to explicitly represent management activities and simulate the system response resulting from these activities. For example, if grazing is a desired management activity for evaluating the watershed response, the model should be formulated to represent the grazing activity through changing model parameters which could be altered according to the intensity of grazing. (5) The uncertainties due to varying climatic and spatial input should be considered. That is, the simulation must consider variations in both mean values and extreme events. This requires a probabilistic approach to describe the stochastic structure of model inputs. (6) The model should be developed within the constraints of available data. Models intended for practical applications should not have requirements for data that are excessively difficult, costly, or time consuming to collect or acquire. If a large quantity of data is required, an effective data storage and retrieval system is necessary. (7) The model should be oriented for use by management personnel. The perspective of the user must be foremost in all model development work. Models intended for use by managers must readily fit into the specific decision-making processes and situations for which they are to be employed, if information resources are to be generated efficiently and used effectively. For usable models to be designed and implemented, developers must be in effective communication with target users throughout the development process. Involving users in model design helps insure that the researcher has full knowledge of the decision-making environment, the actual problems managers face, and the user's perception of the situation being modeled. The model will be more relevant and the user will better trust its validity and capabilies. (8) The model could be easily transferred to several levels of accuracy and resolution. Models operable at several levels of accuracy and resolution will be required in order to provide the full range of tools needed for land and river management. Providing usable and realistic models and guidelines for use by field level managers in many cases will first require developing and testing relatively complex process models. Once the processes involved are thoroughly understood and the sensitive parameters identified, these models can then be regionalized and generalized to provide more simplified models and guidelines for field users. (9) The model computer software system should adopt the modular approach. Adopting the modular approach offers an opportunity to build a coordinated nucleus of standardized system components for use in a wide spectrum of watershed and river systems. This nucleus would be made up of components that provide common interfaces for information transfers between modules that are necessary for storage and retrieval, analysis, and display. Modular systems also have the advantage that individual components can be updated or replaced as needed without disrupting other components of the system. It appears that generalized all-purpose models are expensive to develop, difficult to use and control, and have large data requirements--all of which tend to detract from their field usability. (10) The modeler should provide a useful documentation of computer programming including: i) the system level flow chart showing how modules and files are connected, ii) flowchart of each module, iii) description of each file, iv) narrative descriptions showing how the system is implemented, v) definitions of all variables in each module, and vi) comments on each program or file that show purpose of the code.

11.6 MULTI-STEP DEVELOPMENT FOR ANALYSIS

The development of a model for analysis usually involves the following steps: spatial design, temporal design, model formulation, mathematical solution, model calibration, parameter sensitivity analysis, qualitative examination with physical significance, model simplification, regionalization and generalization, validation, testing and refinement under operational field applications and documentation.

The spatial and temporal designs of a watershed or river system are both necessary to provide a realistic representation of the space-time structure for the simulation models. The watershed geometry, topography, vegetation distribution, and soil distribution would allow either manual or computer-oriented determination of hydrologic response units that constitute the spatial design of a small watershed. A computer program for automatically segmenting a watershed and grouping into desired response units was developed by Simons and Li (1975). The river and its tributaries, their locations, and the location of all pertinent gaging stations, structure, and confluences would allow the spatial design of a large river basin to be developed. The spatial design must also consider the purposes of the study. The temporal design of a system can be made using the historic hydrologic record of the watershed or river basin. This record includes historic maximum, minimum and mean precipitation, temperature, moisture contents and water flows, river stages, precipitation patterns, flow volumes, and the effect of man's activities, such as reservoir construction, on the hydrologic record. The temporal design should be in accordance with the spatial design. Therefore, only those records pertinent to areas included in the spatial design need be analyzed. The temporal design is used to generate the system input for evaluating the future system response.

After the spatial and temporal designs have been made, the governing physical processes of the system response are not difficult to identify. Then, a series of partial or differential equations can be applied to represent the governing processes. This formulates the conceptual system of the model. The model formulation should consider the criteria of a useful mathematical model established earlier. For simulating the dynamic response of water and river systems, perhaps the most important governing equations include the continuity equation for water, the continuity equation for sediment, and the momentum equation (or the energy equation). These three equations can be solved simultaneously or can be approximated by solving the water continuity equation and the momentum equation, first, then refined by using the sediment continuity equation. The second approach is usually applicable because the movement of sediment is much slower than that of water. The numerical solution of these three equations can proceed in two directions. Either an attempt can be made to convert the original system of partial differential equations into an equivalent system of ordinary differential equations by using the method of characteristics (Chang and Richards, 1971), or one can replace the partial derivatives in the original system with quotients of finite-differences by using the explicit method or the implicit method (Amein, 1968, Amein and Fang, 1970, and Chen, 1973). Amein (1968) found that the implicit method is faster and more accurate than other finite difference methods when applied to flood routing in river systems. Chen (1973) modified Amein and Fang's method to derive an unconditionally stable linear implicit method. These implicit methods require downstream boundary conditions and are usually applicable only in the low Froude number flows. The other group of methods is developed by simplifying the momentum equation. The most common simplified method is the kinematic wave approximation. This method assumes that the friction slope is equal to the channel bed slope which neglects the dynamic effects due to pressure, local and convective accelerations. The kinematic wave approximation can be solved either by an analytical or a numerical method. Unfortunately, the analytical solution is often restricted for practical applications because of the formulation of kinematic "shockwave." The "shock" is represented by the intersections of characteristics in the space and time characteristic plane (Kibler and Woolhiser, 1970, and Li et al., 1976a). In the analytical solution, water is conceptually routed along a particular characteristic path which is independent of the other characteristic paths. This additional assumption can cause the formation of "shock." Numerical solutions have not been developed that consider this additional assumption. Besides, numerical solutions usually provide artificial "damping" and "smoothing" effects. Thus, there is usually no kinematic wave "shock" problem when a numerical solution is used. However, if the dynamic effect is significant, the kinematic wave assumption is no longer valid. For watershed runoff problems, the dynamic effects are generally negligible, therefore, the kinematic wave approximation is usually limited to the watershed system. Some of the available numerical methods for kinematic wave approximation are given by Kibler and Woolhiser (1970) and Li et al. (1975). The method developed by Li et al. (1975) is unconditionally stable and is the simplest of those cited. In general, flood routing methods can be categorized into two groups. One group of methods is applicable to watershed flow routing. The other group is suitable for the channel flood routing. Because of the limitations of these methods, until recently there has been no generalized method for prediction of flood movement in both watershed and channel systems. Recently R. K. Simons (1977) developed such a generalized flood routing scheme. For simplicity, known discharge methods are often employed. Examples of these methods are HEC-2 and HEC-6 computer programs developed by the Corps of Engineers (Hydrologic Engineering Center, 1975, 1976).

In the mathematical modeling of system responses, the calibration of model parameters has often relied on an optimization scheme. The dependency on the optimization technique may be reduced if the model is formulated considering physical significance. For the flood routing problem, the parameters describing flow resistance are usually unknown, but their ranges are known from measured data. However, the optimum values of the parameters that reproduce correct model response are usually not known. Hence, model calibration is a necessity. The simplest calibration technique is the trial and error method. Except for models that contain parameters with very narrow searching ranges, the trial and error procedure is inefficient. An efficient procedure is apparently needed for the model calibration. There are many optimization techniques available for the purpose of model calibration. However, the usefulness of a particular optimization technique is very dependent on the formulation of the model being calibrated. Rosenbrock's (1960) optimization technique is usually recommended for finding the optimum set of parameters because it is, by far, the most promising and efficient method for fitting a hydrologic model. Some modification of Rosenbrock's method has been made by Simons and Li (1976) to increase the efficiency of the method.

After development, the model should be examined by a parameter sensitivity analysis. This sensitivity analysis facilitates model parameter calibration, identifies data needs and provides useful information for model simplifcation. Another important evaluation is to examine the model qualitatively using physical significance as a guide. That is, the trends and extremes simulated by the model should be examined to assure that they are meaningful considering physical significance and field experience.

Simplification is a step backward from the more complicated process models that deal with time and space. In general, the more complicated time-space models solve finite difference formulations of the various processes at each time-space point. The simplified model retreats from this approach and averages the processes over both time and space. For most cases, however, the complex procedure provides the better solution. The main disadvantages of the complex models is that they require computer applications and knowledge of the mathematical formulations and assumptions that are often beyond the capability of the average field user. The limitations of regression type or "black box" models and user restrictions imposed by the more complex physical process models made the development of simplified physical process component models necessary. Such simplified models can provide the field user with an easy to use, accurate methodology for estimating system response (Simons, et al., 1977b).

In order to facilitate application of the model, the regionalization of model parameters should be made. This can be achieved by extensive application of the model in various geographical areas. After regionalization has been completed, generalization of the model is possible and it may be applied to various regions. An example of regionalization and generalization is given by the Agricultural Research Service (1975).

The refinement of the model is a continuous process. As more field data becomes available, the model can be improved so that more accurate predictions are possible. The final step involved in model development is the documentation.

11.7 APPLICATIONS

General

Examples of successful applications of mathematical models to various watershed and river systems at the Engieering Research Center, Colorado State University are: (1) a hydrologic and hydraulic study in the Lower Mississippi River (see Holly et al. 1974), (2) a geomorphic study of Pools 24, 25 and 26 in the Upper Mississippi River and Lower Illinois River (see Simons et al., 1975b), (3) a mathematical simulation of water and sediment routing and yields from small watersheds in Beaver Creek Watershed, Arizona (see Simons et al., 1975a), (4) a study of Pool 4 in the Upper Mississippi and Lower Chippewa Rivers (see Simons and Chen, 1976), (5) an application of the Compound Stream Model (see Dass and Simons, 1976), (6) a mathematical simulation of water and sediment yields from roads (see Simons et al., 1976a, 1977a), (7) mathematical models for predicting on-site soil erosion (see Li et al., 1977 and Simons et al., 1977b), (8) a mathematical simulation of infiltration and subsurface flow routing (see Li et al., 1976b), (9) a computer application in mapping potential landslide sites (see Simons et al., 1976b), (10) an investigation of the effects of Chippewa River erosion and silt reduction measures (Simons and Chen, 1977) and (11) design of Bunay-Barry Canal and Mansury-Rayyon-Bagr Canal of Wadizabid Canal System (Simons and Li, 1977).

There are many additional on-going modeling efforts at Colorado State University. The major ones include: (1) Yazoo Basin sedimentation study for Corps of Engineers (principal investigators--D. B. Simons and R. M. Li), (2) Upper Mississippi River two-dimensional mathematical model study for U.S. Fish and Wildlife Service (principal investigators--D. B. Simons and Y. H. Chen), (3) Developing, testing, simplifying, and coupling and demonstrating the various sediment, pollutant routing component models for U.S. Forest Service (principal investigators--D. B. Simons and R. M. Li), (4) Testing and refinement of the road sediment model for U.S. Forest Service (principal investigators--D. B. Simons and R. M. Li), (5) Development of a generalized planning model for evaluation of alternative forest management practice for Environmental Protection Agency (principal investigators--D. B. Simons and R. M. Li), and (6) Flood plain management study for the State of Colorado (principal investigator--D. B. Simons).

Two of the foregoing studies are discussed in some detail. The first one is a simple but practical problem pertaining to canal design. The second is an example of the analysis of a complex river basin.

Bunay-Barry Canal Design

A general layout of the Wadi Canal system, Yemen Arab Republic, is given in Fig. 11-7. Because the canal system intakes water from the river draining a steep watershed in the arid area, the problems encountered in the Wadi Canal system are unique in many aspects compared to usual canal design problems. The unsteady input of water and sediment to the canals, the high sediment load of the Wadi and the heterogeneous distribution of bed and transported material all contribute to the difficulty of the problems involved. The particular canal considered here is the Bunay-Barry canal which offtakes from Diversion Structure No. 1 and has a design discharge of 11.0 m^3/sec.

The source of water for the canals is the flood flows of the Wadi Zabid that fluctuates to a great extent considering both the discharge and the sediment. In the past the water has been diverted by a temporary earth dike constructed across the Wadi. When a large flow occurred in the Wadi, the dike was destroyed and it was usually impossible to rebuild the dike until the next dry season. Concrete diversion weirs are being constructed to command the remodeled canal system.

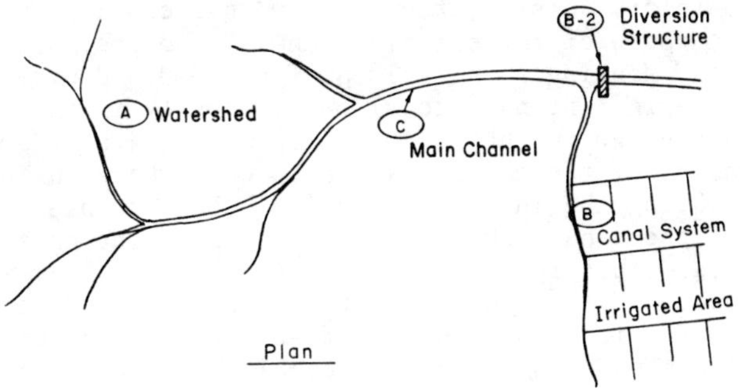

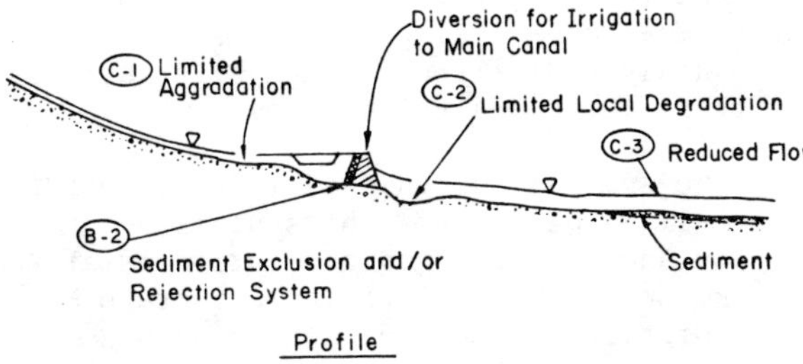

A. Watershed Characteristics
   1. Steep slopes
   2. Small flows
   3. Large velocities
   4. Relatively coarse bed and bank materials
   5. Large sediment transport rates

B. Canal System Characteristics
   1. Designed to operate on a flat gradient to assure stability of banks and beds of the canal system
   2. Sediment delivered to the canal system has been reduced by use of sediment excluders at the diversion dam. Even so, it is difficult to reduce the sediment load to the point where the canal system can carry the sediment without depositing part of it in the canals, laterals, ditches, etc.

C. Effect on Main Channel
   1. Limited aggradation upstream of diversion dam
   2. Limited local degradation downstream of diversion dam
   3. Downstream flow reduced due to diversion for irrigation. This reach will receive more sediment than it normally would from upstream, after the small reservoir upstream of the diversion dam is filled with sediment. Hence, significant aggradation may occur over time downstream of the diversion structure.

<u>Conclusion</u>

In this type of system it is virtually impossible to design a system that will not experience aggradation and require maintenance. However, with a detailed analysis of the systems the problems can be minimized.

Fig. 11-7. Watershed, River and Irrigation Systems

In the initial design, the longitudinal canal slopes are flatter than those of the existing canals. To design the cross section, Manning's n values of 0.033-0.030 were used. The initial design intended to allow a velocity in the canal at design discharge on the order of 1.5 m/sec (4.9 ft/sec). Several reaches of the canals have been constructed using this criteria. During the initial operation, considerable scour of the banks of the canal and some degradation of the canal bed have occurred. This indicates that the existing canal is fairly unstable. It is suspected that the correct Manning roughness coefficient is much less than the value assumed for the initial design. The remodeled canal should have stable banks that require decreasing the canal slope. However, as the bed slope is decreased, it is anticipated that the ability of the canal to carry the sediment load introduced from the Wadi would be impaired. A simple mathematical model has been developed to analyze different design alternatives. The procedures and steps in the analysis follow.

Spatial Design: The spatial design is straightforward. The location map (Fig. 11-8) designates the spatial points of interest. The flow and sediment gaging station is at Maath which is about 1 km upstream of Diversion Structure No. 1. The Roda-Gerbeh Canal offtakes from Diversion Structure No. 1 on the left and Bunay-Barry Canal, the canal being considered in this example, offtakes from Diversion Structure No. 1 on the right.

Flow Duration Analysis (Temporal Design): As mentioned previously the flood waters of the Wadi Zabid fluctuate to a great extent (see Fig. 11-9 for example). The canal considered here is classified in the Group 1 Canals that only intake water during the dry season and the early flood season (November 1 to July 15). This further amplifies the fluctuations of water supply to canals. It is known that except for very small canals, the canals will not operate at their design capacities except for very short intermittent periods. The unsteady water supply to the canals introduces a new type of canal design problem. In order to account for the random nature of the water supply it is necessary to conduct a flow duration analysis.

The operation rule governing water intake influences the water supply to each canal. During the early flood season when Group I Canals are in operation it is assumed that the following rules apply:

1. A base flow of up to 5 m^3/sec will be diverted into the Roda-Gerbeh Canal at Structure No. 1.

2. A radio network in the catchment area will give 8-10 hours advance notice of approaching flood crests and the gates on the Bunay-Barry Canal can be opened.

3. As the flood wave starts to recede the gate openings on Bunay-Barry increase in an effort to maintain this canal at its design discharge as long as possible. Excess flow would be passed on to Structure No. 2 which serves other canals.

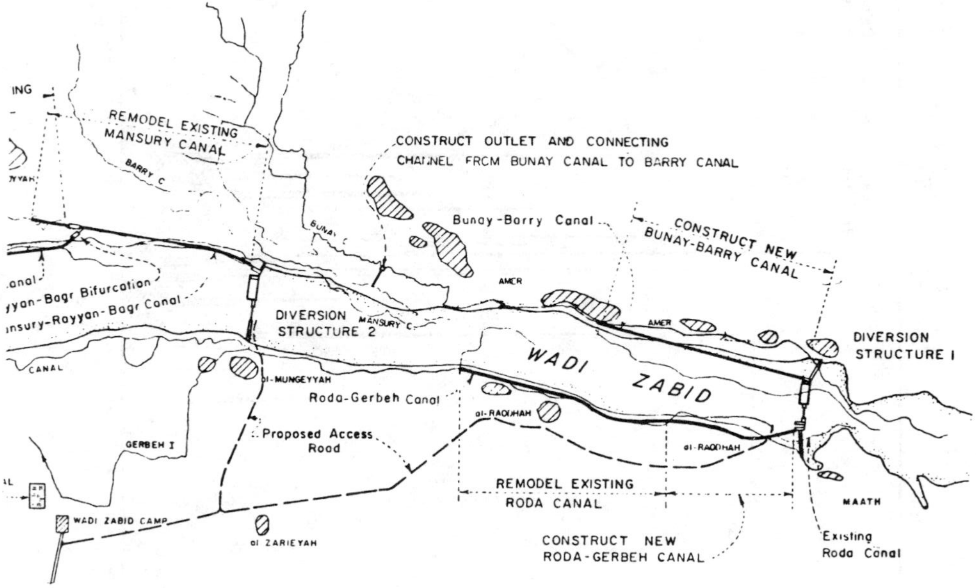

Fig. 11-8. Location Map of Wadi Canal System

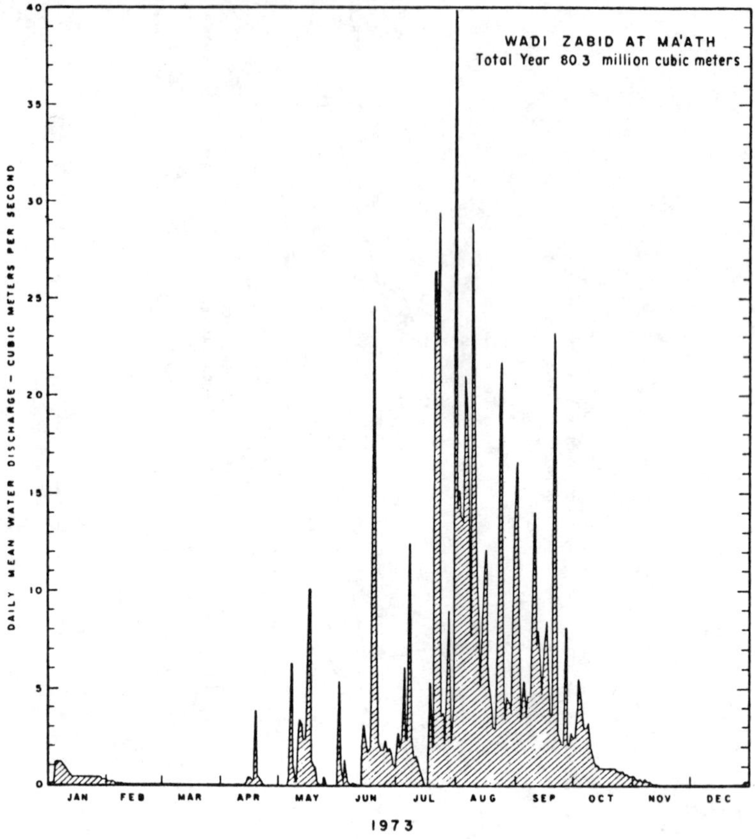

Fig. 11-9. Example of Annual Hydrograph of the Wadi

4. Later in the flood season the Group II and Group III Canals receive their supplies. At this time the headgates on the Bunay-Barry and Mansury-Rayyan-Bagr Canals are closed and the flows are passed through Structures No. 1 and 2 via the sluiceways and over the concrete weirs.

Since water would only be diverted to Bunay-Barry Canal when the discharge at the Wadi is more than 5 m^3/sec during the dry and early flood season, the flow duration of the Wadi at Maath is analyzed considering the above constraints (see Fig. 11-10). The total number of days that water was diverted to Bunay-Barry Canal totaled 141 days for the period from 1969 to 1976. This gives an average of 20.1 days of flow per irrigation season. The flow frequency analysis of water and sediment supply to Bunay-Barry Canal from the Wadi is presented in Table 11-1. This table considers the operation rule, canal design capacity, and the flow duration analysis.

Table 11-1. Flow Frequency Analysis for Bunay-Barry Canal

Discharge (m^3/sec) Range	Average	Concentration (ppm)	Probability
0-1	0.5	1,890	0.220
1-2	1.5	2,590	0.213
2-3	2.5	3,220	0.163
3-4	3.5	3,920	0.113
4-5	4.5	4,500	0.078
5-7	6.0	5,600	0.064
7-10	8.5	7,350	0.050
11**	11.0	9,450	0.035
11	11.0	13,300	0.043
11	11.0	19,600	0.014
11	11.0	25,200	0.007

*0.7 of the concentration of the Wadi at Maath
**Design discharge

It is assumed that the flow of water into the Canal would carry 70 percent of the suspended sediment concentration of the Wadi. The measured suspended sediment concentrations of the Wadi at Maath for different discharges is given in Fig. 11-11. A line fitted by eye was used to determine the suspended sediment concentrations presented in Table 11-1.

Canal Design (Model Formulation, Mathematical Solution, Model Calibration, and Application of Model: The canal should be designed either to transport the sediment through it or to have both stable banks and bed for the design discharge. If the primary objective is to transport the sediment inflow, the canal design bed slope must be fairly steep since the banks would erode unless stabilized. Conversely, if the primary objective of the design is to have both stable banks and bed, the design

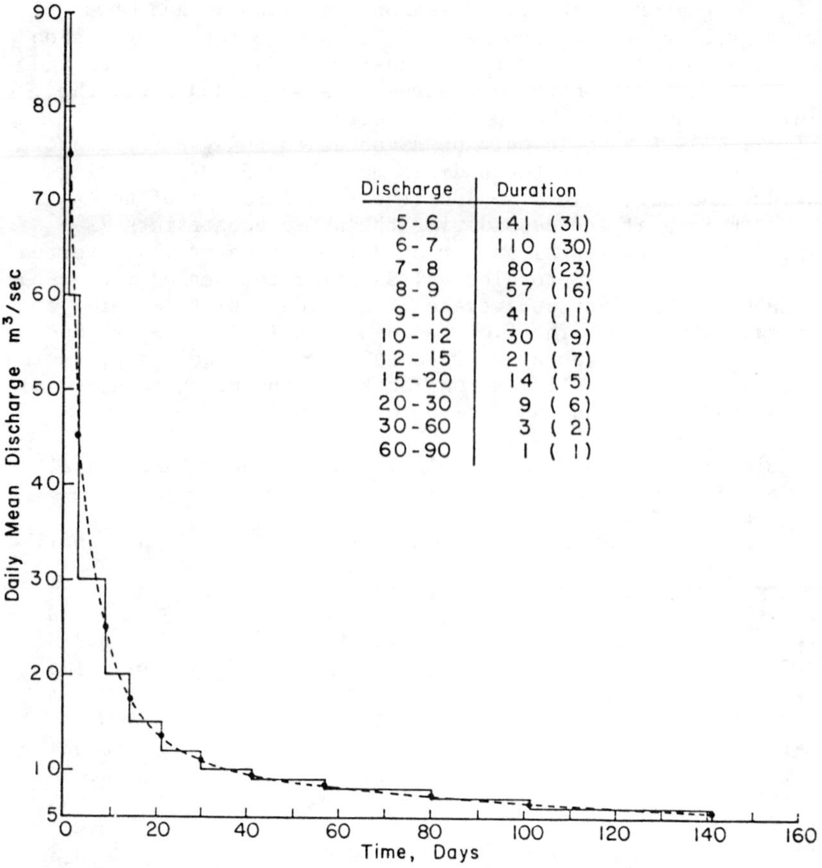

Fig. 11-10. Flow Duration Analysis of the Wadi at Maath for Potential Water Supply to Bunay-Barry and Mansury-Rayyan-Bagr Canals

bed slope must be flatter. This would greatly reduce the capability of the canal to transport any measurable bed material load. This design alternative would be attractive if the sediment concentration in the water supplied to the canal was relatively small.

As mentioned earlier, the water and sediment inflows to the canal fluctuate greatly with time. Conventional canal design methods using the design discharge only is not valid, especially if the primary objective is to transport the sediment load introduced at the headworks. This section describes the canal design considering the different alternatives.

The design discharge is 11 m^3/sec, side slope of the trapezoidal section is 1.5:1, the bottom width is 5.3 m and Manning's n is 0.02.

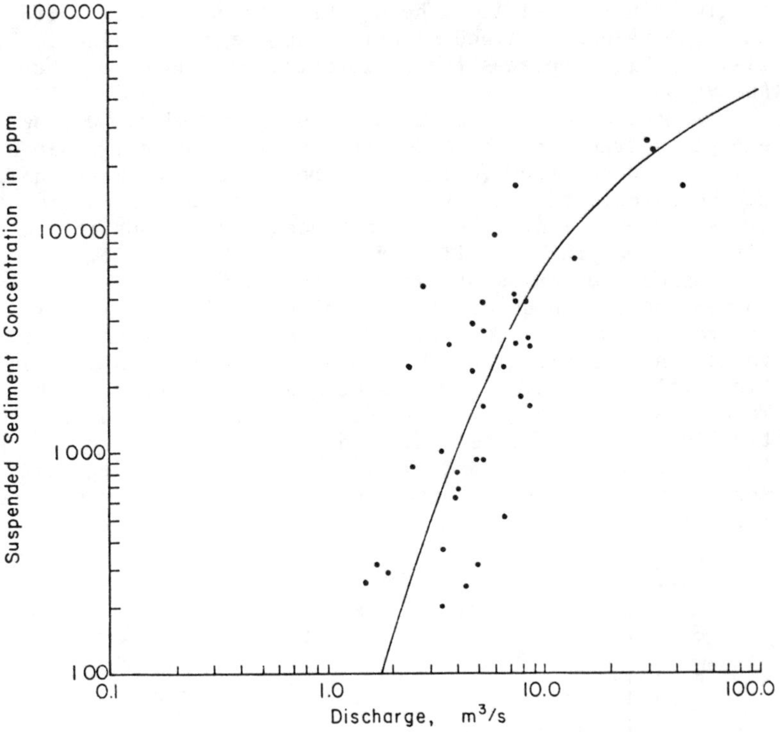

Fig. 11-11. Measured Suspended Sediment Concentration of the Wadi at Maath

The sediment transport rates are very important for estimating the erosional and/or depositional potential of canals. The measured suspended sediment concentration of the Wadi at Maath is shown in Fig. 11-4. Assuming that the bed load is only 15 percent of the suspended load (according to the study by Sabol, 1977) the total sediment concentration can be estimated.

An equation for estimating total bed material load was derived by a combination of the empirical formulas utilizing the bed load function proposed by TESCO-VIZITORV-VITUKI (1971) and the Einstein suspended load procedure (Einstein, 1950).

Due to the small discharges involved in the measurements of suspended sediment concentrations, it is assumed that the average diameter of transported bed load particles is 0.25 mm (average of fine and median sands). With an assumed value of the von Karman constant (for clear water, 0.4), the total sediment load can be determined for various flow conditions. For best results, it was determined that the von Karman constant should be 0.53 for this system. A comparison of the calculated and measured results considering total sediment concentration is given in Fig. 11-12. The measured suspended concentrations were the values obtained by utilizing the fitted line in Fig. 11-11. This completes the calibration procedure for the analysis.

If the primary concern is to design a bed slope steep enough to transport the sediment inflows, the design slopes for different discharges (with the corresponding sediment inflows) can be determined using Manning's equation and the sediment transport equation. The design slopes for various inflow conditions shown in Table 11-1 are given in Fig. 11-13.

Since the inflow of water to the canal is highly variable, conventional methods considering only the design discharge are not valid. Hence, a new method of design that accounts for the variation of inflows and the long-term degradation and aggradation cycle is developed. This method is derived by using Manning's equation, the sediment transport equation, the continuity equation for sediment and the flow frequency analysis. The design bed slope that would minimize the net degradation and aggradation of the channel bed is determined by solving the following minimization problem

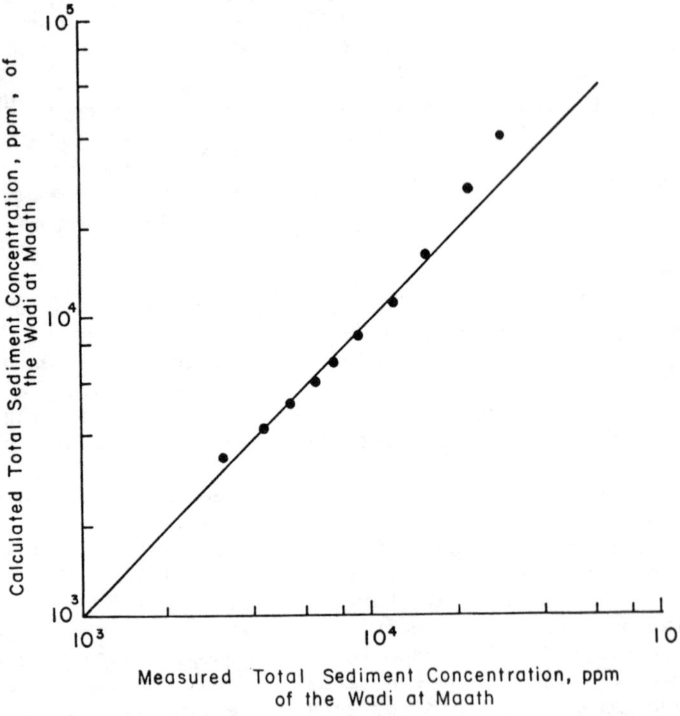

Fig. 11-12. Comparison of Calculated and Measured Sediment Concentration of the Wadi at Maath

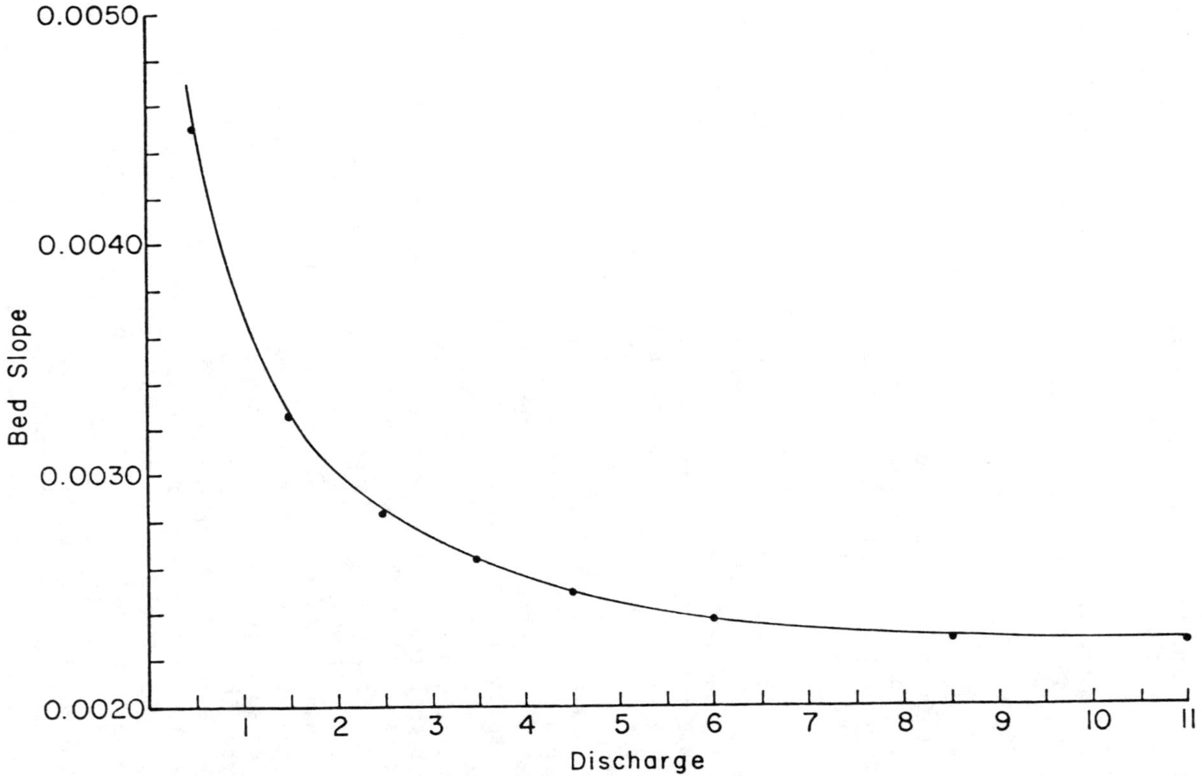

Fig. 11-13. Design Slopes Considering Sediment Transport for Various Inflow Conditions of Bunay-Barry Canal

$$\underset{S}{\text{Min}} \sum P_i \, (C_i - I_i)$$

where P_i is the probability of the ith flow range, C_i is the canal sediment transport capacity during the ith flow condition, and I_i is the sediment inflow rate during the ith flow condition.

A computer program has been developed to solve the above problem. The optimum bed slope for the Bunay-Barry Canal which minimizes the net degradation and aggradation of the canal bed is 2.59 meters per 1000 meters. The depth of flow and mean velocity at the design discharge are respectively 0.81 m and 1.96 m/sec.

It is important to note that the above design would minimize the net degradation and aggradation on a seasonal basis but would not eliminate all canal maintenance. It is expected that some aggradation of coarser particles in the head reach and at the intake would tend to plug the system and increase the bed slope locally near the headworks if no maintenance dredging occurred. The prediction of the local response of the canal in the head reach can be made by conducting a more detailed analysis considering routing sediment through the canals by size fraction. The above design also indicates high velocity will result for this design condition. Undoubtedly, the banks would be fairly unstable if no stabilization measures were used. The banks of both canals would be stable if they were protected by an armor layer composed of particles having a size of 25 mm (median diameter of the Wadi bed) or larger. This size of bank protection material was determined using the method recommended by E. W. Lane and the U.S. Bureau of Reclamation (see Simons and Senturk, 1977).

If the primary interest is to design a bed slope flat enough that the banks would be stable without treatment, it would have a different design slope and associate problems.

The bank stability is dependent on the size of armoring material that would develop as a natural consequence of erosion of the banks. Judging from the parent soil condition (mostly clay and silt) and the transported material in the Wadi, it is very conservative to assume an armor size of at least 4.0 mm. However, in reality, an armor size of 5 mm could develop on the banks considering characteristics of the transported material. Applying the same method recommended by E. W. Lane and the U.S. Bureau of Reclamation (Simons and Senturk, 1977), and Manning's equation, the appropriate bed slopes that would assure stable banks for various discharges and armor sizes are determined (see Fig. 11-14).

Considering canal bank stability, it is desirable to use the design discharge because it is the largest and most erosive discharge that would flow in the system. The hydraulic design conditions, considering that banks will be stable if armored

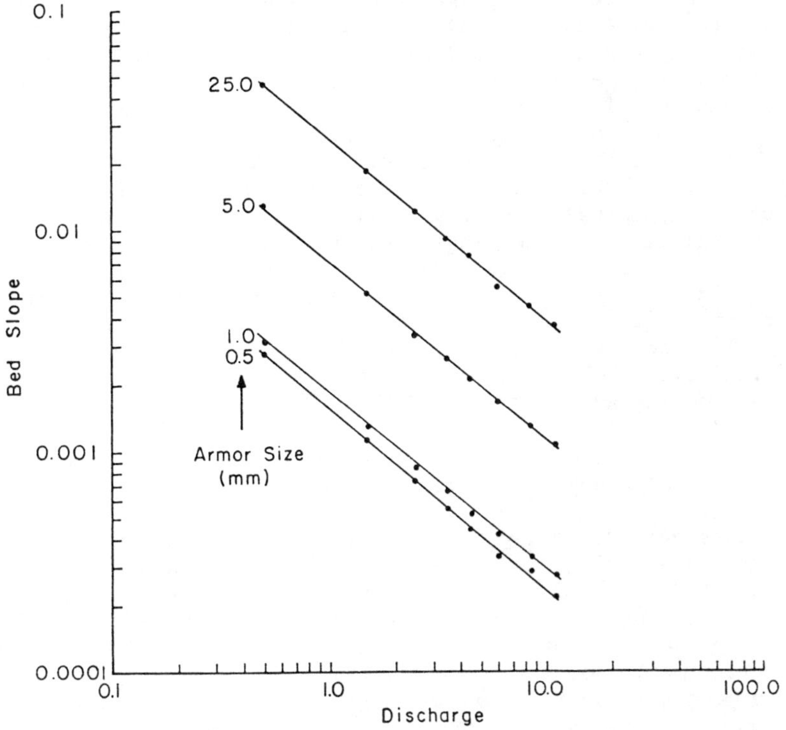

Fig. 11-14. Design Slopes Considering Bed Stability for Various Discharges of Bunay-Barry Canal

with 5.0 mm material are: the design slope is 1 m per 1000 m, the flow depth is 1.02 m, and the mean velocity is 1.46 m/sec.

As mentioned earlier, the flat bed slopes required to assure basic stability of the canal would induce a large amount of deposition because of virtually limited sediment transport capacity. It is determined that the above design would induce about 41,300 metric tons of deposition of sediment in Bunay-Barry Canal (assuming 20.1 days of flow in the canal per irrigation season). This analysis does not account for the adjustment of bed gradient during the depositional process. The actual deposition volume would be somewhat less than the above estimated values. Nevertheless, the estimated deposition volume is certainly large enough to indicate the difficulty of maintaining the canal. Such a design alternative is feasible only if adequate sediment excluders, ejectors, and/or sediment traps are utilized to remove excess sediment at the diversion works or in the head reach of the canal. If a sediment trap is employed, the design volume should be approximately 26,000 m^3.

Considering the optimum canal design it is advisable to select an alternative that would minimize changes in the present design and minimize future maintenance costs. Because the bank material is predominantly formed of silt and clay, the banks would be highly unstable if a steep slope without bank stabilization was used. Riprapping the banks for the entire channel system is probably not feasible. Thus, the main concern is the stability of the canal. It is recommended that the final design should be in accordance with the criteria considering bank stability. Accordingly, the recommended longitudinal slope for the Bunay-Barry Canal is 1.0 m per 1000 m. However, a sediment trap to remove excessive sediment should be constructed with the capacity indicated in the head reach of the canal. Because the assumed armor size of 5 mm is not a certainty, it is strongly suggested that riprapping the head reach, using an armor size of about 25 mm or larger, should be considered. This size of material can be easily obtained from the Wadi bed since the canal head reach parallels the Wadi for some distance downstream of the diversion structure.

Discussion: Utilizing the above example, the usefulness of the model considering the criteria established in Section 11.5 is evaluated. (1) The above design considered both short and long term canal responses. The bank stability and the need to riprap the banks of the head reach is a short term measure, and the estimation of size of armor that will develop in the future, the long-term degradation and aggradation, and future maintenance requirements are for long term considerations. (2) The spatial design presented covers only the canal. For better results one should conduct a comprehensive system analysis that routes the sediment considering particle size through the Wadi and the canal system. (3) The method presented here can be applied in any area where the design parameters are properly measured or calibrated. (4) The above method of analysis is sensitive to management activities. The consideration of both sediment transport and bank stability in the analysis reflects this sensitivity. (5) The flow duration analysis identifies the randomness of the water and sediment input and is an indication of the uncertainties of input. (6) The method of analysis was developed within the constraints of available data. (7) The method is sufficiently simple, so that it can be easily applied. (8) The current model could be extended to complicated systems that involve the routing of water and sediment for the whole system. Conversely, the design alternatives can be easily modified by utilizing the information established in this analysis such as in Figs. 11-13 and 11-14. (9) The developed computer program is made up of many subroutines, it has modules for considering optimization, canal hydraulics, sediment transport, and bank stability. (10) The documentation is simple if it is necessary.

Yazoo Basin Sedimentation Study

The Yazoo River Basin occupies nearly one quarter of the state of Mississippi (see Fig. 11-15). The Engineering Research Center, Colorado State University is currently conducting a comprehensive study for the Corps of Engineers to evaluate various development plans. The plans for improvement include construction of a cross-country channel in Tallahatchie, Lefore and Humphreys Counties to provide additional channel capacity to pass flood waters from the upper basin. The study involves a systems analysis of the main channel and its tributaries in such a way that water and sediment is routed from the watershed and through the main channel to determine the effectiveness of the system considering flood control, navigation and the location of aggradation and degradation in the main channel and its tributaries and methods of minimizing these and related problems. The analysis provides a means of evaluating various design alternatives to determine the best plan for development. The first alternative considered involves dredging sediments from the main channel. From the model one could determine the effects of channel enlargement on the flow line, on reduced velocities, on sediment deposition rates, etc. Since the first alternative may not be the most satisfactory one and since developing the channel as indicated in the first alternative may lower the flow line and cause additional sediment flow to the main channel from the tributary system, the next alternative involves an analysis of the system with control structures located on the tributaries just upstream of the point where major tributaries enter the Yazoo main stem. These structures would prevent head cutting and the discharge of excess sediments into the main channel. Another alternative is to utilize various techniques to further minimize sediments contributed from the watershed to the main channel. For example, control structures such as check dams could be designed, spaced and located to further reduce the sediment inflow from the tributaries to the main channel. The analysis could be carried out to determine the optimum design considering control structures and dredging to minimize maintenance problems in the main channel after it has been modified to meet the requirements dictated by flood control needs, and navigation requirement.

For this analysis, every step identified in Section 11.6 is important. A schematic diagram showing the development of the model is given in Fig. 11-16.

Overall System Design: As shown in Fig. 11-15, the Yazoo River Basin is composed of a main stem and numerous tributaries. The model is designed to simulate the river system as a whole. In the Yazoo River Basin, tributaries to the main stem can be divided into controlled, uncontrolled, and point source type streams. Controlled tributaries are those such as the Yalobusha which has large storage reservoirs located on them. These

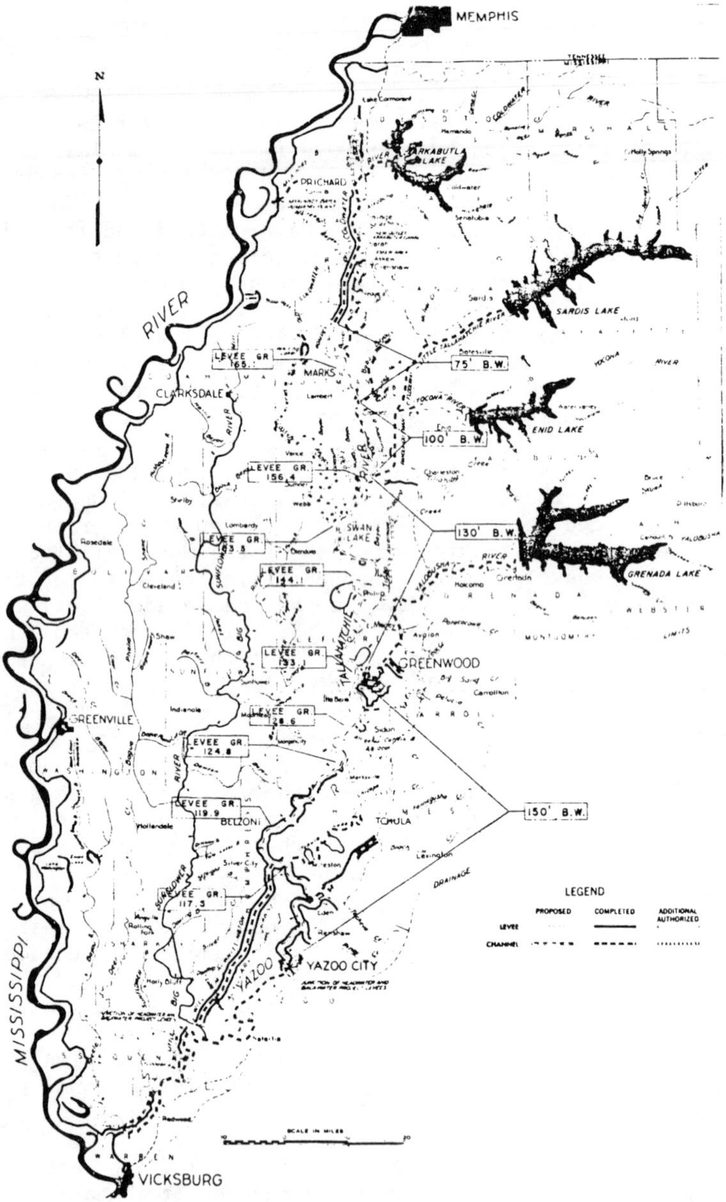

Fig. 11-15. Location Map of Yazoo Basin, Mississippi

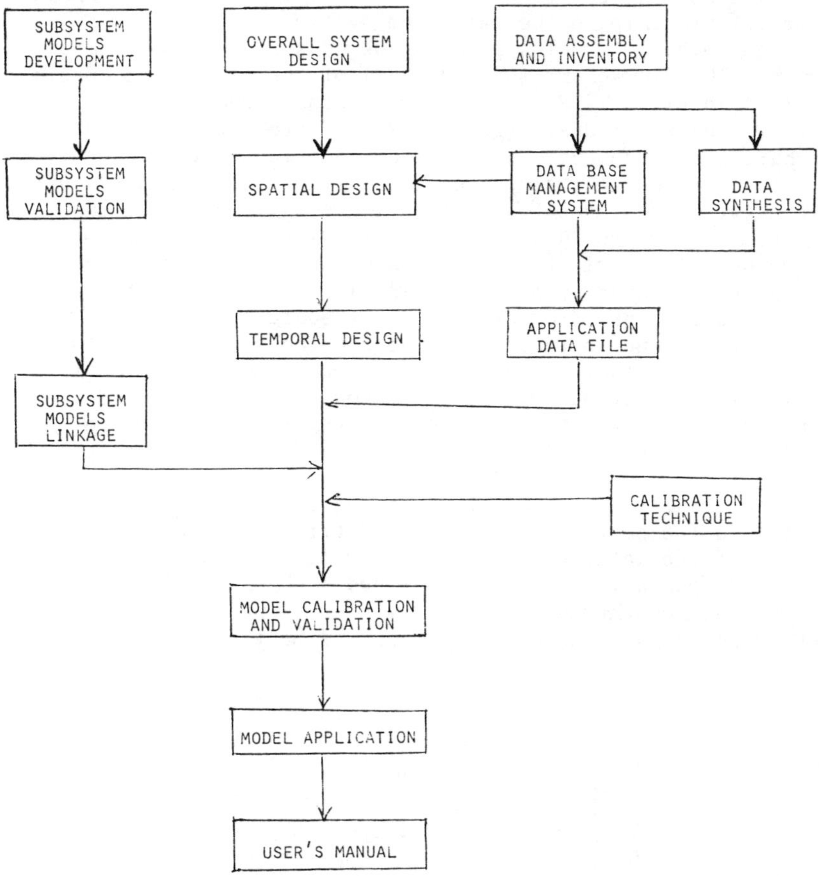

Fig. 11-16. Step of Development for Yazoo Basin Sedimentation Study

include: Arkabutla, Sardis, Enid, and Grenada Reservoirs. These controlled tributaries usually play signficant roles in the system. Uncontrolled tributaries are generally smaller than the controlled type and do not have large storage reservoirs on them. Nevertheless, they are important in the analysis. An example of this type is Big Sand Creek. The third type of tributary, point source, is generally smaller than the other two and can be considered as a single point input. In these tributaries, the potential effect of sediment inflow on the main stem response is minimal. However, in order to conserve the flow continuity, they should be considered in the analysis as point source inputs. An example of this type is Piney Creek. The classification of tributaries was made by examining system response (see Spatial Design). These three types of tributaries are shown schematically in Fig. 11-17. This figure also shows their relationship to the main stem system.

The computer model system can be constructed in a similar manner. The flow chart for the model of the Yazoo River Basin is shown in Fig. 11-18. The desired output from the main stem model is the water discharge, the water level, the sediment discharge, and the aggradation or degradation of the channel. These outputs will be simulated for different alternatives of river system development and operating conditions. The input required can be provided by the output from the controlled and uncontrolled tributary models and the point source models. The input for these types of models will consist of water and sediment discharges from primary input sources and lateral inflows from land surfaces. Reliable methods are needed to generate these required input data if measured data are not available. The tributary and point source models require input from several components. Key components are the watershed model for the uncontrolled tributaries and point source inputs, and reservoir operation rules for the controlled input. The watershed model will simulate water and sediment supply being input into the system by uncontrolled tributaries and point sources if the watershed boundary is well defined. If the watershed boundary is not well defined such as Bobo Bayou, a method based on channel geometry is used. The reservoir operation component will handle water and sediment supply aided by

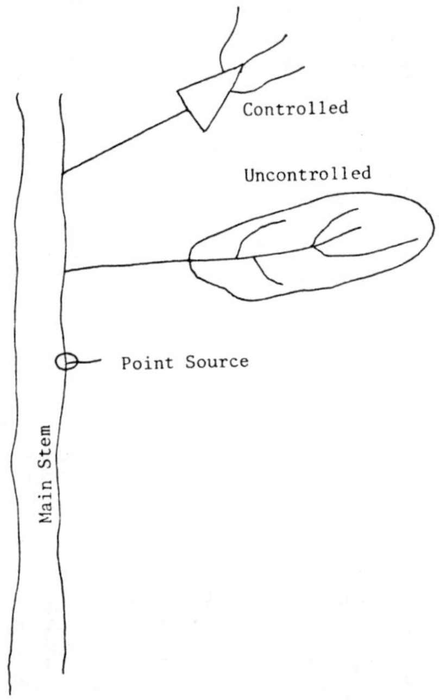

Fig. 11-17. Schematic River System

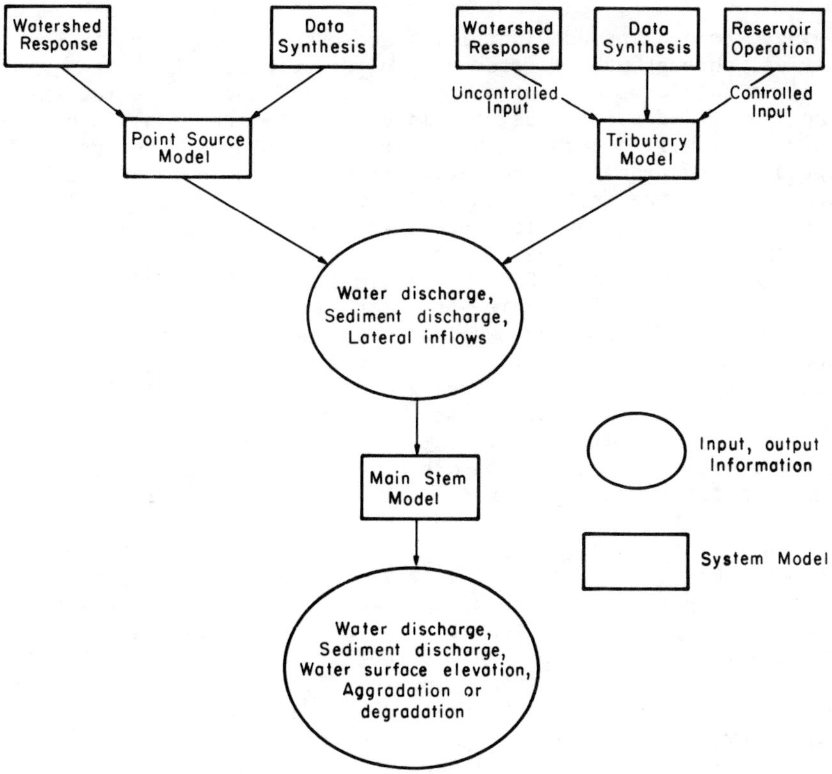

Fig. 11-18. Linkage of Yazoo River Model

operation of the reservoirs on controlled tributaries. The third component, data synthesis, is used to synthesize other necessary data from tributaries, where little or no records exist, for example, sediment discharge, cross sections, stage, and water discharge. This method is also used to fill in missing data in the more complete records. Also, the data synthesis component is used to generate various temporal and spatial data that may be required in the main river system. These situations will then be evaluated to determine the corresponding system response. Similarly, the point source models are composed of data synthesis and watershed response component models.

The structuring of the Yazoo River Basin model as shown in Fig. 11-18, enables a sequential, logical flow from one compo- to the next. In this manner the model system is easier to understand and use. After the system can realistically simulate the river response, management studies will be conducted in order to determine optimal means of controlling changes in bed elevation along the main stem. The major means of control now being examined include inhibiting sediment inflows from tributaries, through use of grade control structures, check dams, as well as bank erosion control measures. In order to coordinate the use of different subsystem models, a command model is needed. This model will determine the order of computation, select options, and transfer information from one module to another according to the instructions and the logic of the physical system. The framework for the command model has been developed. Two application packages have been developed for different levels of resolution. One is for simulating the flood and sediment movement in a three-hour time resolution which involves the linkage of the watershed model, the unsteady flow tributary model, and the unsteady flow main stem model (see Fig. 11-19). This package is particularly useful in predicting the flood movement in the Yazoo River System. Another package is for simulating the sediment movement only and assumes that the water hydrographs are known (known discharge solution, see Fig. 11-20). This simplified package involves the use of water balance estimations and the known discharge uncoupled model for sediment routing. This package can be used to evaluate the system response using a larger time increment such as one or two days or even longer. Each linked application package consists of the following six main elements: data input, initialization, design alternative selection, hydrograph generation, simulation which determines the system response, and an output or summary part. For validation runs, the hydrograph generation is replaced by the measured and synthesized flow records.

The subsequent problems encountered in the conceptual system design are the determination of locations of main tributaries and point sources and the generating of hydrographs for the simulation of future response.

Spatial design: Spatial and temporal designs of the Yazoo River system are necessary to provide a realistic representation of the space-time structure for simulation models. The river and its tributaries, their locations and the location of all pertinent gaging stations, structures, and confluences allow the spatial design to be developed. The spatial design is based on the potential contribution to the bed elevation changes. By applying the sediment continuity equation, a sediment transport equation (Colby's method is used in this study), and a set of typical flow conditions, the bed elevation changes along the main stem between each two neighboring confluences were determined. Also, the percentage changes contributed by the tributaries were determined considering the ratio of sediment transport rates between the main stem and its tributaries. By summing all of the changes in the sediment storage volume (product of bed elevation change, wetted perimeter, and the space increment between two neighboring confluences), and relating this total change to the individual change, the percentage contributed by each tributary to the change in bed elevation in the main stem is determined. After ranking the potential contributions according to the computed percentages

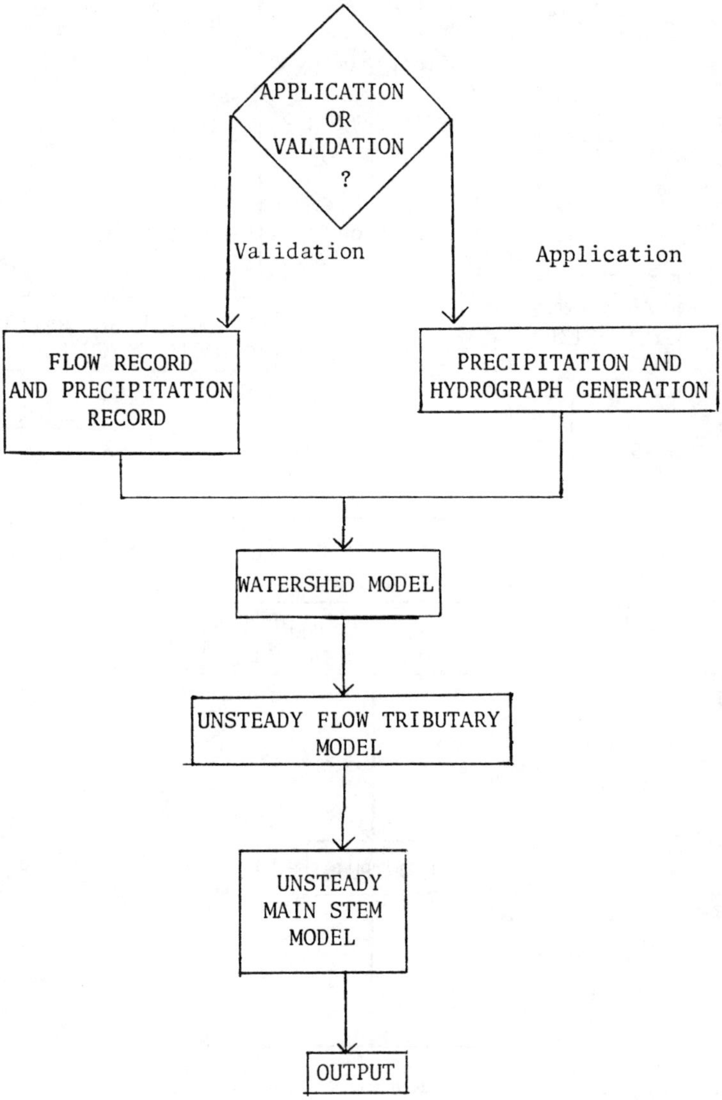

Fig. 11-19. Unsteady Flow Simulation

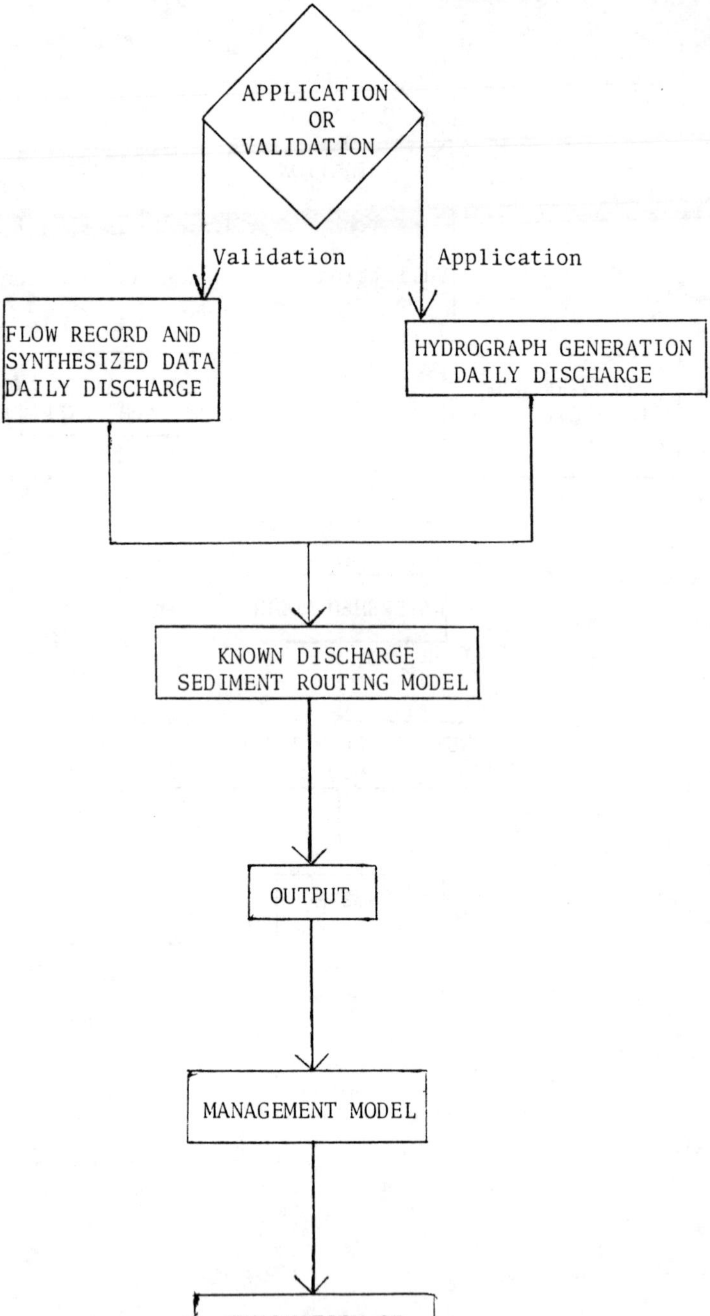

Fig. 11-20. Known Discharge Simulation

the determination of important tributaries (either controlled or uncontrolled) and point source input can be sorted out. The abstracted system compared to the complicated system (Fig. 11-15) is given in Fig. 11-21. There are 31 important input points to the Yazoo Basin and only 9 of the tributaries are considered to be important for the analysis. They are Arkabutla Canal, Little Tallahatchie River, Yacono River, Tillatoba Creek, Yalobusha River, Potococowa Creek, Teoc Creek Big Sand Creek and Pelucia Creek. Other input points are considered point source inputs.

Temporal design: Temporal design of the system is determined by the historic hydrologic record of the basin. This record includes historic maximums, minimums and mean flows, river stages, precipitation patterns and volumes, and the effect of man's activities, such as reservoir construction, on the hydrologic record. Analysis of the historic hydrologic record should be in accordance with the spatial design. Therefore, only those records pertinent to drainages included in the spatial design need be analyzed. The temporal design is a typical time series analysis.

Model simulation is dependent on the type of input record used. The record may consist of historic flows or synthesized data based on the historic record. Numerous statistical techniques are available for extending historic records forward in time for simulation. A typical time-series analysis will be employed to generate 50 years of simulation (11 years recorded plus 39 years of generated data). This time-series analysis considers the general trend, the cyclic component and the random nature ("white noise") of the discharge data.

Subsystem Models: The key subsystem models which have been developed for analysis include: the watershed model, the unsteady flow tributary model, the unsteady main stem flow model, and the known discharge sediment routing model. The watershed model utilizes a simplified representation of the geometry of each drainage considering only two sloping planes and one channel. The water routing is accomplished by an analytical solution to the kinematic wave approximation and the excess rainfall estimation is based on an explicit solution of Green and Ampt's approach (see Chapter 9). If a larger time increment is used such as one or two days, the routing component can be skipped. The unsteady flow tributary model is used to route water and sediment from the watershed, through the tributary and into the main stem. The routing is accomplished by using finite difference approximations to solve the water continuity, sediment continuity, and kinematic wave approximation to the momentum equation (see Chapter 9). The unsteady flow main stem model is for simulating the dynamics of flood and sediment movement in the main stem using a relatively short time increment such as two hours. The method applies finite difference approximations to solve the equations for water continuity, sediment continuity and the full dynamic equations utilizing a double sweep algorithm (see Chapter 10).

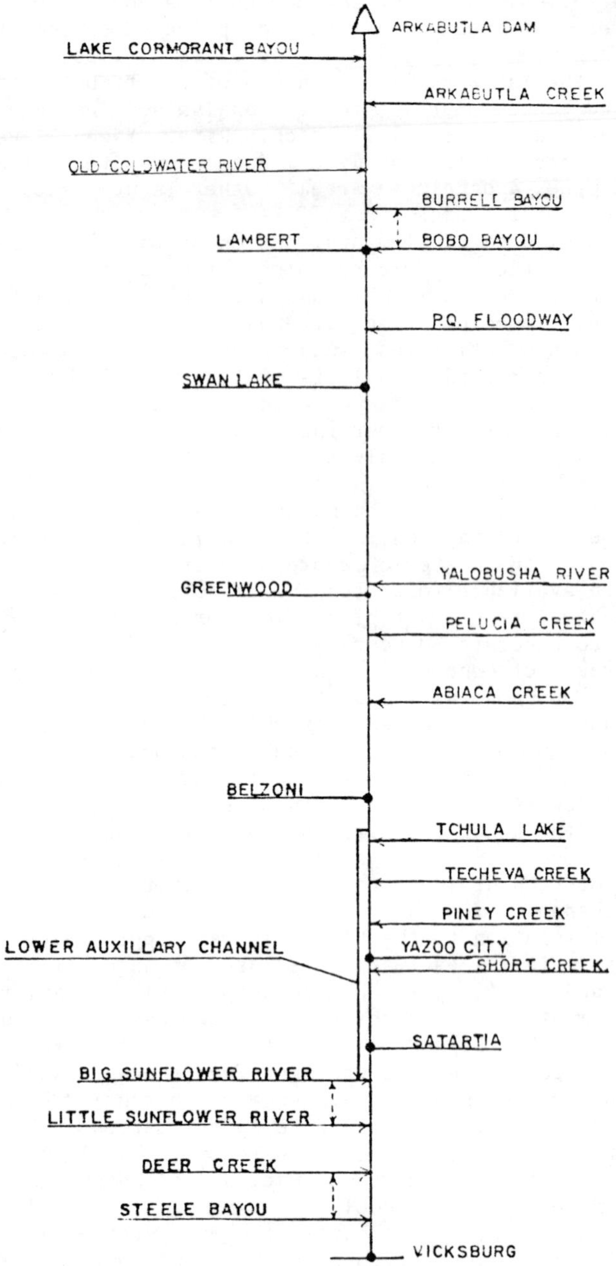

Fig. 11-21. Schematic of Yazoo River Basin System

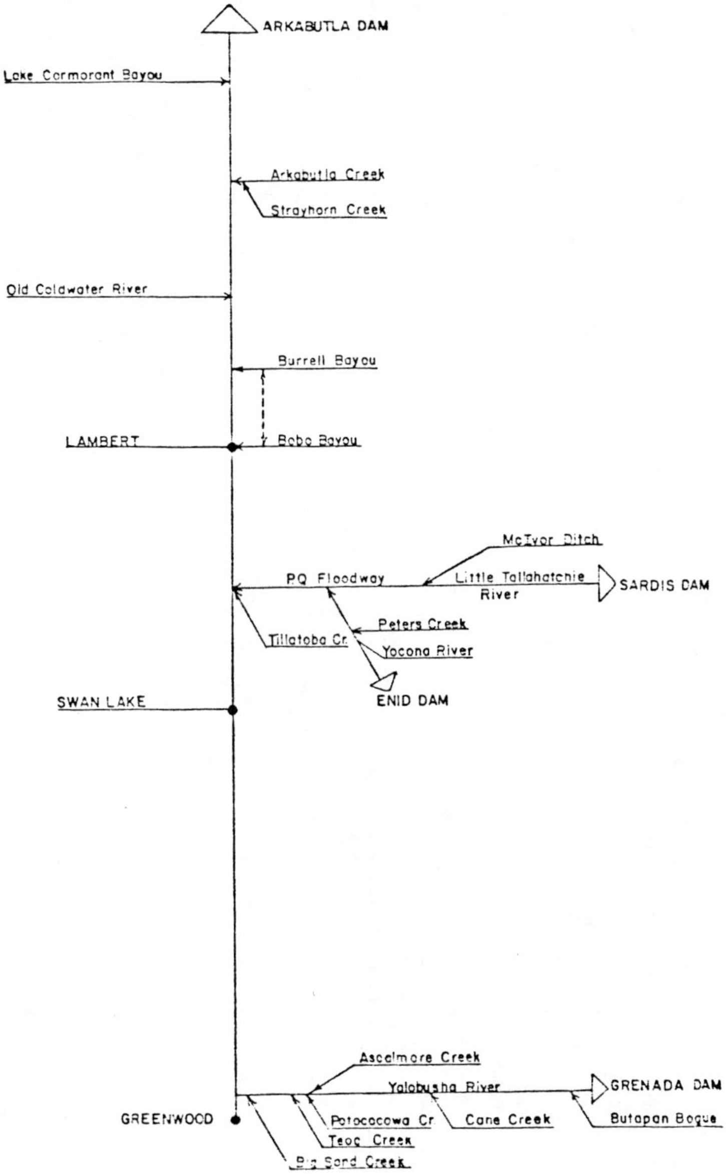

Fig. 11-21 (Continued). Schematic of Yazoo River Basin System

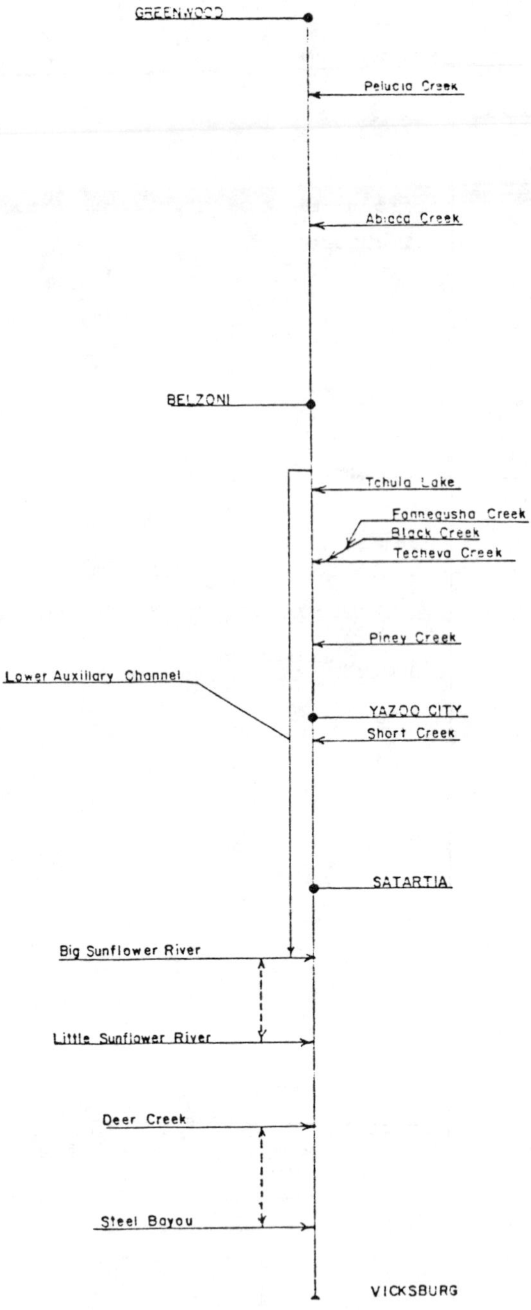

Fig. 11-21 (Continued). Schematic of Yazoo River Basin System

The known discharge sediment routing model is useful to evaluate the system response for a larger time increment. It determines backwater profile first and simulates the sediment movement using a finite difference scheme. The theories used are similar to the HEC-6 Model (see Chapter 8) but the model formulation and mathematical solution are different.

Data Synthesis: As pointed out in Section 11.5, a useful model should be developed considering the constraints of available data. For large basin analysis (like the Yazoo) it is virtually impossible to have a complete data base for analysis. A large amount of data must be synthesized from the secondary data base. As mentioned earlier, some discharges, stages, cross sections and sediment discharges must be synthesized. The stages are synthesized utilizing stage-discharge relations. Data on cross sections in the secondary tributaries are sporadic. They are synthesized by using relationships between channel geometry and top width. Four methods for computing bed material discharge have been examined using available data. The methods are (1) the modified Einstein procedure (Colby and Hembree, 1955), (2) Yang's formula (Yang, 1976), (3) Colby's method (Colby, 1964) and (4) Toffaleti's method (1969). The modified Einstein procedure is used as a basis for comparison. The comparison results show that Colby's method gives a good estimate of the bed material discharges for the channels in the basin. In order to fully utilize the measured sediment data, a set of sediment transport equations has been developed for the basin utilizing the computed results from the Modified Einstein Method. Three approaches will be used to synthesize hydrographs from ungaged sites. They include the watershed model, channel geometry, and flow continuity. The watershed model was discussed earlier. There are several tributaries that are amenable to water discharge estimation by the channel geometry method. These drainages are located primarily in the bottom lands and have small stream gradients. During high flow in the larger downstream channels into which these tributaries empty the drainages may act as distributary networks rather than contributary channels. Because watershed boundaries of these drainages are not well defined discharges for these tributaries can only be meaningfully determined using the channel geometry comparisons. Such comparisons are based on several channel characteristics. Assuming that precipitation input, moisture content, and flow resistance are the same for nearby watersheds and channels, the discharge of an ungaged channel can be estimated by using Manning's equation based on available channel geometry and the discharge record of a nearby gaged channel. The ungaged discharge is estimated by scaling the nearby gaged discharge using the ratio of computed bank-full discharges of the two channels determined from Manning's equation. The third approach is based on flow continuity. This method is simply to use upstream and downstream records to determine the input from a particular source.

Data storage and retrieval system: The development of a data base and data management system for the analysis is necessary since both collected data and the results of the analysis involve large quantities of information. This system should be capable of storing and managing all information from the data acquisition phase and the data analysis phase. The efficiency of a data storage and retrieval system is very important since it affects the availability and the quality of the input data of the simulation models used in the study of the basin. Therefore, the developed data base must reflect the logical structure of the watershed and the river channel system of the Yazoo River Basin. The specific data base scheme for each data category was worked out to ease the access, retrieval, or updating of the data elements in the data bank. A data management system has been established along with the data base to assist users in the retrieval and reformating of the data to obtain the desired information for simulation and analysis. The data management system is of a modular type and utilizes an overlay structure to increase its flexibility in changing component elements and reducing computer memory requirements. The general structure of this system is outlined in Fig. 11-22.

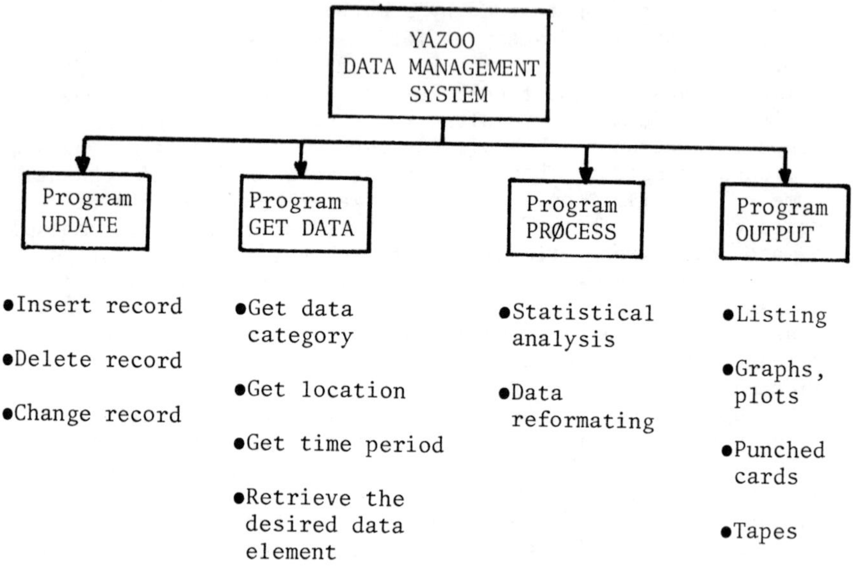

Fig. 11-22. Overall Structure of the Yazoo Data Management System.

11.8 SUMMARY

This chapter describes a brief literature review, an overview of some case studies, and the criteria of a useful mathematical model for simulating watershed and river system response and indicates the advantages and disadvantages of various existing models. Multi-step development of a model is discussed. This includes spatial design, temporal design, model formulation, mathematical solution, model calibration, parameter sensitivity analysis, qualitative examination with physical significance, model simplification, regionalization and generalization, validation, testing and refinement under operational field applications and documentation. In addition, several modeling efforts and applications to various watershed and river systems at the Engineering Research Center, Colorado State University, are identified.

11.9 REFERENCES

Agricultural Research Service, USDA, 1975. Control of water pollution from cropland. Vol. I-A, Manual for Guideline Development, November.

Amein, M., 1968. An implicit method for numerical flood routing. Water Resources Research, AGU, Vol. 4, No. 4, Aug., pp. 719-726.

Amein, M., and Fang, C. S., 1970. Implicit flood routing in natural channels. J. Hyd. Div., ASCE, Vol. 96, No. HY12, Proc. Paper 7773, Dec., pp. 2481-2500.

Chang, F. F. M. and Richards, D. L., 1971. Deposition of sediment in transient flow. J. Hyd. Div., ASCE, Vol. 97, No. HY6, Proc. Paper 8191, June, pp. 837-849.

Chen, Y. H., 1973. Mathematical modeling of water and sediment routing in natural channels. Ph.D. Dissertation, Civil Eng. Dept., Colorado State University, Fort Collins.

Colby, B. R., 1964. Practical computation of bed-material discharge. J. of the Hydraulics Div., ASCE, Vol. 90, No. HY2.

Colby, B. R. and Hembree, C. H., 1955. Computation of total sediment discharge, Niobrara River near Cody, Nebraska. U.S. Geol. Survey, Water Supply Paper 1357.

Dass, P. and Simons, D. B., 1976. Mathematical model to simulate channel deformation. Symposium on Inland Waterways for Navigation, Flood Control and Water Diversions, ASCE, Vol. 1, Fort Collins, Colorado, August 10-12, pp. 1-15.

Einstein, H. A., 1950. The bed load function for sediment transportation in open channel flows. USDA, Technical Bulletin, No. 1026.

Holly, F. M., Simons, D. B., and Chen, Y. H., 1974. Hydrologic study, Saint Maruice towhead, coal fired station, Gulf State Utilities Company. Prepared for Stone and Webster Engineering Corp., Boston, Massachusetts, July.

Hydrologic Engineering Center, Corps of Engineers, U.S. Army, 1975. Water surface profiles. Vol. 6, Hydrologic Engineering Methods for Water Resources Development.

Hydrologic Engineering Center, Corps of Engineers, U.S. Army 1976. HEC-6 scour and deposition in rivers and reservoirs. Users Manual.

Kibler, D. F. and Woolhiser, D. A., 1970. The kinematic cascade as a hydrologic model. Hydrology Paper No. 39, Colorado State University, Fort Collins,

Li, R. M., Simons, D. B., and Eggert, K. G., 1976b. Process model of infiltration and subsurface flow routing. Prepared for USDA Forest Service, Rocky Mountain Forest and Range Experiment Station, Flagstaff, Arizona, December.

Li, R. M., Simons, D. B., Shiao, L. S., and Chen, Y. H., 1976a. Kinematic wave approximation for flood routing. Rivers 76 Symposium on Inland Waterways for Navigation, Flood Control, and Water Diversions, Colorado State University, Vol. 1, pp. 377-398.

Li, R. M., and Simons, D. B., and Simons, R. K., 1977. A mathematical model for evaluating on-site soil erosion. Prepared for USDA Forest Service, Rocky Mountain Forest and Range Experiment Station, Flagstaff, Arizona, February.

Li, R. M., Simons, D. B., and Stevens, M. A., 1975. Nonlinear kinematic wave approximation for water routing. Water Resources Research, Vol. 11, No. 2, April, pp. 245-252.

Rosenbrock, H. H., 1960. An automatic method for finding the greatest or least value of a function. The Computer Journal, Vol. 3, pp. 175-184.

Sabol, G. V., 1977. Preliminary report on the estimation of the total fluvial sediment transported by the Wadi Mawr. Yemen Arab Republic, April.

Simons, D. B. and Chen, Y. H., 1976. A mathematical model study of pool 4 in the Upper Mississippi and Lower Chippewa Rivers. Prepared for the U.S. Fish and Wildlife Service, Twin Cities, Minnesota.

Simons, D. B. and Chen, Y. H., 1977. Investigation of the effects of Chippewa River erosion and silt reduction. Prepared for U.S. Army Engineer District, St. Paul, Minnesota.

Simons, D. B. and Li, R. M., 1975. Watershed segmentation by a digital computer for mathematical modeling of watershed response. Prepared for USDA Forest Service, Rocky Mountain Forest and Range Experiment Station, Flagstaff, Arizona, December.

Simons, D. B. and Li, R. M., 1976. Procedure for estimating model parameters of a mathematical model. Prepared for USDA Forest Service, Rocky Mountain Forest and Range Experiment Station, Flagstaff, Arizona, April.

Simons, D. B. and Li, R. M., 1977. Design of Bunay-Barry Canal and Mansury-Rayyan-Bagr Canal of Wadi Zabid Canal System. Report prepared for Tipton and Kalmbach, Inc., Denver, Colorado, July.

Simons, D. B., Li, R. M., and Shiao, L. Y., 1976a. Preliminary procedural guide for estimating water and sediment yield from roads in forest. Prepared for USDA Forest Service, Rocky Mountain Forest and Range Experiment Station, Flagstaff, Arizona, November.

Simons, D. B., Li, R. M., and Shiao, L. Y., 1977a. Formulation of road sediment model. Prepared for USDA Forest Service, Rocky Mountain Forest and Range Experiment Station, Flagstaff, Arizona, March.

Simons, D. B., Li, R. M., and Stevens, M. A., 1975a. Development of models for predicting water and sediment routing and yield from storms on small watersheds. Prepared for USDA Forest Service, Rocky Mountain Forest and Range Experiment Station, Flagstaff, Arizona.

Simons, D. B., Li, R. M., and Ward, T. J., 1977b. Simple procedural method for estimating on-site soil erosion. Prepared for USDA Forest Service, Rocky Mountain Forest and Range Experiment Station, Flagstaff, Arizona, February.

Simons, D. B., Schumm, S. A., Stevens, M. A., Chen, Y. H., and Lagasse, P. F., 1975b. A geomorphic study of pools 24, 25 and 26 in Upper Mississippi and Lower Illinois Rivers. Prepared for the Waterways Experiment Station, Vicksburg, Mississippi.

Simons, D. B., and Senturk, F., 1977. Sediment transport technology. Water Resource Publications, Fort Collins, Colorado.

Simons, D. B., Li, R. M., and Shiao, L. Y., 1976a. Preliminary procedural guide for estimating water and sediment yield from roads in forest. Prepared for USDA Forest Service, Rocky Mountain Forest and Range Experiment Station, Flagstaff, Arizona, November.

Simons, R. K., 1977. A modified kinematic approximation for water and sediment routing. M.S. Thesis, Civil Engineering Department, Colorado State University, Fort Collins.

Sobol, G. V., 1977. Preliminary report on the estimation of the total fluvial sediment transported by the Wadi Mawr, Yemen Arab Republic. April.

TESCO-VIZITERV-VITUKI, 1971. Sediment transportation. Budapest.

Toffaleti, F. B., 1969. Definitive computations of sand discharge in rivers. J. of the Hydraulics Div., ASCE, Vol. 95, No. HY1, January, pp. 225-246.

U.S. Forest Service, 1976. Non-point water quality modeling in wildland management: a state-of-the-art assessment. USDA Forest Service Interagency Agreement No. EPA-IAG-05-0660, Washington, D.C.

Yang, C. T., 1976. Minimum unit stream power and fluvial hydraulics. J. of the Hydraulics Division, Proceedings, ASCE, Vol. 102, No. HY7, Paper 12238, pp. 919-934.

Chapter 12

WATER QUALITY MODEL FOR AGRICULTURAL RUNOFF

by

Anthony S. Donigian, Jr., Senior Research Engineer
Hydrocomp Inc.

and

Norman H. Crawford, President
Hydrocomp Inc.

Chapter 12

WATER QUALITY MODEL FOR AGRICULTURAL RUNOFF

12.1 INTRODUCTION

The development of models to simulate the water quality impact of nonpoint source pollutants is receiving considerable attention by the engineering and scientific community. One of the major reasons for this interest was the passage of the Federal Water Pollution Control Act Amendments of 1972, specifically requiring the evaluation of the contribution of nonpoint source pollution to overall water quality. This chapter describes a modeling effort whose goal is the simulation of water quality resulting from agricultural lands. The beginnings of this research modeling effort date from 1971 when the U.S. Environmental Protection Agency, through the direction of the Environmental Research Laboratory in Athens, Georgia (ERL-Athens) sponsored the development and initial testing of the Pesticide Transport and Runoff (PTR) Model (Crawford and Donigian 1973). The Agricultural Runoff Management (ARM) Model discussed by Donigian and Crawford (1976) and Donigian, et al. (1977) is the combined result of further model testing and refinement, algorithm modifications, and inclusion of additional capabilities not present in the PTR Model.

This chapter is a summary of the model description and testing results included in the reports mentioned above. The ultimate goal of the ARM Model development work is to establish a method to evaluate the efficacy of management practices for controlling the loss of sediment, pesticides, nutrients, and other nonpoint pollutants from agricultural lands.

The guiding philosophy of the modeling effort is to represent, in mathematical form, the physical processes occurring in the transport of nonpoint source pollutants. The hydrologic and water quality related processes occurring on the land surface (and in the soil profile) are continuous in nature; hence, continuous simulation is critical to the accurate representation of these physical processes. Although nonpoint source pollution from the land surface takes place only during runoff-producing events, the status of the soil moisture and the pollutant prior to the event is a major determinant of the amount of runoff and pollutants that can reach the stream during the event. In turn, the soil moisture and pollutant status prior to the event is the result of processes which occur between events. Cultivation and tillage practices, pesticide and fertilizer applications, pesticide degradation and nutrient transformations, all critically affect the mass of pollutant that can enter the aquatic environment during a runoff-producing event. Models that simulate only single events cannot accurately evaluate agricultural land management practices since between-event processes are ignored. Although all between-event processes cannot be quantitatively described at the present state of technology, continuous simulation provides a sound framework for their approximation and for further research into their quantification.

When modeling nonpoint source pollution, the above stated philosophy is joined by the fact that the transport mechanisms of such pollutants are universal. Whether the pollutants originate from pervious or impervious lands, from agricultural or urban areas, or from natural or developed lands, the major transport modes of runoff and sediment loss are the same. (Wind transport may be significant in some areas, but its importance relative to runoff and sediment loss is usually small.) In this way, the simulation of nonpoint source pollution is analogous to a three-layered pyramid. The basic foundation of the pyramid is the hydrology of the watershed. Without accurate simulation of runoff, modeling nonpoint pollutants is practically impossible. Sediment loss simulation, the second layer of the pyramid, follows in sequence the hydrologic modeling. Although highly complex and variable in nature, sediment modeling provides the other critical transport mechanism. The pinnacle or final layer of the pyramid is the interaction of various pollutants with sediment loss and runoff, resulting in the overall transport simulation of nonpoint source pollutants.

The general goals of the research effort described in this chapter are (1) to utilize the most advanced state of present technology in the simulation of nonpoint source pollutants, and (2) to delineate critical areas for further research and investigation. The final version of the ARM Model is designed for general applicability throughout the United States and for use by state and local agencies for the water quality evalutation of agricultural land management practices.

12.2 THE AGRICULTURAL RUNOFF MANAGEMENT (ARM) MODEL

The ARM Model simulates runoff (including snow accumulation and melt), sediment, pesticides, and nutrient contributions to stream channels from both surface and subsurface sources. No channel routing procedures are included. Thus, the model is applicable to watersheds that are small enough that channel processes and transformations can be assumed negligible. Although the limiting area will vary with climatic and topographic characteristics, watersheds greater than 2 to 5 sq km are approaching the upper limit of applicability of the ARM Model. Channel processes will significantly affect the water quality resulting from larger watersheds.

Figure 12-1 demonstrates the general structure and operation of the ARM Model. The major components of the model individually simulate the hydrologic response (LANDS) of the watershed, sediment production (SEDT), pesticide adsorption/desorption (ADSRB), pesticide degradation (DEGRAD), and nutrient transformations (NUTRNT). The executive routine, MAIN, controls

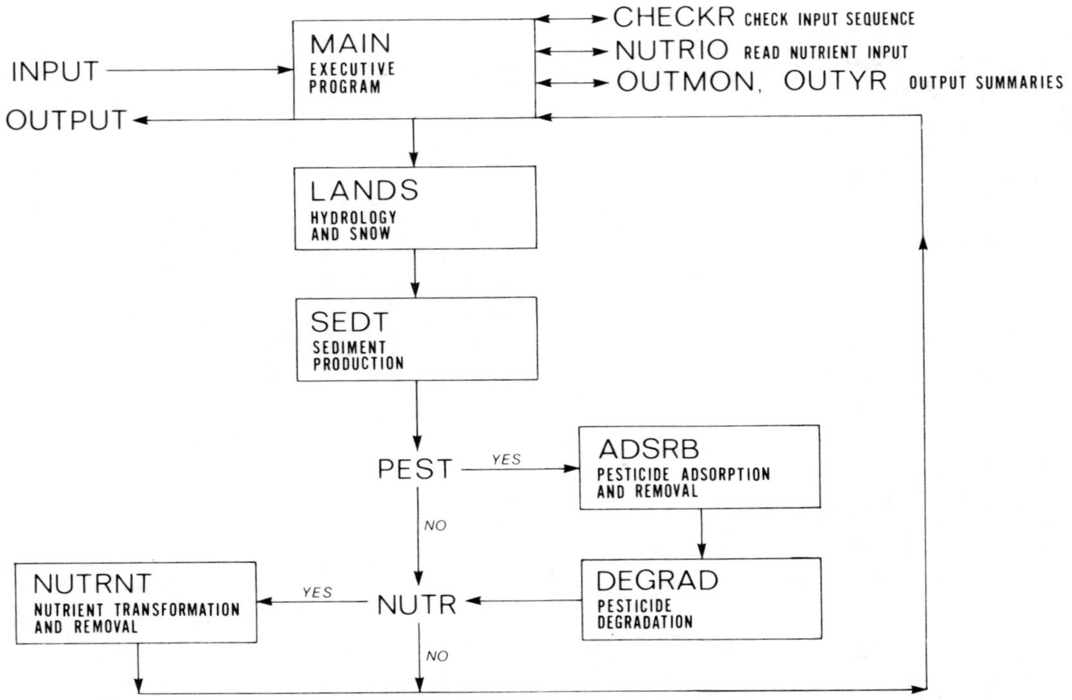

Fig. 12-1. ARM Model Structure and Operation

the overall execution of the program; calling subroutines at proper intervals, transferring information between routines, and performing the necessary input and output functions. Table 12-1 describes the functions of each of the ARM Model components.

In order to simulate vertical movement and transformations of pesticides and nutrients in the soil profile, specific soil zones (and depths) are established so that the total soil mass in each zone can be specified. Total soil mass is a necessary ingredient in the pesticide adsorption/desorption reactions and nutrient transformations. Figure 12-2 depicts the zones and depths assumed in the ARM Model. The depths of the surface and upper soil zones are specified by the model input parameters, SZDPTH and UZDPTH, respectively. The upper zone depth corresponds to the depth of incorporaton of soil-incorporated chemicals. It also indicates the depth used to calculate the mass

Table 12-1. ARM Model Components

Major Program	Component Subroutine	Function
MAIN		Master program and executive control routine
	CHECKR, CHECKS	Checks input parameter errors
	BLOCK DATA	Data initialization for common variables
	NUTRIO	Reads and checks nutrient input data
	OUTMON	Prints monthly output summaries
	OUTYR	Prints yearly output summaries
LANDS		Performs hydrologic simulation and snowmelt calculations
SEDT		Performs sheet erosion simulation
	ERDBUG	Outputs to the printer erosion files written to disk (for error checking)
ADSRB		Performs pesticide soil adsorption/desorption simulation
	DSPTN	Performs desorption calculations
DEGRAD		Performs pesticide degradation simulation
NUTRNT		Performs nutrient simulation
	TRANS	Performs nutrient transformations

Fig. 12-2. Assumed Soil Depths for Pesticide and Nutrient Storage

of soil in the upper zone whether agricultural chemicals are soil-incorporated or surface applied. The depths of the surface and lower zones are important because the active surface zone is crucial to the washoff and degradation of agricultural chemicals, while the extent of the lower zone determines to what degree soluble pollutants will contaminate the groundwater. The zonal depths will vary with the geology and topography of the watershed. The relative specification of the soil depths indicated in Fig. 12-2 has been satisfactory for the watersheds tested to date.

The transport and vertical movement of pesticides and nutrients, as conceived in the ARM Model, is indicated in Fig. 12-3. Pollutant contributions to the stream can occur from the surface zone, the upper zone, and the groundwater zone. Surface runoff is the major transport mechanism carrying dissolved chemicals, pesticide particles, or sediment and adsorbed chemicals. The interflow component of runoff can transport dissolved pesticides or nutrients occurring in the upper zone. Vertical chemical movement between the soil zones is the result of infiltrating and percolating water. From the surface, upper, and lower zones, uptake and transformation of nutrients and degradation of pesticides is allowed. On the watersheds tested, the groundwater zone has been considered a sink for deep percolating chemicals since the groundwater flow contribution has been negligible. However, on larger watersheds this contribution could be significant.

Hydrology

To comprehend the movement of pesticides and nutrients in the ARM Model, one must have a basic understanding of the hydrology subprogram, LANDS. A flowchart of LANDS is shown in Fig. 12-4. The mathematical foundation of LANDS was originally derived from the Stanford Watershed Model (Crawford and Linsley, 1966) and has been presented with minor variations in numerous subsequent publications. For this reason, the algorithms will not be fully described here. The major parameters of the LANDS subprogram are defined in Table 12-2. These parameters are identical to those in the PTR Model and also in Hydrocomp Simulation Programming (Hydrocomp, 1976). In brief, the LANDS subprogram simulates the hydrologic response of the watershed to inputs of precipitation and evaporation. LANDS simulates runoff continuously through a set of mathematical functions derived from theoretical and empirical evidence. It is basically a moisture accounting procedure on the land surface for water in each major component of the hydrologic cycle. The parameters (Table 12-2) within the mathematical functions are used to characterize the land surface and soil profile characteristics of the watershed. These parameters must be selected, tested, and modified when LANDS is applied to a new watershed. Calibration is the process whereby the parameters are modified

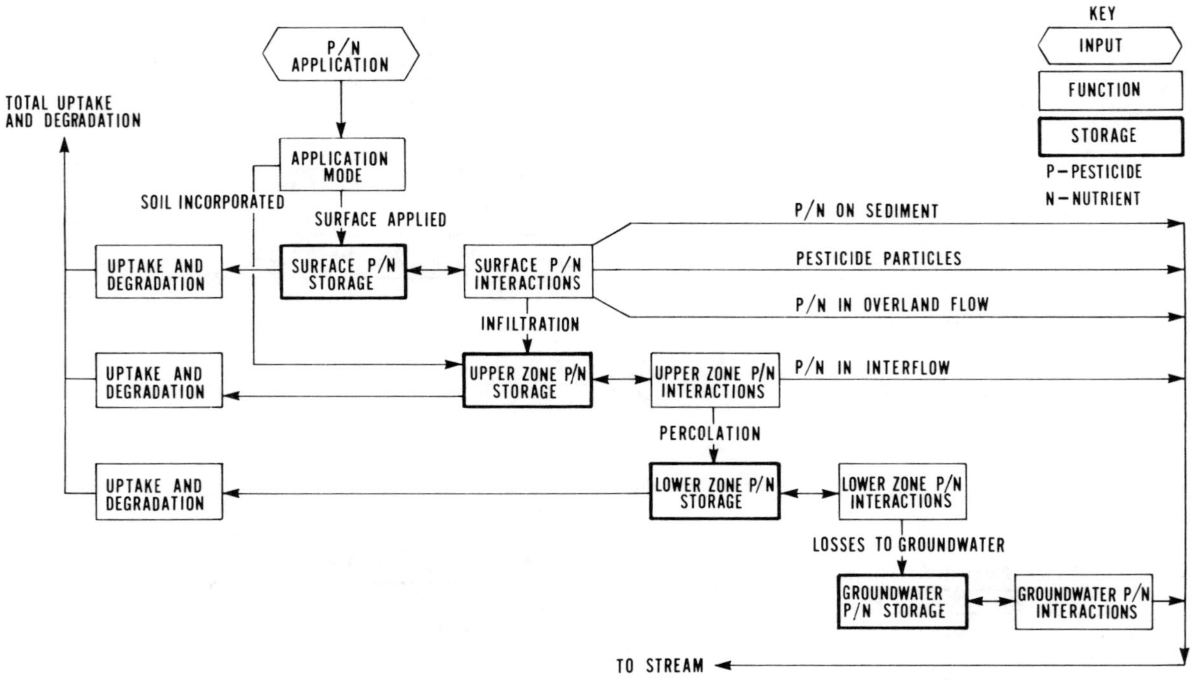

Fig. 12-3. Pesticide and Nutrient Movement in the ARM Model

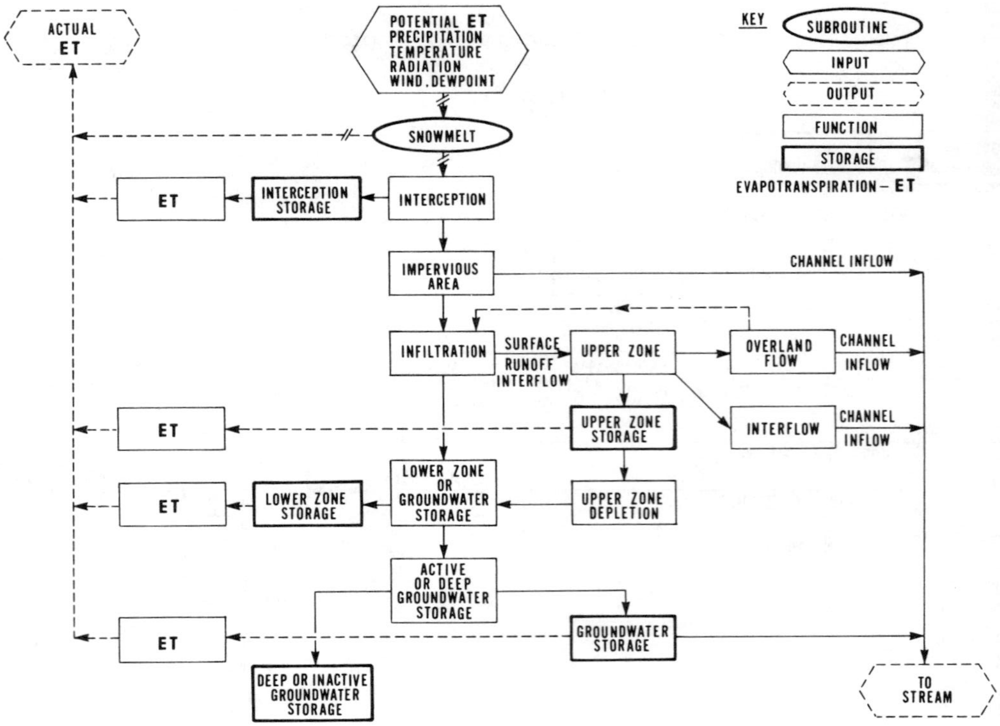

Fig. 12-4. LANDS Simulation

as a result of a comparison of simulated and recorded runoff data for the watershed.

Modifications to the Stanford and Hydrocomp Simulation Programming (HSP) versions of the LANDS algorithms have been discussed in the PTR Model Report. The present version of the LANDS subprogram of the ARM Model includes these modifications to simulate the areal variation in agricultural chemical concentrations on the land surface.

Areal Zone Concept: The HSP LANDS module was modified for pesticide transport to accommodate the expected areal variation in pesticide concentration over the land surface. It is generally accepted in hydrology that infiltration is time and area dependent. Infiltration capacity will vary even within small watersheds with reasonably homogeneous soil characteristics. This areal variation in infiltration results in source

Table 12-2. Hydrologic Model (LANDS) Parameters

A	A fraction representing the impervious area in a watershed.
EPXM	The interception storage parameter, related to vegetal cover density.
UZSN	The nominal upper zone soil moisture storage paramter.
LZSN	The nominal lower zone soil moisture storage parameter.
K3	Index to actual evaporation (a function of vegetal cover).
K24L, K24EL	Parameters controlling the loss of water from groundwater storage. K24L is the fraction of groundwater recharge that percolates to deep groundwater tables. K24EL is the fraction of the segment area where shallow water tables put groundwater within reach of vegetation.
INFIL	This parameter is a function of soil characteristics defining the infiltration characteristics of the watershed.
INTER	This parameter defines the interflow characteristics of the watershed.
L	Length of overland flow plane.
SS	Average overland flow slope.
NN	Manning's "n" for overland flow.
IRC, KK24	The interflow and groundwater recession parameters.
KV	The parameter KV is used to allow a variable recession rate for groundwater discharge.

areas, or zones, with low infiltration capacity within the watershed, contributing a large component of overland flow. Since overland flow and sediment loss are the major mechanisms of pesticide transport to the watercourse, the low infiltration source areas will also experience a greater loss of pesticide than the remainder of the watershed. Consequently, the pesticide concentration on the land surface will vary, in spite of an initially uniform application. The pesticide concentration within the soil profile will also vary as a function of the volume of infiltration. Obviously, the extent of pesticide areal variation depends upon the solubility and transport characteristics of the specific pesticide applied and upon topographic and watershed characteristics. Natural hydrologic conditions and watershed characteristics are sufficiently non-uniform to justify the above described mechanisms leading to areal variations in infiltration and pesticide concentrations.

HSP LANDS employs a cumulative frequency distribution of infiltration capacity to account for the areal variation. Fig. 12-5a graphically presents the infiltration function of HSP LANDS. A mean infiltration capacity, f, is calculated and a linear approximation to the actual cumulative distribution is assumed. Interflow is determined as a function of infiltration and lower zone moisture storage. It is evaluated in Fig. 12-5a as a second linear cumulative distribution denoted by f(c-1). Since the X-axis is unity (100 percent of watershed area), the area of each wedge in Fig. 12-5a represents the portion of the moisture supply allocated to each component. During any time interval, the available moisture supply is distributed to surface detention, interflow detention, and infiltration. Overland flow and interflow are determined as losses from surface detention and interflow detention, respectively. Lower zone moisture storage and groundwater components are derived from infiltration component.

The LANDS subprogram of the ARM Model employs the same infiltration function as HSP LANDS, with one modification: the watershed is divided into five zones, each representing 20 percent of the total area. The zonal division is based on infiltration capacity. Schematically, Fig. 12-5b shows that zone 1 will infiltrate much less water than zone 5. Conversely, zone 5 will provide less overland flow than zone 1. Thus, the areal variation in infiltration capacity is approximated. Zones with lower infiltration capacity will serve as the major source areas for overland flow, sediment, and pesticide loss. Generally, zones with high infiltration will contain more pesticide in the soil profile because of the greater amount of infiltrated water.

Conceptually, the zones are not necessarily concentric, continuous, or contiguous. Each is connected directly to the stream channel by the overland flow plane. As with any simulation model, this source zone concept is an approximation. It is an attempt to portray mechanisms which are known to occur, but are impossible to simulate in detail.

Fig. 12-5a. Cumulative Frequency Distribution of Infiltration Capacity Showing Infiltrated Volumes, Interflow and Surface Detention

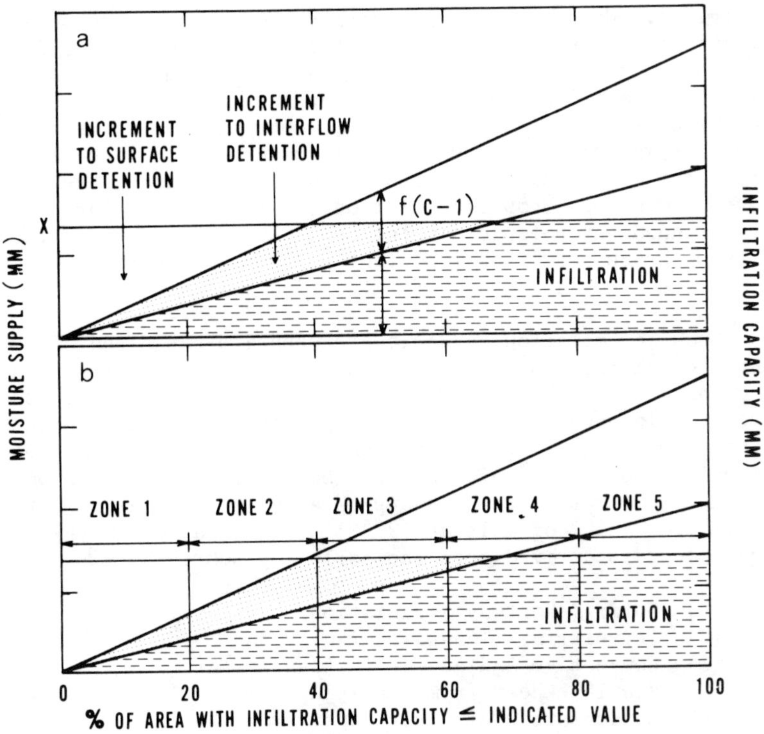

Fig. 12-5b. Source-Zones Superimposed on the Infiltration Capacity Function

Fig. 12-5. Infiltration Capacity and Areal Source-Zone Functions

Snow Accumulation and Melt Simulation: In the simulation of water quality processes, the mechanisms of snow accumulation and melt are often neglected. The stated reasons for this omission generally pertain to an assumed minor influence on water quality, the extensive data requirements, and the extreme complexity of the component processes. Obviously, in the southern latitudes of the United States and at many coastal locations, snow accumulation during winter months is often negligible. However, considering its location in a temperate climatic zone, over 50 percent of the continental United States experiences significant snow accumulation. In many areas streamflow contributions from melting snow continue throughout the spring and well into the summer. For many urban areas, the supply of water during the critical summer period is entirely a function of the extent of snow accumulation during the previous winter. Snow accumulation and melt is a major factor in continuous hydrologic simulation. Thus, the consideration of these processes is an important part of any hydrologic model which is to provide a basis for the simulation of water quality processes.

Snow accumulation and melt are separate but often concurrent mechanisms. The initial snow accumulation is largely a function of air (and atmospheric) temperature at the time of precipitation; whereas, snowmelt is an energy transfer process in the form of heat between the snowpack and its environment. Basically, 80 cal/cm^2 of heat must be supplied to obtain one centimeter of water from a snowpack at 0 °C (203 cal/cm^2 or 750 Btu/cm^2 for 1 in. of melt at 32 °F). This heat or energy requirement is derived from the following sources:

(1) Solar (shortwave) radiation
(2) Terrestrial (longwave) radiation
(3) Convective and advective transfer of sensible heat from overlying air
(4) Condensation of water vapor from the air
(5) Heat conduction from soil and surroundings
(6) Heat content of precipitation

The complexity of the snowmelt process is due to the many factors that influence the contributions from each of the above energy sources. Figure 12-6 conceptually indicates the factors and processes involved in snow accumulation and melt on a watershed.

The snowmelt reaching the land surface results from complex interactions between the melt components, climatic conditions, and snowpack characteristics. For the most part, the snowpack behaves like a moisture reservoir gradually releasing its storage. However, the combination of extreme climatic conditions and snowpack characteristics can lead to abnormally high liquid moisture holding capacity and sudden release of melt in relatively short time periods. The objective of snow accumulation and melt simulation is to approximate the physical

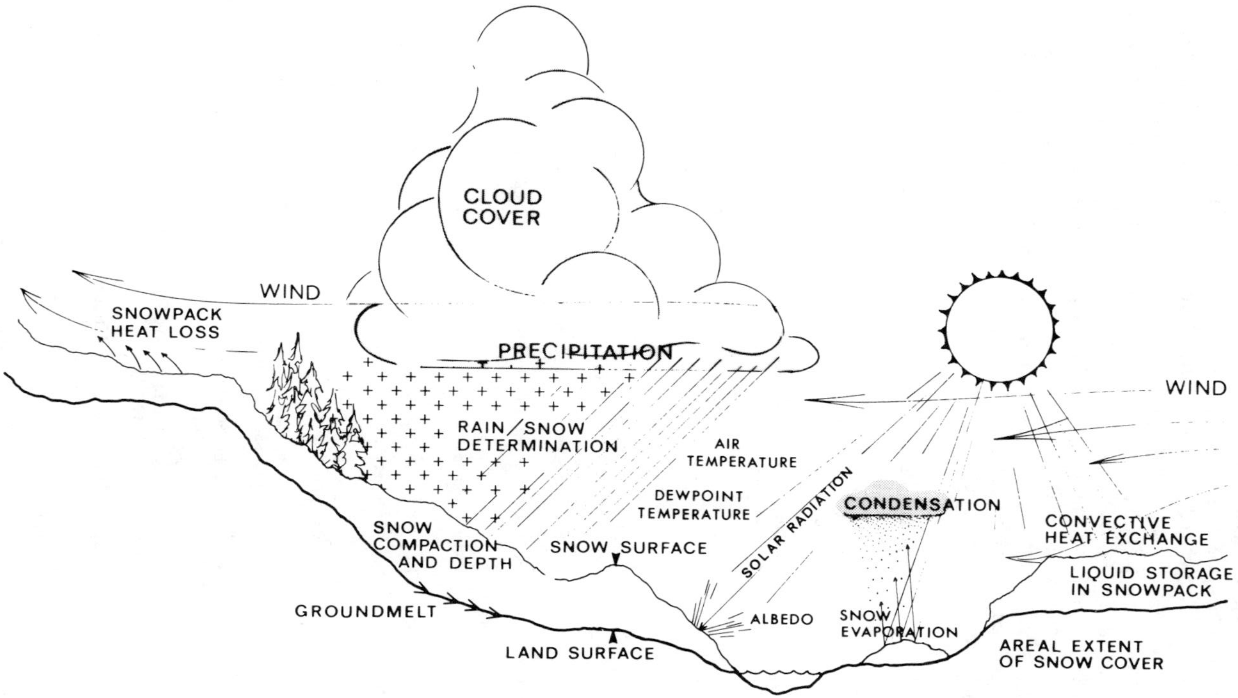

Fig. 12-6. Snow Accumulation and Melt Processes

processes and the interactions in order to evaluate the timing and volume of melt water released from the snowpack. The algorithms used in simulating the processes shown in Fig. 12-6 are based on extensive work by the Corps of Engineers (1956), Anderson and Crawford (1964), and Anderson (1968). Empirical relationships are employed when quantitative descriptions of the process are not available. The snowmelt parameters are listed in Table 12-3. The algorithms are identical to those employed in HSP and have demonstrated reasonably successful results on numerous watersheds. A flowchart of the snowmelt routine is shown in Fig. 12-7, and the individual algorithms are described in the ARM Model (Donigian and Crawford 1976) and HSP (Hydrocomp 1976) reports.

Table 12-3. Snowmelt Parameters

RADCON:	Parameter to adjust theoretical solar radiation melt equations to field conditions
CCFAC:	Parameter to adjust theoretical condensation and convection melt equation to field conditions
EVAPSN:	Parameter to adjust theoretical snow evaporation to field conditions
MELEV:	Mean elevation of the watershed
ELDIF:	Elevation difference between the temperature station and the midpoint of the watershed
TSNOW:	Wet-bulb air temperature below which snowfall occurs
MPACK:	Water equivalent of the snowpack required for complete coverage of the watershed
DGM:	Daily groundmelt
WC:	Maximum water content of the snow
IDNS:	Index density of new snow at 0°F
SCF:	Snow correction factor to compensate for deficiencies in the gage during snowfall
PETMAX:	Temperature below which input potential evapotranspiration is reduced by 50 percent
PETMIN:	Temperature below which input potential evapotranspiration is reduced to zero
PETMUL:	Potential evapotranspiration multiplier to adjust observed daily input values
WMUL:	Wind multiplier to adjust observed daily wind values
RMUL:	Solar radiation multiplier to adjust observed daily solar radiation values
F:	Fraction of watershed with forest cover
KUGI:	Index to the extent of undergrowth in forested areas

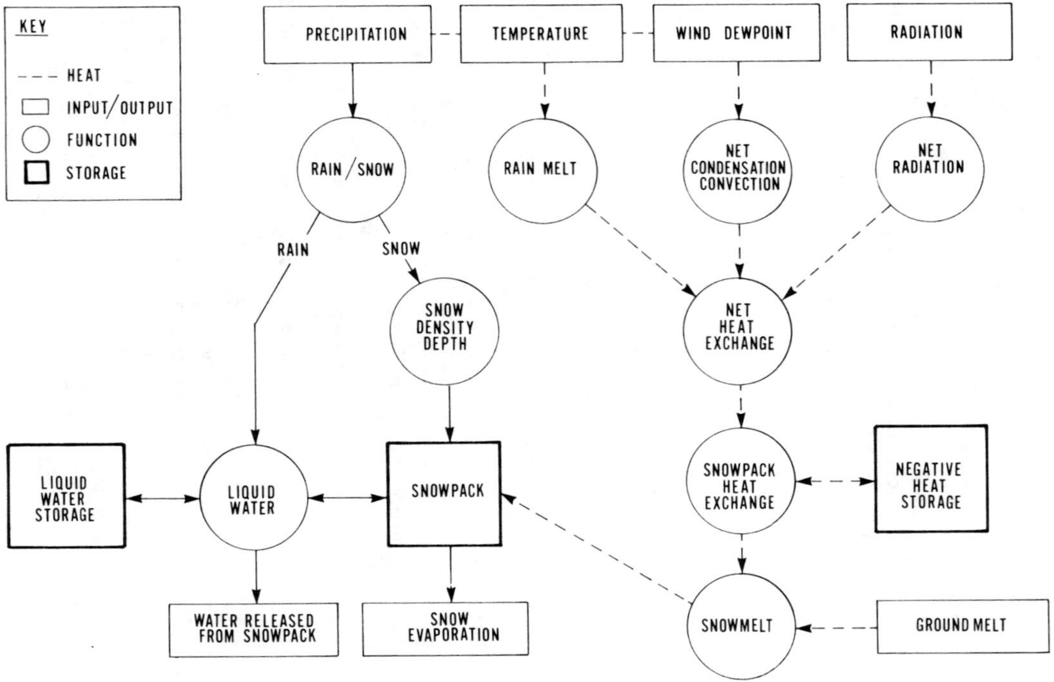

Fig. 12-7. Snowmelt Simulation

Sediment Loss

The algorithms for simulating soil loss, or erosion, were initially derived from research by Negev at Stanford University (1967) and have been subsequently influenced by the work of Meyer and Wischmeier (1969), Onstad and Foster (1975), and Fleming and Fahmy (1973).

Although Negev simulated the entire spectrum of the erosion process, only sheet and rill erosion were included in the ARM Model. The two component processes of sheet and rill erosion pertain to (1) detachment of soil fines (silt and clay fraction) by raindrop impact, and (2) pick-up and transport of soil fines by overland flow. These processes are represented in the ARM Model by the following algorithms:

Soil fines detachment:

$$RER(t) = (1 - COVER(T))*KRER*PR(t)^{JRER} \qquad (12\text{-}1)$$

Soil fines transport

$$SER(t) = \begin{cases} KSER*OVQ(t)^{JSER}, \text{ for } SER(t) \leq SRER(t) & (12\text{-}2) \\ SRER(t), \text{ for } SER(t) > SRER(t) & (12\text{-}3) \end{cases}$$

where:

RER(t)	=	soil fines detached during time interval t, tonnes/ha
COVER(T)	=	fraction of vegetal cover as a function of time, T, within the growing season
KRER	=	detachment coefficient for soil properties
PR(t)	=	precipitation during the time interval, mm
JRER	=	exponent for soil detachment
SER(t)	=	transport of fines by overland flow, tonnes/ha
JSER	=	exponent for fines transport by overland flow
KSER	=	coefficient of transport
SRER(t)	=	reservoir of soil fines at the beginning of the time interval, t, tonnes/ha
OVQ(t)	=	overland flow occurring during the time interval, t, mm
F	=	fraction of overland flow reaching the stream during the time interval, t
ERSN(t)	=	sediment loss to the stream during the time interval, t, tonnes/ha

In the operation of the algorithms, the soil detachment (RER) during each time (5 or 15 min.) interval is calculated by Eq. (12-1) and added to the total fines storage or reservoir (SRER). Next, the total transport capacity of the overland flow (SER) is determined by Eq. (12-2). Sediment is assumed to be transported at capacity of sufficient fines are available, otherwise the amount of fines in transport is limited by the fines storage, SRER (Eq. 12-2). The sediment loss to the waterway in the time interval is calculated in Eq. (12-3) by the fraction of total overland flow that reaches the stream. A land surface flow routing technique determines the overland flow contribution to the stream in each time interval. After the fines storage (SRER) is reduced by the actual sediment loss to the stream (ERSN), the algorithms are ready for simulation of the next time interval. Thus, the sediment that doesn't reach the stream is returned to the fines storage and is available for transport in the next time interval. The methodology attempts to represent the major processes of importance in soil erosion so that the impact of land management practices (tillage, terracing, mulching) can be specified by their effects on the sediment parameters.

Since land cover by growing crops and crop residues has a major impact on sediment loss, the variability in the land surface cover is explicity represented in the ARM Model by the function shown in Fig. 12-8. Monthly cover values assumed to occur on the first day of the month are specified by the user. Cover on any day is determined by linear interpolation between the monthly values.

The timing and severity of tillage operations also have a controlling effect on the sediment loss from an agricultural watershed. The effect of tillage on sediment processes is to increase the mass of soil fines available for transport and produce a reasonably uniform distribution of fines across the watershed. Consequently, the ARM Model allows the user to specify the dates of tillage, planting, or other land-surface disturbing operations. For each of these dates the user must specify a new detached soil fines storage (SRERTL) resulting from the operation. At the beginning of each tillage day the ARM Model resets the fines storage in each of the areal zones to the new value, resulting in a uniform fines distribution across the watershed. The amount of fines storage produced by different tillage operations is related to the depth and extent of the operation, and soil characteristics. Further study is

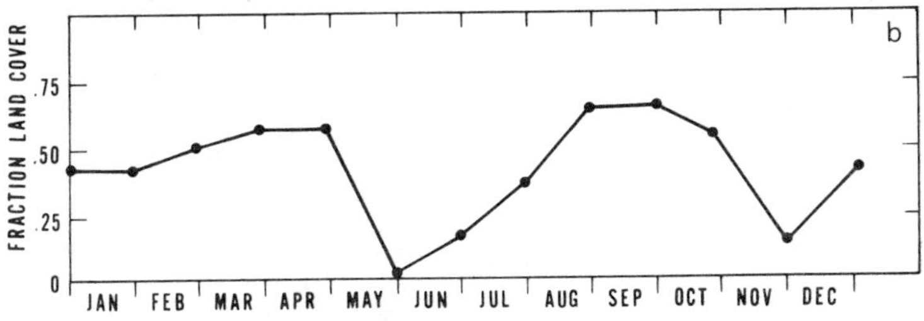

Fig. 12-8. Land Cover Algorithm in the ARM Model

needed to develop guidelines for the specification of fines storage as affected by tillage and other agricultural management operations.

To account for the settling and compaction of the soil surface following tillage, the SCMPAC parameter has been included in the ARM Model to decrease on each nonrain day the amount of fines available for transport. SCMPAC is essentially a first-order decay parameter that reduces the soil fines storage in the following manner:

$$SRER(T) = SRER(T-1)*(1.0-SCMPAC) \qquad (12\text{-}4)$$

where SRER(T) = soil fines storage on day T, tonnes/ha
SRER(T-1) = soil fines storage on day T-1, tonnes/ha
SCMPAC = soil compaction factor, day^{-1}

On the watershed the decrease in soil fines is due to natural settling and aggregation of particles and compaction by rainfall and agricultural operations. These processes are complex functions of soil characteristics, meteorologic conditions, and agricultural practices. The use of the SCMPAC parameter is a simple approximation of the overall impact of these processes. Future research should consider more detailed methods of representation.

The sediment production parameters of the ARM Model are listed in Table 12-4. The present version of the sediment loss algorithms is a stepping stone on the continuing path of model development. With additional testing, refining and retesting, a greater understanding of the erosion process and methods for its simulation will evolve.

Table 12-4. Sediment Production Parameters

COVPMO	Fraction of land cover on a monthly basis (12 values)
TIMTIL	Time when soil is tilled (Julian day, i.e., day of the year, e.g., January 1 is 1, December 31 is 365 or 366, etc.), (12 dates).
YRTIL	Corresponding year (last two digits only) for TIMTIL (12 values).
SRERTL	Find deposits produced by tillage corresponding to TIMTIL and YRTIL (12 values).
JRER	Exponent of rainfall intensity in soil splash equation.
KRER	Coefficient in soil splash equation.
JSER	Exponent of overland flow in sediment washoff equation.
KSER	Coefficient in sediment washoff equation.
SRERI	Initial detached soil fines deposit.
SCMPAC	Soil compaction factor, day^{-1}.

Pesticide

Adsorption/desorption: Once the hydrology and sediment production of a watershed have been simulated, the process of pesticide adsorption/desorption onto sediment particles determines the amount of pesticide loss which will occur. This process establishes the division of available pesticide between the water and sediment phases, and thus specifies the amounts of pesticide transported in solution and on sediment. The algorithm employed to simulate this process in the ARM Model is as follows:

$$X/M = KC^{(1/N)} + F/M \tag{12-5}$$

where:
- X/M = pesticide adsorbed per unit soil, μg/gm
- F/M = pesticide adsorbed in permanent fixed state per unit soil. F/M is less than or equal to FP/M, where FP/M is the permanent fixed capacity of soil in μg/gm for pesticide. This can be approximated by the cation or anion exchange capacity for that particular soil type.
- C = equilibrium pesticide concentration in solution, mg/l
- N = exponent
- K = coefficient

Basically this algorithm is comprised of an empirical term, F/M, plus the standard Freundlich single-valued (SV) adsorption/desorption isotherm (Fig. 12-9a). The empirical term, F/M, accounts for pesticides which are permanently adsorbed to soil particles and will not desorb under repeated washing. As indicated in Fig. 12-9a, the available pesticide must exceed the capacity of the soil to permanently adsorb pesticides before the adsorption/desorption equilibrium is operative. Thus the pesticide concentration on soil particles must exceed FP/M before the equilibrium soil and solution pesticide concentrations are evaluated by the Freundlich curve. An in-depth description and discussion of the underlying assumptions is presented in the PTR Model report.

A major conclusion of the PTR Model development work was that the above algorithm did not adequately represent the division of pesticides between the sediment and solution phases. This was especially true for pesticides which are transported by both sediment and surface runoff, i.e., soluble pesticides which also adsorb onto soil particles. Research has indicated that the assumption of single-valued adsorption/desorption (Fig. 12-9a) is not valid for many pesticides (Davidson and McDougal, 1973). In these cases, the adsorption and desorption processes would follow different curves as indicated in Fig. 12-9b. Although a controlled laboratory experiment cannot hope to duplicate the vagaries of nature

Fig. 12-9a. Single-Valued Adsorption/Desorption Algorithm

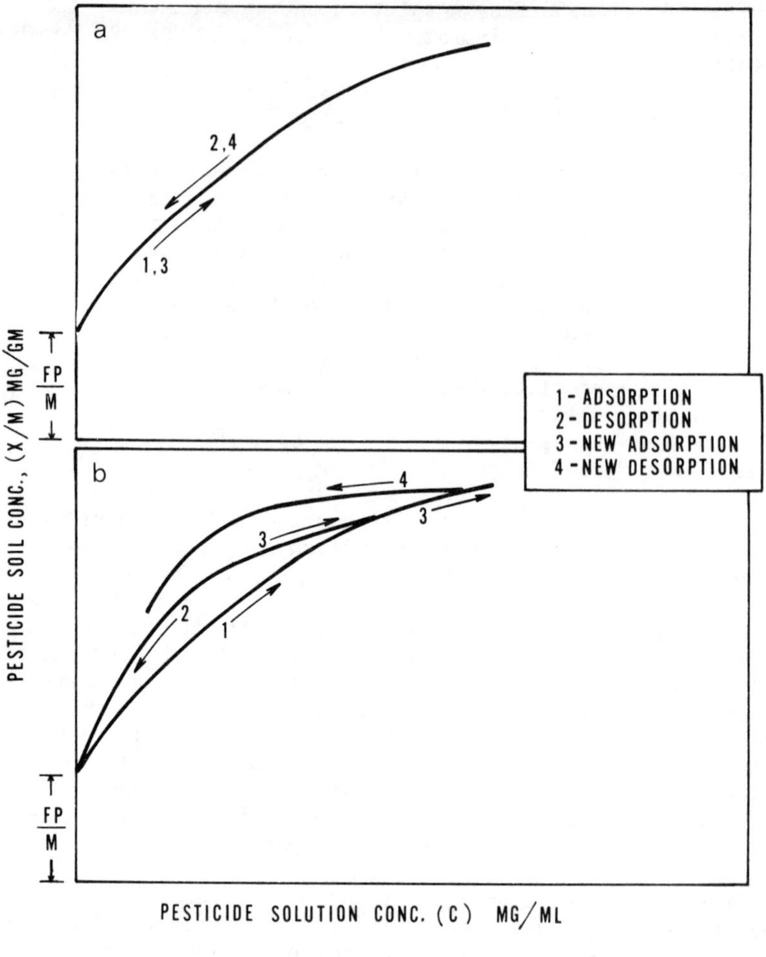

Fig. 12-9b. Non-Single-Valued Adsorption/Desorption Algorithm

Fig. 12-9. Adsorption/Desorption Algorithms in the ARM Model present in a field situation, the basic mechanisms should be similar in both circumstances. Since field data have been inconclusive, the present version of the ARM Model allows the user to specify the use of either single-valued (SV) as in Fig. 12-9a or non-single-valued adsorption/desorption as in Fig. 12-9b). Table 12-5 defines the pesticide simulation parameters in the ARM Model. The DESORP parameter indicates the adsorption/desorption function to be used. The NSV algorithm (Fig. 12-9b) utilizes the above SV algorithm (path No. 1) as a base from which different desorption curves are calculated. The form of the desorption curve is identical to Eq. (12-5) except that K and N values are replaced by K' and N' respectively. The prime denotes the desorption process. The user specifies the N' value as an input parameter (NP), and the ARM Model calculates K' from the following expression based on work by Davidson, Mansell, and Baker (1973).

$$K' = K^{(N/N')} S_{max}^{(1-N/N')} \qquad (12\text{-}6)$$

Table 12-5. Pesticide Simulation Parameters

APMODE	Application mode, SURF-surface applied, SOIL-soil applied
DESORP	NO-single-valued adsorption/desorption used, YES-non-single-valued adsorption/desorption algorithm used
PSSZ	Initial pesticide storage in surface area
PSUZ	Initial pesticide storage in upper zone
PSLZ	Initial pesticide storage in lower zone
PSGZ	Initial pesticide storage in groundwater zone
TIMAP	Time of pesticide application (Julian day) (12 values)
YEARAP	Year of pesticide application (last two digits only)(12 values)
SSTR	Pesticide application for entire watershed (12 values)
CMAX	Maximum solubility of pesticide in water
DD	Permanent fixed adsorption capacity
K	Coefficient in Freundlich adsorption equation
N	Exponent in Freundlich adsorption equation
NP	Exponent in Freundlich desorption equation
DDG	Julian day when KDG(1) begins (max. of 12 values)
YDG	Corresponding year in which KDG applies
KDG	Pesticide decay rate (per day) (max. 12 values)

where: K' = desorption coefficient
K = adsorption coefficient
N' = desorption exponent
N = adsorption exponent
S_{max} = solution pesticide concentration prior to initiating desorption

When the desorption process is initiated, the maximum attained solution concentration S_{max}, is utilized with K, N, and N' to calculate a value of K'. As desorption continues (path No. 2), the model continues to use the K' and N' values to calculate the soil and solution concentrations. When re-adsorption is initiated (path No. 3), the model follows the desorption curve back to the junction with the SV adsorption curve, and continues on this curve until desorption again occurs. At the new occurrence of desorption, a new K' is calculated resulting in a new desorption curve (path No. 4). The process is continued indefinitely producing a series of desorption curves emanating from the base SV adsorption curve.

Pesticide Attenuation: The attenuation processes of degradation and volatilization of pesticides are critical to the simulation of pesticide loss since these mechanisms control the mass of chemical available for transport at any time following application. Highly volatile or degradable pesticides can be reduced to insignificant levels after only one month of exposure in the field. On the other hand, non-volatile or non-degradable pesticides can continue to contribute to stream pollution months, or possibly years, after the initial application. In addition, volatilization and degradation, by microbial, chemical, or photochemical means, often accounts for the great majority of the applied pesticide removed from the soil environment; surface runoff and erosion removal of pesticides is generally a small fraction of the total application amount.

The ARM Model includes a step-wise first-order pesticide attenuation function that allows the use of different degradation rates for separate time periods following application. The function calculates the combined degradation of pesticides by volatilization, microbial degradation, and other attenuation mechanisms. This approach was chosen after evaluating both simpler and more sophisticated degradation models (Donigian, et al., 1977). Because of the critical impact of pesticide attenuation, additional research is needed to develop predictive models that can represent the effects of soil and environmental conditions on pesticide persistence in the soil.

Nutrients

Water pollution from agricultural land has been increasing due to greater use of machinery and chemicals to improve crop yields. Chemicals are applied to prevent unwanted plants (herbicides) and animals (pesticides), and to increase available plant nutrients (fertilizers). After application, herbicides persist in the soil until they are degraded to less harmful compounds or are removed from the soil by washoff or leaching. Fertilizers on the other hand are applied as a supplement to nutrients present in the soil profile. Plants do not absorb all the applied fertilizer. Typically, only 5 to 10 percent of the applied phosphorus and about 50 percent of the applied nitrogen is recovered in the crop. The remaining nutrients can be retained in the soil in unavailable forms or lost by volatilization, leaching, and surface washoff. Although greater fertilizer application will improve crop yields, it will increase nutrients in the soil available for contamination of streams and groundwaters.

Methods for nutrient control can be investigated and developed through costly field experiments or through the use of a mathematical model of the important processes occurring on and in the soil profile.

Nutrient simulation in the ARM Model attempts to predict nutrient losses from erosion, surface washoff, leaching, and biological conversion. With testing and calibration the model could be used to develop fertilizer management plans to maximize fertilizer efficiency and minimize the water quality impact of fertilizer use.

Nutrient Cycles

Nitrogen: Many nitrogen compounds are indigenous to the soil and undergo chemical and biological transformations of importance to crop production and pollution control. A general nitrogen cycle for agricultural lands is depicted in Fig. 12-10. Most soil nitrogen is in the organic form as decaying plant residues and rather resistant soil humus (Stevenson, 1965). Organic nitrogen can be broken down to ammonia through the process of mineralization, also called ammonification. Ammonia is usually strongly adsorbed to soil surfaces and can undergo nitrification to nitrite and nitrate. Nitrite is rapidly converted to nitrate which is the most common form of the mobile nitrogen compounds. Dissolved nitrates can be removed by overland flow and interflow, and leached to groundwater. Biologically, nitrate can be absorbed by plants, reduced anaerobically to various nitrogen gases and immobilized by microorganisms in the presence of nitrogen-deficient organic material. Nitrogen absorbed by plants is often lost from the soil through harvesting. Nitrogen input to the soil occurs by a number of pathways including rainfall, plant residues, dry fall of dust and dirt, biological fixation of atmospheric nitrogen, and direct application of fertilizer nitrogen. Although the soil nitrogen cycle is quite complex, the major pathways can be sufficiently quantified to allow mathematical simulation.

Phosphorus: While phosphorus does not exist in as many forms as nitrogen, phosphorus compounds undergo transformations important to agriculture as shown in Fig. 12-11. Organic

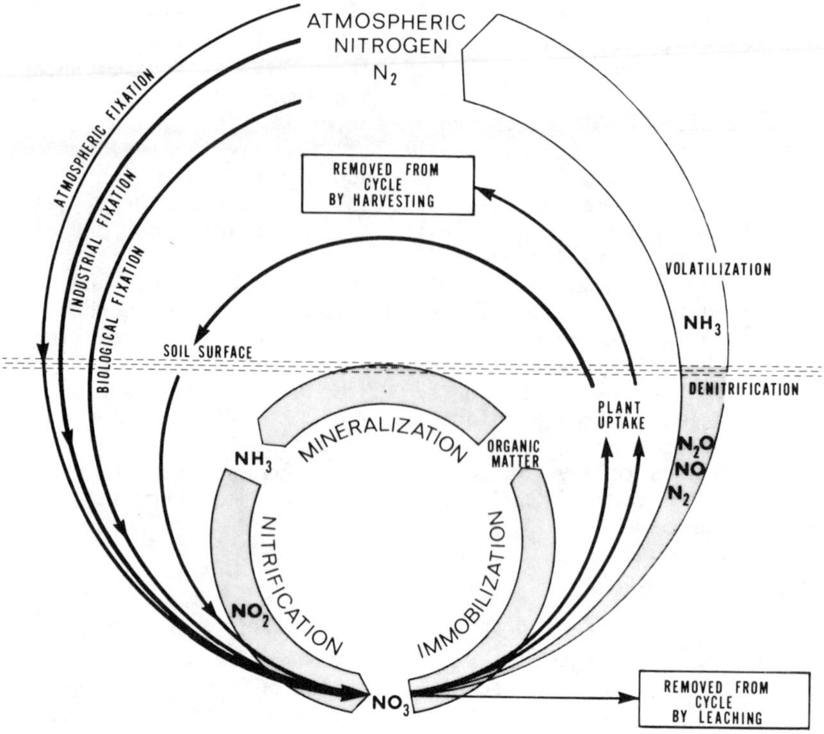

Fig. 12-10. Nitrogen Cycle

phosphorus can be mineralized to inorganic phosphates and under special circumstances, the reaction can be reversed to immobilization of inorganic phosphates to organic phosphorus. Inorganic phosphates are either strongly adsorbed to clay particles, or present as insoluble calcium, magnesium, iron or aluminum phosphates. Soluble phosphate concentration rarely exceeds 0.2 mg/l. Thus, the major mechanism for the loss of phosphorus compounds is soil erosion (Loehr 1974).

A number of models have been developed recently to predict nutrient washoff from agricultural lands. Models in which actual soil processes were considered are discussed below.

A complex watershed model for irrigated land was developed by Dutt, Shaffer, and Moore (1972) at the University of Arizona. The model includes procedures for calculating moisture flow and chemical and biological nutrient reactions. The nitrogen transformation rates were developed by regression analysis on data from arid regions.

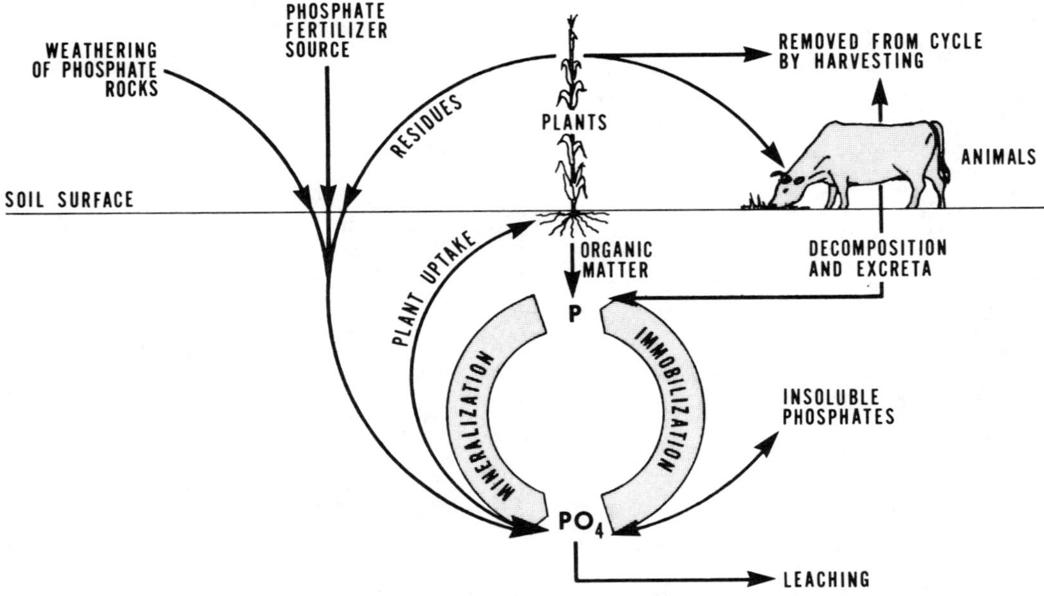

Fig. 12-11. Phosphorus Cycle

Hagin and Amberger (1974) have developed a computer model for predicting nitrogen and phosphorus movement and transformations on agricultural land. They used the IBM Continuous System Modeling Program (CSMP), for simulating ecological processes and transport phenomena in the soil. The model includes mineralization and immobilization of nitrogen, nitrification, denitrification, sediment washoff of phosphate and transport of oxygen and heat.

The Agricultural Research Service has developed the Agricultural Chemical Transport Model (ACTMO) which includes hydrologic, sediment, and chemical transport simulation (Frere, et al., 1975). The nitrogen simulation considers mineralization of organic nitrogen to nitrate, plant uptake of nitrate, and nitrate removal by overland flow and leaching.

A preliminary model of nitrogen transformations in agricultural soils was reported by Mehran and Tanji (1974). They developed a complex nitrogen transformation model for batch reactors assuming all reactions proceed by first-order kinetics. Through adjustment of reaction rates, the model was able to reproduce data collected in four different laboratory experiments.

Algorithm Descriptions: In the ARM Model all chemical and biological reactions are represented by first-order kinetics. The rate of a first order reaction is proportional to the amount of the reactant; the proportionality factor is the rate constant.

With first-order kinetics, the biological conversion of compound A to compound B with reaction rate constant k can be expressed as

$$A \xrightarrow{k} B \qquad (12\text{-}7)$$

The rate of this reaction is expressed in terms of the rate of change in A and B with time or

$$-\frac{d}{dt}\{A\} = \frac{d}{dt}\{B\} = k\{A\} \qquad (12\text{-}8)$$

Solution of the differential equation for A and B yields

$$A = A_o e^{-kt} \qquad (12\text{-}9)$$

$$B = A_o\ (1 - e^{-kt}) \qquad (12\text{-}10)$$

where: A = initial amount of compound A at time t = 0

In chemical reactions there is also a backward reaction of B going to A, expressed as

$$A \underset{k_b}{\overset{k_f}{\rightleftarrows}} B \qquad (12\text{-}11)$$

and

$$-\frac{d}{dt}\{A\} = \frac{d}{dt}\{B\} = k_f\{A\} - k_b\{B\} \qquad (12\text{-}12)$$

where: k_f = forward rate constant

k_b = backward rate constant

At equilibrium when the rate of change in concentration is zero, Eq. (12-10) becomes

$$0 = k_f\{A\} - k_b\{B\} \qquad (12\text{-}13)$$

On solving for A, a linear relationship is obtained between A and B at equilibrium

$$\{A\} = \frac{k_b}{k_f}\{B\} \qquad (12\text{-}14)$$

Chemical reactions that proceed rapidly can be viewed as instantaneously obtaining equilibrium or quickly approaching equilibrium with rapid forward and backward reaction rates. Modeling of adsorption-desorption chemical reactions with first-order kinetics produces a linear relationship between adsorbed and dissolved compounds at equilibrium. This is a simplification of the equilibrium relationship defined by more complex methods such as the Freundlich and Langmuir equations. A general discussion of adsorption-desorption reaction kinetics is given by Oddson, et al. (1970).

In chemical and biological reactions, an increase in temperature will cause an increase in the reaction rate for a certain temperature range. Reaction rates can be adjusted for different temperatures by a simplification of the Arrhenius Equation (Sawyer and McCarthy 1967):

$$k_T = k_{35}\Theta^{(T-35)} \qquad (12\text{-}15)$$

where: k_T = reaction rate at temperature T

k_{35} = reaction rate at 35°C

Θ = temperature correction coefficient

T = temperature in degrees Celsius

Typically biological reaction rates will double with each 10 Celsius degree rise in temperature. This corresponds to Θ=1.07. For nutrient transformations in the ARM Model, the reaction rates are modified for temperatures less than 35°C. At temperatures of 35°C or greater, the reaction rates are assumed to remain constant. In this temperature range the assumption of a constant temperature correction coefficient, Θ, is doubtful, and different bacterial species demonstrate widely varying behavior. Each nutrient reaction rate requires its own temperature correction coefficient. Soil temperatures are calculated in the ARM Model with regression equations that relate the temperature in each soil layer to the air temperature (Donigian, et al, 1977).

Nitrogen Transformations: Six different forms of nitrogen and eight reaction rates are used to represent nitrogen transformations in the soil. Fig. 12-12a is a diagram of the nitrogen forms and their interaction. Table 12-6 presents the resulting system of coupled differential equations. The reaction rate equations for the specific transformations are developed below. The original version of the nutrient model (Donigian and Crawford 1976) has been modified as described by Donigian, et al. (1977). The modifications pertain to the plant uptake function, the use of nitrite and nitrate in combined form (as opposed to separate nitrite and nitrate components), and the use of concentration as opposed to mass per hectare in the nutrient transformations.

Mineralization and immobilization: These processes are difficult to measure independently. Researchers usually report only the net amount of mineralization or immobilization. The basic mechanisms occurring in the soil can be visualized as:

$$\text{Organic -N} \xrightarrow{\text{mineralization}} \text{NH4, NO3} \xrightarrow[\text{(uptake)}]{\text{microbial immobilization}} \text{protein complexes}$$

There is net mineralization when mineralization exceeds microbial uptake, and net immobilization when uptake exceeds mineralization. The amount of organic nitrogen in the soil far exceeds other nitrogen forms. Mineralization, even at the slow rate, can have considerable impact on the amount of inorganic nitrogen available for plant uptake and leaching. The most significant studies to date on quantifying the rate of organic nitrogen mineralization have been done by Stanford and others (1972, 1973, 1974) at the Agriculture Research Service, Beltsville, Maryland.

The mineralization rate equation used in the ARM Model is

$$-\frac{d}{dt}\{\text{ORG-N}\} = \text{KAM}\{\text{ORG-N}\}\Theta_{\text{KAM}}^{(T-35)} \qquad (12\text{-}16)$$

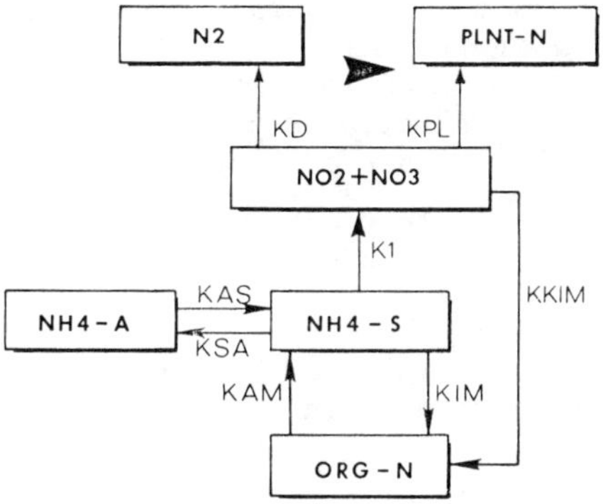

A. Nitrogen transformations in ARM model

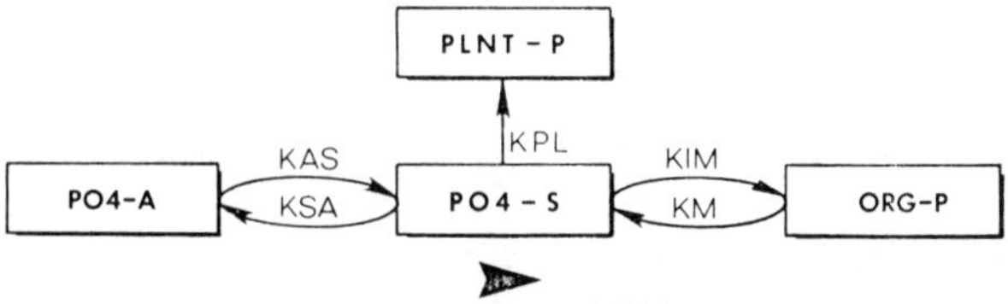

B. Phosphorus transformations in ARM model

Fig. 12-12. Nutrient Transformations in the ARM Model

where: ORG-N = organic nitrogen concentration, ppm
KAM = mineralization rate constant at 35°C per day
Θ_{KAM} = temperature correction coefficient for mineralization
T = soil temperature, °C

The decrease in organic nitrogen will result in an increase of ammonia as shown in Fig. 12-12a. At this time corrections for oxygen and moisture levels are not included, except for inhibition of the reaction at low moisture levels.

The ARM Model represents immobilization as potentially removing ammonia and nitrate according to the following equations.

$$-\frac{d}{dt}\{NH4\text{-}S\} = KIM\{NH4\text{-}S\}\Theta_{KIM}^{(T-35)} \qquad (12\text{-}17)$$

$$-\frac{d}{dt}\{NO3 + NO2\} = KKIM\{NO3 + NO2\}\Theta_{KKIM}^{(T-35)} \qquad (12\text{-}18)$$

where: KIM, KKIM = immobilization rate constants at 35°C per day

NH4-S, NO3+NO2 = ammonia in solution and nitrate/nitrite concentrations, ppm

Θ_{KIM}, Θ_{KKIM} = temperature correction coefficients

T = temperature, °C

Table 12-6. Coupled System of Differential Equations For Nitrogen Transformations

Organic Nitrogen:

$$\frac{d}{dt}\{\text{ORG-N}\} = \text{KIM}\,\{\text{NH4-S}\} + \text{KKIM}\,\{\text{NO3+NO2}\} - \text{KAM}\,\{\text{ORG}\}$$

Solution Ammonia:

$$\frac{d}{dt}\{\text{NH4-S}\} = \text{KAM}\,\{\text{ORG-N}\} - (\text{KSA} + \text{K1} + \text{KIM})\,\{\text{NH4-S}\} + \text{KAS}\,\{\text{NH4-A}\}$$

Adsorbed Ammonia:

$$\frac{d}{dt}\{\text{NH4-A}\} = \text{KSA}\,\{\text{NH4-S}\} - (\text{KAS})\,\{\text{NH4-A}\}$$

Nitrate:

$$\frac{d}{dt}\{\text{NO3+NO2}\} = \text{K1}\,\{\text{NH4-5}\} - (\text{KD} + \text{KKIM} + \text{KPL})\,\{\text{NO3+NO2}\}$$

Nitrogen Gas:

$$\frac{d}{dt}\{\text{N2}\} = \text{KD}\,\{\text{NO3+NO2}\}$$

Plant Nitrogen:

$$\frac{d}{dt}\{\text{PLNT-N}\} = \text{KPL}\,\{\text{NO3+NO2}\}$$

Nitrification: Nitrification is a two-step process in which ammonia is oxidized first to nitrite and then to nitrate. This is an important soil reaction because a largely immobile form of nitrogen, ammonia, is converted to a highly mobile form, nitrate, which may be absorbed by plants, lost by leaching and denitrification, or removed by surface runoff. Alexander (1965) provided a good description of the nitrification process and an overview of current research. Quantification of the nitrification process can be approached from a simplification of the works of A. D. McLaren (1970) of U.C. Berkeley who has published many articles relating nitrogen transformations to enzyme kinetics and bacterial growth dynamics.

In the ARM Model nitrification was originally represented as a two-step process but has been changed to a one-step mechanism assuming the rapid and immediate oxidation of nitrite to nitrate (Donigian, et al., 1977). The oxidation of solution ammonia to nitrate is simulated as:

$$-\frac{d}{dt}\{NH4\text{-}S\} = K1\ \{NH4\text{-}S\}\Theta_{K1}^{(T-35)} \qquad (12\text{-}19)$$

where: NH4-S = solution ammonia concentration, ppm
K1 = nitrification rate constant at 35°C per day
Θ_{K1} = temperature correction coefficient
T = temperature, °C

Denitrification: Until recently prediction of denitrification rates has not been possible although the mechanisms have been known for some time. Denitrification is favored in wet, poorly aerated soils that have sufficient decomposable organic matter. The tremendous increase in the use of nitrogen fertilizers and the possibility of losing over 30 percent of the applied nitrogen through denitrification has sparked recent interest in quantifying and predicting these losses (Broadbent and Clark 1965).

Denitrification is represented in the ARM Model as a single step process reduction of nitrate plus nitrite to nitrogen gas. The rate equation is:

$$-\frac{d}{dt}\{NO2 + NO3\} = KD\ \{NO2 + NO3\}\Theta_{KD}^{(T-35)} \qquad (12\text{-}20)$$

where: NO3 + NO2 = nitrate plus nitrite concentration, ppm
KD = denitrification rate constant at 35°C per day
Θ_{KD} = temperature correction coefficient
T = temperature, °C

In spite of the importance of oxygen level on the denitrification rate, it was not possible to include a correction for oxygen level because it is not simulated in the present model.

Thus, at this time, the denitrification reactions are either turned on or turned off all the time depending on the value of KD and the moisture level. Future work will attempt to include oxygen uptake and diffusion in the soil and allow for internal adjustment of denitrification rate as a function of oxygen level.

Plant Uptake: The primary mechanism for removal of nitrogen from agricultural land is through plant uptake. Viets (1965) provided a general review of nitrogen uptake by plants. Van der Honert and Hooymons (1955) showed that the rate of nitrate uptake was a first-order reaction at nitrate concentrations less than 5 mg/l and a zero-order reaction at higher concentrations. The effect of temperature and pH on the rate of uptake was also discussed.

The ARM Model represents plant uptake of nitrates according to the following equation:

$$\frac{d}{dt}\{PLNT\text{-}N\} = KPL\ \{NO3 + NO2\}\Theta_{KPL}^{(T-35)} \qquad (12\text{-}21)$$

where: PLNT-N = nitrogen concentration taken up by plants, ppm
NO3+NO2 = nitrate/nitrite concentration, ppm
KPL = plant uptake rate constant, per day
Θ_{KPL} = temperature correction coefficient
T = temperature, °C

The uptake rate KPL varies with crop growth and is calculated as

$$KPL = KPL_{max} * UPTK \qquad (12\text{-}22)$$

where: KPL_{max} = maximum rate of plant uptake, day^{-1}
UPTK = average fraction of maximum uptake occurring during a month

A KPL_{max} value and twelve monthly UPTK values are input by the user for the upper soil layers (surface and upper zones) and the lower zone to allow flexibility in specifying the time distribution, origin, and rate of nutrient uptake by plants.

All plant nitrogen is assumed to be removed during harvesting. Future work will need to evaluate the extent to which plant nitrogen contributes to soil nitrogen in the form of plant residues remaining on the watershed. However, application of organic nitrogen forms at harvesting can be used to approximate the impact of crop residues with the model.

Ammonia

Adsorption/desorption: Ammonia can exist in three different forms in the soil: dissolved in water, adsorbed to surfaces of soil particles and fixed inside crystal lattices. Mortland and Wolcott (1965) discussed the various ammonia complexes with clays but did not present a general theory to allow prediction of the different forms. Instead of developing a complex model for specific soil types and conditions, a much simpler approach was used that might represent a much broader range of soils. The ARM Model assumes that two forms of ammonia exist in the soil: the adsorbed ammonia attached to the soil particles, and dissolved ammonia which moves with the soil water. The rate of transfer from one type to the other is governed by first-order reactions. These reactions can be represented by

$$\{NH4\text{-}S\} \underset{KAS}{\overset{KSA}{\rightleftharpoons}} \{NH4\text{-}A\} \tag{12-23}$$

and the rate equations are

$$-\frac{d}{dt}\{NH4\text{-}S\} = \frac{d}{dt}\{NH4\text{-}A\} = KSA\{NH4\text{-}S\}\Theta_{KSA}^{(T-35)} - KAS\{NH4\text{-}A\}\Theta_{KAS}^{(T-35)} \tag{12-24}$$

where: NH4-S = ammonia concentration in solution, ppm
NH4-A = ammonia concentration adsorbed to soil, ppm
KSA = first-order rate constant for adsorption reaction at 35°C, per day
KAS = first-order rate constant for desorption reaction at 35°C, per day
Θ_{KSA}, Θ_{KAS} = temperature correction coefficients
T = temperature, °C

Usually very little ammonia is in solution; most is adsorbed to soil particle surfaces. This would correspond to an adsorption reaction rate much greater than the desorption rate or KSA>KAS.

Phosphorus Transformations: Phosphorus was assumed to exist in only four forms: organic phosphorus, solid phosphate compounds, dissolved phosphates, and phosphorus absorbed by plants. The reactions of mineralization-immobilization, adsorption-desorption, and plant uptake are modeled as first-order rates. A diagram of the phosphorus cycle as represented by the ARM Model is given in Fig. 12-12b, and Table 12-7 contains the system of coupled differential equations.

<u>Mineralization and immobilization</u>: Organic phosphorus is not as important in the phosphorus cycle as organic nitrogen is in the nitrogen cycle. Larsen (1967) reviewed the literature on soil phosphorus and did not present any general findings on mineralization and immobilization rates. In the ARM Model, phosphorus mineralization and immobilization mechanisms were assumed to be similar to the corresponding nitrogen processes. Thus, they are represented as

$$-\frac{d}{dt}\{ORG\text{-}P\} = KM\ \{ORG\text{-}P\}\Theta_{KM}^{(T-35)} \qquad (12\text{-}25)$$

$$-\frac{d}{dt}\{PO4\text{-}S\} = KIM\ \{PO4\text{-}S\}\Theta_{KIM}^{(T-35)} \qquad (12\text{-}26)$$

where: ORG-P = organic phosphorus concentration, ppm
PO4-S = phosphate concentration in solution, ppm
KM = first-order mineralization rate at 35°C per day
KIM = first-order immobilization rate at 35°C per day
Θ_{KM}, Θ_{KIM} = temperature correction coefficients
T = temperature, °C

Table 12-7. Coupled System of Differential Equations For Phosphorus Transformations

Organic Phosphorus:

$$\frac{d}{dt}\{ORG\text{-}P\} = -\ KM\ \{ORG\text{-}P\} + KIM\ \{PO4\text{-}S\}$$

Solution Phosphate:

$$\frac{d}{dt}\{PO4\text{-}S\} = KM\ \{ORG\text{-}P\} - (KIM + KSA + KPL)\ \{PO4\text{-}S\} + KAS\ \{PO4\text{-}A\}$$

Adsorbed and Combined Phosphate:

$$\frac{d}{dt}\{PO4\text{-}A\} = KSA\ \{PO4\text{-}S\} - KAS\ \{PO4\text{-}A\}$$

Plant Phosphorus:

$$\frac{d}{dt}\{PLNT\text{-}P\} = KPL\ \{PO4\text{-}S\}$$

Soil organic phosphorus is assumed to be insoluble and only leaves the watershed with the eroded sediment.

Adsorption/desorption: Organic phosphorus mineralization results in the release of inorganic phosphates which can remain in the soil solution; precipitate as sparingly soluble salts of calcium, magnesium, aluminum, or iron phosphates; or adsorb onto the surface of clay or calcium carbonate soil particles. The model represents these three forms of phosphate in two categories; that is, phosphates in solution, and phosphates in solid form including both adsorbed and precipitated forms. Solid phosphates will be referred to as adsorbed phosphates, and the transfer between solution and adsorbed phosphates is modeled by adsorption and desorption reactions:

$$\{PO4\text{-}S\} \underset{KSA}{\overset{KSA}{\rightleftarrows}} \{PO4\text{-}A\} \qquad (12\text{-}27)$$

The resulting rate expressions are

$$-\frac{d}{dt}\{PO4\text{-}S\} = \frac{d}{dt}\{PO4\text{-}A\} = KSA\{PO4\text{-}S\}\Theta_{KSA}^{(T-35)} - KAS\{PO4\text{-}A\}\Theta_{KAS}^{(T-35)} \qquad (12\text{-}28)$$

where: PO4-S = phosphate concentration in solution, ppm
PO4-A = phosphate adsorbed concentration, ppm
KSA = first-order rate constant for adsorption at 35°C, per day
KAS = first-order rate constant for desorption at 35°C, per day
Θ_{KSA}, Θ_{KAS} = temperature correction coefficients
T = temperature, °C

Plant uptake: Fried et al. (1957) studied the rate of phosphate uptake by plant roots under laboratory conditions and found that the absorption rate was approximately proportional to the solution concentration, thus a first order mechanism. Van der Honert, et al. (1955) showed that phosphate uptake was a first-order reaction up to 1.0 mg PO_4/1. Since soil solutions rarely exceed this concentration, a first-order uptake mechanism is a reasonable assumption.

The rate expression used in the ARM Model is

$$\frac{d}{dt}\{PLN\text{-}P\} = KPL\ \{PO4\text{-}S\}\Theta_{KPL}^{(T-35)} \qquad (12\text{-}29)$$

where: PLNT-P = plant phosphorus concentration, ppm
PO4-S = phosphates concentration in solution, ppm

KPL = first-order adsorption rate
Θ_{KPL} = temperature correction coefficient
T = temperature, °C

The monthly variability in plant uptake is represented in changes to KPL as discussed above for nitrogen uptake.

The nutrient model assumes that plant phosphorus can be removed from the watershed only by harvesting. This assumption is valid for plants, such as grain crops, that contain phosphorus largely in the portion harvested. Moreover, the conversion of phosphorus in plant residues to soil organic phosphorus is a slow process especially in dry, cold regions. However, in warm, humid areas and where substantial plant residues remain on the watershed, the conversion of plant phosphorus to soil organic phosphorus may be significant. Further development of the nutrient model will need to evaluate the importance of this process and possibly allow for its simulation.

Nutrient Model Assumptions and Summary

The nutrient model required many assumptions in its development. A review of these assumptions is essential to a full understanding of the model. The assumption of first-order kinetics is generally valid for chemical and biological reactions when the reactants are not in high concentrations. From the literature cited, it appears conditions existing in the soil are such that first-order kinetics is a reasonable assumption. Temperature correction of reaction rates using a simplified form of the Arrhenius equation is flexible and can closely approximate changes in rates reported in the literature. The reaction rates were assumed to be constant for temperatures greater than 35°C because the behavior of chemical and biological reactions is not well defined at high temperatures.

The environmental factors of pH, moisture, oxygen, and organic matter are not directly taken into account for reaction rate modification. Soil pH is relatively constant due to the high buffering capacity of the soil itself. Any pH correction could be done when the reaction rates are input to the model. Reaction rates are inhibited at low moisture levels because biological activity is dependent on soil moisture, but no other adjustment is performed in the present version of the model. Oxygen levels in the soil are needed to determine if oxidative processes like mineralization and nitrification, or reductive processes, like denitrification, will occur. Organic matter in the soil can deplete the oxygen in the soil and accelerate the rate of denitrification.

Some of the limitations in the nutrient model due to neglecting pH, moisture, oxygen, and organic matter can be circumvented by having separate reaction rates for each of the four soil layers. For example, the denitrification rate could be set to zero in the surface and upper zone because they are usually well aerated. Likewise the denitrification rate would be close to zero in the groundwater zone because of low organic content. Thus, the input of four values, one for each soil layer, is a temporary correction for soil properties and environmental factors at different depths.

12.3 MODEL RESULTS

Data Collection and Analysis Programs

The ARM Model development effort is supported by an extensive data collection and analysis program sponsored by the U.S. Environmental Protection Agency's Environmental Research Laboratory in Athens, Georgia (ERL-Athens). Test sites located in Georgia and Michigan have been instrumented for continuous monitoring and sampling of runoff and sediment. Collected samples are refrigerated on site and later analyzed for pesticide and nutrient content. In addition, meteorologic conditions are continuously monitored and soil core samples are taken and analyzed immediately following application and periodically throughout the growing season. Table 12-8 presents pertinent characteristics of the test watersheds. The individual programs in Georgia and Michigan are described in the ARM Model report (Donigian and Crawford, 1976).

During the current study testing efforts were concentrated on the P2 watershed (Fig. 12-13) in Georgia and the P6 watershed (Fig. 12-14) in Michigan. Both P2 and P6 are natural nonterraced watersheds. Data were available for the 1973 through 1975 growing seasons. P2 and P6 were monitored and analyzed for various pesticides and nutrient forms.

Runoff and Sediment Simulation Results

The P2 and P6 watersheds were selected for simulation testing of the ARM Model because both pesticides and nutrients had been applied and then sampled in the runoff and the soil profile. Both are nonterraced watersheds and P6 is the larger of the Michigan watersheds. Also the data collected on these watersheds contained a minimum of inconsistencies and errors. This combination of requirements provided the best conditions in which to test all components of the ARM Model. Model testing for runoff and sediment loss was performed on the P2 watershed for the period of record, May 1973 through September 1975. Data for the P6 watershed prior to May 1974 was questionable due to uncontrolled seepage under the outflow weir. This was corrected in April 1974.

Simulation of hydrology and sediment processes must be sufficiently accurate to allow evaluation of the pesticide and nutrient simulation results. Pesticides and nutrients are transported from the watershed either in solution in the runoff water or adsorbed onto eroded sediment particles. Thus,

Table 12-8. Watersheds for ARM Model Testing

Watershed Designation	Location	Owner/ Operator	Area (ha)	Mean Elevation (m above msl)	Soils	Conservation
P1	Watkinsville, Georgia	USDA/EPA	2.70	238	Cecil sandy loam	Non terraced grass waterway installed 11/74
P2	Watkinsville, Georgia	USDA/EPA	1.30	231	Cecil sandy loam	non terraced
P3	Watkinsville, Georgia	USDA/EPA	1.26	239	Cecil sandy loam	terraced with grass waterway
P4	Watkinsville, Georgia	USDA/EPA	1.38	239	Cecil sandy loam	terraced with grass waterway
P6	East Lansing, Michigan	Michigan State Univ.	0.80	272	Spinks loamy sand, also Traverse/Hillsdale sandy loam	non terraced
P7	East Lansing, Michigan	Michigan State Univ.	0.55	271	Spinks loamy Sand, also Tuscola Sandy loam	non terraced

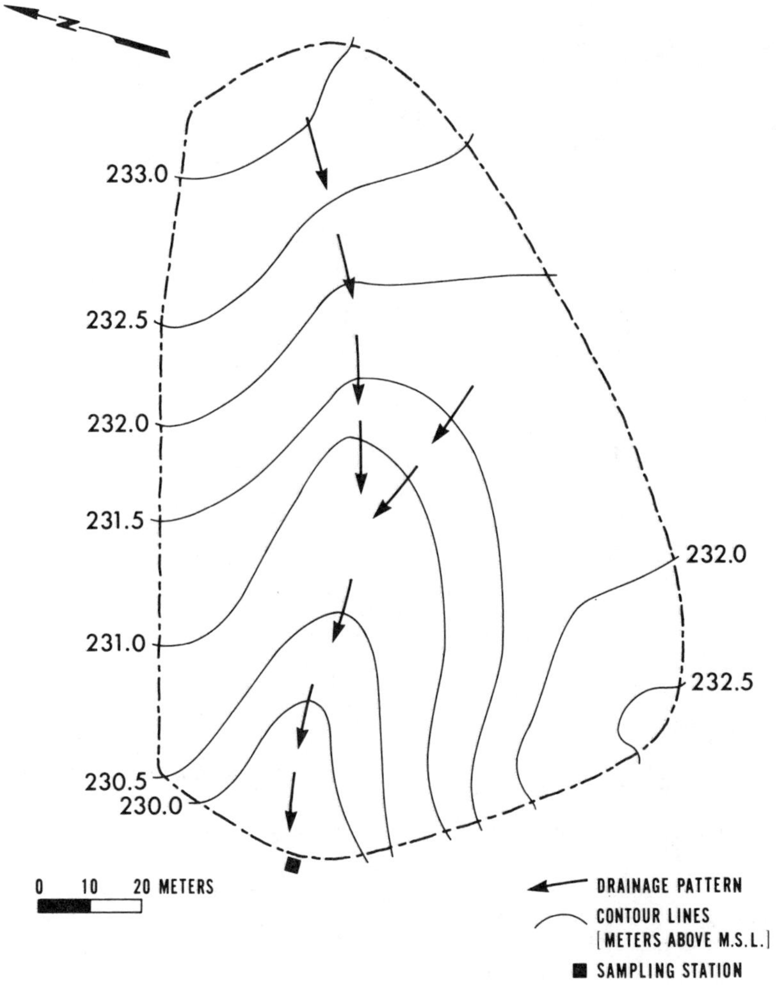

Fig. 12-13. P2 Watershed Watkinsville, Georgia (1.3 ha)

pesticide and nutrient runoff results are dependent on the simulated runoff and sediment loss. Similarly, the simulated vertical movement of pesticides and nutrients in the soil profile depends on the model's representation of infiltration, percolation, and soil moisture processes.

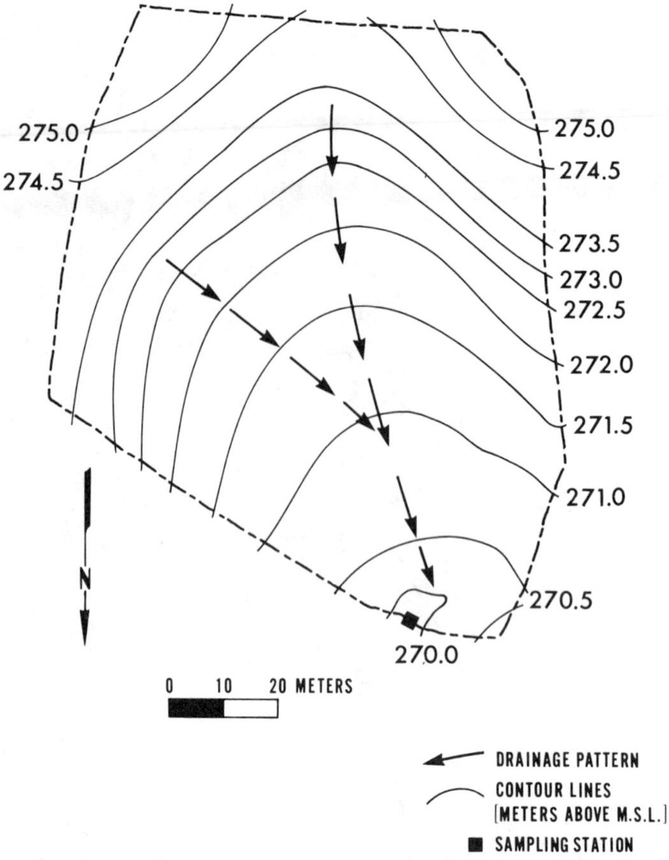

Fig. 12-14. P6 Watershed East Lansing, Michigan (0.8 ha)

Although the agreement between simulated and recorded runoff and sediment loss discussed below is generally good, discrepancies do exist. Some are a result of problems or errors in data collection and analysis. Such obvious errors must be considered when evaluating simulation results. Others are due to limitations of the hydrology and sediment algorithms of the model. Recommendations on how these limitations can be overcome and where additional research is necessary will be discussed at the end of this section.

P2 Watershed Simulation: A monthly volume comparison of simulated and recorded runoff on the P2 watershed (Fig. 12-15) shows that the simulation of runoff is good for the important summer months. In the winter months (December through March) there is the tendency to over-simulate runoff.

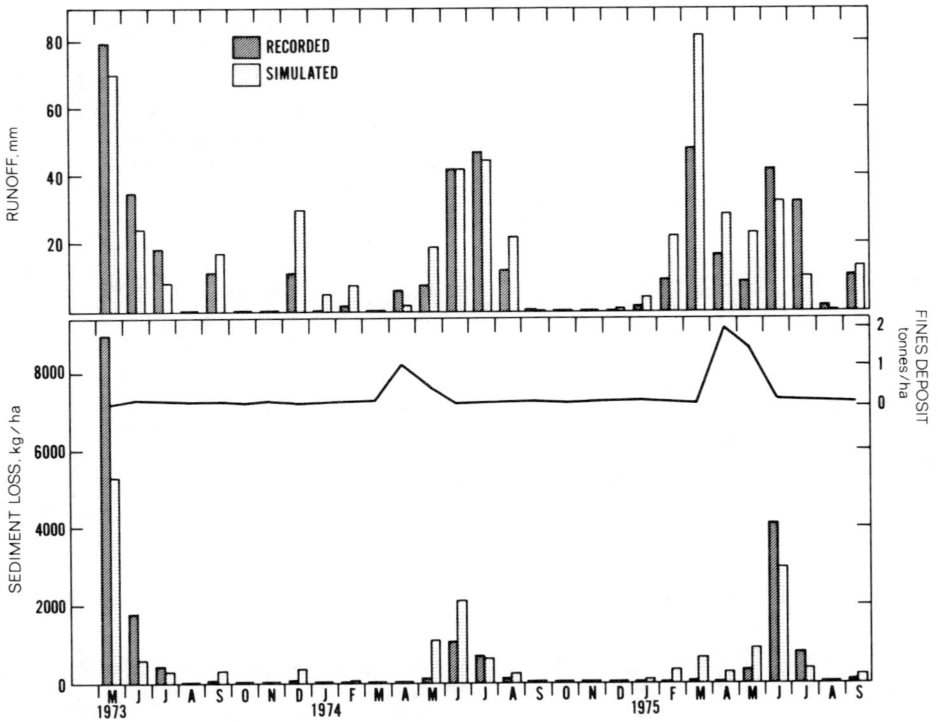

Fig. 12-15. Monthly Runoff and Sediment Loss From the P2 Watershed

The 1974 recorded sediment loss is lower than the other years. In 1974 tillage and planting occurred in late April and the first major sediment producing event occurred about 60 days later. In 1973 and 1975, tillage and planting was in mid-May with the first major event about 15-20 days later. Thus, the extra lag time and intervening changes on the watershed in 1974 appears to be the major factor for the lower sediment loss.

Sediment loss as represented in the model, results from rainfall detachment of soil particles and transport of these particles or fines to the stream by surface runoff. In addition, production of soil fines by tillage operations significantly contributes to the total fines available for transport. As shown in Fig. 12-15, the major sediment loss occurs within 2 to 3 months following tillage operations. Changes in crop canopy and settling and compaction of the soil surface are the major processes occurring in this intervening period that determine the sediment-producing impact of the first event following tillage. The ARM Model accepts 12 monthly crop cover values for the first day of each month; crop cover on any day is determined by linear interpolation. Thus for multi-year simulation runs the crop cover is the same for corresponding days in each year of the simulation. For the P2 watersheds, the average cover calculated by this procedure underestimates the crop cover in 1974 and overestimates the cover in 1973 and 1975. This effect contributes to the differences in Fig. 12-15. Separate calibration runs with different cover values for each year would improve the agreement between the simulated and recorded sediment loss.

The individual hydrographs and sediment graphs for the P2 watershed show good agreement between simulated and recorded results for the major storms during the summers of all three years (Figs. 12-16, 12-17). However, the effect of tillage in dampening the peak of the recorded hydrograph and retarding the overland flow is noticable in the storm of June 11, 1975 (Fig. 12-17). The hydrologic algorithms included in the ARM Model do not handle this man-made phenomenon. In general, an accurate runoff simulation will produce reasonable sediment simulation results on the P2 watershed.

P6 Watershed Simulation: Like the P2 watershed, the monthly runoff and sediment loss simulation for the P6 watershed is generally quite good (Fig. 12-18). However, problems with the simulation of snowmelt and runoff during the months of February and March 1975 are evident. These problems, which result in an under simulation of runoff during these months, are attributed to difficulties in properly simulating snowmelt. During winter periods frozen ground conditions severely decrease infiltration rates resulting in greater runoff than would otherwise occur. Also, temperature gradients in the soil profile can result in upward movement of moisture that would be available for runoff during the thaw periods.

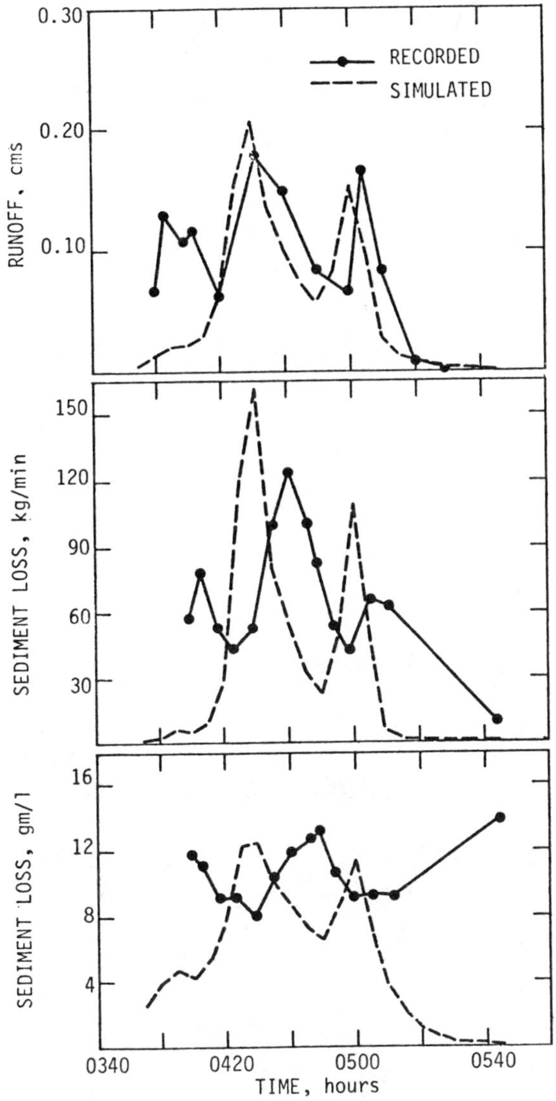

Fig. 12-16. Runoff and Sediment Loss From the P2 Watershed for the Storm of May 28, 1973

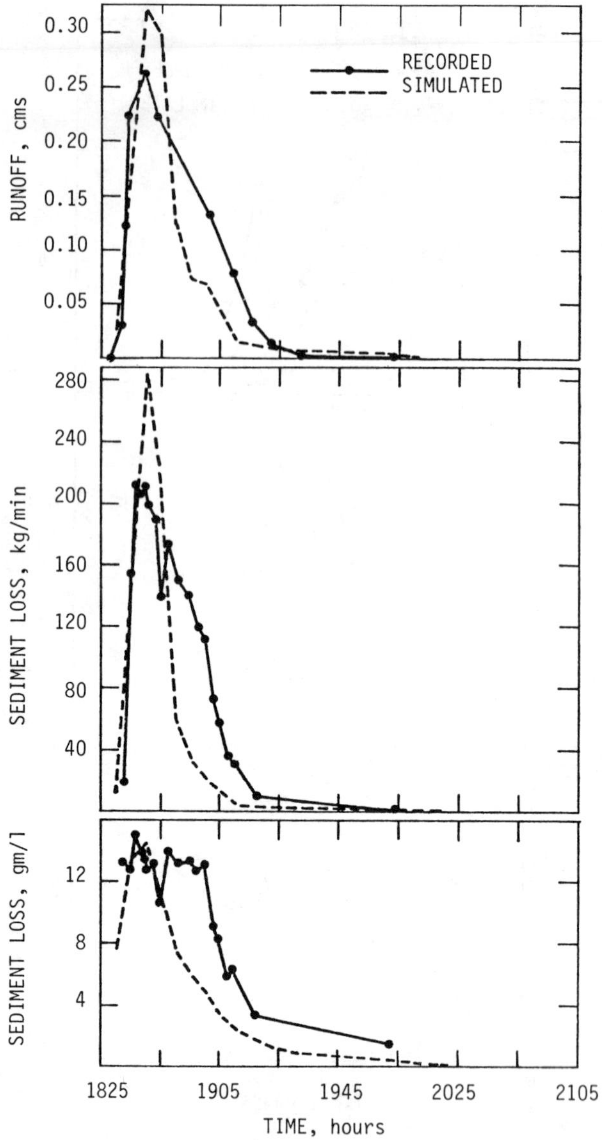

Fig. 12-17. Runoff and Sediment Loss From the P2 Watershed for the Storm of June 11, 1975

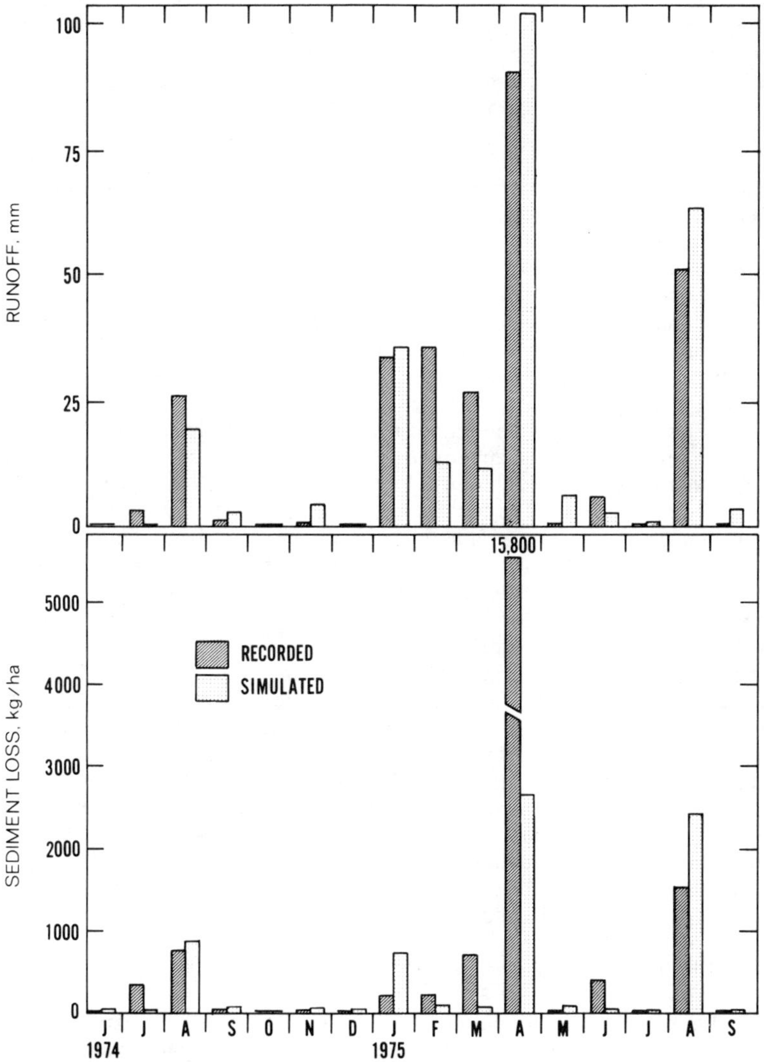

Fig. 12-18. Monthly Runoff and Sediment Loss From the P6 Watershed

Except for April 1975, simulated and recorded results compared well for monthly sediment loss on the P6 watershed, as shown in Fig. 12-18. The April discrepancy is due entirely to simulation of the April 18 storm (Fig. 12-19). Runoff during this storm is simulated quite well while sediment is grossly under-simulated. The sediment concentrations appear to be low because the sediment algorithms under-simulated the sediment fines available for transport. Freeze-thaw cycles during the winter could have produced fines which were washed from the watershed during this first major spring storm. Also, detachment or scour by runoff, which is not presently simulated, likely contributed to the high recorded sediment concentrations especially with the saturated surface soil conditions existing on the watershed. Detachment by runoff is a significant mechanism to some degree on most watersheds. Its incorporation and simulation in the ARM Model would allow better representation of the highly variable nature of erosion, and should be the next major improvement to the sediment algorithms. Except for the April 1975 storm the simulated and recorded runoff and sediment loss (Fig. 12-20) for individual storm events is generally quite good.

A discussion of the runoff and sediment loss simulation is not complete without mention of mass removal and concentration graphs when viewing sediment loss results. Concentration (measured in mg/l) fluctuates according to the amount of runoff water available. It is an indication of the importance of the possible pollutant (in this case sediment) in the waterway and directly influences the quality of the receiving waters. This type of measurement is important in streams containing water year round, as the organisms in the streams are sensitive to changes in concentrations of pollutants and particularly toxic substances. However, the streams in the P2 and P6 watersheds exist only during storm events. They receive no groundwater and therefore are simply dry waterways when surface runoff and interflow are not occurring. In cases like these, mass removal (kg/min) is a better indicator of the effects of watershed management and the pollutant loss. Mass removal also shows a direct connection between the pollutant loss and its transport mechanism, runoff. Thus, it is a good indication of the ability to simulate the combined runoff-sediment loss process.

Pesticide Runoff Simulation Results

The pesticides paraquat (1, 1'-dimethyl-4, 4-bipyridinium ion) and atrazine (2-chloro-4-ethylamino-6-isoproplyamino-s-triazine) were applied to both the P2 and P6 watersheds during the 1973-1975 growing seasons. These pesticides have been simulated by the ARM Model and the amount of pesticide runoff compared with recorded data.

Paraquat is a highly ionic herbicide that rapidly and irreversibly adsorbs onto sediment particles. Thus, the amount of paraquat washed off a watershed is directly related to the

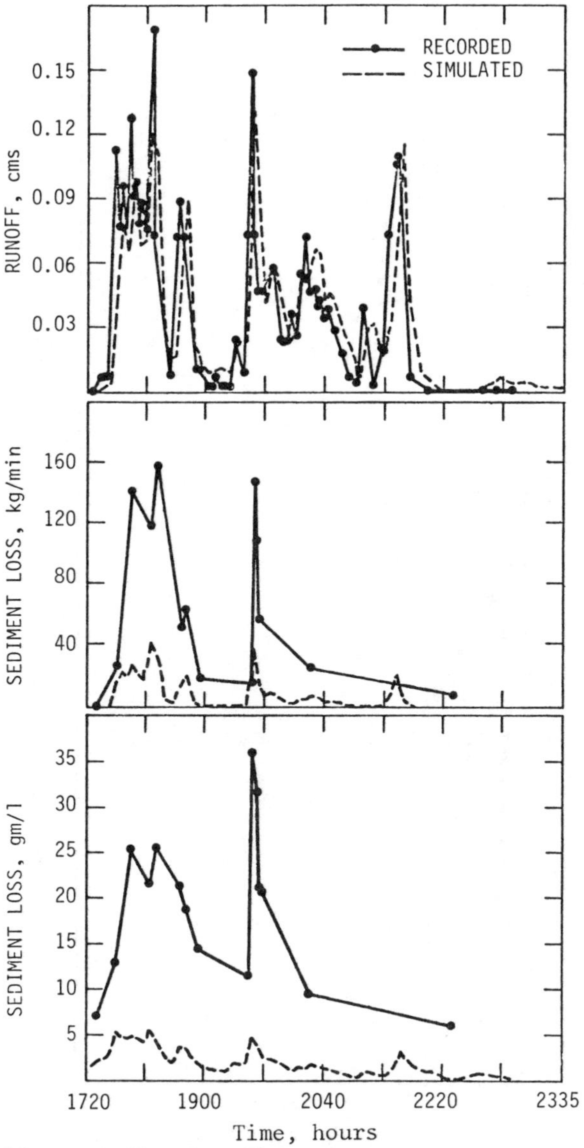

Fig. 12-19. Runoff and sediment Loss From the P6 Watershed for the Storm of April 18, 1975

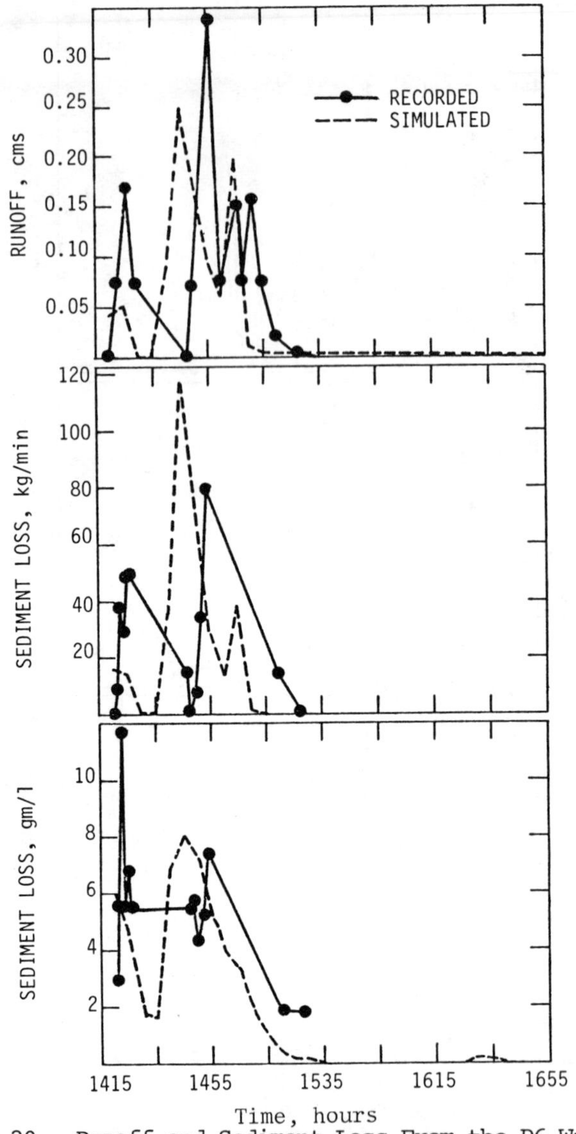

Fig. 12-20. Runoff and Sediment Loss From the P6 Watershed for the Storm of August 20, 1975

amount of sediment eroded. By comparison, atrizine both dissolves in water (35 ppm) and adsorbs on sediment. Thus, the study of atrazine allows the comparison of the single-valued (SV) and nonsingle-valued (NSV) adsorption/desorption pesticide functions. Evaluations of these functions is important because the majority of pesticides are transported by both runoff and sediment. The division between the water and sediment phase is critical to the evaluation of the potential pollutional impact of different pesticides. Highly soluble pesticides will infiltrate deeper into the soil profile than less soluble ones. Soluble pesticides will be affected by practices reducing runoff while soil erosion control will limit the washoff of pesticides transported on sediment particles. Thus, determining the ability of the SV and NSV algorithms to correctly divide a pesticide between its adsorbed and solution phase coupled with an accurate degradation approach is essential to a good simulation of pesticide runoff.

Monthly Simulation Results: The comparison of monthly simulated and recorded runoff amounts of paraquat on the P2 watershed (Fig. 12-21 and Table 12-9) is good. Paraquat, as noted above, is not found in solution and, therefore, is transported on sediment. Paraquat simulation is high compared to the recorded data in May and June, 1974. A review of the P2 sediment loss simulation will show that these months are oversimulated for sediment loss. Thus, a better simulation of sediment removed on the P2 watershed will result in closer agreement between simulated and recorded paraquat removal for this period.

The P2 atrazine results also show a good comparison between simulated and recorded monthly runoff values (Fig. 12-21 and Table 12-9). Figure 12-21 also shows the results of using both the SV and NSV algorithms for atrizine simulation. In addition, a breakdown is given for the recorded atrizine transported in solution and on sediment. Atrazine runoff results for SV and NSV algorithms differ little except in the first summer; May and June of of 1973. However, for the three summers when atrazine was applied on the P2 watershed, the NSV results generally look better than the SV results in terms of removal in solution, which is the major removal mechanism. Little atrazine is removed on sediment (less than .15 percent of the total amount of atrazine removed). In comparing the NSV and SV simulation of atrazine removal, the NSV results produce more removal, although less than recorded. But because of the small amounts of atrazine measured as removed from the P2 watershed it cannot be said that the results are conclusive. Also, problems arise from the variability in determining the value for K in the Freundlich adsorption/desorption algorithm. Because of the lack of laboratory data evaluating the Freundlich constants for the watershed soils exact values have not been determined. Moreover laboratory determinations will often

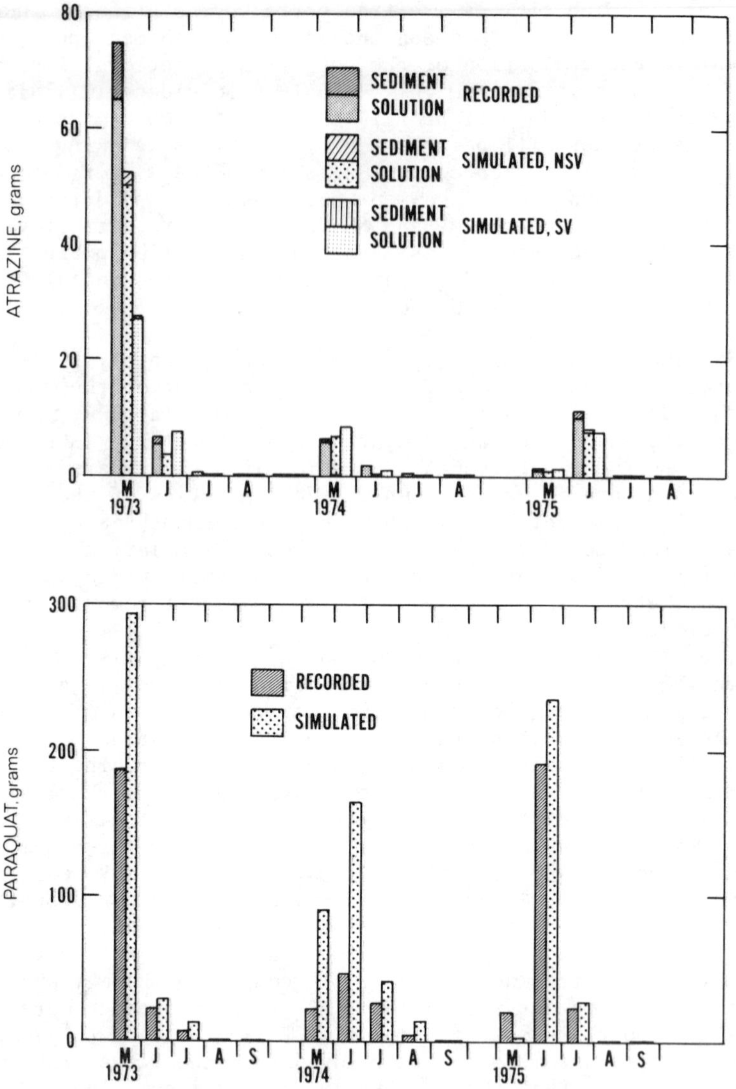

Fig. 12-21. Monthly Pesticide Removal From the P2 Watershed

Table 12-9. Pesticide Runoff Simulation Results for the P2 Watershed

		Atrazine						Paraquat	
		Recorded		NSV Simulation		SV Simulation		Recorded	Simulated
		(sed)	(sol)	(sed)	(sol)	(sed)	(sol)	(sed)	(sed)
1973	May	10	65	2	50	1	27	188	294
	Jun	1	5	0	4	0	8	23	29
	Jul	0	1	0	0	0	0	7	14
	Aug	0	0	0	0	0	0	0	0
	Sep	0	0	0	0	0	0	0	13
	Total	11	71	2	54	1	35	218	350
1974	May	1	6	0	7	0	9	23	91
	Jun	0	1	0	0	0	1	47	164
	Jul	0	0	0	0	0	0	27	42
	Aug	0	0	0	0	0	0	5	15
	Sep	0	0	0	0	0	0	0	0
	Total	1	7	0	7	0	10	102	312
1975	May	1	1	0	1	0	1	20	34
	Jun	1	10	1	8	0	8	191	237
	Jul	0	0	0	0	0	0	25	27
	Aug	0	0	0	0	0	0	5	15
	Sep	0	0	0	0	0	0	0	0
	Total	2	11	1	9	0	9	236	312

notes: NSV - nonsingle-valued adsorption/desorption
SV - single-valued adsorption/desorption simulation
sol - solution form
sed - sediment adsorbed form

produce a range of values. This variability in K hinders evaluation of the relative merits of using either the SV or NSV algorithm.

A further complication, but one which tends to support the concept of NSV adsorption/desorption, is the relative change in partitioning of runoff atrizine between the adsorbed and solution forms during the growing season. Recorded data have shown that the tendency to adsorb onto sediment increases with time after application. NSV adsorption/desorption will produce the same phenomenon, but it is not clear that this is the mechanism responsible for these observations in the field data. More laboratory research into this question is necessary to verify this observation and determine the physical and chemical reasons behind it.

The simulation of paraquat and atrazine on the P6 watershed (Fig. 12-22 and Table 12-10) was not as good as the P2 pesticide simulation. Paraquat runoff results are acceptable, except for the month of August, 1975. Sediment removal for this month is high; paraquat removal is higher. However, it is difficult to determine where the problem lies. Review of paraquat data collected from soil core samples and discussion with EPA personnel familiar with the sampling program on the P6 watershed indicate that the recorded data must be viewed with caution. Uncertainty in the accuracy of this data makes evaluation of the pesticide results difficult.

The monthly atrazine runoff results on the P6 watershed show even less resemblance between the simulated and recorded values than was the case for paraquat. The accuracy of the recorded data can be questioned, but other problems are also evident. Runoff is over-simulated in May 1974 because the hydrologic impact of plowing and planting operations is not represented in the model. The storm on May 29, 1974 occurred eight days after application and removed all of the atrazine for the month of May. For this storm there was a large amount of atrazine available for runoff in the surface zone and it was simulated as running off in solution during this storm. Atrazine was again applied in November, 1974. This application was followed by cold weather and snow. The lowered temperatures decreased the rate of biological degradation (although there is no soil core data to confirm this hypothesis) and more atrazine was then available for runoff during warm melt periods (January, February, March, and April of 1975) than simulated. A biologically oriented degradation approach which uses soil temperatures would be better able to hande this problem of pesticide winter application and degradation.

Storm Event Simulation: An analysis of individual pesticide runoff events on the P2 watershed soon after application shows that they match recorded results very well (Figs. 12-23 to 12-25). The paraquat simulation (Fig. 12-9) matches the sediment simulation for the storm of May 28, 1973, as one would expect. Atrazine removal by sediment for this event is low

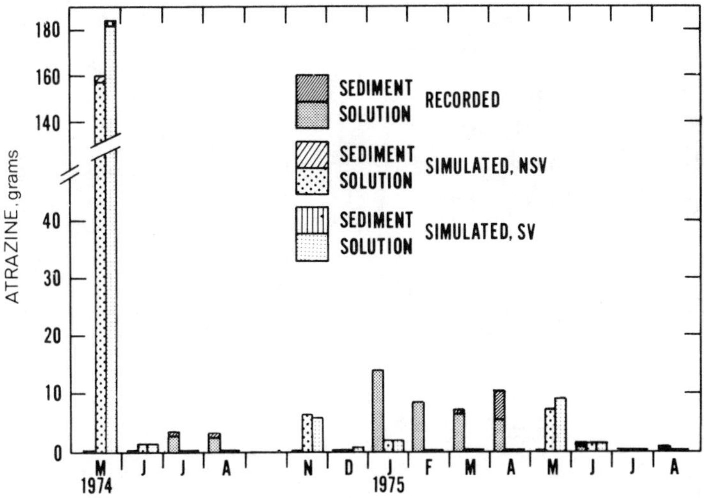

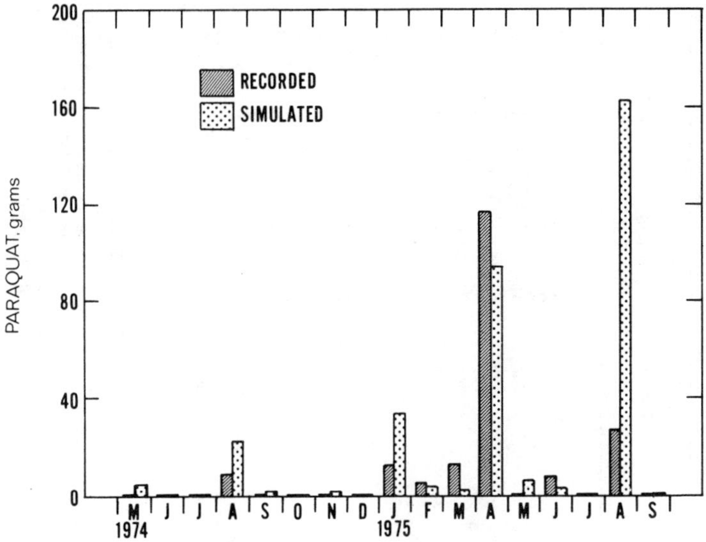

Fig. 12-22. Monthly Pesticide Removal From the P6 Watershed

Table 12-10. Pesticide Runoff Simulation for the P6 Watershed

		Atrazine						Paraquat	
		Recorded		NSV Simulation		SV Simulation		Recorded	Simulated
		(sed)	(sol)	(sed)	(sol)	(sed)	(sol)	(sed)	(sed)
1974	May	0	0	3	137	3	152	0	5
	Jun	0	0	0	1	0	1	0	0
	Jul	1	3	0	1	0	1	0	0
	Aug	2	1	0	0	0	0	9	23
	Sep	0	0	0	0	0	0	0	2
	Oct	0	0	0	0	0	0	0	0
	Nov	0	0	0	0	0	0	0	2
	Dec	0	0	0	0	0	0	0	0
Total		3	4	3	139	3	154	9	32
1975	Jan	0	14	0	2	0	2	12	34
	Feb	0	9	0	0	0	0	6	4
	Mar	1	6	0	0	0	0	13	3
	Apr	5	5	0	0	0	0	117	94
	May	0	0	0	0	0	0	0	6
	Jun	0	1	0	1	0	1	8	4
	Jul	0	0	0	0	0	0	0	0
	Aug	0	1	0	0	0	0	27	163
	Sep	0	0	0	0	0	0	0	1
		6	36	0	3	0	3	183	309

notes: NSV - nonsingle-valued adsorption/desorption
SV - single-valued adsorption/desorption simulation
Sol - solution form
Sed - sediment adsorbed form

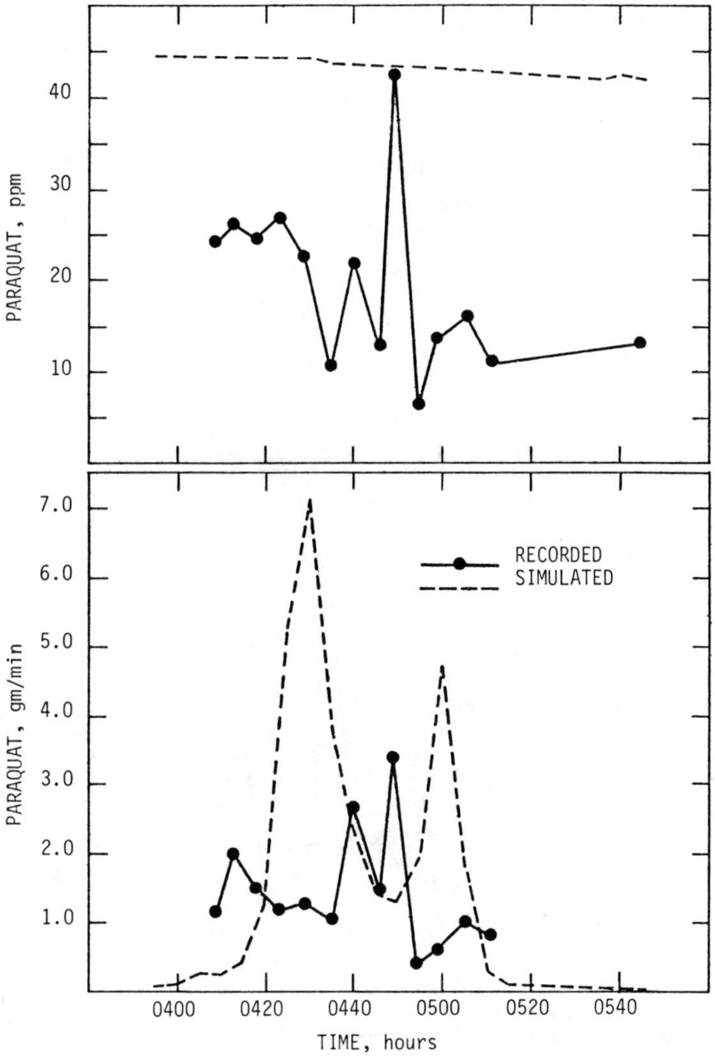

Fig. 12-23. Paraquat Removal From the P2 Watershed for the Storm of May 28, 1973

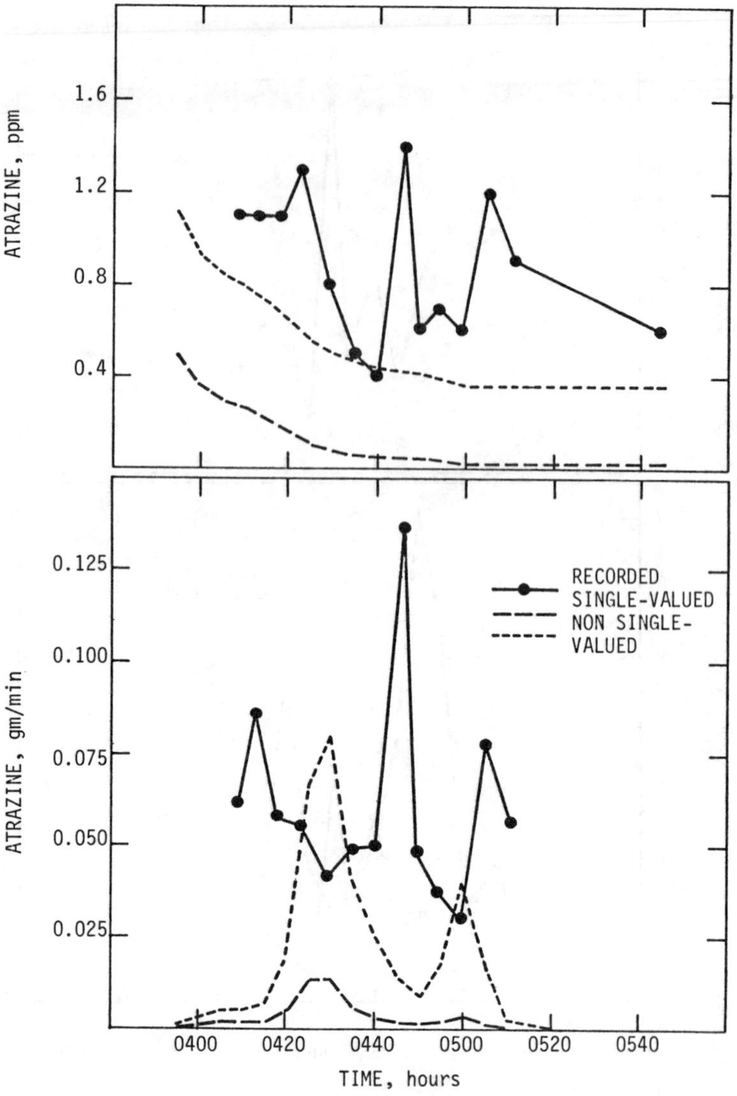

Fig. 12-24. Atrazine Removal on Sediment From the P2 Watershed for the Storm of May 28, 1973

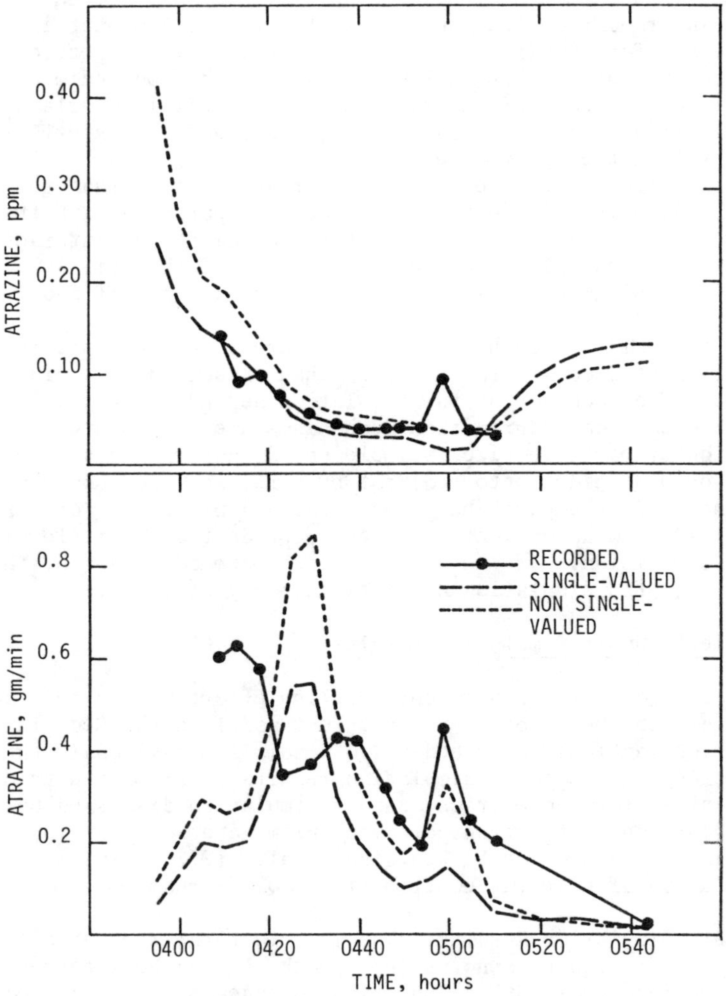

Fig. 12-25. Atrazine Removal in Solution From the P2 Watershed for the Storm of May 28, 1973

(Fig. 12-24), but still within the range of accuracy required. But even with the use of NSV adsorption/desorption the amount of atrazine removed is not as great as was recorded. Simulation of atrazine removed in solution is excellent however.

It can be seen at later events after application (Fig. 12-26) that the simulated amount of atrazine is not as great as that recorded. The reason for the small amount of atrazine runoff is that only a relatively small amount is available for runoff. Atrazine (and other soluble pesticides) can be removed from the watershed by transport on sediment particles eroded from the surface zone, runoff in overland flow interacting with the surface layer, and runoff in interflow coming from the upper zone.

Solution pesticides show greater variability than the adsorbed forms and are not simulated as accurately with the ARM Model. A better understanding of the movement and interactions of soluble pesticides in the upper layers of the soil is needed to better simulate their transport by surface runoff and interflow.

Pesticides attached to sediment are relatively stable and are simulated considerably better than dissolved pesticide forms. The accuracy of the simulation depends on how well the surface zone and degradation mechanisms are represented. If erosion by particle size is implemented, adsorption and desorption of pesticides onto sediment by particle size should also be included. Research has demonstrated a preference of certain pesticides to adsorb onto sediment dependent on particle size and composition. If the ARM Model can take advantage of this knowledge the simulation of pesticides may be improved.

Nutrient Runoff Simulation Results

The goal of the nutrient modeling effort was to test and evaluate the behavior of the nutrient model on the Georgia (P2) and Michigan (P6) watersheds. To accomplish this goal, the hydrology and sediment simulation results provided the transport mechanisms for the nutrient runoff simulation discussed below. The coupling of the transport mechanisms and soil nutrient storages (see Section B, Donigian et al., 1977) results in the simulation of nutrient contributions to a stream.

Monthly Simulation Results: The monthly total removal of nitrogen (N) and phosphorus (P) for the P2 and P6 watersheds, divided into sediment and solution components, is presented in Figs. 12-27 and 12-28, respectively. The forms of nitrogen on sediment for both watersheds are illustrated in Fig. 12-29, while the forms of nitrogen in solution are presented in Fig. 12-30. Figure 12-31 provides the simulated and recorded phosphorus in solution for both watersheds. The corresponding values for these graphs are given in Tables 12-11 and 12-12.

The nutrient related parameters for the P2 watershed were calibrated on both the 1974 and 1975 summer periods. This

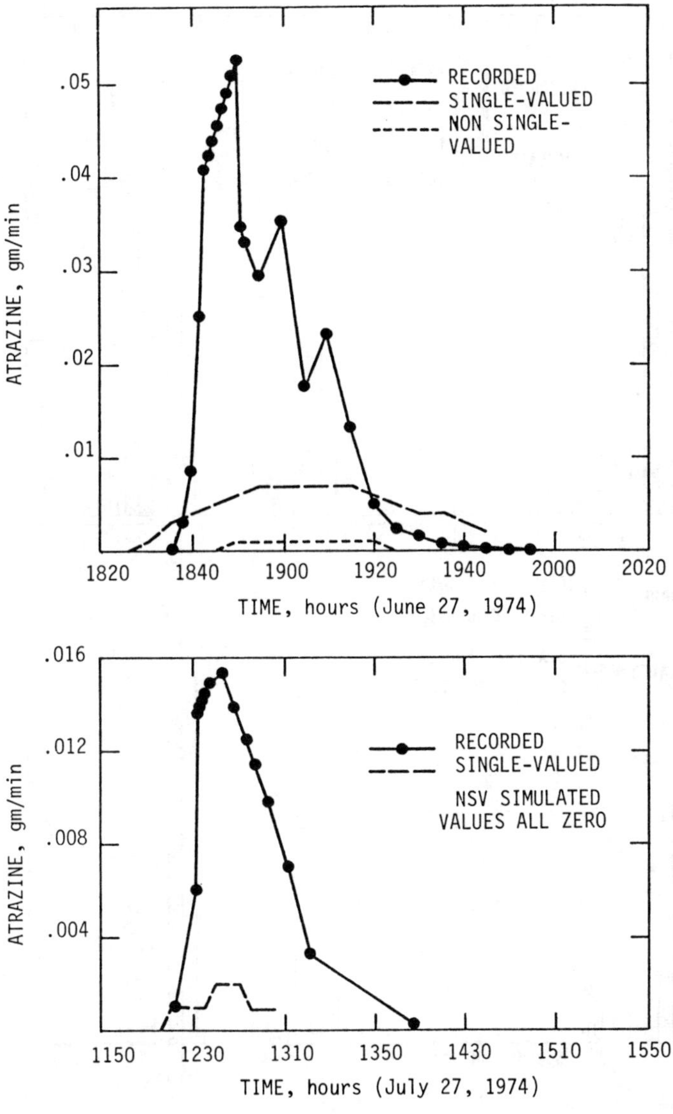

Fig. 12-26. Atrazine Removal in Solution From the P2 Watershed for the Storms of June 27, and July 1974

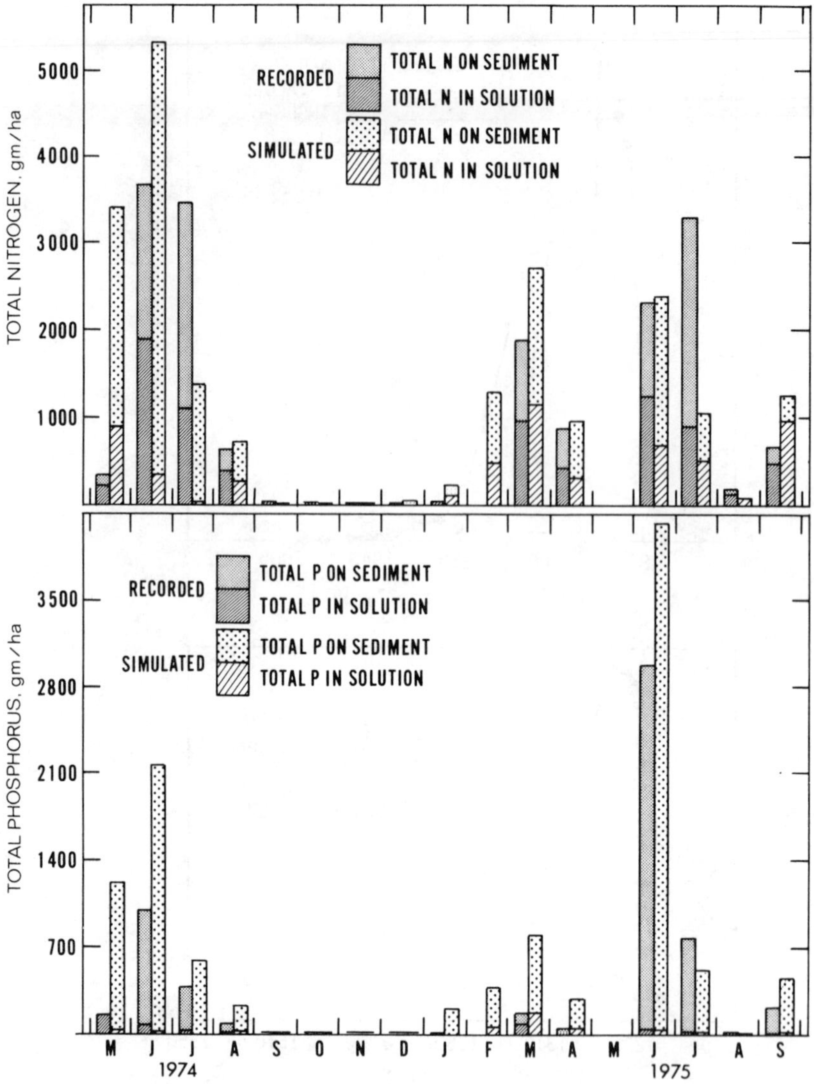

Fig. 12-27. Monthly Total Nitrogen and Phosphorus Runoff From the P2 Watershed

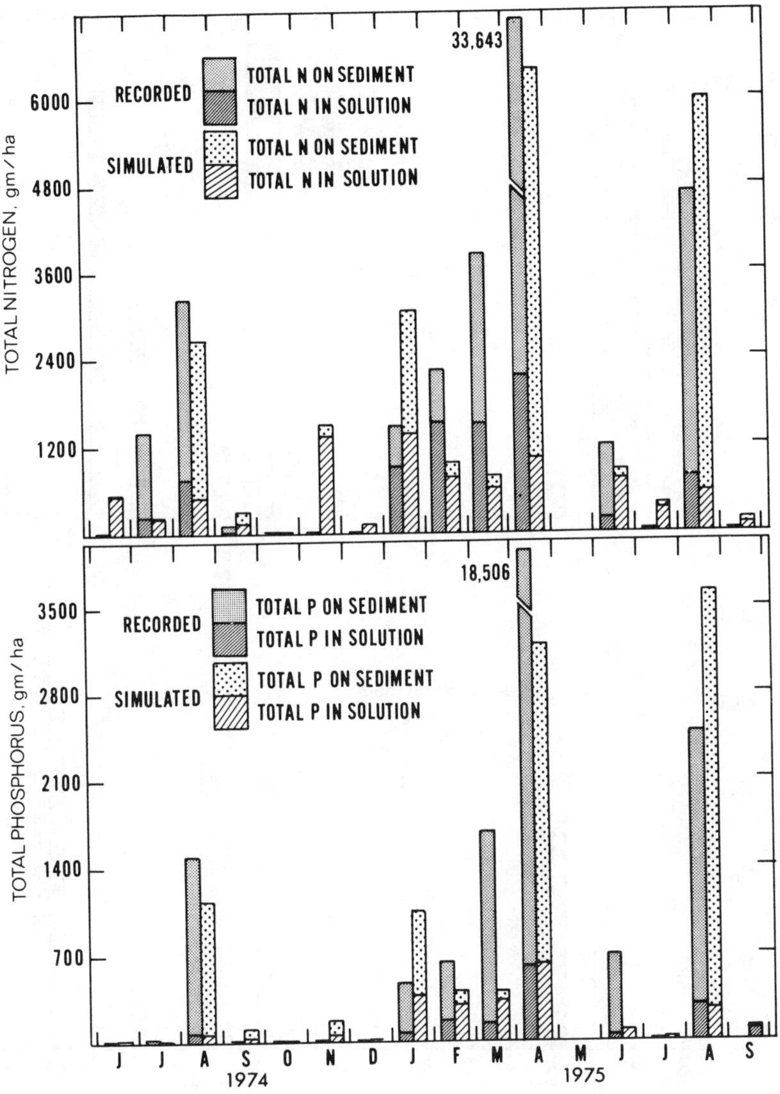

Fig. 12-28. Monthly Total Nitrogen and Phosphorus Runoff From the P6 Watershed

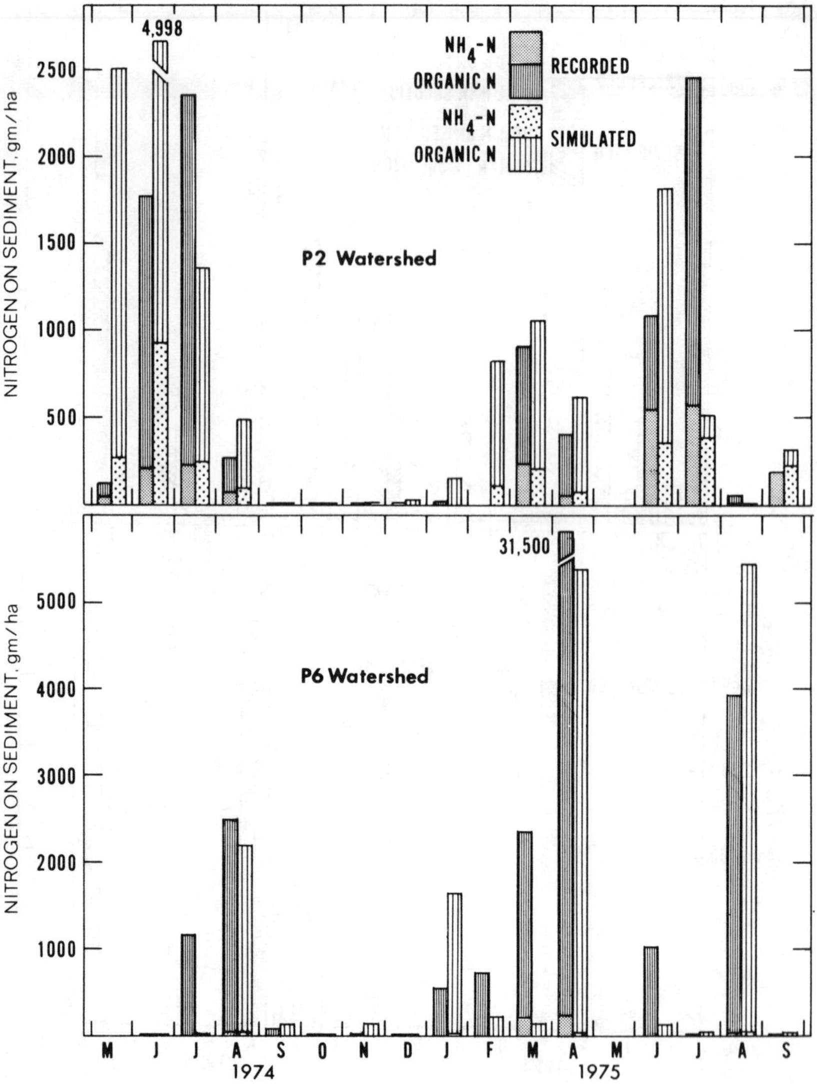

Fig. 12-29. Monthly Organic N and NH_4 on Sediment From the P2 and P6 Watersheds

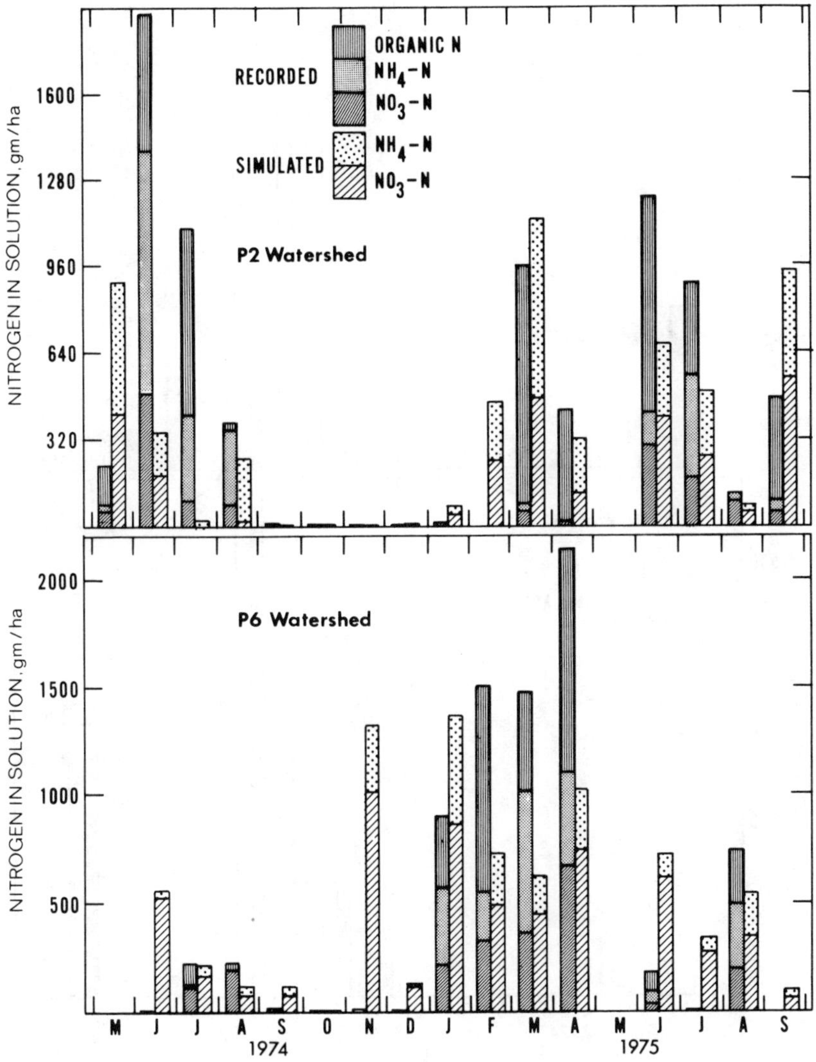

Fig. 12-30. Monthly NH_4, NO_3, and Organic N Removal in Solution From the P2 and P6 Watershed

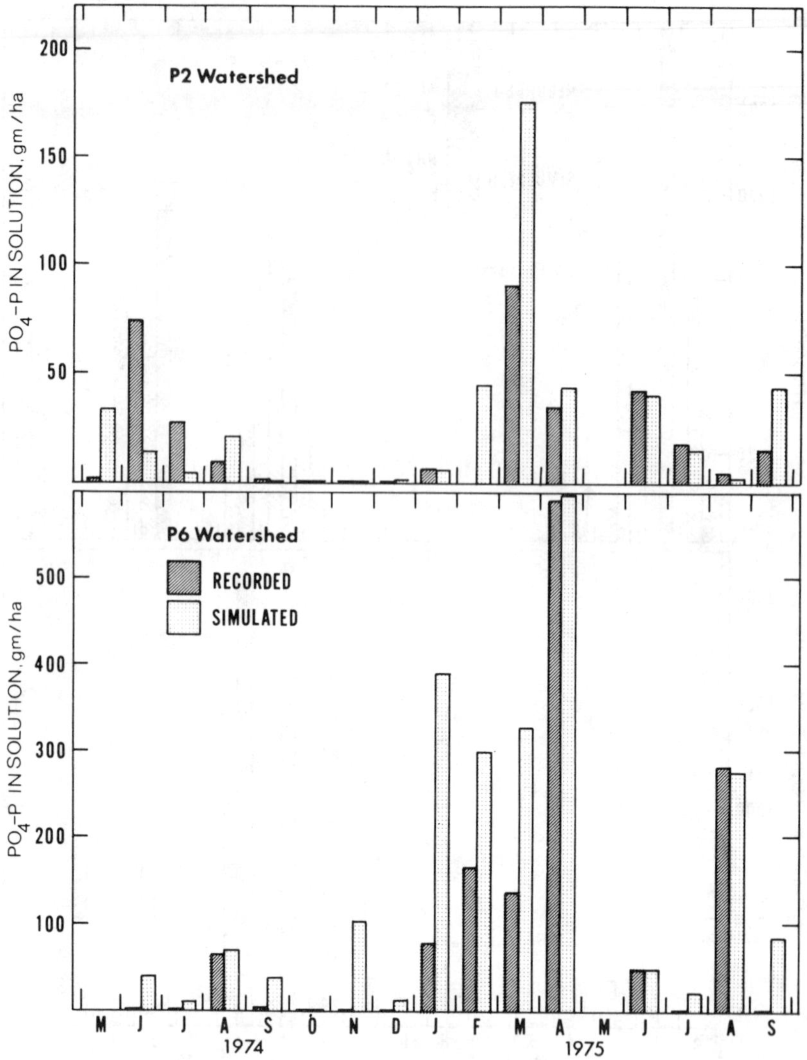

Fig. 12-31. Monthly Total PO_4 Removal in Solution From the P2 and P6 Watersheds

Table 12-11. Nutrient Results for the P2 Watershed

Month		Runoff	Sediment	NO_3-N	NH_4-N		Total N		PO_4-P	Total P
				(Sol)	(Sol)	(Sed)	(Sol)	(Sed)	(Sol)	(Sed)
		(mm)	(kg/ha)	(gm/ha)						
1974 May	Rec	7.6	102.1	57.2	18.2	42.1	222.1	123.8	2.3	150.9
	Sim	18.3	1145.3	437.1	464.0	276.8	901.1	2510.6	33.6	1189.4
June	Rec	41.9	968.0	487.7	901.7	214.4[b]	1897.2	1778.4[b]	74.3	922.5[a]
	Sim	41.4	2138.1	187.2	159.2	918.8	346.4	4997.7	13.4	2155.9
July	Rec	47.0	684.6	100.8[b]	304.4	225.1	1098.0[b]	2361.8	27.2	359.5
	Sim	44.0	600.6	0.0	20.2	239.9	20.2	1346.3	3.4	583.4
Aug	Rec	12.2	104.3	82.4	271.5[c]	69.2	379.5	258.8	10.0	58.8
	Sim	18.7	224.0	21.3	224.0	85.2	245.3	493.1	21.3	213.7
Sept	Rec	0.6	0.9	8.9	0.0	4.8	10.2	12.2	1.0	1.5
	Sim	0.1	0.0	0.0	0.0	0.0	0.0	0.0	0.0	0.0
Oct	Rec	0.0	0.0	0.0	0.0	0.0	0.0	0.0	0.0	0.0
	Sim	0.0	0.0	0.0	0.0	0.0	0.0	0.0	0.0	0.0
Nov	Rec	0.0	0.0	0.0	0.0	0.0	0.0	0.0	0.0	0.0
	Sim	0.0	0.0	0.0	0.0	0.0	0.0	0.0	0.0	0.0
Dec	Rec	0.1	0.0	1.2	0.2	0.0	5.6	0.0	0.3	0.0
	Sim	0.6	11.2	0.0	3.4	4.5	7.9	26.8	1.1	11.1
1975 Jan	Rec	1.5	3.4	3.0	5.0	2.5	14.6[c]	21.4	6.0	1.2[b]
	Sim	3.1	65.0	46.0	29.1	21.3	75.1	151.1	6.7	197.1

Table 12-11. (Continued)

Feb	Rec	9.3	13.2	-	-	-	-	-	-	-
	Sim	20.4	351.9	23.9	216.3	112.1	456.2	821.1	54.9	326.4
Mar	Rec	48.1	59.3	58.5	25.1	233.3	954.8	908.8	90.3	82.4
	Sim	78.5	340.7	473.0	661.3	206.2	1,134.3	1,561.4	176.0	623.8
Apr	Rec	16.5	22.3	16.5	0.0	56.2	432.5	411.1	34.3	18.0
	Sim	27.9	271.2	124.4	192.8	75.1	317.2	614.5	53.8	247.2
May	Rec	8.9	340.7	67.2	43.7	83.5	172.3	117.2[b]	8.1	218.6
	Sim	22.6	889.8	-	-	-	-	-	-	-
Plant uptake and initial storages were reset for the simulation run beginning June 10.										
June	Rec	41.9	4,124.6	298.4	114.0	547.5	1,214.3	1,083.3	42.5	2,933.5
	Sim	30.0	3,117.5	397.9	273.5	357.5	671.4	1,695.9	40.3	4,083.7
July	Rec	32.3	973.1	183.7	397.8	568.8[c]	893.2	2,465.5[c]	17.5	778.3
	Sim	9.8	401.2	254.4	237.6	384.4	492.0	548.4	14.6	400.8
Aug	Rec	1.6	11.2	103.2	18.3[c]	11.6	124.2[c]	46.6	4.7	16.6
	Sim	0.2	2.3	54.9	21.3	1.1	76.2	1.1	2.2	1.1
Sept	Rec	10.6	90.9	58.3	34.6	183.8[c]	476.8	183.8[c]	15.0	221.5
	Sim	12.4	248.8	552.6	385.6	218.6	938.2	317.7	43.7	419.9
TOTAL except Feb. & May '75	Rec	261.9	7,144.7	1,459.8	2,159.3	2,271.4	7,723.0	9,655.5	325.4	5,544.7
	Sim	285.0	8,565.9	2,548.8	2789.4	2,787.4	5,225.3	14,264.6	410.1	10,127.1

a. Total value is an estimate because of missing samples.
b. Excludes certain minor storm events that were not sampled.
c. questionable value

Sol - solution form
Sed - sediment adsorbed form

Table 12-12. Nutrient Results for the P6 Watershed

Month		Runoff (mm)	Sediment (kg/ha)	NO_3-N	NH_4-N (Sol)	NH_4-N (Sed)	Total N (Sol)	Total N (Sed)	PO_4-P (Sol)	Avail P (Sed)	Total P (Sed)
				(gm/ha)							
1974 June	Rec	0.0	0.	0.0	0.0	0.0	0.0	0.0	0.0	0.0	0.0
	Sim	0.3	2.	527.6	28.1	0.0	555.7	5.1	32.6	1.1	2.3
July	Rec	3.0	389.	116.8	0.1	0.3	220.9	1177.4	0.3	1.6	31.0
	Sim	0.5	9.	166.1	44.9	0.0	211.0	18.7	9.0	2.2	8.5
Aug	Rec	25.8	796.	196.3	123.9	63.0	726.3	2506.3	67.4	220.3	1424.5
	Sim	19.8	928.	389.5	86.4	64.0	475.9	2196.8	68.5	208.8	1064.3
Sep	Rec	0.8	35.	4.3	16.6	2.4	22.1	99.6	2.8	10.9	16.3
	Sim	2.7	72.	86.4	37.0	3.4	123.4	153.8	38.2	15.7	76.4
Oct	Rec	0.0	0.	0.0	0.0	0.0	0.0	0.0	0.0	0.0	0.0
	Sim	0.0	0.	0.0	0.0	0.0	0.0	0.0	0.0	0.0	0.0
Nov	Rec	0.6	10.	4.5	5.1	0.5	11.4	22.5	1.3	3.3	4.4
	Sim	4.3	78.	1107.9	209.9	3.4	1317.8	156.5	103.3	16.8	75.8
Dec	Rec	0.0	0.	0.0	0.0	0.0	0.0	0.0	0.0	0.0	0.0
	Sim	0.2	0.	120.1	0.0	0.0	120.9	0.0	12.3	0.0	0.0
1975 Jan	Rec	33.2	248.	218.4	359.5	-	897.3	573.4	77.4	-	410.6[a]
	Sim	35.3	753.	862.1	504.0	31.4	1366.2	1668.2	390.6	390.6	661.0
Feb	Rec	35.1	244.	330.1	664.8	-	1505.4	727.4[a]	165.5	-	474.4[a]
	Sim	12.6	101.	490.5	241.3	2.3	731.8	212.6	299.7	21.3	99.5
Mar	Rec	26.5	723.	364.9	647.3	-	1479.9	2370.0[a]	139.5	-	1558.5[a]
	Sim	11.5	72.	451.2	177.4	0.0	628.6	148.0	217.8	14.6	69.6
Apr	Rec	88.7	12,315.	674.	625.	243.[b]	2133.	31510.	590.	1378.[b]	16,538.
	Sim	100.3	2,640.	745.	280.	14.	1025.	5390.	594.9	509.6	2,604.9
May	Rec	0.1	3.	1.5	1.0	-	4.0	5.9	0.4	0.4	3.8
	Sim	0.6	91.6	-	-	-	-	-	-	-	-

Table 12-12. (Continued)

		Plant uptake, initial storages, and soil temperature regression constants were reset for the simulation beginning June 1									
June	Rec	5.8	410.	40.5	60.6	12.9	178.1	1035.4[a]	49.0	41.1	644.5[a]
	Sim	2.6	54.	617.4	113.3	1.1	730.7	127.0	47.1	4.5	78.8
July	Rec	0.1	2.	0.9	0.8	-	2.5	8.3	0.4	0.3	4.1
	Sim	0.7	13.	275.0	68.5	0.0	343.5	28.3	18.0	1.1	18.7
Aug.	Rec	50.0	1629.	208.0[a]	290.8[a]	36.4[a]	747.4[a]	3959.6[a]	284.9[a]	99.9[a]	2191.6[a]
	Sim	62.6	2329.	354.7	191.9	56.1	546.6	5430.0	275.0	202.1	3334.3
Sept	Rec	0.0	0.	0.0	0.0	0.0	0.0	0.0	0.0	0.0	0.0
	Sim	3.1	18.	68.5	41.5	0.0	100.0	37.4	83.1	1.2	22.8
TOTAL	Rec	269.5	16,801.	2,158.7	2,794.5	-	7,924.3	43,989.9	1,390.8	-	23,297.9
Except May	Sim	256.5	7,069.	6,262.0	2,024.2	175.7	8,277.1	15,5 2.4	2,300.1	1,389.6	8,116.9

a. Total value is an estimate because of missing samples
b. Excludes certain minor storm events that were not sampled.

Sol - Solution form
Sed - Sediment adsorbed forr

provided a single set of reaction rates that were subsequently used for both summer periods. Following the calibration, the entire period from April 1974 through April 1975 was simulated to evaluate the model behavior during the nonsummer period. The P6 watershed was calibrated only on the summer 1974 period. The calibrated nutrient reaction rates were then used for the 1975 summer period and provided reasonable results. As on the P2 watershed, a single simulation run was then made from May 1974 to May 1975 to evaluate the nonsummer simulation. The nutrient storages, monthly plant uptake fractions, and soil temperature regression constants were then reset on June 1, 1975 for the summer 1975 conditions.

Some general comments can be made on the simulation of Total P and Total N, as presented in Figs. 12-27 and 12-28 and Tables 12-11 and 12-12. Generally the accuracy of the simulation results for total nutrients corresponds well with that for runoff and sediment removal. For example, when sediment was under-simulated on the P6 watershed for April 1975 so was Total N and Total P. For some of the monthly results this relationship to runoff and sediment simulation was not as strong. Moreover, in some cases runoff and sediment loss are over-simulated and nutrients are under-simulated. Examples of this behavior occur on the P2 watershed for the months of July 1974 and June 1975. In July 1974 runoff is high, sediment removal is low, but Total N removal is low and Total P removal is high. In June 1975, both runoff and sediment simulation is low while Total N and Total P removal are high. A similar event occurs on the P6 watershed where during August 1974 runoff is under-simulated, sediment loss over-simulated, yet both Total N and Total P are under-simulated. These seeming inconsistencies are due to particular solution or sediment nutrient forms comprising most of the observed nutrient loss and corresponding discrepancies in simulating the associated transport mechanism.

In general, the simulation of monthly Total N and Total P for the entire test period gives reasonable results as an initial test of the nutrient algorithms. For the entire simulation period the Total N removal on the P2 watershed is simulated within 13 percent of the recorded value, while Total P is over-simulated by 80 percent. The over-simulation was partially caused by the over-simulation of sediment loss. For the P6 watershed, with the exclusion of April 1975, the Total N and Total P removals for the entire simulation period are good: within 6 percent of the recorded values.

The sediment portion of the runoff for both watersheds carries more nitrogen and phosphorus than does the solution component of the runoff. However, the solution nitrogen does at times contribute to much of the Total N removal. Phosphorus in solution is a smaller portion of the Total P washoff on the P2 watershed than on P6. This is likely caused by less phosphorus fertilizer applied on P2.

The sediment portion of the runoff for both watersheds carries more nitrogen and phosphorus than does the solution component of the runoff. However, the solution nitrogen does at times contribute to much of the Total N removal. Phosphorus in solution is a smaller portion of the Total P washoff in the P2 watershed than on P6. This is likely caused by less phosphorus fertilizer applied on P2.

The sediment nutrients are either organic or inorganic forms. Measurements of inorganic phosphorus on sediment were not made except for some analysis of Available P (Table 12-12). It is assumed that most of the phosphorus associated with the sediment was in the organic form. Most of the simulated phosphorus is organic. Thus, the results presented and discussed for Total P can be considered to be similar to the simulation of the Organic P content of the sediment.

Breakdown of the forms of Total N attached to the sediment are shown in Fig. 12-29. Much more sediment associated nitrogen is organic than inorganic. Like phosphorus, the Organic N simulation is similar to the Total N results. The accuracy of the simulation depends on the accuracy of the sediment simulation. however, the simulation results for the inorganic fraction, NH_4-N, varies. The largest portion of NH_4-N is found in the summer after spring fertilization. Generally, the simulation of NH_4-N is reasonably good, particularly when the sediment simulation is good.

The solution nutrient simulation results like the sediment results are strongly influenced by the simulation of the transport mode. The runoff simulation influences the amount of nutrient removed in solution. Most of the soluble nutrient removal comes from the interflow component of runoff. Hence, the interflow hydrology parameter, INTER, has a major impact on the solute runoff simulation. For example, INTER for P2 was reduced from 0.70 to 0.59 for the nutrient simulation. This resulted in a minor difference in the runoff simulation (less than 3 percent increase) during the period of May 1, 1974 through September 30, 1974. However, this reduction in INTER decreased NO_3-N in solution from 4.52 kg/ha to 0.64 kg/ha during this period.

The simulated and recorded forms of nitrogen in solution are shown in Fig. 12-30. The nutrient model does not simulate Organic N in solution even though the recorded results show that there can be large amounts of Organic N relative to inorganic forms. The large amounts of Organic N found in the water phase of the runoff may have been actually attached to sediment particles that passed the filtering process in the laboratory analysis. Organics may also have occurred in solution by the disassociation of sediment organics in the runoff samples.

NO_3-N and NH_4-N in solution are simulated by the model. The overall result of the NO_3-N simulation is to over-simulate NO_3-N on the P6 watershed and slightly over-simulate it on P2. The over-simulation of P6 NO_3-N runoff cannot be directly attributed to the runoff since it is under-simulated. For P6, the simulation of NO_3-N is worse than average during the months of June and November 1974 and June and July 1975. The simulated soil storages of NO_3-N are higher than recorded in November and lower than observed in both Junes, but the July storage is close to the observed values. However, the soil storage and runoff results do not account for the over-simulation of NO_3 for the three summer months. This situation indicates a need to further study the the transport mechanisms. The influence of the interflow parameter on NO_3-N runoff simulation appears to be more important than the soil storage simulation. The November over-simulation of NO_3-N can be explained by the over-simulation of both runoff and soil storage.

The simulation of NO_3-N for the P2 watershed generally follows the hydrology simulation, except for some of the summer months. The effect of the over-simulation of soil NO_3-N simulation indicates that further study of the hydrology simulation for the P2 watershed and the transport mechanisms is needed.

The simulation of NH_4-N in solution for the entire simulation period is good. The recorded NH_4-N in solution for both watersheds is slightly greater than the NO_3-N. The results are generally better for NH_4-N than NO_3-N on both of the watersheds, although the NH_4-N simulation results generally follow the NO_3-N trends. However, for the winter period on the P2 watershed, NH_4-N is over-simulated; very little NH_4-N is recorded in the P2 runoff during this period. The over-simulation is perhaps due to over-mineralization of organics to NH_4-N in the soils during this period. Much more NH_4-N is recorded for the Michigan P6 watershed during the winter because of fertilization in November and the possibility of NH_4-N in the snow.

Monthly PO_4-P in solution is shown in Fig. 12-31. Much more PO_4-P is in solution from the P6 watershed than from P2 mainly because more fertilizer was applied to P6: 224 kg/ha compared to 54 kg/ha. Although some inconsistencies exist, the overall simulation of solution PO_4-P from the P2 watershed is good, particularly in view of the slight over-simulation of runoff. The P6 simulation of PO_4-P is high, mostly because of the over-simulation of the winter period. However, the P6 summer months are over-simulated when the recorded values are less than 3 gm/ha. Otherwise the three major summer runoff months of August 1974 and June and August 1975 for P6 are well simulated for PO_4-P in solution.

The nutrient model at present simulates monthly results of nutrients on sediment better than nutrients in solution. The value of a more accurate sediment than solution nutrient simulation will depend on whether short term or long term effects are most important. Soluble inorganic nutrients has a more immediate impact on water quality than the sediment-associated nutrients since they are more readily used by algae than either the organic or sediment forms. Thus, soluble nutrients can directly contribute to accelerated eutrophication. However, over the longer term organics and sediment-associate inorganics can decompose and release inorganics in solution available for stimulation of aquatic growth.

Storm Event Simulation: Analysis of storm event simulation results is needed to determine how well the model is representing the separate solution and sediment nutrients in relation to their transport components. Figures 12-32 to 12-35 show results for the August 27, 1974 storm on the P6 watershed. The results are presented in both concentration and mass removal (gm/min) units. Also, each graph pertains either to the solution or sediment transport mode of the specific component.

Some general conclusions can be made concerning all the individual storm event results. The summer nutrient simulation is generally better than the winter. The poor winter nutrient simulation is partially due to inaccurate runoff, sediment, and soil temperature simulation. Nutrient simulation on the P6 watershed is also generally better than on P2. This is true for both the sediment and solution nutrient forms. The poor sediment nutrient simulation on P2 is because the concentration of nutrients varied from one storm event to the next more than it did for P6, for which the model simulates the concentration as remaining fairly stable.

The less accurate soluble nutrient simulation on the P2 watershed compared to P6 is because interflow is a smaller portion of total runoff and is more variable. Consequently, soluble nutrients transported mostly in interflow demonstrate the same highly variable nature as the simulated interflow component.

NO_3-N is the most difficult solution component to simulate. This is particularly noticeable in the P6 storm results where large variations occur. NO_3-N is difficult because it is affected by many transformations as well as being soluble and thereby easily transported by and in the water. The other nutrients simulated in solution, NH_4-N and PO_4-P, do not undergo as many transformations and are also found in adsorbed form. Since the adsorbed form is usually more stable than the solution form, NH_4-N and PO_4-P are simulated somewhat better than NO_3-N.

All the solution nutrient results including NO_3-N are closer to the recorded values in mass per minute than in concentration (ppm). Since nutrients are not toxic to aquatic life in these small quantities, the simulation of total mass of nutrients in the runoff is considered more crucial than concentration. Moreover, eutrophication which occurs most readily in lakes is dependent more on the total nutrient input than on localized storm concentrations. Thus, simulation of the total mass loading of nutrients should be the primary goal of nutrient modeling in the runoff. The simulation of nutrient loadings with the ARM nutrient algorithms meets this goal since mass loadings are simulated better than instantaneous concentrations.

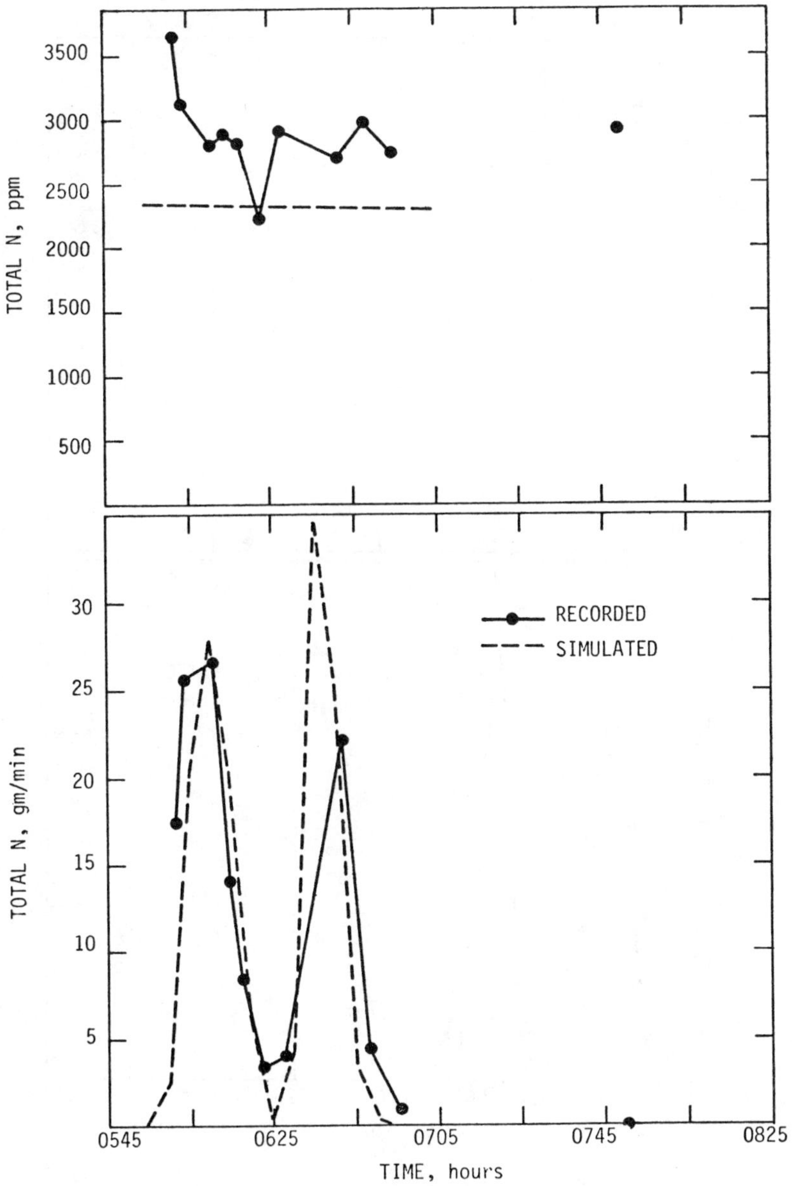

Fig. 12-32. Total N Removal on Sediment From the P6 Watershed for the Storm of August 27, 1974

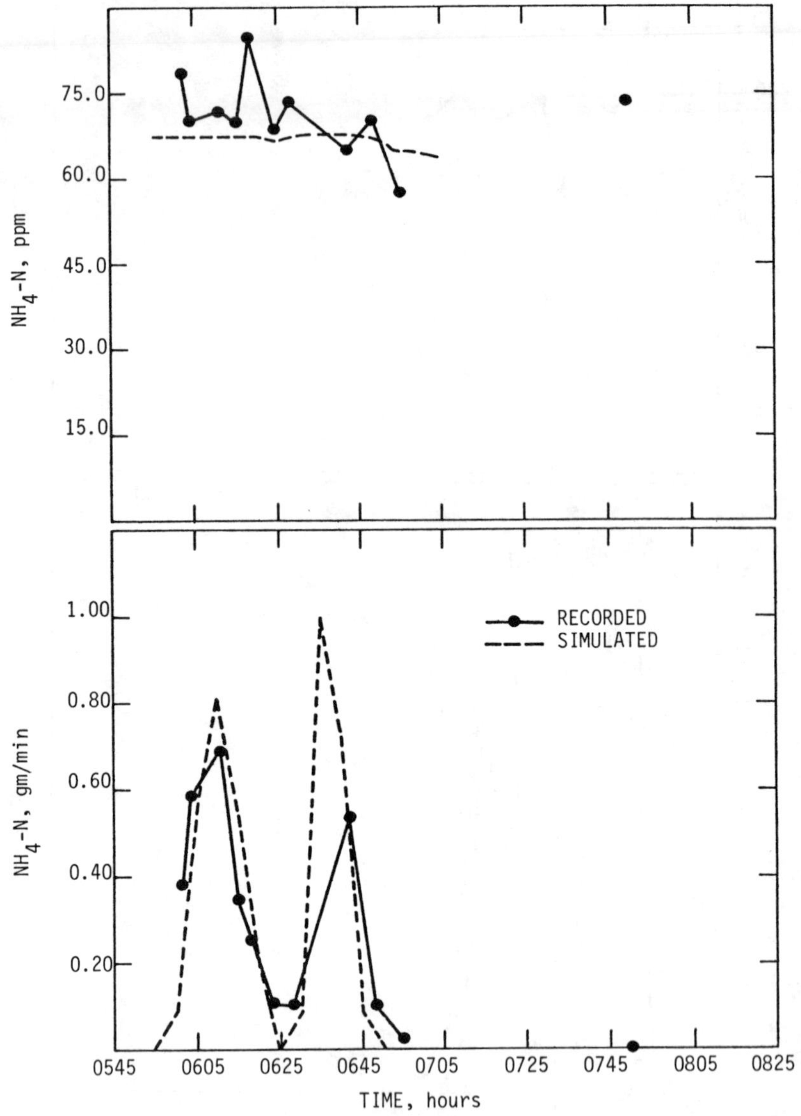

Fig. 12-33. NH_4 Removal on Sediment From the P6 Watershed for the Storm of August 27, 1974

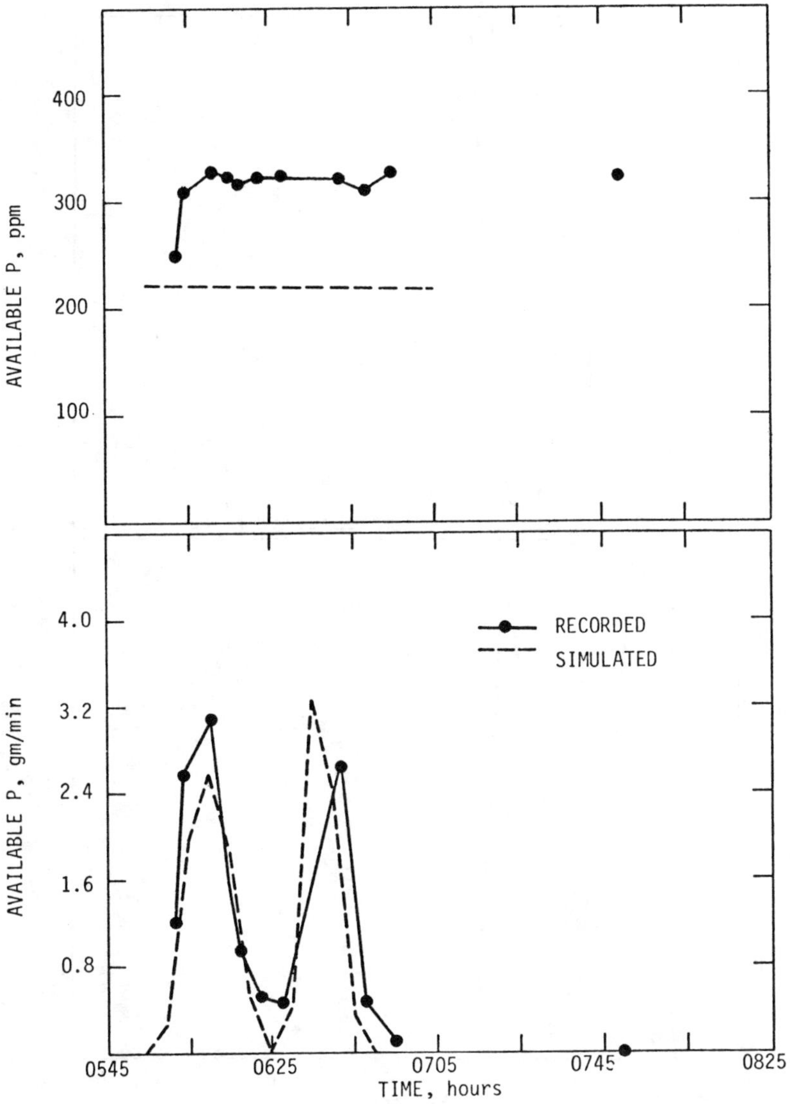

Fig. 12-34. Available P Removal on Sediment From the P6 Watershed for the Storm of August 27, 1974

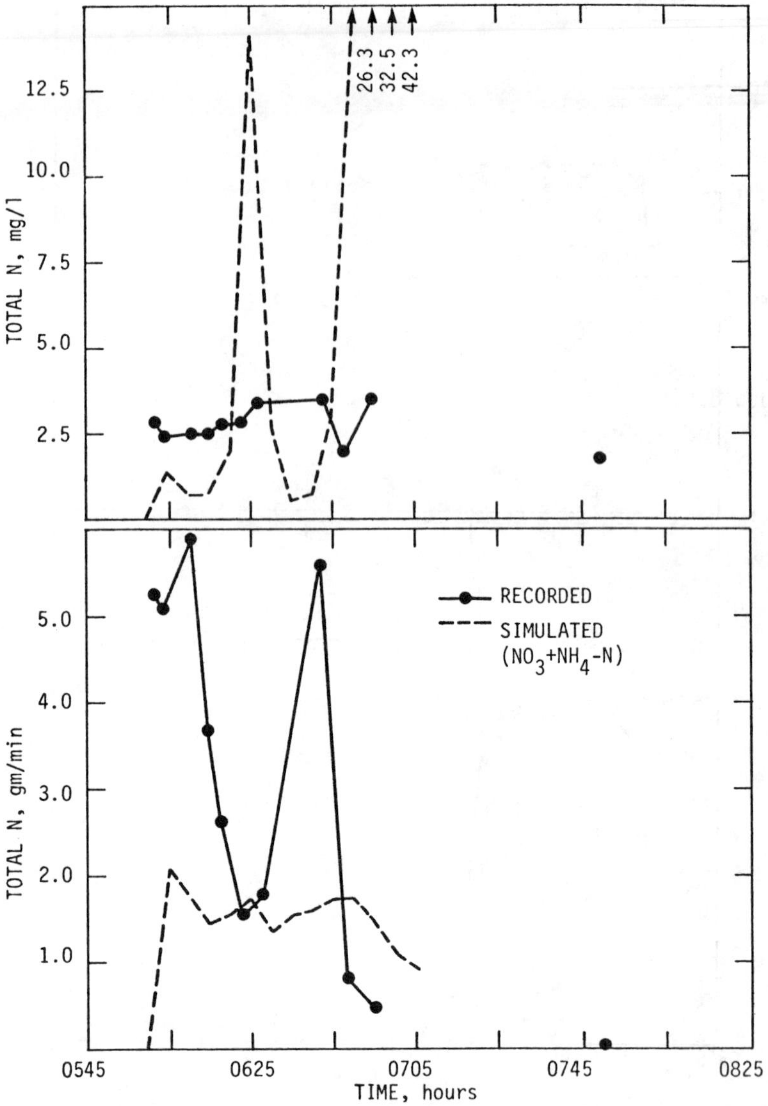

Fig. 12-35. Total N Removal in Solution From the P6 Watershed for the Storm of August 27, 1974

Summary

The nutrient model provides a reasonable framework for simulation of nutrient runoff. The results show that the model can be calibrated to a variety of situations. The soil storage simulation provides a reasonably accurate representation of the soil nutrients which are the sources for nutrient components in runoff. Nutrient simulation is found to rely heavily on the hydrology and sediment simulation.

Within the present framework of the nutrient model, the accuracy of the nutrient simulation is a direct function of the ability to correctly simulate runoff and sediment loss. Overall, the ARM nutrient model is able with proper calibration to give reasonable estimates of nutrient mass in runoff. However, further study is needed in the area of nutrient transport through surface and subsurface pathways. This involves the study of surface and upper zone relationships, especially soil moisture, surface runoff, and interflow simulation. Even though the nutrients on sediment are simulated better than those in solution, the relationships of nutrient components to erosion processes should be studied. The behavior and transport of nutrients at the soil surface is an important topic for future research. With this additional knowledge, the ARM Model could more accurately simulate nutrients in the soil and in the runoff.

12.4 APPLICATIONS AND FUTURE RESEARCH

Model Applications in Environmental Planning

The Agricultural Runoff Management Model represents runoff, sediment, pesticide, and nutrient processes on agricultural lands. Model testing has shown that the soil and transport mechanisms that determine the quantity and quality of agricultural runoff can be simulated.

Applications of the model in environmental planning are:

1. Generation of extended continuous time series of runoff quantity and quality that correspond to present or planned agricultural use. These time series can be:
   a. statistically summarized to show the frequency of a particular concentration of a water quality variable as it leaves agricultural lands,
   b. summed to show monthly or annual total mass loadings into rivers and lakes,
   c. used as input to stream flow quality models that will predict continuous quality variables in rivers and lakes,
   d. used in economic evaluation of alternative environmental plans, as outlined by Donigian and Linsley (1976),
   e. used in "worst case" risk analysis for runoff of toxic substances;

2. Short term forecasts (1 day to 6 months) of the effects of current pesticide or nutrient applications. The ARM Model is a first attempt to continuously represent interdependent complex processes. Much has been learned about the processes that are critical to agricultural runoff quality. Further research is indicated for many model components, and additional field testing is needed. Future research is discussed below.

Future Research Requirements

1. Further testing of all portions of the ARM Model is needed to demonstrate the capabilities of the model and to delineate and confirm topics for future research. Past testing has shown that the model can represent the mechanisms determining the quantity and quality of agricultural runoff in two different climatic and geographic regions. Testing in other areas is needed to strengthen confidence in the model and demonstrate its general applicability.
2. Model testing on larger watersheds, in the general range of 50 to 500 hectares, is needed to evaluate the model's representation of processes in the lower soil zones and chemical contributions to groundwater. In this way, simulation of the movement of soluble chemicals through subsurface pathways to a stream or waterbody can be investigated.
3. The impact of tillage operations on runoff and sediment production should be quantified and incorporated into the ARM Model. This is especially important because the first events following pesticide and fertilizer applications are the most significant for agricultural chemical runoff.
4. The relationship of different agricultural land management practices to the ARM Model parameters should be investigated and quantified if the model is to be used to evaluate methods of controlling agricultural nonpoint pollution. The extent to which specific practices can or cannot be simulated with the ARM Model, and the corresponding changes in model parameters needs to be delineated. Such work will likely produce additional recommendations for model modifications to accomodate agricultural practices common to different regions of the country.
5. Future testing of the pesticide and nutrient simulation algorithms should evaluate the extent to which laboratory measurements and literature values of model parameters can be used to simulate pesticide and nutrient runoff without calibration on recorded data. Although the model algorithms have been designed to use commonly measured values, calibration of parameters has been used to evaluate the pesticide and nutrient algorithms after initial parameter values had been obtained from the literature. General use

of the ARM Model will be severely limited if each application requires calibration of certain pesticide and nutrient parameters on recorded data.

6. To improve the reliability and accuracy of the ARM Model the following recommendations are extended:
   a. Better definition of moisture, temperature, and chemical behavior in the surface soil layer (e.g., 0-1 cm) is needed since the processes in this zone directly affect the quality of surface runoff.
   b. In conjuction with (a), more accurate representation of the entire soil profile and intermediate moisture transfer is required for detailed simulation of vertical chemical movement.
   c. Refinement of erosion algorithms should concentrate on inclusion of detachment and scour by overland flow, consideration of gully erosion, and erosion simulation as a function of particle size. This work could significantly improve the sediment simulation and help to interface the ARM Model output with instream sediment transport models.
   d. If particle size erosion is added to the ARM Model, the pesticide and nutrient algorithms for sediment related processes should be re-evaluated to incorporate the dependence on particle size.
   e. Predictive models are needed for pesticide attenuation processes such as chemical, photochemical, and biological degradation and volatilization. Ideally, such models should consider the impact of soil environmental conditions (e.g., moisture, temperature, oxygen), be capable of application in various regions of the country, and utilize parameters easily measured in a laboratory.
   f. Evaluation and quantification of the impact of moisture and oxygen conditions on the soil nutrient reaction rates is needed. If this impact is significant, relationships should be developed and added to the ARM Model nutrient algorithms.
   g. Additional nutrient processes such as nitrogen fixation, ammonia uptake by plants, and the impact of nutrient contributions in precipitation, should be evaluated for their effect and possible inclusion in the model.
7. To promote the general use of the ARM Model for investigation, evaluation, and management of agricultural runoff, the following topics should be considered:
   a. Guidelines for parameter evaluation, calibration, and analysis of simulation results are needed to assist the user in model application. This is especially lacking for the pesticide and nutrient parameters, and may require research to evaluate adsorption parameters, nutrient reaction rates, and nutrient uptake by various crops under field conditions.

b. User workshops and assistance programs should be developed to attract and initiate potential model users. Some technical background and training is necessary for effective use of the ARM Model. Also, the establishment of the model on a central, nationwide computer network would then make the model available to users across the country without the need for separate source codes for each user. This would considerably simplify model updates and technology transfer with advances in the state of the art.

Acknowledgement

The authors gratefully acknowledge the financial support of the U.S. Environmental Protection Agency, Office of Research and Development. Coordination and direction was provided by the Environmental Research Laboratory in Athens, Georgia.

12.5 REFERENCES

Alexander, M., 1965. Nitrification. In: Soil nitrogen. M.V. Bartholomew and F.E. Clark (eds.), Madison, Wis., Am. Soc. Agron. Agronomy Monograph No. 10.

Anderson, E.A., and N.H. Crawford, 1964. The synthesis of continuous snowmelt runoff hydrographs on a digital computer. Department of Civil Engineering, Stanford University. Stanford, California. Technical Report No. 36.

Anderson, E.A., 1968. Development and testing of snow pack energy balance equations. Water Resour. Res. 4(1):19-37.

Broadbent, F.E., and F. Clark, 1965. Denitrification. In: Soil nitrogen. W.V. Bartholomew and F.E. Clark (eds.), Madison, Wis., Am. Soc. Agron. Agronomy Monograph No. 10.

Crawford, N.H., and A.S. Donigian, Jr., 1973. Pesticide transport and runoff model for agricultural lands. Office of Research and Development, U.S. Environmental Protection Agency, Washington, D.C. EPA 660/2-74-013.

Crawford, N.H., and R.K. Linsley, 1966. Digital simulation in hydrology: Stanford watershed model IV. Department of Civil Engineering, Stanford University. Stanford, California. Technical Report No. 39.

Davidson, J.M., and J.R. McDougal, 1973. Experimental and predicted movement of three herbicides in a water-saturated soil. J. Environ. Qual. 2(4):428-433.

Davidson, J.M., R.S. Mansell, and D.R. Baker, 1973. Herbicide distributions within a soil profile and their dependence upon adsorption-desorption. Soil Crop Sci. Soc. Florida Proc.

Donigian, A.S., Jr., et al., 1977. Agricultural runoff management (ARM) model. Version II: Refinement and testing.

Environmental Research Laboratory, Office of Research and Development. U.S. Environmental Protection Agency. Athens, Georgia. EPA 600/3-77-098.

Donigian, A.S., Jr., and N.H. Crawford, 1976. Modeling pesticides and nutrients on agricultural lands. Environmental Research Laboratory, Athens, Georgia. EPA 600/2-7-76-043.

Donigian, A.S., Jr., and R.K. Linsley, 1976. The use of continuous simulation in the evaluation of water quality management plans. Office of Water Research and Technology, U.S. Department of the Interior. Contract No. 14-31-0001-5215.

Dutt, G.R., M.T. Shaffer, and W.J. Moore, 1972. Computer simulation model of dynamic bio-physiochemical processes in soils. University of Arizona, Department of Soils, Water and Engineering Agricultural Experiment Station. Tucson, Arizona. Technical Bulletin 196.

Fleming, G., and M. Fahmy, 1973. Some mathematical concepts for simulating the water and sediment systems of natural watershed areas. Department of Civil Engineering, Strathclyde University. Glasgow, Scotland. Report HO-73-26.

Frere, M.H., C.A. Onstad, and H.N. Holtan, 1975. ACTMO, an agricultural chemical transport model. U.S. Department of Agriculture, Agricultural Research Service, Hyattsville, Maryland. ARS-H-3.

Fried, M., et al., 1957. Kinetics of phosphate uptake in the soil-plant system. Soil Sci. 84(6):427-437.

Hagin, J. and A. Amberger. 1974. Contribution of fertilizers and manures to the N and P load of waters. A computer simulation. Report submitted to the Deutsche Forschungs Gemeinschaft. 123 p.

Hydrocomp Inc., 1976. Hydrocomp Simulation Programming: Operations manual, 2nd ed., Hydrocomp Inc. Palo Alto, Calif.

Larsen, S., 1967. Soil phosphorus. Advan. Agron. 19:151-210.

Loehr, R.C., 1974. Agricultural waste management: Problems, processes and approaches. New York, Academic Press.

McLaren, A.D., 1970. Temporal and vectorial reactions of nitrogen in soil: A review. Can. J. Soil Sci. 50(2):97-109.

Mehran, M., and K.K. Tanji, 1974. Computer modeling of nitrogen transformations in soils. J. Environ. Qual. 3(4):391-395.

Meyer, L.D., and W.H. Wischmeier, 1969. Mathematical simulation of the process of soil erosion by water. Trans. Am. Soc. Agric. Eng. 12(6):754-758.

Mortland, M.M., and A.R. Wolcott, 1965. Sorption of inorganic nitrogen compounds by soil materials. In: Soil nitrogen. W.V. Bartholomew and R.E. Clark (eds.), Madison, Wis., Am. Soc. Agron. Agronomy Monograph No. 10.

Negev, M.A., 1967. Sediment model on a digital computer. Department of Civil Engineering, Stanford University. Stanford, California. Technical Report No. 76.

Oddson, J.K., L. Letey, and L.V. Weeks, 1970. Predicted distribution of organic chemicals in solution and adsorbed as a function of position and time for various chemicals and soil properties. Soil Sci. Soc. Amer. 34:412-417.

Onstad, C.A., and G.R. Foster, 1975. Erosion modeling on a watershed. Trans. Am. Soc. Agri. Eng. 18(2):288-292.

Sawyer, C.N. and P.L. McCarty, 1967. Chemistry for sanitary engineers, 2nd ed. New York, McGraw-Hill Book Company.

Stanford, G., and S.J. Smith, 1972. Nitrogen mineralization potential in soil. Soil Sci. Soc. Amer. Proc. 36:465-472.

Stanford, G., M.H. Frere, and D.E. Schwaninger, 1973. Temperature coefficient of soil nitrogen mineralization. Soil Sci. 115:321-323.

Stanford, G., and E. Epstein, 1974. Nitrogen mineralization--water relations in soils. Soil Sci. Soc. Amer. Proc. 38:103-107.

Stevenson, F.J., 1965. Origin and distribution of nitrogen in soil. In: Soil nitrogen. W.V. Bartholomew and F.E. Clark (eds.), Madison, Wis., Am. Soc. Agron. Agronomy Monograph No. 10.

U.S. Army Corps of Engineers, 1956. Snow hydrology, summary report of the snow investigations. North Pacific Division. Portland, Oregon.

Van den Honert, T.H., and J.J.M. Hooymons, 1955. On the adsorption by maize in water culture. Acta Bot Neerlandica 43:376-384.

Viets, F.G., 1965. The plant's need for and use of nitrogen. In: Soil nitrogen. W.V. Bartholomew and F.E. Clark (eds.), Madison, Wis., Am. Soc. Agron. Agronomy Monograph No. 10.

Chapter 13

UNSTEADY FLOW MATHEMATICAL MODELING TECHNIQUES

by

Ben Chie Yen, Professor of Civil Engineering, University of Illinois at Urbana-Champaign, Urbana, Illinois 61801

Chapter 13

UNSTEADY FLOW MATHEMATICAL MODELING TECHNIQUES

13.1 INTRODUCTION

River models can be classified according to the tools used as physical models, analog models and mathematical models. The mathematical models can again be classified as hydraulic models and hydrologic models; the former consider the mechanics of the flow, whereas the latter utilize no more than the continuity relationship of the flow and usually treat the river as a lumped system. Mathematical models can also be classified according to the solution techniques as analytical models, graphic models and numerical models. Analytical models are those with explicit analytical solutions. Graphical models solve the problem graphically. Numerical models are those requiring numerical procedures to obtain solutions.

Mathematical hydraulic models can be one-, two- or three-dimensional. One-dimensional models use cross-sectional average flow parameters and consider the variations of the parameters in both the longitudinal and transverse directions. However, some two-dimensional models consider the variations in the longitudinal and vertical directions while assuming homogeneous condition along the transverse direction. Three-dimensional models use "point" flow parameters and consider the longitudinal and transverse, as well as vertical, directions. Obviously three-dimensional modeling is more complicated than two- and one-dimensional modeling.

Hydraulic models can be classified according to the time variation of the flow as steady flow models and unsteady flow models. They can also be classified according to the geometric boundary condition as fixed boundary models and loose boundary models. Fixed boundaries can be rigid or flexible; the latter rarely occur in rivers. Loose boundaries usually involve erosion, deposition, and transport of cohesive and noncohesive sediment.

Mathematical modeling of steady open-channel flow has been discussed extensively in the literature (e.g., see Chow, 1959; Henderson, 1966). Modeling of loose boundary flow has been discussed in Chapter 9. Three- and two-dimensional modeling for rivers is still at the developmental stage. Therefore, in this chapter the discussion is centered on mathematical hydraulic, fixed boundary, one-dimensional, unsteady flow models, although many of the discussions may also be valid for other cases.

13.2 BASIC FLOW EQUATIONS

The basic flow equations mathematically describing unsteady river flow are the continuity equation, derived from the principle of conservation of mass, and the momentum equation from Newton's second law, the principle of conservation of linear momentum. Jean-Claude Barre de Saint-Venant (1871) derived these equations more than a century ago, long before the techniques to solve them became available. Because the Saint-Venant equations are often used improperly and many engineers do not realize that they are not exact, it is worthwhile to discuss here the assumptions involved in them.

One rigorous approach to derive the unsteady open-channel flow equations is by integrating the "point" form continuity equation and Navior-Stokes equation, which is the "point" form momentum equation. Integration over the depth yields depth-averaged continuity and momentum equations which are useful for two-dimensional modeling. Integration over the flow cross section yields the unified general one-dimensional continuity and momentum equations. Although these one- and two-dimensional flow equations have been derived for the more general case of nonhomogeneous fluids (Yen, 1973, 1975), only the case of incompressible homogeneous fluid is discussed in the following.

Conceivably, the exact form of the flow equations depends on the coordinate system used. Unfortunately, attention to this point was not given in the past, resulting in confusion and errors in expressing the flow equations. Open-channel flow problems can be formulated by using either the "natural" coordinate system or the "gravity-oriented" coordinate system. Both are orthogonal coordinate systems. As shown in Fig. 13-1,

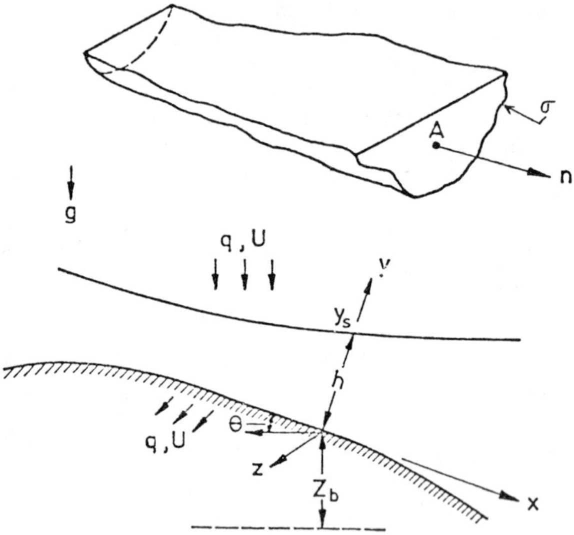

Figure 13-1. Natural Coordinates with Depth Measured Normal to Channel Bed

in the natural coordinate system the x-axis is along the longitudinal direction of the channel. The flow depth and cross section are taken normal to the x direction, i.e., not vertical.

Conversely, in the gravity-oriented coordinates (Fig. 13-2), the x-axis is along the horizontal direction whereas the flow

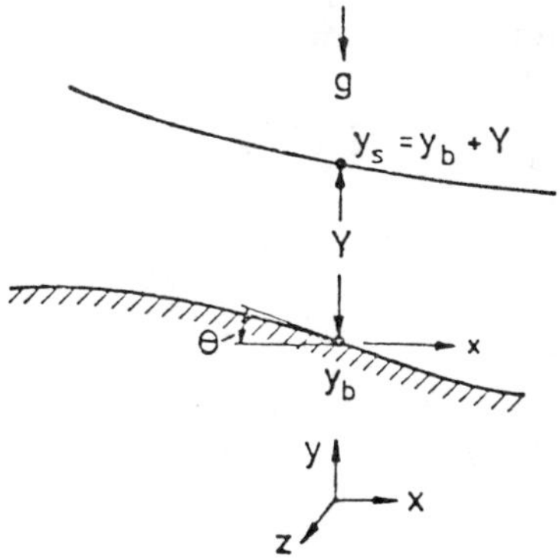

Figure 13-2. Gravity-Oriented Coordinates with Depth Measured Vertically

depth and cross section are taken vertically normal to x. In solving problems, which of these two coordinate systems should be used depends on the nature of the problem. For prismatic channels with constant bed slope the natural coordinate system is usually preferred, whereas for irregular channels with the bed and water surface profiles given in elevations the gravity-oriented coordinate system is preferred.

Two other non-orthogonal coordinate systems have often been used mistakenly for open-channel flows. One designates x horizontally while taking the flow depth and cross section normal to the channel bed, not vertically. The other defines x along the longitudinal direction of the channel, not horizontally, while taking the flow depth and cross section vertically. Yen (1975) showed that the flow equations for these two non-orthogonal coordinate systems are rather complicated and unsuitable for practical uses.

The point-form continuity equation for an incompressible fluid is (Rouse, 1959):

$$\frac{\partial u_i}{\partial x_i} = 0 \qquad (13\text{-}1)$$

in which t is time and u_i is the instantaneous local velocity component of the fluid along the x_i direction where i = 1, 2, and 3 (i.e., x, y, and z). Repetition of the subscript i in a term implies summation over the three possible orthogonal coordinate directions. The one-dimensional open-channel flow continuity equation is derived through integration of Eq. (13-1) over the flow cross section as described previously (see Yen, 1973 for details):

$$\frac{\partial A}{\partial t} + \frac{\partial Q}{\partial x} = \int_\sigma q \, d\sigma \qquad (13\text{-}2)$$

in which A is the flow cross sectional area; Q is the discharge through A; σ is the perimeter bounding A; and q is the time rate of lateral flow per unit length of σ, having a dimension of length/time, and being positive for lateral inflow and negative for outflow. Equation (13-2) is valid for both natural and gravity-oriented coordinates. The integration of q at the right hand side of Eq. (13-2) permits consideration of time and spatial variations of the lateral flow, as for the case of variable rainfall and evaporation on the free surface.

The point-form momentum equation for the mean motion (averaged over turbulence) of a homogeneous incompressible fluid is

$$\frac{\partial \bar{u}_i}{\partial t} + \frac{\partial}{\partial x_j}(\bar{u}_i \bar{u}_j) = \bar{F}_i - \frac{\partial}{\partial x_i}(\bar{p}/\rho) + \frac{\partial}{\partial x_j}(\bar{\tau}_{ij}) \qquad (13\text{-}3)$$

in which F_i is the component of the body force per unit mass along the x_i direction, p is the local pressure intensity, ρ is the fluid density, and

$$\bar{\tau}_{ij} = \mu\left(\frac{\partial \bar{u}_i}{\partial x_j} + \frac{\partial \bar{u}_j}{\partial x_i}\right) - \rho \, \overline{u'_i u'_j} \qquad (13\text{-}4)$$

where μ is the dynamic viscosity of the fluid. The bar above a term indicates time averaging over turbulence, and the prime on u indicates the turbulence fluctuation with respect to the mean. Integrated over the cross section A, the one-dimensional momentum equation for gravity-oriented coordinates with depth Y measured vertically is (Yen, 1975):

$$\frac{1}{gA}\frac{\partial Q}{\partial t} + \frac{1}{gA}\frac{\partial}{\partial x}\left(\frac{\beta}{A}Q^2\right) + \frac{\partial}{\partial x}(k\,Y) + (k-k')\frac{Y}{A}\frac{\partial A}{\partial x}$$

$$= S_o - S_f + \frac{1}{\gamma A}\frac{\partial T}{\partial x} + \frac{1}{gA}\int_\sigma q\, U_x \, d\sigma \qquad (13\text{-}5)$$

and for natural coordinates with depth h measured normal to x,

$$\frac{1}{gA}\frac{\partial Q}{\partial t} + \frac{1}{gA}\frac{\partial}{\partial x}\left(\frac{\beta}{A}Q^2\right) + \frac{\partial}{\partial x}(Kh\cos\theta) + (K-K')h\cos\frac{1}{A}\frac{\partial A}{\partial x}$$

$$= S_o - S_f + \frac{1}{\gamma A}\frac{\partial T}{\partial x} + \frac{1}{gA}\int_\sigma q\, U_x \, d\sigma \qquad (13\text{-}6)$$

In Eqs. (13-5) and (13-6), g is the gravitational acceleration; S_o is the channel bottom slope, which is $\sin\theta$ for natural coordinates and $\tan\theta$ for gravity-oriented coordinates, where θ is the angle between the channel bottom and a horizontal plane (Fig. 13-1 and 13-2); S_f is the friction slope; γ is the specific weight of the fluid; U_x is the x-component velocity of the lateral flow when joining the channel flow; k, k', K, and K' are pressure distribution correction factors; β is the momentum flux correction factor given by

$$\beta = \frac{A}{Q^2} \int_A \bar{u}^2 \, dA \qquad (13\text{-}7)$$

in which u is the x-component of the local (point) velocity; and T represents the force acting normal on A due to internal stresses:

$$T = \int_A \left(2\mu \frac{\partial \bar{u}}{\partial x} - \rho \overline{u'^2}\right) dA \qquad (13\text{-}8)$$

Derivation of Eq. (13-5) and (13-6) can be found elsewhere (Yen, 1973, 1975). The value of β is equal to unity if the x-component velocity distribution over A is uniform. The pressure distribution correction factors for the gravity oriented coordinates are defined as

$$k = \frac{1}{\gamma A Y} \int_A [\bar{p} + \gamma(y - y_b)] \, dA \qquad (13\text{-}9)$$

$$k' = \left(\gamma Y \frac{\partial A}{\partial x}\right)^{-1} \int_\sigma \left[\bar{p} \frac{\partial \bar{r}}{\partial x} + \gamma(y - y_b) \frac{\partial \bar{r}}{\partial x} + \overline{p' \frac{\partial r'}{\partial x}}\right] d\sigma \qquad (13\text{-}10)$$

where y and y_b are the elevations of the point and the channel bed, respectively, above a reference datum; and r is the normal displacement of σ with respect to space or time projected on a plane parallel to A, being positive outward. The boundary σ may fluctuate, e.g., due to wavy surface. Correspondingly, for the natural coordinates,

$$K = \frac{1}{\gamma A h \cos\theta} \int_A (\bar{p} + \gamma y \cos\theta) dA \qquad (13\text{-}11)$$

$$K' = \left(\gamma h \cos\theta \frac{\partial A}{\partial x}\right)^{-1} \int_\sigma \left[(\bar{p} + \gamma y \cos\theta) \frac{\partial \bar{r}}{\partial x} + \overline{p' \frac{\partial r'}{\partial x}}\right] d\sigma \qquad (13\text{-}12)$$

Special values of the pressure correction factors k, K, k', and K' are given in Table 13-1.

Equations (13-2) and (13-5) or (13-6) are the unified exact one-dimensional equations for open-channel flow. The flow can be turbulent or laminar, rotational or irrotational, unsteady or steady, uniform or nonuniform, subcritical or supercritical, gradually or rapidly varied, and with or without lateral discharges. The channel can have arbitrary cross sectional shape and alignment with fixed or loose and impervious or pervious bed. But they are also extremely difficult to

Table 13-1. Special Values of Pressure Correction Factors for Incompressible Homogeneous Fluids

Correction Factor	Coordinate System	Special Value	Condition
k	Gravity	1	Constant piezometric pressure distribution over A
K	Natural	1	Same as above
K	Natural	1	Hydrostatic pressure distribution over A
k	Gravity	$(1+\cos^2\theta)/2$	Hydrostatic pressure distribution over rectangular cross section
k'	Gravity	1	Constant piezometric pressure distribution over nonfluctuating A and σ
K'	Natural	1	Same as above
K'	Natural	1	Hydrostatic pressure distribution over nonfluctuating A and σ

solve because of the temporal and spatial variations of the correction factors β, k, k', K, and K'. However, in most cases the variations of these correction factors are either small or predictable, making the solution (numerically or analytically) possible. In fact, from an engineering viewpoint, one big advantage that the cross sectional averaged one-dimensional equations have is the possibility of obtaining a solution by treating these correction factors as constants or prescribed expressions, whereas the solutions for the corresponding point-form continuity and momentum equations are either impossible or costly. Nevertheless, acceptable assumptions are necessary to facilitate possible solutions of one-dimensional equations.

For gradually varied flows, the momentum equations can be simplified by the following assumptions:

(a) The pressure distribution correction factors k, k', or K, K' are each equal to unity, corresponding to the conditions listed in Table 13-1.

(b) The gradient of the force due to internal stresses, $\partial T/\partial x$, is relatively small and negligible.

Accordingly, Eqs. (13-5) and (13-6) can be reduced for gravity-oriented and natural coordinates, respectively, as

$$\frac{1}{gA}\frac{\partial Q}{\partial t} + \frac{1}{gA}\frac{\partial}{\partial x}\left(\frac{\beta}{A}Q^2\right) + \frac{\partial Y}{\partial x} \qquad -S_o + S_f = \frac{1}{gA}\int_\sigma q\, U_x\, d\sigma \tag{13-13}$$

dynamic wave

quasi-steady dynamic wave

diffusion wave

kinematic wave

$$\frac{1}{gA}\frac{\partial Q}{\partial t} + \frac{1}{gA}\frac{\partial}{\partial x}\left(\frac{\beta}{A}Q^2\right) + \frac{\partial}{\partial x}(h\cos\theta) - S_o + S_f = \frac{1}{gA}\int_\sigma q\, U_x\, d\sigma \tag{13-14}$$

These equations can also be expressed in terms of the cross sectional averaged velocity $V = Q/A$ instead of Q. With the additional assumption that $\beta = 1$, the simplified momentum equations for gravity-oriented and natural coordinates are, respectively,

$$\frac{1}{g}\frac{\partial V}{\partial t} + \frac{V}{g}\frac{\partial V}{\partial x} + \frac{\partial Y}{\partial x} \qquad - S_o + S_f = \frac{1}{gA}\int_\sigma (U_x - V)q\, d\sigma \tag{13-15}$$

dynamic wave

quasi-steady dynamic wave

diffusion wave

kinematic wave

$$\frac{1}{g}\frac{\partial V}{\partial t} + \frac{V}{g}\frac{\partial V}{\partial x} + \frac{\partial}{\partial x}(h\cos\theta) - S_o + S_f = \frac{1}{gA}\int_\sigma (U_x - V)q\, d\sigma \tag{13-16}$$

It should be noted here that in Eqs. (13-13) and (13-15), $(\partial Y/\partial x) - S_o = (\partial y_s/\partial x)$, where y_s is the water surface elevation. Use of the water surface gradient avoids the determination of the channel slope S_o, which is often difficult to determine locally for rivers. It should also be noted that V in Eqs. (13-15) and (13-16) is along the x-direction; i.e., V is horizontal in Eq. (13-15) and inclined along the channel longitudinally in Eq. (13-16).

The continuity equation that is usually used together with Eqs. (13-15) or (13-16) to solve problems is

$$\frac{\partial Y}{\partial t} + D \frac{\partial V}{\partial x} + V \frac{\partial Y}{\partial x} = \frac{1}{B} \int_{\sigma} q \, d\sigma \qquad (13\text{-}17)$$

for gravity-oriented coordinates, and

$$\frac{\partial h}{\partial t} + D \frac{\partial V}{\partial x} + V \frac{\partial h}{\partial x} = \frac{1}{B} \int_{\sigma} q \, d\sigma \qquad (13\text{-}18)$$

for natural coordinates, where B is the water surface width and $D = A/B$ is called hydraulic mean depth. Equations (13-17) and (13-18) can be derived from Eq. (13-2) assuming prismatic rigid boundary channel.* Derivation of the above equations can be found elsewhere (Yen, 1971) and is not repeated here.

Equations (13-2) and (13-13) or (13-14), or Eqs. (13-15) and (13-17), or Eqs. (13-16) and (13-18) are commonly referred to as the Saint-Venant equations. They are also called shallow water wave equations. They are often termed as "complete" unsteady flow dynamic wave equations in the sense that they contain all the major influential dynamic terms representing the gradually varied unsteady flow. However, they are by no means exact, as has been revealed by the assumptions discussed previously. They are invalid when the assumptions, particularly that on pressure distribution, are seriously violated. Thus, the Saint-Venant equations are unreliable when applied to the initial stage of dam break problems, channel control sections with highly curvilinear flow, supercritical flow with roll waves, and flow with the Froude number near unity (hydraulic jump and hydraulic drop).

Mathematically, the Saint-Venant equations are a pair of first order partial differential equations of hyperbolic type. Analytical solutions can be obtained only for a few special cases for which the equations are considerably simplified, and often with linearization. Some graphical methods have been proposed in the past for solving these equations, but they have seldom been used since the advent of digital computers. Nearly all the solutions are now sought numerically. Numerical models can be grouped into finite difference models and finite element models. Application of finite element methods to unsteady open-channel flow problems is still in its infant stage. At present it appears that finite element methods may offer some advantages when the boundary conditions are complicated and the fluid is highly nonhomogeneous. For many river flow routing problems, finite difference is still the method of solution.

When applying the Saint-Venant equations to simulate unsteady flow in a channel, the friction slope, S_f, is usually estimated by using Manning's formula

*otherwise add a term containing $V(\partial A/\partial x)_{h}$ = constant.

$$S_f = \frac{n^2V^2}{C_n} R^{-4/3} = \frac{n^2Q^2}{C_nA^2} R^{-4/3} \qquad (13\text{-}19)$$

in which n is Manning's roughness factor; R is hydraulic radius, which is equal to A divided by the wetted perimeter; and $C_n = 1$ for SI units and 2.22 for English units. It may also be estimated by using the Darcy-Weisbach formula

$$S_f = \frac{f}{8gR} V^2 = \frac{f}{8gR} \frac{Q^2}{A^2} \qquad (13\text{-}20)$$

in which f is Weisbach's resistance coefficient; or by Chezy's formula

$$S_f = \frac{V^2}{C^2R} = \frac{Q^2}{C^2RA^2} \qquad (13\text{-}21)$$

in which C is Chezy's coefficient. Rigorously speaking, the values of n, f, and C for unsteady nonuniform flow have not been established. They are functions of flow unsteadiness, nonuniformity, the Reynolds and Froude numbers, and channel boundary roughness conditions including bed forms for alluvial channels (Rouse, 1965; Yen, 1971). In general, the friction slope is different from the dissipated energy gradient, the total-head gradient, the hydraulic gradient, and the channel bottom slope. Only for steady uniform flow without lateral flow are these gradients equal to one another. However, it appears that the steady uniform flow values of n, f, and C which can be found in standard references provide good approximations to gradually varied unsteady flow corresponding to the conditions for which the Saint-Venant equations are valid.

Although the Saint-Venant equations are not exact, it has been found that from an engineering viewpoint they are sufficiently reliable and adequate to model gradually varied unsteady flow as long as the assumptions on pressure distribution and $\partial T/\partial x$ are satisfied. As an example, solutions of the Saint-Venant equations [Eqs. (13-16) and (13-18)] using five different numerical schemes are compared with experimental results obtained by Ackers and Harrison (1964) at the British Hydraulics Research Station (HRS) at Wallingford, England, and by Yevjevich and Barnes (1970) at Colorado State University (CSU), and the comparisons for two experiments summarized in Table 13-2 are shown in Figs. 13-3 and 13-4.

13.3 BOUNDARY AND INITIAL CONDITIONS

Solution of an unsteady river flow problem obviously depends on the initial and boundary conditions imposed on the river flow. In other words, when using the Saint-Venant equa-

Table 13-2. Experimental Data Adopted for Comparison With Theoretical Solutions for Open-Channel Flow in Single Pipe

Channel and Inflow Properties	CSU Run No. 19	HRS Run No. 115
Pipe Diameter, D, ft	3.0	0.25
Length, L, ft	822	300
Slope, S_o	0.00099	0.0010
Roughness	f = 0.012	k = 0.00004 ft.
Base Flow, Q_b, cfs	2.60	0.0189
Peak Inflow, $Q_p(0)$, cfs	26.17	0.0532
Duration of Inflow Hydrograph, t_i, sec	89.0	36.7
Shape of Inflow Hydrograph	Symmetric, triangular	Symmetric, trapezoidal with peak inflow for 3.3 sec.

tions to model river flow, two initial conditions and two boundary conditions must be specified in order to obtain a unique solution that describes the flow.

Initial Conditions

The initial conditions specify the flow conditions at the initial time (t=0) of computation of the unsteady flow. The two initial conditions for the Saint-Venant equations are the velocity, $V(x,0)$, or discharge, $Q(x,0)$, paired with the area, $A(x,0)$, or depth, $Y(x,0)$ or $h(x,0)$, specified for the entire channel length at the initial time $t=0$. However, if an initial dry-bed condition is specified with V or $Q=0$ and A or $h=0$ at $t=0$, a numerical singularity is generated. For such a case one can assume a non-zero but small and negligible initial depth so that the computations can proceed. After all, the Saint-Venant equations are unreliable for dry bed because under such conditions the Weber number (interfacial, surface tension) effect is important, which is not accounted for in the Saint-Venant equations.

Boundary Conditions

In addition to the initial conditions, two boundary conditions are required in order to yield a unique solution of the

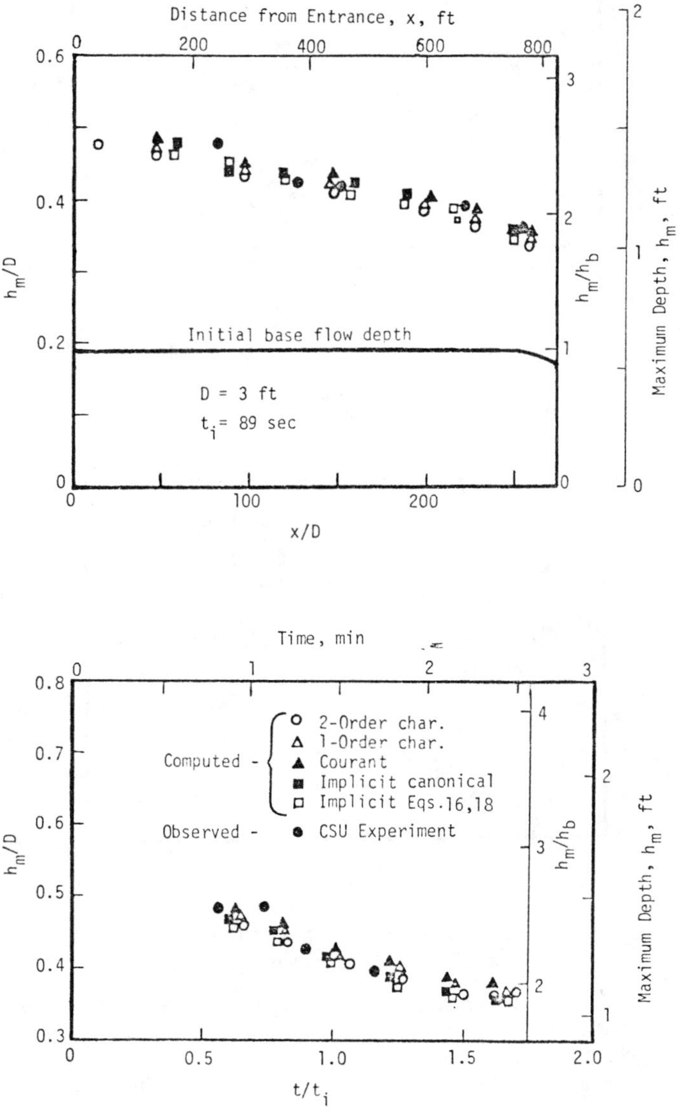

Figure 13-3. Comparison of Theoretical Solutions with CSU Experimental Data

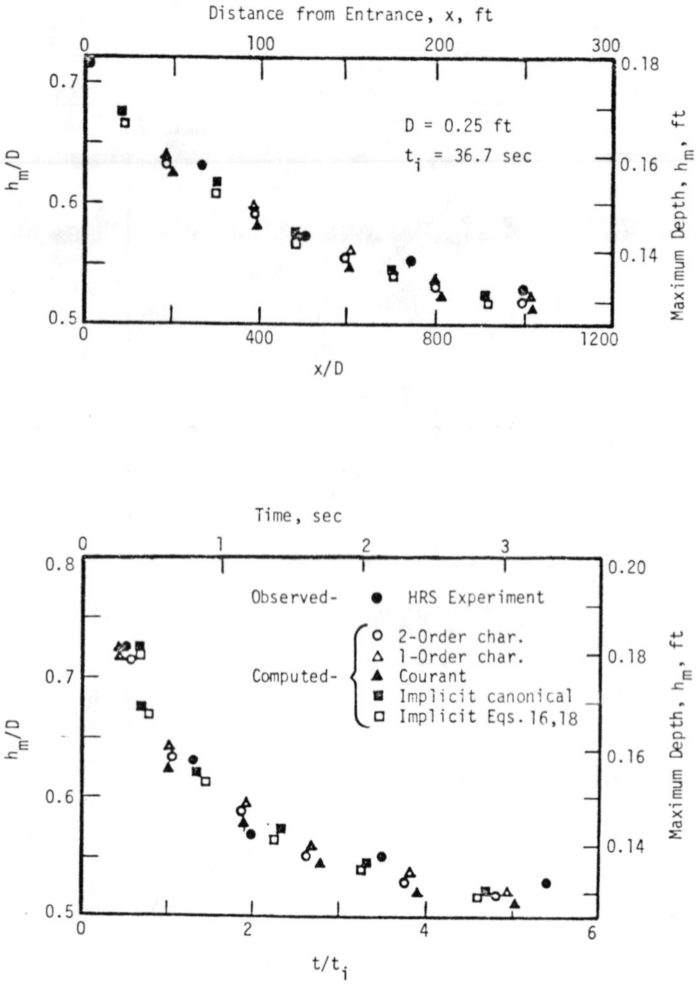

Figure 13-4. Comparison of Theoretical Solutions with HRS Experimental Data

Saint-Venant equations. The boundary conditions specify the time variations of the discharge, velocity, depth, or area of boundary locations. For a supercritical flow, both boundary conditions must be specified at the upstream boundary of the channel. For a subcritical flow, one boundary condition must be specified at the upstream end of the channel whereas the other must be at the downstream end.

In routing of unsteady flow in a river network, the computation usually proceeds downstream from the upstream end, station by station and channel by channel. For a supercritical flow there is no downstream backwater effect and both boundary conditions are specified at the upstream end. They are either known or computed in previous computations. Hence, there are no computational difficulties provided the river junction characteristics are known.

Conversely, for a subcritical flow, the upstream boundary condition reflects the effect of the flow upstream, whereas the downstream boundary condition reflects the downstream backwater effect. For a river network, the downstream boundary condition of a channel is actually by itself an unknown and a part of the solution being sought. Therefore, the numerical solution for this channel cannot be obtained until its downstream boundary condition is specified. This unknown downstream boundary in turn provides the upstream boundary condition for the immediate downstream channel. This situation of unknown boundary conditions repeats channel by channel for all the internal channels of a river network, until the "outlet" of the network where the downstream condition is specified.

Properly specified downstream boundary condition is of great importance in seeking a realistic solution for a subcritical flow. As an example, Sevuk and Yen (1973b) demonstrate four vastly different results for an open-channel network of identical inflow hydrographs but different downstream boundary conditions (Figure 13-5). Liggett and Cunge (1975) discussed the significance of properly posed boundary conditions in seeking solution of the Saint-Venant equations. Further discussion on boundary conditions as related to solution of unsteady flow in river networks will be given later.

Channel Junctions

The precise hydraulic description of the flow at channel junctions is rather complicated and difficult because of the high degree of flow mixing, separation, turbulence, and energy loss. Yet correct representation of the junction hydraulics is important in realistic and reliable computation of flow in river networks. In addition to the continuity relationship, the dynamic relationship can be represented by either the energy or the momentum equations. In applications, the momentum equations are rarely used because they are vector equations and the pressure acting on the junction boundaries is usually difficult to describe. The energy equation is usually expressed in a simplified form of one of the following equations:

$$\Sigma Q_i \left(\frac{V_i^2}{2g} + y_i + Z_i \right) = Q_o \left(\frac{V_o^2}{2g} + y_o + Z_o + h_f \right) \qquad (13\text{-}22)$$

$$\frac{V_i^2}{2g} + y_i + Z_i = \frac{V_o^2}{2g} + y_o + Z_o + h_{fi} \qquad (13\text{-}23)$$

in which Q is flow into or from the junction with a velocity V; y is depth of flow; Z is elevation of channel bed at the

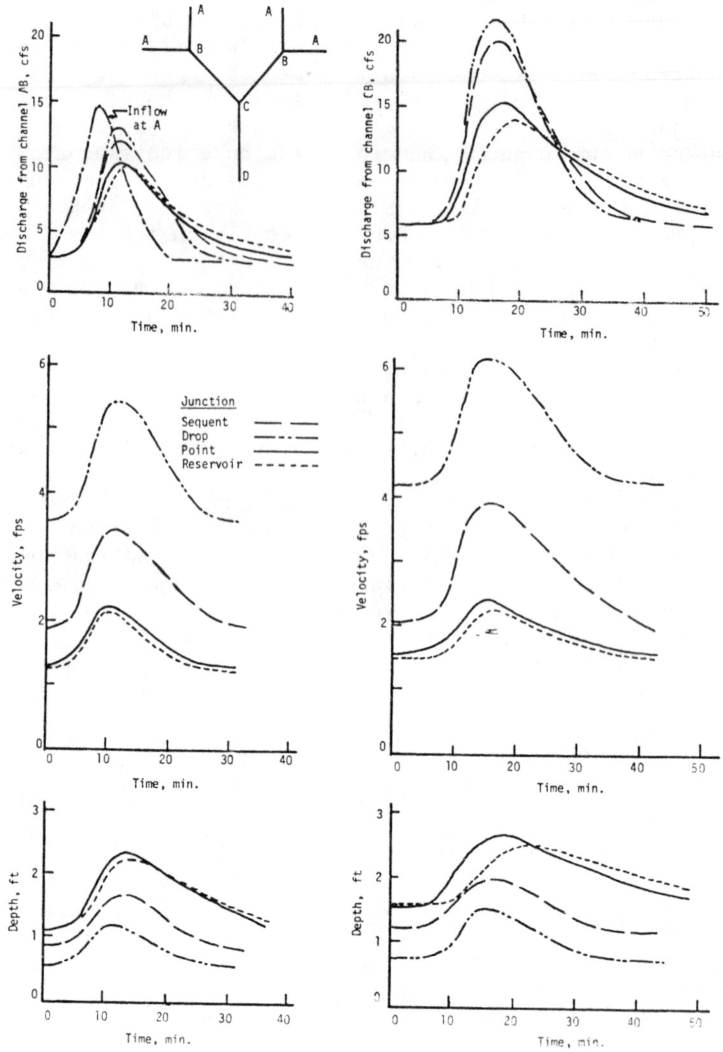

(a) Flow Entering BC

(b) Flow Entering CD

Figure 13-5. Effect of Junction Condition

junction; and h_f is loss of energy head. The subscript i indicates the i-th inflow channel at the junction; the subscript o represents the outflow channel; and h_{fi} is not generally available and hence assumptions are necessary to facilitate solutions. For practical purposes, the junctions can be classified into point type and reservoir type depending on whether the junction storage capacity is negligible relative to the volume of the flow.

(a) Point-type junction. For junctions with insignificant storage capacity, the junction can be considered as a point-type junction which is assumed to be represented by a single confluence point without storage. The net discharge into the junction is therefore zero at all times. Hence

$$\Sigma Q_i = Q_o \tag{13-24}$$

A typical point-type junction with two inflow channels and one outflow channel is shown schematically in Fig. 13-6, for which

$$Q_1 + Q_2 = Q_o \tag{13-25}$$

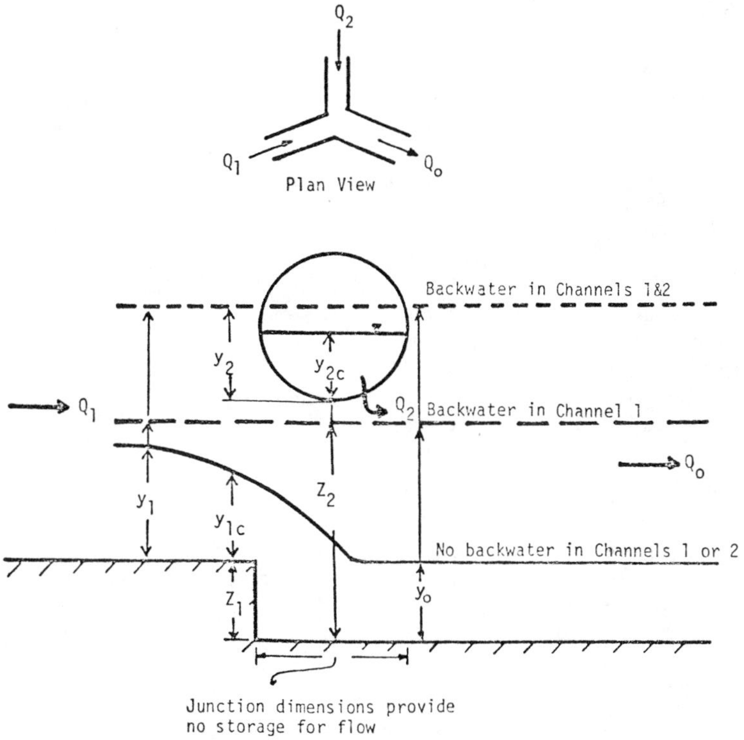

Figure 13-6. Point-Type Junction

For subcritical flow in the inflowing channels, the flow discharges freely into the junction only when a free-fall exists over a nonsubmerged drop at the end of the channel. Otherwise, the subcritical flow in the inflowing channel is subject to backwater effect from the junction. Since the junction is considered as a point, the energy compatibility condition can be represented by a common water surface at the junction for all the joining channels. Thus, by referring to Fig. 13-6,

$$y_i = y_{ic} \qquad \text{if } Z_i + y_{ic} > y_o + z_o \tag{13-26a}$$

$$y_i + Z_i = y_o + Z_o \qquad \text{otherwise} \tag{13-26b}$$

in which y_i = depth of flow (measured vertically) of the i-th inflowing channel at the junction; y_{ic} = critical depth corresponding to the instantaneous flow rate Q_i; Z_i = height of the drop of the i-th inflowing channel; and y_o and Z_o = depth and drop, respectively, of the outflowing channel. Flow in the outflow channel may be either subcritical or supercritical. In the latter case, y_o in Eq. (13-26a) is equal to the critical flow depth, y_{oc}, corresponding to the instantaneous flow rate Q_o.

Flow in the inflow channels can also be supercritical, discharging freely into the junction, provided the flow at the downstream end of the channel is not submerged by the backwater in the junction. For such case, the discharge of the inflowing channels into the junction can be computed without considering the flow condition in the junction. Subsequently, the discharge into the outflowing channel, Q_o, can be computed directly by using Eq. (13-24).

(b) Reservoir-type junction. The reservoir-type junction has a relatively large storage capacity in comparison to the flow. Consequently, it can be assumed to behave like a reservoir with a horizontal water surface and capable of adsorbing and dissipating all the kinetic energy of the inflows. The net discharge into the junction is equal to the time rate of change of storage in the junction, i.e.,

$$\Sigma Q_i - Q_o = \frac{ds}{dt} \tag{13-27}$$

in which s = water stored in the junction, and other symbols are as defined in Eq. (13-24). The depth of water, H, in the junction is assumed equal to the specific energy of the flow at the entrance of the outflowing channel, i.e.,

$$H = y_o + \frac{V_o^2}{2g} + Z_o \tag{13-28}$$

A typical three-way reservoir-type junction is shown schematically in Figure 13-7.

Since the kinetic energy of the inflows is assumed lost at the junction, for subcritical flow in the inflow channels,

$$Z_i + y_i = H \qquad \text{if } Z_i + y_{ic} < H \tag{13-29a}$$

$$y_i = y_{ic} \qquad \text{otherwise} \tag{13-29b}$$

If the flow in the outflow channel is supercritical, critical flow condition exists at its entrance and hence H in Eq. (13-29a) should be replaced by the minimum specific energy corresponding to the instantaneous flow rate Q_o.

As in the case of point-type junctions, supercritical flow in the inflowing channels discharges freely into the reservoir-type junction provided the inflow is not submerged by the backwater from the junction, and the discharge from the inflowing channels into the junction can be computed without considering the existing flow conditions in the junction or outflowing channel.

13.4 APPROXIMATIONS TO SAINT-VENANT EQUATIONS

Because Saint-Venant equations are rather complicated and it is not an easy task to obtain their solutions for unsteady river flows, various approximations to these equations have been proposed to provide simpler but acceptable solutions. From a hydraulic viewpoint, these approximations can be classified, according to the terms of the momentum equation considered, as quasi-steady dynamic wave, diffusion wave, and kinematic wave approximations as shown in Eqs. (13-13) and (13-14) or (13-15) and (13-16). If cnly the continuity equation is considered and the momentum equation is ignored, the approximation is a hydrologic routing model (e.g., see Chow, 1964) and is not discussed here.

Kinematic Wave Approximation

The kinematic wave approximation is the simplest but also the least accurate model of the three approximations to the Saint-Venant equations. It retains only the two slope terms of the momentum equation (any one of Eqs. (13-13), (13-14), (13-15) or (13-16) and ignores the inertial and pressure terms [the first three and last terms in Eqs. (13-13), (13-14), (13-15) or (13-16)]; i.e.,

$$S_o = S_f \tag{13-30}$$

where S_f can be approximated by the Manning [Eq. (13-19)], Darcy-Weisbach [Eq. (13-20)], or Chezy [Eq. (13-21)] formula as discussed previously.

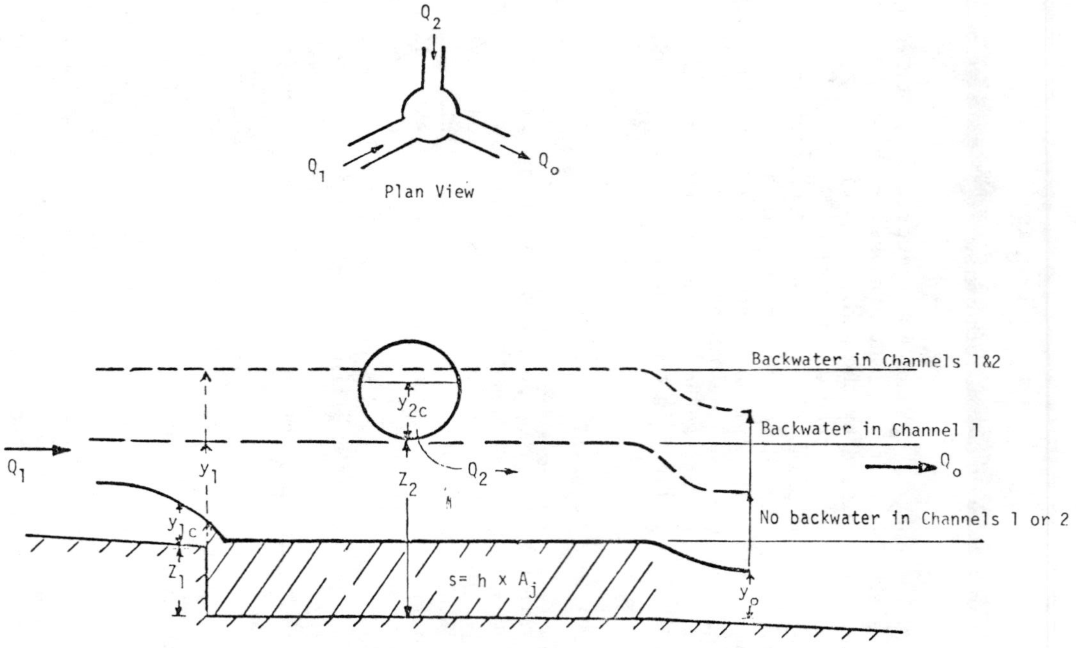

Figure 13-7. Reservoir-Type Junction

Equation (13-30) together with the continuity equation [any one of Eqs. (13-2), (13-17), or (13-18)] forms a nonlinear kinematic wave model for which, except for special cases, solutions are obtained numerically. However, because of the simplification of the momentum equation, only one boundary condition is required instead of two, as in the case of the Saint-Venant equations. Usually this single boundary condition is supplied by the flow condition at the channel upstream. For subcritical flow, elimination of the downstream boundary condition also eliminates the mechanism used to account for the downstream backwater effect. Consequently, solution for a river network by using the nonlinear kinematic wave model can be accomplished in a cascading manner, solving the flow channel by channel and junction by junction, starting from the most upstream channels and proceeding downstream in sequence, no matter whether the flow is supercritical or subcritical. Since the downstream backwater effect cannot be accounted for, the nonlinear kinematic wave model is unreliable for subcritical flow when the downstream backwater effect is important. Ironically, kinematic wave approximation is supposed to be relatively more accurate when the inertial effect is relatively small, i.e., subcritical flow with Froude number not close to unity.

In the kinematic wave approximation, neglecting the inertial and pressure terms in the momentum equation also eliminates the mechanism for flood wave attenuation. Theoretically, the continuity equation permits translation and some distortion of the flood hydrograph as the flood propagates downstream, but there is no peak attenuation (Lighthill and Whitham, 1955). However, in seeking solutions by using finite difference numerical procedures, inevitably numerical attenuation is introduced. This numerical attenuation usually acts advantageously in the same direction as the actual flood wave attenuation, and it is often misunderstood by some engineers who, mistakenly thinking that small computational steps will give more accurate results, use reduced sizes of computational steps in the hope of getting more attenuation. In fact, they are incurring more computational expense to obtain a solution that is closer to what the kinematic wave equations represent--no attenuation--and usually farther from the actual physical condition.

Because of its relative simplicity and computational economy, the nonlinear kinematic wave approximation has become very popular in the last decade. Indeed, it is a useful and powerful approximation for many unsteady flow problems, providing sufficient accuracy from the practical viewpoint. However, one should always keep in mind the assumptions and limitations associated with the approximation so that it will not be used improperly. Several models have been proposed either to simplify and improve the computational aspects of the nonlinear kinematic wave approximation (e.g., Muskingum-Cunge method), or to provide some remedy for the downstream backwater effect (e.g., Storm Water Management Model). However, these variations of the nonlinear kinematic wave approximation are not discussed here.

Diffusion Wave Approximation

The next higher level approximation is the diffusion wave approximation, which incorporates the pressure term [third term in Eqs. (13-13), (13-14), (13-15) or (13-16)] in addition to the two slope terms in the momentum equation, i.e.,

$$\frac{\partial Y}{\partial x} = S_o - S_f \tag{13-31}$$

for gravity-oriented coordinates, and

$$\frac{\partial}{\partial x} (h \cos\theta) = S_o - S_f \tag{13-32}$$

for natural coordinates. The simplified momentum equation [Eq. (13-31) or (13-32)] is combined with the continuity equation [any one of Eqs. (13-2), (13-17) or (13-18)] to form the nonlinear diffusion wave model.

Inclusion of the pressure term in the diffusion wave model substantially improves the solution accuracy. It permits peak attenuation in addition to distortion and translation of the hydrograph. It provides a means of accounting for the downstream backwater effect, if any. However, to obtain a unique numerical solution, it requires that two boundary conditions be specified, as in the case of the Saint-Venant equations. Therefore, it also requires simultaneous or iterative numerical solutions and is more complicated than the nonlinear kinematic wave approximation. Nevertheless, through careful formulation it is possible to reduce computational expense considerably (Akan and Yen, 1977).

Quasi-Steady Dynamic Wave Approximation

The quasi-steady dynamic wave approximation neglects only the local acceleration term [first term in Eqs. (13-13), (13-14), (13-15) or (13-16)] and considers all others terms of the momentum equation. This simplified momentum equation is coupled with the continuity equation to form the nonlinear quasi-steady dynamic wave model. It accounts for the downstream backwater effect and permits peak attenuation, distortion and translation of the hydrograph, just as for diffusion and full dynamic wave models. It also requires that two boundary conditions be specified for the solution to be unique. Its numerical solution procedure is nearly as complicated as that for the Saint-Venant equations.

For gradually varied unsteady river flows, except for highly nonuniform channels (such as rapid expansion or contraction), the local and convective acceleration terms (first two terms in the momentum equation) are usually of the same order of magnitude but have opposite signs. Therefore, neglecting only one of them gives worse results than neglecting both. In other words, the quasi-steady dynamic wave approximation gives less satisfactory results than the diffusion wave approximation, and hence the latter is preferred not only because it is simpler, but also because it is more accurate.

Comparison of Approximate Models

In order to provide an idea of the relative accuracy of these approximations, their solutions can be compared with the solutions of the Saint-Venant equations. An example comparison is made by considering a rather critical condition of a flood wave routed through a long (4000 ft) channel. A four-point noncentral, implicit, finite difference numerical scheme is adopted and identical time and space increments are used in the computation for all the cases. A typical result for an inflow hydrograph having a duration t_i = 40 min routed through the channel having a circular cross section of diameter D = 6 ft and slope S_o = 0.0006 over a constant baseflow Q_b = 20 cfs is shown in Figure 13-8. The computed results show that the nonlinear kinematic wave model is the least accurate one, and the diffusion wave approximation is better than the quasi-steady dynamic wave approximation, as discussed previously. Note that there is numerical attenuation for the kinematic wave approximation. This attenuation will be smaller if a finer computational grid is used. Further discussions related to comparison of the different approximations have been presented by Gunaratnam and Perkins (1970), Bettess and Price (1976), and Grijsen and Vreugdenhil (1976).

13.5 FINITE DIFFERENCE SOLUTION SCHEMES

The nonlinear flow routing models, each of which is formed by the continuity equation coupled with the momentum equation or its simplified form, are mathematically a set of first order quasi-linear hyperbolic partial differential equations. Only for special cases can these equations be further simplified to yield analytical solutions. Usually, solutions can be obtained only numerically, with appropriately specified initial and boundary conditions. In the finite difference numerical methods a river channel is divided into a number of reaches and the continuous time is divided into time steps, which together form the computational grids (x-t plane). The differentials in the flow equations are approximated by the corresponding finite differences of the parameters between the computational grid points so that the differential equations are approximated by algebraic equations and then solved numerically. Many finite difference numerical schemes have been proposed to solve these equations. They can be classified into the following three groups.

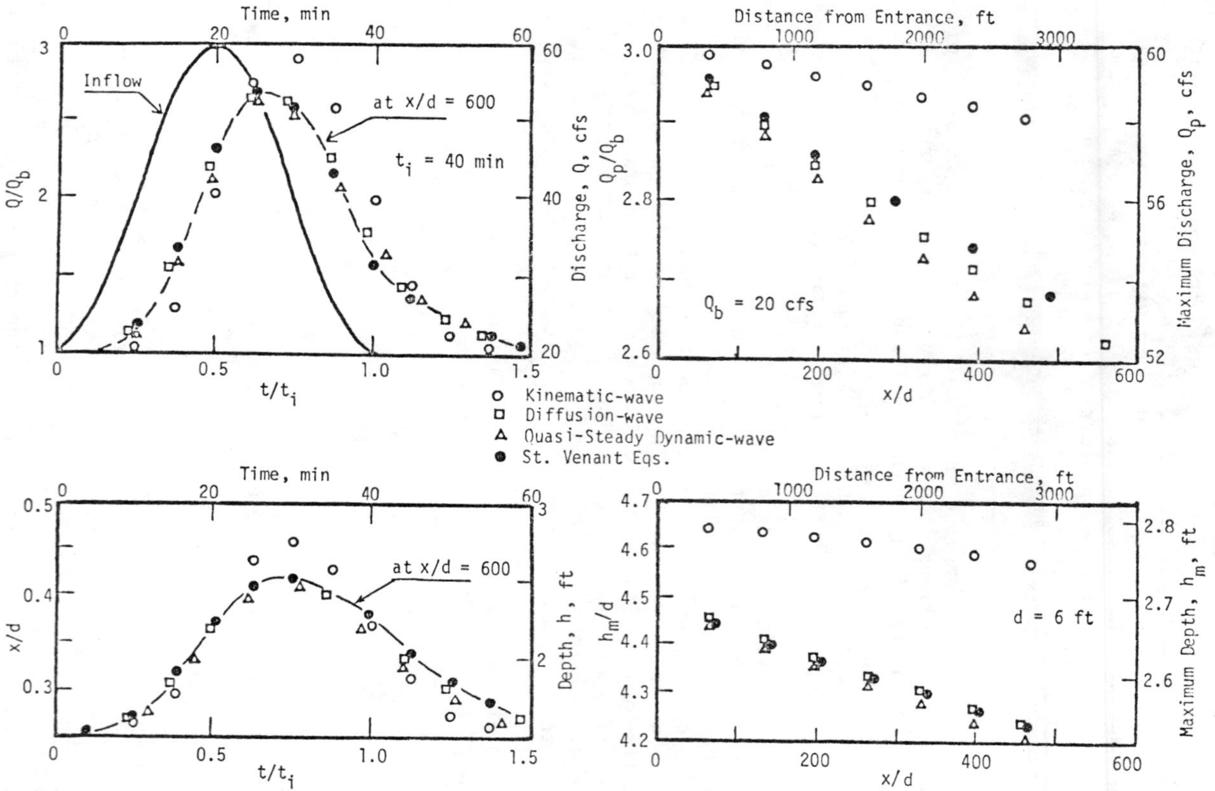

Figure 13-8. Comparison of Hydraulic Routing Methods

(a) Explicit schemes. The explicit schemes express the unknown parameters explicitly as functions of known quantities, and solve them directly. They are relatively easy to understand, easy to formulate and easy to program, but they are also computationally highly inefficient because of numerical stability problems. To minimize the numerical instability, the computational grids are usually selected to satisfy the Conrant criterion,

$$\frac{\Delta x}{\Delta t} \geq V + (gA/B)^{1/2} \tag{13-33}$$

in which Δx and Δt are computational space and time intervals, respectively; other terms are as defined previously. Thus, in order to ensure numerical stability, for a given reach Δx, Δt is so small that the computation becomes very costly. Therefore, explicit schemes are only useful in flood routing of short duration events such as flash floods.

(b) Implicit schemes. The implicit schemes express the unknown parameters implicitly in simultaneous algebraic equations and then solve them using an appropriate solution technique. They are relatively much more difficult to formulate and program, but if done properly and carefully they can be computationally very efficient and stable. The finite difference computational grid sizes Δx and Δt can be chosen independently.

(c) Method of characteristics. The method of characteristics solves two sets of "characteristic" equations, each set consisting of a pair of ordinary differential equations. These equations are transformed mathematically from the Saint-Venant equations. For example, transformation of Eqs. (13-15) and (13-17) with no lateral flow ($q = 0$) yields the following characteristic equations:

$$dV + (gB/A)^{1/2}\, dY + g(S_f - S_o)\, dt = 0 \tag{13-34a}$$

$$dx = [V + (gA/B)^{1/2}]\, dt \tag{13-34b}$$

$$dV - (gB/A)^{1/2}\, dY + g\,(S_f - S_o)\, dt = 0 \tag{13-35a}$$

$$dx = [V - (gA/B)^{1/2}]\, dt \tag{13-35b}$$

The symbols are all as defined previously. Equation (13-34) is known as the forward characteristic, whereas Eq. (13-35) is the backward characteristic. The characteristic equations are usually solved numerically using finite differences to approximate differentials. They may be expressed explicitly or implicitly using rectangular space-time grids or characteristics grids.

There are many different schemes for each of the three groups just mentioned, depending on how many grid points are used in expressing the finite differences to approximate the differentials, how (relative locations) these grid points are chosen, and how the computational grid is formulated. Although there are clearly inferior schemes that one should avoid, which of the many schemes should be used to solve an unsteady flow problem depends on the nature of the problem, the accuracy required, and the computational facilities available. Those interested in finite difference solution techniques of hyperbolic partial differential equations should refer to references on this topic (e.g., Ames, 1969; Forsythe and Warsow, 1960; Mahmood and Yevjevich, 1975; Remson et al., 1971; Richtmeyer and Morton, 1967; Smith, 1965). Some comparison of the finite difference schemes to unsteady river flow problems can be found, e.g., in Amein and Fang (1969), Baltzer and Lai (1968), Bettes and Price (1976), Price (1974), Sevuk and Yen (1973a), Strelkoff (1970), and Yevjevich and Barnes (1970). For gradually varied river flows with the current computer capabilities, usually the implicit schemes and the method of characteristics are preferred to the explicit schemes.

13.6 RIVER NETWORK SOLUTION TECHNIQUES

A river network consists of a number of river channels joined at a number of junctions. These junctions pose additional difficulties in routing unsteady flow through a river network because they behave as internal boundary conditions, as discussed previously in Section 13.3, Boundary and Initial Conditions. In other words, the flow conditions at the junctions are part of the solution being sought. From a practical viewpoint, the junctions can be treated as either point-type or reservoir-type as discussed previously. There are four methods to facilitate solution of unsteady flow in river networks as follows.

Simultaneous Solution Method

When the diffusion wave, quasi-steady dynamic wave or dynamic wave (Saint-Venant equations) models are applied to a river channel, two boundary conditions in addition to the initial conditions are required to yield the unique solution. For a subcritical flow, one boundary condition should be at the upstream end of the channel, whereas the other should be at the downstream end. The latter, which provides through the junction the upstream boundary condition for the immediately following channel, is actually an unknown and a part of the solution being sought. In using the implicit numerical scheme, the flow in an interior channel can only be expressed in terms of the unknown boundary conditions at its upstream and downstream ends. This situation repeats until the last channel of the network is considered, for which the downstream boundary condition at the outlet of the network is specified. Thus, the solution can be obtained only by solving all of the flow equations at the channel and junctions simultaneously. For example, for a Y-shaped network having three channels and a point-type junction, with three computational stations for each channel, there are 20 simultaneous nonlinear algebraic equations to be solved for the 20 unknowns. The simultaneous solution is computed for the entire network for each time step.

The simultaneous solution method is exact in the sense of solving all of the difference equations for the entire network. But it is also very costly and requires a large digital computer when the network is large, making the method impractical.

One-Sweep Explicit Solution Method

By applying the direct explicit numerical schemes to the river network, it is possible to solve for the junction flow conditions explicitly before proceeding to solve for the downstream channel for the same time step. In this method, the unknown junction and channel flow parameters are expressed explicitly in terms of known quantities of the network at a previous time and of upstream points at the present time. Thus, simultaneous solution for the entire network is avoided. The solution proceeds starting from the upstream end towards the downstream end for the entire network for the present time step in sequence in a one-sweep manner before the computation advances to the next time step. Examples can be found in Stoker (1957), Dronkers (1964), and Balloffet (1969). This method bears the drawback of explicit schemes concerning computational stability problems.

One variation of this method is the Lax-Wendroff scheme adopted by Murota et al. (1973). Another variation is to solve for the junction flow conditions explicitly using the junction flow equations for each of the joining channels (e.g., Harris, 1968; Larson et al, 1971; Hsie et al, 1974). This approach is an improvement over the direct explicit schemes in view of computational stability, but the computational time requirement is still severe.

Overlapping Segment Method

The overlapping segment method is a single step, successive iteration technique which is demonstrated schematically in Fig. 13-9. The river network is considered to be formed by a number of overlapping segments. Each segment is formed by a junction together with all the channels joining it. Thus, except for the most upstream and downstream channels, each (interior) channel belongs to two segments--as a downstream channel for one segment and then as an upstream channel for the other segment, i.e., "overlapped." Each segment is solved as a unit. The flow equations are first applied to each of the branches of the most upstream segment for which the upstream boundary condition is known, and solved numerically with appropriate junction equations. If the flow is subcritical and the boundary condition

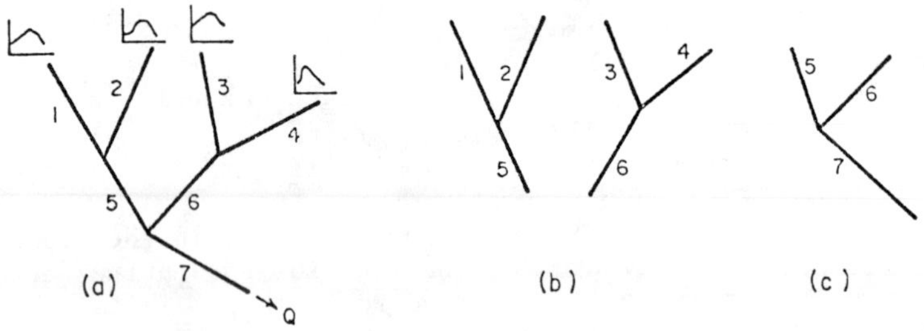

Figure 13-9. Method of Overlapping Segments

at the lower end of the downstream channel of the segment is unknown, the forward or backward differences, depending on the numerical scheme, are used as an approximate substitution. Simultaneous numerical solution is obtained for all the channels and junctions of the segment for each time step, repeating until the entire flow duration is completed. For example, for the network shown in Fig. 13-9a, solutions are first obtained for the two segments shown in Fig. 19-9b. Since the downstream boundary condition of the segment is assumed, the solution for the downstream channel is discarded, whereas the solutions for the upstream channels are retained. The computation now proceeds to the next immediate downstream segment (e.g., the segment shown in Fig. 13-9c). The upstream channels of this new segment were the downstream channels of each of the preceding segments for which solutions have already been obtained. The inflows into this new segment are given by the outflow from the junctions of the preceding segments. With the inflows known, the solution for this new segment can be obtained. This procedure is repeated successively, segment by segment, going downstream until the entire river network is solved. For the last (most downstream) segment of the network, the prescribed boundary condition at its downstream end is used.

The method of overlapping segments greatly reduces the necessary computer size and time requirements when solving for large river networks. It accounts for downstream backwater effect and simulates reversal flow, if it occurs. Its accuracy and practical usefulness have been demonstrated by Sevuk (1973) and Yen and Akan (1976).

Solution by the overlapping segment method accounts for the downstream backwater effects of subcritical flow only for the adjacent upstream channels of the junction, but cannot reflect the backwater effect from the junction to channels farther upstream if such case occurs. However, by considering the length to depth ratio of natural river branches, the effect of backwater beyond the immediate upstream branches is small, and hence imposes no significant error in routing of river flows. For the rare case of two junctions being closely located, the overlapping segment method can be modified to include the short branch between the junctions as the internal branch and to use longer branches as the upstream and downstream channels. It should also be mentioned that the overlapping segment method, as well as the two previously described methods, can be modified to account for divided channels, i.e., loop networks in addition to tree-type networks.

Cascade Method

In the nonlinear kinematic wave approximation, only one boundary condition is required for routing the flow through a channel. Usually the upstream boundary condition is specified. Since the downstream boundary condition is not required and the downstream backwater effect is not accounted for, the solution of the upstream channel is not affected by the downstream channel. At the junctions, only the continuity relationship [Eqs. (13-24) or (13-27)] is needed, and the dynamic condition [Eqs. (13-26) or (13-29)] is ignored. Consequently, solution of flow can be obtained in a cascading manner, solving first for the most upstream channels individually over the entire flow hydrograph duration. Computation then proceeds and is repeated for the next downstream channel. Thus, the solution is obtained channel by channel individually and sequentially, moving downstream until the entire network is solved. No simultaneous or network iterative solution is required. This method is relatively simple and inexpensive, but it is inaccurate if the downstream backwater effect is important.

The cascade method can be used for diffusion wave, quasi-steady dynamic wave, and dynamic wave models without causing additional error if the flow is entirely in supercritical regime. However, such a case rarely occurs in natural rivers. Moreover, it is not unusual that an engineer, not realizing the importance of the downstream boundary condition, unknowingly applies the cascade method to diffusion wave or dynamic wave models simply by carrying on the computations up to, and perhaps beyond, the length of the channel.

Using example networks, Yen and Akan (1976) compared the reliability of the methods of simultaneous solution, overlapping segments, and cascade. A typical example for the network shown in Fig. 13-10 is reproduced in Fig. 13-11, which shows the former two methods giving identical results, whereas the sequential cascading method gives significant errors.

13.7 CONCLUDING REMARKS

A prerequisite for mathematical modeling of river flows is the use of the correct and appropriate flow equations. In addition, the initial and boundary conditions should be correctly and adequately specified. There exist a number of approximate models and numerical solution schemes for mathematical modeling. Which model and scheme is the best for a particular problem depends on the nature of the problem, the accuracy required, the computational facilities available, and the computational costs.

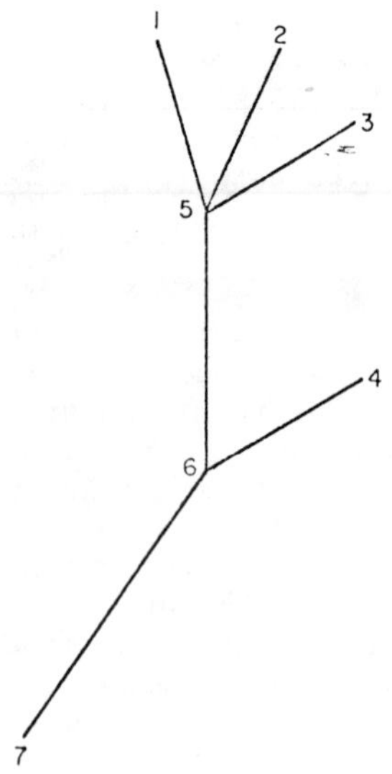

Figure 13-10. Example network

There is no universally superior model or scheme. However, usually for gradually varied river flow, the implicit schemes and method of characteristics are superior to the explicit schemes, and the diffusion wave or dynamic wave models are preferred if the downstream backwater effect is important, using the overlapping segment method if the river network is large. The nonlinear kinematic wave model has the advantage of simplicity and is useful for subcritical flow in river network with negligible downstream backwater effects.

13.8 NOTATION

A = cross-sectional area
B = free surface width of channel cross section
C = Chezy's resistance factor
D = hydraulic depth, A/B; also, diameter of circular channel
F = body force per unit mass
f = Weisbach resistance coefficient
g = gravitational acceleration
H = specific energy head
h = depth of flow, measured normal to bed along y-direction
h_f = loss of energy head
i,j = indices
K = piezometric pressure distribution correction factor for equations using natural coordinates
k = piezometric pressure distribution correction factor for equations using gravity oriented coordinates

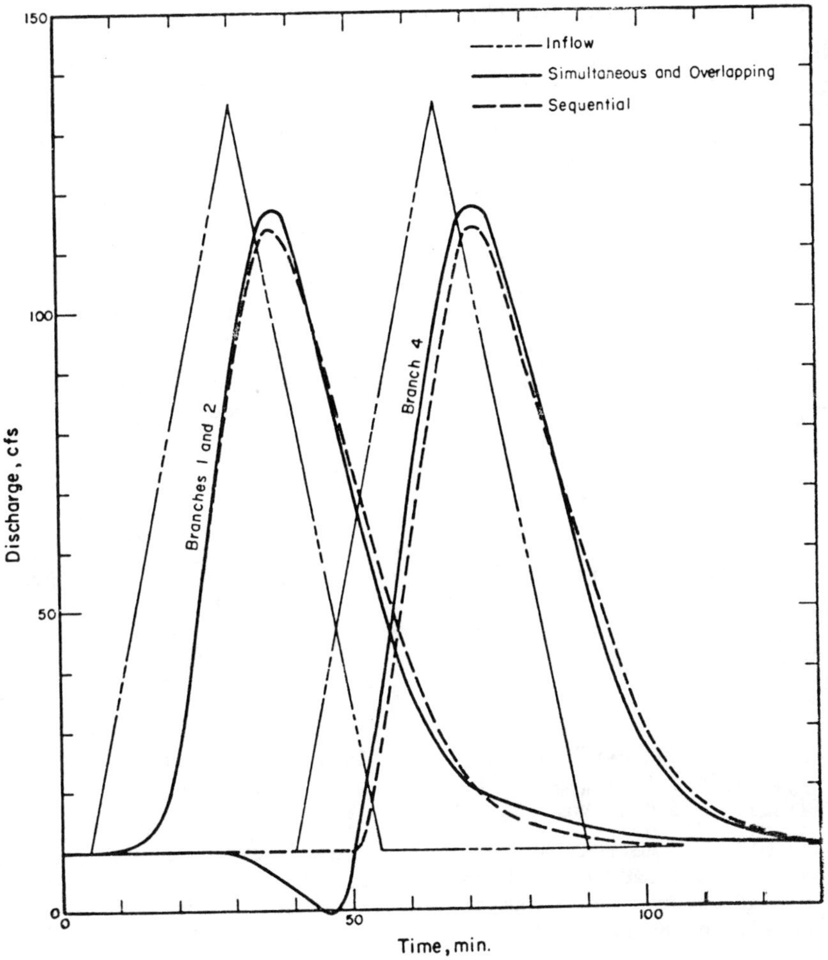

Figure 13-11. Comparison of Network Solution Methods

K' = ambient piezometric pressure correction factor for natural coordinates

k' = ambient piezometric pressure correction factor for gravity oriented coordinates

n = Manning's roughness factor

p = pressure intensity

Q = discharge

q = time rate of lateral flow per unit length of circumference σ

R = hydraulic radius

r = normal displacement of σ with respect to space or time projected on a plane parallel to A

S_f = friction slope

S_o = channel bottom slope, = $-\partial Z_b/\partial x$ or $-\partial y_b/\partial x$
s = storage
T = force due to internal stresses acting on A
t = time
U_x = x-component of velocity of lateral flow when joining channel flow
u = x-component of local velocity, = u_x
u_i = local velocity component along x_i-direction
V = average flow velocity across cross section, = Q/A
x = longitudinal coordinate
x_i = coordinate along i-th direction
Y = depth of flow, measured vertically
y = coordinate perpendicular to x on a vertical plane
y_b,Z = elevation of channel bottom with respect to a horizontal reference datum
β = momentum flux correction factor
γ = specific weight of fluid
θ = angle between channel bottom and horizontal plane
μ = dynamic viscosity of fluid
ρ = mass density of fluid
σ = perimeter bounding cross-sectional area A
$\overline{\tau}_{ij}$ = temporal mean stresses as defined in Eq. 4.

13.9 REFERENCES

Ackers, P., and Harrison, A. J. M., 1964. Attenuation of flood waves in part-full pipes. Proceedings, Institute of Civil Engineers, London, Vol. 28, pp. 361-383.

Akan, A. O., and Yen, B. C., 1977. A nonlinear diffusion-wave model for unsteady open-channel flow. Proceedings, 17th Congress of International Association of Hydraulic Research, Baden-Baden, Germany, Vol. 2, pp. 181-190.

Amein, M., and Fang, C. S., 1969. Streamflow routing with applications to North Carolina rivers. Report No. 17, Water Resources Research Institute, University of North Carolina, 72 p.

Ames, W. F., 1969. Numerical methods for partial differential equations. Barnes and Noble, Inc., New York.

Balloffet, A., 1969. One-dimensional analysis of floods and tides in open channels. Journal Hydraulics Division, ASCE, Vol. 95, No. HY4, pp. 1429-1451, July.

Baltzer, R. A., and Lai, C., 1968. Computer simulation of unsteady flow in waterways. Journal Hydraulics Division, ASCE, Vol. 94, No. HY4, pp. 1083-1117, July.

Bettess, R., and Price, R. K., 1976. Comparison of numerical methods for routing flow along a pipe. Report No. IT 162, Hydraulics Research Station, Wallingford, England.

Chow, V. T., 1959. Open-channel hydraulics. McGraw-Hill Book Company, New York.

Chow, V. T., ed., 1964. Handbook of Applied Hydrology. McGraw-Hill Book Company, New York.

Dronkers, J. J., 1964. Tidal hydraulics. North-Holland Publ. Co., Amsterdam, pp. 203-204.

Forsythe, G. E., and Wasow, W. R., 1960. Finite difference methods for partial differential equations. John Wiley and Sons, New York.

Grijsen, J. G., and Vreugdenhil, 1976. Numerical representation of flood waves in rivers. Publication No. 165, Delft Hydraulics Lab, Delft.

Gunaratnam, D. J., and Perkins, F. E., 1970. Numerical solution of unsteady flows in open channels. Report No. 127, R. M. Parsons Lab for Water Resources and Hydrodynamics, M.I.T., 260 pp.

Harris, G. S., 1968. Development of a computer program to route runoff in the Minneapolis-St. Paul interceptor sewers. Memo No. M121, St. Anthony Falls Hydraulics Lab, University of Minnesota, December.

Henderson, F. A., 1966. Open channel flow. MacMillan Company, New York.

Hsie, C. H., Chow, V.T., and Yen, B.C., 1974. The evaluation of a hydrodynamic watershed model (IHW Model IV), Civil Engineering Studies Hydraulic Eng. Ser. No. 28, Dept. of Civil Engineering, University of Illinois at Urbana-Champaign.

Larson, C. L., Wei, T. C., and Bowers, C. E., 1971. Numerical routing of flood hydrographs through open channel junctions. Water Resources Research Center Bulletin No. 40, University of Minnesota, August.

Liggett, J. A., and Cunge, J. A., 1975. Numerical methods of solution of the unsteady flow equations. Chapter 4 of Unsteady Flow in Open Channels, ed. by K. Mahmood and V. Yevjevich, Water Resources Publications, Ft. Collins, Colo.

Lighthill, M. J. and Whitham, G. B., 1955. On Kinematic Waves: I. flood movements in long rivers. Proceedings, Royal Soc. (London), Ser. A., Vol. 229, No. 1178, pp. 281-316, May.

Mahmood, K. and Yevjevich, V., ed., 1975. Unsteady Flow in Open Channels, Vol. 1. Water Resources Publications, Ft. Collins, Colo.

Murota, A., Kanda, T., and Eto, T., 1973. Flood routing for urban river network. Proceedings, IAHR International Symposium on River Mechanics, Bangkok, Thailand, Vol. 3, pp. 339-350, January.

Price, R. K., 1974. Comparison of four numerical methods of flood routing. Journal Hydraulics Division, ASCE, Vol. 100, No. HY7, pp. 879-899, July.

Remson, I., Hornberger, G. M., and Molz, F. J., 1971. Numerical methods in subsurface hydrology. Wiley-Interscience, New York.

Richtmeyer, R. D. and Morton, K. W., 1967. Difference methods for initial value problems. Wiley-Interscience, New York.

Rouse, H., ed., 1959. Advanced Mechanics of Fluids. John Wiley and Sons, New York, p. 36.

Rouse, H., 1965. Critical analysis of open-channel resistance. Journal Hydraulics Division, ASCE, Vol. 91, No. HY4, pp. 1-25, July.

de Saint-Venant, Jean-Claude Barré, 1871. Théorie du movement non-permanent des eaux avec application aux crues des rivières et à l'introduction des marées dans leur lit. Acad. Sci. (Paris) Comptes Rendus, Vol. 73, pp. 148-154, 237-240.

Sevuk, A. S., 1973. Unsteady flow in sewer networks. Ph.D. Thesis, Dept. of Civil Engineering, University of Illinois at Urbana-Champaign.

Sevuk, A. S. and Ye-, B. C., 1973a. A comparative study on flood routing computation. Proceedings, IAHR Intn'l Symposium on River Mechanics, Bangkok, Thailand, Vol. 3, pp. 275-290.

Sevuk, A. S. and Yen, B. C., 1973b. Comparison of four approaches in routing flood wave through junction. Proceedings, 15th Congress of Interntl. Assoc. Hydraulic Research, Istanbul, Turkey, Vol. 5, pp. 169-172, September.

Smith, G. D., Numerical solution of partial differential equations. Oxford University Press, Oxford, 1965.

Stoker, J. J., 1957. Water waves. Interscience, New York, p. 496.

Strelkoff, T., 1970. Numerical solution of Saint-Venant equations. Journal Hydraulics Division, ASCE, Vol. 96, No. HY1, pp. 223-252, January.

Yen, B. C., 1971. Spatially varied open-channel flow equations. Research report No. 51, Water Resources Center, University of Illinois at Urbana-Champaign, 63 p., December. (Available as PB209532 from US NTIS, Springfield, Va.)

Yen, B. C., 1973. Open-channel flow equations revisited. Journal Engineering Mechanics Division, ASCE, Vol. 99, No. EM5, pp. 979-1009, October.

Yen, B. C., 1975. Further study on open-channel flow equations. Sonderforschungbereich 80 Report, No. SFB80/T/49, University of Karlsruhe, 46 p., April.

Yen, B. C., and Akan, A. O., 1976. Flood routing through river junctions. River 76, Proceedings of Symposium on Inland Waterways for Navigation, Flood Control and Water Diversions, ASCE, Vol. 1, pp. 212-231.

Yevjevich, V., and Barnes, A. H., 1970. Flood routing through storm drains, Parts I to IV. Hydrology Papers, No. 43-46, Colorado State University.

Chapter 14

MODELING OF ICE IN RIVERS

by

George D. Ashton, Chief of Snow and Ice Branch, Cold Regions Research and Engineering Laboratory, Hanover, New Hampshire

Chapter 14

MODELING OF ICE IN RIVERS

14.1 INTRODUCTION

Ice problems are increasingly being attacked using a variety of modeling techniques. The present paper summarizes the nature of river ice phenomena, and describes various means by which ice behavior is modeled. Attention is not confined only to physical models but includes analytical and numerical models which have been used to make predictions of ice behavior. At the outset it is noted that the study, understanding, and management of river ice is in its infancy and the techniques used to model its behavior are relatively unsophisticated when compared, for example, with sediment transport models. This situation, however, allows those concerned with river ice to have at their disposal the tools and techniques developed for other subjects and they may readily be applied to river ice problems <u>if</u> the essence of the particular behavior of river ice is understood qualitatively. This qualitative understanding is most often gained by actual observation of ice phenomena and no amount of laboratory and analytical study can substitute for field observation. With this proviso stated, let us begin.

We choose for present purposes to classify the problems of river ice into the areas of formation, evolution, and breakup of ice covers; hydraulics associated with the presence of ice; thermal effects; and ice forces. All interact with the others, of course, and necessarily there is overlap. It is also noted that most of the problems of river ice are associated with short, intense periods of activity which occur between longer periods of more or less steady state. It is these short periods which are the crux of most design problems and hence the ones with which we are most concerned.

14.2 BACKGROUND LITERATURE

There is considerable literature on the subject of engineering problems of freshwater ice although it is widely dispersed and in many cases somewhat obscure. General treatises exist by Barnes (1928), Altberg (1936), Hoyt (1913), Michel (1970, 1971) and Starosolszky (1969). Various symposia on the general subject have been held, most notably the ice symposia of the IAHR (1970, 1972, 1974, 1975) and of POAC (1971, 1973), and these contain numerous short papers both theoretical and practical. There are bibliographies which provide extensive cross referencing by subject and include the extensive Russian literature (CRREL, 1951-1976).

14.3 FORMATION, EVOLUTION, AND BREAKUP

Formation

The initial formation of ice on rivers is generally in one of three modes: formation of a uniform sheet of ice, formation of an accumulation of ice floes, and formation of frazil. The processes and evolution are summarized schematically in Fig. 14-1.

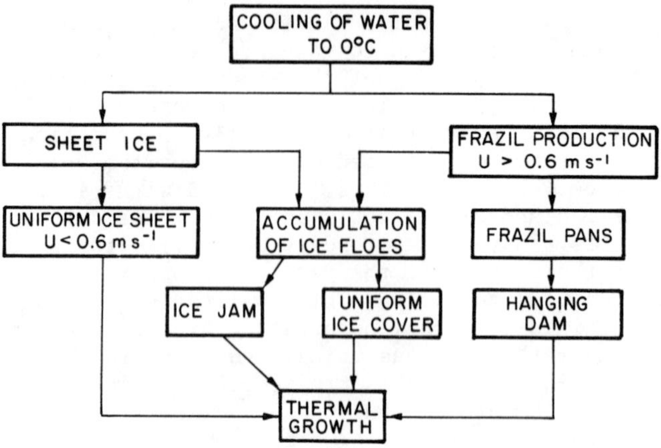

Fig. 14-1. Processes of Formation of Initial Ice Cover

The obvious initial condition is that the water must be cooled to 0°C and continue to lose heat to the atmosphere. A later section will discuss energy budgets associated with the evaluation of the heat loss. When the surface velocity is less than about 0.6 m s^{-1} a uniform sheet ice cover generally forms and in narrow rivers rapidly bridges across the river surface. In wider rivers the sheet ice floes are transported downstream and eventually form an ice cover by accumulation of the floes. The subsequent ice cover may be a uniform thickness accumulation or an irregular accumulation which may significantly block the flow cross section. Both are termed ice jams although the usual connotation of an "ice jam" is the latter.

The mechanics of accumulation of ice floes has been subjected to considerable study, experiment, and field observation including those of Michel (1971), Pariset and Hausser (1961), Cartier (1959), Uzuner and Kennedy (1972, 1976), Ashton (1974a, 1974b) and Larsen (1975). The problem is essentially of two parts: the behavior of individual floes upon arrival at the upstream edge of the accumulation (or other surface obstacle such as an ice boom) and the subsequent thickening of the accumulation by deposition of floes or by thickening. We first consider the behavior of individual floes. Figure 14-2 presents a definition sketch.

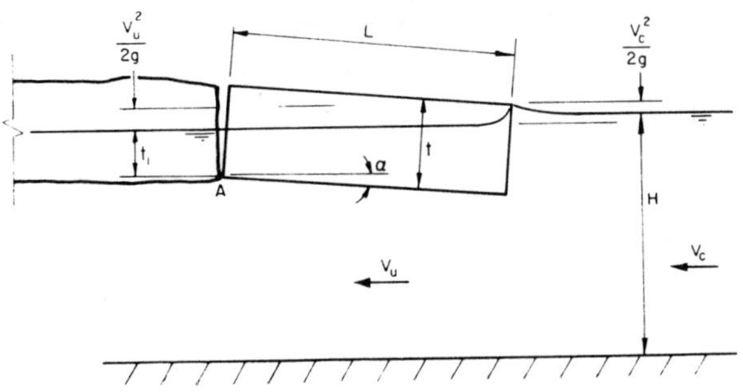

Fig. 14-2. Definition Sketch for Entrainment of Individual Floes

The critical velocity V_c for underturning and submergence can be predicted by a simple equilibrium analysis (Ashton, 1974a) utilizing a "no-spill" condition that entrainment occurs when the stagnation water level exceeds the top edge of the floe with the resulting threshold condition

$$\frac{V_c}{\left[g\ t_i\left(1 - \frac{\rho_i}{\rho}\right)\right]^{1/2}} = \frac{2\left(1 - \frac{t_i}{h}\right)}{\left[5 - 3\left(1 - \frac{t_i}{H}\right)^2\right]^{1/2}} \qquad (14\text{-}1)$$

where g is gravity, t_i is the floe thickness, and ρ_i and ρ are the densities of the ice and water, respectively. Figures 14-3 and 14-4 compare the experimental data with the prediction of Eq. (14-1) (including data for densities other than that of ice). In Fig. 14-3 the length parameter used in the densimetric Froude number is floe thickness while in Fig. 14-4 the flow depth H is used as the length parameter. The two figures are equivalent. In Fig. 14-4 the usual Froude number equivalent is noted on the right-hand ordinate using $\rho_i/\rho = 0.916$ in Eq. (14-1). The results in this figure suggest that accumulation of ice floes by simple juxtaposition cannot occur at Froude numbers $(F = V_c/(gH)^{1/2})$ greater than about 0.13. It does not mean that an ice jam of the catastrophic type necessarily forms at such Froude numbers.

For very small values of t_i/H and t_i/L there is some uncertainty in the appropriateness of the relationship of Eq. (14-1) (see Larsen, 1975) and this has a bearing on model scales which are selected using the densimetric Froude criterion. There are also complications which arise when thin blocks of other material than ice are used to simulate ice. One known difficulty arises from different surface tension interactions

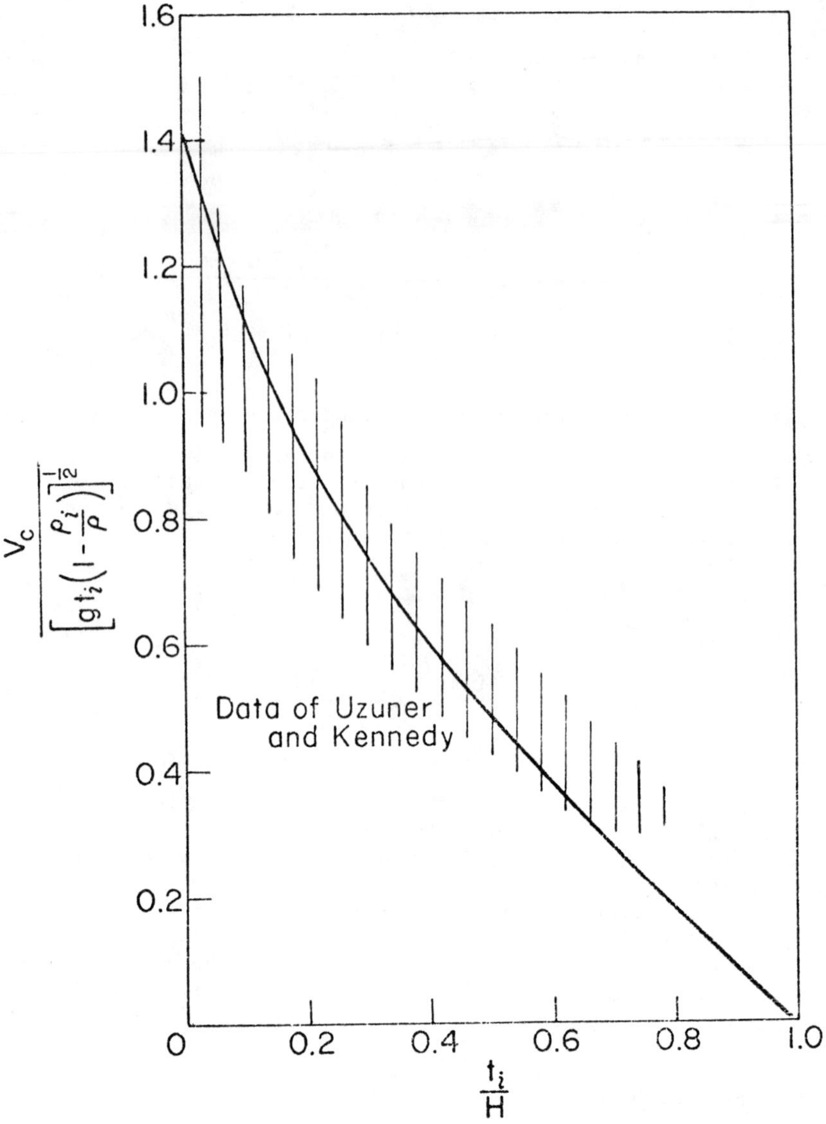

Fig. 14-3. Single Floe Entrainment Threshold with Floe Thickness Froude Criterion. Solid Line is Analysis of Ashton (1974a)

between materials which are non-wetting and ice (J. C. Tatinclaux, private communication) and this has most effect on very thin floes.

There has also been some study by Filippov (1974) on the transport distance of a floe after submergence. The data of

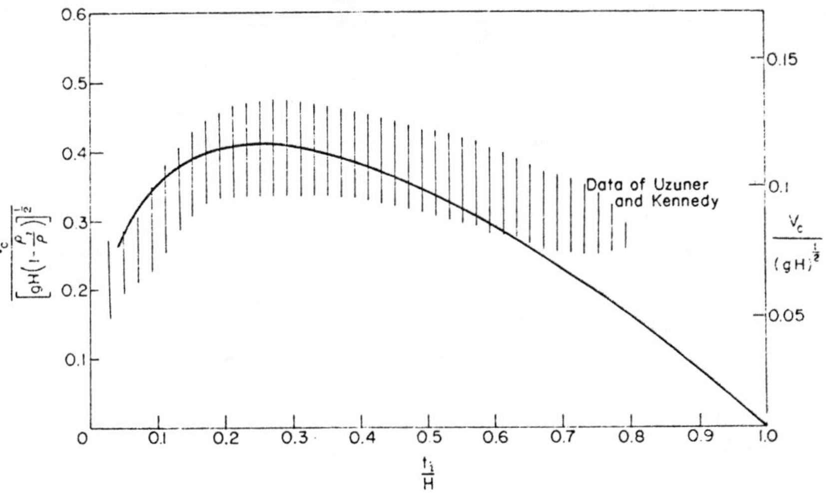

Fig. 14-4. Single Floe Entrainment Threshold with Depth Froude Criterion. Solid Line is Analysis of Ashton (1974a)

Filippov has been collapsed by a dimensional argument by the present author and is presented in Fig. 14-5 where V is the flow velocity and L the floe length.

At a threshold velocity greater than that of Eq. (14-1) the ice accumulates in an equilibrium thickness first described by Pariset and Hausser (1961) and of the form given by:

$$\frac{V}{\left[2g\ t_j\left(1 - \frac{\rho_i}{\rho}\right)\right]^{1/2}} = \left(1 - \frac{t_j}{H}\right) \qquad (14\text{-}2)$$

where t_j is the thickness of the accumulation. Figure 14-6 presents a definition sketch applicable to Eq. (14-2). Substituting $\rho_i/\rho = 0.916$ yields, at $t/h = 1/3$, a maximum depth Froude number of 0.16. Again, it is debatable whether a Froude number of this order or greater is a sufficient condition for formation of the catastrophic type jam. There are several analyses of the evolution of an ice jam accumulation, most notably those of Pariset and Hausser (1961), Pariset et al. (1966), Uzuner and Kennedy (1976), and Tatinclaux (1977, unpublished). In any case it is clear that proper modeling of ice accumulation processes should satisfy both the densimetric Froude similarity associated with the floe characteristics and the Froude similarity of the flow itself. Fortunately, plastic is available which has the correct value of ρ_i/ρ of that of ice; and if there are no thermal or mechanical property-related

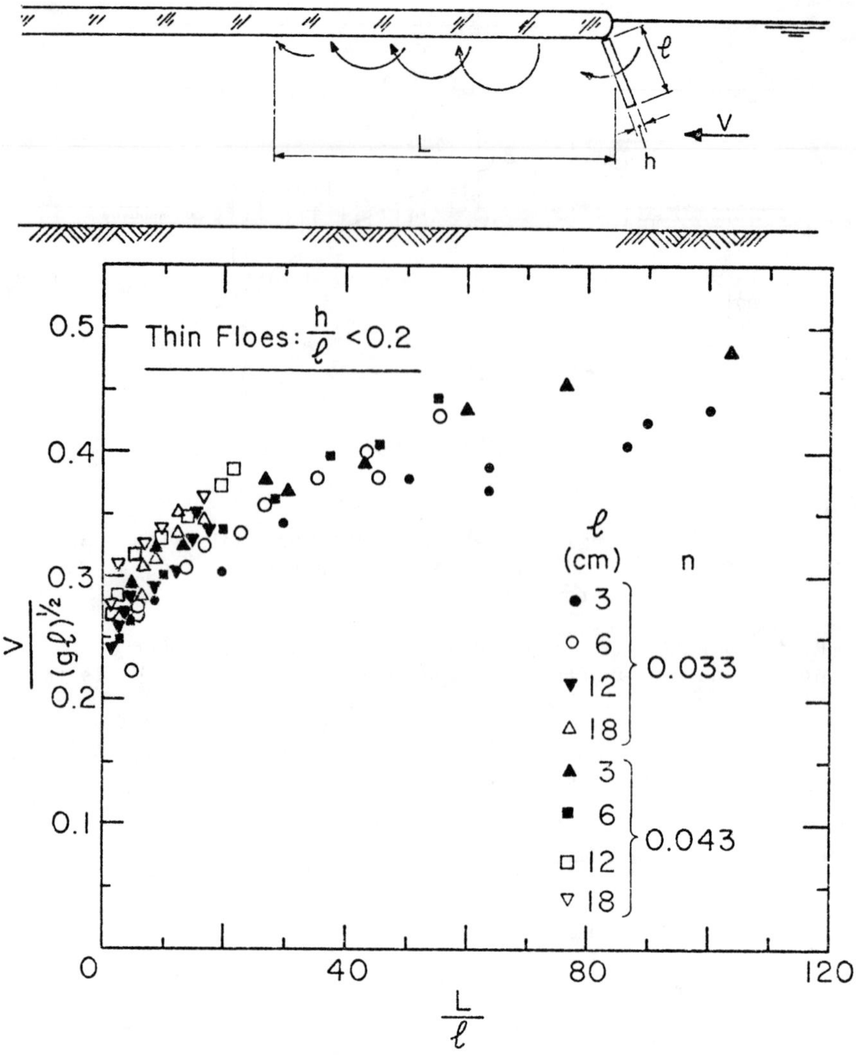

Fig. 14-5. Transport Distance of a Floe after Submergence at the Upstream Edge of an Ice Cover (Interpretation of Data of Filippov (1974))

effects involved in the phenomena being modeled, adequate hydraulic models may be used.

Frazil

Frazil ice forms in turbulent flows when the water surface is open to the atmosphere and the heat loss is sufficient to decrease the water temperature below 0°C. The initial formation is generally in the form of small disc-shaped crystals

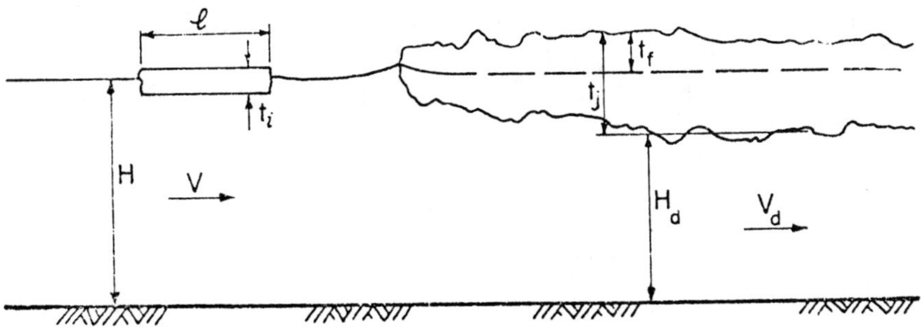

Fig. 14-6. Ice Jam Definition Sketch

with diameters of the order of 5 mm and these grow to platelets 10 to 15 mm in their maximum dimension. In Fig. 14-7 is presented a diagram showing the evolution of frazil downstream from its point of formation (adapted from Michel, 1971). The initial formation is at the surface; the crystals are subsequently entrained by the turbulent flow, grow in size, and flocculate; the flocs rise to the surface, freeze together in pans which are roughly hemispherical in shape; and subsequently form an ice cover by accumulation process described earlier for ice floes. Under certain conditions which are not well defined the frazil may be transported considerable distances and deposited beneath an ice cover to form a "hanging dam" which may cause severe blockage of the flow cross section, or

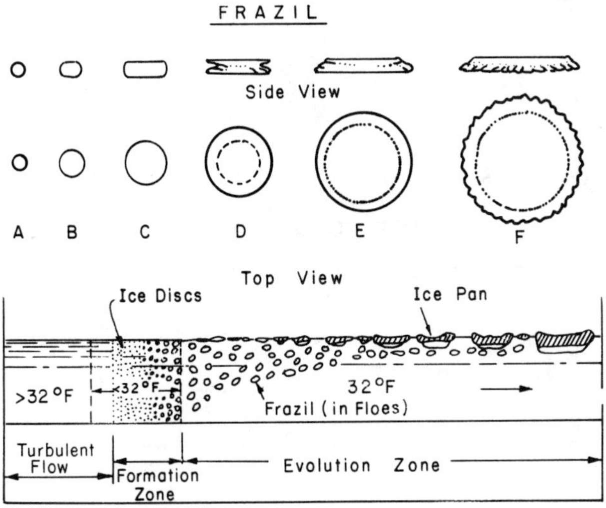

Fig. 14-7. Diagram Illustrating Frazil Evolution (after Michel, 1971)

agglomerate on trashracks thus blocking the flow into the intakes. Field and laboratory experience (Carstens, 1970; Hanley and Michel, 1977) suggest that a surface flow velocity greater than about 0.6 m s^{-1} is required to cause sufficient turbulence to entrain the frazil away from the surface. If less than 0.6 m s^{-1} generally a sheet ice cover forms.

In Fig. 14-8 is presented a typical time-temperature sequence associated with frazil formation. It is important to understand that this temperature evolution is that of a parcel of fluid moving with the flow and there may be a constant production of frazil in a given reach of a river, particularly at rapids section. If the surface area A of open water contributing to frazil production is known, the total production of frazil may be estimated by calculating the energy loss, dividing by the product of the heat of fusion and the ice density, and integrating over time. Generally the energy loss is expressed as a product of a surface heat transfer coefficient h_a applied to the difference between the below-freezing air temperature T_a and the water temperature T_w which may be taken as 0°C. Thus

$$\forall_f = \int_A \int_o^t \frac{h_a(T_w - T_a)}{\rho_i \lambda} \qquad (14\text{-}3)$$

where $\forall_f$ is the volume of frazil produced, and t is time. If detailed energy budget calculations are not available, the present writer has found that $h_a = 24\ \mathrm{W\ m^{-2}\ {}^\circ C^{-1}}$ is a reasonable value for rough estimates.

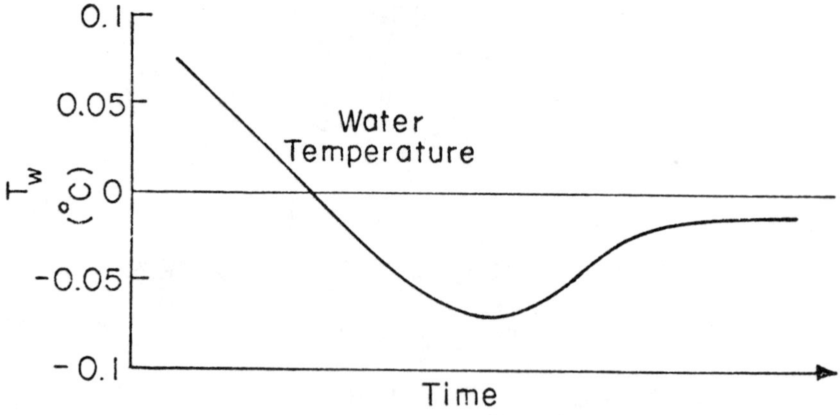

Fig. 14-8. Typical Temperature Time Sequence of Water Undergoing Supercooling and Subsequent Frazil Formation (Adapted from Carstens, 1966)

A number of papers treat various facets of the problem of frazil ice including those of Barnes (1928), Altberg (1936), Devik (1942), Williams (1959), Carstens (1966, 1970), Michel (1963, 1971), and Dean (1977). Those who would seek to model frazil problems in the laboratory without actually using refrigeration to produce frazil itself should consider the problem as an upside-down sediment transport phenomena but are cautioned that the cohesive properties are highly variable depending upon whether the frazil is in its active or passive state. Frazil particles are in their active (or growing state) when the water is supercooled and in their passive state when the matrix water is at 0°C (or above). Only recently have quantitative values of the properties of deposited frazil become available. These include concentration (up to 10^6 m^{-3} particles), porosity (on the order of 0.6) permeability (on the order of $k = 15 \times 10^{-10}$ m^2), or in situ shear strength (ranging from 5×10^3 N m^{-3} to 5×10^4 N m^{-2}) (see Dean, 1977).

Evolution of Ice Cover Thickness

Once an initial ice cover has been formed either as a relatively thin accumulation or as an initial sheet of ice, further thickening results from loss of heat to the atmosphere. The usual simplified analysis of this problem (the so-called Stefan problem) assumes a linear temperature distribution in the ice, the bottom surface to be at the melting point T_m (0°C) and the top surface to be at the ambient air temperature T_a. A simple heat conduction analysis then yields

$$q_i = -\rho_i \lambda \frac{d\eta}{d\Theta} = \frac{k_i}{\eta}(T_a - T_m) \qquad (14\text{-}4)$$

where q_i is the heat transfer through the ice cover of thickness η, Θ is time, and k_i is the thermal conductivity of the ice. Integration, then yields

$$\eta = \left(\frac{2ki}{\rho_i \lambda}\right)^{1/2} [(T_a - T_m)\Theta]^{1/2} \qquad (14\text{-}5)$$

Eq. (14-5) is of the same form as the often used practice of relating the thickening to accumulated degree-days freezing in the form

$$t = \alpha(\Delta T\Theta)^{1\ 2} \qquad (14\text{-}6)$$

and evaluation of the Eq. (14-5) term for α results in $\alpha = 0.000121$ M $s^{1\ 2}$ $°C^{-1\ 2}$ (conveniently equal to 1.0 inches

$\text{day}^{-1}\ {}^2\ {}^{\circ}\text{F}^{-1}\ {}^2$ in the English system). Generally to "fit" actual data α is taken to be 0.6 to 0.7 its theoretical value given by Eq. (14-5) which is to be expected since the derivation of Eq. (14-5) assumed the air temperature equals the top surface temperature and snow layers (which insulate the ice cover) were neglected. A relatively simple compromise is to introduce a heat transfer coefficient h_a at the ice surface and predict the thickening process by step-wise solution of

$$\rho_i \lambda \frac{d\eta}{d\Theta} = \frac{-(T_a - T_m)}{\frac{\eta_i}{k_i} + \frac{\eta_s}{k_s} + \frac{1}{h_a}} \tag{14-7}$$

using daily average air temperatures (η_s and k_s are the thickness and thermal conductivity of any snow layer that exists). h_a is of the order of 24 W m^{-2} °C (corresponding to a wind speed of about 4.5 m s^{-1}) see for example Adams, French, and Kingery (1960).

The above analysis has been subjected to a multitude of variations such as heat transfer to the bottom side, consideration of specific heat capacity effects, and a variety of boundary conditions. Most of these analyses are of dubious practical value since one seldom has at hand the necessary input data to use them. Generally, the ice cover on rivers is thinner than on adjacent lakes subjected to the same weather because of earlier initial formation and heat transfer to the underside by forced convection (see the later section on thermal effects).

Deterioration of Ice Covers

While the thickening process of ice covers is reasonably well understood and predictions limited primarily by the detailed effort required to accurately assess the various components of the energy budget, the deterioration process is only imperfectly understood. It is clear that melting results from heat transfer to the surfaces of the ice cover. However, at the undersurface the forced turbulent convection heat transfer requires unusually accurate knowledge of the water temperature since significant melting occurs at flow temperatures on the order of 0.05°C. There is also some uncertainty in the appropriate value of the heat transfer coefficient in light of the meager data available on wide aspect ratio turbulent channel flow heat transfer. Until such data is available the most practical method is to use the available empirical pipe flow formulas with appropriate consideration of the hydraulic radius. At the top surface of the ice, the problem is even more complex since first melting results in pools of water, changes in albedo, irregular drainage of the melt water through cracks in the ice, varying absorption of radiation with depth, in short, nearly every complication one could imagine. Further, since melting at the top results in an isothermal ice slab, the path of energy to the water below to be applied to the undersurface is complicated. Added to the complications is the fact that added heat from tributary or groundwater contributions enters at the sides and hence must be convected to the center regions by lateral dispersion. It is common during the melt season for near bank portions to melt completely while the center regions undergo little melting.

The turbulent flow of water by the underside of the ice cover, if the water is above 0°C, also results in the formation of a wavy relief pattern called ice ripples. The ripples are dune-shaped features with crests perpendicular to the velocity vector. They have the intriguing property of a wave length inversely proportional to the velocity with a typical value of 0.3 m wave length corresponding to 0.3 m s^{-1} velocity. A non-dimensional plot of most known field and laboratory data is presented in Fig. 14-9. The ripples have been studied in some depth (Ashton and Kennedy, 1972; Ashton, 1972; Hsu, 1973; Carey, 1966, 1967a) and their behavior is reasonably well documented.

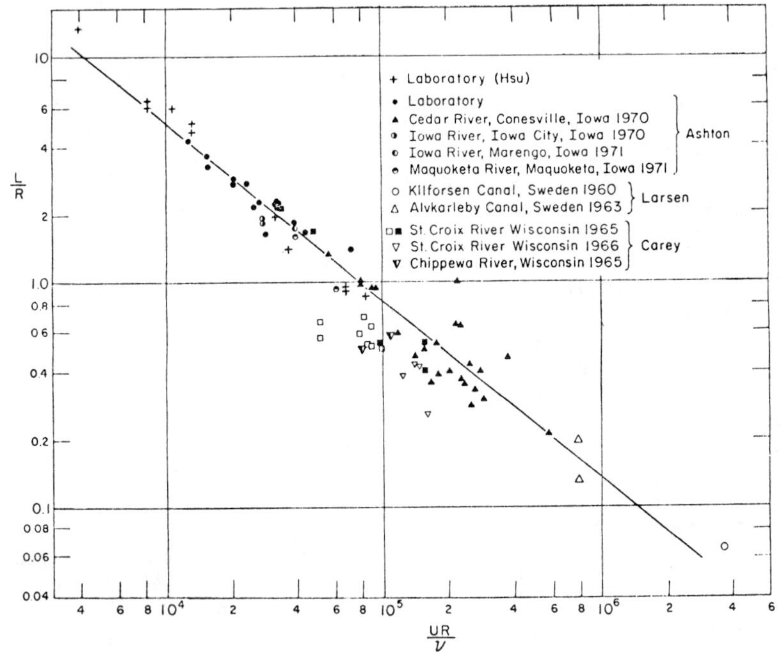

Fig. 14-9. Non-Dimensional Representation of Ice Ripple Length-Velocity Relationship (Adapted from Ashton, 1972b)

Finally, as the ice cover approaches isothermal condition at 0°C, a phenomenon called rotting, or candling, occurs. It is generally conceded that candling is an ice grain boundary melting phenomenon but there is little information on the rate of progression of the candling through the ice cover. Candling may originate at either the top or the bottom sides of the ice cover (or both) and its major practical impact results from the associated weakening of the ice cover which, in turn, may result in the breakup due to hydrodynamic forces when the ice thickness has been but little reduced by direct convective melting. It is related to the absorption of radiation (Koren'kov, 1970; Frankenstein, 1961) by the ice cover but the detailed mechanism is unclear.

Breakup

There are two major problems associated with breakup. When does it occur? Where and under what conditions do resulting catastrophic ice jams occur?

The prediction of breakup time is not only complicated by the deterioration problems discussed above but also by the difficulty of predicting the time of occurrence of the increased flows which usually initiate the breakup. It is generally conceded that significantly increased flows will cause the ice to break up. It is common practice in the management of controlled discharges from hydroelectric plants to minimize the discharge variation when there is an ice cover downstream with potential for damaging jams (Burgi and Johnson, 1971). Once breakup occurs, it is almost inevitable that an ice jam will form somewhere downstream. The critical need is to be able to predict where jamming will occur and the resulting water levels that will be associated with the jam. To date, our knowledge is largely descriptive in nature. Michel (1971) and Deslauriers (1968) have distinguished the simple ice jam, in which the accumulation of ice still permits flow beneath, from the dry ice jam in which the accumulation extends to the bottom and effectively blocks the flow cross section. The simple ice jam is the most amenable to analysis to date (see the earlier section on ice jams). The dry ice jam, which is most often of catastrophic type, has received little analytical treatment. Indeed, observation of the formation and evolution of such a jam is such a formidable sight as to make any students of river ice themselves head for high ground to wait for the event to run its course. No doubt this will be the case for some years to come in spite of the considerable importance of the problem. The author's own view is that it may yield to an analysis incorporating the dynamics of granular media flow principles with unsteady kinematical wave analysis necessary to account for the storage and release of water as such jams form, break up, and reform downstream. The goal of such analyses should be the ability to establish anticipated water levels, a means of warning those who would be flooded (stage rises of 1 m hr^{-1} are not unusual), and a framework by which channel modifications may be assessed for their benefit in reducing jamming potential, or causing the jam to form in a reach where damage would be minimal.

There are measures which are taken before breakup periods to reduce the jamming potential at locations which historically have experienced severe jams (Moor and Watson, 1971). These include weakening of the ice cover by dusting (Cook and Wade, 1968), or mechanical means. Measures taken after a jam has formed include physical removal of ice from the channel section, blasting, dislodgment of ice by icebreakers, and other means intended either to allow the water behind the jam to recede or to reduce the potential obstruction presented to a later breakup.

The investigator seeking to contribute to the solution of the catastrophic jam formation is directed to the large descriptive literature on jamming such as Chizov (1974), Michel (1971), IAHR (1970, 1972, 1974, 1975), Korzhavin (1971), Williams (1973), and Bolsenga (1968), the latter of which presents a considerable, although uncritical, review of the literature.

14.4 HYDRAULICS OF RIVERS WITH ICE

The hydraulics of rivers with ice is distinguished from the usual open channel warm weather hydraulics by the effects which ice has in changing the flow from open channel to closed conduit flow, by the effects of blockage of the flow cross section by ice, and by the unsteady effects associated with the storage and release of water as the ice cover forms or dissipates. The major deficiencies in our knowledge of river ice hydraulics center around quantitative knowledge of the friction factor, storage-discharge relationships, dispersion processes as modified by the ice cover, effects on sediment transport, and design of intakes for ice conditions.

Friction Factor

The friction factor of a river is of two parts: The friction factor directly associated with the underice surface itself and the effect on the overall composite roughness of an ice covered river reach. The friction factor is known to change through the winter season with a general trend of higher values early in the season associated with irregular ice formation and the presence of frazil; a decrease over the winter season generally attributed to a decrease in the amount of frazil present in the flow (and this represents the decrease in blockage effects more than actual decrease in surface friction factor); and finally an increase associated with the formation of relief features on the underside of the ice cover.

Significant papers reporting quantitative data on friction factor of the ice cover f_i include the work of Nezhikovskiy (1964) which clearly shows a trend of decreasing f_i over the winter season, the work of Larsen (1973) and Tesaker (1970) which provides data on f_i associated with a sheet ice cover, and the work of Carey (1966, 1967a, 1967b) which provides good data for f_i associated with the formation of ice ripples. Early in a winter season when the ice cover is thickening, the undersurface is generally quite planar and very smooth. With the onset of thaw, the warm water flowing under the ice cover results in the formation of the ice ripples which attain maximum amplitude-wavelength ratios of about 0.06 (Larsen, 1973). As a rule of thumb, the present author has found that the first day of average air temperature above 0°C after a long colder period is the time at which ice ripples are initiated. They may, of course, occur during the cold periods if the flow temperature is above 0°C as is often the case downstream from an impoundment or a source of artificial thermal effluent.

There also is little good data on the friction factor of an accumulated ice cover where the underside is a jumbled surface of angular blocks and slabs. Such data would be useful assessing the forces which cause thickening of ice jams by shoving and piling (Kennedy, 1975).

The composite roughness of an ice covered channel can be assessed if there is knowledge of the bottom roughness f_b and the ice roughness f_i. Most analyses of this sort divide the flow into two sections, one associated with f_i and one with f_b. Uzuner (1975) has summarized the gist of the various approaches and concluded those of Larsen (1969) and of Hancu (1967) to be the most complete and fundamentally sound. The use of the Chezy and Manning roughness coefficients is also in wide use but conversion of the Darcy-Weisbach f equivalents is straightforward.

Stage Discharge Relationships

In principle, if the composite roughness of an ice covered channel is known, then conventional backwater calculations, appropriately modified for the displacement effect of the ice cover, may be used to determine the stage-discharge relationship at a given river site. The converse is not true, however, since stage-discharge relationships developed by repeated gaging of a river during open surface conditions are not appropriate for ice covered conditions. For equivalent discharges the stage is necessarily higher with an ice cover and, where adequate gaging history during winter is available, a separate stage-discharge relationship may be developed. Even then, the accuracy is poorer than for equivalent open water conditions due to a multitude of complications. These include the temporary storage and release of water resulting from imposing or removing the resistance effects of the ice cover during initial freeze-over or thaw, difficulties arising from blockage effects of frazil deposits, variations of f_i described above, and the generally more meager data base associated with the hazards and complications of gaging during winter conditions. The actual gaging techniques in common usage in North America are discussed by USGS (1965) and by Cook and Cerny (1968). Except in very narrow channels the ice cover is sufficiently flexible to accommodate gradual changes in stage; hence, while the flow is under the pressure of the ice cover it is essentially open channel flow since the ice cover is usually floating in hydrostatic equilibrium except near bank areas. For very wide, shallow rivers grounding of the ice cover may occur; it is sometimes difficult to even find the flow sections, much less obtain representative velocity profiles. The extreme case of the latter is the formation of icings or naleds (from Russian word) which are sometimes massive accumulations of ice due to overflow of the ice cover and repeated freezing of the overflowing water (Carey, 1973).

It is noted also that the actual freezing and thawing of the ice cover extracts and adds water to the flow. In some cases this temporary storage is significant in the management of discharge releases for hydroelectric power production.

Dispersion and Diffusion

There are few measurements of the longitudinal and lateral dispersion coefficient associated with flow in an ice covered river, largely because of the difficulty of field measurements during periods of ice cover. The lack of knowledge of the lateral dispersion coefficient is the most critical of the two and bears importantly on the disposition of thermal effluents discharged into ice covered river systems. The effect of an ice cover is to change the flow from open channel flow to closed conduit flow (in the sense of adding another boundary, not in the sense of a pressure conduit) and one would expect the lateral mixing coefficient $E = ku_*R$ (where k is a constant, u_* is the shear velocity, and R the hydraulic radius) to be accordingly smaller by factor of $2(2)^{1/2}$ for equal depths (Engmann and Kellerhals, 1974).

The only field measurement lateral dispersion under an ice cover known to the writer is that of Engmann and Kellerhalls (1974) who, using dye tracing techniques, found the dispersion coefficient to be somewhat less than theoretically expected.

Perhaps the most obvious use of a dispersion coefficient is the simulation or prediction of width of open water downstream from a point discharge of thermal effluent. The problem is complicated by the coupling that must exist between the open channel dispersion in the open water areas and the closed channel dispersion beneath the adjacent ice cover. Field measurements of the thermal structure at these sites would provide useful data on which to test analytical simulation schemes for the prediction of the associated ice cover suppression. Paily, Macagno, and Kennedy (1974) have included longitudinal dispersion in their theoretical treatment of the ice suppression downstream from an abrupt release of thermal effluent uniformly distributed across the channel, but do not consider the time lag associated with the melting of the ice cover.

Sediment Transport

There has been virtually no documentation on the effects of ice covers and frazil ice on sediment transport in rivers. There often is a noticeable decrease in turbidity during the freeze-up period (Osterkamp, 1975). The imposition of the ice cover on a river reach with a fixed slope, while increasing the total flow resistance, results in a decrease average shear stress per unit area of boundary as may be shown by a straightforward analysis (W. Sayre, unpublished notes). In turn, using this result in any of various bedload analyses leads to the result that bedload movement can be expected to decrease relative to open channel conditions for equivalent water discharges. Similarly, there is little reason to expect dramatic difference in the suspended sediment load although order of magnitude decreases have been observed on the Yukon during initial freeze-up (Osterkamp, private communication). Nevertheless, work needs to be performed to quantitatively evaluate the differences.

It is often conjectured that the presence of ice in a river may result in greater scour depths than for corresponding open water discharges. Observations after the passage of ice jams often report cases of extreme scouring, although it is seldom clear whether the bed material removal was due to the scouring action of the water or due to ice scouring. The problem in either case is intimately tied up with the nature of the accumulation process and we seem to be far from having the ability to predict ice accumulation in sufficient detail to perform meaningful design predictions. Studies of before and after bed morphology would do much to clarify the question. It is known that movement of ice floes down a river can severely degrade riprap bank protection, particularly if it is poorly placed originally, and cause other damage to shore areas and structures.

Design of Intakes for Ice Conditions

Intake design for ice conditions has two major problems, choice of location and detailed design to avoid clogging by ice. The most comprehensive survey of the general problem is that of Hayes (1974) which summarizes North American experience. The essence of the location problem is the upside down counterpart of design of intakes to avoid sediment intake. Unfortunately, the two requirements are often at odds with each other.

The other major problem associated with intakes is clogging of the trashracks, particularly by frazil ice. The usual solution is to heat the trashrack bars, thus preventing adherence. The experience on heat requirements has been summarized by Logan (1974), although the heat transfer relationships used therein are relatively crude. Frazil can and does clog intakes at depths at least as great as 15 m. Where possible, measures should be taken to prevent the frazil supply by inducing an early complete ice cover upstream thus reducing the heat loss to the atmosphere.

14.5 THERMAL EFFECTS

Thermal effects associated with river ice may be considered from two viewpoints; the natural thermal regime of an ice infested river, and the modification of that regime by imposed thermal effluents. It is noted at the outset that most of the heat transfer techniques applied to rivers during non-ice conditions at the water surface are applicable during winter conditions with but slight modification. The energy budget analysis involves calculation of the components due to radiation, convection, advection, conduction, and evaporation. The added complication due to ice is the required energy to melt or freeze the ice and account of the time lag involved. The most comprehensive treatments of the energy budget of an open river surface in an otherwise ice-covered river are those of Dingman, Weeks, and Yen (1968), Pivavarov (1973), and Paily, Macagno, and Kennedy (1974). The difficulties with the radiation component are associated with the absorption and albedo effects of the ice or snow cover. Difficulties with the convective effects are due largely to the extreme air-water temperature differences which sometimes induce convective effects not often experienced during non-ice periods. Associated with these effects is the coupled effect of evaporation and to date the best treatment of evaporation during cold conditions is that of Rimsha and Donchenko (1957) and the so-called Russian winter equation, although the resulting numerical values are not markedly different from those of Ryan, Harleman and Stolzenbach (1974) which is based on fairly extensive experimentation and theory not motivated by cold conditions.

Natural Thermal Regimes

The natural thermal regime of an ice covered river is characterized by water temperatures exceedingly close to 0°C and measurements of it require exceedingly accurate thermometry. In Figs. 14-10 and 14-11 are typical results of measurements of the water temperatures of a natural river of medium size (discharge about 45 $m^3 s^{-1}$) when the river was entirely ice covered. Figure 14-11 presents the vertical temperature profiles corresponding to the data of 27 January in Fig. 14-10 and it may clearly be seen that vertical temperature gradients are negligible. The lateral variations in Fig. 14-10 are typical; namely, a tendency for higher temperatures nearer the bank

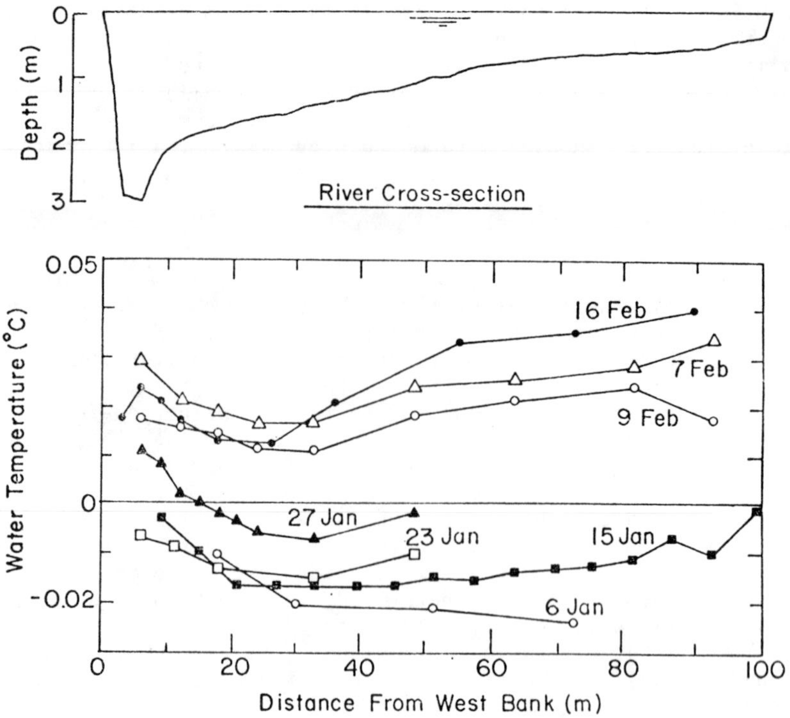

Fig. 14-10. Lateral Distribution of Mid-Depth Water Temperatures (Cedar River, Iowa, 1970; Adapted from Ashton and Kennedy, 1970)

areas which results from the fact that thermal input in a natural river derives from lateral inflow of groundwater and tributary contributions which occur near the shores. It is noted that the sub-zero temperatures are not considered to represent supercooling but rather a depression of the freezing point by dissolved solids. These very small temperature differences are significant since water temperatures as small as 0.05°C may cause rapid melting at the undersurface, a fact easily concluded by application of existing empirical correlations of turbulent heat transfer data for closed conduit flow.

Effects of Thermal Modification

Understanding and modeling of the effects of imposed thermal loads on a river during periods of ice conditions requires appreciation of the path of the energy from discharge into the river to its ultimate residence in the atmosphere. Figure 14-12 presents a schematic diagram of the components of that path.

The near-field mixing and entrainment can be treated similarly as during warm-water conditions although there are

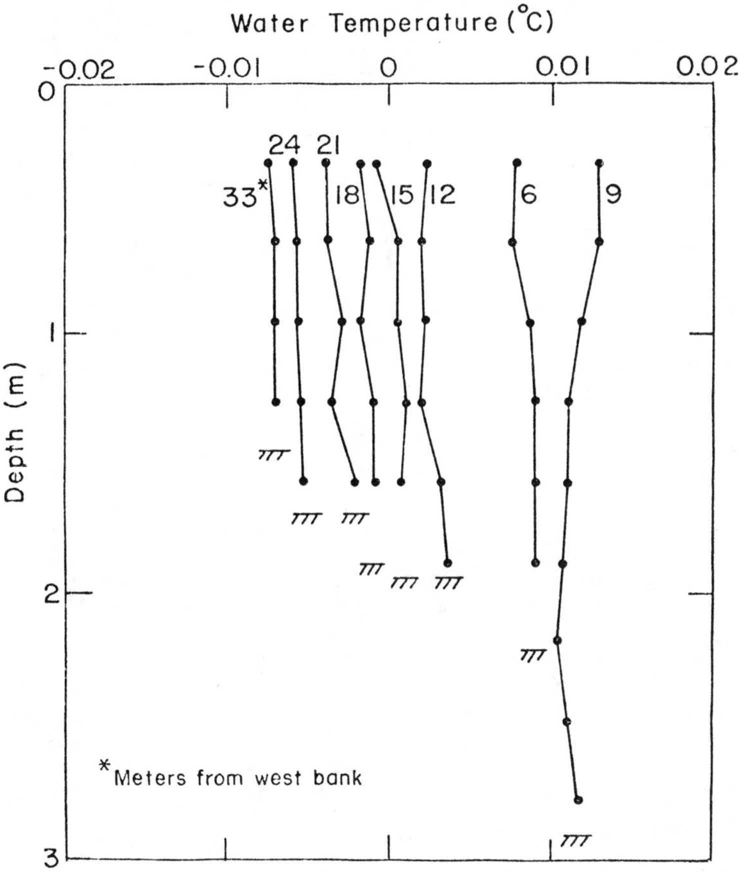

Fig. 14-11. Vertical Temperature Profiles in a Natural River during Ice Cover (Cedar River, Iowa, 1970; Adapted from Ashton and Kennedy, 1970)

some complications associated with the density variations of water from 0°C to 8°C. Next, vertical mixing and entrainment occur followed by horizontal mixing and longitudinal mixing. These processes are primarily dilution.

The actual transfer to the atmosphere occurs through the water surface if the ice cover has been completely suppressed. If not, heat transfer occurs to the undersurface of the ice cover, either causing melting or retarding further thickening. It may also melt frazil ice which is deposited under the ice cover, and there is some transfer to and from the bottom sediments although this is generally small in rivers (as contrasted with lakes where seasonal storage and release of heat from the bottom is significant for winter energy budgets).

With the picture drawn above of the disposition of thermal effluents, we now examine the present state-of-the-art of

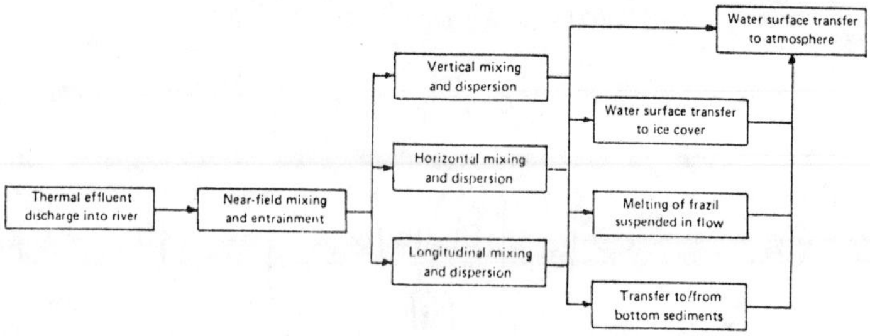

Fig. 14-12. Path of Thermal Energy Imposed on a River

predicting such effects. Nearly all treatments assume an initial complete mixing over the entire cross section of flow and do not consider unsteady effects such as the storage of energy in the ice cover in the form of latent heat of fusion, or of weather variations. The classic paper treating the uniform mixing, steady-state problem is that of Dingman, Weeks, and Yen (1968). Paily, Macagno, and Kennedy (1974) modified that method to include longitudinal dispersion and an instantaneous beginning of the thermal effluent but did not consider the time lag required to melt the ice or the effect of changing weather conditions.

The decrease in flow temperature downstream beneath the ice cover has been treated by Baines (1961) and the present writer with minor differences in choice of the turbulent heat transfer coefficient applicable to the ice-water interface. Typical numerical results are presented in Fig. 14-13 where the downstream temperature T_∞ (relative to the melting temperature T_m) is normalized by initial temperature $T_{\infty}.0$. Example curves have been drawn for various flow depths D and a reference mean velocity of $U = 1\ m\ s^{-1}$. The effect of velocity is not great since, while higher velocities increase the rate of heat transfer to the ice cover, the parcel of water which is cooling is transported proportionately further.

What is needed is a general numerical scheme incorporating all the elements of Fig. 14-12, although there is some question (Weeks and Dingman, 1972) as to whether a good scheme can ever be sufficiently detailed to handle the many special situations which occur at actual sites of interest. Perhaps the best that can be expected in the near future is inclusion of the lateral mixing effect together with an unsteady (weather-driven) numerical simulation of the melting and freezing of the ice cover.

The magnitude of energy required to fully suppress an ice cover can be estimated by multiplying a reasonable value of h_a by the surface area and air-water temperature difference. As an example, with $h_a = 25\ W\ m^{-2}\ °C^{-1}$ a width of 400 m, and -10°C air temperature, 100 MW per km of river reach are required to fully suppress the ice cover.

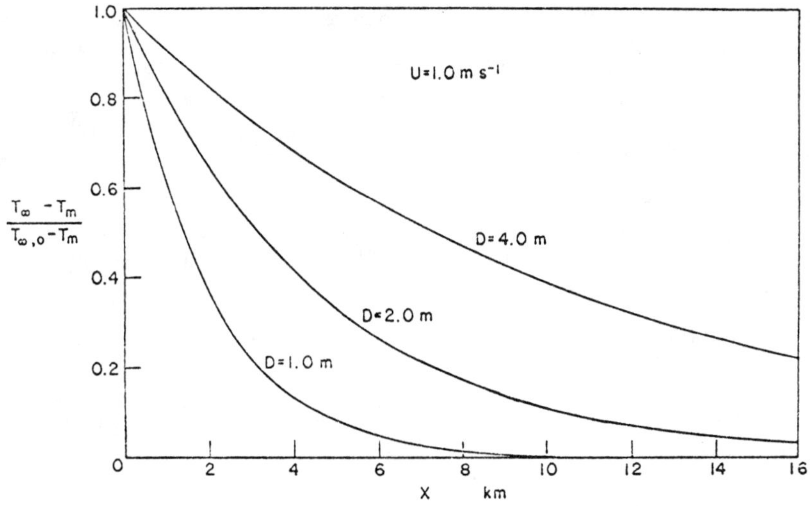

Fig. 14-13. Downstream Attenuation of Water Temperature beneath an Ice Cover

14.6 ICE FORCES

The subject of forces exerted by river ice is too extensive to treat completely in this review. Instead, attention herein will center on the role of fluid mechanics in the problem of ice forces. No attempt will be made to review the effects of the mechanical properties, the various guidelines and codes available to the designer, or the many experimental studies which involve essentially static loading situations in which fluid mechanics plays little role. To some degree this avoids the ultimate pragmatic problem since most designs of structures to resist ice forces utilize the philosophy that the ice must fail rather than the structure, and the ice failure depends largely on dimensions and strength of the ice. Fortunately, there have recently appeared good summaries of the general problem such as those by Korzhavin (1971), Michel (1970), Kerr (1976), and Neill (1976). The various symposia (IAHR, 1970, 1972, 1974, 1975) also contain numerous papers on various aspects of the problem. It is noted, however, that the mechanical properties are not independent of the hydraulic history of the ice formation and affect not only thickness but also the ice crystallography (Michel and Ramseier, 1971).

Neill (1976) has distinguished four principal modes of ice action on fixed piles and piers which are: impact of moving sheets and floes, static pressures due to expansion or contraction of the ice sheet, slow pressure from ice accumulations and vertical movements of the ice cover due to fluctuating water levels. The first and third have elements of fluid mechanics which are of some importance. Clearly the impact of a moving ice floe or sheet derives its momentum from the hydrodynamic forces of wind and water acting upon it and, upon impact of the floe, the added mass of water carried with the floe has an effect on the force levels. Again, it must be emphasized that the prudent designer plans for the ice to fail rather than the structure and the limiting force is determined by the mechanical properties and dimensions of the ice rather than the hydrodynamic forces.

The forces of accumulated ice involve fluid mechanical interactions both in the sense of being the driving force which must be resisted and also in the effects of buoyancy exerted on the individual pieces.

14.7 ACKNOWLEDGMENT

These notes have been prepared and adapted from a general review paper prepared for the Annual Review of Fluid Mechanics (Ashton, 1978, in press). Appreciation is also expressed to the many river ice researchers and engineers who have contributed to the writer's understanding and views on the phenomena of river ice behavior through discussion and criticism of the writer's own efforts.

14.8 REFERENCES

Adams, C. M., French, D. N., Kingery, W. D., 1960. Solidification of sea ice. J. Glaciol., 3:745-769.

Altberg, W. J., 1936. Twenty years of work in the domain of underwater ice formation (1915-1935). Proc. Int. Un. Geodesy and Geophysics, Int. Assn. Sci. Hydrology, pp. 373-407.

Ashton, G. D., 1972. Field implications of ice ripples. In IAHR 1972, pp. 123-129.

Ashton, G. D., 1974a. Froude criterion for ice block stability. J. Glaciol., 13:307-313.

Ashton, G. D., 1974b. Entrainment of ice blocks - secondary influences. In IAHR 1974, Paper A-11, pp. 83-89.

Ashton, G. D., Kennedy, J. F., 1970. Temperature and flow conditions during the formation of river ice. In IAHR 1970, Paper 2.4, 12 p.

Ashton, G. D., Kennedy, J. F., 1972. Ripples on underside of river ice covers. J. Hyd. Div., ASCE, 98:1603-1624.

Baines, W. D., 1961. On the transfer of heat from a river to an ice sheet. Trans. Eng. Inst. Canada, 5:27-32.

Barnes, H. T., 1928. Ice engineering. Montreal: Renouf., 364 p.

Bolsenga, S. J., 1968. River ice jams - a literature review. U.S. Lake Survey, Detroit, MI, Res. Rept. 5-5, 587 p.

Burgi, P. H., Johnson, P. L., 1971. Ice formation - a review of the literature and Bureau of Reclamation experience. Bur. of Reclam., Rep. REC-ERC-71-8, 27 p.

Carey, K. L., 1966. Observed configuration and computed roughness of the underside of river ice. St. Croix River, Wisconsin, USGS Prof. Paper 550-B, pp. B192-B198.

Carey, K. L., 1967a. The underside of river ice. St. Croix River, Wisconsin, USGS Prof. Paper 575-C, pp. C-195-C199.

Carey, K. L., 1967b. Analytical approaches to computation of discharge of an ice-covered stream. USGS Prof. Paper 575-C, pp. C200-C207.

Carey, K. L., 1973. Icings developed from surface water and groundwater. Cold Reg. Res. Eng. Lab., Hanover, N.H., Monograph III-D3, 37 p.

Carstens, T., 1966. Experiments with supercooling and ice formation in flowing water. Geofysiske Publikasjoner, 26:1-18.

Carstens, T., 1970. Heat exchanges and frazil formation. In IAHR 1970, Paper 2.11, 17 p.

Cartier, L., 1959. Field investigations on some ice cover problems in an experimental canal. Proc. Eastern Snow Conf., pp. 45-47.

Chizov, A. N., ed., 1974. Investigation and calculations of ice jams. Cold Reg. Res. Engl. Lab., Hanover, N.H., Draft Translation 473, 106 p.

Cold Reg. Res. Eng. Lab., 1951-1976. Bibliography on cold regions science and technology. CRREL Rep. 12 (annual vols.) Hanover, N.H.

Cook, R. E., Cerny, E. E., 1968. Patterns of backwater and discharge on small ice-affected streams. USGS Water Supply Paper 1892, pp. 114-125.

Cook, R. G., Wade, M. D., 1968. Successful ice dusting at Fairbanks, Alaska, 1966. J. Hyd. Div., ASCE, 94:31-41.

Dean, A. M., 1977. Remote sensing of accumulated frazil and brash ice in the St. Lawrence River. Cold Reg. Res. Eng. Lab., Hanover, N.H., Rep. 77-78, 19 p.

Deslauriers, C. E., 1968. Ice break-up in rivers. Proc. Conf. Ice Pressure against Structures, Natl. Res. Council Canada Tech. Memo. 92, pp. 211-292.

Devik, O., 1942. Supercooling and ice formation in open water. Geofysisk Publikasjoner, 13:1-10.

Dingman, S. L., Weeks, W. F., Yen, Y. C., 1968. The effects of thermal pollution on river ice conditions. Wat. Res. Res., 4:349-362.

Engmann, J. E. O, Kellerhals, R., 1974. Transverse mixing in an ice-covered river. Wat. Res. Res., 10:775-784.

Filippov, A. M., 1974. Modeling the movement of ice floes drawn in under ice cover. Transactions of State Hydrologic Inst. Leningrad, pp. 72-80.

Frankenstein, G. E., 1961. Strength data on lake ice, II. Cold Reg. Res. Eng. Lab., Hanover, N.H., Tech. Rep. No. 80, 18 p.

Hancu, S., 1967. Modelarea hidraulica in curenti de aer sub presiune. Editura Academiei Republicii Socialiste Romania (text in Romanian).

Hanley, T. O. D, Michel, B., 1977. Laboratory formation of border ice and frazil slush. Can. J. Civ. Eng., in press.

Hayes, R. B., 1974. Design and operation of shallow river diversions in cold regions. Bur. Recl. Rep. REC-ERC-74-19, 42 p.

Hoyt, W. G., 1913. The effects of ice on stream flow. USGS Water Supply Paper 337, 77 p.

Hsu, K., 1973. Spectral evolution of ice ripples. Ph.D. Thesis, Univ. Iowa, 147 p.

Int. Assn. Hyd. Res., 1970. Symposium on ice and its action on hydraulic structures. Reykjavik, Iceland, 64 Papers.

Int. Assn. Hyd. Res., 1972. Symposium on ice and its action on hydraulic structures. Leningrad, USSR, 3 Vol. and Summary Vol.

Int. Assn. Hyd. Res., 1974. Symposium on river and ice. Budapest, Hungary, Proceeding Volumes.

Int. Assn. Hyd. Res., 1975. Proc. third international symposium on ice problems. Hanover, N.H., 627 p.

Kennedy, J. F., 1975. Ice-jam mechanics. In IAHR 1975, pp. 143-164.

Kerr, A. D., 1976. The bearing capacity of floating ice plates subjected to static or quasi-static loads. J. Glaciol., 17:229-268.

Koren'kov, V. A., 1970. Experimental research findings on decrease of river ice strength in spring. In IAHR 1970, Paper 5.4, 4 p.

Korzhavin, K. N., 1971. Action of ice on engineering structures. Cold Reg. Res. Eng. Lab., Hanover, N.H., Draft Translation 260, 319 p.

Larsen, P. A., 1969. Head losses caused by an ice cover on open channels. J. Boston Soc. Civil Eng., 56:45-67.

Larsen, P., 1973. Hydraulic roughness of ice covers. J. Hyd. Div. ASCE, 99:111-119.

Larsen, P., 1975. Notes on the stability of floating ice blocks. In IAHR 1975, pp. 305-314.

Logan, T. H., 1974. The prevention of frazil ice clogging of water intakes by application of heat. U.S. Bur. Recl. Rep. REC-ERC-74-15, 21 p.

Michel, B., 1963. Theory of formation and deposit of frazil ice. Proc. Eastern Snow Conf., 8:129-149.

Michel, B., 1970. Ice pressure on engineering structures. Cold Reg. Res. Eng. Lab., Hanover, N.H., Monograph III-Blb, 71 p.

Michel, B., 1971. Winter regime of rivers and lakes. Cold Reg. Res. Eng. Lab., Hanover, N.H., Monograph III-Bla, 131 p.

Michel, B., Ramseier, R. O., 1971. Classification of river and lake ice. Can. Geotech. J., 8:36-45.

Moor, J. H., Watson, C. H., 1971. Field tests of ice jam prevention techniques. J. Hyd. Div. ASCE, 97:777-789.

Neill, C. R., 1976. Design of ice forces on piers and piles: an assessment of design guidelines in the light of recent research. Can. J. Civ. Eng., 3:305-341.

Nezhikovskiy, R. A., 1964. Coefficients of roughness of bottom surface of slush-ice cover. Soviet Hydrology: Selected Papers, 1964: 127-150.

Osterkamp, T. E., 1975. Observations of Tanana River Ice. In IAHR 1975, pp. 201-209.

Paily, P. P., Macagno, E. O., Kennedy, J. F., 1974. Winter-regime thermal response of heated streams. J. Hyd. Div. ASCE, 100:531-551.

Pariset, E., Hausser, R., 1961. Formation and evolution of ice covers on rivers. Trans. Eng. Inst. Canada, 5:41-49.

Pariset, E., Hausser, R., Gagnon, A., 1966. Formation of ice covers and ice jams in rivers. J. Hyd. Div. ASCE, 92:1-24.

Pivovarov, A. A., 1973. Thermal conditions in freezing lakes and rivers. New York: J. Wiley and Sons, 136 p.

POAC, 1971. Port and ocean engineering under Arctic conditions. 1st Conf., Trondheim, Norway.

POAC, 1973. Port and ocean engineering under Arctic conditions. 2nd Conf., Reykjavik, Iceland.

Rimsha, V. A., Donchenko, R. V., 1957. The investigation of heat loss from free water surfaces in wintertime. Tr. Leningrad Gos, Gidrol. Inst., 65:54-83 (in Russian).

Ryan, P. J., Harleman, D. R. F., Stolzenbach, K. D., 1974. Surface heat loss from cooling ponds. Wat. Res. Res., 10:930-938.

Starosolszky, O., 1969. Ice in hydraulic engineering. Norwegian Inst. Tech., Trondheim, Norway, Rep. No. 70-1, 165 p.

Tesaker, E., 1970. Measurements of ice roughness and the effect of ice cover on water levels in three Norwegian rivers. In IAHR 1970, Paper 3.4, 9 p.

USGS, 1965. Discharge measurements at gaging stations. Book 1, Chap. 11, 67 p.

Uzuner, M. S., 1975. The composite roughness of ice-covered streams. J. Hydraul. Res., 13:79-102.

Uzuner, M. S., Kennedy, J. F., 1972. Stability of floating ice blocks. J. Hyd. Div. ASCE, 98:2117-2133.

Uzuner, M. S., Kennedy, J. F., 1976. Theoretical model of river ice jams. J. Hyd. Div. ASCE, 102:1365-1383.

Weeks, W. F., Dingman, S. L., 1972. Thermal modification of river ice covers: progress and problems. IHD/UNESCO/WMO Symposia on Role of Snow and Ice in Hydrology, Banff, Alberta, 2:1427-1435.

Williams, G. P., 1959. Frazil ice - a review of its properties with a selected bibliography. Eng. J. (Canada), 42:55-60.

Williams, G. P., 1973. Seminar on ice jams in Canada. Natl. Res. Council Canada, Tech. Memo. No. 107, 182 p.

Chapter 15

SHORE-ATTACHED THERMAL PLUMES IN RIVERS

by

W.W. Sayre, Professor, Division of Energy Engineering, and Research Engineer, Institute of Hydraulic Research, The University of Iowa

and

R. Caro-Cordero, Hydraulic Engineer, Development Analysis Associates, Inc., Cambridge, Massachusetts

Chapter 15

SHORE-ATTACHED THERMAL PLUMES IN RIVERS

15.1 INTRODUCTION

Currently applicable guidelines of the Environmental Protection Agency (EPA) imply that only closed-cycle evaporative cooling systems will meet the best available practicable control technology and best available technology requirements for the disposal of waste heat from steam-electric generating stations. However, the 1972 amendments also state that if it can be demonstrated that EPA limitations on thermal discharges into rivers are more stringent than is necessary to protect the propagation of fish and wildlife, then EPA may permit less stringent control on a case-by-case basis.

A recent study by Paily et al. (1976) indicates that there is sufficient heat transfer and assimilation capacity at average river flows to accommodate the waste heat load from most of the power plants projected until 1993 along the Upper Mississippi River and the Missouri River above the southern Iowa border, without violating the presently applicable stream temperature standards. Except at times of low river flow, when supplementary closed-cycle cooling at some sites would be required, this could be accomplished with once-through cooling systems which are much more economical and entail significantly less consumptive use of water than closed-cycle systems. Assuming that several other large rivers are also in this category, a strong case can be made for the continued use of once-through cooling systems, often in conjunction with closed-cycle systems, for plants that are spaced along rivers at intervals that would preclude a significant cumulative buildup of excess temperature, provided it can be demonstrated that compliance with the applicable mixing zone standards will be maintained.

Prior demonstration of compliance with mixing zone standards requires prediction of the temperature-rise distribution in the region of the river affected by the thermal discharge, commonly called the thermal plume. This chapter is concerned with the description of a mathematical model for predicting shore-attached thermal plumes and its application to reaches of the Missouri River downstream from the Cooper and Fort Calhoun Nuclear Power Stations. It draws heavily on the results of an investigation (Caro-Cordero and Sayre, 1977) conducted for the Iowa Energy Policy Council.

15.2 DESCRIPTION OF MODEL

Requirements of Model

In recent years several analytical and numerical models have been developed for predicting configurations of thermal plumes resulting from the discharge of heated surface and/or submerged jets into ambient cross flows as in a river. Several of these have been reviewed in detail in publications resulting from investigations at Argonne National Laboratory (Dunn, et al., 1975), and MIT (Jirka, et al., 1975). For one reason or another none of these models in their present state of development is suitable for predicting thermal plumes in fast-flowing streams such as the Missouri River. For discharges that enter the river with an appreciable cross-channel momentum component, most of them assume bell-shaped excess temperature and velocity profiles that are symmetrical about a predicted trajectory or plume centerline that carries well out into the channel. In addition they assume an idealized uniform ambient flow and do not adequately account for the mixing mechanisms associated with the ambient flow. In contrast, the transverse temperature profiles in thermal plumes which have been measured in the Missouri River downstream from the Fort Calhoun, Cooper and North Omaha power stations are not bell-shaped; they are invariably attached to the near shore, with the maximum temperatures in the plume occurring at or near the shore line. Following a brief period of initial mixing, the mixing process and the plume configuration appear to be governed mainly by the turbulence and shifting transverse distribution of the ambient flow. In addition, temperature measurements taken over the depth of flow indicate that buoyancy effects are minimal. Even with initial dimensionless density differences as large as $\Delta\rho/\rho \approx 0.003$, an essentially uniform temperature-rise distribution over the depth of flow is rapidly achieved due to the high level of turbulence in the ambient river flow.

Evidently in fast-flowing, highly turbulent streams such as the Missouri River, the near-field mixing and entrainment mechanisms are dominant only in a small region close to the outfall structure. After leaving this region, the advective and diffusive mechanisms associated with the ambient flow become dominant in determining the configuration of the thermal plume. In comparison with more tranquil streams of comparable size, the near-field zone is much smaller in the Missouri River and the transition to the far-field zone occurs more rapidly and abruptly. Consequently what is needed for predicting thermal plumes in rivers such as the Missouri is a model which focuses correspondingly less on the near-field mechanisms and in greater detail and with fewer restrictions on the far-field mechanisms.

Analytical Background

The model adopted for this presentation is based on the depth-averaged, steady-state continuity and convection-diffusion equations

$$\frac{\partial}{\partial x}(h_3 d\bar{u}^d) + \frac{\partial}{\partial z}(h_1 d\bar{w}^d) = 0 \ , \qquad (15\text{-}1)$$

and

$$h_3 d\bar{u}^d \frac{\partial T}{\partial x} + h_1 d\bar{w}^d \frac{\partial T}{\partial z} = \frac{\partial}{\partial z} \left(\frac{h_1}{h_3} dE_z \frac{\partial T}{\partial z}\right) , \qquad (15\text{-}2)$$

in a meandering coordinate system for natural channels. A definition sketch of this coordinate system, which is described in greater detail by Yotsukura and Sayre (1976), is shown in Fig. 15-1. In Eqs. (15-1) and (15-2) x and z represent distances along the longitudinal and transverse coordinate axes; d is the local depth; $\bar{u}^d$ and $\bar{w}^d$ are depth-averaged values of the local time-averaged velocities in the direction of the longitudinal and transverse coordinate surfaces; T is the depth-averaged value of the local temperature rise above ambient due to the heated discharge; E_z is an overall transverse mixing coefficient due to the combined effects of turbulent diffusion and advection by bend-generated secondary flow; and h_1 and h_3 are metric coefficients (or scaling factors) which relate horizontal distances along the curved longitudinal and transverse coordinate surfaces to the corresponding distances along the x and z axes. The metric coefficients correct for the effect of curvature; if the coordinate surfaces are straight and parallel to their respective axes, h_1 and h_3 assume a value of unity.

Following Yotsukura and Sayre (1976), Eq. (15-2) is now transformed to a more convenient form for use in natural channels by introducing the cumulative discharge,

$$q_c = \int_{z_o}^{z} h_3\, d\, \bar{u}^d\, dz, \qquad (15\text{-}3)$$

where z_o is the transverse coordinate at the bank where the outfall is located, as a new variable to replace z. This a accomplished by first integrating Eq. (15-1) to obtain

$$h_1\, d\, \bar{w}^d = -\frac{\partial q_c}{\partial x} \qquad (15\text{-}4)$$

and then substituting Eq. (15-4) into Eq. (15-2) and applying the chain rule of partial derivatives to obtain

$$\frac{\partial T}{\partial x} = \frac{\partial}{\partial q_c} \left(h_1 d^2\, \bar{u}^d\, E_z \frac{\partial T}{\partial q_c}\right) , \qquad (15\text{-}5)$$

or, in terms of the normalized cumulative discharge $p = q_c/Q$, where Q is the total river discharge,

$$\frac{\partial T}{\partial x} = \frac{\partial}{\partial p}\left(\frac{h_1 d^2 \bar{u}^d E_z}{Q^2}\frac{\partial T}{\partial p}\right) . \tag{15-5a}$$

No generality has been lost and no restrictions have been added in the transformation from Eq. (15-2) to Eq. (15-5). In particular, the capability of representing the transverse mixing process (in a depth-averaged sense) in an irregular channel, wherein the transverse distribution of flow shifts from one section to the next, is retained. Solutions to Eq. (15-5) are obtained in the x - q_c domain, however Eq. (15-3) can be used to transform them back into the x - z domain.

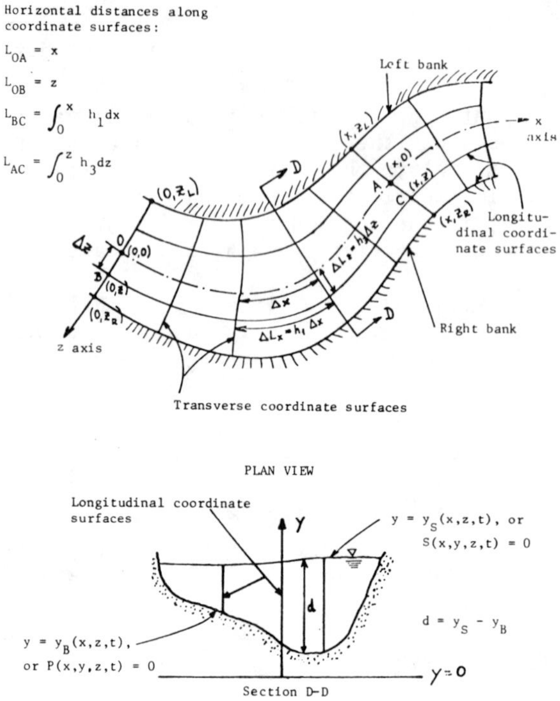

Fig. 15-1. Meandering Coordinate System for Natural Channels

For the condition that

$$h_1 d^2 \bar{u}^d E_z = Q^2 D \tag{15-6}$$

where D = constant, Eq. (15-5a) assumes the form of a simple one-dimensional diffusion equation

$$\frac{\partial T}{\partial x} = D \frac{\partial^2 T}{\partial p^2} , \tag{15-7}$$

for which solutions corresponding to many different initial and boundary conditions are well known. Because D includes various mechanisms in addition to turbulent diffusion and is consequently not a true diffusion coefficient in a physical sense, and to avoid confusing it with the transverse mixing coefficient E_z, we shall call D a transverse diffusion factor.

Although there is no reason to expect that the quantity on the lefthand side of Eq. (15-6) remains constant in either the transverse or longitudinal direction, Yotsukura and Cobb (1972) have shown that solutions of Eq. (15-5a) are not very sensitive to variations in the transverse distribution of this quantity provided that its average value,

$$Q^2 D = \int_0^1 h_1 d^2 \bar{u}^d E_z dp , \tag{15-8}$$

remains the same. Allowing the value of D to change in the longitudinal direction from one subreach to the next along the total reach of interest presents no particular problem. Solutions of Eq. (15-7) have been found to represent the transverse mixing process in the Missouri River (Sayre and Yeh, 1973; Paily, 1975), and other rivers (Jackman and Yotsukura, 1977) quite well.

Form of Solution

The particular form of the solution and its application adopted herein to represent shore-attached thermal plumes in rivers is a refinement and extension of techniques used by Sayre (1975, 1973) and Paily (1975).

Using the principle of superposition, the general form of the solution to Eq. (15-7) can be expressed as

$$T(p,x) = \int_0^1 T_I(p',0)\, f_p(p - p';x)\, dp' , \tag{15-9}$$

where $T_I(p',0)$ is the temperature-rise distribution with respect to the dummy variable p' at x = 0 (or a short distance downstream) following initial near-field mixing; and $f_p(p - p';x)$ is the solution of Eq. (15-7) for a continuous source of unit strength ($Q_o T_o/Q = 1$) concentrated at p = p', x = 0. As a first approximation let us assume that, following initial near-field mixing, T_I is uniformly distributed from p' = 0 to p' = P as shown in Fig. 15-2, i.e.

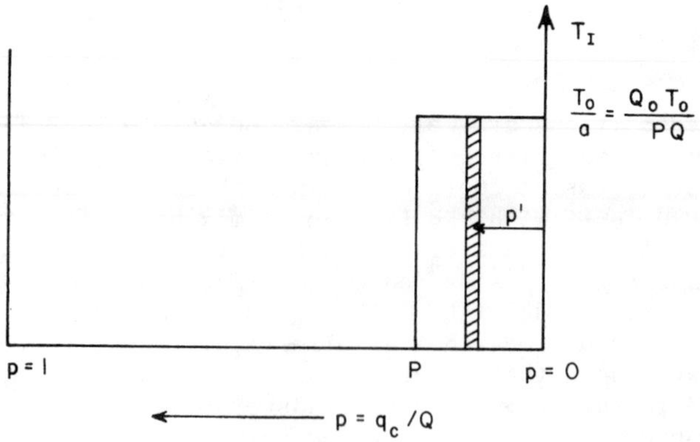

Fig. 15-2. Definition Sketch for Temperature Distribution Following Initial Near-Field Mixing

$$T_I(p',0) = \frac{Q_o T_o}{PQ} = \frac{T_o}{a}, \quad 0 \leq p' \leq P$$

$$= 0 \quad , \qquad P < p' \leq 1 \qquad (15\text{-}10)$$

where Q_o is the volumetric discharge rate of the condenser cooling water into the river; T_o is the temperature rise of the condenser cooling water at the point of discharge; and $P = aQ_o/Q$ is the fraction of the total river discharge with which the cooling water is mixed following dilution by a factor of $\underline{a}$ due to initial near-field mixing. Turning now to the second term inside the integral on the righthand side of Eq. (15-9) and neglecting for the moment the reflections from the channel banks,

$$f_p(p - p';x) = \frac{1}{2\sqrt{\pi Dx}} \exp\left[-\frac{(p - p')^2}{4Dx}\right], \quad -\infty < p - p' < \infty$$

$$= \frac{1}{\sigma_p}\frac{1}{\sqrt{2\pi}} \exp\left[-\frac{s^2}{2}\right], \qquad -\infty < s < \infty \qquad (15\text{-}11)$$

where $\sigma_p = \sqrt{2Dx}$; $s = \dfrac{p - p'}{\sigma_p}$ is the standardized normal variable; and $\sigma_p f_p(p - p';x) = f_s(s)$ is the standardized probability density function of the normal probability law. Substituting into Eq. (15-9), we get

$$T(p,x) = \frac{Q_oT_o}{PQ} \int_0^P f_p(p - p';x)\, dp'$$

$$= \frac{Q_oT_o}{PQ} \int_{\frac{p - P}{\sigma_p}}^{\frac{p}{\sigma p}} f_s(s)\, ds$$

$$= \frac{Q_oT_o}{PQ} [F_s(\frac{p}{\sigma_p}) - F_s(\frac{p - P}{\sigma_p})] , \qquad (15\text{-}12)$$

where F_s ( ) is the cumulative distribution function of the normal probability law. Finally, after including reflection from the channel banks, we get

$$\frac{PQ}{Q_oT_o} T(p,x) = F_s(\frac{p + P}{\sigma_p}) - F_s(\frac{p - P}{\sigma_p})$$

$$+ \sum_{n=1}^{\infty} \Big\{F_s(\frac{2n - p + P}{\sigma_p}) - F_s(\frac{2n - p - P}{\sigma_p})$$

$$+ F_s(\frac{2n + p + P}{\sigma_p}) - F_s(\frac{2n + p - P}{\sigma_p})\Big\} , \qquad (15\text{-}13)$$

hereinafter called the model. As $P \to 0$, i.e., for a concentrated source at $p' = 0$, Eq. (15-13) reduces to

$$\frac{Q}{Q_oT_o} T(p,x) = \frac{2}{\sigma_p} \Big\{f_s(\frac{p}{\sigma_p}) + \sum_{n=1}^{\infty} f_s(\frac{2n - p}{\sigma_p}) + f_s(\frac{2n + p}{\sigma_p})\Big\}. \qquad (15\text{-}13a)$$

Upstream of where the thermal plume encounters the far shore, i.e., $x \leq 0.08/D$, the first two terms on the righthand side of Eq. (15-13) and the first term on the righthand side of Eq. (15-13a) are sufficient.

The solution of Eq. (15-13) for $p = 0$, that is, for the maximum temperature rise which occurs along the near shore, is shown for various values of P in Fig. 15-3. Note that uniform mixing across the channel is achieved at about $Dx = 0.5$ for all values of P that are shown.

Parameters of the Model

Representation of thermal plumes by Eq. (15-13) requires specification of the parameters P (or a), for the initial near-

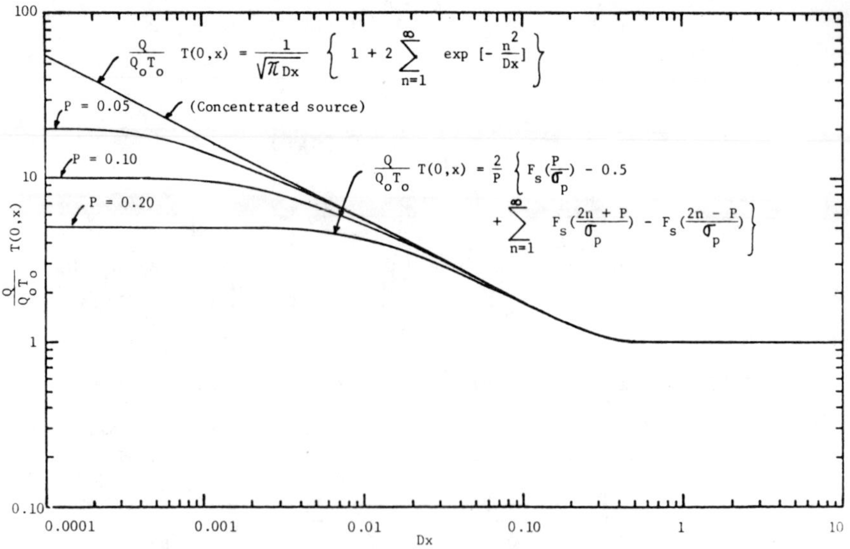

Fig. 15-3. Attenuation of Maximum Temperature Rise Along Near Shore

field dilution, and D, the transverse diffusion factor for mixing in the far field due to mechanisms associated with ambient flow. If the value of D varies along the reach from 0 to x_n, the average value of D for the reach

$$\overline{D}^{x_n} = \frac{1}{x_n}\int_0^{x} D dx \qquad (15\text{-}14)$$

or

$$\overline{D}^{x_n} = \frac{1}{x_n}\sum_{i=1}^{n} D_i \Delta x_i \qquad (15\text{-}14a)$$

is used in evaluating $T(p,x_n)$. In Eq. (15-14a) the reach is divided into n subreaches numbered i - 1,2,3,...,n, of length $\Delta x_i = x_i - x_{i-1}$ wherein D_i is the value of D which remains constant in a subreach, but varies from one subreach to the next. Consequently, in evaluating Eq. (15-13) when D varies along the reach, σ_p is defined by

$$\sigma_p = \sqrt{2\overline{D}^{x_n} x_n}. \qquad (15\text{-}15)$$

The transverse diffusion factor D is related to the overall transverse mixing coefficient E_z by Eq. (15-8). Assuming that E_z does not vary across the channel, Eq. (15-8) can be put in the form

$$D = \frac{E_z}{\bar{U}^A B^2} \; \frac{\overline{h_1 d^2 \bar{u}^d}^Q}{(\bar{d}^B)^2 \bar{U}^A} , \qquad (15\text{-}8a)$$

where $\bar{U}^A$ is the cross-sectional average velocity, B is the channel width at the water surface, $\bar{d}^B$ is the width-averaged depth, and

$$\overline{(\quad)}^Q = \int_0^1 (\quad)\, dp$$

denotes an average for the total flow taken with respect to the cumulative discharge. In Eq. (15-8a), the ratio $E_z/\bar{U}^A B^2$ can be interpreted as the reciprocal of a length scale for cross-sectional mixing, and the ratio $\overline{h_1 d^2 \bar{u}^d}^Q/(\bar{d}^B)^2\, \bar{U}^A$ is a dimensionless factor which varies from 1 to about 2.5, depending on channel shape, curvature and transverse distribution of flow.

Based on the results of earlier transverse mixing experiments by Yotsukura et al. (1970) and Sayre and Yeh (1973) with dye tracer in the Missouri River, Yotsukura and Sayre (1976) reported the semi-empirical relationship

$$\frac{\bar{E}_z^L}{\bar{d}^B U_*} = 0.4 \left[\frac{B}{\bar{d}^B} \frac{\bar{U}^A}{U_*} \frac{\bar{d}^B}{r_c}\right]^2 \qquad (15\text{-}16)$$

for the average value of E_z along a channel bend. In Eq. (15-16) r_c is the radius of curvature of the channel centerline, and $U_* = \sqrt{g \bar{d}^B S}$ is the shear velocity, where S is the energy slope. There is some theoretical basis for Eq. (15-16) in that Fischer (1969) derived a relationship of the form

$$\frac{E_z}{\bar{d}^B U_*} \propto \left(\frac{\bar{u}^d}{U_*} \frac{\bar{d}^B}{r_c}\right)^2 ,$$

using a radial velocity distribution function for fully-developed secondary flow in the central portion of an idealized curved channel. Assuming that $\bar{E}_z^L = E_z$, Eq. (15-16) can be substituted into Eq. (15-8a) to obtain

$$BD = 0.4 \, \frac{\overline{h_1 d^2 \bar{u}^d}^Q}{(\bar{d}^B)^2 \bar{U}^A} \, \frac{B}{\bar{d}^B} \, \frac{\bar{U}^A}{U_*} \left(\frac{\bar{d}^B}{r_c}\right)^2 . \qquad (15\text{-}17)$$

In Fig. 15-4 BD is plotted as a function of $\frac{B}{\bar{d}^B}\frac{\bar{U}^A}{U_*}(\frac{\bar{d}^B}{r_c})^2$. Data obtained in curved laboratory flumes with a uniform rectangular cross section is also shown. The reasons why the laboratory and Missouri River data do not follow the same curve is not known. Assuming that the flow across the rectangular laboratory flumes would be more evenly distributed than the flow across the river, inclusion of the quantity $\frac{\overline{h_1 d^{2-d} u^{d}}}{(\bar{d}^B)^2 \bar{U}^A}$ in the abscissa would cause a larger shift to the right for the river points than for the laboratory points. But the difference in shift would not be nearly enough to account for the 7-fold difference.

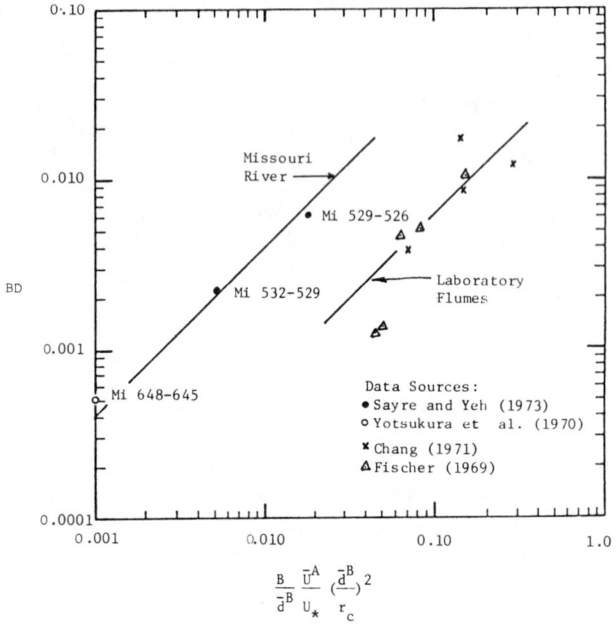

Fig. 15-4 Variation of Transverse Diffusion Factor with Bulk Channel and Flow Properties

Discussion of the parameters of the model and their relationship to plant discharge and ambient flow and channel properties will be resumed in section 15.3.

Zone of Passage

One of the standards for limiting thermal discharges into rivers that is required by several environmental agencies relates to the zone-of-passage. According to this requirement the region wherein the temperature rise exceeds some critical value T_c, usually 5°F, is restricted to a specified mixing zone that is defined in terms of either percent of cross-sectional area or percent of river discharge. Thus at any cross section, the mixing zone corresponds to either the area enclosed by, or the volume flux through, the T_c isotherm. Going downstream, as the thermal effluent mixes with the river water, the mixing zone grows, reaches a maximum, and then shrinks (assuming sufficient river flow to provide the required dilution). A convenient expression for $p_{m_{max}}$, the maximum fractional discharge passing through the mixing zone, can be obtained from Eq. (15-13a). A definition sketch for the mixing zone with respect to discharge is shown in Fig. 15-5. If, as usual, the environmental regulations restrict $p_{m_{max}}$ to 0.25, only the first term in Eq. (15-13a) need be considered. Solving for p_m, we get

$$p_m = \sqrt{2}\, \sigma_p \left[-\ln(\sigma_p \sqrt{\pi/2}\; QT_c/Q_oT_o)\right]^{\frac{1}{2}}$$

$$= 2\sqrt{DX} \left[-\ln(\sqrt{\pi DX}\; QT_c/Q_oT_o)\right]^{\frac{1}{2}}. \quad (15\text{-}18)$$

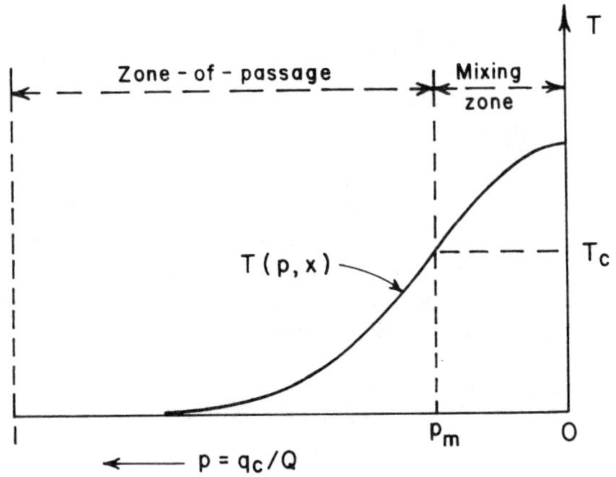

Fig. 15.5 Definition Sketch for Mixing Zone with Respect to Discharge

Setting $\frac{dp_m}{dx} = 0$ according to the classical maximization procedure of elementary calculus leads to

$$p_{m_{max}} = \underbrace{(2/\pi e)^{\frac{1}{2}}}_{0.4839} \frac{Q_o T_o}{Q T_c} . \qquad (15\text{-}19)$$

For the initial distribution specified in Eq. (15-10), the same procedure would result in somewhat larger values of $p_{m_{max}}$ that tend to increase with P, assuming that $T_o/a > T_c$. For $P \lesssim 0.10$, the increase over the value given by Eq. (15-19) is very slight. Expressions for $p_{m_{max}}$ corresponding to some other distribution (with respect to p) functions for T(p,x) are:

$$\text{Rectangular:} \quad p_{m_{max}} = \frac{Q_o T_o}{Q T_c} \qquad (15\text{-}19a)$$

$$\text{Triangular:} \quad p_{m_{max}} = \frac{1}{2} \frac{Q_o T_o}{Q T_c} \qquad (15\text{-}19b)$$

$$\text{Exponential:} \quad p_{m_{max}} = \frac{1}{e} \frac{Q_o T_o}{Q T_c} \qquad (15\text{-}19c)$$

where e is the base of the natural logarithm.

Limitations and Extensions of Model

Dunn et al. (1975) identify five basic physical processes in the dispersion of thermal plumes. They are: jet entrainment, crossflow interaction, buoyant spreading, ambient-flow mixing mechanisms, and surface heat exchange. Although the model as described in the preceding sections focuses primarily on the ambient-flow mixing mechanisms, it can be adapted within limits to account for the other mechanisms.

Jet Entrainment and Buoyant Spreading in the Near Field: The model is clearly incapable of simulating jet entrainment and near-field buoyant spreading in any detailed fashion. However, the parameters a and D_1 can be adjusted to account for the approximate combined effects of these mechanisms on the transverse distribution of depth-averaged temperature in the first subreach. Sayre (1975) and Paily (1975) applied a rudimentary form of this approach by increasing the value of D to account for jet-induced mixing when predicting thermal plumes in the Missouri River. In general, the applicability of the model varies inversely with the size of the near-field zone relative to the size of the river. In cases where the characteristics of the plant discharge, outfall structure, and ambient flow combine to produce a near-field region that occupies a substantial portion of the river channel, the model would be less appropriate than for cases where the near-field is confined to a small region near the outfall structure.

Buoyant Spreading in the Transition Zone: In the transition zone between the near and far fields, sometimes called the intermediate field, buoyant spreading occurs when there is a significant residual density difference between the ambient and plume water following completion of the jet-entrainment stage. This can be accounted for by increasing the value of D_2 in the second subreach. According to results obtained by Yeh (1974) in laboratory flume experiments, the total excess variance (in the z domain) attributable to buoyant surface spreading is on the order of

$$\Delta\sigma_z^2 \simeq 0.5\ bd\left(\frac{\Delta\rho}{\rho S}\right)^{1.5} \qquad (15\text{-}20)$$

for $b/d \gtrsim 1$, where, for our purposes, $b = aQ_o/d\bar{u}^d$ is the effective source width following initial dilution in the jet-entrainment stage, ρ is the mass density of the ambient river water, and $\Delta\rho$ is a representative residual density difference between the ambient and plume water following the initial dilution. For $\Delta\rho/\rho S > 9$, buoyancy-driven secondary circulation may be strong enough to drive the plume away from the near shore. Yeh also found the length of the reach over which buoyant surface spreading adds significantly to the transverse spreading due to turbulent diffusion to be equal to or less than

$$L_B = \frac{\bar{d}^2\bar{u}^d}{E_z} \ . \qquad (15\text{-}21)$$

In natural rivers, factors such as bend-generated secondary circulation and transverse bottom slope may to a substantial degree either counteract or supplement the buoyant spreading. These factors were not present in the experiments on which Eqs. (15-20) and (15-21) are based.

Crossflow Interaction: The mechanism whereby a jet is deflected downstream from its initial direction by the ambient flow, as its cross-channel momentum is diffused and dissipated, is not included in the model as presented herein. However, the model can be made to account for crossflow interaction by incorporating an appropriate equation for computing the trajectory of the plume centerline. Sayre (1975) used modified forms of the trajectory equations of Carter et al. (1973) in conjunction with a diffusion model similar to Eq. (15-13a) for predicting thermal plumes downstream from the Iatan Station in the Missouri River. The x axis was taken as the predicted plume centerline, and the river banks were treated as reflecting boundaries.

For the shore-attached thermal plumes considered herein, the near shore corresponds to the plume centerline so that trajectory equations need not be incorporated. According to the observations of Jirka et al. (1975) the plume resulting from a surface jet discharged at a 90° angle into the ambient flow is attached to the near shore if

$$\frac{U_a}{U_o} > 0.05\left(\frac{d}{H_{max}}\right)^{3/2} \qquad (15\text{-}22)$$

where

$$H_{max} = 0.35\ \mathbb{F}_{DO}\left(\frac{H_o}{B_o}\right)^{1/4}\sqrt{H_o B_o} \qquad (15\text{-}23)$$

is the estimated maximum depth beneath the water surface to which the plume penetrates; U_a is the average ambient velocity in the region of the river channel near the discharge channel; U_o is the average velocity of the discharge-channel flow where it enters the river; and H_o and B_o are respectively the depth and width of the discharge channel, and $\mathbb{F}_{DO} = U_o/\sqrt{gH_o\Delta\rho_o/\rho}$ is the densimetric Froude number at the point of discharge. In applying Eq. (15-22), H_{max} is assumed to be equal to d, a representative local river depth in the vicinity of the outfall, if Eq. (15-23) indicates $H_{max} > d$.

<u>Surface Heat Exchange</u>: In most situations where application of the present model is appropriate, i.e., in the zone downstream from the near field where the plume is mixing across the river, surface heat exchange need not be considered. Computations by Paily (1975), Sayre (1975), and Sayre and Yeh (1973), for reaches up to ten miles in length downstream from various Missouri River power stations, indicate that the reduction in computed temperature rises due to surface heat exchange rarely exceeds 10 percent.

However, under extreme conditions of a cool, dry, windy day, together with a high ambient river temperature, all combined with a low river discharge, the reduction at a downstream distance of ten miles could be as much as 30 percent. Thus there are situations when it may be advisable to take surface heat loss into account. A simple approximate procedure for computing heat loss for the case of a steady waste heat discharge $Q_o T_o$ results from the addition of a heat sink term to Eq. (15-7), which then becomes

$$\frac{\partial T}{\partial x} = D \frac{\partial^2 T}{\partial p^2} - \frac{K}{\bar{d}\bar{U}} T \tag{15-24}$$

where K is a surface exchange coefficient for excess heat, and $\bar{d}$ and $\bar{U}$ are respectively the average depth and average velocity along the path of the plume (not in general equal to the cross-sectional average depth and velocity). The solution of Eq. (15-24) is equal to the solution of Eq. (15-7), for the same initial conditions, multiplied by the decay factor $\exp\left[-\frac{K}{\bar{d}\bar{U}} x\right]$, that is,

$$T(p,x) = T'(p,x) \exp\left[-\frac{K}{\bar{d}\bar{U}} x\right] \tag{15-25}$$

where T'(p,x) is given by Eq. (15-13). The approximate analysis leading to Eq. (15-25) does not take into account the variation of depth and velocity across and along the path of the plume. It implicitly assumes that the time-of-travel from 0 to x is the same for all water particles. Consequently it tends to somewhat underpredict the heat loss, particularly towards the outer edge of the plume.

A more exact procedure, which takes into account the variation of d and $\bar{U}^d$ by dividing the channel into a number of subchannels as well as subreaches, has been presented by Jackman and Yotsukura (1977).

15.3 APPLICATION TO THERMAL PLUMES IN THE MISSOURI RIVER

A detailed description of the application of the model to reaches of the Missouri River downstream from the Fort Calhoun and Cooper Nuclear Power Stations, including comparisons with twelve sets of thermal plume data, has been reported by Caro-Cordero and Sayre (1977). This section summarizes that report, with attention focused on, but not limited to, the Cooper Station plumes and study reach.

Representation of Channel and Flow

A map of the Cooper Station study reach is shown in Fig. 15-6. It is limited to the 4,000-ft length for which thermal plume data is available. The channel is constrained by a system of rock revetments, dikes, and jetties to a width of

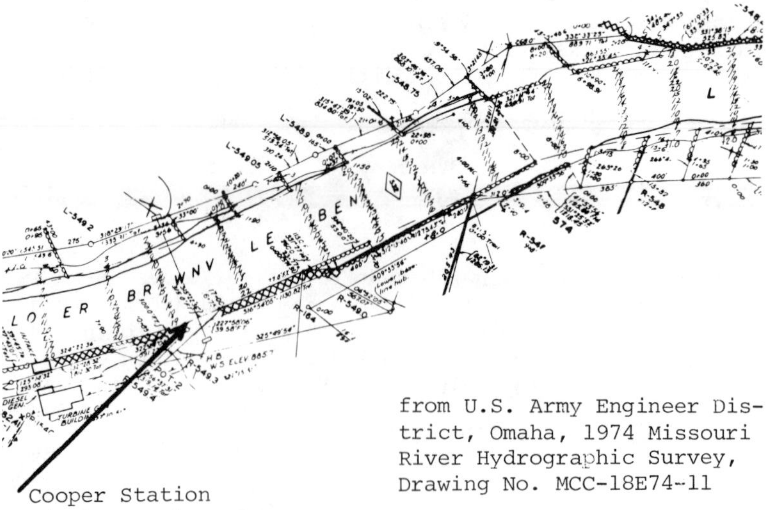

Fig. 15-6. Cooper Station Study Reach

about 700 feet, and an alignment consisting of the transition between two smooth bends. The bottom consists predominantly of fine sand. The slope of the channel is about 0.00021. During the navigation season, which lasts from April through November, the river discharge is maintained at not less than about 35,000 cfs. During the remaining months, the discharge may drop to considerably lower levels, particularly during ice jams. In an extensively studied reach near Omaha, the U.S. Army Engineer District, Omaha (1969) found that a bed configuration consisting of large dunes and bars is typical for the spring and summer months. This generally gives way to a much smoother plane-bed configuration during the fall. Depths of flow vary considerably with discharge, bed configuration, and from place to place in the channel. For a river discharge of 35,000 cfs, average and thalweg channel depths are approximately 13 and 20 feet, respectively, in the Cooper Station study reach.

Taking full advantage of Eq. (15-13) to model thermal plumes requires a detailed representation of channel and ambient flow properties, and how they vary across and along the study reach at the river discharges for which plumes are to be modeled. Sufficiently detailed information on channel geometry is available from Corps of Engineers hydrographic survey sheets. As shown in Fig. 15-6, they provide data on channel alignment and sets of depth soundings taken across the river at approximately 500-ft longitudinal intervals. Fig. 15-7, which is based on hydrographic survey data taken at approximately the same cross section of the Fort Calhoun study reach in 1963, 1967, and 1972, illustrates how the cross-sectional shape can change with time, and for different discharges.

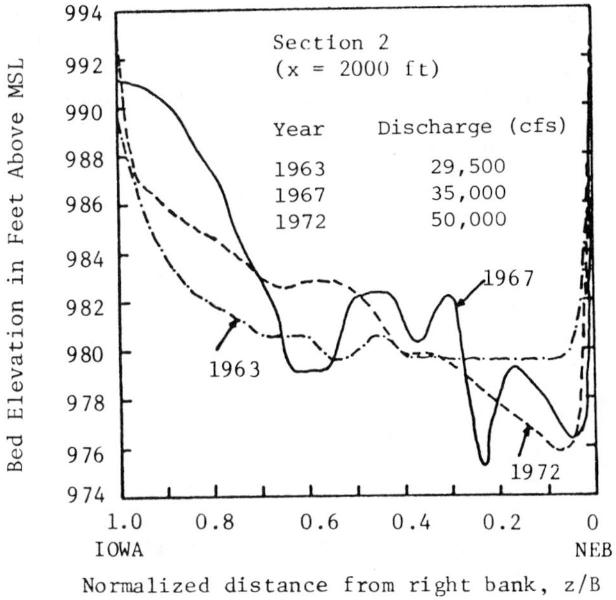

Fig. 15-7. Comparison of Cross-Sectional Shapes in the Fort Calhoun Study Reach for Different Hydrographic Surveys

For the Cooper Station study reach, hydrographic survey data for 1974, as well as for the other three years, was examined. In addition, the NALCO report (Paily, 1976) from which the thermal plume data for Cooper Station was obtained, contained limited depth-sounding data for the cross sections where temperature measurements were obtained. This data was helpful because the water depths were measured simultaneously with the temperatures; however, depths were not measured at enough points to adequately define the cross-sectional shapes. For this reason, and also because the 1974 hydrographic survey data tended to agree better with the NALCO data than did the hydrographic survey data for the other years, the 1974 hydrographic survey data was used to represent the river channel in modeling the thermal plumes from the Cooper Station.

The river discharge at the time of 1974 survey was 37,500 cfs. The stage-discharge relationship in Fig. 15-8a was used to adjust the water surface level either upward or downward for the cases in which the river discharge differed appreciably from this amount. Since there is no gaging station in the Cooper Station study reach, the stage-discharge curves for the nearest upstream and downstream USGS gaging stations at Nebraska City (Mile 562.6) and Rulo (Mile 498.0) were used to construct the curve for the Cooper site. Except for a displacement in elevation, the curves for Nebraska City and Rulo were found to be virtually identical in the range from 20,000 to 60,000 cfs. Therefore it was assumed that the shape of the curve would be the same for the Cooper site also. The stage-discharge combination observed in the 1974 hydrographic survey was used to fix the elevation of the stage-discharge curve for the Cooper site.

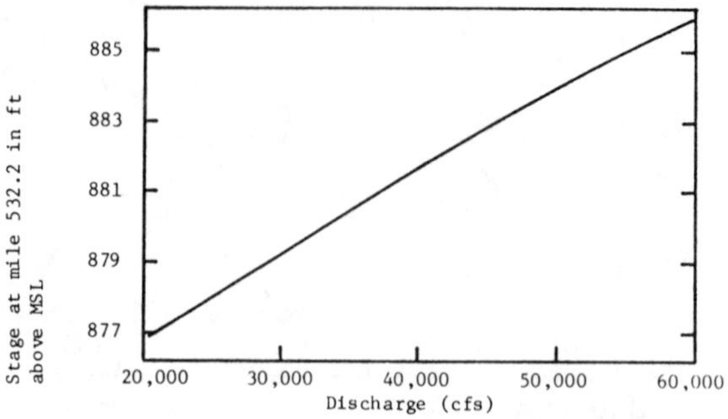

(a) Approximate stage-discharge relationship at Mile 532.2

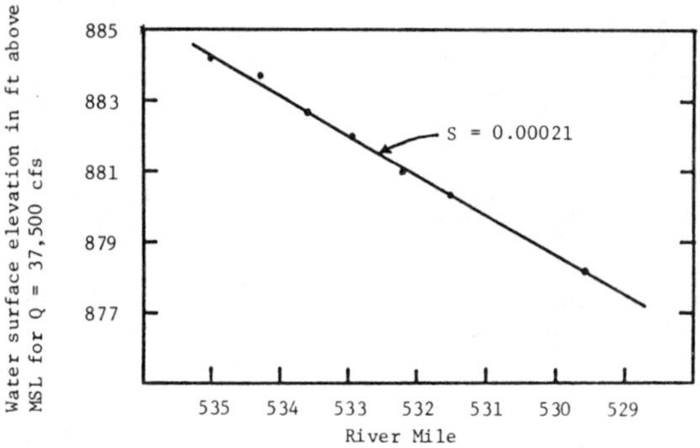

(b) Water surface profile according to 1974 hydrographic survey

Fig. 15-8. Stage Information for Cooper Station Study Reach

The Cooper Station study reach is divided into three subreaches, numbered from 1 to 3, terminating at control sections numbered correspondingly, where the temperatures were measured. The location of the control sections, both in river miles and distances downstream from the station outfall, and data describing the average cross-sectional properties and flow conditions for each subreach in both the Fort Calhoun and

Cooper Station study reaches are listed in Table 15-1. These properties were computed for a composite cross section obtained by averaging the depth-sounding data for the upstream end, middle, and downstream end of each subreach, giving double weight to the set in the middle of the subreach. Transverse bed profiles obtained by this method for the Cooper Station study reach are shown in Fig. 15-9.

Table 15-1. Background Data Describing Average Cross-Sectional Properties of Subreaches

River Discharge	Subreach Number	Downstream Control Section: River Mile	Downstream Control Section: Distance from Outfall	Channel Width	Average Depth	Average Velocity	Shear Velocity	Radius of Bend
Q	i		x	B	$\bar{d}^B$	$\bar{U}^A$	U_*	r_c
cfs		mile	ft	ft	ft	ft/sec	ft/sec	ft
(1)	(2)	(3)	(4)	(5)	(6)	(7)	(8)	(9)
A. Fort Calhoun Station								
35,000	1	645.7	1,000	700	9.87	5.07	0.246	12,000
	2	645.5	2,000	735	10.24	4.65	0.250	12,000
	3	645.3	3,000	690	9.18	5.53	0.237	12,000
	4	645.1	4,000	680	8.71	5.91	0.231	12,000
	5	644.9	5,000	700	10.21	4.90	0.250	12,000
	Average			701	9.64	5.18	0.242	
18,500	1	645.7	1,000	640	6.71	4.31	0.203	12,000
	2	645.5	2,000	650	7.37	3.86	0.212	12,000
	3	645.3	3,000	610	6.10	4.97	0.193	12,000
	4	645.1	4,000	620	5.37	5.56	0.181	12,000
	5	644.9	5,000	630	7.14	4.11	0.209	12,000
	Average			630	6.54	4.56	0.200	
B. Cooper Station								
60,000	1	532.20	800	700	17.60	4.87	0.345	7,700
	2	531.93	2,200	680	18.43	4.79	0.353	7,700
	3	531.59	4,000	670	18.19	4.92	0.351	6,400
	Average			680	18.16	4.86	0.351	
37,500	1	532.20	800	680	13.19	4.18	0.299	7,700
	2	531.93	2,200	670	13.79	4.06	0.306	7,700
	3	531.59	4,000	660	13.53	4.20	0.303	6.400
	Average			668	13.55	4.14	0.303	
22,000	1	532.20	800	660	9.82	3.39	0.258	7,700
	2	531.93	2,200	660	10.23	3.26	0.263	7,700
	3	531.59	4,000	640	9.98	3.44	0.260	6,400
	Average			651	10.04	3.37	0.261	

Note: Averages are weighted by the lengths of the subreaches.

To transfer the temperature rise data back and forth between the z and p domains, the transverse distribution of the normalized cumulative discharge $p = q_c/Q$ is required at the control sections. The cumulative discharge

$$q_c = \int_0^z q \, dz \tag{15-26}$$

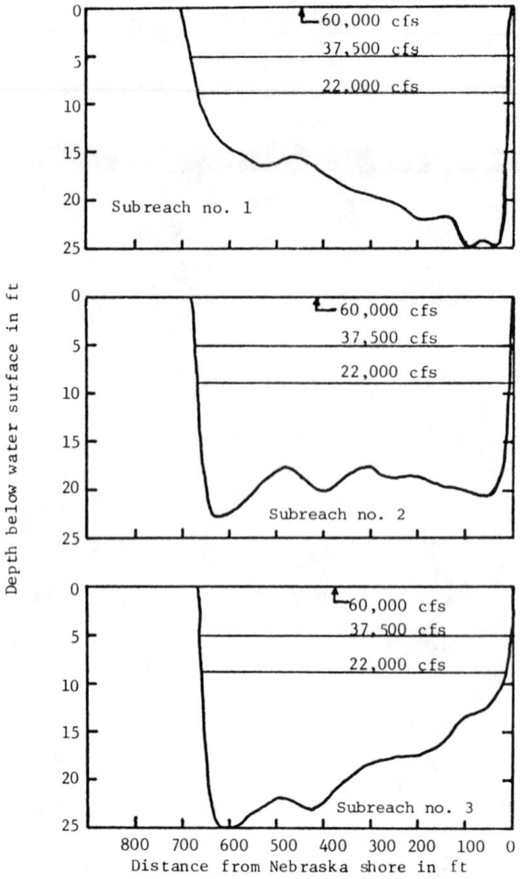

Fig. 15-9. Transverse Bed Profiles for Cooper Station Study Reach

is estimated by numerically integrating the unit discharge

$$q = d\bar{u}^{d} \tag{15-27}$$

synthesized from the variation of d across the section by means of the relationship

$$\frac{q}{\bar{q}^{B}} = b_{o}\left(\frac{d}{\bar{d}^{B}}\right)^{b_{1}} \tag{15-28}$$

which is based on the hypothesis that the transverse distribution of q depends mainly on that of d. Fig. 15-10 shows $q/\bar{q}^B$, determined from sets of velocity measurements taken at three cross sections downstream from Cooper Station, plotted against corresponding values of $d/\bar{d}^B$. The data indicate that $b_o \simeq 1$ and $b_1 \simeq 5/3$, which is consistent with the Manning formula. When using Eq. (15-28) to construct p vs z/B curves at a cross section, the value of b_o must be slightly reduced to make $q_c = Q$ when $z = B$. Although not as precise as methods based on actual velocity-distribution measurements, this method of synthesizing the transverse distribution of q from depth-sounding data has been tested by Sium (1975) against measured data from several rivers, including the Missouri, and found to be reasonably reliable, especially for large alluvial streams.

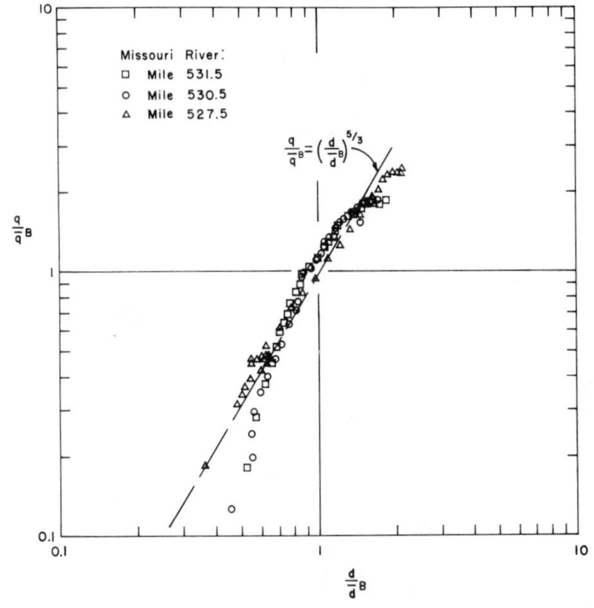

Fig. 15-10. Relative Unit Discharge vs. Relative Depth Relationship for Synthesizing Transverse Distribution of Unit Discharge

Fig. 15-11 illustrates the steps followed in calculating the cumulative discharge distribution according to the above procedure, and also compares synthesized distributions of $q/\bar{q}^B$ and p with those obtained from velocity-distribution measurements. Normalized cumulative discharge curves obtained

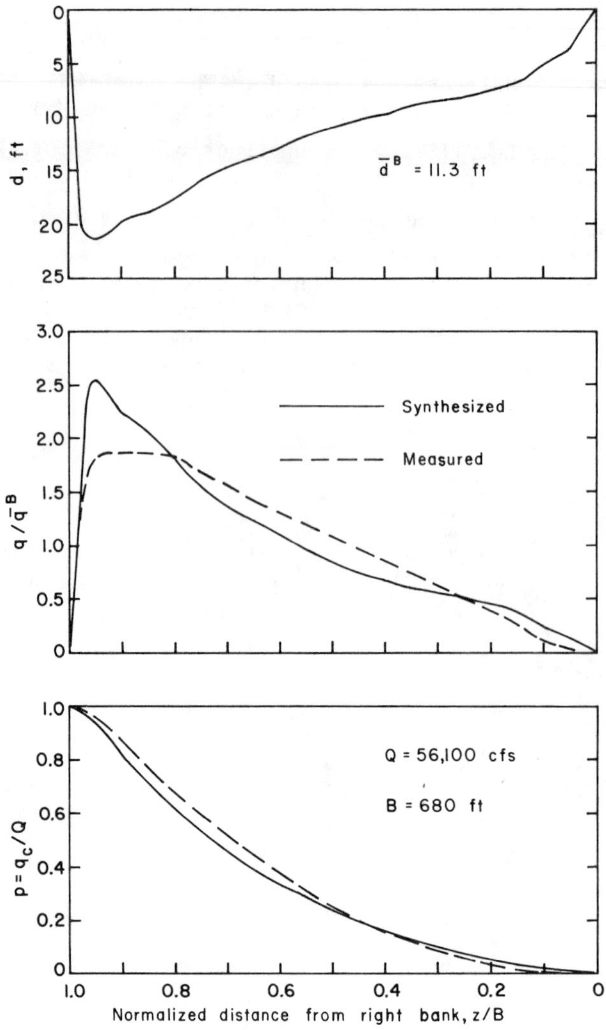

Fig. 15-11. Transverse Distribution of Depth and Normalized Unit and Cumulative Discharge at Mile 531.5, October 11, 1972

by the synthesizing procedure (with $b_1 = 5/3$) for the three control sections in the Cooper Station study reach are shown in Fig. 15-12 for river discharges of 60,000 and 22,000 cfs. On the theory that the transverse distribution of flow has some inertia and does not immediately respond to changes in cross-sectional shape, these curves are based on the average of the $q/\bar{q}^B$ vs z/B distributions for the indicated cross section and the section 500 ft upstream.

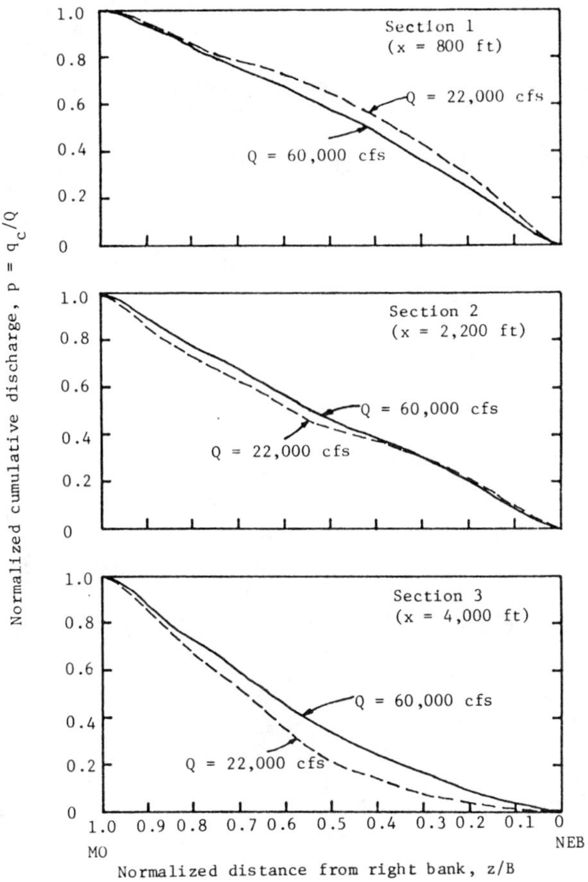

Fig. 15-12. Normalized Cumulative Discharge at Control Sections in Cooper Station Study Reach Synthesized from 1974 Hydrographic Survey

In all of the flow characterization procedures described in this section it is implicitly assumed that no significant change in cross-sectional shape accompanies changes in river discharge and stage, and also that there is a unique stage-discharge relationship. Neither of these assumptions is very good for alluvial channels. With regard to the first assumption, it has already been shown in Fig. 15-7 that the scour and deposition patterns which determine cross-sectional shape are subject to some variation. The relative contributions to these variations from changes in discharge, seasonal effects, randomness, and long-terms trends, are not known. With regard to the second assumption, seasonal shifts in the stage-discharge relation that are evidently associated with temperature-related effects on the bed configuration are known to occur in the Missouri River. Unfortunately, alluvial channel phenomena are not sufficiently well understood to quantitatively evaluate or significantly improve on these assumptions. In the present case, however, the channel shape and transverse flow distribution are certain to retain the general characteristics of larger depths and velocities along the outside of bends and smaller depths and velocities on the inside. Crossings between bends are more apt to be unstable. Although lacking in precision, the characterization of the flow and channel described herein is certainly much better than the common assumption of a straight rectangular channel.

Plant Discharge and Thermal Plume Data

The Cooper and Fort Calhoun Stations both have once-through cooling systems. The outfall for Cooper Station is located at river mile 532.35 and consists of a 120-ft-wide earthen discharge canal oriented at an angle of 24° with the ambient flow. The Fort Calhoun outfall is located at river mile 645.9 and consists of a rectangular concrete discharge tunnel, terminating in a 30-ft-wide by 14.2-ft high opening, oriented at an angle of 57° with the ambient flow. Although the top of the tunnel is submerged when the river discharge exceeds about 25,000 cfs, it is considered as a surface discharge herein because it is not submerged by more than about 2.6 ft at the maximum river discharge considered. At full plant load of 778 MW electrical, the cooling water discharge from the Cooper Station is 1455 cfs with a termperature rise of 18°F. For the 457 MW electrical Fort Calhoun Station the corresponding discharge and temperature rise are 802 cfs and 18°F respectively. Table 15-2 summarizes the plant discharge and background river conditions which existed at the times of the thermal plume surveys.

Samples of thermal plume data are shown in Fig. 15-13 and Fig. 15-14. The Cooper Station data was obtained from a NALCO Environmental Sciences report by Paily (1976), and the Fort Calhoun data from the Omaha Public Power District. The Cooper Station data included water depth at the locations where vertical temperature profiles were measured; the Fort Calhoun data--consisting of surface, mid-depth, and bottom temperature rises--included no information on water depth. The Cooper Station data included enough temperature measurements well away from the plume to adequately define ambient temperature; the Fort Calhoun ambient temperature was determined by upstream monitoring.

None of the twelve sets of thermal plume data showed any evidence of appreciable vertical temperature stratification

Table 15-2. Background River and Plant Discharge Conditons

Station	Date	River		Plant Outfall						Dimensionless Parameters			
		Q	T_a	Q_o	T_o	H_o	B_o	U_o	α	$\frac{U_o}{U_a}$	$\frac{B_o}{H_o}$	$\frac{H_o}{d}$	F_{Do}
		cfs	°F	cfs	°F	ft	ft	ft/sec					
Fort Calhoun	05/03/74	33,000	58.0	802	18.0	14.2	30	1.88	57°	0.33	2.11	0.96	1.97
	08/08/74	36,000	74.5	802	13.3	14.2	30	1.88	57°	0.34	2.11	0.93	1.94
	09/06/74	35,000	67.4	802	15.3	14.2	30	1.88	57°	0.31	2.11	0.95	1.92
	11/25/74	33,500	43.1	802	14.2	14.2	30	1.88	57°	0.34	2.11	0.94	3.31
	12/04/74	18,500	38.5	802	16.8	12.9	30	2.07	57°	0.36	2.33	1.11	4.21
Cooper	08/22/74	36,100	77.6	1088	16.7	5.4	120	1.70	24°	0.45	22.2	0.29	2.43
	10/17/74	37,100	55.1	1116	16.7	5.4	120	1.70	24°	0.44	22.2	0.28	3.17
	01/23/75	21,000	32.2	980	24.6	3.9	120	2.10	24°	0.72	30.8	0.23	7.86
	01/28/75	23,400	32.0	686	36.9	3.8	120	1.50	24°	0.48	31.6	0.26	3.25
	03/20/75	38,800	33.5	744	30.6	7.4	120	0.80	24°	0.20	16.2	0.38	1.46
	08/01/75	63,400	80.4	1470	15.0	5.4	120	2.30	24°	0.53	22.2	0.21	3.43
	12/05/75	57,300	35.6	960	24.1	2.5	120	3.20	24°	0.78	48.0	0.10	11.87

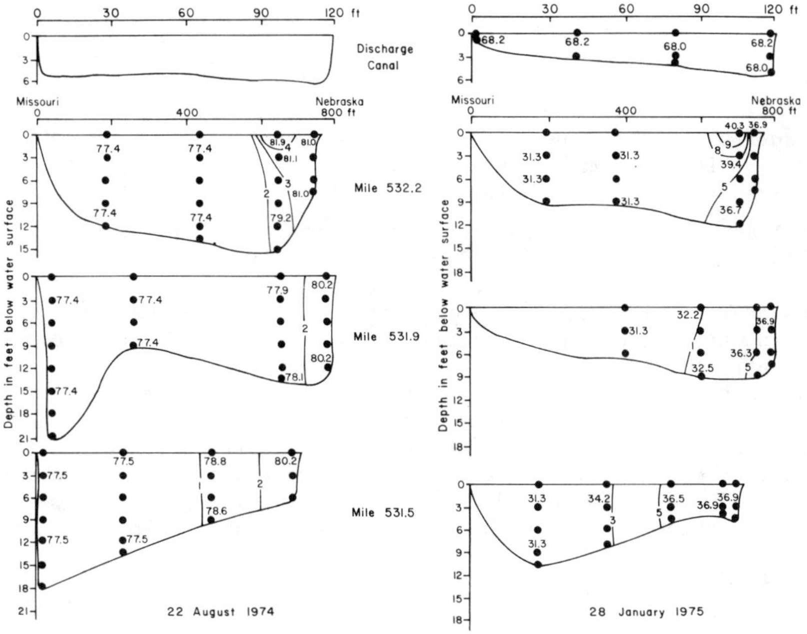

Fig. 15-13. Cross-Sectional Temperature Distributions from Thermal Plume Surveys; Cooper Station Study Reach

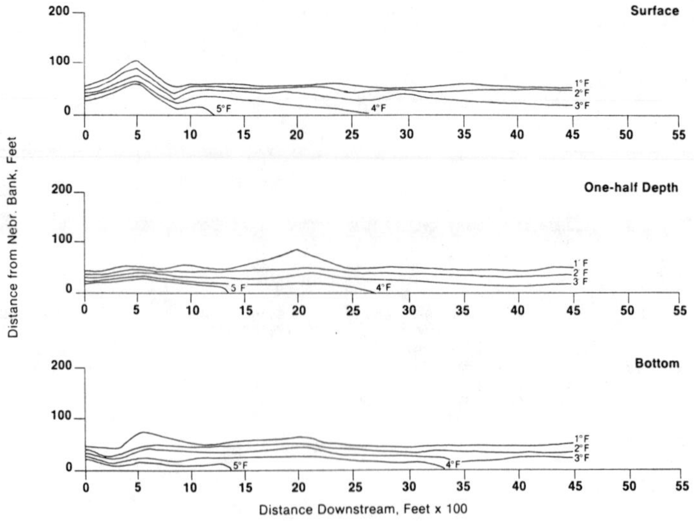

Fig. 15-14. Temperature-Rise Isotherms from Triple-Depth Thermal Plume Surveys; Fort Calhoun Study Reach, November 25, 1974

occurring farther than about 1,000 ft downstream from the outfalls. In both study reaches the plume was definitely shore attached in all cases. The value of U_a/U_o was in every case at least 15 times as great as the critical value for shore attachment computed by Eq. (15-22).

The thermal plume data was prepared for comparison with temperature-rise distributions computed by Eq. (15-13) by: (1) plotting the measured temperature rises as a function of dimensionless distance z/B from the Nebraska shore for each of the control sections and averaging over the depth (giving double weight to the mid-depth temperatures); (2) mapping the measured depth-averaged temperature-rise distributions from the z/B to the p domain by means of the appropriate synthesized p vs z/B distributions; and (3) adjusting the measured T vs p distribution to satisfy the heat-balance relationship

$$Q_oT_o = \int_0^B d\bar{u}^d\ T dz = Q \int_0^1 T dp. \qquad (15\text{-}29)$$

Use of Eq. (15-29) involves the assumption that the percentage of excess heat transferred to the atmosphere over the length of the study reaches is negligible. This assumption is borne out by heat-loss computations using Eq. (15-25), which indicate a heat loss of from one to two percent in both the Cooper and Fort Calhoun study reaches, for a set of meteorological conditions that would produce a much greater than average rate of heat loss.

Of the terms in Eq. (15-29), the transverse distributions of $d\bar{u}^{d}$--i.e., the synthesized p vs z/B relationships-- were considered to be the least reliable. Therefore, the measured T vs p distributions were adjusted to satisfy Eq. (15-29) by multiplying the p values by an adjustment factor (A.F.), defined as

$$\text{A.F.} = \frac{Q_o T_o}{Q \int_0^1 T dp} \quad . \tag{15-30}$$

Values of A.F. were computed for each control section in each survey. They ranged from about 0.2 to about 2.5, and averaged 1.01 and 1.28 respectively for the Cooper and Fort Calhoun study reaches. No consistent trends were evident.

Adjusted p vs z/B distributions are compared with the original synthesized p vs z/B distributions for three of the Cooper Station surveys in Fig. 15-15. For the most part, differences are not much larger than differences between distributions synthesized from different hydrographic surveys for the same cross section. Large deviations, where they occur, are likely due mainly to actual changes in channel cross-sectional shape. In particular, the Cooper Station study reach is located along a reach of the river where the direction of channel curvature reverses. Cross-sectional shapes tend to be unstable in such reaches.

Initial Dilution

Knowledge of the initial dilution factor $a = PQ/Q_o = T_o/T_I$, defined in Eq. (15-10) and Fig. 15-2, is required to specify the initial condition for the model. Values of P were estimated from the adjusted T vs p distribution curves for the control sections at x = 0 as illustrated in Fig. 15-16, where P is the width of the best-fit uniform distribution having the same enclosed area as the curve. Values of P determined from the data in this manner are listed in Table 15-3, together with the corresponding a values, for both the Cooper and Fort Calhoun plume surveys.

Attempts were made to find a general relationship governing the value of a, first by dimensional analysis, and then by correlating the a values in Table 15-3 with existing near-field relationships for predicting the reduction of dimensionless temperature rise, T_s/T_o, along the centerline of thermal plumes resulting from surface jets discharging into an ambient cross flow. A relationship for a obtained by dimensional analysis is

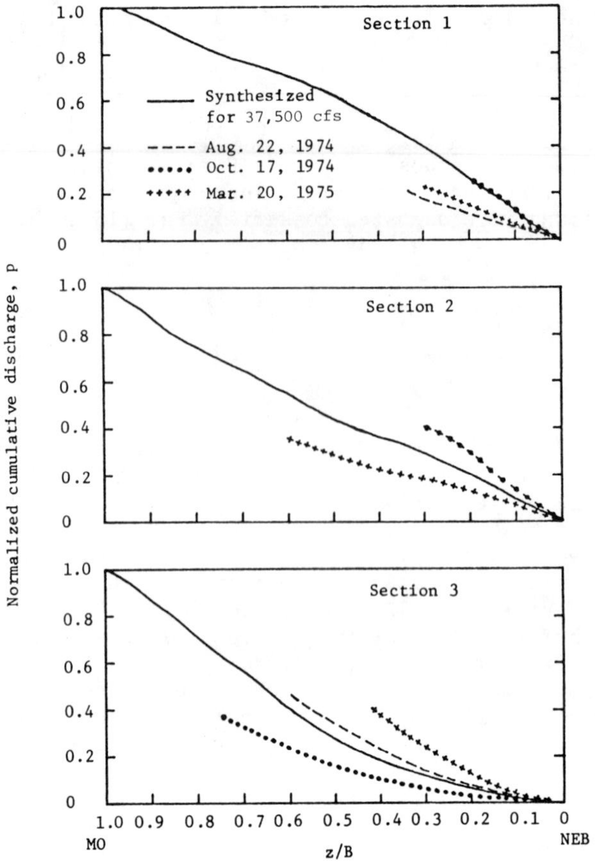

Fig. 15-15. Transverse Distribution of Cumulative Discharge in Cooper Station Study Reach, Showing Adjustments Required by Eq. (15-30), Q = 37,500 cfs

$$a = f\left(\frac{U_o}{U_a},\ \mathbb{F}_{Do},\ \frac{B_o}{H_o},\ \frac{H_o}{d},\ \frac{Q_o}{Q},\ \alpha\right) . \qquad (15\text{-}31)$$

Values of all the dimensionless parameters on the righthand side of Eq. (15-31) are listed in Table 15-2, except for Q_o/Q which is listed in Table 15-3. Comparison of a with these parameters indicates only that: (1) the significantly larger a values for the Fort Calhoun in comparison with the Cooper Station data correlates with the larger discharge angle α, the larger depth ratio H_o/d, and the smaller aspect ratio B_o/H_o; (2) a decreases approximately linearly with increasing Q_o/Q for

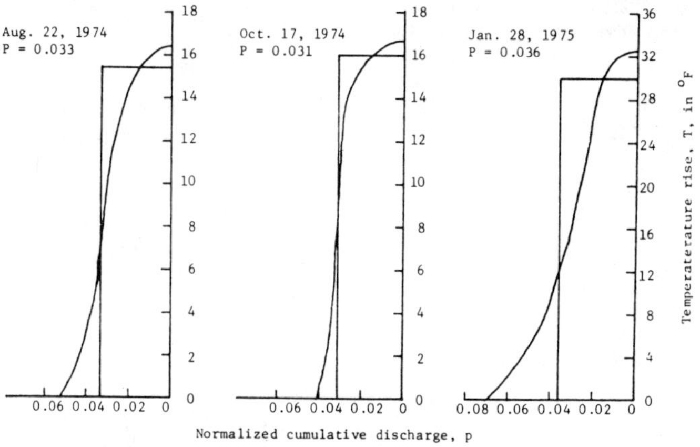

Fig. 15-16. Adjusted Measured T vs p Distributions at Control Section 0 (x = 0), and Fitted Initial Uniform Source Distributions, Cooper Station

Table 15-3. Values of $\bar{a}$ and P Determined from Adjusted T vs p Distribution Curves

Station	Date	P	a	Q_o/Q
Fort Calhoun	05/03/74	0.093	3.83	0.024
	08/08/74	0.054	2.42	0.022
	09/06/74	0.073	3.20	0.023
	11/25/74	0.067	2.79	0.024
	Average	0.072	3.06	
	12/04/74	0.116	2.68	0.043 (1)
Cooper	08/22/74	0.033	1.08	0.030
	10/17/74	0.031	1.04	0.030
	01/23/75	--(2)	--(2)	0.047
	01/28/75	0.036	1.22	0.029
	03/20/75	0.031	1.61	0.019
	08/01/75	0.037	1.60	0.023
	12/05/75	0.033	1.99	0.017
	Average	0.034	1.42	

(1) Thermal plume observed on December 4, 1974 was excluded from the average because it is the only case for which Q differs appreciably from 35,000 cfs.

(2) Available data does not provide an adequate definition of the transverse distribution of T at x = 0.

the Cooper data, which is consistent with the nearly constant P values for Cooper Station in Table 15-3, but no such trend is evident for the Fort Calhoun data since the value of Q_o/Q is nearly the same for four out of the five plumes; and (3) $\underline{a}$ does not vary consistently with either the velocity ratio U_o/U_a, or the densimetric Froude number of the discharge channel $\mathbb{F}_{DO}$.

It is reasonable to expect that $\underline{a}$ should correlate with existing near-field relationships for T_s/T_o, where T_s is the temperature rise at a distance s along the plume centerline from the discharge channel. Both $\underline{a}$ and the reciprocal of T_s/T_o are dilutions, and consequently should be closely related functions of the same dimensionless parameters. This correlation was attempted with values of T_o/T_s computed by the formulas of Shirazi and Davis (1974), Carter and Regier (1974), and Liong (1977), for plant discharge and ambient flow conditions existing at the time of the plume surveys. Except for Liong's formula

$$\frac{T_s}{T_o} = 1.9\ \mathbb{F}_{DO}^{0.06}\left(\frac{H_o}{B_o}\right)^{-0.19}\left(\frac{s}{\sqrt{B_o H_o}}\right)^{-0.64} \qquad (15\text{-}32)$$

which correctly indicates larger $\underline{a}$ values for the Fort Calhoun than for the Cooper Station plumes, the results were not encouraging. This may be due to significant differences between the Missouri River plumes and the data used for developing the formulas, as regards ranges of U_o/U_a and B_o/H_o values. In particular, $U_o/U_a < 1$ for all of the river data, but $U_o/U_a > 1$ for all the data on which the formulas are based. It appears that cases for $U_o/U_a < 1$ have been largely ignored in previous experimental and analytical investigations.

Determination of Transverse Diffusion Factor from Plume Data

At this point, it is assumed that T_o, Q_o and Q are all known, that the river has been represented as a set of subreaches in the x-p domain, and that appropriate values of P or $\underline{a}$ have been selected. It remains only to determine the values of the transverse diffusion factor D_i for the subreaches and the reach-averaged values $\overline{D}^{x_n}$, before the model, Eq. (15-13), can be used to compute thermal plumes. Relationships such as Fig. 15-4 need further verification and refinement before they can be used to predict D with confidence. Achievement of this goal will require the determination of D from many sets of thermal plume data, obtained for a wide range of river flow and geometry and plant discharge conditions. This section is concerned with the application of a stepwise, trial and error, least-squares procedure to three sets of thermal plume data.

Going downstream from one subreach to the next, the procedure for the n-th subreach consists in essence of: selecting trial values of D_n, computing an average value $\bar{D}^{x_n}$ for the reach 0 to x_n for each trial value of D_n, computing T vs p distribution curves at the downstream end of the subreach for each trial value of D_n; comparing the computed distributions with a measured distribution; and finally, determing the best-fit value for D_n. Distributions of $T(p,x_n)$ are computed directly by Eq. (15-13) with σ_p evaluated by Eq. (15-15). Eq. (15-14a) provides the necessary linkage between successive subreaches so that the mixing which has occurred in all n subreaches is properly accounted for.

A more precise description of the procedure is as follows:
For the first subreach (n=1):

1. Using Fig. 15-4 as a guide, select several (about 5) trial values of D_1 for the first subreach that bracket the value of D indicated by Fig. 15-4.

2. Using Eq. (15-13) with $\sigma_p = \sqrt{2D_1x_1}$, compute distribution of $T(p,x_1)$ with respect to p at downstream end of first subreach $(x=x_1)$ for each trial value of D_1. Compare computed and measured distributions and determine the best-fit value of D_1 using least-squares or equivalent criterion.

For succeeding subreaches $(n\geq 2)$:

1. Select trial values of D_n for n-th subreach, using values of D_i determined in subreaches upstream in addition to Fig. 15-4 as a guide.

2. For each trial value of D_n, compute the average value of D for the reach from 0 to x_n according to:

$$\bar{D}^{x_n} = \frac{1}{x_n} \sum_{i=1}^{n} D_i \Delta x_i \qquad (15\text{-}14a)$$

where $\Delta x_i = x_i - x_{i-1}$ and $x_0 = 0$.

3. Compute distributions of $T(p,x_n)$ with respect to p at downstream end of n-th subreach $(x=x_n)$, using Eqs. (15-13) and (15-15), for each trial value of D_n. Compare with measured distribution as in step 2 for first subreach, and determine best-fit value of D_n.

4. Continue to next subreach using updated feedback to assist in selecting trial values of D_{n+1}.

A FORTRAN program which includes this procedure, essentially as outlined above, is listed in Caro-Cordero and Sayre (1977). One procedural difference is that they used a virtual source method instead of Eq. (15-14a) to provide continuity between subreaches. However, the two methods lead to identical results.

Table 15-4 shows the results obtained in applying the procedure to two thermal plumes in the Cooper Station study reach, and one in the Fort Calhoun study reach. The average P values listed in Table 15-3 were used to define the initial conditions. Two sets of results, labeled (a) and (b), or (a) and (c), are listed for each plume. In sets (a), the values of D_i are the best-fit values for each subreach of each plume. They are regarded as 'local' values. Some combining is incorporated into sets (b) and (c). For the Cooper Station plumes, the results for sets (b) are the best-fit common values of D_i for sets grouped according to river discharge. The 8/22/74 plume, for which Q = 36,100 cfs, is grouped together with the 10/17/74 and 3/20/75 plumes, for which Q = 37,100 and 38,800 cfs respectively; the 1/28/75 plume for which Q = 23,400 cfs, is grouped together with the 1/23/75 plume for which Q = 21,000 cfs. For the Fort Calhoun plume, the results for set (c) represent the best-fit single value for the entire study reach. Because of the strong dependence of D_1 on P (or $\bar{a}$), the first subreach is not included in set (c). The set (c) method of combining subreaches was not attempted for the Cooper Station plumes, because the decrease in D going from subreach 2 to subreach 3 is too consistent and pronounced to ignore. In contrast, the variation in D going from one subreach to the next appears to be entirely random when comparing the Fort Calhoun plumes.

Comparison of Computed and Measured Temperature-Rise Distributions

Comparisons between the best-fit computed and measured T vs p distributions are shown in Figs. 15-17 and 15-18 for the plumes and conditions listed in Table 15-4. The designations (a), (b), and (c) for the computed distribution curves correspond to the sets of D values in Table 15-4.

For the most part there is good agreement between the computed and measured distributions, at least when the measured distributions are adjusted by Eq. (15-30) to satisfy the heat-balance relationship given in Eq. (15-29). If velocity measurements and depth soundings are obtained concurrently with the temperature measurements, this adjustment procedure should not be necessary.

Table 15-4. Best-Fit Values of Transverse Diffusion Factor D for Three Thermal Plume Surveys

Station	Date	Q cfs	P	a	i	x_i ft	$D_i \times 10^6$ ft^{-1}	$\bar{D}^{x_n} \times 10^6$ ft^{-1}
(1)	(2)	(3)	(4)	(5)	(6)	(7)	(8)	(9)
Cooper	8/22/74	36,100	0.034	1.13	1	800	5.1	5.1
		(a)			2	2200	4.6	4.8
					3	4000	1.0	3.1
			0.034	1.13	1	800	4.2	4.2
		(b)			2	2200	4.2	4.2
					3	4000	1.1	2.8
Cooper	1/28/75	23,400	0.034	1.16	1	800	4.4	4.4
		(a)			2	2200	5.8	5.3
					3	4000	0.9	3.3
			0.034	1.16	1	800	4.2	4.2
		(b)			2	2200	3.5	3.8
					3	4000	2.3	3.1
Fort Calhoun	11/25/74	33,500	0.072	3.01	1	1000	0.6	0.6
		(a)			2	2000	0.3	0.45
					3	3000	0.2	0.37
					4	4000	1.4	0.63
			0.072	3.01	2	2000	0.5	0.5
		(c)			3	3000	0.5	0.5
					4	4000	0.5	0.5

Figs. 15-17 and 15-18 also show surprisingly good agreement between curves computed with the different sets of D values. Significant differences tend to be restricted to a small region close to the shore. For subreaches 3 and 4 of the Fort Calhoun study reach, the difference between D values is nearly threefold. This indicates that the computed distributions do not respond very rapidly to changes in D over short subreaches, and justifies the set (c) method of combining subreaches. This would amount to the use of longer subreaches corresponding, say, to a complete bend, or a relatively straight reach between two bends. The insensitivity to variation in D over short subreach lengths increases as the distance downstream from the source increases.

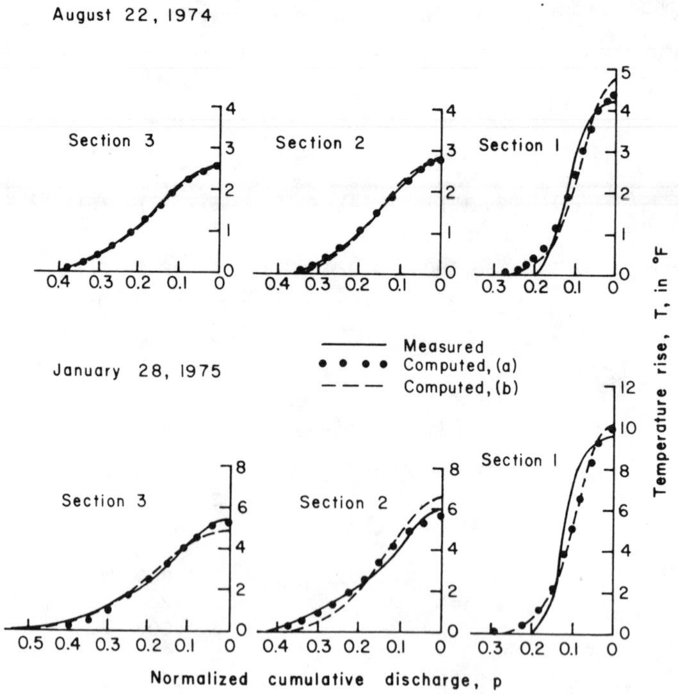

Fig. 15-17 Comparison Between Computed and Measured Transverse Distributions of Temperature Rise, Cooper Station

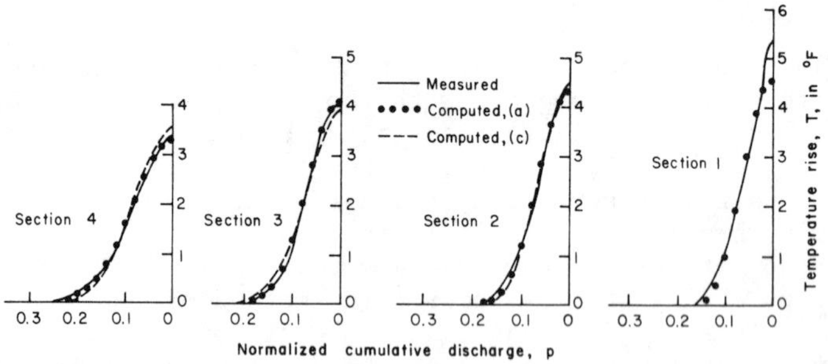

Fig. 15-18 Comparison Between Computed and Measured Transverse Distributions of Temperature Rise, Fort Calhoun Station, November 25, 1974

The Fort Calhoun plumes were also investigated for interdependence between D and $\underline{a}$ in the first subreach by computing a set of five plumes with $\underline{a} = 1.8$. Comparing with set (a) and averaging results for all five plumes, the 1.7-fold reduction in $\underline{a}$ is accompanied by a 1.6-fold increase in D for the first subreach, but no appreciable change in D for the subreaches farther downstream.

An examination of the results for any effect of buoyant spreading on D yielded negative results. Application of Eq. (15-21) indicates that the region of buoyant spreading should have been confined within the first subreach in all cases. Since no evidence could be found that the ratio D_1/D_2 correlates with the parameter $\Delta\rho/\rho S$ from Eq. (15-20), it is concluded that any buoyant-spreading effects which may have existed would have been masked by the interdependence between D and $\underline{a}$ in the first subreach. However, since no correlation between $\underline{a}$ and $\mathbb{F}_{DO}$ could be found, it is unlikely that buoyancy affected the transverse mixing even in the first subreach.

In Figs. 15-17 and 15-18 temperature rise is plotted versus the normalized cumulative discharge, p. If desired, the p vs z/B plots can be used to transform the temperature-rise profiles from the p domain back to the z domain, as shown in Fig. 15-19. Note that, depending on the shape of the p vs z/B curve, the temperature-rise profiles may appear to be considerably wider (or narrower) in one domain than in the other. As the thalweg meanders back and forth between one side of the channel and the other, the plume when viewed in the z domain would appear to go through expansion and contraction cycles.

General Observations Concerning the Transverse Diffusion Factor

Concerning the relationship of D to bulk channel and flow properties, the results in Table 15-4 are not directly comparable to the relationship given in Fig. 15-4. This is because D in Fig. 15-4 represents an average taken over an entire bend length, whereas the D values in Table 15-4 are either 'local' values for a short subreach, or at most an entire study reach, both of which are considerably shorter than one bend length. The experimental results of Chang (1971) with a neutrally-buoyant salt tracer and a fluorescent dye tracer in two meandering laboratory flumes, and of Sayre and Yeh (1973) with a fluorescent dye tracer in a six-mile meandering reach of the Missouri River downstream from the Cooper Station, both show that there is a considerable variation, but with a quite consistent pattern, in the value of the transverse mixing coefficient E_z going around a bend. A similar pattern of variation is expected for D.

Fig. 15-20 shows Chang's results for two experiments with a dye tracer in a large rectangular flume. His 'local' E_z values were obtained by a simulation and least-squares fit

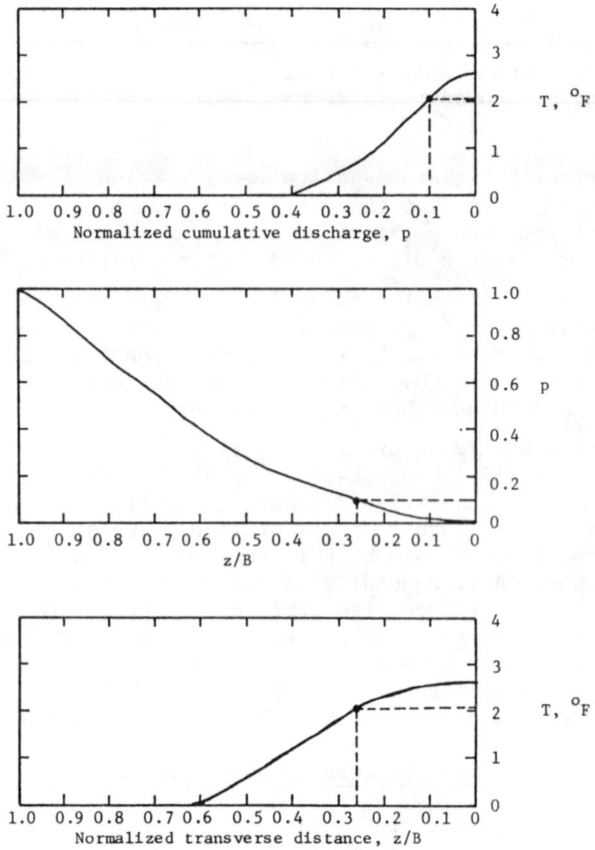

Fig. 15-19. Transformation of Transverse Temperature-Rise Profiles from the p Domain to the z Domain; Cooper Station Study Reach, Section 3, August 22, 1974

method similar to that used for obtaining the D values in Table 15-4. The decrease in D which was observed to occur between subreaches 2 and 3 in the Cooper Station study reach, which is about where the direction of curvature in channel alignment changes, is consistent with the decrease in E_z which occurs between reaches B-C and C-D in the flume. The information in Fig. 15-20 is now used to obtain estimates of the bend-averaged diffusion factor $\overline{D}^L$ by multiplying values of D by the ratio $\overline{E}_z^L/E_z$ for corresponding reaches, as shown in Table 15-5. In applying this procedure, reach A-B of the flume is taken to

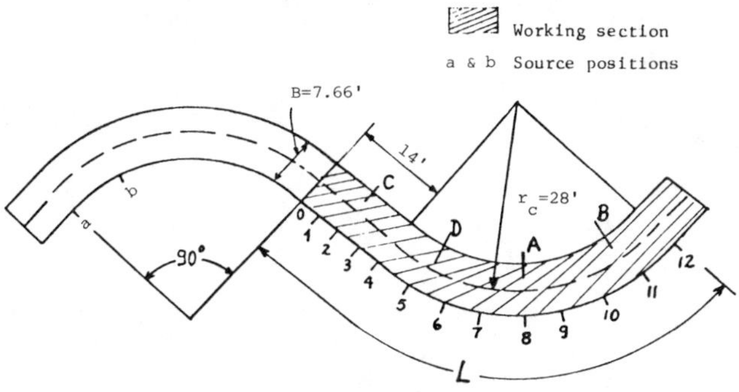

(a) Plan view of flume

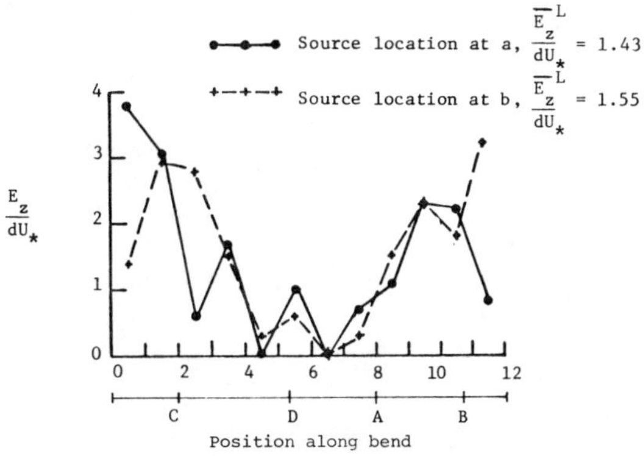

(b) Longitudinal variation of transverse mixing coefficient

Fig. 15-20. Variation of Transverse Mixing Coefficient Along a Bend in Chang's Flume Experiments

correspond to the Fort Calhoun study reach, reach B-C to subreaches 1 and 2 of the Cooper Station study reach, and reach C-D to subreach 3 of the Cooper reach. The values of D in column 5 of Table 15-5 correspond to sets (b) and (c) in Table 15-4.

Columns (8) and (9) in Table 15-5 represent the plume data in terms of the parameters plotted in Fig. 15-4. Because of what appears to be a fairly strong dependence on river discharge, and because the ratio of the ordinate to the abscissa values for the curve representing the river data in Fig. 15-4 is simply a constant (equal to 0.4), the data in columns (8) and (9) is plotted in a different format, as shown in Fig. 15-21. The ordinate, which is the ratio of column (8) to column (9), is plotted as a function of river discharge. An appropriate scaling factor for Q is not apparent. The key to the symbols is given in Table 15-5. Data from the earlier dye plume studies of Sayre and Yeh (1973) and Yotsukura et al. (1970) are included for comparison.

Table 15-5. Summary of Results for Transverse Diffusion Factor

Station (1)	Subreaches (2)	Date (3)	Q cfs (4)	$D \times 10^6$ ft^{-1} (5)	$\bar{E}_z^L/E_z$ (6)	$\bar{D}^L \times 10^6$ ft^{-1} (7)	$B\bar{D}^L$ (8)	$\frac{B}{\bar{d}^B} \frac{\bar{U}^A}{U_*} (\frac{\bar{d}^B}{r_c})^2$ (9)	$\frac{\bar{D}^L \bar{d}^B}{\bar{U}^A/U_*} \left(\frac{r_c}{\bar{d}^B}\right)^2$ (10)
Fort Calhoun ▲	2-4	08/08/74	36,000	0.8	0.8	0.64	0.00045	0.0010	0.45
		09/06/74	35,000	0.8	0.8	0.64	0.00045	0.0010	0.45
		11/25/74	33,500	0.5	0.8	0.40	0.00028	0.0010	0.28
	2-5	12/04/74	18,500	0.9	0.8	0.72	0.00045	0.00065	0.69
Cooper Station ●	1 & 2	01/23/75 01/28/75	22,000	3.8	0.6	2.3	0.0015	0.0014	1.07
		08/22/74 10/17/74 03/20/75	37,500	4.2	0.6	2.5	0.0017	0.0021	0.81
		08/01/75 12/05/75	60,000	2.6	0.6	1.6	0.0011	0.0029	0.38
Cooper Station ○	3	01/23/75 01/28/75	22,000	2.3	1.4	3.2	0.0020	0.0021	0.95
		08/22/74 10/17/74 03/20/75	37,500	1.1	1.4	1.5	0.00099	0.0030	0.33
		08/01/75 12/05/75	60,000	0.6	1.4	0.8	0.00054	0.0042	0.13
Sayre and Yeh (1973)									
○ Mile 532-529			56,100			3.3	0.0023	0.0051	0.45
○ Mile 529-526			56,100			8.6	0.0062	0.0182	0.34
Yotsukura et al. (1970)									
△ Mile 648-645			34,100			0.76	0.00051	0.0010	0.51

The dimensionless factor $\frac{h_1 \overline{d^2 \bar{u}^{\bar{d}Q}}}{(\bar{d}^B)^2 \bar{U}^A}$, included in Eqs. (15-8a) and (15-17) on which Fig. 15-21 is based, is not included in Fig. 15-21. The metric coefficient h_1 is excluded because it remains within the bounds $0.95 \le h_1 \le 1.05$ throughout both test reaches, and it is troublesome to evaluate. The quantity $\frac{\overline{d^2 \bar{u}^{\bar{d}Q}}}{(\bar{d}^B)^2 \bar{U}^A}$, which depends on channel shape and the transverse distribution of flow, was evaluated for several of the subreaches, both for the entire cross section and for just that part occupied by the plume. Since it was found to be reasonably constant, with an average value of about 1.2, and its inclusion did not appear to reduce the scatter, it too is excluded from further consideration. It's inclusion would move all the points in Fig. 15-21 downward by somewhat varying amounts.

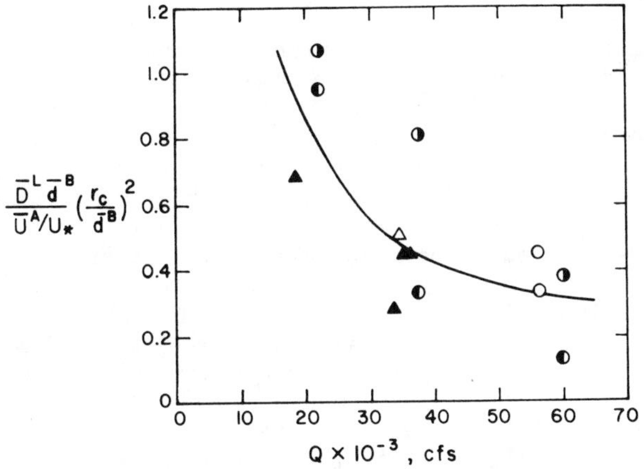

Fig. 15-21. Variation of Estimated Bend-Averaged Transverse Diffusion Factor with Missouri River Discharge

In the opinion of the writers the scatter is primarily due to the rough approximation involved in estimating $\overline{D}^L$ by applying the conversion factor $\overline{E}_z^L/E_z$ based on data from a rectangular flume. Thus any significant improvement of the relationship in Fig. 15-21 is likely to require additional data from study reaches that are long enough to include at least one river bend from beginning to end, and/or more precise information on how D varies going around an alluvial-channel bend.

Another factor in connection with Fig. 15-21 that should be investigated further is the assumption that changes in cross-sectional properties with respect to river discharge can be accounted for simply by raising and lowering the water surface, as in Fig. 15-9, in accordance with a stage-discharge relationship. Insofar as the values of $\overline{U}^A/U_*$ and $\overline{d}^B$ computed for Q = 60,000 cfs and 22,000 cfs in the Cooper Station study reach and for Q = 18,500 cfs in the Fort Calhoun study reach are based on this assumption, it has a bearing on Fig. 15-21.

Nonetheless, it is believed that Fig. 15-21 is a significant improvement over Fig. 15-4, for the Missouri River at any rate, and that the quantity $\frac{\overline{D}^L \overline{d}^B}{U^A/U_*}\left(\frac{r_c}{\overline{d}^B}\right)^2$ is a significant dimensionless form of the transverse diffusion factor for meandering channels. Final verification, and the determination of a suitable scaling factor for Q, will require the analysis of plumes in other meandering rivers with characteristics, particularly size, that differ from those of the Missouri River.

15.4 SUMMARY

In relatively fast-flowing turbulent rivers thermal plumes originating from steam-electric generating stations typically remain attached to the near shore, with the maximum temperatures in the plume occurring at or near the bank, in contrast to more tranquil rivers wherein plumes tend to be more or less symmetrical about a centerline trajectory that carries some distance out into the channel. In comparison with the latter type of plume, shore-attached plumes typically pass rather quickly through the near-field stage into the far-field stage wherein the plume configuration is governed mainly by the mixing mechanisms associated with the ambient flow. The model described in this chapter is particularly suited for representing shore-attached thermal plumes because, in contrast to the commonly used integral-type models, it focuses correspondingly less on the near-field jet-entrainment and crossflow interaction mechanisms, and in greater detail and with fewer restrictions on the far-field mechanisms.

The model consists of an analytical solution of the transformed two-dimensional convection-diffusion equation in which transverse distance is replaced by cumulative discharge. It takes into account the longitudinal variation of cross-sectional shape and size and transverse distribution of velocity that typify natural rivers. The parameters of the model are the initial dilution factor and the transverse diffusion factor, which in general varies along the channel from one reach to the next.

The model is applied to twelve sets of thermal plume data obtained in the Missouri River downstream from the Cooper and Fort Calhoun Nuclear Power Stations. In keeping with the instructional objectives of the Institutes on River Mechanics, aspects pertaining to application are presented in considerable detail.

Analysis of the thermal plume data indicates that the initial dilution factor tends to increase with discharge angle, depth of discharge channel relative to local river depth and width of discharge channel, and possibly with the ratio of river to plant-effluent discharge. The densimetric Froude number of the discharge channel and the ratio of plant discharge velocity to local river velocity (< 1 in all cases considered), appear to be less important. A dimensionless form of the bend-averaged transverse diffusion factor, $\frac{\bar{D}^L \bar{d}^B}{\bar{U}^A/U_*}\left(\frac{r_c}{\bar{d}^B}\right)^2$, is found to decrease by a factor of about three as the Missouri River discharge increases from 20,000 cfs to 60,000 cfs.

15.5 NOTATION

$\underline{a}$	Initial dilution factor, equal to PQ/Q_o. (Underlined in text to avoid confusion with the article a)
A.F.	Adjustment factor, i.e., ratio of computed total excess heat flux to area under measured distribution curve for excess heat flux
b	Effective source width following initial dilution, ft
b_o, b_1	Coefficient and exponent in Eq. (15-28)
B	Channel width at the water surface, ft
B_o	Width of the power-plant discharge channel where it enters the river, ft
d	Local depth of river, ft
$\bar{d}$	Average depth of river along path of plume, ft
$\bar{d}^B$	Width-averaged river depth, ft
D	Transverse diffusion factor, ft^{-1}
$\bar{D}^L$	Transverse diffusion factor averaged over the length of a bend, ft^{-1}
$\bar{D}^{x_n}$	Average value of D for the reach 0 to x_n, ft^{-1}
E_z	Overall transverse mixing coefficient, ft^2/sec
$\bar{E}_z^L$	Overall transverse mixing coefficient averaged over the length of a bend, ft^2/sec
$f_s(\ )$	Standardized normal probability density function
$F_s(\ )$	Normal cumulative distribution function
$\mathbb{F}_{DO}$	Densimetric Froude number of plant discharge, equal to $\dfrac{U_o}{\sqrt{gH_o\Delta\rho_o/\rho}}$
g	Gravitational acceleration, ft/sec^2
h_1, h_3	Metric coefficients for orthogonal curvilinear coordinate system
H_o	Depth of plant discharge channel where it enters the river, ft
H_{max}	Estimated maximum depth beneath the surface to which the plume penetrates, ft
i	Index number for subreaches
K	Surface exchange coefficient for excess heat, ft/sec
L	Length of a bend in a meandering channel, ft
L_B	Length of region of buoyant spreading, ft
n	Number of bank reflections; index number for the last of a sequence of subreaches
p	Cumulative discharge normalized by total river discharge, equal to q_c/Q
p'	Transverse displacement from the origin in the p domain
p_m	Fraction of river discharge for which $T > T_c$
P	Fraction of total river discharge with which the plant discharge is mixed following initial dilution

q	Local river discharge per unit width, ft^2/sec
q_c	Cumulative river discharge, cfs
$\bar{q}^B$	River discharge per unit width averaged across width of channel, ft^2/sec
Q	Total river discharge, cfs
Q_o	Volumetric discharge from the power plant into the river, cfs
r_c	Radius of curvature of the river centerline, ft
s	Distance along the centerline trajectory of the thermal plume from the discharge channel; also the standardized variable of the normal probability density function
S	Energy gradient
T	Depth-averaged value of the local temperature rise above ambient, °F
T_a	Ambient temperature, °F
T_c	Critical temperature rise, usually 5°F
T_I	Temperature-rise in the vicinity of the outfall following initial mixing, °F
T_o	Temperature rise of the condenser cooling water at the point of discharge, °F
T_s	Temperature rise at the centerline of a thermal plume at a distance s from the discharge channel, °F
$\bar{u}^d$	Local depth-averaged velocity of flow in the longitudinal direction, ft/sec
$\bar{U}$	Average velocity of river along path of plume, ft/sec
$\bar{U}^A$	Cross sectional average velocity of river flow, ft/sec
U_o	Average velocity of plant discharge where it enters the river, ft/sec
U_a	Representative ambient velocity in the vicinity of the discharge channel, ft/sec
U_*	Shear Velocity $= \sqrt{g\bar{d}^B S}$, ft/sec
$\bar{w}^d$	Local depth-averaged velocity of flow in the transverse direction, ft/sec
x, z	Distances in longitudinal and transverse directions respectively, ft
z_o	Transverse coordinate distance at $x = 0$ from the x axis to the bank where the outfall is located, ft
α	Angle of discharge channel with respect to ambient flow, in degrees
ρ	Density of ambient river water, slugs/ft^3
$\Delta\rho$	Residual density difference between ambient river and plume water following initial dilution, slugs/ft^3
$\Delta\rho_o$	Density difference between ambient river and plant discharge water, slugs/ft^3

σ_p Standard deviation of transverse distribution in the p domain

$\overline{(\ )}^A$ Denotes averaging over some quantity A, where A may be cross-sectional area, channel width, local depth, length of a bend, or river discharge

15.6 REFERENCES

Caro-Cordero, R., and Sayre, W.W., 1977. Mixing of Power-Plant Heated Effluents with the Missouri River, IIHR Report No. 203, Iowa Institute of Hydraulic Research, The University of Iowa.

Carter, H.H., and Regier, R., 1974. The Three Dimensional Heated Surface Jet in a Cross Flow. Technical Report 88, Chesapeake Bay Institute, The John Hopkins University.

Carter, H.H., Schiemer, E.W., and Regier, R., 1973. The Buoyant Surface Jet Discharging Normal to an Ambient Flow of Various Depths. Technical Report 81, Chesapeake Bay Institute, The Johns Hopkins University.

Chang, Y.C., 1971. Lateral Mixing in Meandering Channels. Ph.D. Dissertation, Department of Mechanics and Hydraulics, The University of Iowa.

Dunn, W.E., Policastro, A.J., and Paddock, R.A., 1975. Surface Thermal Plumes: Evaluation of Mathematical Models for the Near and Complete Field. Report No. ANL/WR-75-3 (in two parts), Water Resources Research Program, Argonne National Laboratory.

Fischer, H.B., 1969. The Effect of Bends on Dispersion in Streams. Water Resources Research, Vol. 5, No. 2, pp. 496-506.

Jackman, A.P., and Yotsukura, N., 1977. Thermal Loading of Natural Streams. U.S. Geological Survey Professional Paper 991.

Jirka, G.H., Abraham, G., and Harleman, D.R.F., 1975. An Assessment of Techniques for Hydrothermal Prediction. MIT Technical Report No. 203, R.M. Parsons Laboratory for Water Resources and Hydrodynamics.

Liong, S.Y., 1977. Effects of Vertical Distortion on Thermal-Hydraulic Modeling of Surface Discharges. Ph.D. Dissertation, Mechanics and Hydraulics Program, The University of Iowa.

Paily, P.P., 1976. Thermal Plume Surveys in the Missouri River Near the Cooper Nuclear Station, April 1974 - December 1975. Report to Nebraska Public Power District from NALCO Environmental Sciences.

Paily, P.P., 1975. An Investigation of the Thermal Plume Characteristics for the Fort Calhoun Power Station Outfall. IIHR Limited Distribution Report No. 30, Iowa Institute of Hydraulic Research, The University of Iowa.

Paily, P.P., Su, T.Y., Giaquinta, A.R., and Kennedy, J.F., 1976. Thermal Regimes of the Upper Mississippi and Missouri Rivers. IIHR Report No. 182, Iowa Institute of Hydraulic Research, The University of Iowa.

Sayre, W.W., 1975. Investigation of Surface Jet Thermal Outfall for Iatan Steam Electric Generating Station. IIHR Report No. 167, Iowa Institute of Hydraulic Research, The University of Iowa.

Sayre, W.W., 1973. Natural Mixing Processes in Rivers. Chapter 6 in Environmental Impact on Rivers (River Mechanics III), Edited and published by H.W. Shen, Fort Collins, Colorado.

Sayre, W.W., and Yeh, T.P., 1973. Transverse Mixing Characteristics of the Missouri River Downstream from the Cooper Nuclear Station. IIHR Report No. 145, Iowa Institute of Hydraulic Research, The University of Iowa.

Shirazi, M.A., and Davis, L.R., 1974. Workbook of Thermal Plume Prediction, Vol. 2, Surface Discharge. National Environmental Research Center, Office of Research and Development, U.S. Environmental Protection Agency.

Sium, Ogbazghi, 1975. Transverse Flow Distribution in Natural Streams as Influenced by Cross-Sectional Shape. M.S. Thesis, Mechanics and Hydraulics Program, The University of Iowa.

U.S. Army Engineer District, Omaha, Corps of Engineers, 1969. Missouri River Channel Regime Studies. M.R.D. Sediment Series No. 13B.

Yeh, Tso-Ping, 1974. Transverse Mixing of Heated Effluents in Open Channel Flow. Ph.D. Dissertation, Department of Mechanics and Hydraulics, The University of Iowa.

Yotsukura, N., and Cobb, E.D., 1972. Transverse Diffusion of Solutes in Natural Streams. U.S. Geological Survey Professional Paper 582-C.

Yotsukura, N., Fischer, H.B., and Sayre, W.W., 1970. Measurement of Mixing Characteristics of the Missouri River Between Sioux City, Iowa, and Plattsmouth, Nebraska. U.S. Geological Survey Water Supply Paper 1899-G.

Yotsukura, N., and Sayre, W.W., 1976. Transverse Mixing in Natural Channels. Water Resources Research, Vol. 12, No. 4, pp. 695-704.

Chapter 16

DATA COLLECTION PLANNING AND SURVEY OF WATER QUALITY MODELS

by

T. G. Sanders, Assistant Professor of Civil Engineering,
Colorado State University, Fort Collins,
Colorado

Chapter 16

DATA COLLECTION PLANNING AND SURVEY OF WATER QUALITY MODELS

16.1 INTRODUCTION

Water quality data which is both accurate and readily available in sufficient quantity is a necessary prerequisite in the formulation and verification of water quality models. Unfortunately, in many instances the costs are becoming prohibitive to collect our own data. Therefore, much of the data which is used in modeling and will be used in the future will be data which is collected and disseminated by public agencies. And although there will be an abundance of data available, the actual value or utility of this data for model development as well as other uses have been questionable because rational criteria in specifying: sampling frequency, station location, and data analysis are nonexistant in the planning and design of data collection networks.

Experience indicates there is a need for the development of rational design criteria to be applied in planning a river quality data collection network derived from fundamental concepts of: 1) the hydrodynamic mixing properties of a stream; and 2) the stochastic nature of the water quality data to be collected and analyzed. Too often, irrespective of the objectives of the sampling agency, too much data has been collected from monitoring networks which lacked consistent and definitive design criteria. Without a justifiable basis, in many cases water quality samples are assumed to be representative of the total river flow and the frequency of sampling is assumed to be adequate for taking into account the variability of the parameter concentrations. Without an understanding of the hydrodynamic mixing properties of a stream to determine a zone of complete mixing, the degree to which the water sample is representative of actual stream conditions is questionable. Similarly, without an understanding of the deterministic as well as stochastic variation of these concentrations, the estimation and quantification of ambient concentrations and trends as required in the 1972 Federal Water Pollution Control Act Amendments is questionable, not to mention the utility of the data for other purposes. Therefore, it is the purpose of this chapter to discuss rational design criteria to be applied in the planning of a water quality monitoring network. These criteria are a basis for specifying sampling station location and sampling frequency for a river quality data collection network having the primary capability of assessing ambient water quality concentrations and determining yearly trends, as well as being the major source of data for developing and verifying water quality models.

During the last two decades the need for large amounts of accurate reliable water quality data has increased immensely both in the public and private sectors. In 1956, when the

United States Public Health Service was the major federal agency responsible for collecting water quality data, the data was needed to assess the level of quality of the nations rivers and provide a data base for "development of comprehensive water resource programs," (USPHS, 1963). By 1965 with the passage of the water quality act additional water quality data was required to first establish and later enforce stream standards. In 1972 the Federal Water Pollution Control Act Amendments specified that each state submit an annual report beginning January 1, 1975 defining the water quality of all navigable waters.

The need for more water quality data in the future will certainly not diminish. In order to justify the implementation of increased pollution control programs proceeding toward a zero discharge goal by 1985, quantitative water quality data should exist assessing the impact of implementing "best practical control technology" by 1977 and "best available control technology" by 1983. Furthermore the economic ramifications of this legislation can be investigated by relating the change in water quality to the cost of the abatement procedures.

More data will be needed in the private sector as well but because of the large commitment by many states and federal agencies in the collection and dissemination of water quality data, it is not a question of quantity per se but of quality. Given the overwhelming amount of water quality data that is available now, it is and has been of questionable utility.

Wolman (1971) found in his investigations of the changes in water quality that although the concentrations of many pollutants should have increased due to the increased industrial expansion and population, the water quality data available did not conclusively verify this assertion. It may be construed from his work that in many cases the amount and type of data were insufficient to formulate valid conclusions; however, most of the water quality legislation since 1956 has specified or at least implied that data should be collected so that the effectiveness as well as the justification for pollution abatement laws and programs might be quantitatively evaluated. Data obtained from water quality surveillance networks were to be used to determine existing quality, to assess trends and to monitor compliance with stream standards, but due to the ineffectiveness of monitoring networks these objectives generally have not been met (Sanders, 1974).

It appears that much of the readily available data has been inadequate for many of its intended uses, therefore, it may also be quite inadequate for constructing and verifying water quality models. In the development of water quality models possibly too much emphasis has been placed on the solution procedures, analytical, numerical, etc., and computer application without or with very little appreciation as to the validity and appropriateness of the available raw data used to estimate model parameters. It is not improbable that many water quality models derived from fundamental relationships which did not represent the process was due to inappropriate and/or non-representive data used to calculate process parameters rather than an incorrect model.

An example of using inappropriate data in the construction of a very simple water quality model is the use of instantaneous grab sample, time series data for relating the concentration of a water quality parameter to mean daily stream flow. A flow-weighted, average daily concentration would be more appropriate. The data may also be non-representative if it were collected at a single point on the river where mixing is incomplete--the water quality parameter concentration being a function of the sampling location within the lateral transect.

Knowledge of the lab analysis procedures in the determination of the water quality parameter concentrations is important as well. Without proper control of the number of analyses of the sample, it is possible that the random variation of the estimated sample mean concentration is equal to or exceeds the variation of the water quality parameter concentrations in nature.

16.2 MONITORING AND DATA COLLECTION ACTIVITIES

Since 1956, national water quality surveillance and developing a data base have been the responsibility of the United States Public Health Service, Department of Health, Education and Welfare; Water Pollution Control Administration, Department of Health, Education, and Welfare; Water Pollution Control Administration, Department of the Interior; and Office of Data Coordination of the United States Geological Survey (USGS). At the state level in Massachusetts, for example, responsibility has shifted from the Department of Public Health to the Water Pollution Control Division of the Department of Natural Resources. Although the responsibility of collection and disseminating water quality data has varied, the primary cause of the general data deficiency from both federal and state monitoring activities is the lack of adequate and comprehensive design criteria in the planning stage for locating the sampling stations, and determining the frequency of sampling.

Even without design criteria the size of the national surveillance system, nevertheless, continues to increase. In 1957 the National Surveillance Network included 50 sampling stations (USPHS, 1963). By 1963, the number of sampling stations increased to 128. Because of this apparent inadequate national coverage, a proposal has been made to include all federal, state, and local monitoring activities into a national surveillance network. If approximately 6,000 monitoring stations at municipal water treatment facilities are included, the number of sampling stations will increase to over 7,600. However, if national coverage is to be "adequate", over 10,000 stations will be required (Sayers, 1971).

The decision to use the data collected daily at water treatment facilities will certainly increase the coverage of a national surveillance network with a minimum of capital and operating costs. However, if the new data is not representative of the actual conditions in the river, the data will be of little benefit.

16.3 SAMPLING LOCATION FOR REPRESENTATIVE WATER QUALITY DATA

The location of a sampling station is probably the most critical design factor in the collection of water quality data; if the sample is not representative of the water mass, the frequency of sampling as well as the mode of data interpretation and presentation becomes inconsequential. Nevertheless, criteria to establish station locations for representative sampling has received relatively little attention from both state and federal agencies responsible for the collection and dissemination of water quality data.

There are two levels of design criteria in the discussion of sampling station location which can be considered: 1) macrolocation (Sanders, 1976), the location of a sampling station relative to an entire river basin, and 2) microlocation, the location of the sampling station relative to the nearest upstream outfall. The utility of data collected from a water quality network will to a large extent be dependent on the consideration given to each level in the planning of the data collection network. However, only criteria for locating the microlocation which defines a zone of complete mixing will be addressed as the macrolocation criteria will be a function of the specific objectives of the data collection agency.

Presently, in the majority of monitoring networks reviewed, the definition of only the macrolocation of the sampling stations is even attempted. Kittrell (1969) recommends that even though complete mixing is assumed, sampling at the quarter points should still be done and if too time consuming, he suggests sampling at the midpoint of the river. Velz (1950) indicated that samples should be taken at

> "...some distance downstream where dispersion throughout the section has been reasonably complete. Even then, care should be taken if the stream is sluggish and wide to test for variability, and if detected, it may be necessary to sample at a number of points across the section."

Either complete mixing is assumed near the point of discharge or the zone of complete mixing is estimated primarily by experience (Sanders, 1974). In some applications, subjective judgment or over-simplified assumptions may be adequate, but this is not sufficient for locating permanent water quality sampling stations. The representativesness of a water quality sample is a function of the uniformity of the sample concentrations in a river's cross-sectional area.

> "A sample, by definition, should be representative, for it is presented as evidence of the quality of the body of

water from which it is obtained," (Haney and Schmidt, 1958).

Wherever the concentration of a water quality parameter is independent of depth and lateral location in a river's cross section, the river at that point is completely mixed and can be defined as a microlocation.

The inability to define the zones of complete mixing is not restricted to locating permanent water quality, sampling stations. It has prevented widespread applicability of tracer dilution methods to measure river discharge as..

> "The accuracy of the method critically depends upon complete mixing of the injected solution through the stream cross section before the sampling station is reached..." (Buchannan, 1964).

Without a mathematical definition for estimating zones of complete mixing, the mixing distance has been determined by trial and error: multiple samples are taken at different points downstream from an outfall until the variability of the samples from the same river cross section becomes insignificant. In such a study on the Columbia River, Reeder (1971) discovered that 10 miles below an outfall in Trail, British Columbia, Canada, the effluent was completely mixed. He found that mixing was incomplete 1/2 mile and 3 miles downstream from the outfall. Although this trial and error procedure can establish the zones of complete mixing, its universal application for establishing the microlocation of a sampling station would be questionable in terms of capital expenditure alone not to mention the time and manpower requirements.

Well mixed zones (microlocation) in a river for representative water quality sampling can be defined, given that several assumptions will apply. By assuming that a tracer distribution from an instantaneous point source is Gaussian in both the lateral and vertical transect and applying classical image theory, a theoretical distance from an outfall to a well mixed zone in a straight, uniform river channel can be determined (Sanders, 1974). This mixing distance is a function of: 1) mean stream velocity, 2) location of the point source and 3) the mean lateral and vertical turbulent diffusion coefficients.

The turbulent diffusion coefficients are estimated in a procedure utilizing the recently derived Cleary and Adrian (1973) form of the analytical solution to the unsteady state, three-dimensional advection and dispersion partial differential equation having appropriate initial and boundary conditions. Because the analytical solution of the governing partial differential equation is used and perturbed by adjusting the magnitude of the three turbulent diffusion coefficients so that the predicted concentrations coincide with the measured tracer concentration, these coefficients can be considered best fit estimates in a least squares sense. Other methods are available to calculate the coefficients such as the Method of Moments (Diachishin, 1963; Fischer, 1967) and the Method of Routing (Fischer, 1968). But these apply to only the longitudinal coefficient; the lateral and vertical coefficients, however, can be estimated from equations derived from the stream hydraulic characteristics (Fischer, 1967) and have compared favorably to the coefficients, estimated using the analytical solution.

The mixing properties of a stream can be defined in terms of the mean longitudinal, lateral and vertical turbulent diffusion coefficients. For many years, the longitudinal turbulent diffusion coefficient, frequently called the longitudinal dispersion coefficient, has been estimated from tracer studies utilizing the solution to the one-dimensional advection and dispersion partial differential equation (Feuerstein and Selleck, 1963 and Fischer, 1967). As more longitudinal dispersion measurements became available, the coefficient was equated to functions of the hydraulic characteristics of the river (Fischer, 1968; Sayre and Chang, 1968; and Thackston and Krenkel, 1967). Similarly, the lateral and vertical turbulent diffusion coefficients have been equated to hydraulic characteristics of the river. Using the analytical solution to the three-dimensional, unsteady-state, advection and dispersion partial differential equation, estimates of the lateral and vertical turbulent diffusion coefficients can be estimated from tracer studies as well.

There are several models available being functions of the mixing coefficients which have been shown to apply for predicting a zone of relatively complete mixing. Ruthven (1971) derived an expression for a mixing distance utilizing the solution to the steady-state, two-dimensional, advection and dispersion equation. By assuming that most streams are shallow enough so that complete vertical mixing is assured in a relatively short distance, he established a relationship (Eq. 16-1) from the two-dimensional solution to predict the mixing distance to a point where concentration variation in the cross section does not exceed ten percent.

$$L \geq 0.075 \frac{w^2 u}{D_y} \tag{16-1}$$

where: L = mixing distance
w = width of channel
u = mean stream velocity
D_y = lateral turbulent diffusion coefficient

In a more general relationship, Sayre (1965) estimated in his research that the mixing distance to a point of uniform concentration in the cross section should be calculated by:

$$L = \frac{1}{2\alpha_1^2} \frac{w^2 u}{D_y} \qquad (16\text{-}2)$$

where:

α_1 = constant associated with the location of a point source and the degree of uniformity of the concentration gradient.

Ward (1973) derived an expression for a mixing distance based on the channel geometry and the point of tracer injection relative to the channel midpoint. This was done primarily to find the point in the river where flow measurement by the dilution method is applicable (uniform concentration in the cross section).

$$L = \frac{K_1}{0.02} \frac{w^2}{d} \qquad (16\text{-}3)$$

where:

K_1 = a function of the lateral turbulent diffusion coefficient, the tracer injection point and the width of the straight river.

When estimating flow by the dye dilution method, the United States Geological Survey (Kilpatrick et al., 1970) recommended the use of the following equations to locate sampling points downstream from dye injection points at the center of the stream and near the bank respectively.

$$L = 1.3u \frac{w^2}{d} \qquad (16\text{-}4)$$

$$L = 2.6u \frac{w^2}{d} \qquad (16\text{-}5)$$

It should be noted that the mixing distances are a function of the square of the river width, the lateral turbulent diffusion and a constant.

16.4 THEORETICAL MIXING DISTANCE

Because the tracer distribution initiated from a point source is theoretically a Gaussian distribution in the lateral cross section as well as in vertical cross section (Cleary and Adrian, 1973), the variances of the lateral and vertical concentration distributions are related to the respective turbulent diffusion coefficients D_y and D_z in the form:

$$\sigma_y^2 = 2D_y t \qquad (16\text{-}6)$$

$$\sigma_z^2 = 2D_z t \qquad (16\text{-}7)$$

where:

σ_y^2 = variance in the lateral direction

σ_z^2 = variance in the vertical direction

It should be noted that the derivative with respect to time of Eq. (16-6) and Eq. (16-7) are the equations which are the basis for calculating the coefficients by the change in variances technique.

Classical image theory provides a basis for relating the turbulent diffusion coefficients, the variance of the tracer distributions, and the length of the mixing zone by treating the stream boundaries as reflecting barriers (Cochrane and Adrian, 1970; Diachishin, 1963; Holley, 1972; and Sayre, 1967). The mixing distance may be defined in terms of vertical or lateral mixing distances, the distance downstream from a point source injection where a specified degree of vertical or lateral mixing occurs. These mixing distances are computed by specifying that the standard deviation of the tracer distribution in the lateral direction (square root of the variance) is equal to the half width of the stream and the standard deviation of the tracer distribution in the vertical direction is equal to the half depth of the stream.

With reference to Fig. 16-1, the Gaussian distribution representing the tracer distribution in either the lateral or vertical direction from a point source at midstream and middepth, is drawn so that the respective boundaries coincide with one standard deviation from the mean. There is actually an infinte number of images for each boundary. However, for simplicity only one is drawn. Because the method of images does apply,

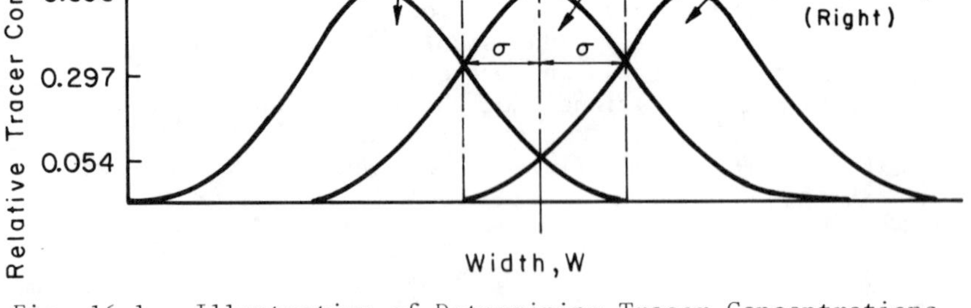

Fig. 16-1. Illustration of Determining Tracer Concentrations Utilizing the Method of Images, Where σ is Equal To the Standard Deviation

two identical distributions can be superimposed on each side of the initial Gaussian curve with their peaks one standard deviation from the boundary. By adding the coordinates of each curve, the sum will indicate the relative concentration distribution. Calculations indicate that when the boundaries are one standard deviation from the center an almost completely uniform concentration distribution in the lateral or vertical direction exists with the deviation between the maximum concentration at the middle and the minimum at the boundary being theoretically no greater than three percent (Sanders, 1974). Theoretically a completely uniform concentration can exist over the cross section if the ratio of the distance from the boundary to the point source divided by the standard deviation of the tracer distribution is slightly less than one (Sayre, 1967); however, for simplicity in the planning of sampling station locations the following definition is recommended: the mixing distance is defined as the longitudinal distance from the point of tracer injection or pollutant discharge in a river to the point downstream where the standard deviation of the theoretical tracer concentration curve equals the lateral (or vertical) distance between the tracer injection point and the farthest boundary.

Replacing t in Eq. (16-6) and Eq. (16-7) by L/u, the mixing distance for complete lateral mixing and the mixing distance for complete vertical mixing can be derived for a point source located equidistant from all boundaries:

$$L_y = \frac{\sigma_y^2 u}{2D_y} \tag{16-8}$$

$$L_z = \frac{\sigma_z^2 u}{2D_z} \tag{16-9}$$

where:

L_y = mixing distance for complete lateral mixing

L_z = mixing distance for complete vertical mixing

Using the same procedure based on the method of images for a point source located at the stream boundary and on the water surface, the maximum theoretical mixing distance for a straight uniform channel can be calculated utilizing Eq. (16-8) and Eq. (16-9) by substituting the entire river width and depth for the standard deviations in the lateral and vertical directions respectively. This will result in a four fold increase in the theoretical mixing distance, from a point source at a midstream and middepth.

Because the theoretical mixing distance for complete vertical mixing of the tracer in most streams will be less than the theoretical mixing distance associated with lateral mixing, the mixing distance for providing a uniform tracer distribution over the entire stream cross section will be equal to the larger of the two mixing distances, that is, the mixing distance associated with lateral mixing.

16.5 ESTIMATION PROCEDURE FOR PARAMETERS OF MIXING DISTANCE MODEL

With reference to Eqs. (16-8) and (16-9), the variables required for estimating mixing distances are: 1) the location of the injection point (relative to the farthest boundary), 2) mean stream velocity, and 3) the turbulent diffusion coefficients, D_y and D_z. The mean stream velocity and the injection point location can be easily determined from the available data, and using a water quality model which defines the movement of a soluble tracer in a stream, estimates of the lateral and vertical turbulent diffusion coefficients can be made as well.

The water quality model used to estimate these coefficients, derived by Cleary and Adrian (1973), is the analytical solution to the three-dimensional, advection and dispersion partial differential equation [Eq. (16-10)] with appropriate initial and boundary conditions. The partial differential equation, initial condition, boundary conditions and solution are presented below in outline form as the complete derivation is available in the above reference.

$$\frac{\partial C}{\partial t} + U\frac{\partial C}{\partial X} = D_x \frac{\partial^2 C}{\partial X^2} + D_y \frac{\partial^2 C}{\partial Y^2} + D_z \frac{\partial^2 C}{\partial Z^2}$$

$$+ g^i_{pt}\, \delta(X-X_1)\delta(Y-Y_1)\delta(Z-Z_1)\delta(t-\tau) \qquad (16\text{-}10)$$

C is the concentration of tracer; D_x, D_y, and D_z are the turbulent diffusion coefficients in the X (longitudinal), Y (lateral) and Z (vertical) directions, respectively; the Dirac delta functions, δ, describe the instantaneous input of mass, g^i_{pt}, for a time of mass release, τ, relative to time, t and coordinate locations (Fig. 16-2) at the point of mass injection, X_1, Y_1, and Z_1; U is the mean velocity.

The following boundary conditions are applicable when the stream's banks, bottom and surface act as reflecting barriers:

$$\frac{\partial C}{\partial Z} = 0 \quad Z = 0 \text{ and } Z = H$$

$$\frac{\partial C}{\partial Y} = 0 \quad Y = 0 \text{ and } Y = W$$

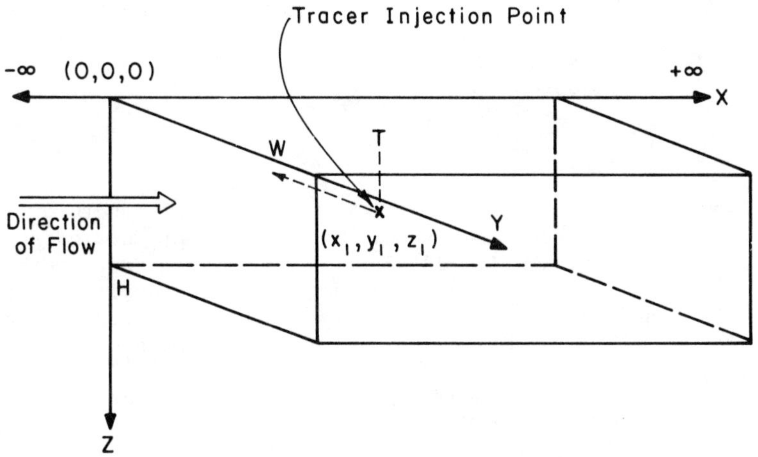

Fig. 16-2. River Coordinate Notation in the Analytical Solution to the Three-Dimensional Advection and Dispersion Partial Differential Equation

where H equals the depth of the stream and W equals the width of the stream. Furthermore, the initial condition is:

$$C(X,Y,Z,0) = 0$$

for:

$$-\infty < X < \infty$$

$$0 \leq Y \leq W$$

$$0 \leq Z \leq H$$

Applying these conditions to Eq. (16-10), Cleary and Adrian (1973) derived the following solution:

$$C(X,Y,Z,t) = \frac{g^{i}_{pt} \exp[-\frac{(X-X_1-Ut)^2}{4\ D_x\ t}]}{WH\ (4\pi D_x t)^{1/2}}$$

$$\{1 + 2 \sum_{n=1}^{\infty} \exp(-\mu_n^2\ D_z t)\ \text{COS}\mu_n Z\ \text{COS}\mu_n Z_1\}$$

$$= \{1 + 2 \sum_{m=1}^{\infty} \exp(-\beta_m^2\ D_y t)\ \text{COS}\beta_m Y\ \text{COS}\beta_m Y_1\} \quad (16\text{-}11)$$

where:

$$\mu_n = \frac{n\pi}{H} \quad n = 1,2\ldots$$

$$\beta_m = \frac{m\pi}{W} \quad m = 1,2\ldots$$

Equation (16-11) is based on the assumption that the turbulent diffusion coefficients and velocity are constant and independent of spatial location. In addition, it is assumed the lateral and vertical velocities are negligible and are equal to zero and that the flow is uniform, steady state and nonstratified. The channel is straight and the water quality parameter is conservative and soluble.

Using the three-dimensional analytical solution which only requires a single point source, the longitudinal, lateral and vertical turbulent diffusion coefficients from the field data can be estimated simultaneously. Ideally a continuous point source is preferred for several reasons, one of which is that once steady-state conditions occur, concentration is independent of time, allowing multiple samples or continuous samples to be taken, thus smoothing out sampling errors by increasing precision. However, in field studies, it is generally impractical and publicly unacceptable to introduce large quantities of tracer material into a stream as is the case with a continuous point source. An indirect economic benefit of using an instantaneous point source is that the sampling equipment could be simplified as no pumps are required to release or collect the tracer. Furthermore, utilizing a point source can simplify the experimental procedure by reducing the amount of equipment, tracer, manpower and time for each run.

In order to obtain the best estimates of the longitudinal, lateral and vertical turbulent diffusion coefficients utilizing Eq. (16-11), the criteria of the minimum sum of the squared deviations can be used.

$$SSD = \sum_{i=1}^{n} (C(X_i,Y_i,Z_i,t_i)_m - C(X_i,Y_i,Z_i,t_i)_T)^2 \qquad (16\text{-}12)$$

SSD = sum of square deviations

n = number of data points

$C(X_i,Y_i,Z_i,t_i)_m$ = tracer concentration measured in the field at point (X_i,Y_i,Z_i) and at time t_i

$C(X_i,Y_i,Z_i,t_i)_T$ = Theoretical tracer concentration predicted by Eq. (16-11) at point (X_i,Y_i,Z_i) and at time t_i

X_i,Y_i,Z_i = longitudinal, lateral and vertical coordinates

t_i = time

An initial estimate for each of the three turbulent diffusion coefficients is made. Using these estimates in Eq. (16-11), theoretical tracer concentrations are determined then subtracted from the respective field concentration; the difference is squared and summed as shown by Eq. (16-12). By systematically comparing the sum of squared deviations, it is possible to find the combination of turbulent diffusion coefficients having the minimum sum, hence the "best estimates" in a least squares sense.

This calculating procedure, commonly known as the gradient method, is one of several methods such as the Newton method, and the Newton-Raphson method which could be used to improve upon the initial estimates of the turbulent diffusion coefficients. Both the Newton and Newton-Raphson methods can converge to the optimum solution more rapidly than the gradient method, because these methods make use of the slopes of the functional relationships between the SSD and each of the turbulent diffusion parameters as a means of moving to the point of minimum SSD. However, depending on the initial guesses of the turbulent diffusion coefficient (how close they are to be best estimates), the final estimates may not be associated with a global minimum of the SSD, but instead, a local minimum; and in many cases, especially when estimating three parameters, the procedures may not converge at all.

An additional impediment associated with both methods is the amount of mathematical calculations in their derivations. The Newton method is basically an iterative method for solving a set of simultaneous nonlinear equations, requiring first and second partial derivatives of the SSD function with respect to D_x, D_y, and D_z.

Considering the complexity of the analytical solution [Eq. (16-11)], and the marginal benefit of using either the Newton or Newton-Raphson method, it was decided that the gradient method would be adequate (Sanders, 1974).

The gradient method is restricted by the step length (this is the magnitude of the increment systematically added to and subtracted from the parameters during each iteration); but by programming the method so that the step length is variable, the speed of convergence is improved and the problem of convergence at a local minimum is diminished. In addition, to add to the likelihood that the global minimum has been found all coefficients should be estimated twice, using the gradient method, but with substantially different initial guesses and step size. Although, theoretically there is a chance that the global minimum has not been attained, even by starting with two different sets of initial guesses, experience with using many different initial estimates for several of the tests displayed continual convergence to the same results (Sanders, 1974).

Table 16-1 lists estimates of the turbulent diffusion coefficients which were best estimates in the least squares sense from the Mill River in Northampton, Massachusetts

(Sanders, 1976). Except for the longitudinal coefficients, very little data is available in the literature for extensive comparison. It was found, however, that the estimates from equations, reported by Fischer (1967) and Holly (1972) compared fairly well with the coefficients estimated using the minimum sum of the squares deviation estimates.

Table 16-1. The Longitudinal (D_x), Lateral (D_y) and Vertical (D_z) Turbulent Diffusion Coefficients Estimated from Data Collected 200 ft Downstream from an Instantaneous Point Source of Rhodamine WT (Sanders, 1974)

Sampler Location	Turbulent Diffusion Coefficients		
Mean Velocity	D_x ft^2/sec (cm^2/s)	D_y ft^2/sec (cm^2/s)	D_z ft^2/sec (cm^2/s)
U = 0.68 ft/sec	0.6	0.10	0.0004
(20.6 cm/s)	(555)	(92.5)	(0.37)
	0.5	0.07	0.0006
	(463)	(64.7)	(0.55)
	0.3	0.07	0.00021
	(278)	(64.7)	(1.94)
U = 0.59 ft/sec	0.5	0.02	0.0004
(17.9 cm/s)	(462)	(18.5)	(0.37)
	0.4	0.03	0.0007
	(370)	(27.8)	(0.65)
	0.3	0.06	0.0004
	(278)	(55.5)	(0.37)
	0.5	0.02	0.0007
	(462)	(18.5)	(0.65)

These equations for estimating the lateral and vertical turbulent diffusion coefficients are functions of easily measured hydraulic characteristics of a stream: the stream depth and shear velocity.

$$D_y = 0.23\ du^* \tag{16-13}$$

$$D_z = \frac{1}{15}\ du^* \tag{16-14}$$

where:

u^* = shear velocity = $\sqrt{gRS_e}$

g = acceleration due to gravity

R = hydraulic radius
S_e = slope of the energy gradient

Substitution of empirical estimates of D_y [Eq. (16-13)] into Eq. (16-8) and D_z [Eq. (16-14)] into Eq. (16-9) should not substantially affect the accuracy of the mixing distance equations.

$$L_y = \frac{\sigma_y^2}{.46d} \frac{u}{u*} \tag{16-15}$$

$$L_z = \frac{7.5\sigma_z^2}{d} \frac{u}{u*} \tag{16-16}$$

Using the empirical estimates of the turbulent diffusion coefficients does, in fact, increase the utility of the mixing distance criteria by requiring estimates only of the stream velocity and the shear velocity, both of which can be determined without going into the field. Average daily river flow for practically every river in the country is available through the USGS and can be used to determine river velocities. Since shear velocity is a function of known or easily estimable parameters (the acceleration due to gravity, the hydraulic radius, and the slope of the energy gradient), it can be determined with a minimum of effort as well.

Application of the microlocation criteria [Eq. (16-15) and Eq. (16-16)] for water quality monitoring stations suggested for the State of Massachusetts is presented in Table 16-2 (Sanders, 1974). The width of the river and the slope of the hydraulic gradient were estimated from USGS contour maps. The mean velocities from time of travel studies were used and, if not available, an average depth was assumed for calculating the velocity using the annual, average daily flow. The acceleration due to gravity was 32.2 ft/sec^2 and the location of the nearest outfall was determined from the Massachusetts Wastewater Discharge Survey (Bayon and Ramsay, 1972).

Although the mixing distances for complete lateral mixing appear exhorbitant, especially on the Connecticut River at Hatfield, Northfield and Springfield, it should be noted that the river is extremely wide and sluggish at these points and that the mixing distance is a function of the square of the distance from the outfall location to the farthest boundary. Ward (1973) found that a meandering river having a width 560 ft and a tracer injected at the midpoint, had a mixing distance of 84,000 ft and that a straight river having a width of 38 ft had a mixing distance of 3820 ft. Certainly the average flow rates, the slope of the energy gradients as well as channel geometries were different for the rivers investigated by Ward

Table 16-2. Application of the Microlocation Criteria for Sampling Stations of the Massachusetts' Portion of the Connecticut River Basin (Sanders, 1974)

River	Location	Order	Q	w	d	U	R	S_e	u*	L_y shore	L_z surface	L_y mid-stream	L_z mid-depth
			ft^3/sec	ft	ft	ft/sec	ft	ft/ft	ft/sec	ft	ft	ft	ft
Conn.	Hatfield	2	13490	850	6	2.64	5.91	0.000146	0.1668	*	713	*	178
Conn.	Northfield	3	10170	650	5	3.12	4.92	0.000146	0.1522	*	771	*	192
Conn.	Springfield	4	16020	1000	8	2.00	7.87	0.000146	0.1924	*	624	*	156
Conn.	Greenfield	4	11650	210	23	2.41	18.86	0.0033	1.4223	7,068	292	1,767	73
Ware	Ware	5	314	125	2	1.25	1.94	0.002	0.3533	60,381	53	15,095	13
Westfield	Knightville	5	318	150	2	1.06	1.95	0.01	0.7920	32,731	20	8,182	5
Millers	Athol	5	611	150	2	2.04	1.95	0.005	0.5600	88,940	54	22,235	13
Deerfield	Charlemont	5	880	190	2	2.32	1.96	0.01	0.7942	114,419	43	28,604	10

*The mixing distance exceed the distance to dams or other physical characteristics of the river which are not taken into account in the mixing distance criteria.

as compared to the rivers in the Connecticut River Basin but the magnitude of mixing distance calculations appear to be consistent.

With the exception of the sampling station at Greenfield, a zone of complete mixing is at least 6 miles downstream from an outfall located on the shore. For the Connecticut River at Hatfield the zone of complete mixing, with all parameters remaining constant, is in excess of 700 miles. Even if the point source were located at middepth and midstream the mixing distance would exceed 190 miles. Therefore, more than one sampling point in the lateral transverse would be necessary to determine a concentration representative of the stream cross section. At every station vertical concentration uniformity is completed in less than a thousand feet, and indicates at least theoretically that multiple depth sampling is not required.

It should be noted that the presence of a dam downstream would be the limit to which the mixing distance criteria would apply. Particularly in the Connecticut River Basin which has an abundance of hydropower dams, however, the water quality parameter distribution immediately downstream from the dam would be completely mixed.

In addition to the assumption that no temperature or concentration stratification exists and that the flow of the pollutant source is negligible compared to the flow of the river, it should be emphasized that the mixing distances calculated are derived from mean estimates of the flow, width and depth and that the rivers are assumed to be straight without bends which cause secondary lateral currents. Obviously the parameters will not remain constant and none of the rivers are straight throughout the entire mixing distance. Subsequently, the actual mixing distances will, most probably, be much less than predicted if concentration and temperature stratification are not significant and there exists no other source of the water quality parameter. As an example, with information received from the Water Survey of Canada (Tofte, 1973), the mixing distance on the Columbia River at Trail, B.C. for complete lateral mixing is calculated using Eq. (16-15) to be 37 miles for a source located on the shore and 9.3 miles for a source located at midstream. As previously noted, Reeder (1971) found from actual field measurements that complete mixing existed 10 miles downstream from the outfall at Trail, B.C. The difference between the theoretical and actual mixing distances is not unexpected as the precise location of the pollutant outfall was unavailable and the Columbia River below Trail was not straight, had several islands separating flow and combined with several smaller rivers prior to the measured point of complete mixing (Sanders, 1974).

16.6 PROCEDURE TO ESTIMATE SIGNIFICANT CONCENTRATION VARIATIONS

If a water quality sampling station is located at a point on a river within the calculated zone of complete mixing, the question of the number and location of sampling points in the river's cross section to provide a representative sample is minor. In this case when the water quality parameter concentrations are independent of depth and width (completely mixed), multiple sampling points will improve the accuracy of the parameter concentration by decreasing the confidence interval of the mean for a given level of significance but will not improve the representativeness of the data. It should be emphasized, however, that the concentration of water quality parameters having momentum vectors such as sediment will generally be a function of depth, except in extremely turbulent mixing and will not be considered here.

Whenever a stretch of river which must be sampled does not include a zone of complete mixing as is the case of the majority of the sampling stations in the Massachusetts' portion of the Connecticut River Basin, the problem of obtaining a representative sample still remains. If the location of the sampling station does not exceed either L_y or L_z, the number and location of sampling points in both the vertical and lateral directions must be considered. Cognizant of the problem of obtaining a representative sample in a study of the assimilative capacity of the Ohio River, the sampling agency divided the river's cross section into three sections having equal area and sampled at their respective midpoints (Ward, 1973). Although a river can be completely mixed vertically, Montgomery and Hart (1974) suggest sampling at least 300 mm below the water surface and 300 mm from the bottom. These locations are recommended to prevent sampling floating solids and bottom sediment regardless of the depth.

The variation of parameter concentrations in a river's cross section can be sizeable. The concentration of phenols, for example, were reported to vary in the Mississippi River at St. Louis from 50 ppb at each shore to approximately zero at midstream (Haney and Schmidt, 1958). There was also significant variation of BOD and dissolved solids. Colston (1974) found in a study of urban runoff that the majority of all pollutants studied exhibited higher concentrations near the bottom of the stream than near the surface. Only the 5-day BOD and Kjeldahl nitrogen concentrations were the same at both locations. The rest of the pollutants including chemical oxygen demand (COD), total organic carbon (TOC), total solids (TS), suspended solids (SS) as well as conservative parameters such as magnesium, manganese, lead and zinc had higher concentration near the stream bottom. Calcium alone had a higher concentration at the surface. However, it is not known whether the sampling points were within the theoretical zone of complete vertical mixing and whether there existed benthic deposits which could affect vertical as well as lateral concentration distributions.

Using a standard statistical procedure analysis of variance, the water quality concentration variation within a river cross section can be analyzed to determine if the variation is significant requiring more than one sampling point or insignificant establishing a relatively, completely mixed zone. This procedure will be applicable for points in a river which are at a distance less than L_y or L_z from an upstream outfall or at points in a river where data is unavailable for estimating zones of complete mixing. Once a level of significance which is a function of allowable variability is established by the sampling agency, the question of variability can be quantitatively analyzed. As long as the concentration variability in time is negligible relative to variation between locations, one way analysis of variance is applicable. However, if the variability of the parameter during the time interval when the samples are taken is significant, an additional dimension in the analysis of variance should be considered.

Under most circumstances, complete mixing will exist in the vertical direction, and the parameter concentrations will vary with time. Therefore a two-way analysis of variance would be required to assess the variability of the parameter concentrations in the lateral cross section. If the variability of the parameter concentration in time were not taken into account in the analysis, the possibility would exist that the variability of the parameter at different locations in the lateral transect may be the result of the time variability and not spatial variability.

In a two-way analysis of variance the assumptions are made that there is a variability of the data in time and location, and that the magnitudes of the respective variabilities are independent and additive. That is,

$$S_{TOT} = S_E + S_T + S_L \qquad (16\text{-}17)$$

where:

$$S_{TOT} = \sum_{i=1}^{n_i} \sum_{j=1}^{n_j} (x_{ij} - \bar{x})^2, \text{ total variation}$$

$$S_T = \sum_{j=1}^{n_j} \sum_{i=1}^{n_i} (\bar{x}_{i\cdot} - \bar{x})^2, \text{ variation due to time}$$

$$S_L = \sum_{i=1}^{n_i} \sum_{j=1}^{n_j} (\bar{x}_{\cdot j} - \bar{x})^2, \text{ variation due to location}$$

S_E = unaccounted random variation

$$\bar{x}_{i\cdot} = \frac{1}{n_j} \sum_{j=1}^{n_j} x_{ij}$$

$$\bar{x}_{\cdot j} = \frac{1}{n_i} \sum_{i=1}^{n_i} x_{ij}$$

n_i = number of data in each column

n_j = number of data in each row

The mean squared deviation with respect to time and location divided by the mean squared error are independent random variables having an F distribution with $(n_i - 1)$, $(n_i - 1)(n_j - 1)$ and $(n_j - 1)$, $(n_i - 1)(n_j - 1)$ degrees of freedom respectively (Guttman and Wilks, 1965). Therefore, using an Anova table (Table 16-3) and the null hypotheses H_o, that the parameter concentration sampled at different times and locations are taken from the same population having an identical mean and variance can be tested against the alternate hypotheses H_1, that the parameter concentrations sampled at different locations and at different times in a river's cross section are drawn from different populations having different means and variances. The hypotheses which test the equality of the variances are:

$$H_o: \frac{\sigma_L^2}{\sigma_E^2} = 1, \quad \frac{\sigma_T^2}{\sigma_E^2} = 1$$
$$H_1: \frac{\sigma_L^2}{\sigma_E^2} \neq 1, \quad \frac{\sigma_T^2}{\sigma_E^2} \neq 1 \qquad (16\text{-}18)$$

where:

H_o is the null hypothesis

H_1 is the alternate hypothesis

$\sigma_L^2 = S_L/n_j - 1$

$\sigma_T^2 = S_T/n_i - 1$

$\sigma_E^2 = S_E/(n_j - 1)(n_i - 1)$

The sequence of the calculations to test the validity of the null hypothesis is illustrated in Table 16-3. Of primary importance is the calculation of S_E, the unaccounted variability. It is the difference between the total variation and the variation in time and location. The calculated ratio of the mean spatial variability and mean unaccounted variability as well as the ratio of the mean temporal variation and mean unaccounted variability designated in the table as F calculations are estimates of the limiting F constants.

If the F calculation were less than the F constant, a function of the degrees of freedom, level of significance and found in most statistics tables, the ratio of the variances equal one and the null hypothesis is valid [Eq. (16-18)].

Table 16-3. An Anova Table Used in a Two-Way Analysis of Variance

	Degrees of Freedom	Sum of Squares	Mean Sum. Sq.	F Calc.
Rows	$n_i - 1$	S_T	$S_T/n_i - 1$	$\frac{S_T(n_j - 1)}{S_E}$
Columns	$n_j - 1$	S_L	$S_L/n_j - 1$	$\frac{S_L(n_i - 1)}{S_E}$
Error	$(n_i - 1)(n_j - 1)$	$S_E = S_{TOT}-S_T-S_L$	$S_E/(n_i-1)(n_j-1)$	
Total	$(n_i)(n_j) - 1$			

However, if the F calculation exceeds the F constant the ratio of the variances is not equal to one and the null hypothesis is not valid. Therefore, the water quality concentrations would not have been derived from the same population and would be a function of the location or time or both.

An example of using a two-way analysis of variance as a method of assessing spatial variation for determining the adequacy of one sampling point to obtain a representative sample in the lateral direction is presented using data collected and analyzed from the Connecticut River (Sanders, 1974). Grab samples were collected every five minutes for a period of 2.5 hours at six different sampling locations. Separated by approximately 120 feet in the lateral direction, the sampling points were 2.3 miles downstream from the nearest outfall (Deerfield Sewage Treatment Plant). This distance exceeded the mixing distance L_z, for uniform concentrations in the vertical direction but was far less than, L_y, the mixing distance for concentration uniformity in the lateral direction. The analysis of variance should detect that concentration variations are significantly different at a specified level of significance in the lateral cross section as complete mixing would not occur 2.3 miles downstream from the outfall. The outfall was located within 20 feet of the western bank of the river and subsequently the parameter concentration should have been higher downstream at the west bank and decreased as one moved due east. The flow at the upstream outfall was insignificant compared to the average flow of the Connecticut River, 0.676 MGD and 8700 MGD respectively.

Sodium was chosen as the parameter to assess the spatial variability in the river's cross section. It is generally not considered a pollutant, but it is conservative, and ubiquitous both naturally and in sewage. Most significant is its ability to be detected with high precision at concentrations of less than 1 mg/l with the Perkin Elmer atomic absorption spectrophotometer Model 303. In addition the sodium concentration in the Connecticut River has been reported upstream from the outfall in the range of 10 mg/l, easily detected by the spectrophotometer.

The depth of the Connecticut River on the day of the sampling was approximately four feet at the predetermined sampling location, which eliminated the necessity for using the depth samplers. Each person sampled approximately 1.5 feet from the water surface: this being the length of an extended arm from the fingers to the elbow. The sampling points (boats) were on a line perpendicular to the western bank. The distances from the western bank measured using a sextant are listed in Table 16-4.

The absorbance which is linearly related to the concentration of the sodium for each of the 186 grab samples are listed in Table 16-4. It is apparent that the concentration at the western side of the river is higher than at the middle or eastern

Table 16-4. The Absorbance Which is Linearly Related to the Sodium Concentration of the 31 Grab Samples at Each of Six Sampling Points on the Connecticut River. Location 1 is Located near the Western Shore of the River and Location 6 near the Eastern Shore. The Samples Were Taken at Five Minute Intervals

Absorbance (percent)

	Location					
Sample Number	1	2	3	4	5	6
1	0.3242	0.2924	0.2976	0.3116	0.3098	0.3080
2	0.2660	0.2388	0.2269	0.2299	0.2262	2.2306
3	0.2644	0.2381	0.2366	0.2351	0.2373	0.2381
4	0.2993	0.2774	0.2790	0.2741	0.2725	0.2725
5	0.2993	0.2676	0.2676	0.2720	0.2668	0.2741
6	0.2636	0.2472	0.2381	0.2381	0.2306	0.2366
7	0.2725	0.2534	0.2472	0.2426	0.2388	0.2403
8	0.2725	0.2441	0.2441	0.2457	0.2503	0.2441
9	0.2660	0.2487	0.2457	0.2503	0.2518	0.2426
10	0.2628	0.2373	0.2472	0.2381	0.2426	0.2426
11	0.2660	0.2457	0.2441	0.2441	0.2381	0.2472
12	0.2692	0.2457	0.2411	0.2411	0.2472	0.2518
13	0.2628	0.2441	0.2403	0.2457	0.2426	0.2441
14	0.2557	0.2411	0.2487	0.2503	0.2441	0.2472
15	0.2628	0.2381	0.2373	0.2381	0.2233	0.2248
16	0.2596	0.2411	0.2381	0.2457	0.2457	0.2396
17	0.2596	0.2496	0.2449	0.2396	0.2426	0.2472
18	0.2636	0.2441	0.2373	0.2351	0.2351	0.2240
19	0.2549	0.2451	0.2381	0.2336	0.2441	0.2441
20	0.2534	0.2336	0.2218	0.2306	0.2336	0.2218
21	0.2549	0.2291	0.2248	0.2351	0.2403	0.2336
22	0.2708	0.2464	0.2291	0.2291	0.2358	0.2336
23	0.2596	0.2457	0.2262	0.2306	0.2314	0.2336
24	0.2457	0.2262	0.2336	0.2248	0.2351	0.2411
25	0.2426	0.2336	0.2226	0.2175	0.3182	0.2306
26	0.2403	0.2218	0.2204	0.2211	0.2262	0.2321
27	0.2226	0.2076	0.2104	0.2013	0.2007	0.2132
28	0.2218	0.1952	0.1911	0.1878	0.1871	0.1952
29	0.2676	0.2472	0.2426	0.2480	0.2396	0.2425
30	0.2765	0.2487	0.2464	0.2457	0.2457	0.2518
31	0.2596	0.2441	0.2381	0.2336	0.2381	0.2449
Mean	0.2632	0.2419	0.2389	0.2389	0.2394	0.2413
Distance From West Bank (ft)	155	247	347	4	595	820

side. And it is also apparent that the sodium absorbance appears to be related to the time of the sampling. The absorbance at every location tended to decrease and then increase during the sampling period. The time dependency, whether it was actual temporal variation, or the result of minor electronic variability of the instrument, could not be ignored. Thus it was necessary to use the two-way analysis of variance to assess the spatial variation in the lateral direction.

Table 16-5 is the Anova table for the two-way analysis of variance of the sodium absorbance.

Table 16-5. An Anova Table of a Two-Way Analysis of Variance of Sodium Absorbance from Six Lateral Sampling Locations in the Connecticut River

	Degrees of Freedom	Sum of Sq. Deviation	Mean Sum of Sq. Deviation	F Calc.	F Constant
Rows	30	0.065	0.002	79.475	1.54
Columns	5	0.014	0.003	103.194	2.27
Error	150	0.004	0.000		
Total	185	0.083			

As indicated by the comparison of the F calculated to the F constant, there is substantial variation of the parameter absorbances both with time (rows) and location (columns). Since by definition of two-way analysis of variance, the variation with time is independent and does not effect the variation with location, the results indicate that the cross section is not completely mixed in the lateral direction: more than one sampling point is needed. The mean absorbance was the highest near the western shore (Location 1) and decreased moving east to the middle of the river (Location 4). However, moving from the middle of the river to the eastern most shore the sodium absorbance increased. It was assumed that the sewage treatment plant located on the east bank 3 miles upstream had such a low flow, less than 0.25 MGD, that it would not affect the downstream concentration. Apparently, the assumption was unfounded. Although the difference of the sodium concentration in the lateral transverse is negligible, it is nevertheless, statistically significant. Other water quality parameters which include total organic carbon, chlorides, BOD, etc., quite probably do vary in the cross section and would be of sufficient magnitude that the concentration differences may not be negligible.

In order to obtain a representative mean parameter concentration at this particular sampling location on the

Connecticut River, more than one sampling point would be required. Parameter concentrations at just the midpoint of the river would not suffice as the samples would not take into account the sewage effluents. Similarly, water quality samples at either shore would not be representative as the effluent concentrations of the sewage treatment plants would have a significant effect on the magnitude of the water quality concentrations. Therefore, it would appear that sampling at the third points of the lateral transverse would be the minimum number of sample locations to adequately define mean concentrations representative of the river's cross section at a location which is not within a zone of complete mixing (Sanders, 1974).

16.7 FREQUENCY OF SAMPLING

Sampling frequency is a very important parameter which must be considered in the planning of a water quality monitoring network which is to provide data for use in the determination and verification of water quality models as well as for estimating ambient water quality and determining long-term trends. A large portion of the costs of operating a monitoring network is directly related to the frequency of sampling. In addition, the reliability and utility of water quality data derived from a monitoring network is likewise related to the frequency of sampling. Addressing this anomaly Quimpo et al. (1970) summarized the significance of sampling frequency and stated that:

> "On the one hand, by sampling too often, the information obtained is redundant and thus expensive, and on the other hand, sampling too infrequently bypasses some information necessitating an extended period of observation."

Significant as sampling frequency is to detecting stream standard violations, maintaining effluent standards, estimating temporal changes in water quality, and defining the natural water quality variations very few quantitative criteria exist which designates appropriate sampling frequencies which can be applied for the planning of a water quality monitoring and data collection network. In many cases professional judgment and cost constraints provide the basis for sampling frequencies. Because sampling agencies will differ substantially in the expertise of their personnel in sampling as well as the amount of money available for monitoring, sampling frequencies to obtain similar data, i.e., estimating ambient water quality characteristics, will differ depending on the agency.

Responsibility for gathering hydrologic data is divided among several federal agencies. The National Weather Service has primary responsibility for gathering precipitation, evaporation and wind data. The Soil Conservation Service and other federal agencies collect snowfall data. The United States Geological Survey (USGS) is the primary federal agency responsible for the collection and dissemination of river flow data. It also collects water quality data on a daily basis at many sampling stations (Rainwater and Thatcher, 1960). Historically the sample collection frequency of various agencies has differed widely. For example, at 730 sampling stations dispersed throughout the United States in 1966, the Federal Water Pollution Control Administration collected water quality samples every week (Gunnerson, 1967). The State of New York's Department of Environmental Conservation collected one sample per month during the period 1965-1967 (State of New York, Department of Environmental Conservation, 1967). In 1973, at the majority of the water quality sampling stations in Canada, samples were collected monthly (Inland Waters Directorate, 1973). Various sampling frequencies were found in Europe as well: in 1972 British authorities collected a 24-hour composite sample every eight days (Briggs, 1972) and in the Ruhr area of Germany in 1967, the water quality management authorities sampled once or twice per year. That specific objectives of the water quality sampling agencies mentioned may vary (which in itself may account for the different sampling frequencies) does not belie the fact that quantitative criteria establishing sampling frequencies were not mentioned or were arbitrarily selected by the various sampling authorities and because the sampling frequencies do vary the utility of the water quality data will likewise vary.

As recently as 1973, the inventory and monitoring section of EPA did not have standard procedures to establish sampling frequencies (Personal Communication, 1973, Walsh, W. J.). In a report by the Office of Water Data Coordination of the USGS "Recommended Methods for Water Data Acquisition" (1973) little mention was made of sampling frequencies except to "collect an appropriate volume of a representative sample." In a government report suggesting sampling frequencies "A Practical Guide to Water Quality Studies of Streams," Kittrell (1969) points out that sampling once a day is sufficient to establish water quality for a monthly average. He noted that samples must be collected every day at different times so that variation of parameter concentrations during the day may be taken into account. To what extent the average monthly parameter concentration is dependent on sampling frequency or how the accuracy is improved using daily samples versus, for example, weekly samples, is not mentioned. However, Kittrell (1969) has attempted to establish and apply a sampling frequency based on the real or perceived variability of the data and the adequacy of the mean derived from the data. Mongomery and Hart (1974) found that systematic sampling was better than random sampling and that sampling should not be multiples of a week. Otherwise, the parameter variations which have a weekly cycle would go undetected.

In an investigation of the accuracy of a mean and standard deviation using systematic sampling (equispaced sampling) data versus randomly sampled data, Rainwater and Avrett (1962) state that in many cases systematic sampling produces more accurate estimates of the annual mean and standard deviation. Time series data of hydrologic events (river flow, etc.), are not random but in general highly correlated to the preceding event. Therefore in a series of water quality data which are highly correlated the sample average from equispaced samples would tend to be a better estimate of the true mean than the average of an equal number of random events. To verify this assertion, the variance of the annual mean and standard deviation of a time series were compared to the variances of the identical statistics of the same time series by using an estimate of the number of independent observations in the time series rather than the actual number of observations. The number of random or independent observations was equal to the number of observations minus one degree of freedom divided by twice the number of sequential sign changes of the data (an observation less than the median followed by an observation greater than the median constitutes a sequential sign change). A mathematical expression for the number of independent observations is:

$$\left[\frac{n-1}{2m}\right] n \qquad (16\text{-}19)$$

where:

n = number of observations
m = number of sequential sign changes

Although empirical in nature, the results indicated that the estimated error (standard deviation) derived from random error formulas which took into account the number of independent observations was greater than the actual error calculated from the data. If the number of sequential sign changes were greater than one-half the number of observations of a time series, there would be no advantage of using equispaced sampling instead of an equal number of random samples. However, for this to occur, sampling intervals of three months or greater are required (Rainwater and Avrett, 1962).

Sampling criteria have been established for water quality monitoring in the State of California and have been derived, not from the variability of the parameter, but, instead, from river flow and river basin characteristics which in some respect are functionally related to the variability of the water quality parameter. Pomeroy and Orlob (1967) defined sampling frequencies as functions of the drainage area and the ratio of the maximum flow to the minimum flow. Without citing quantitative data, it is recommended that large watersheds with a drainage area greater than 1000 square miles should be sampled at least 12 times a year, whereas a watershed which drains 10 square miles should be sampled twice a week. A wild stream which has a ratio of maximum flow to minimum flow greater than 100 should be sampled weekly and a well-regulated river denoted by a ratio of less than 10 should have minimum surveillance, presumably once a month (Fig. 16-3).

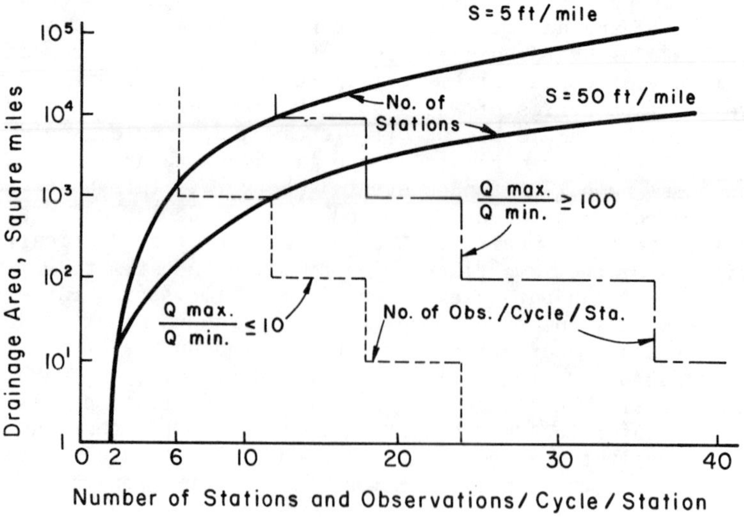

Fig. 16-3. The Minimum Number of Sampling Stations and Sampling Observations for Water Quality Characterization of a River (Pomeroy and Orlob, 1967)

Although not yet applied to present monitoring networks, methods have been developed which may be used to determine sampling frequencies based on the variability of the water quality parameters. Other methods available establish sampling frequencies based on the periodicities of parameter fluctuations, the capacity to detect a stream standards violation and the information content of the annual mean and variance.

In a study on the Potomac River at Washington, D.C. Gunnerson (1966) found that sampling intervals of 2-3 hours can adequately characterize the variation of dissolved oxygen in the Potomac estuary. It was found by spectral analysis techniques and plotting the variance of the data versus sampling interval (lag) that the dissolved oxygen had natural periodicities denoted by peaks on a graph of 12 hours, 24 hours, and 14 days. These natural periodicities were related to the semidiurnal tides, photosynthesis and the lunar fortnightly tide respectively. A similar analysis of specific conductance was made. However, only one natural periodicity was found - semidiurnal, a result of the tides. Theoretically the time interval between samples which define a periodic function having a period of 12 hours would have been 6 hours (which is the Nyquist frequency) but in practice sampling intervals less than one-half the periodicity were required to adequately describe the natural fluctuations of parameter concentrations.

Wastler (1963) concluded from his investigation of BOD (biochemical oxygen demand) and DO (dissolved oxygen) concentration variations in the Potomac River at the confluence of the Anacostia River (Fig. 16-4), that spectral analysis techniques can be a valuable tool in the interpretation of equally spaced time series data and detecting periodic fluctuations.

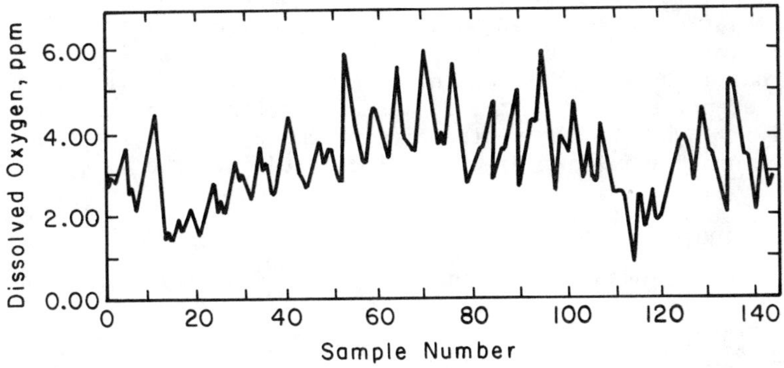

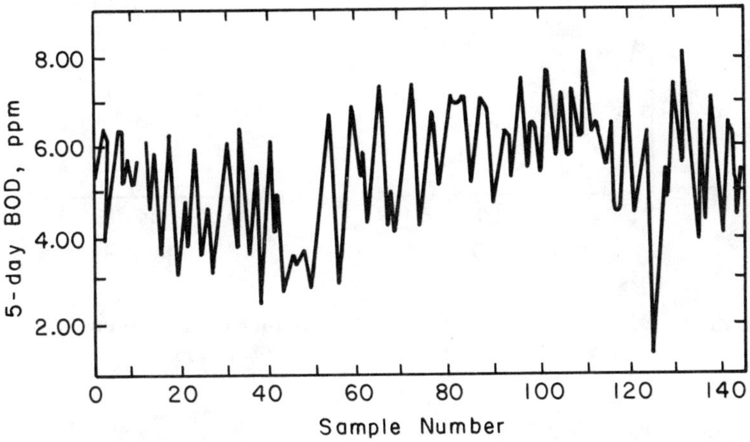

Fig. 16-4. Dissolved Oxygen and Biochemical Oxygen Demand Obtained from the Potomac Estuary Taken at 4-Hour Intervals (Wastler, 1963)

However, he noted that to fully utilize the potential of spectral analysis techniques several "rules of thumb" should be considered. They are:

1. The time series record must be fairly long - 100 or more data.
2. No missing data, and that if data are missing, interpolation must be done which will not affect the spectral density plots as long as the missing data

are "widely scattered" and represent no more than 5% of the total amount of data.

3. The time interval between samples must be approximately 1/3 the length of the shortest periodicity expected.
4. The number of lags which can be investigated is equal to approximately 10% of the number of available data.

Wastler established that both DO and BOD exhibited periodicites of 12 and 24 hours (Fig. 16-5). The 12 hour periodicities were a result of tidal action. However, the diurnal periodicity of the DO was a function of photosynthetic activity, whereas the periodicity of the BOD was a function of the diurnal variation of waste discharged into the river.

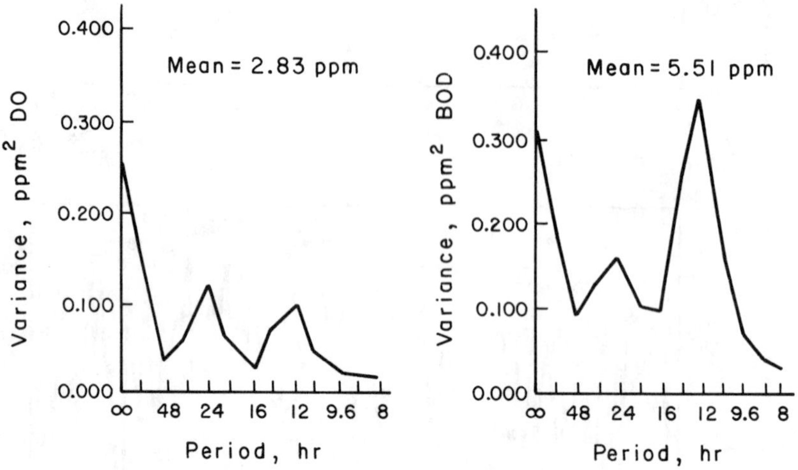

Fig. 16-5. Dissolved Oxygen and Biochemical Oxygen Demand Spectra from Data Plotted in Fig. 16-4 (Wastler, 1963)

Gunnerson (1967) and Wastler (1963) have shown that by using spectral analysis techniques a minimum number of samples required during a period of time to detect and effectively evaluate the parameter variations can be determined. For example, when sampling for DO or specific conductance concentrations in the Potomac River, a sampling frequency greater than 4 times a day is required. In non-tidal rivers, for parameters which exhibit diurnal periodicities sampling frequencies could be decreased to 3 times a day. It appears that DO, BOD, specific conductance and copper (Montgomery and Hart, 1974) exhibit diurnal fluctuations requiring several samples per day or a 24-hour composite, but the periodicities, if they exist at all, of the majority of the other water quality parameters remain to be determined.

There is a question as to the applicability of the spectral analysis technique by itself to define sampling frequencies for a water quality monitoring network other than to determine periodicities of water quality parameters from a set of equispaced time series data and specifying a sampling interval of one-third the period. It may be concluded, however, that a minimum of three equispaced samples a year are necessary to characterize a parameter concentration having an annual cycle. Although the resulting annual mean parameter would hardly seem adequate for determination of annual trends or use in a water quality model, the four month time interval between samples would represent the maximum time interval between samples. Montgomery and Hart (1974) and Pomeroy and Orlob (1967) have recommended that six equispaced samples per year be required to characterize parameter concentrations having an annual cycle.

Sampling frequency can also be specified quantitatively as a function of the information content of the data. This information content I, as defined by Quimpo and Yang (1970), is the effective number of independent observation N_1 divided by the actual number of observations N_2 and is, in fact, a measure of the correlation of the time series data.

$$I = \frac{N_1}{N_2} \tag{16-20}$$

For a highly correlated time series in which each observation is functionally related to the preceding observation the effective number of independent observations would be very low and the relatively new information content of the data would be low. For a time series having no correlation among successive observations the number of independent observations would equal the number of actual observations and the relative information content would be equal to one.

Bayley and Hammersley (1946) defined two effective numbers: one is associated with the variance of the sample mean and the other is associated with the variance of the sample variance.

$$\mathrm{Var}(\bar{x}) = \frac{\sigma^2}{n_b{}^*} \tag{16-21}$$

$$\mathrm{Var}(s^2) = \frac{2\sigma^4}{(n_v{}^* - 1)} \tag{16-22}$$

where:

x_i = an equally spaced time series observation

$\bar{x} = \frac{1}{n} \sum_{i=1}^{n} (x_i)$

σ^2 = variance of x_i (theoretical)

n_b^* = effective number of independent observations associated with the variablity of the annual mean

n_v^* = effective number of independent observations associated with the variability of the annual variance

$s^2 = \frac{1}{n-1} \sum_{i=1}^{n} (x_i - \bar{x})^2$

n = number of data

Defining the autocorrelation coefficeint $\rho(j)$ as:

$$\rho(j) = \frac{E(x_i x_j)}{\sigma^2} \qquad (16\text{-}23)$$

where:

$\rho(j)$ = autocorrelation coefficient of the variable x_i separated by j time units;

$E(x_i x_j)$ = the expected value of the product $x_i x_j$

It was found that the effective numbers may be defined as a function of the summation of the autocorrelation coefficients of the time series.

$$n_b^* = \frac{n}{\sum_{j=0}^{n-1} \rho(j)} \qquad (16\text{-}24)$$

and

$$n_v^* = \frac{n}{\sum_{j=0}^{n-1} \rho^2(j)} \qquad (16\text{-}25)$$

Expanding the series in the denominator:

$$\sum_{i=0}^{n-1} \rho(j) = \rho(0) + \rho(1) + \rho(2) + \rho(3) \ldots \rho(N-1) \qquad (16\text{-}26)$$

and since: $\rho(k) = \rho(1)^k$ for a 1st order Markov process and $\rho(0) = 1$, Eq. (16-26) can be simplified to:

$$\sum_{i=0}^{n-1} \rho(j) = 1 + \rho(1) + \rho(1)^2 + \rho(2)^3 \ldots + \rho(1)^{n-1} \qquad (16\text{-}27)$$

Inserting Eq. (16-27) into Eq. (16-24), multiplying both the numerator and denominator by $(1-\rho(1))$ and subtracting, Eq. (16-24) becomes:

$$\frac{n_b^*}{n} = \frac{1-\rho(1)}{1-\rho(1)^n} \qquad (16\text{-}28)$$

Similarly Eq. (16-25) can be simplified, Quimpo and Yang (1970)

$$\frac{n_v^*}{n} = \frac{1-\rho(1)^2}{1-\rho(1)^{2N}} \qquad (16\text{-}29)$$

As long as the water quality time series can be assumed to be a first order Markov process Eq. (16-28) may be used to estimate the information content for various sampling intervals. Equation (16-24) is less restrictive and may also be used to establish the relationship of the information content to sampling frequency.

Utilizing Eq. (16-29) which is a function of the number of independent observations used to estimate the variability of the sample variance Quimpo and Yang (1970) determined the information content of stream flow, time-series, data whose different sampling intervals ranged from 1 to 20 days (Fig. 16-6). The mean river flow was calculated for each sampling

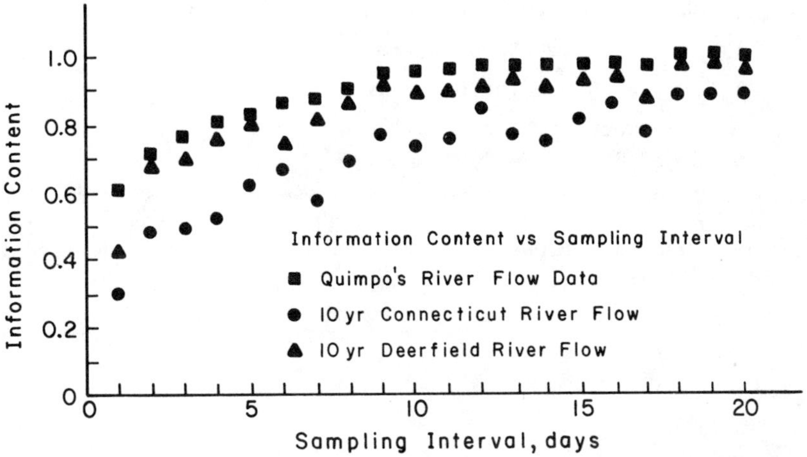

Fig. 16-6. Information Content of the Annual Variance as a Function of Sampling Frequency for Flow Data from Quimpo, the Connecticut River and the Deerfield River (Sanders, 1974)

period rather than using the data separated by the sampling interval. In other words, the mean river flow for a three day period was used instead of the river flow every three days. The use of the time interval means would have a smoothing effect in a plot of information content versus time interval; the effect of cyclic flow variations would be less noticeable in the calculation of the information content. With reference to Fig. 16-6, the curve which defines information content versus sampling frequency of Quimpo's data is smoother than the curves from the Connecticut and Deerfield Rivers' data which were not averaged. The decrease of information at the 7 and 14 day time intervals illustrates the effect of the weekly flow regulation for power production. Quimpo's data indicated that for an information content $I_2 = .80$ the sampling interval for stream flow was every 4 days and for $I_2 = .90$ the sampling interval was 9 days while the intervals for the same information content of the Connecticut and Deerfield River flows were 5 and 11 days, and 8 and 16 days respectively.

Using the information content defined as the reciprocal of the variance of the mean of the random variable, $\bar{\varepsilon}_t$, of a time series fitted with a first order Markov model, Knisel and Yevjevich (1967) found that as the time interval increases, the time series of the flow averaged over the time intervals represented a decrease of information. An example of a 1st order Markov process is:

$$Y_t = r\, Y_{t-1} + \varepsilon_t \qquad (16\text{-}30)$$

where:

Y_t = an observation at time t

Y_{t-1} = the preceding observation

r = calculated correlation coefficient

ε_t = random variable

To be more specific, in terms of information, it is "more advantageous" to analyze mean daily flows rather than mean weekly or monthly flows. Information content "I" is defined in Eq. (16-31) as:

$$I = \frac{N}{s_t^2} = \frac{T}{s_t^2 \Delta T} \qquad (16\text{-}31)$$

where:

N = number of data

s_t^2 = variance of the random shock, ε_t

T = length of record

ΔT = time interval between sample averages.

The apparent discrepancy between information content as used by Quimpo and that of Knisel is due to the disparity in the definitions. Quimpo (1968) related information as a function of the effective number of random data divided by the actual number of data. On the other hand, Knisel and Yevjevich (1967) and Dawdy et al. (1972) define information as the reciprocal of the variance of the mean random variable. As the time interval between observations increases, the residual mean variance increases, thus decreasing the information content of the time series.

The selection of a sampling frequencies for a water quality monitoring network can be derived from the criteria discussed: the periodicities of water quality parameters, and information content.

Specifying sampling frequencies to detect harmonic variations of data, Gunnerson (1966) and Wastler (1963) recommend an interval of one-third the periodicity of the harmonic: this would indicate at least three samples a year or one every 4 months are needed for an estimate of the annual mean. Using the level of new information as the basis for specifying sampling frequencies, Quimpo (1968) found that river temperature sampled daily had an information content of less than 0.2 and that data sampled every two weeks had an information content of 0.9. From estimates of Knisel and Yevejevich's (1967) data, information content of daily flow data was approximately twice that of flow averaged over a two week interval.

Although each of the aforementioned criteria provide a rational basis for specifying sampling frequencies, each is limited. Use of spectral analysis techniques discussed by Gunnerson (1966) and Wastler (1963) indicates the maximum sampling intervals necessary in the detection of harmonics but do not indicate the accuracy of statistics which result from the data. Information content, as criteria to specify sampling frequencies, is inadequate because both Quimpo and Yang (1970) and Knisel and Yevjevich (1967) definitions of information content do not give an indication of the variability or precision of an annual statistic calculated from the time series, data sampled at the different time intervals. The frequency of sampling a water quality parameter can be derived as a function of its random variability, more specifically, sampling frequency can be specified according to the expected half-width of the confidence interval of the annual mean parameter concentration.

In a series of random events, the accuracy of the mean is a function of the number of sample observations. As the number of samples increases, the expected confidence interval of the mean decreases at a rate inversely proportional to the square root of the number of samples (Mace, 1964). Therefore, for a specified confidence interval of the mean, a sampling frequency (number of samples per year) can be determined. Unfortunately, the majority of hydrological time series data including daily river flow are not random but highly correlated, nonstationary and therefore inappropriate for standard statistical analysis. However, after transforming the time series data by removing the correlation and nonstationarity, the remaining residual data can be assumed to be random, identically distributed and amenable to standard statistical analysis.

In order to estimate the relationship between sampling frequency and the confidence interval of the mean, the Students "t" statistic can be utilized. The variable t defined by Eq. (16-32) has a Students "t" distribution if the observations x_i; $i = 1,2,\ldots n$ are stationary, independent, identically distributed, and Gaussian.

$$t = \frac{\bar{x} - \mu}{S/\sqrt{n}} \qquad (16\text{-}32)$$

where:

$\bar{x}$ = calculated sample mean

μ = theoretical or population mean = $E[x]$

S^2 = the sample variance of $x_i = \sum_{i=1}^{n} \frac{(x_i - \bar{x})^2}{n-1}$

n = number of independent observations.

For a given level of significance, the variable t will be contained in a confidence interval bounded by known constants. That is, the probability that the random variable t is contained within the known constants is equal to the level of significance $(1-\alpha)$ and the probability that the variable t is not contained within the constants is equal to α. Using common statistical notation, this can be written as:

$$P_r\{t_{\alpha/2} < \frac{\bar{x}-\mu}{S/\sqrt{n}} < t_{1-\alpha/2}\} = 1 - \alpha \qquad (16\text{-}33)$$

where:

$t_{1-\alpha/2}$, $t_{\alpha/2}$ = constants defined from the Students "t" distribution associated with the level of significance and the number of samples

$1 - \alpha$ = specified level of significance

Having established the distribution and knowing that $t_{1-\alpha/2} = -t_{\alpha/2}$, the confidence interval of the population mean is:

$$\bar{x} + \frac{t_{\alpha/2}S}{\sqrt{n}} > \mu > \bar{x} - \frac{t_{\alpha/2}S}{\sqrt{n}} \qquad (16\text{-}34)$$

and the expected width of the confidence interval of the mean (Mace, 1964) is:

$$\text{Confidence Interval} = \frac{2\, t_{\alpha/2}S}{\sqrt{n}} = 2R \qquad (16\text{-}35)$$

where:

S = standard deviation of x_i

R = one-half the expected confidence interval of the population mean

Equation (16-35) can be applied directly to specify the number of replicate laboratory analyses for each water quality sample, this is due to the fact that any aliquot from a discrete well mixed water sample can be considered a sample from an identical, independent population. The water quality parameter concentration of the aliquot (as long as it is conservative or does not change during the period of analysis) will be random, uncorrelated, identically distributed and can be assumed Gaussian. Subsequently, the number of sample analyses for estimating a mean concentration can be determined as a function of the confidence interval of the mean, using the Students "t" distribution. The unknown parameter, which either must be calculated or known, is the standard deviation (square root of the variance). This parameter is a function of the procedure used to determine the water quality, parameter concentrations and should be estimated for each water quality parameter.

Because there are standard procedures (Standard Methods, 1965) to be used for the analysis of water and because the variation of some of these procedures are known, estimates of the standard deviation are available without performing 30-40 chemical analyses of the water sample. If, however, the standard deviations are not published, which is the case for many water quality parameters, the standard deviations must be calculated preferably in the labs and by persons who will be performing the routine analyses.

Using the mercuric nitrate methodology for determining chloride concentrations, as an example, the standard deviation of samples having a concentration of 241 mg/l is 3.5 mg/l (Standard Methods, 1965). Therefore, a plot of the one-half width of the confidence interval of the mean versus the number of sample analyses can be drawn utilizing Eq. (16-35) (Fig. 16-7). The mean chlorides concentration at a 95% level of significance analyzing only two samples will have an expected half-width of the confidence interval of approximately 31 mg/l. However, by analyzing four samples, the half-width of the confidence interval of the mean will decrease to 5.6 mg/l: a six-fold increase in the accuracy of the mean concentration.

Although the estimation of the standard deviation and the plot of one-half the confidence interval versus the number of analyses is very time consuming, it will enable the establishment of an allowable error. As a result, the specification of the accuracy of each sample mean will ultimately increase the validity and utility of the subsequent data. Clark (1950) made the observation:

> "...it hardly seems logical to apply a refined mathematical treatment to data which are not within say, 5% of the most probable value to begin with."

That is to say that model development, analysis of variance or

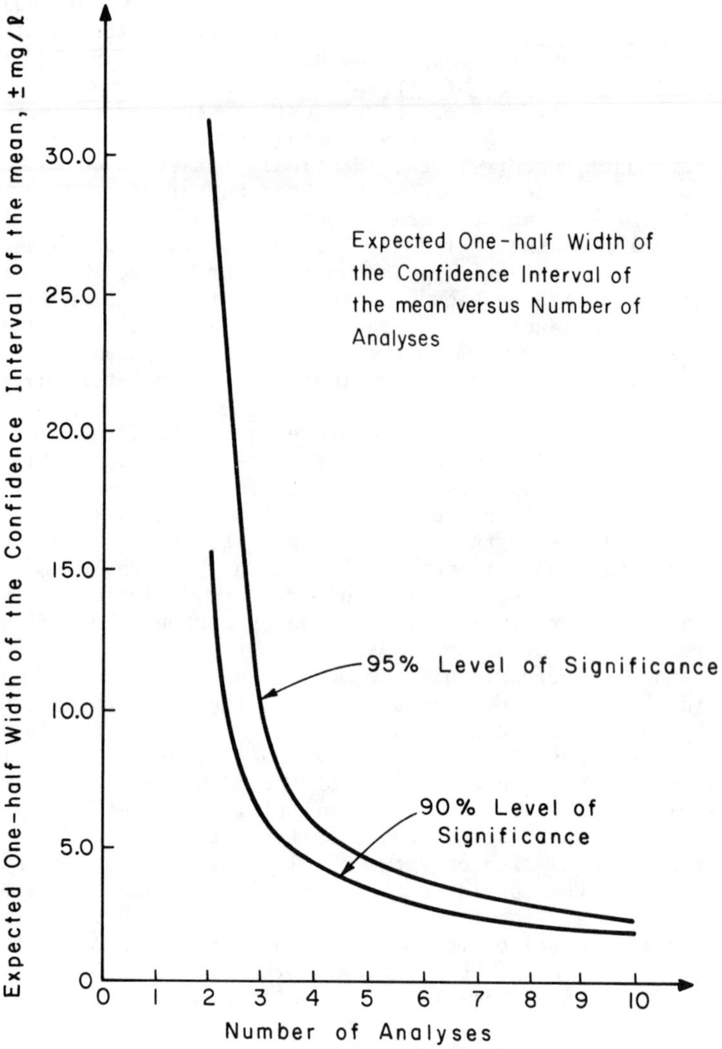

Fig. 16-7. The Expected One-Half Width of the Confidence Interval of Mean Chloride Concentration Versus the Number of Sample Analyses Using the Mercuric Nitrate Method

even trend analysis is ludicrous if the data from each water quality sample is not accurate within certain limits.

If the variability of the water quality parameter concentrations due to laboratory analysis is not known and minimized within reasonable limits, quantitative use of the data to model the spatial and temporal variation of the parameter concentrations may not be possible. The variability of parameter concentrations within a river's cross section as well as the variability in time may not exceed the variability of the mean as a result of the laboratory analysis, at a specified level of significance.

The confidence interval of the mean is, therefore, a function of the standard deviation of the observed data, the constant from the Students "t" distribution and the square root of the number of data. A similar application of the Students "t" distribution serves as the basis for the specification of sampling frequencies as well, using a quantitative definition of the confidence interval of the annual mean.

It would seem logical that the derivation and development of appropriate sampling frequencies for a water quality monitoring and data collection network should be derived from water quality data: dissolved solids, total organic carbon, biochemical oxygen demand, conductivity, etc. However, Sanders (1974) found that records of water quality data collected daily of sufficient length are available at very few sampling stations in the United States and were nonexistent in the Connecticut River Basin. Therefore, because an abundance of daily river flow data existed at many guaging stations in the Connecticut River Basin, and because river flow has been found to be highly correlated with many water quality parameters (suspended solids dissolved solids, specific conductance, etc.), river flow data were used to establish sampling frequencies for the water quality monitoring network. It should be noted, however, as water quality data are accumulated the procedures used in the determination of sampling frequencies can be readily applied to water quality data, especially those variables which are not necessarily related to river flow.

16.8 RIVER FLOW MODEL

Before the relationship between the sampling frequency and the confidence interval of the annual mean statistic can be determined, the daily river flow data must be modeled to isolate a residual which is random, independent, identically distributed, and assumed Gaussian.

The data utilized in the derivation of the mathematical model to estimate daily stream flow were 10 years of daily flow observations at the USGS gauging station (01170500) on the Connecticut River near Montague City, Massachusetts. The average flow for the length of record (68 years) was 13,490 cubic feet per second (cfs) with a maximum discharge of 236,000 cfs and a minimum discharge of 215 cfs. Like the majority of

USGS river gauging stations the flow at Montague City is measured continuously and averaged daily (Sanders, 1974).

Because the Connecticut River flow appeared to be non-stationary over the ten years of record as the annual mean was time dependent, a trend component was subtracted from each observation prior to the derivation of a mathematical model of the flow. Thus it was initially assumed that an adequate model for daily river flow was:

$$Q_t = b_t + c_t + v_t \qquad (16\text{-}36)$$

where:

Q_t = daily river flow

b_t = periodic, deterministic component

c_t = linear, long-term trend

v_t = stochastic component

The trend was assumed to be linear in time and was estimated as the slope of the line of best fit for the 10-year flow data using linear regression. Each trend component c_t was equal to the slope (trend) multiplied by the time differential from the origin of the trend.

$$c_t = \text{slope } (t - t_o) \qquad (16\text{-}37)$$

where:

slope = slope of the line of best fit

t_o = time origin of the trend

t = time.

The mathematical model to estimate river flow and determine the random residual is based upon the parametric time series models developed by Box and Jenkins (1970). In addition, the parametric time series was fitted to the logarithm of the flow as previous research (Benson, 1972; McMichael and Hunter, 1972) indicated that logarithms of the river flow could be more appropriately modelled by Gaussian processes than could untransformed data.

Although parametric models may be used to model logarithms of river flow, it has been found that a better fit of the model to flow data can result if the periodic, deterministic component is removed from the trend-free observations prior to estimating the parameters of the time series model (McMichael and Hunter, 1972). Thus the next step is to suggest that

$$w_t = \text{Log}(Q_t - c_t) - d_t \qquad (16\text{-}38)$$

where:

w_t = a stationary (perhaps correlated) sequence of observations

d_t = deterministic component of log river flow related to the annual river flow cycle and the weekly hydropower production cycle.

A review of the river flow data at Montague City revealed a noticeable annual cycle as well as a weekly cycle: the former was a result of summer low flow and spring freshets, the natural annual cycle; the latter was a result of the highly regulated flow for power production during the week with ponding on the weekend above the gauging station (Fig. 16-8).

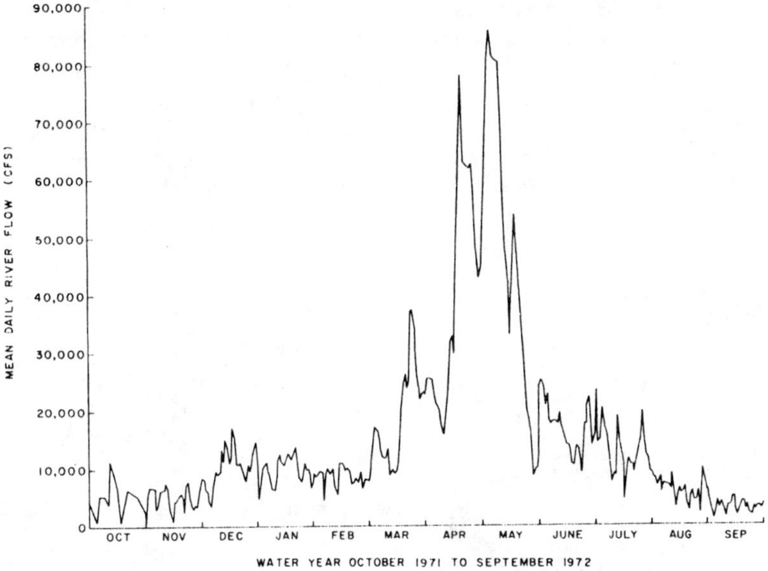

Fig. 16-8. Mean Daily Discharge of the Connecticut River at Montague City, Massachusetts

In order to account for both the annual and weekly cycles in the flow model, a cosine function was utilized (McMichael and Hunter, 1972).

$$d_t = A_1 + A_2\cos(\frac{2\pi}{7} t + \omega_1) + A_3\cos(\frac{2\pi}{365} t + \omega_2) \qquad (16\text{-}39)$$

where:

A_1 = constant

A_2 = amplitude of the weekly cycle

A_3 = amplitude of the annual cycle

ω_1 = the phase angle of the weekly cycle from the beginning of the water year October 1

ω_2 = the phase angle of the annual cycle from the beginning of the water year October 1.

$$A_1 = \frac{1}{n} \sum_{t=1}^{n} \log(Q_t - c_t) \tag{16-40}$$

where:

n = number of data

Q_t = averaged flow for day t (cfs)

The remaining parameters of the periodic deterministic component, Eq. (16-39), were estimated by the Newton technique from a function which minimized the sum of the squared deviations between the predicted and actual flows. The variance of the log of the flow minus the stationary component was reduced 33.5 percent with the addition of the deterministic component of the model. However, it was apparent from inspection of $\{w_t\}$ that they were not random but were serially correlated, thereby justifying the formulation of a parametric times series model.

An example of a parametric time-series model is an autoregressive model of the first order which is simply a first-order Markov model (Box and Jenkins, 1970).

$$(1 - \phi B)w_t = a_t \tag{16-41}$$

where:

w_t = a correlated variable at time t

ϕ = a constant equal to the correlation coefficient

B = a linear lag operator defined as: $Bw_t = w_{t-1}$.

Equation (16-41) can be rewritten as:

$$w_t - \phi w_{t-1} = a_t \tag{16-42}$$

A moving average model of the first order is denoted by Eq. (16-43):

$$w_t = (1 - \theta B)a_t \tag{16-43}$$

where:

θ = constant.

The combination of Eqs. (16-41) and (16-43) is a first order autoregressive moving average model.

$$(1 - \phi B)w_t = (1 - \theta B)a_t \tag{16-44}$$

or

$$w_t = \phi w_{t-1} + a_t - \theta a_{t-1} \tag{16-45}$$

The first model investigated was first order autoregressive [Eq. (16-41)] as the autocorrelation decreased almost exponentially as the lag increased. The constant ϕ, being the only unknown in Eq. (16-41) (besides σ_a^2, the population variance, which is implicit) was determined utilizing the minimum sum of squared deviations methodology. It should be noted that ϕ is, in fact, the correlation coefficient between w_t and w_{t-1}.

The model was developed further as a generalization: a moving average component [Eq. (16-43)] was introduced to improve the fit of the model to the data by decreasing the residual variance and because the residual variance was decreased with the addition of the moving average component, the final form of the parametric time series model used was:

$$w_t = \phi w_{t-1} + a_t - \theta a_{t-1} \tag{16-45}$$

For a sequence of time series data, whether serial correlation and stationarity exist or not, a mean and variance can be calculated. However, only if the time series data is stationary, uncorrelated, and Gaussian can the variance be utilized to determine confidence intervals of the mean at a given level of significance. The variance of the log Connecticut River flow (log Q_t) for example, is a function of the amplitude of the weekly and annual harmonics and the trend as well as the inherent randomness of the flow data. Therefore, a large portion of the calculated variance of the log of the daily mean flow calculated from data having different sampling intervals should be independent of the sampling frequency because the trend component and the periodic component are both deterministic. Gilroy and Steel (1972) found that the parameters estimating the amplitude of functions defining river temperature as well as the variance of the temperature were independent of the sampling frequency. With reference to Table 16-6, the magnitude of the weekly and yearly harmonics remained approximately the same except for the annual amplitude calculated from data having a seven day sampling interval; the sign changes were a result of the change of the phase angles as the time interval between samples increased. The inordinate increase of the annual amplitude for the seven day sampling interval was a result of systematic sampling which did not include the weekend low flows.

16.9 FORMULATION OF THE SAMPLING FREQUENCY MODEL

Sanders (1974) recommended that eight monitoring stations be established in the Massachusetts portion of the Connecticut River Basin (Fig. 16-9). Therefore, river flow data from the USGS gauging station in closest proximity to each monitoring station were analyzed to determine the relationship between sampling frequency and the confidence interval, 2R, of the annual, mean residuals, $\bar{a}_t$. And because the confidence interval is symmetric about the mean, the sampling frequency criterion is derived as a function of the half-width of the confidence interval, R, of the mean residuals. The river and location of the gauging stations were: 1) Connecticut River at

Table 16-6. The Amplitude and Phase Angles of the Weekly and Yearly Harmonics Estimated from Flow Data Collected at Montague City at Different Sampling Intervals

Sampling Interval Days	Amplitude of the Weekly Harmonic	Amplitude of the Annual Harmonic	Phase-Angle of the Weekly Harmonic	Phase-Angle of the Annual Harmonic
	A_1	A_2	ω_1	ω_2
1	0.608	-0.227	3.052	1.631
2	0.610	-0.233	-3.213	27.666
3	-0.613	0.233	-0.035	0.307
4	-0.623	0.221	-0.049	1.175
5	-0.593	0.214	-0.023	2.075
6	-0.612	-0.194	0.004	-0.054
7	-0.717	-48.780	-0.029	1.575
8	-0.593	-0.241	0.039	1.669
9	-0.655	0.268	0.075	52.814
10	-0.581	0.218	0.043	0.344
15	-0.612	-0.234	0.172	1.771
20	-0.596	-0.254	0.158	0.171
30	-0.569	-0.183	0.407	-3.408

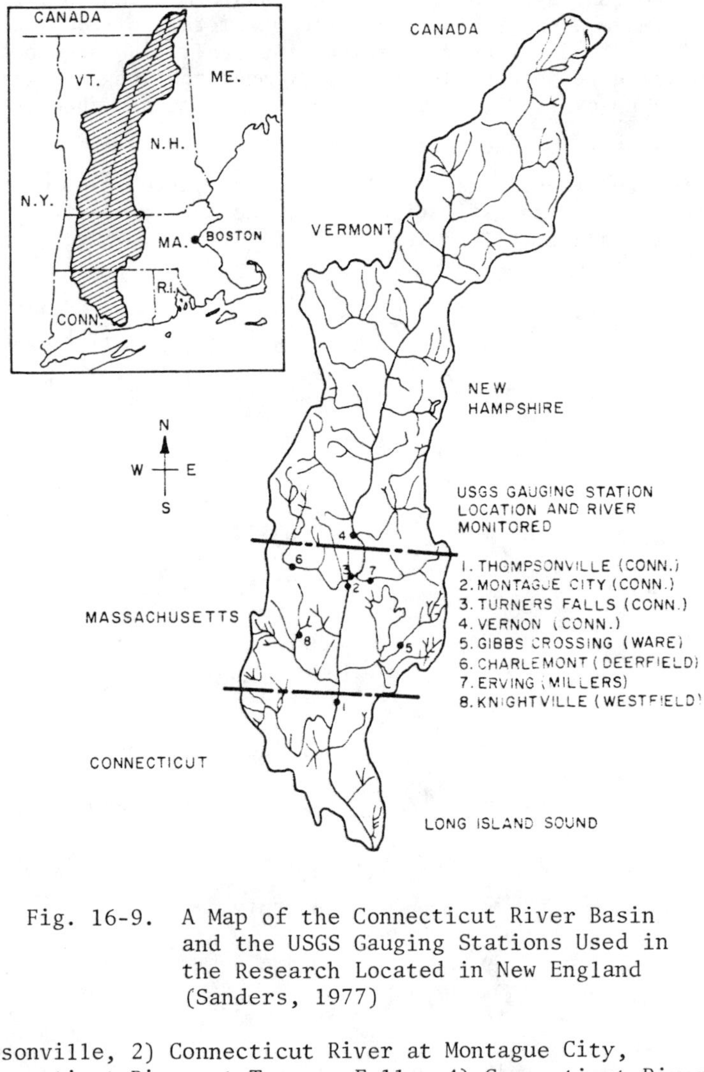

Fig. 16-9. A Map of the Connecticut River Basin and the USGS Gauging Stations Used in the Research Located in New England (Sanders, 1977)

Thompsonville, 2) Connecticut River at Montague City, 3) Connecticut River at Turners Falls, 4) Connecticut River at Vernon, 5) Ware River at Gibbs Crossing, 6) Deerfield River at Charlemont, 7) Millers River at Erving, and 8) Westfield River at Knightville (Fig. 16-9). Annual mean river flow varied from a high of 16,020 cfs at Thompsonville, Connecticut for the Connecticut River to a low of 314 cfs at Gibbs Crossing, Massachusetts for the Ware River.

To determine the relationship of sampling frequency and sample variance, S_a^2, of the residual component $\{a_t\}$ of log river flow at the gauging stations, Eq. (16-45) was fitted to each set of time series data having sampling intervals varying from 1 to 180 days. Utilizing the transformed mean daily flow data from each gauging station, the parameters of Eq. (16-45) were determined for every sampling interval. The sequence of the model development was to first remove the long-term trend c_t (if any existed), convert to natural logs and then estimate the parameters of the deterministic component d_t which included the weekly and annual harmonics (Table 16-6). This was followed by an estimation of ϕ, the autoregressive parameter, and finally an estimate of θ, the moving average parameter (Table 16-7).

Table 16-7. The Magnitude of ϕ and θ for the Time Series Model of the Connecticut River Flow at Montague City.

Sampling Interval Days	ϕ	θ
1	0.83534	-0.09263
2	0.72275	-0.25028
3	0.71956	-0.16318
4	0.68420	-0.12638
5	0.61584	-0.16017
6	0.59428	-0.04523
7	0.67401	-0.07662
8	0.57012	-0.04975
9	0.48675	-0.08175
10	0.51343	-0.05057
15	0.4273	0.00128
20	0.34941	0.04191
30	0.17405	-0.00324

A plot of the sample variance of the residuals versus the sampling interval of the Connecticut River flow at Montague City (Fig. 16-10) indicates that σ_a^2 (which is estimated by S_a^2) approaches a limiting value asymptotically. This

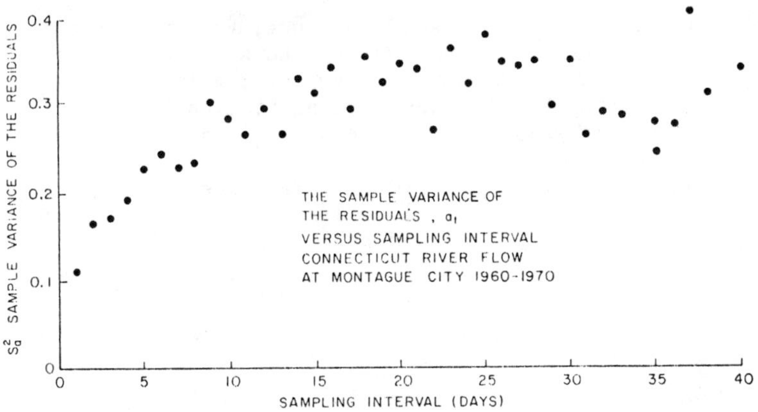

Fig. 16-10. The Sample Variance, S_a^2, of the Residuals Using 10 Years of Flow Data for the Connecticut River at Montague City Versus the Sampling Interval

dependence of the residual variance on the time interval up to 12 days is a result of the correlated nature of the time series. However, once the sampling interval exceeded approximately 12 days the residual variance stabilized, which indicated the predicted and necessary independence with sampling interval, more specifically, a constant variance. From each set of flow data the residual variance for sampling intervals greater than 11 days appeared to fluctuate about a constant residual variance which can be shown to be an estimate of $Var(w_t)$, the variance of the stochastic component of the transformed flows, $\{w_t\}$. This is in fact equal to the variance of the trend-free log flow, $Var[\log(Q_t - c_t)]$. Therefore, R, not only defines the expected one-half width of the confidence interval of the mean residuals it also defines the expected one-half width of the confidence interval of the trend-free, annual mean log flow.

The mean of the residual variances were calculated for each gauging station and are summarized in Table 16-8. To be consistent with the presumption of independence of the variance with time, only those residual variances from time series data having sampling intervals greater than twelve days were used to calculate the mean.

After determining the mean residual variance at each sampling station, the standard deviation was calculated (the square root of the mean residual variance) and used to estimate the relationship between the expected one-half width of the confidence interval of the annual mean log flow, R, and

Table 16-8. The Mean Residual Variance, Variance of the Mean Residual Variance and Range of the Mean at the 95 Percent Level of Significance for the Flow at the Sampling Station Locations in Massachusetts (Sanders, 1974)

	Drainage Area (mi^2)	Mean Residual Variance S_a^2	Variance of S_a^2	One-Half Confidence Interval of S_a^2 (95% Level of Significance)
Ware River @ Gibbs Crossing	199	0.27200	0.00861	0.0264
Conn. River @ Thompsonville	9661	0.27866	0.00205	0.0129
Deerfield River @ Charlemont	362	0.30420	0.00810	0.0255
Conn. River @ Montague City	7865	0.35096	0.00643	0.0228
Millers River @ Erving	375	0.37036	0.00852	0.0262
Conn. River @ Vernon	6266	0.42422	0.01713	0.0311
Westfield River @ Knightville	162	0.43388	0.00990	0.0282
Conn. River @ Turners Falls	7163	0.48906	0.03059	0.0496

sampling frequency. Inserting the standard deviation into Eq. (16-35) along with the value of the Students "t" constant, R associated with a given number of data (365/sampling interval) can be calculated. A plot of the results shows the hyperbolic relationship between sampling frequency and the magnitude of R at a 95 percent level of significance (Fig. 16-11). The half-width of the expected confidence interval of the annual mean log flow for the Ware River (199 sq mi drainage area) was the smallest while R for the Connecticut River at Turners Falls (7164 sq mi drainage area) was the largest. Both are highly regulated by dams, but the regulation of the former is primarily for flood control and the latter for power production.

Figure 16-11 can be used to estimate the magnitude of the confidence intervals for different sampling intervals. For

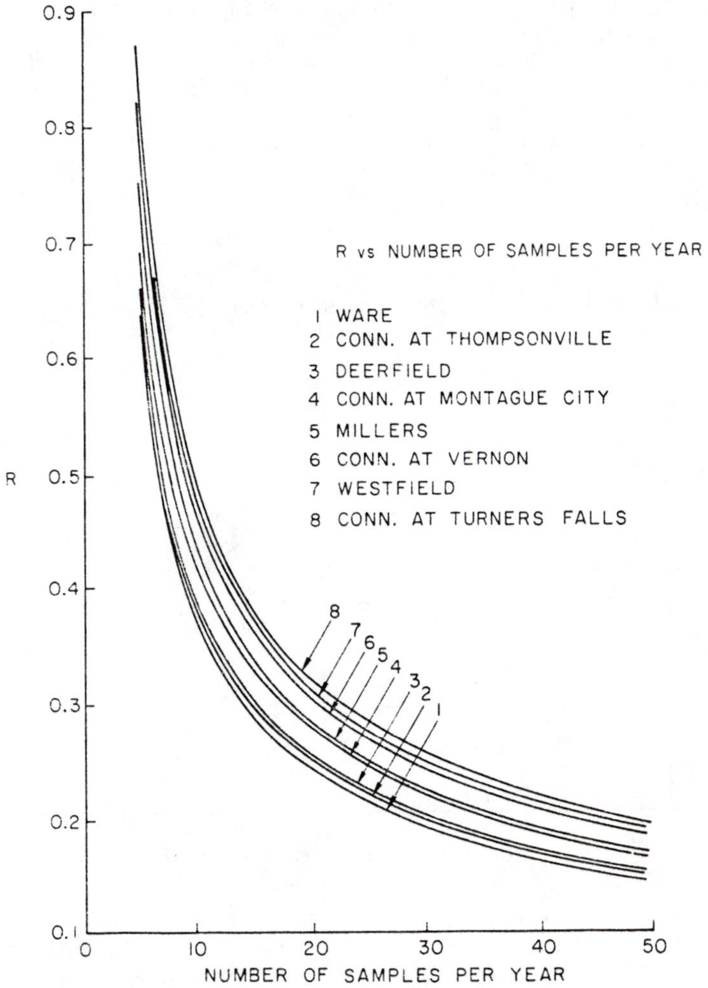

Fig. 16-11. The Magnitude of the Expected Half Width of the Confidence Interval of the Annual Mean Log Flow Versus the Number of Samples Per Year for Each Sampling Station in the Massachusetts Portion of the Connecticut River Basin (Sanders, 1974)

example, sampling the flow of the Ware River three times a year (once every four months), the expected confidence interval of the mean annual log flow is nine times the magnitude of the confidence interval obtained from a sampling interval of seven days. However, sampling once a month versus once every seven days only increases the expected confidence interval by 2.3 times. It should be noted that the level of significance will affect the magnitude of the confidence interval as well. At 95 percent level of significance and sampling every two weeks, the expected one-half width of the confidence interval for the Deerfield mean log flow is 0.225 (Fig. 16-12) whereas for a 90 percent level of significance the half width is 0.190. The confidence interval is smaller for the 90 percent level because it has a 10 percent chance of being exceeded.

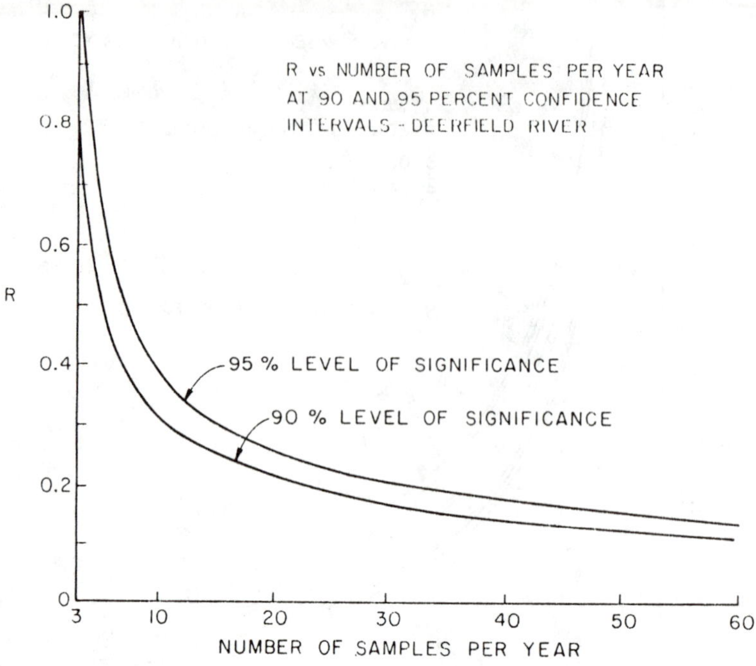

Fig. 16-12. The Expected Half-Width of the Confidence Interval of the Annual Mean Log Flow Versus the Number of Samples Per Year for the Deerfield River at 90 Percent and 95 Percent Levels of Significance

Rainwater and Avrett (1962) investigated the standard errors of an annual mean from data sampled at various frequencies. The standard error is defined as the standard deviation of the sample means. The sampling frequencies studied were once a day, once a week, bi-weekly and monthly because these frequencies were the most commonly used by sampling agencies. Their results indicated that there is an increase in the error of the annual mean dissolved solids concentration as the time between samples increases; the error of the annual average concentration using daily samples was between 10-20 percent of the error of the annual average using monthly samples. And the errors using weekly and bi-weekly samples were between 25-40 percent and 40-65 percent, respectively, of the error using monthly samples. The expected one-half width of the confidence interval of the annual mean log flow of the Ware River and Connecticut River at Turners Falls sampling weekly and bi-weekly were similar in magnitude to the actual errors measured by Rainwater and Avrett (1962), being approximately 44 percent and 64 percent, respectively of the expected half-width of the confidence interval from sampling monthly.

16.10 APPLICATION OF FREQUENCY CRITERION

It has been a common practice to have a uniform sampling frequency within a river basin. As a result, the random variability of the water quality parameters at the different sampling locations is not considered. Using this criterion, the annual means would vary in accuracies depending on the random variability of the water quality parameter. However, specifying sampling frequencies on the basis of equality of the estimated half-width of the confidence interval of the annual mean log flows, the rivers with large drainage basins or their flow highly regulated for flood control would be sampled less frequently than the unregulated streams with a smaller drainage basin--a savings in expenditure for the collection and analysis of samples as well as for data storage and interpretation while the expected confidence interval of the respective annual mean log flows would remain the same.

It can be arbitrarily decided that a maximum of 28 water quality samples per year could be collected at each sampling station. Also, because the majority of the rivers in the Connecticut River Basin are regulated and possess a weekly cycle, the specified frequencies shall not be even multiples of seven. This requires that sampling will be necessary during several weekends per year. Furthermore, due to the diurnal fluctuations of several water quality parameters and the daily cycle of most pollutant discharges into rivers, the sample will be a 24-hour, flow-weighted, composite.

The station with the greatest expected one-half width of the confidence interval of the annual mean log flow, in this case 0.28 at Turners Falls, would be sampled at the maximum sampling frequency of 28 samples per year while the remaining stations would be sampled proportionately less but would have identical expected confidence intervals (Fig. 16-11). The sampling frequencies for each monitoring station location are calculated and summarized in Table 16-9. Use of the sampling frequency criterion is not restricted by having an initial estimate of the maximum number of samples which can be collected at a given station. The sampling frequencies can be determined at each station, given the total number of samples to be taken per year in a basin or the specification of an expected confidence interval width of the annual mean log flow. A relative measure for accuracy could be specified, as well,

Table 16-9. The Proposed Sampling Interval, Sampling Frequency and the Expected One Half Width of the Confidence Interval of the Annual Mean Log Flow for the Eight Sampling Stations in the Massachusetts Portion of the Connecticut River Basin (Sanders, 1977)

	Sampling Interval (days)	Sampling Frequency Per Year	Half-Width of Expected Confidence Interval of Annual Mean Log Flow
Ware River @ Gibbs Crossing	23	16	0.28
Conn. River @ Thompsonville	22	17	0.28
Dearfield River @ Charlemont	20	18	0.28
Conn. River @ Montague City	19	20	0.28
Millers River @ Erving	17	22	0.28
Conn. River @ Vernon	16	23	0.28
Westfield River @ Knightville	15	25	0.28
Conn. River @ Turners Falls	13	28	0.28

using a specified one-half width of the confidence interval divided by the expected mean log flow.

At each station, with the exception of Charlemont and Turners Falls, the sampling interval was calculated by dividing 365 by the number of samples per year and rounding off to the higher integer value. The sampling intervals for the Deerfield River at Charlemont and the Connecticut River at Turners Falls were 21 days and 14 days, respectively and, as both are multiples of seven, the intervals were decreased by one day so that the weekend low flow would not be neglected in the calculation of the annual statistic. Instead of collecting and analyzing 224 samples per year which would be required for a network having uniform sampling frequencies, only 169 samples per year are required--a 25 percent decrease in the number of samples while maintaining a uniform confidence interval for the annual statistics.

The proposed sampling frequency criterion and methodology to determine sampling frequencies are applicable to any water quality parameter. Being a function of the random variability of the data at each proposed sampling station and derived from a statistical analysis of this variability, the criterion establishes sampling frequencies based on an expected confidence interval of the annual statistic. When sampling river flow or possibly coliform concentration, the annual mean log should be utilized as the governing annual statistic. However, when sampling other water quality parameters, the annual arithmetic or geometric means may be more appropriate. The choice of a statistic will depend on the variation of the water quality parameter and the capacity of the predictive model to generate independent, stationary, uncorrelated, and Gaussian residuals.

16.11 REFERENCES

__________, 1962. "The National Water Quality Network 1962." Presented at the Meeting of Interstate Commission on Potomac River Basin, Hagerston, Maryland, May 25, 1962.

__________, 1963. "USPHS Water Pollution Surveillance System--Annual Compilation of Data, October 1, 1962 - September 30, 1963." PHS Publication No. 663 (Revised) (1963 Edition), USDHEW, PHS, Division of Water Supply and Pollution Control.

__________, 1965. Standard Methods for the Examination of Water and Wastewater. 12th Edition, American Public Health Association, Inc., New York.

__________, 1972. "Recommended Methods for Water Data Acquisition." Preliminary Report of the Federal Interagency Work Group on Designation of Standards for Water Data Acquisition, U.S. Department of Interior, Geological Survey.

Bayley, G. V. and J. M. Hammersley, 1946. "The Effective Number of Independent Observations in an Autocorrelated Time Series." J. Roy. Statist. Soc., 8(1-B), pp. 184-197.

Bayon, E. J. and J. F. Ramsay, 1972. "Wastewater Discharge Survey Connecticut River Basin." Publication of Massachusetts Water Resources Commission, Division of Water Pollution Control, August, 209 pages.

Box, G. E. P. and G. M. Jenkins, 1970. Time Series Analysis Forecasting and Control. Holden-Day, San Francisco, 542 pages.

Briggs, R., 1972. "Water Quality Monitoring Networks--Practice in Great Britain." Casebook on Hydrological Network Design Practice, World Meteorological Organization, WMO No. 324.

Buchanan, T. J., 1964. "Time of Travel of Soluble Contaminants in Streams." Journal of the Sanitary Engineering Division, ASCE, Vol. 90, SAJ, June, pp. 1-12.

Clark, R. N., 1950. Discussion of "Sampling for Stream Pollution." By Beltz, C. J., Journal of Sewage and Industrial Wastes, FWPCA, Vol. 22, No. 5, May, pp. 683-684.

Cleary, R. W. and D. D. Adrian, 1973. "New Analytical Solutions for Dye Diffusion Modeling." Journal of the Environmental Engineering Division, ASCE, Vol. 99, No. EE3, June, pp. 213-227.

Cochrane, David, 1970. "Application of Image Theory to Mixing in Streams." Thesis required for partial fulfillment of the requirements for the degree of Master of Science, Civil Engineering, University of Massachusetts, Amherst, Mass., July.

Colston, N. V., 1974. "Characterization of Urban Land Runoff." Paper presented at the ASCE National Water Resources Conference, Los Angeles, January, 29 pages.

Dawdy, D. R., M. E. Moss and N. C. Matalas, 1972. "Application of Systems Analysis to Network Design." Casebook on Hydrological Network Design Practice, World Meteorological Organization, WMO, No. 324.

Diachishin, A. N., 1963. "Dye Dispersion Studies." Journal of the Sanitary Engineering Division, ASCE, Vol. 89, SAI, January, pp. 29-48.

Diachishin, A. N., 1963. "Waste Disposal in Tidal Waters." Journal of the Sanitary Engineering Division, ASCE, Vol. 89, No. SA4, August.

Feuerstein, D. L. and R. E. Selleck, 1963. "Flourescent Tracers for Dispersion Measurements." Journal of the Sanitary Engineering Division, ASCE, Vol. 89, No. SA4, August.

Fischer, H. B., 1967. "The Mechanics of Dispersion in Natural Streams." Journal of the Hydraulics Division, ASCE, Vol. 93, No. 6, November, pp. 187-216.

Fischer, H. B., 1968. "Dispersion Predictions in Natural Streams." Journal of the Sanitary Engineering Division, ASCE, Vol. 94, October, SA5, p. 927-943.

Gilroy, E. J. and T. D. Steele, 1972. "An Analysis of Sampling-Frequency Alternatives for Fitting a Daily Stream Temperature Model." International Symposium on Uncertainties in Hydrologic and Water Resource Systems, Proceedings, Vol. II, pp. 594-608.

Gunnerson, C. G., 1966. "Optimizing Sampling Intervals in Tidal Estuaries." Journal of the Sanitary Engineering Division, ASCE, Vol. 92, No. SA2, April, pp. 103-123.

Gunnerson, C. G., 1967. "Stream Flow and Quality in the Columbia River Basin." Journal of the Sanitary Engineering Division, ASCE, Vol. 93, No. SA6, December, pp. 1-16.

Guttman, I. and S. S. Wilks, 1965. Introductory Engineering Statistics. John Wiley and Sons, Inc., New York.

Haney, P. D. and J. Schmidt, 1958. "Representative Sampling and Analytical Methods in Stream Studies." Sewage and Industrial Wastes, Vol. 30, No. 6, June, pp. 812-820.

Holley, Edward R., 1972. Discussion of "Dispersion in Natural Streams." by Bansal, Mahendla K., ASCE Hydraulic Division, Vol. 98, No. HY6, June.

Inland Waters Directorate, Water Quality Branch, 1973. "Index of Water Quality Stations." Ottawa, Canada.

Kilpatrick, F. A., L. A. Martens and J. R. Wilson, 1970. "Measurement of Time of Travel and Dispersion by Dye Tracing." Techniques of Water Resources of the United States Geological Survey, Book 3, Chapter A9.

Kittrell, F. W., 1969. "A Practical Guide to Water Quality Studies of Streams." U.S. Dept. of Interior, FWPCA, CWR-5, 135 pages.

Knisel, W. G., Jr. and V. M. Yevjevich, 1967. "The Statistical Measure of Hydrologic Time Series." Proc. International Hydrol. Symposium, 1, Fort Collins, Colorado.

Mace, A. C., 1964. Sample Size Determination, New York: Reinhold Publishing Co., pp. 35-36.

McMichael, F. C. and J. S. Hunter, 1972. "Stochastic Modeling of Temperature and Flow in Rivers." Water Resources Research, Vol. 8, No. 1, February, pp. 87-98.

Montgomery, H. A. C. and I. C. Hart, 1974. "The Design of Sampling Programmes for Rivers and Effluents." Water Pollution Control, Vol. 33, No. 1, pp. 77-101.

Pomeroy, R. D. and G. T. Orlob, 1967. "Problems of Setting Standards and of Surveillance for Water Quality Control." California State Water Quality Control Board Publication No. 36, Sacramento, California, May.

Quimpo, R. G., 1968. "Stochastic Analysis of Daily River Flows." Journal of the Hydraulics Division, ASCE, Vol. 94, No. HY1, January, pp. 43-57.

Quimpo, R. G. and Jing-Yea Yang, 1970. "Sampling Considerations in Discharge and Temperature Measurement." Water Resources Research, Vol. 6, No. 6, December, pp. 1771-1774.

Rainwater, R. H. and J. R. Avrett, 1962. "Error Inference in Systematic-Sample Statistics in Stream Quality Studies." Journal of the American Water Works Association, June, pp. 757-768.

Rainwater, F. H. and L. L. Thatcher, 1960. "Methods for Collection and Analysis of Water Samples." Water Supply Paper No. 1454, U.S. Geological Survey, Washington, D.C.

Reeder, S. W., 1971. "Cross-Sectional Study of the Effects of Smelter Waste Water Disposal on Water Quality of the Columbia River Downstream From Trail British Columbia." Technical Bulletin No. 39, Department of Energy, Mines and Resources, Inland Waters Branch, Ottowa, Canada.

Ruthuen, P. M., 1971. "The Dispersion of a Decaying Effluent Discharged Continuously into a Uniformly Flowing Stream." Water Research, Pergamon Press, London, Vol. 5, No. 6, June, pp. 343-352.

Sanders, T. G., 1974. "Rational Design Criteria for a River Quality Monitoring Network." Submitted to the Graduate School of the University of Massachusetts in partial fulfillment of the requirements for the degree of Doctor of Philosophy, August, 229 pages.

Sanders, T. G., D. D. Adrian and B. B. Berger, 1976. "Designing a River Basin Sampling System." Water Resources Research Center Publication No. 62, University of Massachusetts, Amherst.

Sanders, T. G., D. D. Adrian and J. M. Joyce, 1977. "Mixing Length for Representative Water Quality Sampling." J. Water Pollution Control Federation, V. 49, No. 12, pp. 2467-2478, Dec.

Sanders, T. G. and D. D. Adrian, 1978. "Sampling Frequency for River Quality Monitoring." Water Resources Research, V. 14, No. 4, pp 569-576, August.

Sayers, W. T., 1971. "Water Quality Surveillance: The Federal-State Network." Environmental Science and Technology, February, pp. 114-119.

Sayre, C. and F. M. Chang, 1968. "A Laboratory Investigation of Open-Channel Dispersion Process for Dissolved, Suspended, and Floating Dispersions." Geological Survey Professional Paper 433-E, U.S. Printing Office.

Sayre, W. W., 1965. Discussion of "Canal Discharge Measurements with Radioisotopes." By J. Schuster, Journal of the Hydraulics Division, ASCE, Vol. 91, No. HY6, November, Part 1, pp. 185-189.

State of New York, Department of Environmental Conservation, 1967. "Periodic Report of the Water Quality Surveillance Network 1965 through 1967 Water Years." 390 p.

Thackston, E. L. and P. A. Krenkel, 1967. "Longitudinal Mixing in Natural Streams." Journal of the Sanitary Engineering Division, Vol. 93, No. SA5, October, pp. 67-90.

Velz, D. J., 1950. "Sampling for Effective Evaluation of Stream Pollution." Sewage and Industrial Waste Water Pollution Control Federation Journal, Vol. 22, No. 5, May, pp. 666-689.

Wolman, G. M., 1971. "The Nation's Rivers Problem are Encountered in Appraising Trends in Water and River Quality." Science, American Association for the Advancement of Science, Vol. 174, No. 4012, November 26, 1971, pp. 905-918.

Ward, P. R. B., 1973. "Prediction of Mixing Lengths for River Flow Gaging." Journal of the Hydraulics Division, ASCE, Vol. 99, No. HY7, July.

Ward, R. C., 1973. "Data Acquisition Systems in Water Quality Management." EPA-RG-73-014, Office of Research and Monitoring, U.S. Environmental Protection Agency, May, 259 pages.

Wastler, T. A., 1963. "Application of Spectral Analysis to Stream and Estuary Field Surveys." Public Health Service Publication No. 999-WP-7, Cincinnati, Ohio, November, 31 pages.

Chapter 17

RIVER WATER QUALITY MODEL CONSTRUCTION

by

P. A. Krenkel, Executive Director, Water Research Center, Desert Research Institute, The University of Nevada, Reno, Nevada

and

V. Novotny, Associate Professor, Department of Civil Engineering, Marquette University, Milwaukee, Wisconsin

Chapter 17

RIVER WATER QUALITY MODEL CONSTRUCTION

17.1 INTRODUCTION

With the passage of Public Law 92-500 (Water Pollution Control Amendments of 1972) and subsequent implementation of the planning aspects of that act, the importance of mathematical modeling has come to the forefront of water quality management. While the act purports to be concerned only with water quality, the various planning provisions clearly lead to land use planning. It is obvious that attainment of the objectives of the act are dependent on present and future control of all wastewater discharges including those from so called "non-point sources." The only practical method for delineating present and future water quality conditions is by mathematical simulation, thus the use of mathematical modeling is mandatory.

In addition, while the act dictates the use of effluent standards, in receiving waters where those standards will not result in the desired water quality (water quality limiting cases), allowable discharges must be determined by the use of mathematical models.

One other use of modeling, which will make the process even more important, is for determining where "Best Available Treatment" (BAT) is required (1983 Standards), should the recommendations of the National Water Quality Commission be adopted by the Congress. In short, these recommendations were to defer the 1983 requirements because of extreme cost and probable lack of commensurate benefit, and apply the BAT on a case-by-case basis. Again, mathematical modeling will be an integral part of this process.

17.2 OVERVIEW OF MODELING

A detailed description of all of the complexities of water quality modeling would occupy a treatise. Thus, the purpose of this discussion will be to demonstrate the approach taken in adopting one of the most frequently used available river models (DOSAG 1 - Anon. 1970) to a real situation. Inasmuch as the readers of this material are probably well versed in the art of model construction, the emphasis will be on the parameters involved, their sensitivity and their incorporation into the model. Since this is a deterministic model, the basic process consists of writing mass-balance relationships for all material entering and leaving an infinitesimal volume. As will be later demonstrated, the DOSAG 1 model is one-dimensional and steady-state, thus somewhat simplifying the model construction. For a fundamental description of the procedure followed, the reader is referred to Thomann (1972).

It should be noted that several models are available in varying degrees of complexity. If one is dealing with the

Environmental Protection Agency (EPA), the choice of model may depend on a particular region's preference. The model chosen for discussion herein (DOSAG 1), with some modification, is probably the most frequently used for the purposes previously described. The model is primarily concerned with the oxygen balance in a river system, inasmuch as Biochemical Oxygen Demand (BOD) is still considered to be the single most important measure of water pollution.

Models can be classified according to whether or not they handle conservative or non-conservative substances, where a conservative substance is one only affected by dilution and a non-conservative substance changes its characteristics with time, i.e., BOD, coliform bacteria, temperature. Inasmuch as the conservative substances are relatively simple to model and meaningful models exist for temperature and coliform bacteria, the authors chose to use the DOSAG 1 model for oxygen and use other models where necessary. It is pertinent to note that while increased model accuracy usually implies increased complexity, increased complexity does not necessarily yield increased accuracy.

The range of choice for modeling is quite broad inasmuch as one can choose from a number of existing models with different complexities or devise a model "from scratch." Obviously, using an existing model, where a-plicable, is the preferred option because the cost of development has already been paid. For an excellent discussion of existing models, their capabilities, selection and documentation, the reader is referred to Grimsrud, et al (1976). The following guidelines for the selection and use of models were taken directly from that reference:

1. The first step is to define the problem and determine both what information is needed and what questions need to be answered.
2. Use the simplest method that can provide the answers to your questions.
3. Use the simplest model that will yield adequate accuracy.
4. Do not try to fit the problem to a model, but select a model that fits the problem.
5. Do not confuse complexity with accuracy.
6. Always question whether increased accuracy is worth the increased effort and cost.
7. Do not forget the assumptions underlying the model used, and do not read more significance into the simulation results than is actually there.

17.3 THE BASIC MODEL

The basic classical model was proposed by Streeter and Phelps (1925) and was later summarized and published by Phelps in 1944. It is interesting to note that the fundamentals involved were quite sound and are still used today, with some modifications. The differential equation proposed was:

$$\frac{dD}{dt} = K_1L - K_2D \tag{17-1}$$

where D is the oxygen deficit (C_s-C), C is the oxygen concentration observed, C_s is the oxygen saturation concentration, L is the carbonaceous biochemical oxygen demand, K_1 is the deoxygenation coefficient, K_2 is the reaeration coefficient, and t is time.

Equation 17-1 integrates into the familiar form of the oxygen sag equation, which is:

$$D = \frac{K_1L_o}{K_2 - K_1}(e^{-K_1t} - e^{-K_2t}) + D_oe^{-K_2t} \tag{17-2}$$

where L_o and D_o are the ultimate biochemical oxygen demand and oxygen deficit, respectively.

Streeter and Phelps' approach is limited to only two phenomena, namely, deoxygenation of the water due to the bacterial decomposition of carbonaceous organic matter, and reaeration caused by the oxygen deficit and turbulence. The rate at which the BOD is exerted was assumed to be identical with that observed using the laboratory BOD test and a proportionality was assumed to exist between the reaeration rate and certain hydraulic parameters of flow.

Numerous investigators have subsequently questioned Streeter and Phelps' theory inasmuch as the self-purification process in small streams does not follow the model. A classical illustration of the disagreement of field observations with the theory was published by Kittrell and Kochtitzky (1947). These studies were on the Holston River, Tennessee, which is used as an example in this discussion. Pertinent conclusions of that study were as follows:

1. The laboratory deoxygenation coefficients, K_b, varied approximately with the five-day BOD concentration of the samples collected below all sources of pollution. The range was 0.34 to 0.095 (base 10) for BOD results of 20.6 to 1.89 ppm.

2. The stream deoxygenation coefficients, K_1, were many times greater than the laboratory coefficients below the sources of pollution. Furthermore, as the distance downstream from the sources increased, the relationship was reversed. The range of the stream coefficients observed was 2.26 to 0.017 (base 10) at 20^oC.

3. The rate of reduction of BOD was much more rapid in this stream (the Holston River) than in the usual deep, sluggish stream, the reduction being about seven times as great for 0.5 day and four times as great for one day.

4. The rate of first stage BOD reduction in the stream water followed the normal course of bacterial death rates.

This indicated that the rate was not constant but diminished progressively with time below the entrance of pollution.

5. It was suggested that the rapid reduction in BOD may be due partially to the absorption of organic material in biological slimes on the bed of the stream.

6. The assumption that all deposited material exerted its full potential demand on the dissolved oxygen of the stream water resulted in obviously excessive reaeration coefficients, K_2, that ranged from 15.52 to 5.22 (base 10) at $20^{\circ}C$. It was suggested that portions of the deposited material may have decomposed anaerobically and escaped from the stream as unoxidized gases, without utilizing oxygen from the stream water.

7. A turbulent and relatively shallow river could assimilate more organic pollution without excessive dissolved oxygen depletion than could a deep sluggish stream. Figure 17-1 depicts the relative recovery capacity of the Holston River as compared to a sluggish stream, such as the Ohio River studied by Phelps. ($K_1 \approx 0.25$ per day, base e).

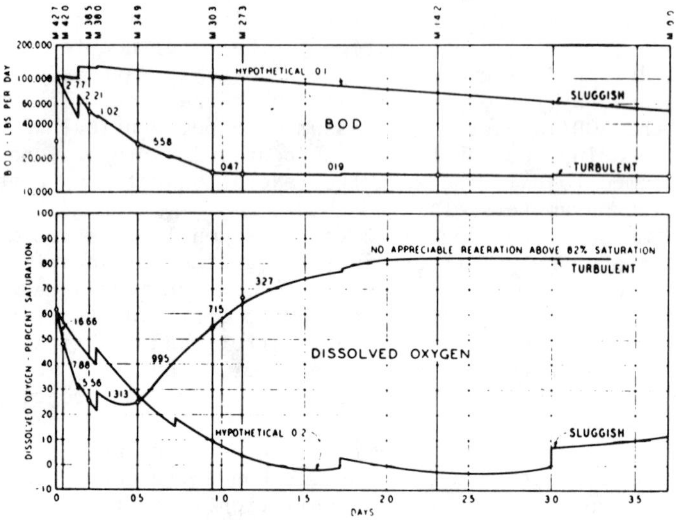

Fig. 17-1. Relative Recovery Capacities of Turbulent and Sluggish Stream (after Kittrell & Kochtitzky, 1947).

17.4 MODIFICATIONS OF THE STREETER-PHELPS THEORY

As previously mentioned, the Streeter and Phelps theory is based on the assumption that there are only two major processes taking place in the self-purification of streams, i.e., BOD and oxygen are being removed along the stretch by the bacterial oxidation of the organic matter, and oxygen is being replaced by reaeration at the water surface. Dobbins (1969) suggested that there are several other factors which contribute to the BOD and oxygen variations in streams as follows:

1. The removal of BOD by sedimentation or absorption.

2. The addition of BOD along the stretch by the scour of bottom deposits or by the diffusion of partly decomposed organic products from the benthal layer into the water above.

3. The addition of BOD along the stretch by local runoff.

4. The removal of oxygen from the water by diffusion into the benthal layer to satisfy the oxygen demand in the aerobic zone of this layer.

5. The removal of oxygen from the water by purging action of gases rising from the benthal layer.

6. The addition of oxygen by photosynthetic action of plankton and fixed plants.

7. The removal of oxygen by the respiration of plankton and attached plants.

8. The continuous redistribution of both BOD and oxygen by longitudinal dispersion.

Additional factors, e.g., effects of biological slimes, were also mentioned by Velz and Gannon (1962).

If unoxidized nitrogen is present (such as ammonia, or unoxidized nitrogen expressed as Total Kjeldahl Nitrogen - TKN) it may also represent a significant sink of oxygen and must be included in the oxygen balance. Summarizing these findings it is possible to state that the variations of water quality in rivers are caused by:

1. Quantitative and qualitative variations in the volume of flowing liquid, such as deoxygenation by the microorganisms present in the water, surface reaeration, denitrification, photosynthesis and respiration of planktonic microorganisms, and, to some degree, sedimentation of suspended organics.

2. External effects, such as those of heterotrophic slimes on BOD and oxygen removal, photosynthetic action of aquatic plants and benthic algae, oxygen demand of benthal deposits, etc.

It should be noted that the components of the oxygen balance systems are coupled, i.e., some outputs from one system represent inputs to another one. The effects of each of the inputs and the magnitudes of the inputs and parameters can best be presented by considering the generalized equation for coupled systems. Thomann (1972) considered the following equations specific to the D.O. problem:

BOD variation:

$$\frac{\partial L}{\partial t} = -\frac{1}{A}\frac{\partial}{\partial X}(QL) - K_r L + L_r\left[\frac{\partial Q}{\partial X}/A\right] + L_{rd} \qquad (17\text{-}3)$$

Nitrogenous oxygen demand variation:

$$\frac{\partial L^N}{\partial t} = -\frac{1}{A}\frac{\partial}{\partial X}(QL^N) - K_N L^N + L_r^N\left[\frac{\partial Q}{\partial X}/A\right] + L_{rd}^N \qquad (17\text{-}4)$$

Dissolved oxygen variation:

$$\frac{\partial C}{\partial t} = -\frac{1}{A}\frac{\partial}{\partial X}(QC) - K_d L - K_N L^N + K_a (C_s - C) - S_B (X_1 t) + P(X_1 t) - R(X_1 t) + C_r \left[\frac{\partial Q}{\partial X}/A\right] \qquad (17\text{-}5)$$

where:

C = dissolved oxygen concentration - mg/l
C_r = dissolved oxygen run-off - mg/l
C_s = dissolved oxygen saturation - mg/l
A = cross-sectional area - m^2 or sq. ft.
Q = river flow - m^3/day or cu. ft/day
L_r = BOD concentration of run-off - mg/l
L_{rd} = rate of addition of BOD along the stretch - mg/l x day
L^N = nitrogenous oxygen demand - mg/l
L_r^N = nitrogenous oxygen demand of run-off - mg/l
L_{rd}^N = rate of addition of N-BOD along the stretch - mg/l x day
S_B = benthal oxygen demand - mg/l x day
P = D.O. produced by plant or phytoplankton photosynthesis - mg/l x day
R = D.O. removed by plant respiration - mg/l x day
K_2 = atmospheric reaeration coefficient - day^{-1}
K_r = overall BOD removal coefficient - day^{-1}
K_d = BOD deoxygenation coefficient - day^{-1}
K_N = nitrification coefficient - day^{-1}
X = distance

The dissolved oxygen system is thus represented by three first order, linear differential equations. It is assumed in using these equations, that the effects of longitudinal dispersion may usually be neglected. However, it has been demonstrated by O'Connor (1961) that in slow-moving, highly mixed streams, such as estuaries, the longitudinal dispersion becomes important and must be included in the balance equations.

Assuming steady state conditions and negligible run-off within the stretch, Eqs. (17-3), (17-4) and (17-5) can be integrated into the following forms:

BOD variation:

$$L = L_o e^{-K_r \frac{x}{u}} + \frac{L_r d}{K_r}\left(1 - e^{-K_r \frac{x}{u}}\right) \qquad (17\text{-}6)$$

Nitrogenous demand variation:

$$L^N = L_o^N e^{-K_N \frac{x}{u}} + \frac{L_{rd}^N}{K_N}\left[1-e^{-K_N \frac{x}{u}}\right] \tag{17-7}$$

and oxygen deficit variation:

$$D = D_o \exp\left[-\left(K_a \frac{x}{u}\right)\right] \qquad \text{(a)}$$

$$+ \left[\frac{K_d}{K_a - K_r}\left\{\exp\left[-\left(K_r \frac{x}{u}\right)\right] - \exp\left[-\left(K_a \frac{x}{u}\right)\right]\right\}\right] L_o \qquad \text{(b)}$$

$$+ \left[\frac{K_N}{K_a - K_N}\left\{\exp\left[-\left(K_N \frac{x}{u}\right)\right] - \exp\left[-\left(K_a \frac{x}{u}\right)\right]\right\}\right] L_o^N \qquad \text{(c)}$$

$$+ \frac{K_d}{K_a K_r}\left\{1-\exp\left[-\left(K_a \frac{x}{u}\right)\right]\right\}$$

$$- \left[\frac{K_d}{(K_a - K_r)K_r}\left\{\exp\left[-\left(K_r \frac{x}{u}\right) - \exp\left[-\left(K_a \frac{x}{u}\right)\right]\right\}\right] L_{rd} \qquad \text{(d)}$$

$$-\left\{1-\exp\left[-\left(K_a \frac{x}{u}\right)\right]\right\} \frac{P}{K_a} \qquad \text{(e)}$$

$$+\left\{1-\exp\left[-\left(K_a \frac{x}{u}\right)\right]\right\} \frac{R}{K_a} \qquad \text{(f)}$$

$$+ 1-\exp\left[-\left(K_a \frac{x}{u}\right)\right] \frac{S_B}{K_a} \qquad \text{(g)}$$

(17-8)

where subscript o denotes the initial conditions (initial concentrations).

The terms in the various parts of the solution are interpreted as:

a. initial value of D.O. deficit
b. deficit due to point source of carbonaceous BOD
c. deficit due to point source of nitrogenous oxygen demand
d. distributed addition of BOD along the stretch
e. reduction of the deficit due to photosynthesis
f. deficit due to algae and plant respiration
g. distributed benthal demand effect

In order to determine the oxygen deficit for different temperatures, flow conditions, etc., several supporting models have been developed for evaluation of the pertinent parameters, as will be subsequently discussed.

17.5 DETERMINATION OF PARAMETERS

Several methods exist for the determination of each parameter of the oxygen balance. In almost all cases, the values of these parameters must be checked against field data which can be obtained either from water quality monitoring stations and/or from water quality survey data. Thus, a sensitivity analysis of the model and verification with field data is a necessary step in developing an adequate water quality model. Using only theoretical equations in a complex water quality system model is always dangerous and can lead even an experienced environmental systems engineer to erroneous conclusions. In addition, the use of specialists in limnology and water chemistry is desirable.

The parameters necessary for the design of an oxygen balance model can be separated into five recognizable categories as follows:

1. Hydrologic parameters
2. Hydraulic parameters
3. Oxygen sinks
4. Oxygen sources
5. Temperature effects

17.6 HYDROLOGIC PARAMETERS

Two major hydrologic factors are required as inputs to the oxygen balance, which are flow and termperature. The design flow is usually taken as the minimum 7-day 10-year flow. For design water temperature, the average temperature of the warmest month (July or August) with a recurrence interval of once in 10 years or 20 years is probably appropriate. In many waste assimilative capacity studies, the waste load is associated with a heat load (cooling water averages). Thus, the effect of the waste water on the thermal regime of the receiving water must be determined, if necessary.

17.7 HYDRAULIC PARAMETERS

The hydraulic parameters are supporting inputs which are related to the reaction rates, detention times, etc. Other hydraulic parameters include velocity, hydraulic radius or depth, slopes and bottom roughness. Stream velocity and/or detention time is usually determined by a dye test as was the case in the Holston River studies. Fluorescent dyes such as Rhodamine WT or B are convenient since the minimum detectable concentrations are less than 1 ppb. Ideally, stretches of the stream with uniform hydraulic characteristics should be selected. In some cases, the velocity can be directly computed using a velocity equation (Mannings', Prandtl Logarithmic or others). Depth - can be determined from cross-sections or computed knowing the width and corresponding flow and velocity. Depth and velocity should be determined for several flows, the results of which can be conveniently plotted and correlated on log-log paper, resulting in the following type relationships:

Depth, $$H = aQ^b \tag{17-9}$$

Velocity, $$U = \frac{X}{t} = cQ^d \tag{17-10}$$

Detention time, $$t = \frac{X}{U} = \frac{X}{c} Q^{-d} \tag{17-11}$$

Width, $$W = eQ^f \tag{17-12}$$

where X = length of the stretch.

Table 17-1 shows typical values of the coefficients. USGS gaging station data can be used for determination of the coefficients for reaches close to the stations.

Table 17-1. Coefficients of the Relation of Depth and Velocity (detention time) with Flow

Stream	Coefficients			
	On depth		On velocity	
	a	b	c	d
1. Impounded waters	H (constant)	0.0	1/A	1.0
2. Ideal channel flow (Mannings formula)	$\frac{1}{(WM)^{3/5} S_e^{3/10}}$	0.6	$\frac{M^{3/5} S_e^{3/10}}{W^{2/5}}$	0.4
3. Tennessee Valley	-	0.48	-	0.48
4. Codorus Creek, PA	-	0.40	-	0.60
5. Holston River, TN				
a) R.M. 142.2 - 131.5		0.32		0.674
b) R.M. 3.5 - 118.4		0.46		0.539
6. Scioto River, Susquehanna Basin, PA		0.30		0.70

A = cross-sectional area
M = Manning's roughness coefficient
W = width
S_e = energy slope

As shown in Table 17-1, b + d ≈ 1.0. Furthermore, natural streams are somewhere between the two extremes of impounded waters and ideal channel flow. Therefore, b may be expected to be within the interval 0 to 0.6 and d within the interval 0.4 to 1.0.

17.8 OXYGEN SINKS

The biochemical deoxygenation rate is related to BOD removal and subsequent oxygen demand brought about by the BOD removal process. Figure 17-2 depicts the processes which take part in the BOD removal in small and medium sized shallow streams.

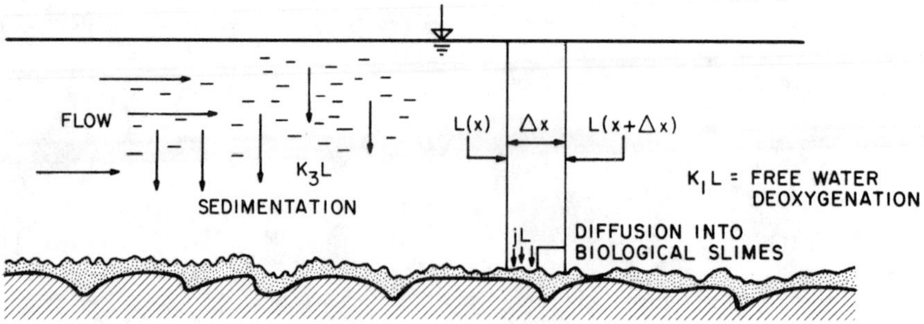

$$U\frac{\Delta t}{\Delta x} = -K_1L - \frac{jL}{H} - K_3L = -(K_1 + B + K_3)L$$

Free Water Deoxygenation; Diffusion into Slimes; Sedimentation; K_r; K_d

Figure 17-2. BOD Removal Processes in a Shallow Stream

The coefficients are defined as follows:

K_r = overall BOD removal coefficient

K_d = overall BOD deoxygenation coefficient

K_1 = deoxygenation coefficient of free flowing water (approximately equal to the laboratory BOD test coefficient)

K_3 = a coefficient describing sedimentation which does not exert an oxygen demand (anaerobic benthal decomposition)

It follows that:

$$K_r = K_d + K_3$$

and

$$K_d = B + K_1$$

where B was defined by Velz and Gannon (1962) as a coefficient describing a boundary effect of absorption by biological slimes attached to a solid bottom (rocks, gravel, etc.). Several authors have attempted to describe the coefficient B by a formula. Since the coefficient, B, is thought to be related to the depth, it would express the ratio of the biological slime volume to the volume of water and possibly to the turbulence intensity, which would account for the diffusion intensity. Bosko (1957, 1966) intuitively proposed the formula:

$$B \approx \eta \frac{U}{H} \tag{17-13}$$

where the coefficient of bottom activity ranges from η = 0.1 for slow flowing, sluggish streams to η = 2.0 (B base e) for rocky mountain streams. Novotny (1969) proposed the formula:

$$B \approx \nu \frac{S_e^{1/4}}{H^{3/4}} \tag{17-14}$$

In the above equations, the depth H is in meters and S_e is in ft/1000 ft., which was based on the hypothesis that the diffusion process between the free-flowing water and bottom slimes is a boundary layer type of diffusion. Thus, assuming that the diffusion rate through the boundary layer controls organic transfer from the free-flowing stream, the coefficient ν is supposed to range from $\nu \approx 0.1$ for streams with moving bottom (sand, mud) to $\nu \approx 3.0$ (B base e) for rocky streams. The possible ranges of the coefficients η and ν are shown in Table 17-2.

Table 17-2. Coefficients η and ν for the Estimation of K_d.

Character of the Stream	Bosko's formula η base e	Novotny's formula ν base e
1. Streams with rocky bottom, higher pollution	2.0	2.0 - 3.0
2. Coarse gravel bottom with large rocks, higher pollution	1.5	1.1 - 2.0
3. Coarse gravel	1.2	0.8 - 1.1
4. Gravel, lower pollution	0.9	0.5 - 0.8
5. Sand, lower pollution	0.6	0.3 - 0.5
6. Slow moving rivers with muddy bottom	0.3	< 0.3

As noted by Kittrell and Kochtitzky (1947) for the Holston River, K_1 is not constant, but decreases with decreasing BOD concentration.

The variation of K_1 (base 10) in the Holston River is shown in Fig. (17-3). One might expect the same behavior to occur with the coefficient B, although to a lesser extent. Thus, the formula of Bosko or Novotny yields an approximation and must be used with caution.

Most of the microorganisms and fungi which take part in the self-purification process are facultative, i.e., they

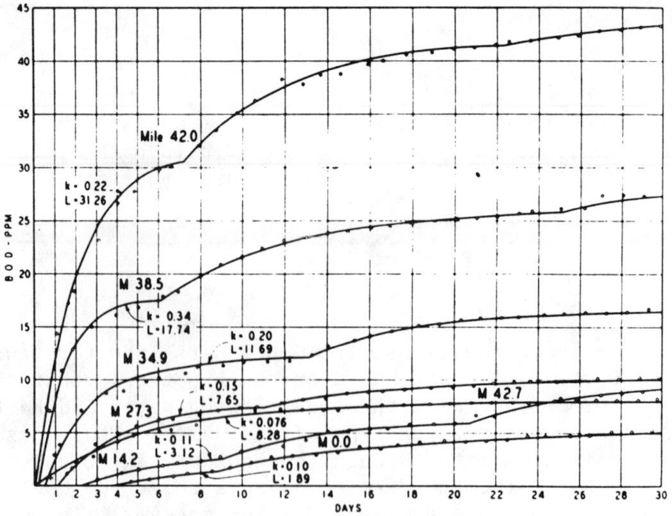

Fig. 17-3. 30-Day B.O.D. Curves (after Kittrell and Kochtitzky, 1947).

metabolize the organics in either an aerobic or anaerobic environment. However, anaerobic metabolism is much slower than aerobic ($K_1 \approx 0.12$ - base e). On the other hand, the most dominant species which can be found in the biological bottom slimes of polluted rivers, *Sphaerotilus natans*, is strictly aerobic. Thus, with low (near zero) oxygen concentrations, one might expect $B \rightarrow 0$. In an anaerobic environment, nitrates and possibly sulphates become the electron acceptors in the BOD removal process, which reduces nitrates to nitrogen gas and sulphates to hydrogen sulfide.

17.9 BENTHAL OXYGEN DEMAND

In polluted streams, the river bottom below a waste water outfall may be covered by highly active biological materials, sludges, biological slimes, algae, etc. The growth and accumulation of these materials result either from deposition of suspended organics and/or from the transfer of soluble organics (BOD) from the flowing water to the biological slimes. The BOD is then metabolized to new cell growth and aerobically or anaerobically decomposed. Deposition of the suspended organics usually occurs during low flows when the velocities are low. Velz (1953) and Imhoff and Fair (1956) state that if the velocity is less than 0.6 to 1.0 fps, deposition takes place, and scouring occurs when the velocity increases above 1.0 - 1.5 fps. Thus, deposited materials may be resuspended and cause a secondary increase of BOD.

Only the upper layer of deposited sludges is aerobic, the remainder of the sludge banks being anaerobic. Imhoff and Fair (1956) have reported that the thickness of the aerobic layer is approximately 5 mm. The gaseous products of the anaerobic decomposition process, CH_4 and H_2S, proceed up through the sludge layer into the overlying water. Typical values of the dissolved oxygen uptake by sludge deposits vary from 1-10 $gm/m^2/day$. Thomann (1972) summarized some of the earlier works on the magnitude of the benthal demand, the results of which are presented in Table 17-3.

Table 17-3. Average Values of Oxygen Uptake Rates of River Bottoms. After Thomann (1972)

Bottom Type and Location	Uptake (gms O_2/m^2 - day) @ 20°	
	Range	Approximate Average
Municipal Sewage Sludge - Outfall Vicinity	2-10.0	7
Municipal Sewage Sludge - Aged Downstream of Outfall	1-2	1.5
Cellulose Fiber Sludge	4-10	7
Exclusive Mud	1-2	1.5
Sandy Bottom	0.2-1.0	0.5
Mineral Soils	0.05-0.1	0.07
River Ivel [1]		2.15
River Hiz [1]		2.9

(1) Anon., Water Poll. Res. Laboratory - England (1967)

In order to obtain S_B as defined by Eq. (17-5), the areal benthal oxygen demand (assuming complete vertical mixing of the overlying water) must be divided by the average depth, or:

$$S_B = \frac{O_2 \text{ uptake (gm/m}^2 \text{ - day)}}{H} \qquad (17\text{-}15)$$

Soluble organic wastes support the growth of heterotrophic bottom microorganisms and fungi. As previously mentioned, filamentous bacteria, such as *Sphaerotilus natans*, are typical. Although *Sphaerotilus* may sometimes become the predominant species, many other bacteria take part in this biological "bottom" activity.

From these considerations, it should be possible to relate the benthal oxygen demand of these aerobic bottom growths to the BOD deoxygenation rate, i.e., to the bottom deoxygenation factor, B. From the general microbiology of biological waste water treatment processes, the following relationship has been accepted for expressing the oxygen uptake rate:

$$O_2 \text{ uptake} = a' S_r + b' X_v \qquad (17\text{-}16)$$

where:

S_r = BOD removed

X_v = cell concentration

a', b' are coefficients

The ranges of a' for activated sludge and trickling filters vary within the interval 0.4 to 0.8. Applying Eq. (17-16), one can write an approximate relationship for benthal oxygen demand as follows:

$$S_B \approx a' BL + b' X_{v/H} \qquad (17\text{-}17)$$

The first term on the righthand side of Eq. (17-17) expresses the oxygen demand due to the BOD removed from the overlying water and the second term expresses the endogenous respiration of the microbiological flora.

As a rough estimate, which also includes a safety factor, it can be assumed that:

$$S_B \approx BL \qquad (17\text{-}18)$$

Thus, it is not necessary to include a separate term for the benthic oxygen demand of the biological slimes in Eqs. (17-5) and (17-8). However, this assumption requires further investigation. For *Sphaerotilus*, values of oxygen demand up to 18 g/m^2/day at 20° C have been reported (anon. 1967).

17.10 NITRIFICATION

In the nitrification processes, the unoxidized organic and ammonia nitrogen is oxidized in two steps. In the first step, ammonia is oxidized to nitrite by *Nitrosomonas* as follows:

$$2NH_4^+ + 3O_2 \rightarrow 2NO_2^- + 2H_2O + 4H^+$$

In the second step, nitrite is oxidized to nitrate by *Nitrobacter* as shown by:

$$2NO_2^- + O_2 \rightarrow 2NO_3^-$$

Theoretically, it takes 4.57 g of oxygen to oxidize 1 g of unoxidized nitrogen; however, somewhat lower values have been observed in practice. In most studies, the accepted oxygen demand of 1 g of unoxidized nitrogen expressed as Total Kjeldahl Nitrogen (TKN) is about 4.33 g of O_2. Nitrifying organisms are strict aerobes, i.e., nitrofication occurs only if sufficient oxygen is present. If the D.O. concentration is depleted below 2.0 to 2.5 mg/l, the reaction rate decreases rapidly. Under septic or anaerobic conditions, denitrification occurs where nitrates become the source of oxygen and the oxidized nitrogen is reduced to nitrogen gas. Thus, under aerobic conditions, the level of total nitrogen remains approximately constant although some nitrogen may be biologically absorbed. However, if oxygen is lacking and denitrification occurs, the total nitrogen may be substantially decreased. Nitrifying organisms are also sensitive to pH and function best over a pH range of 7.0 to 8.0 (Ruane and Krenkel, 1976). Long term BOD studies of a series of the actual river samples can provide information on the magnitude of the nitrification coefficient. A plot of the log of TKN vs. time can also be useful in determining K_N, typical plots of the various forms of N being shown on Figure 17-4 (Bernal-Lopez, Krenkel, & Ruane, 1976).

17.11 ATMOSPHERIC REAERATION

Gas transfer theory demonstrates that stream reaeration by atmospheric oxygen is proportional to the turbulence intensity at the water surface and the ratio of the surface area to the water volume. In the early work of Streeter and Phelps (1925), the reaeration coefficient, K_2, was reported in the form:

$$K_2 = C \frac{U^n}{H^2} \tag{17-19}$$

The magnitude of the reaeration coefficient has subsequently been subjected to intensive investigations ranging from purely theoretical as that of O'Connor and Dobbins (1958) to empirical field investigations as those of Churchill, et al. (1962). These studies usually resulted in a relationship of the type:

$$K_a = \frac{CU^n}{H^m} \tag{17-20}$$

Reported values of the coefficients are summarized in Table 17-4.

Thackston and Krenkel (1969) proposed the following expression for the reaeration coefficient, which assumes that the surface reaeration is proportional to the energy dissipation:

$$K_2 = 0.000125\ (1 + \mathbb{F}^2)\ \frac{U^*}{H} \tag{17-21}$$

where: U^* = friction velocity = $(gHS_e)^{1/2}$

$\mathbb{F}$ = Froude number = $\frac{U}{(gH)^{1/2}}$

g = gravity acceleration

S_e = energy slope

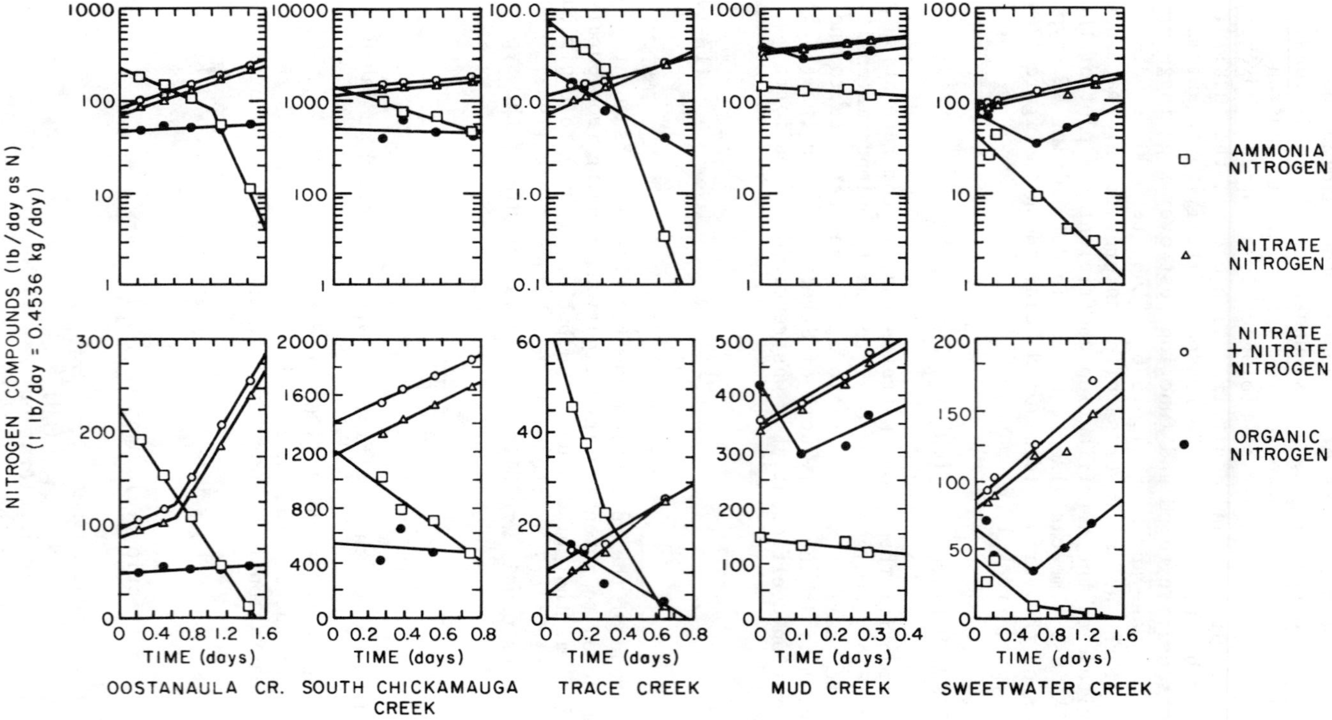

Fig. 17-4. The Variation of Nitrogen Compounds on Five Streams.

Table 17-4. Summary of Coefficients for the Reaeration Equation

$$K_2 = C \frac{U^n}{H^m}$$

Investigator	C	n	u
O'Connor and Dobbins (30)	12.90	0.5	1.5
Owens and Gibbs (34)	9.4	0.67	1.65
Churchill, et al. (31)	11.56	0.969	1.673
Langbein and Durum (35)	7.59	1.0	1.33

u = velocity in fps
H = depth in ft.
K_2 is to the base e

17.12 PHOTOSYNTHESIS AND RESPIRATION OF ALGAE AND GREEN PLANTS

Organisms containing chlorophyll can assimilate atmospheric carbon dioxide and water and liberate oxygen. A simplified photosynthetic reaction can be written as:

$$6CO_2 + 6H_2O \xrightarrow{\text{light}} C_6H_{12}O_6 + 6O_2$$

Plant and algae respiration occurs continuously throughout the day, the photosynthetic release of oxygen during the daytime during bright days being higher than the respiration, thus resulting in a net oxygen gain. The ratio of the photosynthetic oxygen production to the respiration, P/R, has a seasonal pattern. Common ratios of P/R are equal to 1:4. During dull or dark days, the ratio may be less than 1, i.e., respiration is greater than photosynthetic production.

The effect of aquatic plants on D.O. within a particular stream reach at a particular time of day is a function of the plant density, the light intensity, the water depth, the turbidity, and the D.O. concentration. The most important factor is apparently plant density. Oxygen production is proportional to plant density only to a certain limit, then oxygen production decreases with continuously increasing plant density and may even show a net oxygen loss. These phenomena may occur because the plants become so dense that some are shaded by other overlying plants. In such situations, even though high D.O. values are produced during the mid-day hours, extremely low D.O. concentration may occur at night.

Photosynthetic oxygen production is independent of the oxygen deficit; thus, it can result in oxygen supersaturation during a bright sunny day. Photosynthetic oxygen production rates as high as 50 gm O_2/m^2/day have been reported.

The effect of plant photosynthetic activity and respiration on the oxygen balance can be determined using several methods. These include (a) the light and dark bottle techniques, (b) photosynthetic chambers, (c) evaluation of diurnal D.O. measurements, and (d) radiocarbon techniques. Supplementing these techniques are measurements of related variables, such as chlorophyll. The dark and light bottle technique accounts only for plankton and does not measure the influence of attached plants.

Resulting (P-R) curves can usually be represented by a sine curve function as:

$$(P\text{-}R) = C_1 \sin 2\pi t + C_2 \qquad (17\text{-}22)$$

Other functional representations include time series analysis and a Fourier series function. A comparison of a TVA model using Eq. (17-22) and observed values of oxygen is shown on Figure 17-5 (Ruane, 1972).

17.13 TEMPERATURE EFFECTS

Almost all of the reaction rates in the oxygen balance are temperature dependent, the relationships being expressed in the following form:

$$K_T = K_{20}\, \Theta^{(T-20)} \qquad (17\text{-}23)$$

where K_T = reaction rate at temperature T

K_{20} = reaction rate at temperature 20° C

Θ = thermal factor which has the following accepted values

Deoxygenation rates K_r, K_d, K_1	$\Theta = 1.047$
Reaeration rate K_2	$\Theta = 1.025$
Nitrification rate K_N	$\Theta = 1.106$
Benthic oxygen demand S_B	$\Theta = 1.065$
Photosynthesis - Respiration	$\Theta = 1.08$

Temperature also has an effect on the oxygen saturation value, which can be approximated as follows:

$$C_s = 14.652 - 0.41022T + 0.007991\, T^2 - 0.00077774\, T^3 \qquad (17\text{-}24)$$

17.14 DOSAG 1 MODEL

DOSAG 1 (1970), which was developed by EPA and by the Texas Water Development Board, is a steady state model which predicts dissolved oxygen concentrations in streams and canals resulting from a specified set of streamflows, wasteloads and temperature conditions. The model possesses other features such as determination of the stream flow required to maintain a

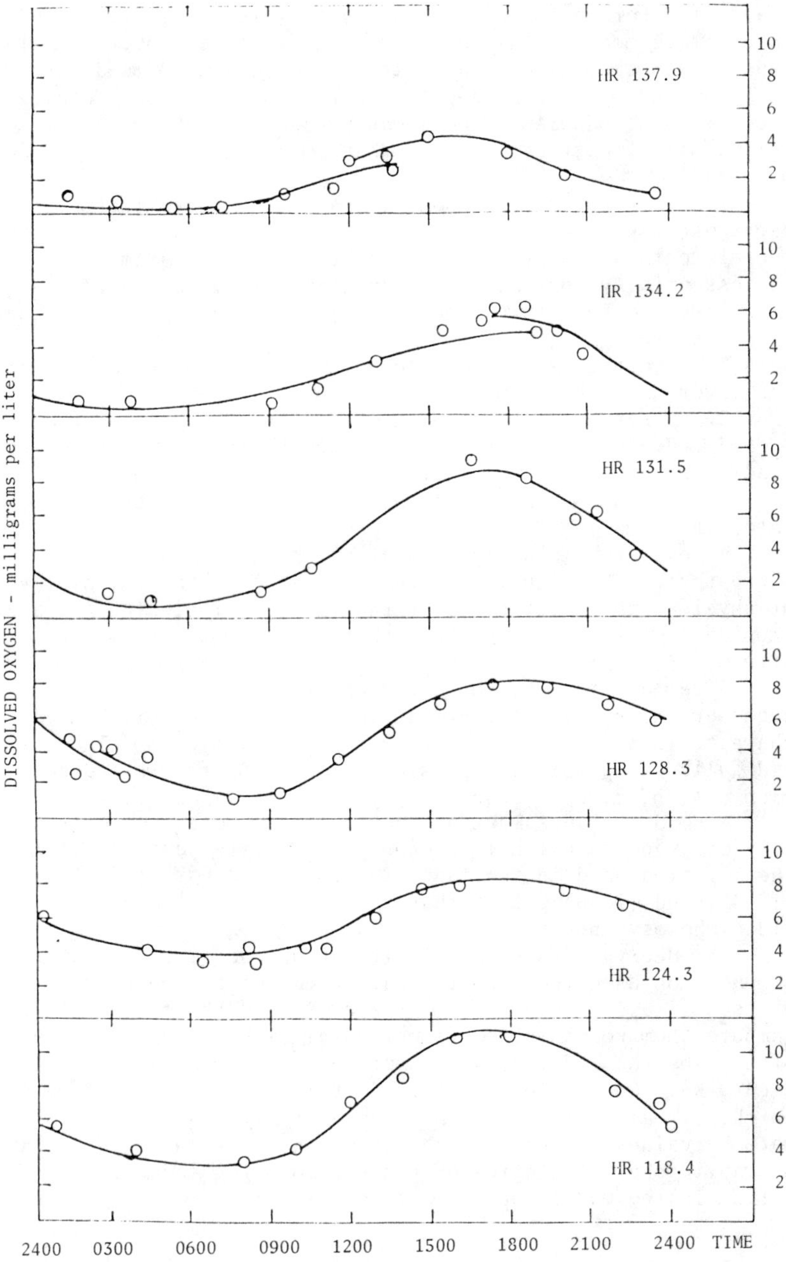

Fig. 17-5. Comparison of Measured D.O. with TVA Computed Values

specific dissolved oxygen goal, and will search the system for available storage to achieve that goal. The model may be used to estimate mean monthly dissolved oxygen levels over a full year. Both carbonaceous and nitrogenous oxygen demands are included and up to five degrees of treatment for both can be specified. The DOSAG 1 model may be adapted to any stream system, a major restriction being that impoundments cannot be modeled.

In order to use the DOSAG 1 Quality Routing Model, the user must take the stream system in question and divide it into the elements which are used as inputs to the program. There are essentially four major elements into which a system must be decomposed so that it can be modeled using this program. These elements are:

1. Junctions - the confluence between two streams within the river basin being modeled.
2. Stretches - the length of a river between junctions.
3. Headwater stretches - the length of a river from its headwater to its first junction with another stream.
4. Reaches - the subunits which comprise a stretch (headwater or normal).

A new reach is designated at any point in the stretch where there is a significant change in the hydraulic, biologic, or physical characteristics of the channel, including the addition of a waste load, or the withdrawal of water from the stream.

After a stream has been represented schematically, it is necessary to specify the hydraulic, physical and biochemical characteristics of each reach in the stream system. The basic hydraulic and physical inputs are flow, temperature and coefficients a, b, c, d as described by Eqs. (17-9) and (17-10).

A La Grangian solution technique is used to solve the dissolved oxygen equation in the DOSAG 1 quality routing model. The solution used in the model is similar to Eqs. (17-6), (17-7), and (17-8) except that the DOSAG 1 model does not include photosynthesis and benthal demand.

The deoxygenation and nitrification coefficients enter the computation as direct input data. Both coefficients are expressed at a standard 20° C temperature and the model will recompute them to a desired temperature. Several options are offered by the model for computation of the reaeration coefficient, K_2. One option is to read it in for each of the reaches in the stream system. The program user may also choose to estimate K_2 values for each reach, based on the known physical and hydraulic characteristics of the stream being modeled. This option in the model applies a general formula for K_2, which is:

$$K_2 = \frac{A_3 V^{B_3}}{H^{C_3}} \qquad (17\text{-}25)$$

where A, B and C are coefficients read as inputs.

Another technique for computing the reaeration coefficient is a direct proportionality between the reaeration coefficient and the stream discharge, or:

$$K_2 = A_4 Q^{B_4} \tag{17-26}$$

The fourth technique available for computing the reaeration coefficient for each reach is based on the investigation by Thackston and Krenkel (1969) which is identical to Eq. (17-21).

The program user may specify any of the four methods described above for the prediction of the reaeration rate coefficient for a given reach.

Waste discharges are entered into the system by specifying a new reach at each location at which a discharge takes place. The model has provisions for withdrawing water at any location within the stream system. In addition, the water is withdrawn from the stream with the quality existing at the location of withdrawal as determined by the model.

17.15 SUMMARY

The ever expanding need for water quality models has been stated and their utilization delineated. The basic model was presented, and the various inputs discussed, including reaeration, photosynthesis, benthic oxygen demand and carbonaceous and nitrogenous oxygen demand. Mathematical descriptions of these phenomena were presented and caveats elucidated. A description of the DOSAG 1 model was given (including its limitations). The application of the principles presented herein will be illustrated in the next chapter.

17.16 REFERENCES

Anon, 1970, DOSAG 1, Simulation of water quality in streams and canals. Texas Water Quality Devel. Board, Austin.

Anon, 1967, Water Pollution Research Lab., Steverage, England.

Bosko, K., 1957, Hydraulicke parametre ako ubasivaatelia samocisticej schopnosh recipientor. Vodohospodorsky casopsis SAV, 5.

Bosko, K., 1966, Discussion to paper I-1, Advances in Water Pollution Research, Vol. III, Pergamon Press.

Churchill, M.A. et al, 1962, The prediction of stream reaeration rates. Advances in Water Poll. Reserach, Vol, I, London, Pergamon Press.

Dobbins, W.E., 1969, BOD and oxygen relationships in streams. Jour. San. Engr., Div. Am. Soc. Civ. Engr., 90.

Grimsrud, G.P., et al., 1976, Evaluation of water quality models, a management guide for planners. ORD, USEPA-600/5-76-004, July, Washington, D.C.

Imhoff, K. and G.M. Fair, 1956, Sewage treatment. Wiley, New York.

Kittrell, F.W. and O.W. Kochtitzky, Jr., 1947, Natural purification characteristics of a shallow turbulent stream. Sewage Works Jour., Vol. XIX, No. 6, November.

Lopez-Bernal, F., P.A. Krenkel, and R.J. Ruane, 1976, Nitrification in free-flowing streams. 8th IAWPR International Conference, Sydney, Australia.

Novotny, V., 1969, Boundary layer effect on the course of the self-purification of small streams. Advances in Water Pollution Research, Pergamon Press.

O'Connor, D.J., 1961, Oxygen balance of an estuary. Trans. Amer. Soc. of Civil Engr., 126.

O'Connor, D.J. and W.E. Dobbins, 1958, Mechanism of research in natural streams. Trans. Amer. Soc. Civ. Engr., v. 123.

Phelps, E.B., 1944, The oxygen balance. Stream Sanitation, Wiley.

Ruane, J., 1972, personal communication.

Ruane, R.J. and P.A. Krenkel, 1975, Nitrification and other factors affecting nitrogen in the Holston River. IAWPR Conference on Nitrogen as a Water Pollutant, Copenhagen, Denmark.

Streeter, A.W. and E.B. Phelps, 1925, A study of the pollution and natural purification of the Ohio River. PHS, Washington.

Thackston, E.L. and P.A. Krenkel, 1969, Reaeration prediction in natural streams. Journal, San. Engr. Div., Am. Soc. Civil Engr., SAI.

Thomann, R.V., 1972, System analysis and water quality management. Environmental Research and Applications, Inc., New York.

Velz, C.J., 1953, Recovery of pollution streams. Sew. Works Journal, 100, 12.

Velz, C.J. and J.J. Gannon, 1962, Biological Extraction and accumulation in stream self-purification. Advances in Water Pollution Research, Pergamon Press.

Chapter 18

BASIC APPROACH TO WATER QUALITY MODELING

by

P. A. Krenkel, Executive Director, Water Research Center, Desert Research Institute, the University of Nevada, Reno, Nevada

and

R. J. Ruane, Program Manager of Special Projects and Research, Water Quality and Ecology Branch, Tennessee Valley Authority, Chattanooga, Tennessee

Chapter 18

BASIC APPROACH TO WATER QUALITY MODELING

18.1 INTRODUCTION

As stated in the previous chapter, computer models are valuable tools that can be used to advantage in making water quality management decisions. Regulatory agencies use mathematical models of the assimilative capacity of receiving streams to set effluent limitations on industrial and municipal wastewater discharges. Industrial wastewater dischargers and water quality planners use models to explore various alternatives to meet water quality objectives. Alternatives often considered include various levels of wastewater treatment, flow augmentation, moving the point of discharge to a larger stream, or artificial instream aeration. These investigations in conjunction with cost benefit analyses can also provide a factual basis for appealing to regulatory authorities for less stringent wastewater effluent limits to avoid unreasonable economic burden imposed by existing effluent limits without commensurate water quality improvement. One of the more important uses of models by both groups includes the determination of the sensitivity of model outputs to estimated reaction coefficients (i.e., K_d, K_2, K_n) and other variables that possibly may have been estimated (i.e., S_B, H, U, H, P, R).

This chapter will present a basic approach used in determining the assimilative capacity of a stream and several illustrating case histories. In the first case DOSAG 1 will be used on a free-flowing river in which both carbonaceous and nitrogenous BOD is oxidized. This case illustrates sensitivity analyses for K_d, K_2, T, H, and necessary levels of wastewater treatment. In the second case, a modified version of the Streeter and Phelps model is applied to a river where photosynthesis and respiration take place. Alternative control measures are considered, including level of wastewater treatment, flow augmentation, artificial instream aeration, temperature, and DO control in the stream above the waste discharge.

18.2 BASIC APPROACH TO WATER QUALITY MODELING

To determine the assimilative capacity of a stream, a series of steps should be followed:

Step 1. The first step is defining the objectives of the study, which includes a determination of the standards for water quality and applicable critical minimal streamflow, the planning period for 10-, 20-, or 30-year wasteload projections, and consideration of the allocation of assimilative capacity for other wastewater discharges.

Step 2. The next step is to contact those who may have existing information so as to minimize unnecessary effort. In many cases the U.S. Geological Survey will provide stream flow information, critical minimum flow determinations, stream geometry, and contour maps. In addition, water quality and time of travel data, and existing computer models can often be obtained from the nearest EPA regional office, state regulatory agencies, regional river control agencies (such as the Tennessee Valley Authority and Corps of Engineers), U.S. Geological Survey, other local industries, area-wide planning agencies, and universities.

Step 3. If information is not available to construct a model, the sensitivity of water quality management decisions to the assimilative capacity of the stream can be estimated using several procedures. If assimilative capacity is found to be an unimportant factor, informed decisions can be made without conducting a detailed stream study.

Three estimating procedures are considered here: (1) Fair, Geyer, and Okun (1968); (2) an approach using two statistically derived equations (Ruane 1971); and (3) an approach using estimated values of U, K_2, K_n, and K_d in the Streeter-Phelps equation.

The Fair, Geyer, and Okun procedure has been used for many years by practicing engineers and is well documented in the literature.

The statistical equations of procedure (2) were derived by multiple regression analysis of 20 assimilative capacity studies of streams in the Tennessee Valley. The following equation was developed on the basis of all 20 streams analyzed:

$$Y = 398{,}700 \frac{(DO_{mix})^{0.951}\, Q^{1.026}\, S^{0.580}}{T^{1.474}\, (DO_{sag})^{1.434}} \tag{18-1}$$

where Y is the assimilative capacity of the stream in pounds per day of 5-day BOD at 20°C (includes upstream BOD); Q is the streamflow, in cubic feet per second, immediately downstream from point of waste discharges; T is the stream temperature, in degrees Celsius; S is the streambed slope, in feet per foot (determined from quadrangle maps with a scale of 1:24,000); DO_{mix} is the DO of the stream in milligrams per liter, after waste mixes with stream; and DO_{sag} is the minimum allowable DO of the stream in milligrams per liter. A correlation coefficient of 92 percent was obtained with this relationship.

The following equation was derived by eliminating from the original 20 streams the 4 streams with excessive atmospheric aeration and the one stream with an excessively high oxidation rate:

$$Y = 10{,}138 \frac{(DO_{mix})^{1.094}\, Q^{0.864}\, S^{0.06}}{T^{1.423}\, (DO_{sag})^{1.474}} \tag{18-2}$$

where the variables are the same as those defined in Eq. (18-1). The correlation coefficient obtained with this relationship was

95 percent. For the latter case, all the computed values were within 50 percent of the observed values (Figure 1).

These equations should be used only if the independent variables fall within the range of the original data. For both equations, the ranges for temperature, DO_{mix}, and DO_{sag}, are 20 to 30°C, 5 to 5 mg/l, and 3 to 5 mg/l, respectively. For Eq. (18-1), the range of slopes was from 0.00037 to 0.00675, and the range of streamflows was from 0.8 to 148 cfs (1.4 to 252 cu m/min). For Eq. (18-2), the range of slopes was from 0.00037 to 0.0062, and the range of streamflows was from 0.8 to 83 cfs (1.4 to 141 cu m/min). Note that the range of slopes for Eq. (18-2) included only one value significantly greater than 0.003, while the four data points that were eliminated from Eq. (18-2) and that represent relatively shallow, rapid streams were for streams having slopes equal to or greater than 0.003.

Elevation above mean sea level was not included in either equation because of its high correlation with slope; however, the effect of elevation on DO saturation remains a part of the estimated values for assimilative capacity because elevation influenced the DO saturation values used in deriving the equations. Elevations for the 20 streams investigated ranged from 300 to 2,300 ft. (90 to 700 m) for Eq. (18-1) and from 300 to 2,000 ft. (90 to 610 m) for Eq. (18-2). The higher values of slope were associated with higher elevations.

Equation (18-2) is recommended for computing assimilative capacity of streams that have a slope less than 0.003. All but one datum point for Eq. (18-2) are for streamflows of less than 25 cfs (42 cu m/min); however, a preliminary comparison of the prediction equation with the results from two larger rivers [155 cfs (264 cu m/min), and 800 cfs (1,360 cu m/min)], suggests that Eq. (18-2) may also be representative of larger streams.

For rapid, shallow streams, use of Eq. (18-1) should probably be considered. Although the method yields results considerably higher than those from Eq. (18-2), these results are lower than the observed data for rapid streams. The results, therefore, should be conservative.

The 5-day BOD values resulting from the above equations are usually converted to ultimate carbonaceous BOD using a deoxygenation coefficient of 0.11 $days^{-1}$ at 20°C (base 10). This ultimate carbonaceous BOD is the allowable stream waste load, and it is assumed that the UOD (ultimate oxygen demand including the nitrogenous demand) of the waste load cannot exceed this value.

If water quality data are not available for a stream, the following assumptions usually yield conservatively low assimilative capacity results: (1) stream temperature is equal to the maximum allowed by water quality standards, (2) DO in the stream above the outfall is 85 percent of saturation concentration, (3) natural 5-day BOD in the stream above the outfall is 1.5 mg/l, and (4) when the wastewater discharge exceeds the

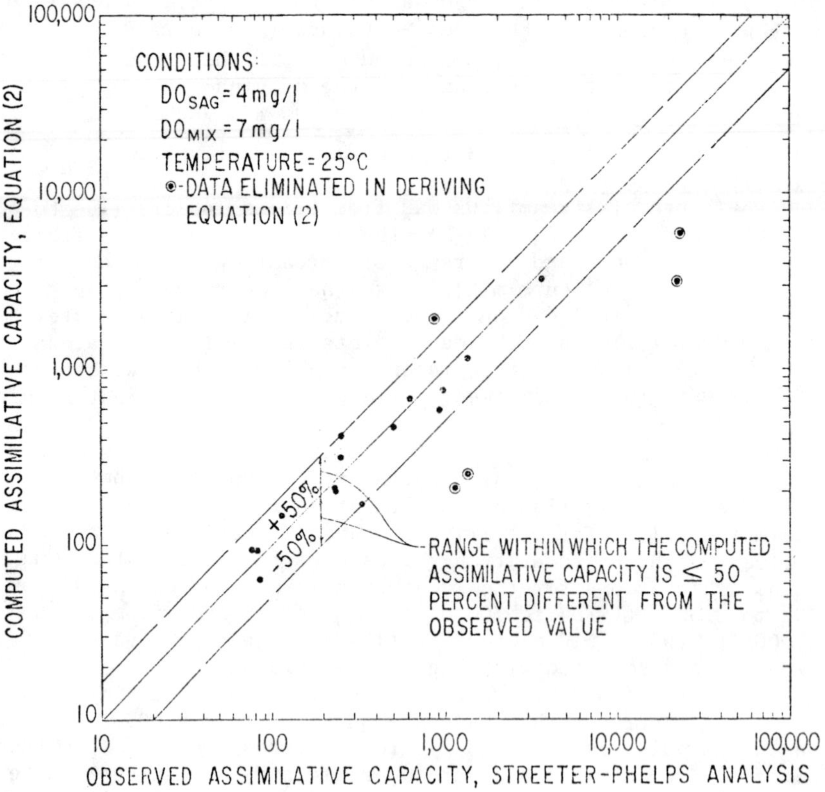

Figure 18-1. Computed Versus Observed Assimilative Capacities for the Streams used to Derive Eq. (18-2).

minimum critical flow in the receiving stream, the DO immediately downstream from the waste discharge is equal to the minimum concentration allowed by water quality standards. Assumptions 2 and 3 are based on the premise that water quality above the outfall is not affected by upstream pollution.

To account for uncertainties in estimating the assimilative capacity of streams, a range of "probable" error is used to test sensitivity to the water quality management decisions. From the results of the statistical analysis, this probable range of error was estimated to be 150 percent. It should be noted that the actual assimilative capacity could be more than 50 percent greater than the calculated value, but is unlikely to be less than 50 percent of the calculated value.

The third method considered here involves the use of the Streeter-Phelps equation, which has been modified to account for nitrification, and coefficients for this equation as estimated on the basis of easily measured hydraulic parameters such as streamflow, slope of the stream bed, and approximate width of the stream.

As stated earlier, streamflow information is available from the U.S. Geological Survey or watershed authorities. For the purposes of estimating stream water quality coefficients, stream bed slopes can be determined from quadrangle maps on which elevation contours are included. The approximate width of a stream of 200 feet or more can also be estimated from quadrangle maps. For smaller streams, a site visit is necessary; however, less than one man-day is generally required to obtain sufficient information.

The estimation of K_d using the Bosko and Novotny equations was discussed in the previous chapter. Bosko's formula required a determination of velocity and depth, and Novotny's equation required slope and depth. Foree (1976) developed empirical relationships for estimating velocity using information on drainage area, the specific discharge, and slope of the stream bed. For $0.30 \leq q < 0.90$ and $5 < A < 250$

$$U = 14 + 5.6\ (q - 0.08)\ S_e^{.5} \tag{18-3}$$

where q = the specific discharge in the stream in cubic feet per second per square mile, A = drainage basin area in square miles, and S_e = slope of stream bed in feet per mile. For $q < 0.30$ and $5 < A < 250$

$$U = 0.40 + 4.1\ qS_e^{0.5} \tag{18-4}$$

The data used to develop these equations were for streams having $0.45 < Q < 4.0$ and ≈ 0 (pool) $< S_e < 41.8$. Foree stated that caution should be used in applying these equations outside the range for which they were developed. Stall and Yang (1970) have also presented relationships between velocity and hydraulic variables.

Using the continuity equation and assuming approximately a rectangular channel section for the stream, depth can be estimated from the following equation:

$$H = \frac{Q}{UW} \tag{18-5}$$

where W = width of the stream.

Foree (1976) also developed a relationship between K_2 and S_e

$$K_2 = 0.30 + 0.19\ S_e^{1.2} \tag{18-6}$$

where K_2 is to the base e at 25°C and S_e is in feet per mile. However, the other expressions for K_2 given in the previous chapter can also be used if velocity and depth can be estimated as shown above. It should be noted that sensitivity analysis on the estimated velocity is especially critical since an error in velocity determination compounds the error in estimating the depth. K_2 is especially sensitive to these errors because for most of the formulas velocity is in the numerator and depth is in the denominator.

Bansal (1976) recently reported a relationship between a dimensionless term including K_n and the Reynolds and Froude numbers

$$\log \left(\frac{K_n H^2}{\nu}\right) = -3.421 + 1.36 \log \left(\frac{\mathbb{R}}{\mathbb{F}}\right) \qquad (18\text{-}7)$$

in which K_n is to the base e and ν is the kinematic viscosity of water.

In using the above approach to model water quality, extreme caution should be exercised. The user should refer to the literature to determine the applicability of the above equations to specific cases and should perform sensitivity analyses on the estimated variables. A considerable amount of research is needed to improve the above relationships for U, K_2 and K_n.

Step 4. As recommended in the preceding chapter, if the approximate analysis in Step 3 indicates that a valid model is needed for the receiving stream, a model should be selected as suggested by Grimsrud et al. (1976).

Step 5. If the information obtained in Step 2 is not sufficient for developing the water quality model, a stream survey for water quality will be necessary. It is preferred that survey data be taken during a summer low-flow, steady state. A time-of-travel study during this period using tracer dyes is also highly recommended. If water quality data does exist on the stream in question, factors that may influence the coefficients should be considered. For example, if the wastewater discharge characteristics have improved since the date of the last survey, it is conceivable that the nitrification rate may have increased and that the carbonaceous BOD oxygenation rate may have decreased. Also if the stream was heavily burdened with macrophytes or attached algae growths, these biological growths may have decreased significantly and have affected the velocity in the stream as well as photosynthesis and respiration.

Step 6. Sensitivity analyses on critical variables that may affect management decisions is again necessary. The use of sensitivity analyses will be discussed in the case histories that are described in the latter part of this chapter.

Step 7. After completion of the sensitivity analyses, the investigator may want to consider the need for additional data or studies to resolve important economic issues related to the high cost considerations for wastewater discharges. For example, two techniques have now been developed for measuring reaeration coefficients directly (Rathbun et al., 1975, and Tsivoglou, 1976).

Step 8. The final step in this process is to make the water quality management decision based on the above investigations and results.

18.3 APPLICATION OF DOSAG-1 MODEL

A schematic diagram of the stream network is shown in Fig. 18-2. The stream system was divided into three stretches:

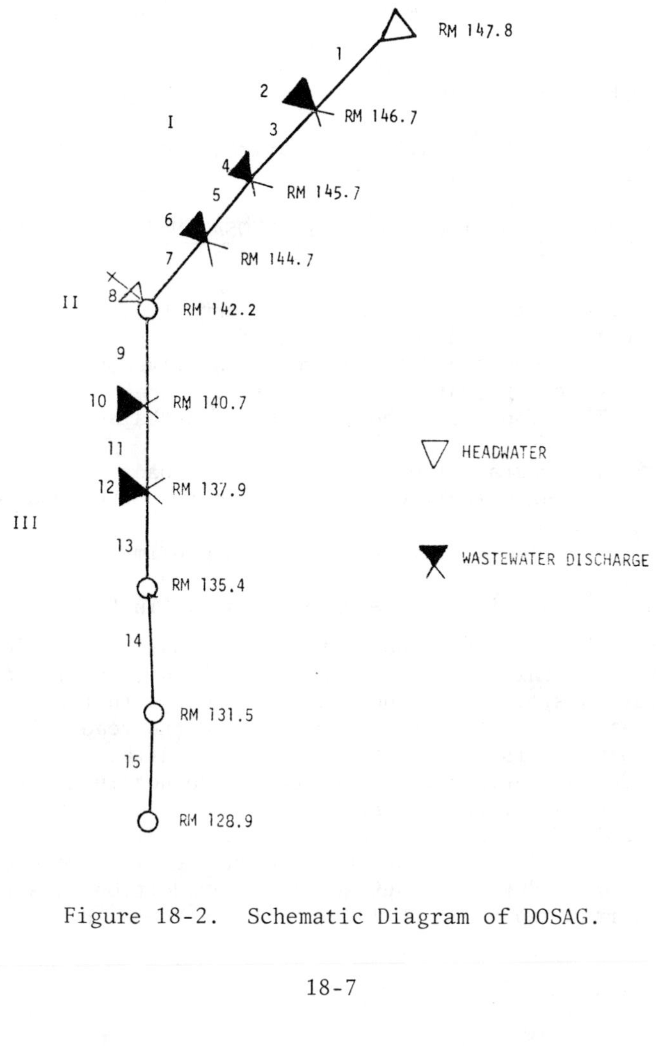

Figure 18-2. Schematic Diagram of DOSAG.

1. Main River R.M. 142.2-147.8
2. Tributory River R.M. 0.0-1.0
3. Main River R.M. 128.9-142.2

The system has ten stream reaches plus five wastewater discharges to give a total of fifteen reaches.

The hydraulic characteristics were determined from survey data, which indicated that the value of the exponent on the velocity equation is 0.674 for river miles above R.M. 131.5 and 0.539 for river miles below R.M. 131.5. The multipliers were determined from known velocities and discharges and from the exponent as described in a previous chapter. The coefficients for depth determination were unknown. Therefore, the exponent on the depth equation was determined using the known velocity exponent from the relationship:

$$b + d = 1 \tag{18-8}$$

and the final values of the multipliers were obtained from the model delineation and verification.

18.4 MODEL VERIFICATION

For the DOSAG-1 model verification, two sets of survey data were available--1946 and 1969.

Delineation and Verification of the DOSAG Model Using the 1946 Data

The 1946 data offer an excellent opportunity for the development of a valid model for the following reasons:

(a) The reported data are representative because of the care with which the data were collected.

(b) The effect of weeds on the oxygen balance was not reported.

(c) The character of the wastewater loads contributed to the river has not significantly differed from the 1946 conditions to date.

(d) The flow conditions closely approximated the critical low flow.

The BOD_5 profile of the river is shown in Fig. 18-3. During the delineation procedure, the deoxygenation coefficients were varied within acceptable ranges until a good fit was obtained. A similar procedure was followed with the dissolved oxygen data. The only parameter by which the reaeration rate can be adjusted is depth. Since the first attempt, using calculated depths, produced results that were not in accord with the observed data, the depths in each reach were varied until a reasonable fit was obtained. The plot finally derived is shown in Fig. 18-4. The depths used in Fig. 18-4 were close to those reported during a 1969 survey. In addition, a survey in 1973 reported depths in RM 142.2 to 131.5 to be in the order of

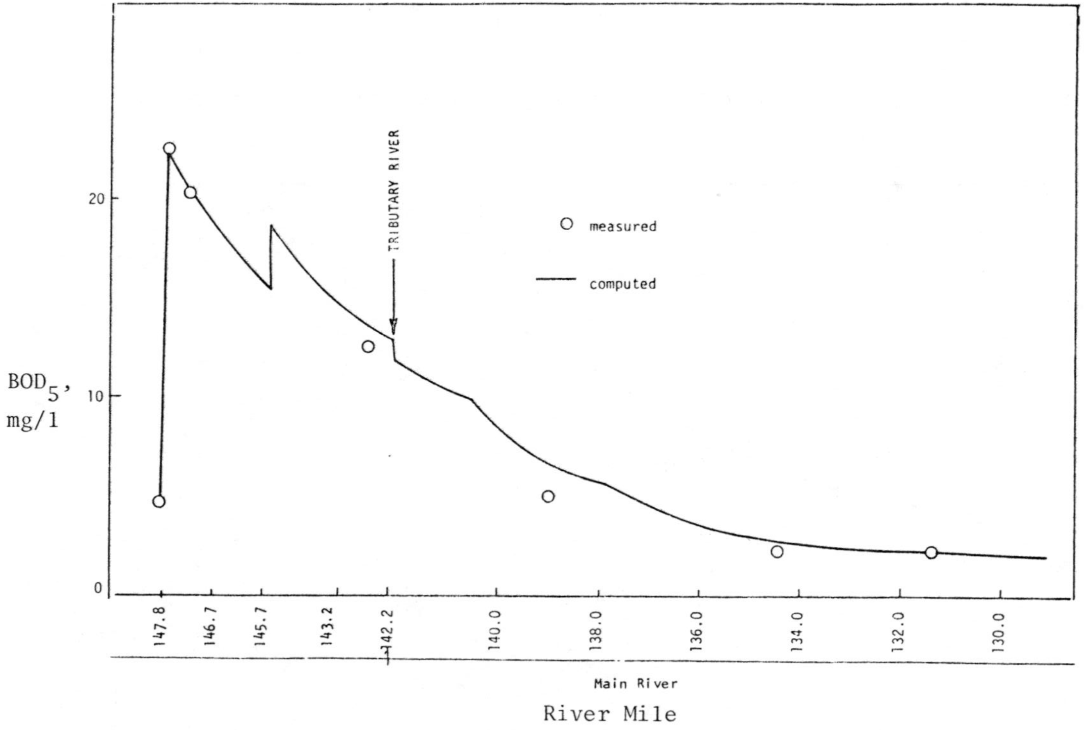

Figure 18-3. Verification of DOSAG Model BOD_5, 1946 Survey (Temperature 23°C.)

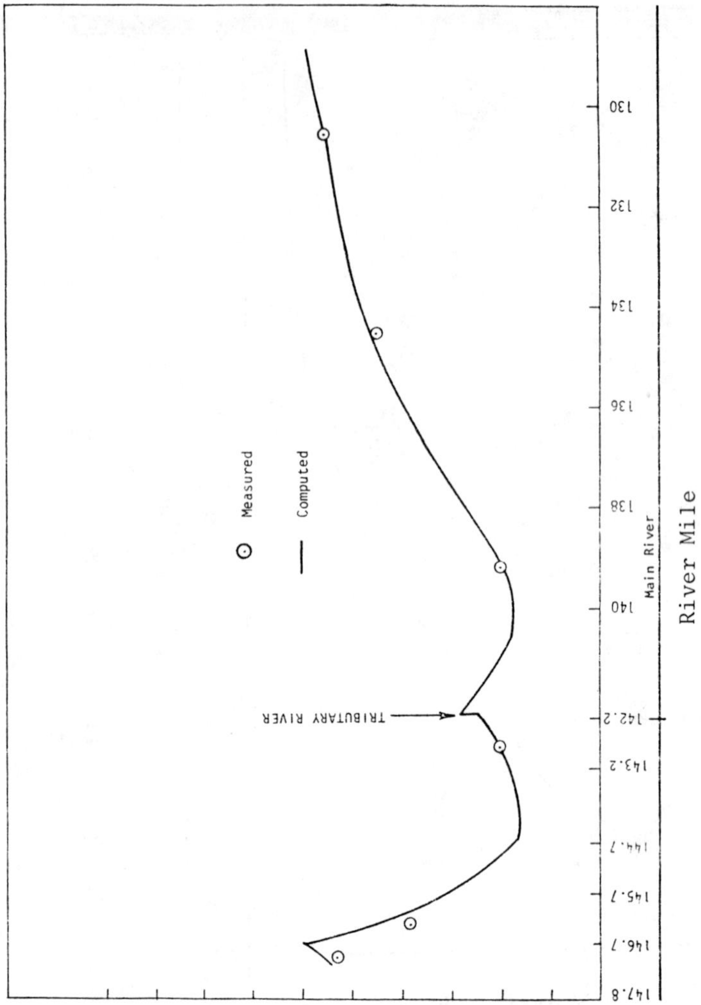

Figure 18-4. Verification of DOSAG Model for D.O. - 1946 Survey.

1 to 2.5 feet, with some locations showing depths below 1 foot.

Verification of the Model Using 1969 Survey Data

The model, as delineated and verified using 1946 survey data, was then compared with the 1969 survey data. The basic coefficients were not changed. The discharges, temperatures, and wastewater loads were adjusted to those reported for the 1969 survey. The computed BOD's, DO values and nitrogenous demand were then plotted and compared with the observed values. The BOD plot (Fig. 18-5) shows excellent agreement, as does the plot of the nitrogenous demand (Fig. 18-6).

As shown in Fig. 18-7, the computed DO profile is somewhat lower than the observed values. This deviation may be explained by noting that the 1969 survey was conducted during bright summer days when the aquatic weeds contributed substantially to the DO variations.

Since the DOSAG-1 model does not take photosynthesis and respiration into account, computed DO represents conditions without the aquatic weeds. With this in mind, the computed oxygen profile can be considered satisfactory.

The verification of the DOSAG-1 model on both the 1946 and 1969 surveys demonstrated that it can be successfully applied (excluding the effects of photosynthesis). Proper care and effort must be devoted to the delineation of the model, selection of adequate parameters, etc. Only then can the model be used for the determination of waste assimilative capacity.

Verification of the Estimated Deoxygenation Coefficients

The results obtained by using the model allow the theoretically determined parameters to be compared with those applied by the model and verified by the two surveys. The results of the comparison of the theoretical deoxygenation, B, with the actual value of $K_d - K_1$ are shown in Table 18-1.

18.5 DETERMINATION OF THE WASTE ASSIMILATIVE CAPACITY

The model was constructed on the 1946 water quality investigation, which was deemed to be the best approximation of water quality conditions that would occur during a low flow period discounting aquatic weeds. The solution matrix was designed with the following vectors:

(1) Wastewater load in the Main River
 50,000, 25,000 and 14,000 lbs of BOD/day

(2) Water temperature
 $26^{\circ}C$, $30^{\circ}C$ and $34^{\circ}C$

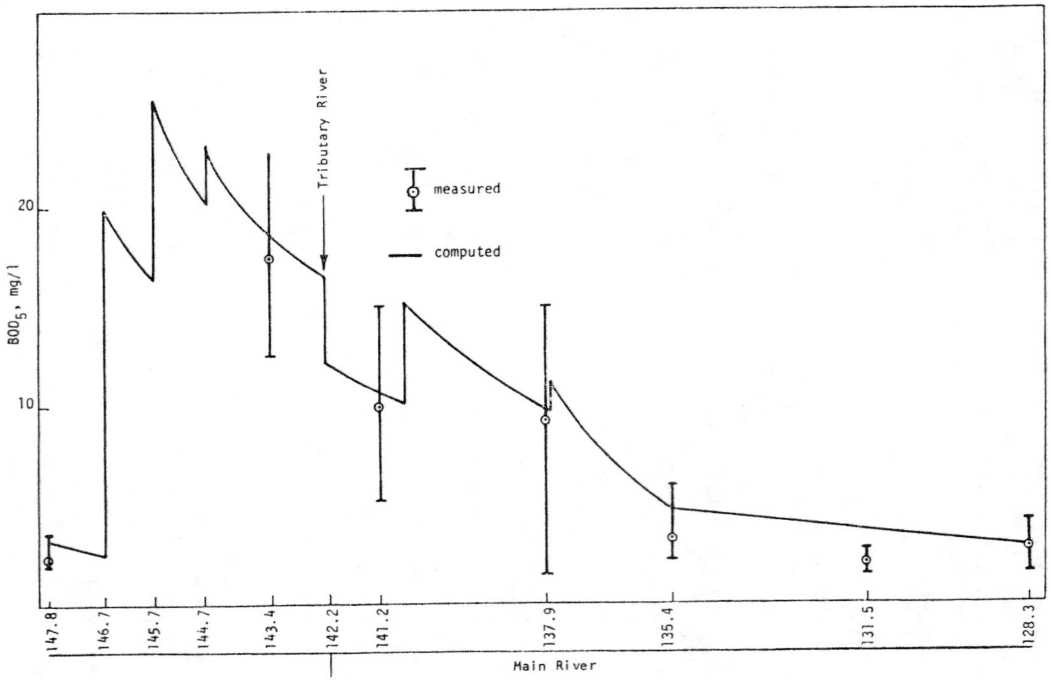

Figure 18-5. Verification of DOSAG Model BOD_5 - 1969 Survey (Temperature 26°C.)

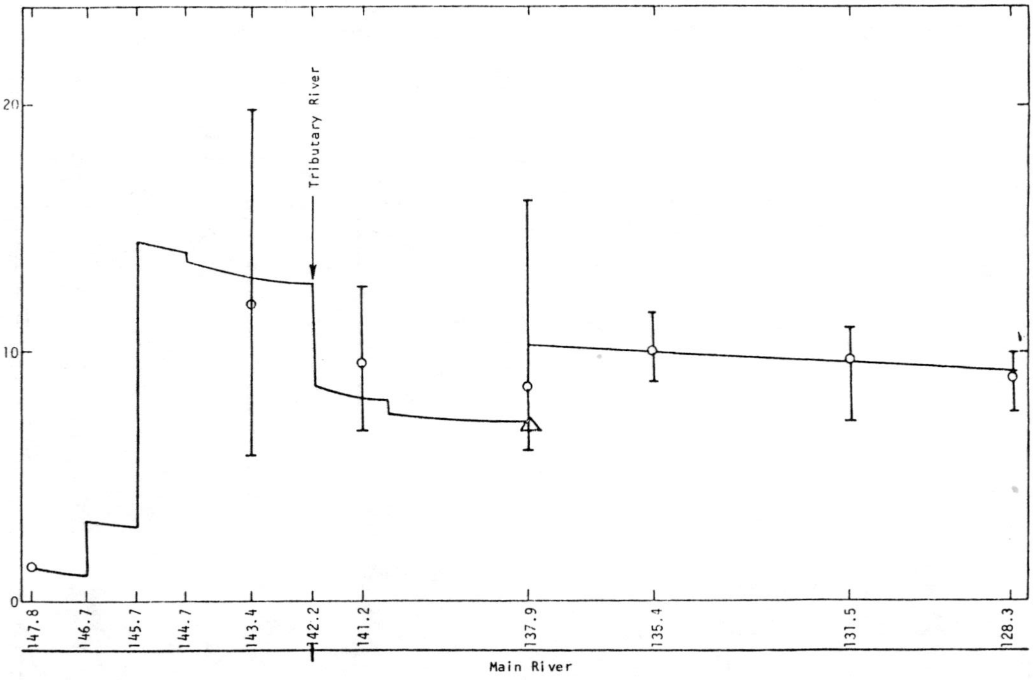

River Mile

Figure 18-6. Verification of DOSAG Model Nitrogenous Demand - 1969 Survey (Temperature 26°C.)

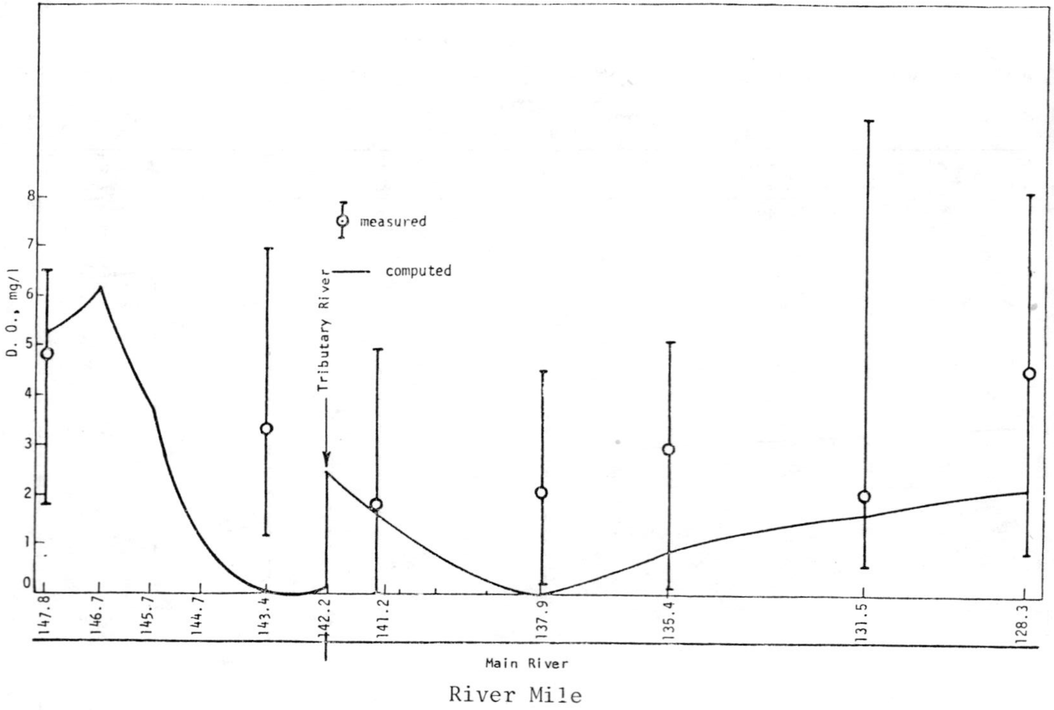

Figure 18-7. Verification of DOSAG Model Dissolved Oxygen - 1969 Survey (Temperature 26°C).

Table 18-1. Comparison of Estimated and Model Deoxygenation Factors B.

Reach RM	Character of Bottom	Velocity fps	Depth ft	Slope per miles	Coefficient		Computed B: Bosko Eg. 4-13	Computed B: Novotny Eg. 4-14	Measured K_d-K_1 (B) days^{-1}
145.7 - 144.7	gravel	0.63	2.5	0.62	1.2	1.0	0.30	1.08	1.0
144.7 - 142.2	sand	0.42	2.5	0.62	0.6	0.4	0.10	0.43	0.25
142.2 - 140.7	gravel rock	1.70	1.9	0.5	1.5	1.5	1.34	1.90	2.12
140.7 - 137.9	gravel rock	1.04	2.4	0.8	1.5	1.5	0.65	1.79	1.60
137.9 - 135.4	gravel rock	0.94	2.4	0.234	1.5	1.5	0.59	1.34	1.14
135.4 - 131.5	gravel lower pollution	1.36	2.3	0.234	0.9	0.6	0.41	0.55	0.55
131.5 - 128.9	gravel lower pollution	0.89	6.1	0.234	0.8	0.5	0.12	0.22	0.15

(3) Deoxygenation coefficient, K_r, in the river

100% of the present

75% of the present

40% of the present, and

$K_r = 0.23\ day^{-1}$

(4) Average depth in the Main River between RM 142.2 and 128.9

1.8 ft, 2.55 ft, and 4.2 ft.

The upstream dissolved oxygen concentration at RM 147.8 was selected to be 5.0 mg/l, which was similar to the reported average concentration of oxygen in the discharge from an upstream dam. It was observed that because of high reaeration occurring between RM 147.8 and 146.7 (the first waste discharge), the dissolved oxygen concentration increased about 1 mg/l.

It was also assumed that the D.O. concentration of all wastewaters discharged would not fall below 5.0 mg/l, as stipulated by the local regulatory authority. The wastewater flows and withdrawals corresponding to 50,000 lbs BOD/day are presented in Table 18-2.

Table 18-2. Wastewater Discharges Corresponding to 50,000 lbs. BOD_5/Day.

Number of Reach	Sewage and Industrial Flows: Flow Rate cfs	Diss. Oxygen mg/l	Carbon BOD mg/l
1	0.0	0.0	0.0
2	573.0	6.3	28.6
3	0.0	0.0	0.0
4	16.1	5.0	390.0
5	-27.1	0.0	0.0
6	26.3	5.0	56.0
7	0.0	0.0	0.0
8	0.0	0.0	0.0
9	0.0	0.0	0.0
10	0.0	0.0	20.0
11	-159.3	0.0	0.0
12	159.3	5.9	22.7
13	0.0	0.0	0.0
14	0.0	0.0	0.0
15	0.0	0.0	0.0

The headwater water quality characteristics were as follows:

Table 18-3. Headwater Water Quality Characteristics.

No. of Headwater	River	Percent DO, Sat	Carbon BOD mg/l	Nitrogen BOD, mg/l
1	Main RM 5.6	65.7	2.3	1.1
2	Trib. RM 1.0	85.0	2.5	0.9

Other BOD loads were computed by the model assuming 50% and 72% waste reduction. Dissolved oxygen profiles were computed and plotted for various BOD loads in the river for temperatures of 26°C and 30°C, and for different reductions of the deoxygenation coefficient, K_r. Two typical plots are shown in Figures 18-8 and 18-9. The Thackston and Krenkel formula for K_2 was used.

A similar analysis was performed for nitrogenous BOD but is not presented here. The information obtained in the various plots similar to Figures 18-8 and 18-9 was extracted and plotted as shown in Figures 18-10 and 18-11. Similar plots were also developed for 25 and 75 percent reduction in K_r. These plots were then used to develop Fig. 18-12, which shows the effect of K_r reduction on the assimilative capacity of the river. It should be noted that for the hypothetical case where $K_r = 0$, the waste assimilative capacity converges to infinity.

When long-term BOD curves of the wastewater effluents with existing treatment were compared to BOD curves of a pilot plant activated sludge effluent, it was concluded that the activated sludge treatment would reduce the deoxygenation coefficient by 50%. This same reduction (greater than 50%) can be expected for the river deoxygenation rate.

The effect of depth on the waste assimilative capacity (DO standard = 3.0 mg/l) is shown in Fig. 18-13. It is significant to note that a deviation in the depth estimation of one foot represents a change in the waste assimilative capacity of 5,000 pounds of BOD per day.

Different formulas for K_2 were evaluated and the resulting assimilative capacity was as follows (assuming K_r is reduced 50 percent):

K_2 Formula	K_2 Day^{-1} Base e	Assimilative Capacity (lbs/day)
Churchill, et al. (1962)	0.95	2300
O'Connor and Dobbins (1958)	1.7	7300
Thackston and Krenkel (1969)	2.4	13500

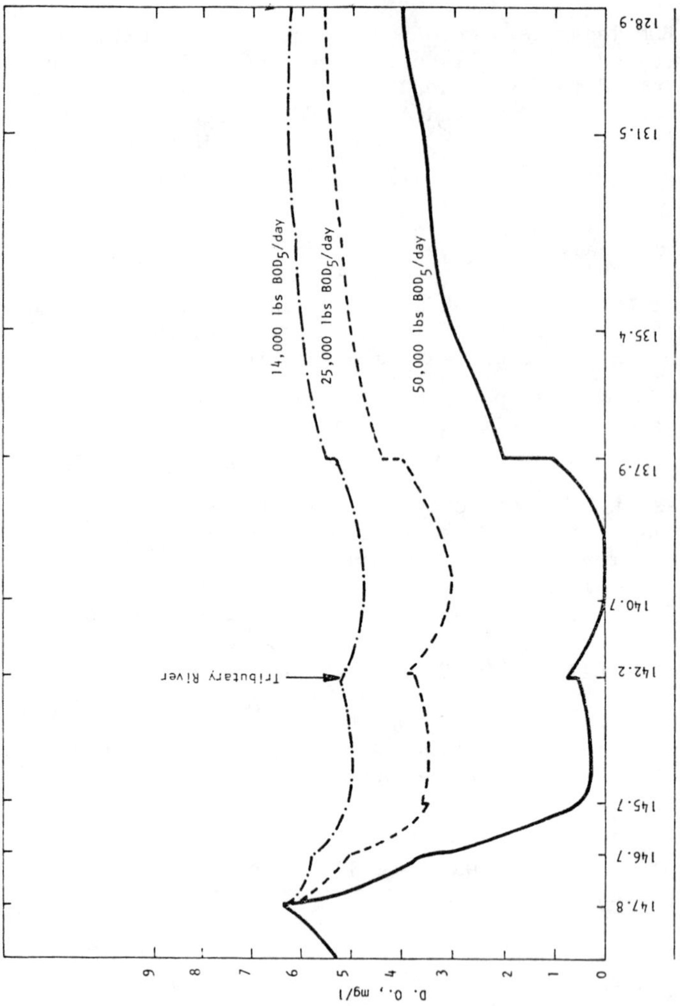

River Mile

Figure 18-8. DO Profiles for Various BOD_5 Loads in the Main River (Temperature 26°C - K_r - Existing).

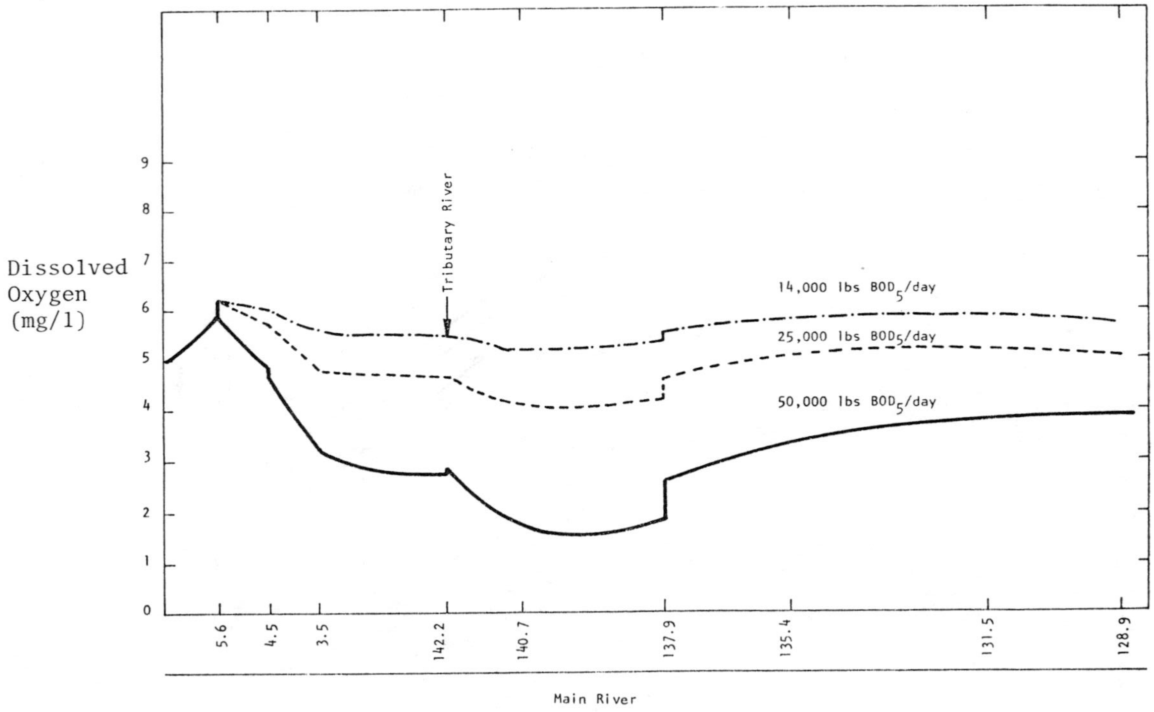

Figure 18-9. DO Profiles for Various BOD_5 Loads in the Main River. (Temperature 30°C - K_r - 60% Reduced).

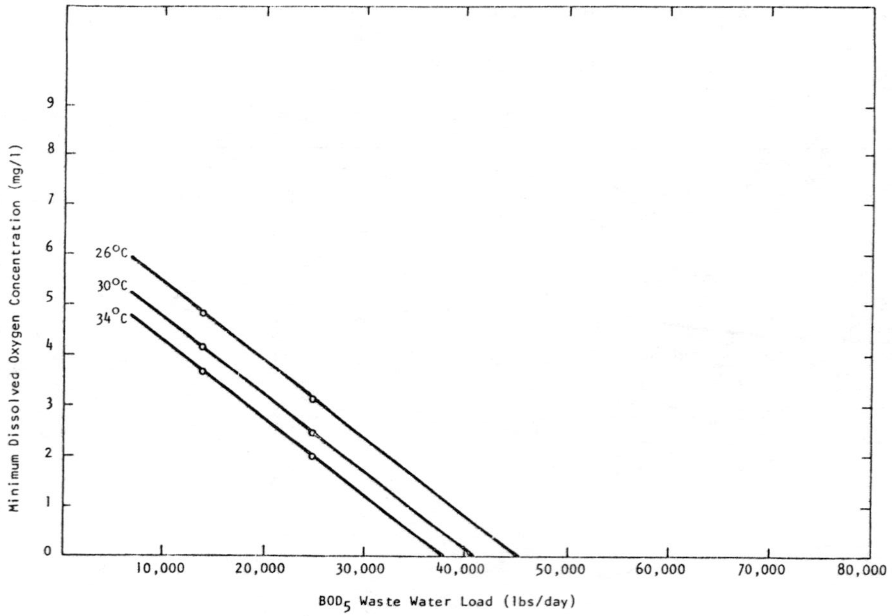

Figure 18-10. Relationship of Minimum Dissolved Oxygen Concentration to BOD_5 Load in the Main River (K_r Existing).

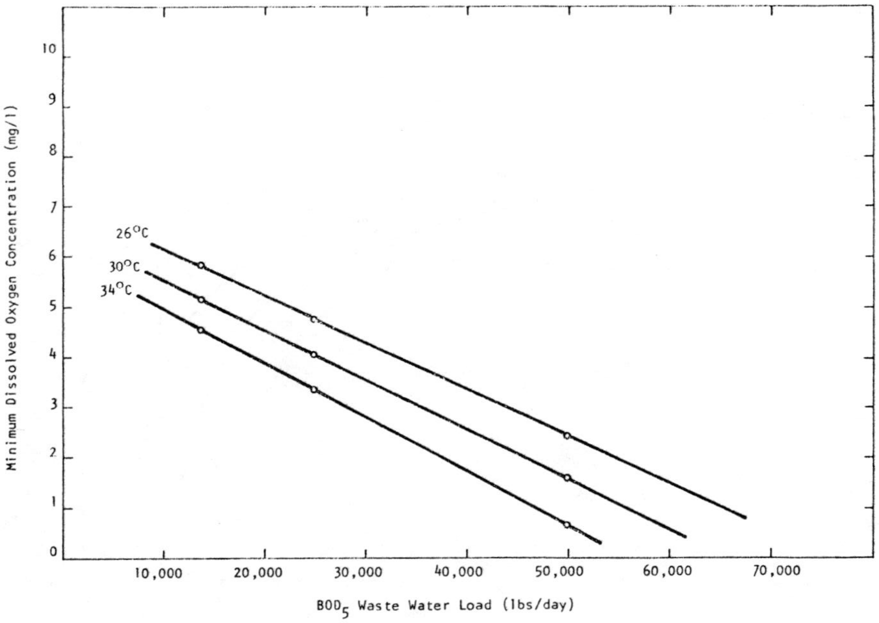

Figure 18-11. Relationship of Minimum Dissolved Oxygen Concentration to BOD_5 Load in the Main River (K_r - 60% Reduced).

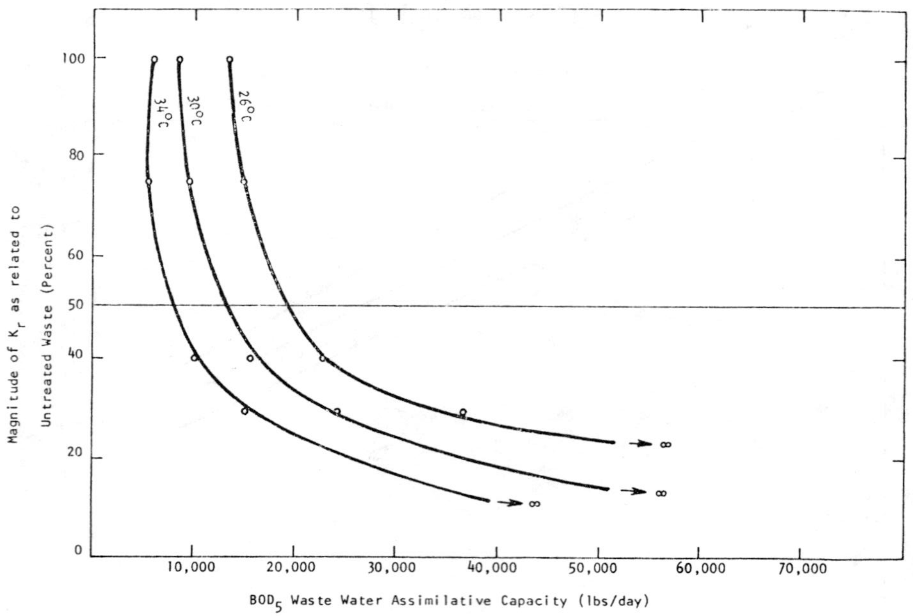

Figure 18-12. Waste Assimilative Capacity Related to Magnitude of K_r for a Minimum DO of 5.0 mg/1.

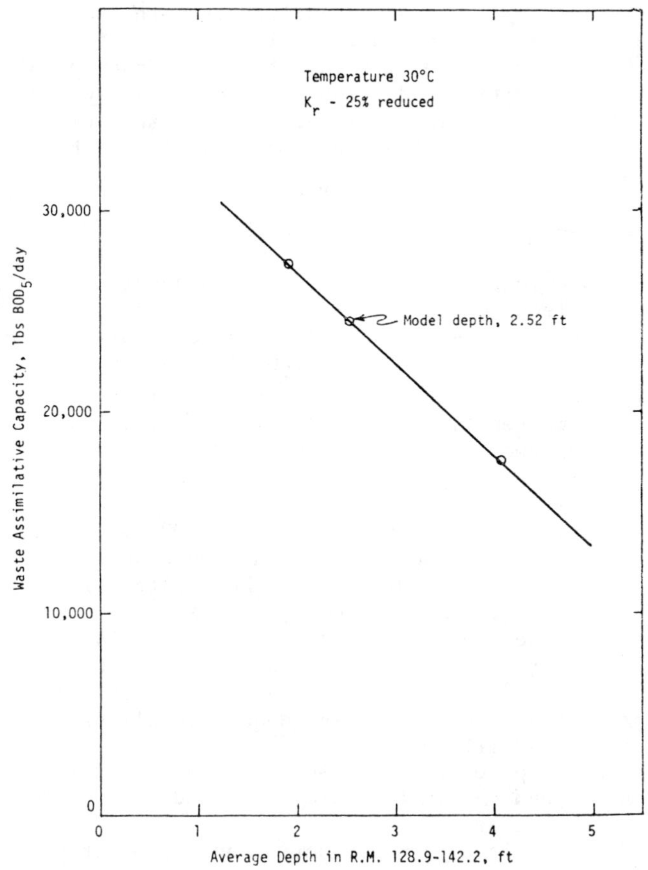

Figure 18-13. Effect of Depth Variation on Waste Assimilative Capacity of Main River.

Best Estimate of the Waste Assimilative Capacity

The estimate of the waste assimilative capacity is based on the following assumptions:

Flow conditions:	750 cfs in Main River above tributary
	800 cfs in Main River after confluence
Temperature:	30^oC in all reaches
Depth:	2.52 ft.
K_r reduction:	50 percent
K_2:	Thackston and Krenkel formula

From Fig. 18-12 the corresponding waste assimilative capacity which would satisfy the minimum DO requirement of 5.0 mg/l of DO in all reaches would be 13,500 lbs BOD/day.

Although depth and K_2 significantly affected the assimilative capacity of the river, the range of assimilative capacity values did not in turn significantly affect the actual DO in the river. This was deduced from plots similar to those in Fig. 18-9. If this line of argument proved unacceptable to regulatory authorities, additional field surveys would have been justified to obtain a more accurate measure of depth, and, possibly, to use one of the inert gas transfer techniques to make an actual determination of K_2.

18.6 MODELING EFFECTS OF AQUATIC WEEDS

This case considers a river in which aquatic weeds significantly affect DO (see Fig. 17-5 in the previous chapter). The river reach is downstream from a large industrial complex that has a total discharge of 100,000 lbs/day of BOD_5 and 50,000 lbs/day of nitrogenous BOD. Factors that have been analyzed and mathematically modeled in the 40-mile river reach include carbonaceous BOD, nitrogenous BOD, benthic oxygen demand, aquatic plant photosynthesis and respiration, and reaeration. Results showed that carbonaceous BOD was significant only for the first 10 miles, nitrogenous BOD was appreciable in the lower 31 miles, benthic oxygen demand was least significant, and aquatic weeds exerted the greatest influence on DO. The mathematical model used was the same as that given in Eq. (17-7) in the previous chapter. The model indicated that aquatic weeds exerted such a tremendous demand that some solution other than secondary waste treatment and low-flow augmentation from upstream reservoirs was needed to alleviate the DO problem in the 40-mile reach.

For the purpose of this discussion, emphasis will be placed on P and R in the terms 7(e) and 7(f) presented in the previous chapter.

Photosynthetic oxygen production and respiration are discussed together because only their net effect was evaluated in this analysis. For all practical purposes, photosynthetic activity occurs only during daylight hours, as expressed in the following equation:

$$CO_2 + H_2O + \text{minor nutrients} \xrightarrow{\text{light}} \text{plant cells} + O_2$$

Plant respiration occurs continuously throughout the day; however, it may vary in magnitude in proportion to the DO concentration. These two factors, P and R, are combined into one factor (P-R) which is actually the net amount of oxygen produced and consumed by the plants.

The effect of aquatic plants on DO within a particular stream reach at a particular time of day is a function of the plant density and distribution, light intensity, water depth, turbidity, temperature, and DO. The most important factors are plant density and distribution. Oxygen production is proportional to plant density only to a certain limit. Increasing plant density results in decreased oxygen production and may even cause net oxygen consumption if the density becomes too great. This occurs because the plants become so dense that some of the plants are shaded by other overlying plants.

Photosynthetic oxygen production is proportional to light intensity. Figure 18-14 illustrates how these two factors affect DO in a river. A high density plant growth will cause low DO concentrations even on bright days. Although high DO values are produced during the afternoon, extremely low DO concentration result at night. Lower density plant growths result in net addition of DO to the stream only on bright days. On dull days, even low density plant growths consume more DO than they produce. Based on these considerations, aquatic plants must be regarded as a disadvantage in any stream. Westlake (1966) reported that net DO production is not only affected by the average plant density, which is the density normally reported for a stream reach, but also by the distribution of the plants within the reach. In other words, even though the average density may be relatively low, if the weeds are concentrated within a small area, the net effect of the weeds may be to consume more DO than produced.

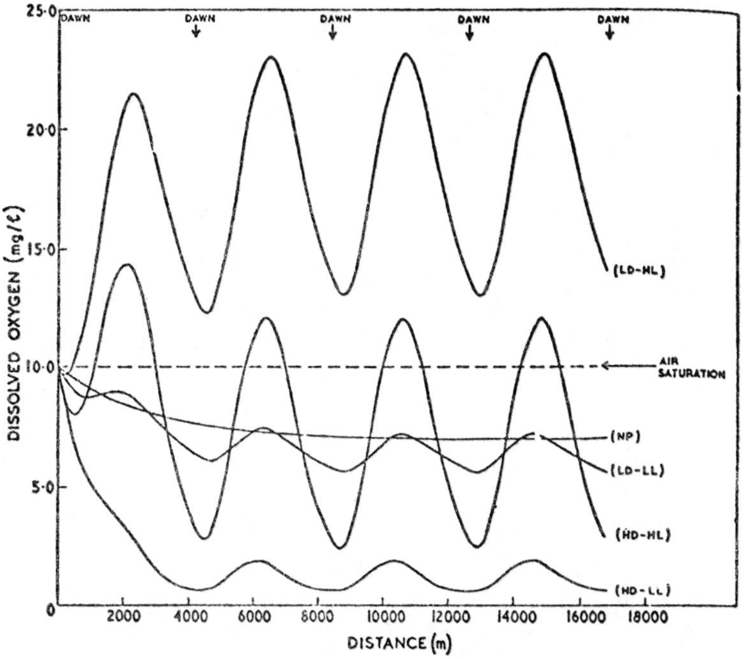

Figure 18-14. Calculated Effect of Differences in Plant Density and Daily Sunlight Distributions on the Distribution of Oxygen in a River. (After Owens, Knowles and Clark, 1969.)

The significance of water depth, turbidity, and temperature effects on photosynthetic oxygen production have not been quantitatively evaluated. Water depth and turbidity influence light intensity. The weed growths in this river were apparently very much influenced by water depth and, as shown in Fig. 18-15, plant density is inversely proportional to water depth. The effect of temperature on plant respiration has been reported by Bedick (1966) and for this analysis it was assumed that temperature affected photosynthetic oxygen production to the same degree. Bodick found that respiration increased by a factor of 2.16 with a change in temperature from 10° to 20°C.

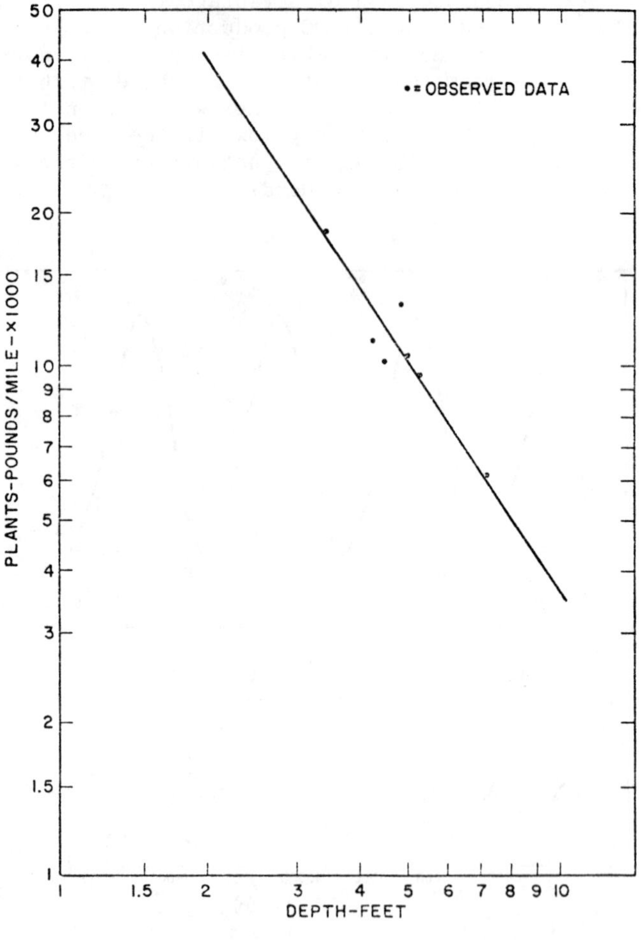

Figure 18-15. Relationship Between Standing Crop of Aquatic Macrophytes and Depth of River Water.

Using this result and assuming that plant respiration and photosynthetic oxygen production follow the van't Hoff-Arrhenius relationship, the following equation describes the effects of temperature on (P-R):

$$(P\text{-}R)_T = (P\text{-}R)_{T_1}\ [1.080^{(T-T_1)}] \qquad (18\text{-}9)$$

where T is the temperature at which (P-R) is wanted and T_1 is the temperature at which (P-R) is known. The factor 2.16 is in close agreement with a factor of 2.3 reported by Westlake for both respiration and photosynthesis and by Talling (1957) for phytoplankton, but is somewhat less than the factor of 2.6 reported by McDonnell and Weeter (1972) for respiration of *E. canadensis* and *P. crispus*.

Respiration by *potamogeton* has been reported by Bedick (1966) to vary with DO concentration according to the following relationship:

$$R_2 = R_1 \left| \frac{0.658 + 0.0693\ (DO_2)}{0.658 + 0.0693\ (DO_1)} \right| \qquad (18\text{-}10)$$

Without incurring significant error, this relationship may be simplified to the following equation:

$$R_2 = R_1 \left| \frac{10 + DO_2}{10 + DO_1} \right| \qquad (18\text{-}11)$$

Similar relationships for other species of aquatic weeds have been developed by Bedick (1966), Owens and Maris (1964), and McDonnell and Weeter (1972).

The effect of plant photosynthetic activity and respiration on DO balance may be determined by one of two methods--the light and dark bottle technique and the analysis of diurnal DO variations in the stream. The light and dark bottle method could not be used on this river because this technique only accounts for plankton and does not measure the influence of attached plants. Other disadvantages of this method have been reported by O'Connor and DiToro (1970), Symons (1966), and Pratt and Berkson (1959). The analysis of the diurnal DO variation in the stream was first proposed by Odum (1956). In flowing streams where the diurnal DO variation differs from station to station, Odum's upstream-downstream method should be used. Basically, this method is a solution of the Streeter-Phelps equation with various terms added to the equation to account for (in addition to [P-R]) benthic oxygen demand and nitrogenous BOD. Hence, (P-R) is determined by trial and error. In this analysis all other variables were either calculated or estimated and (P-R) was calculated to fit the mathematical model to the observed DO conditions of the survey. The resulting (P-R) curves for each stream reach that was analyzed are presented in Fig. 18-16. To simplify calculations, the following sine curve was used to represent the curved portions of (P-R):

$$(P\text{-}R) = C_1 \sin 2\pi t + C_2 \tag{18-12}$$

where C_1 is the amplitude of the sine curve, C_2 is the offset between the axis where (P-R) equals 0 and the sine curve

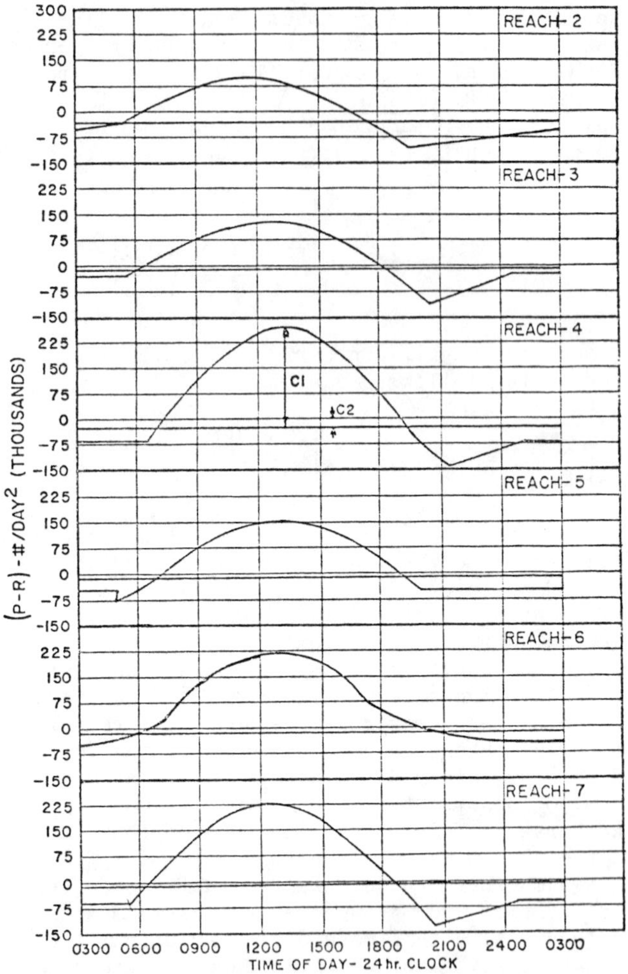

Figure 18-16. Computed (P-R) Curves to Represent River Reaches 2-7.

axis where sin $2\pi t$ equals 0, t is the difference in time between the time of day where sin $2\pi t$ equals 0 at dawn and the time of day in question. This relationship is illustrated in Fig. 18-17 and the values of C_1 and C_2 for each respective reach are shown in Fig. 18-16. In most cases, the sine function was used to represent (P-R) from a period approximately one hour before the curve intersected with its axis and two hours after the curve intersected with the axis. Normally, the maximum respiration rate was reached two hours after the curve intersected with the axis. Following this point, the respiration gradually decreased for a period of approximately four hours and then remained constant for the remainder of the night. Minor deviations from the usual procedure had to be made for Reaches 2, 5, and 6 to obtain better solutions for predicting the DO concentration. The reasons why these deviations were necessary are not well understood. The maximum respiration rate was observed at dusk for most of the reaches, probably because concentrations of dissolved oxygen are higher during this time than during the remainder of the night.

The relative magnitudes of the positive and negative values of (P-R) shown in Fig. 18-16 are primarily a function of plant density. However, as discussed above, the magnitudes of (P-R) are not directly proportional to plant density but may be inversely proportional, depending upon the distribution of the plants within the reach, i.e., whether the plants are concentrated in one area or spread evenly throughout the reach.

Another method for modeling photosynthetic oxygen production and respiration has been proposed by O'Connor and DiToro (1970), who suggest using a Fourier series to represent photosynthetic oxygen production and assume respiration to be constant. The Fourier series, while adequate in most instances, is rather inflexible and difficult to adjust to those cases for which the Fourier series will not fit the observed diurnal DO curves. The major disadvantage with their model used for this river is that plant respiration cannot be treated separately, i.e., the effects of DO concentration on respiration cannot be incorporated very readily. Kartchner et al. (1969) proposed a Fourier series and a time series index approach for representing the dissolved oxygen variation in a stream. These models could be used to represent photosynthetic oxygen production. The time series index approach has the advantage of adequately representing irregular diurnal DO curves. They used these approaches in analyzing data accumulated by water quality monitors.

O'Connor and DiToro (1970) used an unsteady state term for representing photosynthetic oxygen production, not only evaluating the diurnal variation but also the variation of photosynthetic oxygen production within the reach as a function of time. In this analysis it was assumed that (P-R) was constant within the respective reaches and therefore did not vary with time of travel within the reach. Hence, in the integra-

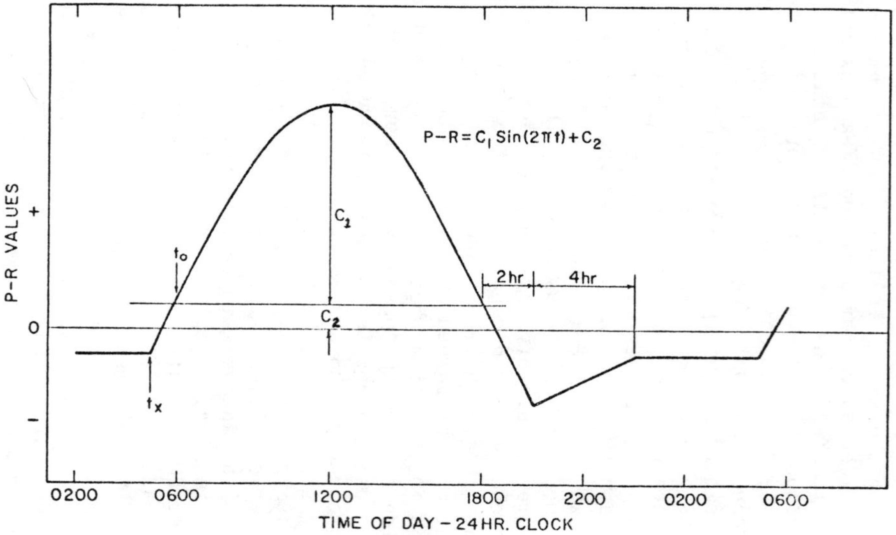

Figure 18-17. Definition of Symbols in Equation (18-12).

tion of the differential equation for DO in a stream reach, (P-R) was treated as a constant. Although (P-R) varies with time and position within each reach, it may be considered as a constant for relatively short intervals of flow time within the diurnal variation without causing significant error.

The peak respiration which occurs around dusk for most of the (P-R) curves determined for the reaches investigated is similar to results obtained by other investigators (Armstrong et al., 1968, and O'Connell and Thomas, 1965). They are not similar to the results reported by other investigators, including Odum (1956), Gunnerson and Bailey (1963), and Odum (1957). A possible reason why the latter investigators did not find increased respiration rates near dusk is that Odum assumed respiration did not vary throughout the day and Gunnerson and Bailey analyzed a diurnal DO curve which varied from only 10 milligrams per litter to 11 milligrams per liter, yielding a DO difference of only 1 milligram per liter throughout the day. Copeland (1965) presented evidence that community metabolism increases as labile organic material is produced during photosynthesis, thus resulting in increased respiration during the photosynthetic period. However, studies by TVA (1962) show that plant respiration does not vary with light intensity. Several investigators, including Gessner and Pannier (1958), Owens and Maris (1964), Bedick (1966), and McDonnell and Weeter (1972), have found plant respiration, or possibly the total biota respiration, to be a function of DO concentration. O'Connell and Thomas (1965) determined (P-R) in the Truckee River using the upstream-downstream technique and a field laboratory "algal chamber." The results for (P-R) were very similar except that peak respiration was measured using the upstream-downstream analysis but not in the "algal chamber" analysis. The authors did not give the DO concentration for the "algal chamber" and did not state whether this DO concentration varied as it did in the river. Therefore, it may be possible that the peak respiration did not occur because the DO was not varied.

The net result of photosynthetic oxygen production and respiration is dependent on streamflow as follows:

$$(P\text{-}R)_2 = (P\text{-}R)_1 \left(\frac{Q_2}{Q_1}\right)^d \qquad (18\text{-}13)$$

where (P-R) is expressed in pounds per day^2, Q is the streamflow, and d is as defined in the previous chapter.

Using the mathematical model for DO that was developed for this river, it was determined that the aquatic weeds exerted such a large respirational demand on the DO that even if no organic or nitrogenous wastes were discharged to the river, the DO would still be below the state standards. Similar problems have been reported by other investigators (McDonnell and Kountz, 1966, and Fisher, 1971). This problem exists even on "bright" days but is especially critical on "dull" days, as indicated by DO data collected during another survey. Using the DO simulation model, it was estimated that the minimum DO would be less than 1 milligram per liter for a minimum flow of 760 cfs, 27°C, and 7 milligrams per liter of DO in the river above the waste discharges, even if no wastes were discharged to the river.

Thus, it was necessary to explore solutions other than waste treatment for increasing DO in the river. Using the model, the following factors and combinations were evaluated to determine their effects on the minimum DO concentration: advanced or tertiary waste treatment, increased DO above the waste discharges, streamflow augmentation, decreased temperature, artificial instream reaeration, and control of the aquatic weed growths.

The first combination of solutions considered were percentage of waste removal, streamflow augmentation, increased upstream DO, and decreased temperature. A range of possible values was selected for each alternative and three values were chosen for each alternative within the respective range. The values considered for each of these alternatives are as follows: percent waste removal--85, 90, and 95; streamflow--760, 1,150, and 1,500; upstream DO--5, 6, and 7; and stream temperature--17°, 21°, and 25°C. The consideration of three values for each of the alternatives required computations for 81 separate DO profiles similar to the one shown in Fig. 18-18. Since the minimum DO is the critical condition for which the alternatives must be designed, this DO concentration was summarized for each of the 81 DO profiles. To best summarize the results of the above calculations, the following linear multiple regression equation was developed between minimum DO and the controllable variables:

$$\begin{aligned}\text{Minimum DO} = &-6.49\\ &+ 0.104\ (\text{percent removal of wastes})\\ &+ 0.00241\ (\text{streamflow})\\ &- 0.253\ (\text{stream temperature})\\ &+ 0.379\ (\text{initial DO concentration})\end{aligned}$$

This relationship has a standard error of estimate equal to 0.2 milligrams per liter and a multiple correlation coefficient equal to .983, which indicates that the relationship is statistically significant at the 1 percent level. The independent variables were shown to be truly independent by using an analysis of moment, covariance, and correlation matrices.

Using the above model for minimum DO as a function of the four independent variables, it may be shown that on "bright" days the state stream standards may be maintained by manipulating only these variables. This is important because these "bright" day conditions are expected to be prevalent and, therefore, an adequate DO level can be obtained without removing the aquatic weeds or using artificial instream reaeration. One combination of values that may be used to attain 3 mg/l DO

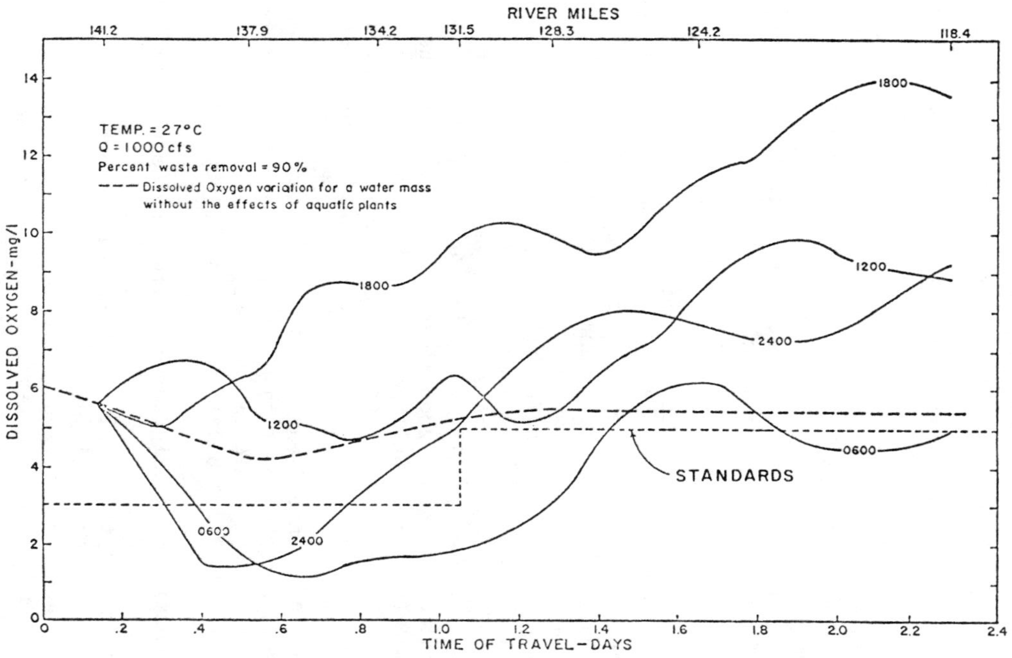

Figure 18-18. Computed D.O. Concentrations for Various Times of Day.

is as follows: 1,500 cfs, 92 percent removal of wastes, 25°C, and 7 milligrams per liter of DO at RM 1.2. Of course, other combinations of these variables may be used to attain 3 milligrams per liter on "bright" days, but it is necessary to use optimization techniques to determine the best combination.

Artificial instream reaeration is the only alternative solution that can be used to maintain state stream standards without increasing the other variables beyond the minimum requirements (85 percent removal of wastes, 760 cfs, and a temperature increase equal to the maximum allowable). In addition, on "dull" days, either artificial instream reaeration or essentially complete removal of aquatic weeds is necessary to meet state stream standards. To determine the cost of instream reaeration, it is necessary to determine (1) the amount of oxygen necessary during critical conditions, and (2) the amount of time the aerators would be required to operate. The latter requirement cannot presently be calculated and can only be evaluated by determining how solar radiation, turbidity, and temperature affect DO in the river.

To estimate the maximum amount of artificial instream reaeration required, the DO model was varied to measure the effects of "dull" days and to determine the amount of artificial instream reaeration required to maintain certain DO concentrations. To simulate DO conditions during critical conditions, the model was altered to fit the minimum DO concentrations observed during "dull" day conditions. To simulate this condition, it was assumed that (P-R) was a constant value equal to some fraction of the respiration rate. It was determined that 0.65 times the observed respiration rates for the remaining downstream reaches yielded a good close approximation of the minimum DO observed under "dull" day conditions.

Artificial instream aeration was included in the model by automatically increasing the DO in the increments of 0.5 milligrams per liter as required to maintain the desired DO level.

This modified DO model was used to determine the required pounds of oxygen and the location of the necessary aerators for maintaining various DO levels under different stream conditions. The results of a typical analysis are plotted in Fig. 18-19. These results are based on a streamflow equal to 760 cfs, DO at RM 1.2 equal to 7 milligrams per liter, percent waste removal equal to 95, and a temperature of 27°C. Each of the variables was manipulated within its respective limits to determine its effect on total pounds of artificial aeration required and on aerator location.

For 85 percent waste reduction, 760 cfs, 27°C, and 5 mg/l DO at SFHRM 1.2, the total amount of artificial instream aeration required was 146,000 pounds of oxygen per day. If the stream temperature was lowered or raised 2°C, the amount of oxygen required would be decreased by 28,000, or increased by 34,000 pounds per day, respectively. The required instream reaeration would decrease by 6,000 pounds per day if the upstream DO was increased from 5 to 7 milligrams per liter and

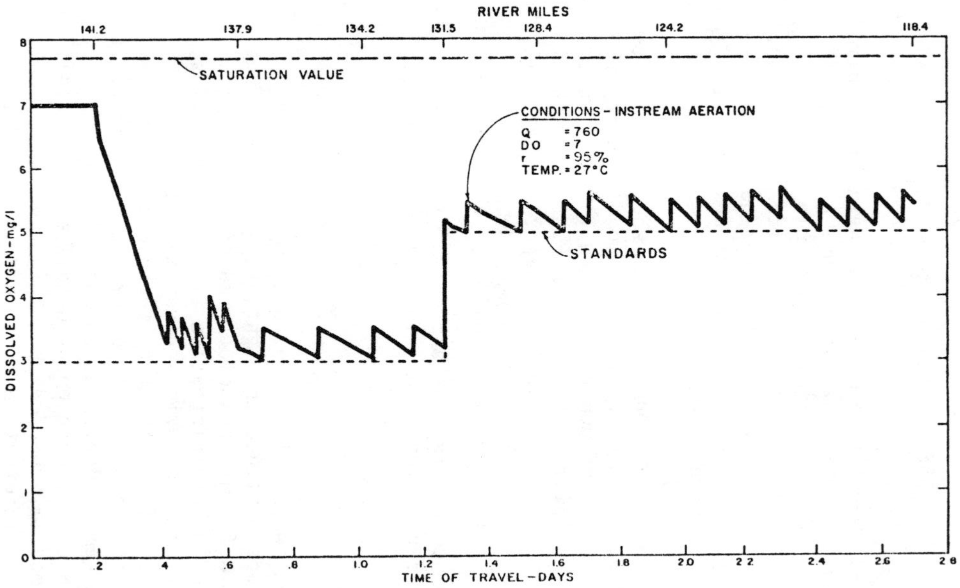

Figure 18-19. Typical Model Output Using Artificial Instream Aeration to Maintain the Specified D.O. Standard Considering the Conditions Stated in the Figure.

would decrease by 11,000 pounds per day if the streamflow was increased from 760 to 1,500 cfs. If percent waste removal was increased to 90 percent, the required instream aeration would be reduced by 15,000 pounds per day. Obviously, many combinations of these variables may be considered to reduce the amount of instream aeration required. For this reason, the following equation was developed to determine the necessary amount of instream aeration to maintain the present state stream standards:

$$I = 73,580 + 15,600T - 3588r - 19.18Q - 5296\ DO_{1.2} \qquad (18\text{-}14)$$

where I is the total amount of supplemental aeration required from RM 1.2 to RM 106.3. This relationship has a standard error of estimate of 1,500 lbs/day and a multiple correlation coefficient equal to 0.997, indicating statistical significance at greater than the 1 percent level. The independent variables were shown to be truly independent by using an analysis of moment, covariance, and correlation matrices.

The DO model was evaluated to determine whether it would simulate expected "natural" DO concentrations when the effects of wastes and (P-R) were eliminated from the model. Using the relatively high benthic oxygen demand measured in the river, the model predicted the lower curve shown in Fig. 18-20. This DO profile is somewhat lower than would be expected "naturally." Most unpolluted streams have a benthic oxygen demand of about 1 gm O_2/m^2/day. When this value was used, the resulting upper DO profile yielded DO concentrations approximately 90 percent of the saturation concentration. This level of DO is normal for large streams.

18.7 SUMMARY

A basic approach to applying water quality models for analysis of streams was outlined in eight steps. These eight steps consist of (1) defining the objectives of the study; (2) obtaining data and other information from municipalities, industries, universities, and various government agencies; (3) the application of three estimating techniques for the assimilated capacity of the stream to determine the sensitivity of water quality management decisions to assimilative capacity so as to define the need for developing detailed models; (4) assuming a detailed model is needed, selecting the appropriate model; (5) assuming the information contained in Step 2 is insufficient, conducting necessary field studies to obtain required data; (6) performing sensitivity analyses on critical variables that may affect management decisions; (7) on the basis of results in Step 6, conducting additional studies to refine variables that may be important to environmental and economic decisions; and (8) deciding best alternatives for water quality management.

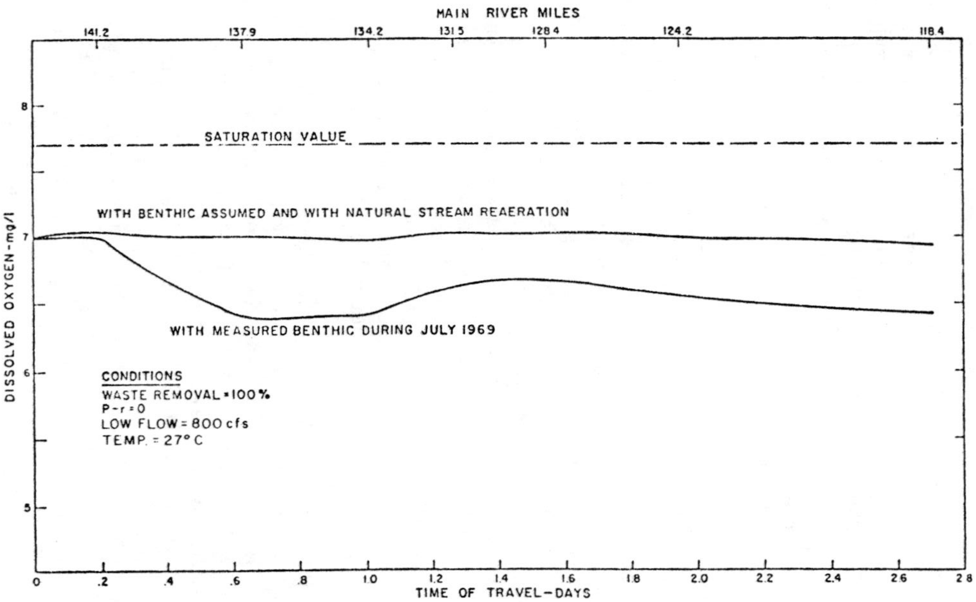

Figure 18-20. Model Output for Conditions of No Wastes Being Discharged to the River and No Aquatic Weeds Present in the River. The Two Curves Illustrated Represent D.O. that Would Occur if Benthic Oxygen Demand Remained High as Was Measured in 1969 or Decreased to Natural Oxygen Demand that has been Measured in Other Streams.

Two illustrative case histories were then presented. The DOSAG-1 model was applied to a case involving both carbonaceous and nitrogenous BOD. This case illustrated sensitivity analyses for K_d, K_2, T, H, and necessary levels of wastewater treatment. A schematic diagram of the stream and wastewater discharge network was presented and the hydraulic characteristics of the stream were also presented. The model was developed on the basis of one set of data and verified using another set of data collected at a later date. An approach was then illustrated for the determination of the waste assimilative capacity under various wastewater loads, water temperature, deoxygenation coefficient, average depth, and various K_2 formulas. Based on analysis of these variables, the best estimate of the waste assimilated capacity was presented along with the rationale for the selected variables.

The second illustrative example involved the development and application of a modified Streeter and Phelps model which incorporated photosynthesis and respiration, and the application of this model to explore various alternative control measures such as increased wastewater treatment, flow augmentation, artificial instream aeration, and temperature and DO control in the stream above the waste discharge. The mechanisms by which aquatic plants affect DO in streams were presented. In addition, the effects of temperature and streamflow quantity on (P-R) and the effect of DO on respiration were presented. Various methods for modeling (P-R) were discussed and a new technique was presented. Using the model that was developed for DO in this case, it was determined that the aquatic weeds exerted such a large respirational demand on the DO that even if organic and nitrogenous waste discharges to the river were eliminated, the DO would still be below the state standards. Thus, the various alternatives for improving DO listed above were considered. A relationship was developed between minimum DO in the river and the four variables of streamflow, temperature, initial DO, and percentage of removal of waste. It was determined that by manipulating these variables it was possible to achieve a minimum DO of 3 mg/l on "bright" days. The only alternatives that could be applied without affecting the minimum limits of the above four variables were artificial instream aeration and removal of the aquatic weeds. In addition, the latter two alternatives were the only feasible alternatives to achieve a reasonable minimum DO level on "dull" days. A relationship was developed between the total of artificial aeration required and levels of wastewater treatment, streamflow, temperature, and DO at the upstream point of wastewater discharge.

18.8 REFERENCES

Armstrong, N.E., and E.F. Gloyna, 1968. Ecological aspects of stream pollution. In Advances in Water Quality Improvement,

Water Resources Symposium No. 1, edited by Gloyna-Eckenfelder, Austin, Texas.

Bansal, M.K., 1976. Nitrification in natural streams. Journal Water Pollution Control Federation, October.

Bedick, T.M., 1966. Effect of oxygen concentration on the respiration of aquatic macrophytes, Master of Science thesis, Pennsylvania State University, December.

Churchill, M.A., et al., 1962. The prediction of stream reaeration rates, Advances in Water Pollution Research, Vol. 1, London, Pergamon Press.

Copeland, B.J., 1965. Evidence for regulation of community metabolism in a marine ecosystem. Ecology, Vol. 46, No. 4.

Fair, G.M., J.C. Geyer and D.A. Okun, 1968. Water purification and wastewater treatment and disposal. In Water and Wastewater Engineering, Vol. 2, John Wiley & Sons, Inc.

Fisher, 1971. The cross-Florida barge canal: a lesson in ecology. Civil Engineering, April.

Foree, E.G., 1976. Reaeration and velocity prediction for small streams, Journal of the Environmental Engineering Division, Vol. 102, No. EE5, October.

Gessner, F. and F. Pannier, 1958. Influence of oxygen tension on respiration of phytoplankton, Limnology and Oceanography, Vol. 3:478-480.

Grimsrud, G.P., et al., 1976. Evaluation of water quality models, a management guide for planners, ORD, USEPA-600/5-76-004, Washington D.C., July.

Gunnerson, C.G. and T.E. Bailey, 1963. Oxygen relationships in the Sacramento River, Journal - Sanitary Engineering Division - ASCE, Vol. 89, SA4, August.

Kartchner, A.D., N. Dixon, and D.W. Hendricks, 1969. Modeling diurnal fluctuations in stream temperature and dissolved oxygen, 24th Annual Purdue Industrial Waste Conference, May.

McDonnell and Kountz, 1966. Algal respiration in a eutrophic environment, Journal Water Pollution Control Federation, May.

McDonnell, A.J., and W.W. Weeter, 1972. Variable respiration in aquatic plant communities, Proceedings of Amer. Soc. of Civil Engineering, Environmental Engineering Division, April.

O'Connell, R.L. and N.A. Thomas, 1965. Effect of benthic algae on stream dissolved oxygen, Journal of Sanitary Engineering Division, ASCE, paper No. 4345, June.

O'Connor, D.J. and D.M. DiToro, 1968. The distribution of dissolved oxygen in a stream with time varying velocity, Water Resources Research, Vol. 4, No. 3, June.

O'Connor, D.J. and W.E. Dobbins, 1958. Mechanism of research in natural streams, Trans. Amer. Soc. Civ. Engr., Vol. 123.

Odum, H.T., 1956. Primary production in flowing waters, Limnology and Oceanography, Vol. 1, No. 2, April.

Odum, H.T., 1957. Trophic structure and productivity of Silver Springs, Florida, Ecological Monographs, 27:55-112, January.

Owens, M., and P.J. Maris, 1964. Some factors affecting the respiration of some aquatic plants, Hydrobiologia, Vol. 23, pp. 533-543, June 30.

Pratt, D.M. and H. Berkson, 1959. Two sources of error in the oxygen light and dark bottle method, Limnology and Oceanography, 4(3):328-334.

Rathbun, R.E., D.J. Schultz, and D. W. Stephens, 1975. Preliminary experiments with a modified tracer technique for measuring stream reaeration coefficients, U.S. Department of the Interior Geological Survey, Open File Report 75-256, Bay St. Louis, Mississippi, June.

Ruane, R.J., 1971. Statistical equation for estimating the assimilative capacity of a stream for BOD (unpublished report), Tennessee Valley Authority, Water Quality Branch.

Stall, J.B. and C.T. Yang, 1970. Hydraulic geometry of 12 selected stream systems of the United States (Research Report No. 32), University of Illinois, Water Resources Center, Urbana, July.

Symons, J.M., W.H. Irwin, R.M. Clark and G.G. Robeck, 1967. Management and measurement of DO in impoundments, Journal-Sanitary Engineering Division - ASCE, Vol. 93, No. SA6, Dec.

Talling, J.F., 1957. Photosynthetic characteristics of some freshwater plankton diatoms in relation to underwater radiation, New Phytologist 56:29-50.

Thackston, E.L. and P.A. Krenkel, 1969. Reaeration prediction in natural streams, Journal, Sanitary Engineering Division, ASCE, SA1.

Tsivoglou, E.C., and L.A. Neal, 1976. Tracer measurement of reaeration: III. Predicting the reaeration capacity of inland streams. Journal Water Pollution Control Federation, December.

Westlake, D.F., 1966. A model for quantitative studies of photosynthesis by higher plants in streams, Air & Wat. Pollution Int. Journal, Vol. 10, pp. 883-896, Pergamon Press.

Chapter 19

APPLICATION OF WATER QUALITY MODELS

by

D. W. Hendricks, Associate Professor of Civil Engineering, Colorado State University, Fort Collins, Colorado

Chapter 19

APPLICATION OF WATER QUALITY MODELS

19.1 INTRODUCTION

The water use locus of a river basin is its stream system. Streams are sources of water supply and, at the same time, they function as sinks for *used water*. Two categories of used water of interest are: (1) point source discharges of treated waste water, i.e., from municipalities and industries, and (2) non-point source flows, i.e., from irrigated agriculture, urban storm water runoff, etc.

The flows of used water will affect stream water quality through the direct effect of the mixing of two waters of different chemical composition, and through the induced effects caused by the reactions in the mixed stream. If the resultant stream water quality is to be maintained within the limits prescribed by law (i.e., stream standards), then the flows of used water must be "managed." The key management questions are: To what extent will ambient stream water quality respond to a proposed level of basin-wide effluent discharge control? And conversely, what level of basin-wide effluent discharge control is required to meet proposed stream standards? These questions may be restated in more poignant "cost effectiveness" terms as: (1) For a given expenditure what will be the results? and (2) For a given result, what will be the cost?

The only way to answer either of these complimentary questions with certainty is to initiate a treatment plant construction program and then monitor stream water quality to empirically ascertain the results. If the results achieved are not as good as desired, then another level of treatment may be imposed. The costs associated with each result should be obtained also. This empirical approach is an important part of the overall management process, but it is not adequate in developing a program initially (i.e., when foresight is needed, vis a vis hindsight). Tools are necessary to provide foresight with respect to both stream water quality response and costs of control. A water quality simulation model is such a tool which can aid in developing the former. Its construction and use is the subject herein.

The development and use of a water quality model involves both scientific-technical knowledge and engineering judgment. One must understand first the basic scientific principles which are operative in governing the water quality behavior of a stream system. Equally important is the largely judgmental process of making the always necessary approximations, or assumptions, about scientific reality, which are consistent with the task required and the time and funds available. This judgmental process extends to the problem of "calibration" of the model in terms of field data, and to interpretation of the model results. Thus the modeling process for engineering purposes is still inherently an *art*.

This paper describes the construction and use of a water quality model. It starts with fundamental theoretical premises, then it outlines the basic structure of a digital computer model, demonstrating in the process how the elements of a real system are adapted to the theory. Finally, the discussion reviews the manner in which the model is queried with questions, and in which the output is interpreted and used.

19.2 CONTEXT OF MODELING

The construction and operation of a water quality model are more than technical exercises. First, one must select the parameters of water quality which are to be modeled. Then the societal context, which the model serves, must be understood and described. Finally, the physical elements of the system require delineation. The following paragraphs review these concerns.

Dimensions of Water Quality

The term "water quality" refers to the condition, or "state," of a given parcel of water. Usually it is described by such measurements and indicators as temperature, coliform count, dissolved oxygen, biochemical oxygen demand (BOD), total dissolved solids, and any cations and anions that may be of interest. Table 19-1 enumerates some of these commonly measured parameters. The list is representative only; a wide

Table 19-1. Common Indicators of Water Quality

TEMPERATURE (°C)
COLIFORM COUNT (organisms per 100 ML)
DISSOLVED OXYGEN (MG/L)
BIOCHEMICAL OXYGEN DEMAND (MG/L)
TOTAL DISSOLVED SOLIDS (MG/L)
CATIONS (MG/L)
SODIUM
POTASSIUM
CALCIUM
ANIONS (MG/L)
CHLORIDE
SULFATE
ORTHOPHOSPHATE
NITRATES
SUSPENDED SOLIDS (MG/L)

range of constituents could be included. The choices used in the model depend upon which ones are of interest. This in turn relates to the purposes for which the water is to be used (e.g., irrigation, municipal water supply, in-stream uses, etc.).

Social Objectives Related to Water Quality

The choice of water quality parameters to be modeled is dependent upon the objectives associated with each category of water use. Table 19-2 illustrates this hierarchal relationship, showing how water uses, objectives, and specific water quality parameters are related. For example, from Table 19-2 one can

Table 19-2. Uses of Water and Water Quality Objectives

CATEGORY OF WATER USE	OBJECTIVES	WATER QUALITY PARAMETERS
DOMESTIC WATER SUPPLY	—HEALTH —PALATABLE —HARDNESS	—TOXIC MATERIALS —HARMFUL ORGANISMS —COLOR, TASTE, ODOR —Ca, Mg, Fe
INDUSTRIAL PROCESS WATER	—CHEMICAL COMPATABILITY WITH PROCESSES	—SPECIFIC CHEMICALS —COLOR
INDUSTRIAL COOLING	—MINIMIZE SCALE —MINIMIZE BIOLOGICAL FOULING	—$CaCO_3$ —ALGAE COUNTS
RECREATION	—HEALTH —NUISANCE —ESTHETICS	—HARMFUL ORGANISMS —NUTRIENTS —ALGAE COUNTS —COLOR
FISHERY MAINTENANCE	—PHYSIOLOGICAL COMPATABILITY	—DISSOLVED OXYGEN —TEMPERATURE
ESTHETIC PRESERVATION	—NUISANCES	—FLOATING MATERIALS —SLUDGE BANKS —GREASE —COLOR
IRRIGATION	—SALT TOLERANCE —TOXICITY	—TOTAL DISSOLVED SOLIDS —TOXIC SUBSTANCES

see that for domestic water supply, "health" is a major objective, and that "toxic materials" and "harmful organisms" are two corresponding parameters used in monitoring. Generally, the parameters monitored are indicators of the suitability of the water for the intended use. Table 19-3 illustrates the idea in a different tabular format. It shows, for example, that a "vector," i.e., a column in Table 19-3, of water quality parameters and limitations is associated with each respective use.

Standards, Criteria, and Management Objectives

Most water uses have specific limits for some or many water quality parameters. They are called *criteria* if set by the technical requirements or needs of the user, and *standards* if codified by law. Domestic use has "drinking water standards"--set by law to protect human health. Table 19-3 shows a few of the criteria used for domestic drinking water; they may be wholly or in part adapted as standards. The lack of criteria under the industrial and irrigation headings is due to the

Table 19-3. Some Standards and Criteria for Selected Water Uses

PARAMETER	USE		
	DOMESTIC	INDUSTRIAL	IRRIGATION
TEMPERATURE (°C)	20	VARIABLE	NONE
COLIFORM COUNT (per 100 ML)	1		
DISSOLVED OXYGEN (100 ML)	SATURATION		
BIOCHEMICAL OXYGEN DEMAND (MG/L)	0		
TOTAL DISSOLVED SOLIDS (MG/L)	500		1000

inappropriateness of absolute criteria for every water quality constituent for every use. Despite such difficulties, the attainment of a particular set of water quality standards associated with certain societal goals is the objective of a water quality management program. Such goals are *normative* (i.e. value-oriented) in character, and hence they may reflect the values of a particular society at a given time.

To protect the ambient water quality for a variety of intake uses (domestic, industrial, irrigation) and in-stream uses (such as recreation, ecological preservation and health), both stream standards and effluent standards have been developed. Stream standards were developed by every state pursuant to the requirements of PL89-234 (the 1965 amendments to the federal water quality act), while effluent standards have been established pursuant to PL92-500 (the 1972 amendments).

Colorado, and most other states, have classified natural waters according to use; the classifications are designated A_1, A_2, B_1, B_2. Some of the water quality characteristics associated with these classifications are shown in Table 19-4. The highest classifications, A_1 and A_2, are for a potable water supply. Prior to the advent of PL92-500, C and D classifications were permitted, which provided for a much wider range of water quality variation.

Both stream standards and effluent standards are used in management of stream water quality; they are complimentary. For example, if the effluent standards are not adequate to achieve the stream standards, they may be more stringent. Whether or not a given effluent standard will result in the achievement of a given stream standard can be investigated through the use of a water quality model. In fact, this question was a principal concern of the National Commission on

Table 19-4. Stream Standards--Classifications by Ambient Water Quality[1/]

CLASS	FECAL COLIFORM (PER 100 ML)	DO (MG/L)	pH	T (°C)
A_1	200	6.0	6.5 - 8.5	20
A_2	200	5.0	6.5 - 9.5	32
B_1	10,000	6.0	6.0 - 9.0	20
B_2	10,000	5.0	6.0 - 9.0	32

[1/]Adopted January 17, 1974 by the Colorado Water Pollution Control Mission.

Water Quality, which was given the charge of ascertaining the feasibility of implementing selected provisions of PL92-500.

Use Systems

The use of water entails a deterioration in water quality. For example, domestic use loads the wastewater with organic matter, suspended matter, bacteria, and mineral constituents. Agricultural use both concentrates and loads the return flow stream with mineral salts. Figure 19-1 illustrates the idea

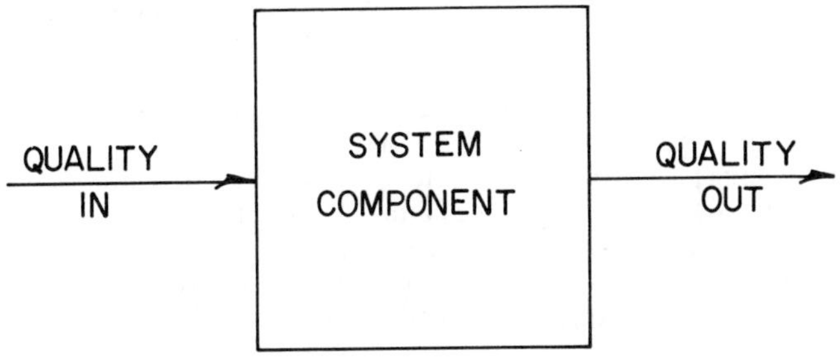

Fig. 19-1. Process of Water Quality Deterioration with Use

that water use by a system component involves the *process* of water quality degradation. Thus the depiction of a water use system within a basin, consisting of transport, use, treatment, and waste discharge, must include consideration of water quality changes.

Figure 19-2 is a simplified schematic of a water resource system; it is a collection of system components. Deterioration of water quality is associated with each use by the system components depicted. The system may be altered internally to meet any effluent discharge standard. This can be done through

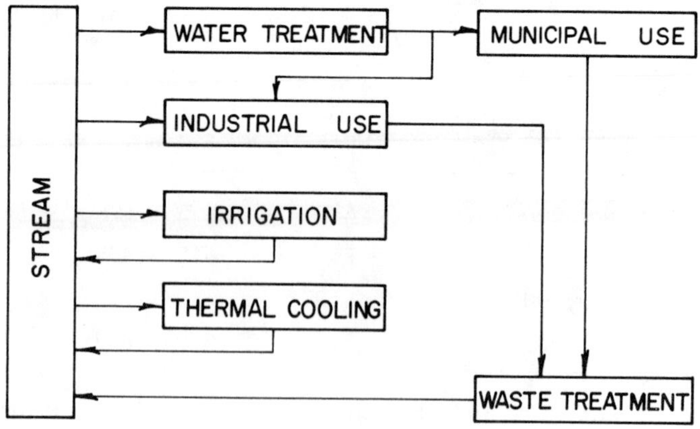

Fig. 19-2. A Water Use Subsystem

inducing modified patterns of use and waste loading, through different wastewater treatment schemes, and through water exchanges between system components in lieu of direct discharge to the stream system to meet any specified water quality effluent discharge limitations.

The Stream Assimilation Subsystem

Ordinarily the receiving water body--usually a river--is the sink for used water from the various use subsystems. The stream is then the *locus* of the system. Man-induced changes are superposed and may disrupt the natural order of the stream, e.g., stream ecology. Thus, stream standards and effluent standards have the purpose of limiting the degree of disruption of this natural order to within acceptable limits.

The hydrologic and water quality portions of the stream subsystem are inseparable. Thus, both flows and water quality must be characterized for each input to the main stream and for each diversion. Figure 19-3 shows schematically some of the

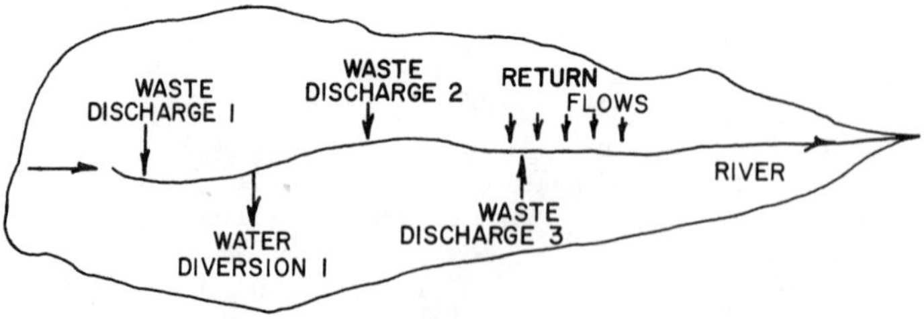

Fig. 19-3. Stream Assimilation Subsystem

flow components of the water use-stream subsystems which must be delineated in order to characterize the assimilation subsystem. These components are stream flows, waste discharges, return flows, and diversions.

Simulation Modeling

A large number of inputs and diversions may occur within any given stream reach. Further, the reactions of the water quality constituents introduced into the stream may be rather complex. The only feasible way to integrate all of these flows and reactions is through the construction of a simulation model, programmed for a high speed computer.

A simulation model provides a predictive capability. Thus one can anticipate the results of a policy, a program, a project, an operating mode, etc., in terms of changes in stream water quality. For example, if a new waste discharge of 1000 kilograms per day is added to the stream as depicted in Fig. 19-4, what might be the effect on the dissolved oxygen levels,

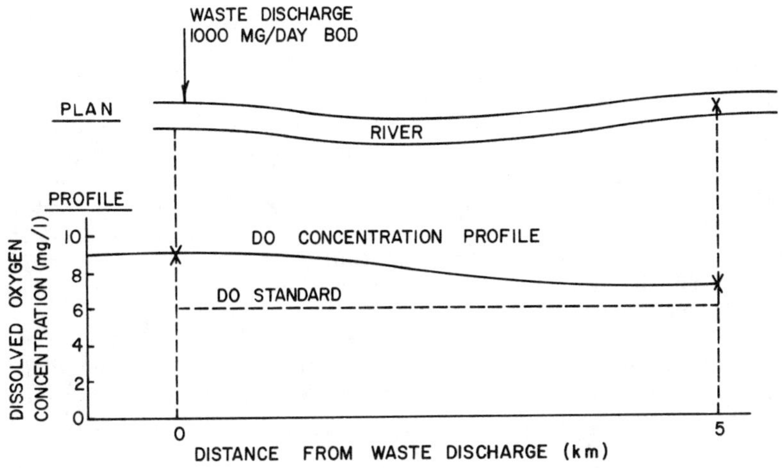

Fig. 19-4. Illustration of the Effect of a Wastewater Discharge on the Dissolved Oxygen Concentration Profile

say five kilometers downstream? The dissolved oxygen (D.O.) concentration profile shown could be predicted by a water quality D.O. simulation model. Comparison of such a profile with the D.O. standard, as shown in Fig. 19-4, tells whether a violation of the standard is likely. On the other hand, if a reservoir is located upstream of the waste discharge, one might ask if reservoir releases could be programmed such that stream standards for dissolved oxygen are not violated. A variety of such questions can be answered through simulation modeling.

The Materials Balance Concept

A rational water quality simulation model must be constructed by a materials balance formulation. The essence of such a model is merely an accounting statement in terms of mass flows in and out of a given fixed element in space, and reactions which occur within. Figure 19-5 illustrates graphically the materials balance principle.

Fig. 19-5. The Difference Between Inputs and Outputs for Any Given Element is Due to Reaction Within

The fixed spatial element upon which the materials balance formulation is constructed may be any arbitrary volume, such as a cube, or a slice of a stream reach. Usually in stream modeling, a slice is used as the volume element. The slice implies that the model constructed is one-dimensional. In practical modeling the thickness of the element may be rather large, i.e., a reach of stream, as depicted in Fig. 19-6. However, the reach length, ΔZ, should be short enough that the constituent in question is approximately homogeneous in concentration within the spatial element. An important assumption is that any input into the spatial element is instantaneously mixed and that the contents of the element have spatial homogeneity within. From this standpoint a cube is a better approximation to homogeneity than a slice. The mathematical formulation of the materials balance based upon the cube would result in a three-dimensional mathematical model; however, the one-dimensional model based on the slice is much easier to handle.

The terms in the materials balance formulation on the fixed spatial element may be classified in two general categories: (1) those representing mass transfer processes, and (2) those representing reactions. Table 19-6 identifies the most significant terms in each general category. It is important to be cognizant of the fact that each term could have influence, depending upon the constituent being modeled and the conditions of the real situation. Thus one can begin to comprehend the complexity of the real system, vis a vis the model representation of it. From this comprehension one can judge how to select the proper terms to be included in a model constructed for engineering purposes. This involves balancing the accuracy needed against model complexity, modeling skill available, feasibility of verification, and other factors unique to the task at hand. The main point is that a model used for engineering purposes usually is a simplification of the real world system. The model is built for a purpose.

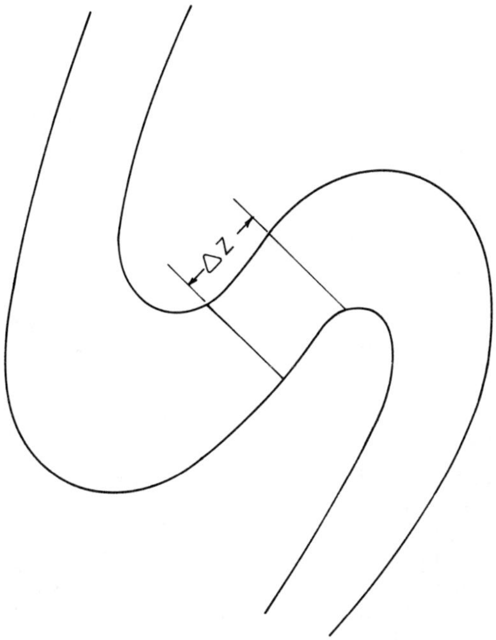

Fig. 19-6. Stream Reach ΔZ About Which a Mathematical Model May be Formulated.

Table 19-5. Examples of Mass Transfer Process and Reaction Terms Relevant to Development of a Water Quality Model

Mass Transfer Processes	Reactions
Convection	Bacterial Metabolism
Turbulent Diffusion	Acid-Base Reactions
Interfacial Molecular Diffusion	Redox Reactions
gas transfer	Chelation
mineral dissolution	Adsorption
ion-exchange	Ion-exchange
molecular adsorption	Gas dissolution
	Gas precipitation
	Mineral dissolution

Mathematics of Materials Balance

Figure 19-7 is a graphic representation of the materials balance principle applied to a slice element, fixed in space.

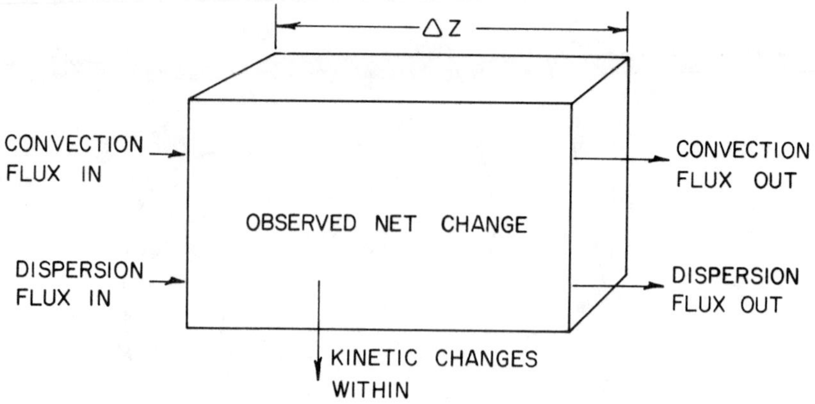

Fig. 19-7. Inputs, Outputs and Changes Within a Slice Element of Water

The transport terms are labeled *convection flux* and *dispersion flux*, respectively. Convection flux includes all material transported by the bulk flow in the stream, which has a mean velocity $\overline{v}$, added or taken from the slice element at the two boundaries. Dispersion flux is that material transport rate superimposed on the convection transport rate by random motion due to the fluid turbulence. The kinetic terms are the rate of change of concentration due to chemical or biological reactions of various sorts, i.e., those listed in the reactions column of Table 19-5. The *observed change* is the net result of all invluences within the element.

The graphic representation of the materials balance principle can be given verbal expression, i.e., the observed changes within the volume element equals the convection flux in minus convection flux out, plus dispersion flux in minus dispersion flux out, plus kinetic changes within. Equation (19-1) is the corresponding equality statement for a volume element fixed in space, as shown in Fig. 19-7.

OBSERVED CHANGES = CONVECTION FLUX IN − CONVECTION FLUX OUT +

DISPERSION FLUX IN − DISPERSION FLUX OUT +

KINETIC CHANGES WITHIN (19-1)

This statement can be formulated mathematically several ways. The essential assumption, as noted previously, is that the contents of the volume elements are homogeneously distributed. Thus the size of the volume element and the resulting form of the mathematical expression depend upon whether it is necessary that the volume element be finite or infinitesimal for the homogeneity assumption to be valid. If the constituent of interest, c, is changing rapidly with distance, i.e., the slice ΔZ must be infinitesimal, then the materials balance expression must be developed in differential form. The mathematical expression for this case is:

$$\left[\left(\frac{\partial(c)}{\partial t}\right)_o = -\bar{V}\frac{\partial c}{\partial z} + D\frac{\partial^2 c}{\partial z^2} + \left(\frac{\partial c}{\partial t}\right)_k\right]_{i,t} \qquad (19\text{-}2)$$

where c = concentration of water quality constituent being modeled (M/L^3), t = time elapsed from which initial conditions were described (T), $\bar{v}$ = mean velocity of flow stream (L/T), z = distance downstream from some initial point (L), D = coefficient of hydraulic dispersion for stream (L/T^2), o = subscript used to designate the *observed* change rate of the constituent concentration within the volume element, k = subscript used to designate the *kinetic* term in the materials balance equation, and i = subscript used to designate a particular volume element. NOTE: The bracket around the overall equation is subscripted in i and t; this means that all terms in the equation are unique to that particular i and t designated.

If, on the other hand, the system can be described macroscopically--that is, the water quality changes are relatively small with distance; i.e., ΔZ is perhaps a *reach* instead of an infinitesimal *slice*, then the appropriate mathematical description is:

$$\left[\left(\frac{\partial(Vc)}{\partial t}\right)_o = Q_e c_e - Q_o c_o + D\frac{\partial^2 c}{\partial z^2}V + V\left(\frac{\partial c}{\partial t}\right)_k\right]_{i,t} \qquad (19\text{-}3)$$

where V = volume of element (L^3); Q_e = flow entering reach volume, i (L^3/T); Q_o = flow leaving reach volume, i (L^3/T), c_e = concentration of water quality constituent being modeled at point of entering reach i (M/L^3), and c_o = concentration of water quality constituent at point of exit of reach i (M/L^3). Equation (19-3) is the basis for stream modeling. It can be expanded to incorporate additional terms as needed to fit the reality of the situation.

Expressions in Materials Balance Equation

To appreciate the generality of Eq. (19-3) and to provide a better intuitive understanding of it, each of the differential and product expressions forming the equation should be examined individually. This is done in the following.

$\left[\frac{\partial(Vc)}{\partial t}\right]_o$ - This differential expression can be expanded to give $V\frac{\partial c}{\partial t} + c\,\frac{\partial V}{\partial t}$. The first differential term is the observed rate of change of c in the volume as a result of all influences on the right side of the equation. The second term in this expansion accounts for volume changes during the time increment dt. The subscript o means *observed*.

$Q_e c_e$ - The product expression for convective transport can be considered a collective term for all inputs into the stream reach ΔZ. It can be expanded into several terms to include hydrologic inputs, discrete waste discharges, and diffuse return flows along the reach ΔZ. These individual terms all are "lumped" to occur at the beginning of the reach.

$Q_o c_o$ - As with the input flux term, the output flux term can be expanded to include stream outflow from the reach, diversions, and channel losses. All of these terms are lumped to occur at the end of the reach.

$D\,\frac{\partial^2 c}{\partial z^2}$ - This is the dispersion transport term caused by the existence of a velocity profile in the channel cross section, and the random motion of turbulence. The dispersion coefficient, D, must be determined by in-situ field tests.

$\left[\frac{\partial c}{\partial t}\right]_k$ - The rate of change of c due to reaction is given by the kinetic term. The subscript k is used as the designation for the *kinetic* term. If several reactions are involved, all must be delineated. For example, if dissolved oxygen is the constituent in question, then kinetic equations associated with biochemical reactions, respiration, photosynthesis, and reaeration must be included. An expression for the photosynthesis term can be quite complex because of the continuous change during daylight hours due to sunshine intensity variation.

All of the terms in Eq. (19-2) and (19-3) are time varying. The flow Q_e is the result of natural hydrologic influences plus diversions of the upstream $i+1$ reach, and returns of the i reach upon which the calculation is being performed. Fig. 19-8 shows how these various flow inputs and flow outputs are aggregated at the head and tail ends of a given reach.

The concentration of c in the reach i is computed as the flow weighted average concentration. Input varies due to mixing from flow after all waste discharges, return flows, and diversions are added to or taken from the streamflow entering reach i. If the constituent is conservative, e.g., the chloride ion, then its concentration in reach i is determined by the simple mixing model. However, if the constituent is reactive, e.g., a sugar compound, then its mass is changed within reach i by reaction. In either case a new concentration, $c_{i,t+\Delta t}$, is calculated for reach i. This calculation utilizes

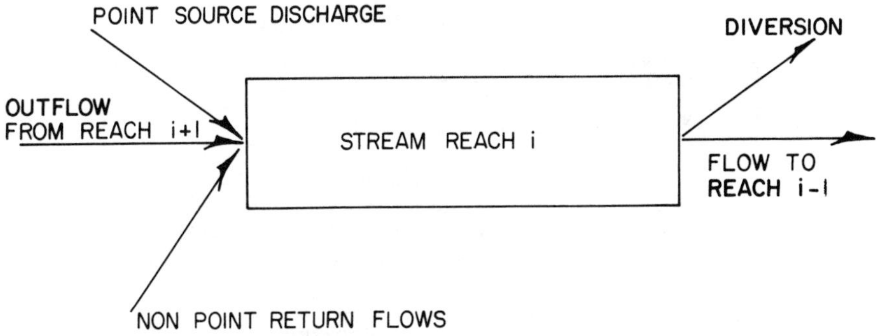

Fig. 19-8. Aggregation of Flows at Ends of Stream Reach i.

Eq. (19-3) operating over a time period Δt. Equation (19-3) is then applied to reach i-1, which becomes reach i. The calculation proceeds in the downstream direction to obtain $t+\Delta t$ concentration values at each i. These values are stored for use as $c_{i,t}$ values during the next iteration, i.e., after time is increased by Δt. This general computation scheme is outlined schematically in Fig. 19-9.

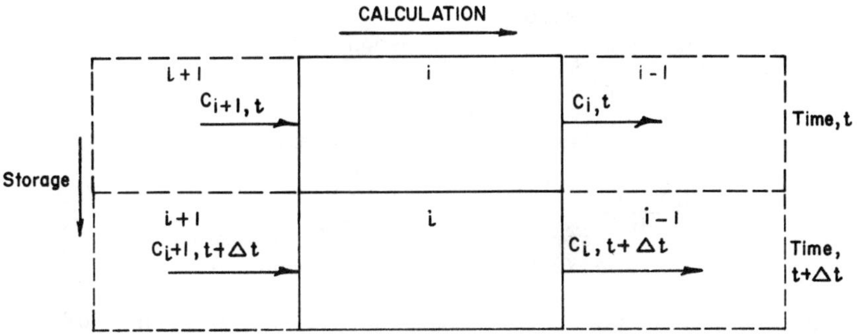

Fig. 19-9. The Flow Entering Slice i Over Time Δt, the Reactive Constituents React Over Time, Δt, and a New Concentration $c_{i,t+\Delta t}$ is Calculated

Reaction Kinetics

The kinetic term $[(\partial c)/(\partial t)]_k$ for reach i describes the resultant rate of change of concentration c in reach i due to the aggregate of all reactions within reach i, which are going on simultaneously. For example, if dissolved oxygen is the constituent in question, the complete description of the reactions and processes involved add up to a complex picture, such as depicted in Fig. 19-10.

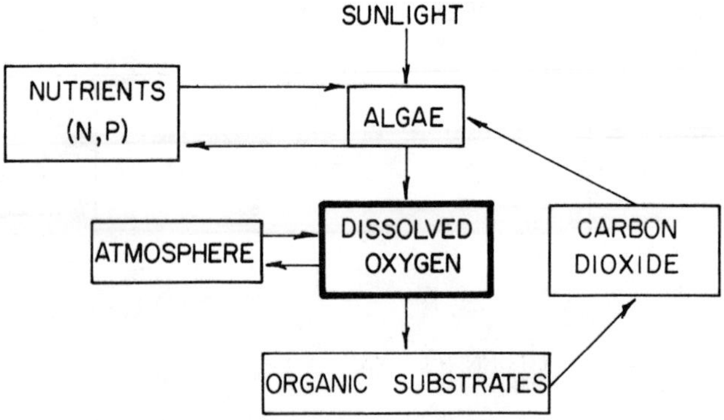

Fig. 19-10. Partial Representation of Reactions and Processes Affecting Dissolved Oxygen Level in a Stream.

Figure 19-10 depicts schematically the addition of oxygen to water by reaeration from the atmosphere and by photosynthesis. At the same time, oxygen is depleted from the water by aerobic biochemical reactions and by respiration. Further, if the water is supersaturated with oxygen, it will be lost to the atmosphere by the same exchange process which causes reaeration in oxygen-deficient waters.

The kinetic statement, corresponding to the reactions is:

$$\left[\frac{dc}{dt}\right]_k = \left[\frac{dD.O.}{dt}\right]_{\text{photosynthesis}} + \left[\frac{dD.O.}{dt}\right]_{\substack{\text{atmospheric}\\ \text{exchange}}} + \left[\frac{dD.O.}{dt}\right]_{\substack{\text{biochemical}\\ \text{reaction}}} \tag{19-4}$$

Equation (19-4) says merely that the net rate of change of oxygen concentration in a given volume element is the sum of the rates of all other reactions and processes which are occurring simultaneously. The individual terms are rather complex, but some kinetic models commonly used for each term are:

$$\left[\frac{dD.O.}{dt}\right]_{\text{photosynthesis}} = k_p \text{ [algae cells]} \tag{19-5}$$

$$\left[\frac{dD.O.}{dt}\right]_{\substack{\text{atmospheric}\\ \text{exchange}}} = k_T \{[D.O.]_s - [D.O.]\} \tag{19-6}$$

$$\left[\frac{dD.O.}{dt}\right]_{\substack{\text{biochemical}\\ \text{reaction}}} = k_B[BOD] \cdot \text{[microbial cells]} \tag{19-7}$$

where:

1. Brackets, e.g., [D.O.], means concentration, i.e., of oxygen, in mg/l.

2. $[D.O.]_s$ is the equilibrium (or saturation) concentration of oxygen with respect to the atmosphere; it is proportional to the partial pressure of oxygen above the water and it is temperature dependent as well.

3. k_p, k_T, k_B are kinetic coefficients.

4. [BOD], biochemical oxygen demand, is the concentration of biochemically reactive organic substrate in mg/l.

5. [microbial cells] is concentration of reacting cells in cells per liter.

6. [algae cells] is the concentration of photosynthesizing algae cells in cells per liter.

Equation (19-5) says that the rate of change of dissolved oxygen due to photosynthesis is first order with respect to the concentration of algae cells; the constant k_p would be dependent upon light intensity. The concentration of carbon dioxide or bicarbonate ion is assumed to be not rate limiting. If it is rate limiting, the equation would be second order with respect to algae cells and carbon dioxide concentration.

Equation (19-6) says that the rate of oxygen exchange between the atmosphere and the water is proportional to the difference between the equilibrium concentration of oxygen with respect to the atmosphere and the existing actual concentration at a given moment. If the latter is less than the former, i.e., $[D.O.] < [D.O.]_s$, due to aerobic biochemical reactions, which consume oxygen, then the exchange process is reaeration. If $[D.O.] > [D.O.]_s$, due to the photosynthesis reaction, then the exchange process is toward the atmosphere. The k_T term is a mass transfer coefficient related to fluid turbulence and molecular diffusion. Similarly, Eq. (19-7) is a second order kinetic equation with respect to organic substrate concentration, expressed as BOD, and the concentration of microbial cells metabolizing the substrate.

The Steady State Approximation

Equations (19-2) and (19-3) are not amenable to analytic solution; they are too complex mathematically. This complexity is accentuated by a variety of boundary conditions (e.g., diversions, tributary streams, waste discharges, etc.) which are different from reach to readh. In order to handle the mathematical complexity, the unique boundary conditions at each reach i, and the possible time varying nature of certain boundary conditions, such as flow, a numerical solution is necessary. This, of course, requires the use of a digital computer. However, Eq. (19-3) can be made less formidable mathematically through the use of Lagrangian coordinates. With this approach the mathematical model is constructed about a volume element which moves with the stream at the velocity of the streamflow. Thus there can be no flow across the boundaries. This assumption ignores the hydraulic dispersion phenomenon. Thus, with no mass fluxes across the boundaries of the element, Eq. (19-3) can be simplified to:

$$\left\{ \left|\frac{\partial c}{\partial t}\right|_0 = \left|\frac{\partial c}{\partial t}\right|_k \right\}_{i,t} \tag{19-8}$$

If the more general "c" term is replaced by "D.O.," for dissolved oxygen, in Eq. (19-8), it becomes the basis for the well-known Streeter-Phelps equation for oxygen sag. Usually, it is expressed in terms of oxygen deficit. The differential equation is, in terms of the formulation of Eq. (19-8):

$$\frac{dD}{dt} = k_1 L - k_2 D \tag{19-9}$$

where D = oxygen deficit, i.e., $[D.O.]_s - [D.O.]$ (mg/l); L = biochemical oxygen demand (mg/l); k_1 = reaction coefficient (sec^{-1}), same as k_B in Eq. (19-7); and k_2 = reaeration coefficient (sec^{-2}), same as k_T in Eq. (19-6).

Equation (19-9) is an ordinary linear differential equation, which has a well-known analytical solution. This solution is not important for the purposes herein, however. The important point is to recognize that Eqs. (19-8) and (19-9) are simplifications of Eq. (19-3). In the former equation one deals with a fixed spatial element, while in the latter equation the orientation is on a *parcel*, or a volume element, which moves with the stream. The former permits flow across the boundaries of the spatial element while the latter does not. The left side of Eq. (19-9) is equivalent to the left side of Eq. (19-8), and so forth for the right sides. It should be noted also that the Streeter-Phelps model leaves out the photosynthesis kinetic term; i.e., it ignores oxygen changes due to photosynthesis.

Computation Scheme

Equation (19-3) is completely general and has no restrictions. It is especially applicable to situations having variable boundary conditions (e.g., diversions or return flows which may vary over time). Because of the necessity to recompute all the variable terms (e.g., concentrations, flows, dispersion coefficient, kinetic coefficients, etc.) at each reach and to keep track of variable boundary conditions, it requires more effort to set up and more computer time to run the corresponding computer program.

Equation (19-9) is applicable only with the assumption of the steady state approximation; i.e., all flows and waste loadings are held constant with time. Because of this it uses less computer time. However, the flows and other boundary conditions can be changed to fit new steady state conditions. The computation proceeds reach by reach as in the general model, but in the steady state model no iterations with respect to time are required.

The application of either Eq. (19-3) or Eq. (19-8) requires that the boundary conditions be delineated and that a computation scheme be devised. Figure 19-11 shows a general layout of a stream system set up for mathematical simulation.

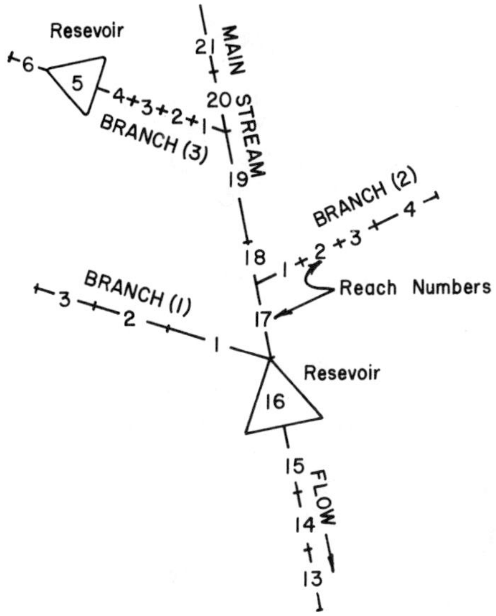

Fig. 19-11. Computation Scheme for Application of Materials Balance Equation (Waddel et al., 1974).

The iteration starts at the top of the main stem and then proceeds reach by reach downstream. When a tributary is reached, the program shifts to the top of the tributary. A reservoir is handled according to whatever instructions are programmed.

If the model is based upon Eq. (19-8) then there is only one pass through the whole system. Equation (19-8) will compute the concentration-distance profile reach by reach along a sequence of n reaches as noted above.

If the model is based upon Eq. (19-3) the computation algorithm is more complex. The computation begins at the top of the main stem and handles tributaries as described above. Eq. (19-3) is applied at reach i, in turn. However, the computation has an iteration, or a "do loop," with respect to time. It computes the concentration in reach i at time $t+\Delta t$, i.e., $c_{i,t+\Delta t}$, by Eq. (19-3). This value is stored for use in the next pass. The computation then proceeds to reach $i-1$. The computation can be seen by noting that the left side of Eq. (19-3) can be expressed:

$$\frac{c_{i,t+\Delta t} - c_{i,t}}{\Delta t}$$

The output of this computation algorithm is a three-dimensional solution which may be expressed as $C(Z)_t$, i.e., concentration distance profile for a given time; or as $C(t)_Z$, i.e., concentration-time history at a given point on the stream. The steady state model output provides only a concentration-distance profile.

An example of a steady state model based upon Eq. (19-8) is the PIONEER I, reported by Waddel et al. (1974). The model can be used for a large number of water quality constituents, since it has programmed in it the commensurate kinetic expressions.

Determination of Model Coefficients

The "model coefficients" include the various kinetic coefficients and a dispersion coefficient. These coefficients can be determined by utilizing either of two approaches: (1) empirical "calibration" of the set of model coefficients as a whole, or (2) independent determination of each coefficient. The "calibration" approach is a trial and error process which involves parametrically assuming values for these coefficients, as a set, until finally there is a match between the concentration profile as measured in the field and the computed profile. This trial and error calibration is then repeated for another set of boundary conditions, i.e. low stream flows vis a vis high flows, which utilizes another corresponding field measured concentration profile. If a set of coefficients can be found such that two or more model computed concentration profiles for different boundary conditions approximate the field measurements, one can say the model is "calibrated." These coefficients can be used then for any other boundary conditions which one may want to impose, i.e., for the purpose of exploring the effects of a proposed effluent standards program on the concentration profile.

Figure 19-12 shows the results of several runs for the PIONEER I model by Waddel et al. (1974) comparing model computed concentration profiles with measured field data. These "runs" for the two constituents shown, i.e., dissolved oxygen and ammonia, involved two different sets of boundary conditions. The runs shown were for the purpose of "veryifying" the model. This means that with the coefficient set determined, the model should predict concentration profiles for other boundary conditions. Evidently the model was successful in this test (from an engineering point of view) and so it can be said to be

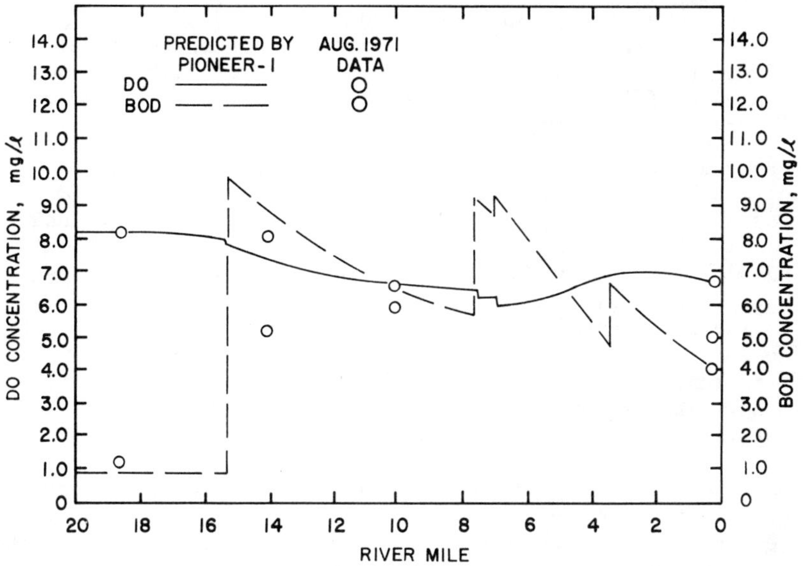

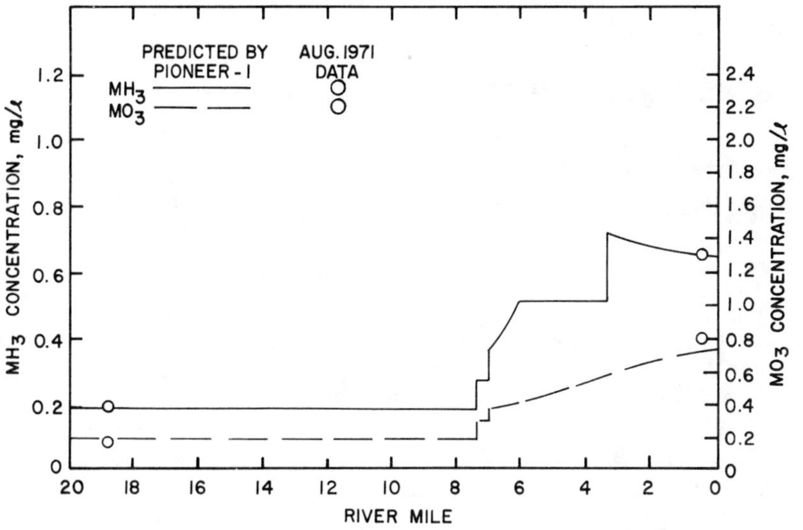

Fig. 19-12. Sample Outputs for the PIONEER I Model (Waddel et al., 1974)

"verified." With the model thus calibrated and verified, new boundary conditions can be imposed as desired.

The other general approach for determining the model coefficients involves independent evaluation of each. The dispersion coefficient, the reaeration coefficient, and the reaction rate coefficient, i.e., D, k_T, and k_B, respectively, have been the subject of a considerable amount of research, while there has been little effort to determine the photosynthesis coefficient, k_p. The latter is dependent upon sunlight and so is subject to hourly variations for each day and daily changes for a given hour; this means that if photosynthesis is considered, a considerable amount of added complexity is introduced into the model. Even though one may be able to find information on some of these coefficients from the literature, it is advisable to base these determinations on in-situ field tests. Dixon et al. (1970) developed a model which incorporated the empiricism of continuous 24-hour field measurements over an annual cycle. However, such tests should employ the theoretical rationale of Eqs. (19-3), (19-4), (19-5), (19-6) and (19-7). The latter three equations assume that dissolved oxygen is the constituent in question. If the constituent of interest is, say, ammonia or nitrate, then of course other kinetic equations would have to be postulated for the respective reactions. These equations can be applied in such a way as to isolate only the coefficient of interest. How this can be done is outlined below.

Outline of Approaches for Determination of Model Coefficients

1. Dispersion coefficient, D. Inject a slug of non-reactive dye into the stream and measure $c(t)_{z=const}$ at some point downstream. Equation (19-3) reduces to the form

$$\left|\frac{\partial c}{\partial t}\right|_o = D\,\frac{\partial^2 C}{\partial z}$$

which has a mathematical solution. The dispersion coefficient D can be calculated as the unknown since the terms in the mathematical solution can be measured.

2. Reaeration coefficient, k_T. The reaeration coefficient is related to stream turbulence by O'Conner and Dobbins (1956), Krenkel and Orlob (1962), and others. The expression given by the former for k_T is: $k_T = (D_L\overline{v})^{1/2}/H^{3/2}$, where D_L is the oxygen film diffusion coefficient (.00008 ft^2/hr at 20°C), $\overline{v}$ is the mean velocity in the stream, and H is the average stream depth.

3. Reaction rate coefficient k_B. A waste input should be located which has about constant discharge. Then for a period in which the flow in the stream is about constant, measure a concentration-distance profile, i.e., $C(Z)_{t=const}$. Thus the left side of Eq. (19-3) is zero. Also the photosynthesis term can be neglected if the field data are obtained between sunset and sunrise. If the D and k_T coefficients have been determined previously, k_B can be calculated. (Note: the "bottle method" is not recommended because conditions in the bottle deviate too much from those of the field.)

4. Photosynthesis coefficient, k_p. As seen in Figs. 19-13 and 19-14, stream dissolved oxygen concentration goes through a daily cycle; the amplitude and base of the cycle varies monthly. Thus if one picks a station, i.e., z = constant, to measure the daily variation in dissolved oxygen, one can measure the "solution" of Eq. (19-3) in the form $C(t)_{z=const}$. Since D, k_T, and k_B presumably have been determined previously, one can calculate k_p for the hour and the month in question. A set of index curves similar to those shown in Fig. 19-14 should be developed.

Definition of Boundary Conditions

The boundary conditions for a simulation model are the sets of limits for variables. The term "boundary condition" has a fairly definite meaning as applied to analytic solutions of mathematical equations. However, it is used in a much broader sense here. Here boundary conditions refer to the sets of limits of the model variables; these are unique to the particular application of the model. Boundary conditions for any water quality simulation model refer to the spatial configuration of all inputs and outputs to and from the main stem, and their respective flows and water quality conditions. Also included are all other initial conditions read into the program by the data deck. The diurnal and annual variations in the ambient stream dissolved oxygen (e.g., Figs. 19-13 and 19-14) as represented by a set of Fourier curves also would constitute a set of boundary conditions. The ambient stream water temperature, or a Fourier curve fit of its diurnal-monthly variation, would be a boundary condition, too.

Development of Boundary Conditions for an Actual Water Quality Model

The model is really no better than its boundary conditions. An example of how boundary conditions are depicted for a stream system is seen in Figs. 19-15, 19-16, 19-17 and 19-18, for the South Platte River near Denver. Figure 19-15 shows the river layout near Denver, depicting the tributaries and the reaches

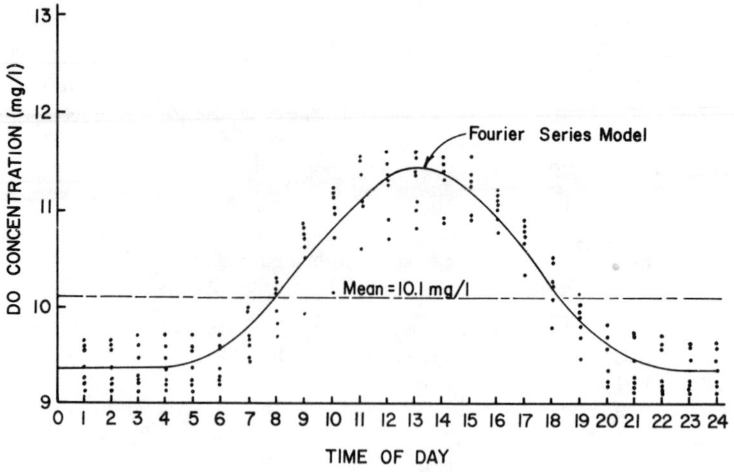

Fig. 19-13. Diurnal Variation in Dissolved Oxygen at a River Station on the Little Bear River in Utah (Dixon et al., 1970)

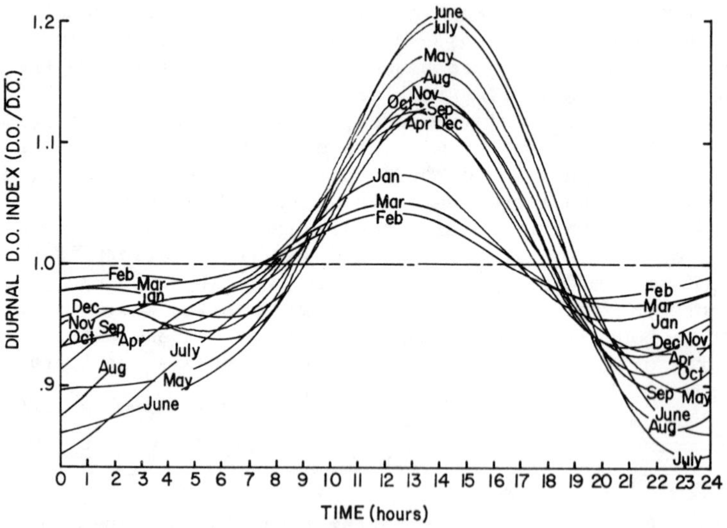

Fig. 19-14. Diurnal Dissolved Oxygen Index Curves for Each Month of the Year (Dixon et al., 1970)

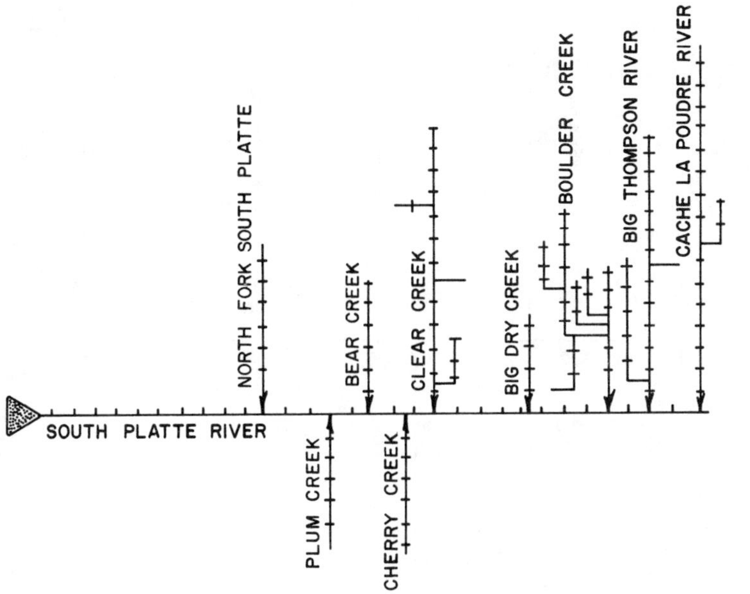

Fig. 19-15. The River Layout for the South Platte Basin Showing Tributaries and Reaches (Waddel et al., 1974).

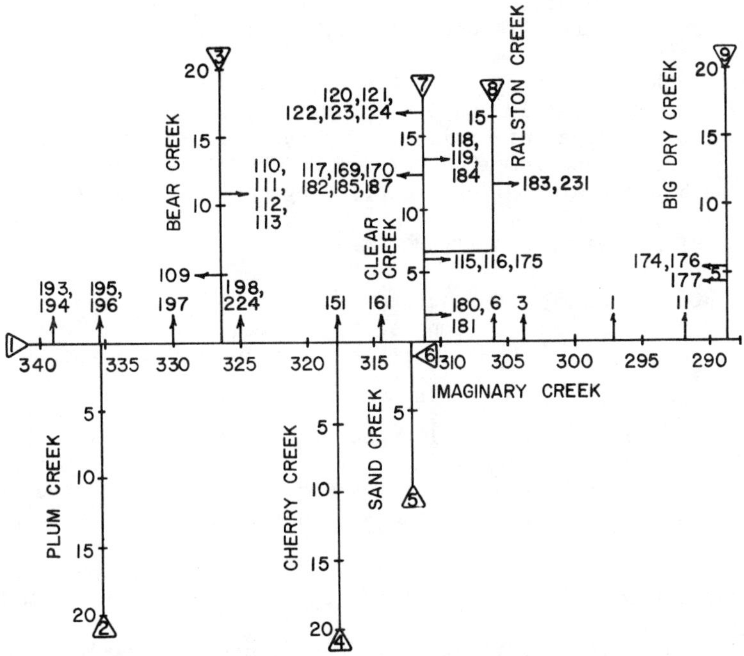

Fig. 19-16. Layout in Denver Vicinity Showing Diversions (Waddel et al., 1974).

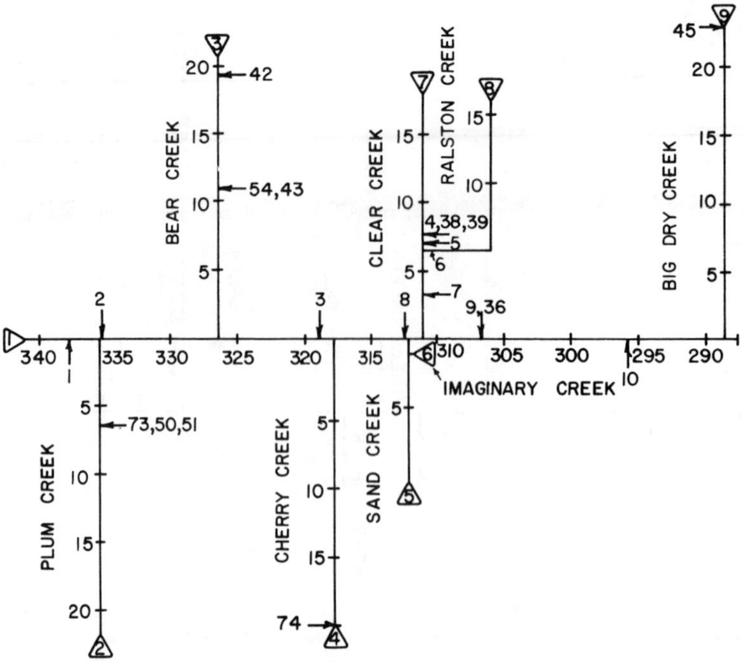

Fig. 19-17. Layout in Denver Vicinity Showing Municipal Water Sources (Waddel et al., 1974)

within both the main stem and the tributaries. Figure 19-16 shows the layout of irrigation diversions, while Fig. 19-17 shows the municipal diversions. Figure 19-18 shows how these data are coded and compiled for the PIONEER I computer program for the South Platte River, reported by Waddel et al. (1974).

19.3 USING A WATER QUALITY MODEL

Once the system is described in terms of its boundary conditions, new ones can be imposed to ascertain their effect on stream water quality. Some of the questions which might be explored include:

1. If the proposed 1977, 1983, 1985 effluent discharge standards are imposed on all present and projected point source discharges in the South Platte basin, what will be the corresponding concentration profiles for various water quality constituents for average river flows, and for the ten year low flows?

2. Is there an optimum location for a new regional wastewater treatment plant with respect to its impact on basin water quality?

3. What is the effect of different stream flow levels from reservoir releases on stream water quality?

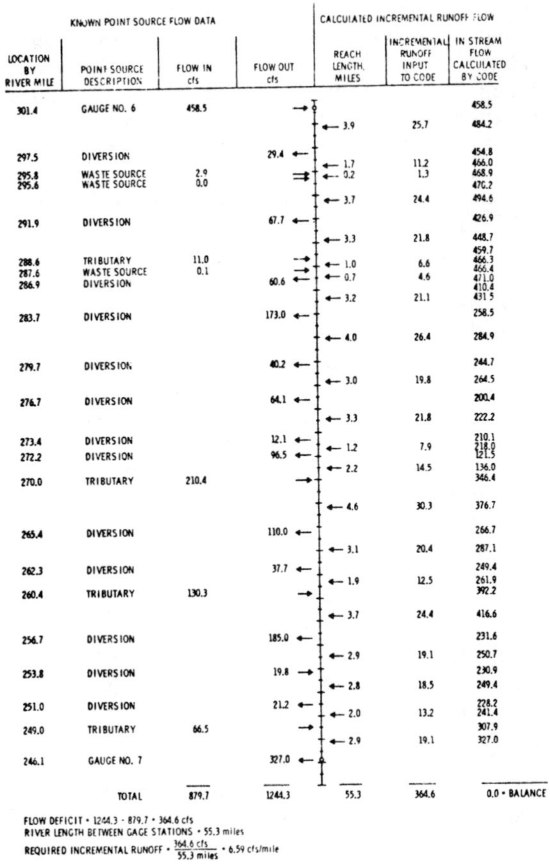
KNOWN POINT SOURCE FLOW DATA
CALCULATED INCREMENTAL RUNOFF FLOW
LOCATION BY RIVER MILE
POINT SOURCE DESCRIPTION
FLOW IN cfs
FLOW OUT cfs
REACH LENGTH, MILES
INCREMENTAL RUNOFF INPUT TO CODE
IN STREAM FLOW CALCULATED BY CODE
301.4 GAUGE NO. 6 458.5 458.5
3.9 25.7 484.2
297.5 DIVERSION 29.4 454.8
1.7 11.2 466.0
295.8 WASTE SOURCE 2.9 0.2 1.3 468.9
295.6 WASTE SOURCE 0.0 470.2
3.7 24.4 494.6
291.9 DIVERSION 67.7 426.9
3.3 21.8 448.7
288.6 TRIBUTARY 11.0 459.7
1.0 6.6 466.3
287.6 WASTE SOURCE 0.1 466.4
0.7 4.6 471.0
286.9 DIVERSION 60.6 410.4
3.2 21.1 431.5
283.7 DIVERSION 173.0 258.5
4.0 26.4 284.9
279.7 DIVERSION 40.2 244.7
3.0 19.8 264.5
276.7 DIVERSION 64.1 200.4
3.3 21.8 222.2
273.4 DIVERSION 12.1 210.1
1.2 7.9 218.0
272.2 DIVERSION 96.5 121.5
2.2 14.5 136.0
270.0 TRIBUTARY 210.4 346.4
4.6 30.3 376.7
265.4 DIVERSION 110.0 266.7
3.1 20.4 287.1
262.3 DIVERSION 37.7 249.4
1.9 12.5 261.9
260.4 TRIBUTARY 130.3 392.2
3.7 24.4 416.6
256.7 DIVERSION 185.0 231.6
2.9 19.1 250.7
253.8 DIVERSION 19.8 230.9
2.8 18.5 249.4
251.0 DIVERSION 21.2 228.2
2.0 13.2 241.4
249.0 TRIBUTARY 66.5 307.9
2.9 19.1 327.0
246.1 GAUGE NO. 7 327.0
TOTAL 879.7 1244.3 55.3 364.6 0.0 = BALANCE
FLOW DEFICIT = 1244.3 - 879.7 = 364.6 cfs
RIVER LENGTH BETWEEN GAGE STATIONS = 55.3 miles
REQUIRED INCREMENTAL RUNOFF = 364.6 cfs / 55.3 miles = 6.59 cfs/mile

Fig. 19-18. Water Balance Calculation Illustrated for a Henderson-Kersey (Waddel et al., 1974)

Many such questions can be formulated. They involve merely the imposition of corresponding boundary conditions on the model.

Samples of Output

The first question above was asked in a 1974 nationwide study by the National Commission on Water Quality pursuant to PL92-500. Contracts were let for 50 such river basin studies throughout the United States. One of the water quality studies authorized by the Commission is reported by Hendricks and Bluestein (1976). The main objective of the study was to determine the effect of imposing 1977, 1983, 1985 effluent discharge standards on stream concentration profiles of various water quality constituents.

A steady state model is adequate to handle the question posed in the above study. The non-steady state model, i.e., Eq. (19-3), would have more resolution than is needed.

The PIONEER I, a steady state model developed for the South Platte River by Battelle Northwest in 1973-74 (Waddel et al., 1974) was used in the study reported by Hendricks and Bluestein (1976). The Hendricks and Bluestein study imposed the 1977, 1983 and 1985 effluent discharge standards for the South Platte River, and also the critical river flow conditions. The hydrologic conditions were imposed for the lower quartile year flows for two representative seasons; August and December conditions were deemed most critical. The seven-day ten-year low flow was used also as an even more critical flow boundary condition for the model. The water quality constituents of interest included: temperature, orthophosphate, nitrate, ammonia, BOD, and dissolved oxygen. Each of these constituents was modeled for the entire length of the South Platte from above Denver to the Nebraska state line. Some thirty runs were made to cover the various combinations of effluent limitations and hydrologic conditions. Some examples of output from the PIONEER I model are given in Figs. 19-19 through 19-25.

Figure 19-19 is the flow profile for the South Platte River for the August-Lower-Quartile-Year flow. Tributary flows, diversions, and return flows account for the flow variability seen. For reference purposes, Denver is at mile 319 and is the reason for the large flow increase at that location.

The dissolved oxygen concentration profile is seen in Fig. 19-20, while the BOD and ammonia concentration profiles are seen in Figs. 19-21 and 19-22, respectively. Again, the marked changes at mile 319 is due to the outfall of the Denver Metro Sewage Treatment Plant.

Figure 19-23 shows the comparison in the dissolved oxygen profiles for August and December. The lower temperature in December is one of the major factors accounting for the higher dissolved oxygen levels for that month. Figure 19-24 is a composite of outputs from several different runs, showing how the different basin wide pollution control policies result in different levels of response in the dissolved oxygen concentration profile. The 1977 profile for secondary treatment shows not too much improvement over 1971 conditions, while a tertiary treatment program in 1983 would result in the dissolved oxygen level meeting standards throughout the stream. The 1985 zero discharge of pollution was programmed into the model literally and the result is seen in the 1985 curve. Figure 19-25 is another example of similar output, but for ammonia.

Interpretation

The outputs shown in these exhibits should not be taken literally. Despite the great amount of hydrologic detail required to construct a reasonable model, there were still many assumptions imposed to make it workable. The model output does,

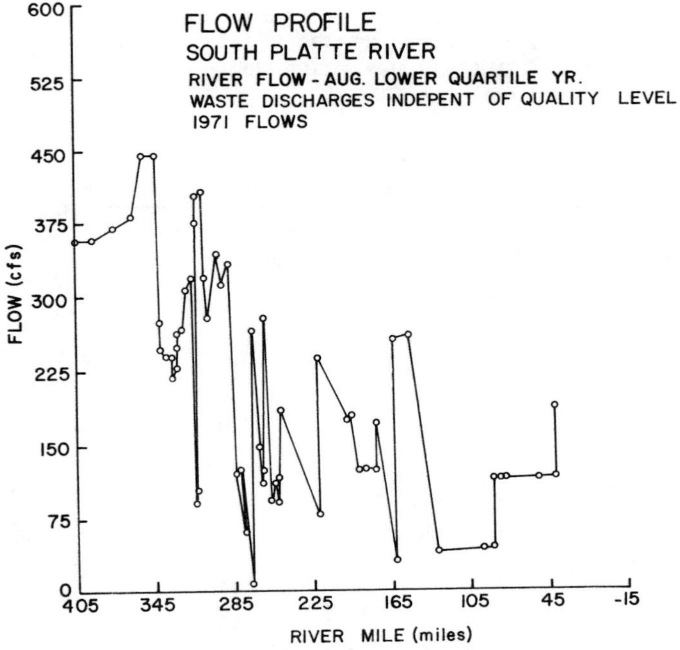

Fig. 19-19. Flow Profile for the South Platte River (Bluestein and Hendricks, 1975)

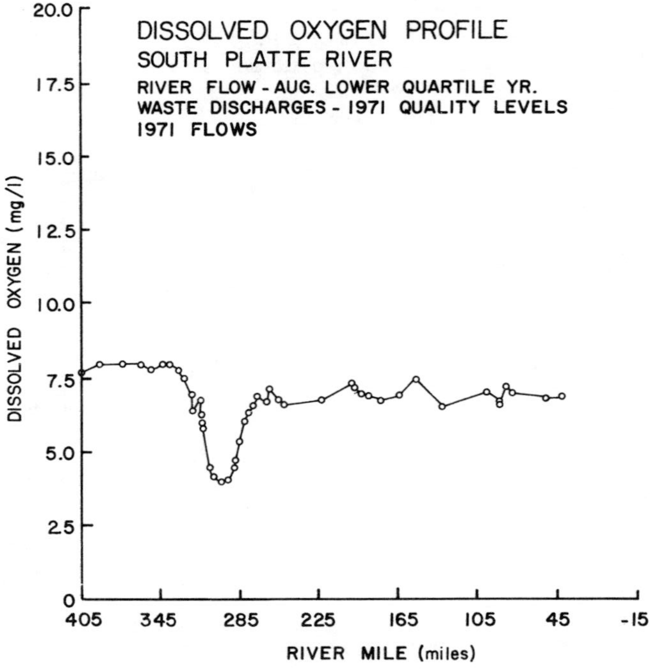

Fig. 19-20. Dissolved Oxygen Concentration Profile South Platte River (Bluestein and Hendricks, 1975)

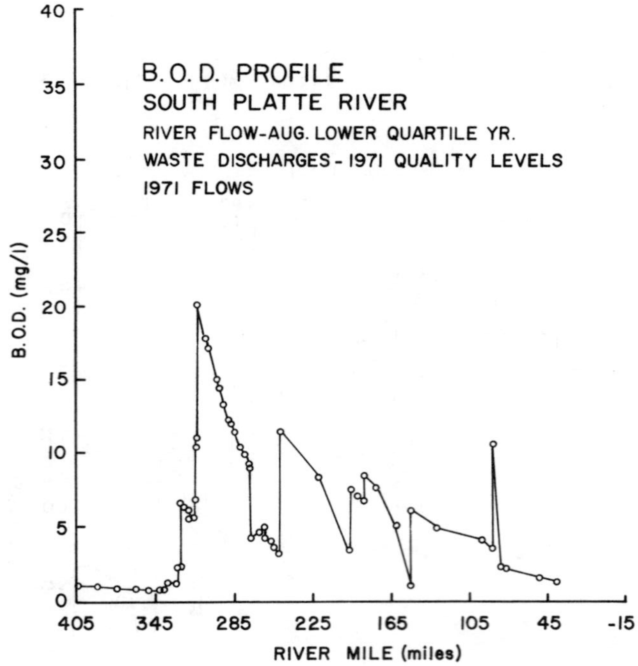

Fig. 19-21. BOD Concentration Profile South Platte River (Bluestein and Hendricks, 1975)

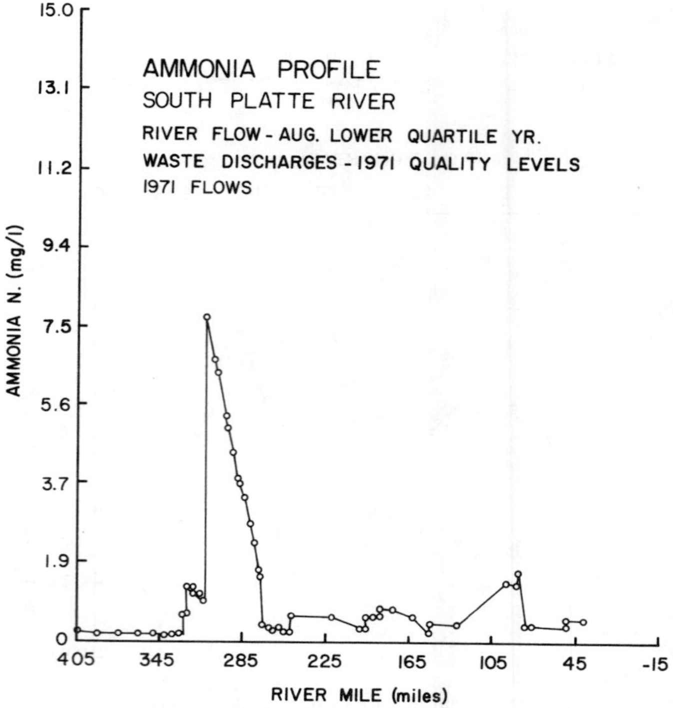

Fig. 19-22. Ammonia Concentration Profile South Platte River (Bluestein and Hendricks, 1975)

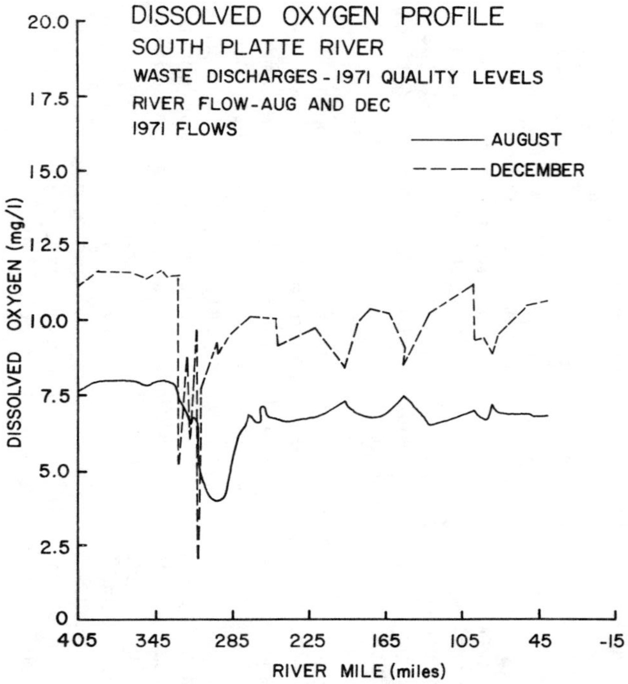

Fig. 19-23. Comparison in Dissolved Oxygen Concentration Profile for August & December, S. Platte River (Bluestein and Hendricks, 1975)

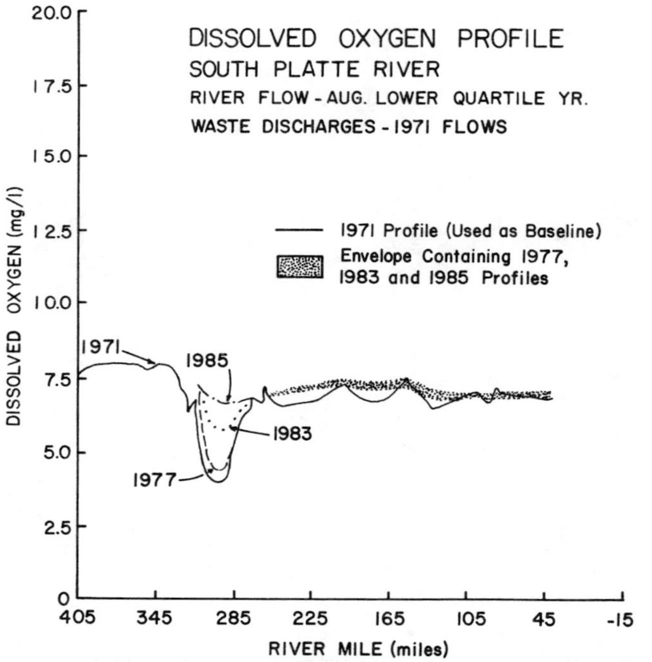

Fig. 19-24. Response of D.O. Concentration Profile, S. Platte River to 1977, 1983, 1985 Effluent Discharge Limitations (Bluestein and Hendricks, 1975)

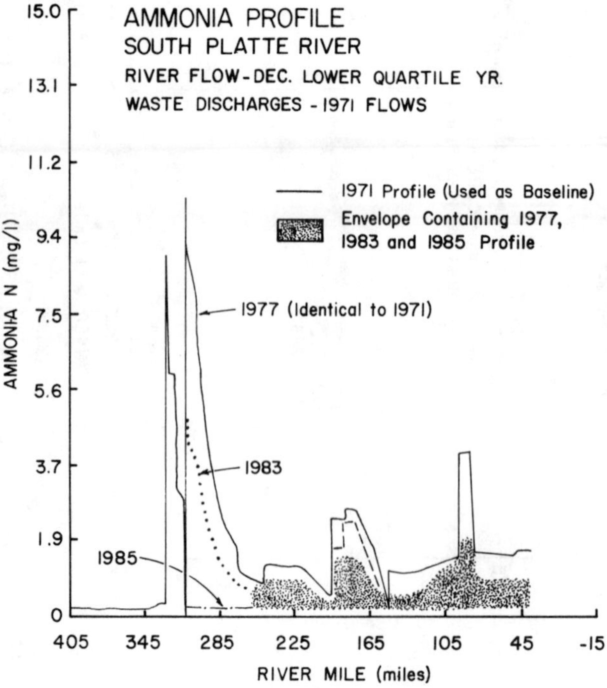

Fig. 19-25. Response of Ammonia Concentration Profile, South Platte River, to 1977, 1983, 1985 Effluent Discharge Limitations (Bluestein and Hendricks, 1975)

however, show trends which can be expected and the probable effects of imposing different conditions on a stream system. The key question involved in the foregoing modeling exercise for the South Platte is: if a huge national investment is undertaken to control waste discharges as indicated by the schedule, will the water quality actually improve sufficiently that the river can be brought to the required standards? This question was explored through the use of the PIONEER I water quality model.

Although output from only the PIONEER model has been shown here, several such models have been developed over the past ten years. Many agencies perfer to develop their own. The same general development approach as outlined in the foregoing sections is required irrespective of the particular model used.

Water quality models are possible because of the advent of large computers which can keep track of large amounts of data. The delineation of boundary conditions, the determination of model constants, and the depiction of the appropriate reactions and the associated kinetic equations are some of the difficulties in water quality modeling. A water quality system is very complex and so in any water quality modeling exercise, a considerable amount of judgment is inherent in development of the most appropriate assumptions.

The mathematical and scientific rigor of model construction provides a delineation of the principles of system behavior. However, as one examines the complexity of the real situation, it becomes clear that some amount of relaxation of the rigor is necessary - unless one has unlimited time and budget. It is here that the engineering philosophy becomes important: the model is seen not as an end, but as a tool to facilitate engineering assessments (e.g., what will be the stream water quality response to a given effluent discharge program?). So the questions are: (1) Can we shorten the time of computation? The answer is yes, by choosing the steady state approximation. (2) Can we simplify boundary conditions? The answer is yes, by "lumping" various flows in a reach, and by assuming a steady flow. (3) Can we approximate some of the reactions and reaction kinetics? (4) Can we reduce field data required?

19.4 REFERENCES

Bluestein, M. H., and Hendricks, D. W., 1975. Biota and water quality of the South Platte River. Past-Present-Projected, Environmental Engineering Technical Report, Department of Civil Engineering, Colorado State University, Fort Collins, Colorado.

Dixon, N. P., Hendricks, D. W., Huber, A. L., and Bagley, J. M., 1970. Developing a hydro-quality simulation model. Utah Water Research Laboratory, Report PRWG67-1, Utah State University, Logan, Utah.

Hendricks, D. W., and Bluestein, M. H., 1976. Response of the South Platte to effluent limitations. Journal of the Environmental Engineering Division, No. EE4, American Society of Civil Engineers, August.

Kartchner, A. D., Dixon, N. P., and Hendricks, D. W., 1969. Modeling diurnal fluctuations in stream temperature and dissolved oxygen, Proceedings of the 24th Annual Purdue Industrial Waste Conference.

Krenkel, P. A., and Orlob, G. T., 1962. Turbulent diffusion and the reaeration coefficient, Journal of the Sanitary Engineering Division, American Society of Civil Engineers (see also Transactions ASCE, Vol. 128, Part III:293-334, 1963).

O'Conner, D. J., and Dobbins, W. E., 1956. The mechanism of reaeration in natural streams, Journal of the Sanitary Engineering Division, American Society of Civil Engineers (see also Transactions ASCE, Vol. 123:641-684, 1958).

Waddel, W. W., Cole, C. R., and Baca, R. G., 1974. A water quality model for the South Platte River basin. Documentation Report, Battelle Pacific Northwest Laboratories, Richland, Washington, April.

Chapter 20

ADDITIONAL REMARKS ON EXTREMAL FLOODS, BASIC EQUATIONS, RIVER CHANNEL PATTERNS, MODELING TECHNIQUES AND RESEARCH NEEDS

by

H. W. Shen, Professor of Civil Engineering, Colorado State University, Fort Collins, Colorado

Chapter 20

ADDITIONAL REMARKS ON EXTREMAL FLOODS, BASIC EQUATIONS, RIVER CHANNEL PATTERNS, MODELING TECHNIQUES AND RESEARCH NEEDS

20.1 INTRODUCTION

Alluvial river behavior is a complex subject. Through centuries of concerted effort, mankind has gradually learned to manage its activities along river systems. Physical modeling of rivers has long served as a powerful tool for prediction of the behavior of individual river reaches, and with the availability of large-scale digital computers, mathematical models have received increasing attention. Although much remains to be learned about alluvial river behavior, mathematical models of river systems can provide us with certain quantitative answers.

Research needs in erosion and sedimentation have been discussed by Einstein (1961), Simons and Gessler (1971), Chiu et al. (1977) and others. This chapter addresses only modeling of rivers and is divided into several sections dealing with the estimation of maximum flood flows, basic flow and sediment equations, stability of river channel patterns, and modeling techniques. Some additional remarks on these various subjects are also included.

20.2 EXTREMAL FLOODS

Perhaps the major hazard for mankind in dealing with rivers is the possibility of overbank flood flows. Historical records indicate that hundreds of thousands of human lives have been lost in a single major catastrophic flood. The following two figures show some of the results of two major floods, one occurring in a large, wide river plain and the other in a narrow mountain canyon. Hoyt and Langbein (1955) have discussed various flood problems from different viewpoints and have also listed major flood events in the United States.

Normally, experimenters have attempted to determine two criteria: a so-called "upper limit" based on some version of maximum probable flood, and a "lower limit" based on extension of previous rainfall and flood records. There is still much to be learned before these criteria can be reliably and accurately determined.

Chapters 2, 3 and 4 of this text discuss available methods for the prediction and analysis of flood flows. However, none of these can determine the correct approach to the derivation of a reliable flood estimate for a given return period at a particular location on a river. Method selection remains largely an art.

Fig. 20-1. Some of the damage caused by the August 1, 1976, Big Thompson flood in Colorado (courtesy Jim Ruff, Colorado State University).

Fig. 20-2. Flooded area of Sioux City, Iowa, April 15, 1952 (courtesy U.S. Army Corps of Engineers).

Maximum Probable Floods

"Maximum probable flood" has been defined as the largest flood for which there is any reasonable expectancy in a particular climatic era. This is not normally meant to be used as a design criterion, but as an estimate of an upper limit. Usually there are two approaches to the determination of the maximum probable flood. The first is to transpose onto the river system being studied an extremely heavy rainfall pattern from a similar region, and the second is to assume full moisture saturation for a certain air temperature (especially useful in tropical areas) and also to assume that a maximum of 60% of this moisture could be precipitation. Actually, such studies should perhaps be conducted by climatologists and/or hydrometeorologists.

Some specific research needs in this area include:

1. how can climatic regions be classified?
2. what is the maximum possible precipitation for each region and for what duration should it be obtained?
3. what is the predominant direction of moisture inflow for a particular river valley?
4. what are the major factors that should be used in the classification of similar climatic regions?
5. what are the variations of precipitation with elevation within each river valley?

and many others. Of course, if snow melt is an important contribution to stream flow in a particular region, its characteristics should be investigated also. Gilman (1964) provides a good brief description of the various aspects of rainfall analysis.

Analysis of Data

The National Environmental Research Council (1975) provides a good analysis of flood studies. Currently the Federal Highway Administration and Colorado State University are preparing a manual entitled "Hydrologic Analysis for Highway Engineers." This manual will be perhaps the most up-to-date reference book on the various techniques of flow analysis. Since flow analysis is such a broad field, it is not feasible even to list all the research needs here; instead, we propose to look at flood frequency analysis from a different angle.

Gumbel Type I. Fisher and Tippett (1928) initially developed extreme value distribution and Gumbel (1958) later expanded their development, which is now known as the Gumbel distributions.

The Type I distribution,

$$F_X(x) = e^{-e^{-\alpha(X_n - V)}} \qquad (20\text{-}1)$$

has been widely accepted for use in the estimation of floods. In this relationship, $F_X(x)$ is the cumulative distribution function, α is a shape parameter and V is a scale parameter. A special graph paper known as "Gumbel paper" has been designed so that one can obtain the flood magnitude X_n for any return period n.

Recently Shen and Bryson (1978) found the relationship:

$$\frac{X_{n_1}}{X_{n_2}} = \frac{\log n_1}{\log n_2} \tag{20-2}$$

where n_1 and n_2 are any two return periods that satisfy the following relationships:

$$\frac{1}{2n_1} << 1 \quad \text{and} \quad \frac{1}{2n_2} << 1 \tag{20-3}$$

Equation (20-2) is extremely useful. It indicates that if the Type I distribution is correct, the magnitude of a 1000-year flood is three times the magnitude of a 10-year flood.

It can be shown that the upper legs of most distributions (such as normal, gamma, etc.) approach the Type I distribution. Research needs in this area include:

1. Under what conditions (geological, topographical and climatic) would the probability distribution of annual flood peaks follow a Type I distribution?

2. For what return period can each of the parent distributions be approximated by a Type I distribution?

3. Since the derivation of the Gumbel distribution requires each flood event to be identical independent distributed from any other event, how can partial duration series be used to increase the usefulness of data collected in a limited time span?

Gumbel Type II. The U.S. Water Resources Council (1967, 1976) has recommended that the log-Pearson Type III distribution be used for flood frequency analysis. A variable, Y, is said to have a log-Pearson Type III distribution if its logarithm has a Pearson Type III (gamma) distribution. Shen and Bryson (1978) have shown that the maximum of several Y's of the log-Pearson Type III distribution will have a Gumbel Type II extreme value distribution. Writing the Gumbel Type II distribution in the following form,

$$F_X(x) = e^{-BX^{-k}} \tag{20-4}$$

Shen and Bryson found that, for $1/(2n) << 1$,

$$\frac{X_{n_1}}{X_{n_2}} = \left(\frac{n_1}{n_2}\right)^{\frac{1}{k}} \qquad (20\text{-}5)$$

This relationship is also rather useful. If k is known, the ratio of the flood magnitudes of the two return periods can be found. Shen and Bryson further found that the values of B (for flood frequency analysis based on data collected in regions of Texas, New Mexico, and Colorado) were usually less than 2, and a Type II distribution would indicate a much greater flood magnitude than if the maximum floods followed a Gumbel Type I distribution. On the other hand, data collected from the long, narrow Cauca River basin in Colombia, where the rainfall is extremely heavy, indicates that the k value is rather large and that the extreme floods estimated from a Gumbel Type I distribution would be larger than those from a Gumbel Type I distribution.

Some of the research needed on the Gumbel Type II distribution includes:

1. How does the parameter k vary with the characteristics of the river basin and its climatic environment?
2. For what return period can each of the parent distributions be approximated by a Type II distribution?
3. Since for a certain range of values of k the mean does not exist, and for another range of values of k the variance does not exist, how should these special situations be dealt with?

20.3 BASIC FLOW AND SEDIMENT EQUATIONS

The basic equations used to study the movement of water and sediment through rivers are:

1. Flow continuity equation
2. Sediment continuity equation
3. Flow momentum equation
4. Sediment transport equation
5. Flow resistance equation
6. Infiltration equation and other losses

Flow Continuity Equation

The flow continuity equation requires no comment if the river does not change its cross sectional area. However, if it does, a change of cross section with time must be included.

Sediment Continuity Equation

The sediment continuity equation for a case of one-dimensional flow can be written as

$$\eta \frac{\partial(WZ)}{\partial t} + \frac{\partial Q_b}{\partial x} = k_1 \qquad (20\text{-}6)$$

where η is the porosity of the sediment deposit, W is the width of the channel, Z is the bed elevation, Q_b is the bed load transport, t is the time, x is the longitudinal distance, and k_1 is the change of sediment deposit on the bed due to the deposition of suspended load and/or lateral sediment inflow.

A critical point in the solution of the above equation is the prediction of the channel width change. Bank stability depends on many factors, such as the bank material, presence of vegetation and other organic matter, fluctuation of the water level, character of the seepage, and so on. This situation would be rather complex if one included the consideration of the change of geomorphic factors such as river meandering and braiding. A great deal of research work is needed in this area.

If one is only interested in the average behavior of a long river reach, such as that of large rivers, the transfer between bed load and suspended load is not important. However, when a relatively small river reach is considered for suitability of fish habitat, one must consider the variation of the sediment scour and deposit in a short reach. In that case, the variation of k_1 with flow and sediment in terms of time and space coordinate systems should be known in order to solve Eq. (20-6).

Flow Momentum Equation

As discussed in Chapter 13, "the complete gradually varied flow equation" (Saint Venant Equation) is also called the dynamic wave equation. There are three reduced versions of this equation. If one neglects the unsteady term by introducing the input of different constant flow discharges at various time steps, it is called the quasi-steady dynamic wave equation or known-discharge solution, as described in Chapter 10. One can further assume that the change in water surface is much greater than the change in velocity head, and the diffusion wave equation is formed. The diffusion wave equation is not commonly used because it is not much simpler to solve than the dynamic wave equation. Finally, the kinematic wave approximation is the simplest, but also the least accurate model of the three approximations to the Saint Venant equation. It retains only the two slope terms--river bottom slope and energy slope. Detailed discussion of this solution is given in Chapters 9, 10 and 13.

Since there are several modifications or simplifications of the complete equation, an obvious need is to determine the applicability and limitations of each simplified version. Woolhiser and Liggett (1967) found that a single dimensionless parameter is a suitable criterion for a choice between the complete equation and the kinematic wave approximation. This is an excellent conceptual paper based on inspectional analysis. However, much more data and detailed analyses are needed to establish the relative accuracy of results using the kinematic wave approximation and the complete equation under various conditions.

Sediment Transport Equation

This is perhaps the weakest link in the mathematical modeling of river systems. Sediment input into a river system is generally limited by either the upstream sediment supply or the capability of a river to transport. For very fine material the capability of a river to transport is enormous, and thus the amount of this material in the river system is limited by upstream supply; this part of the load is called the wash load. On the other hand, for relatively coarse material the amount in the river system is limited by the capability of a river to transport it. This type of material, which can be found readily on the river bed, is called the bed material load.

The understanding of sediment production and yield from a watershed or from farm land is rather meager; this is perhaps the toughest research problem facing us today. The sediment yield from farm land depends on the characteristics and variability of agricultural products, crop growth, topographical conditions, management factors, and many others. The sediment yield from a forested watershed depends on the characteristics and variability of many factors such as rainfall; vegetation, including trees and other organic matter; soil, including cohesive soil and infiltration rate; rain splashes; the antecedent conditions of each storm; and the effects of landslides and other geotectonic activities. None of these factors are easy to study. The "Universal Soil Loss Equation" is the result of perhaps the most comprehensive research program ever undertaken. Shen and Li (1976) summarized many previous studies, and a great deal more research should be carried out.

The sediment transported by a river can be divided into two parts: the suspended bed material load and the bed load. The suspended bed material load is usually obtained by the sum of the product of flow velocity and suspended bed material concentration at different flow depths. Since the flow in a river system is turbulent, many suspended samples must be taken to obtain a reliable mean value. Studies are needed to determine the number of suspended samples necessary to achieve a mean value within a certain error range for different flow and sediment bed configurations. Preliminary tests conducted in the laboratory by the writer indicate that, in most cases, ten samples are needed to achieve a mean within a 10% error for flows over plane bed (except for the heavy suspended sediment concentration near the bed). However, it is believed that for flows over dune beds, the number of samples required to establish a mean value within a 10% error may be greater.

A vertical logarithmic velocity profile is usually reasonably accurate for turbulent flow over a plane bed. For a dune bed, the vertical logarithmic velocity profile for flow is not valid for the regions near the dune crests, and more studies to establish the variation of vertical flow velocity over different regions of a dune bed are needed. It is generally believed that Toffaletti's equation (1969) is the best for the calculation of sediment transport load for large rivers, and Colby's equation (1964) is preferred to estimate the sediment load for smaller rivers; however, much more study is needed due to the availability of more data. A reliable and rugged sampler which can continuously measure suspended sediment concentration in rivers is definitely needed.

The sediment bed load is rather difficult to measure. Bed load may be obtained by any of the following methods:

1. through improved knowledge of the transfer function between bed load and suspended load;
2. by the extension of vertical suspended sediment concentration to include bed load,
3. by direct measurements with an improved trap, and
4. by monitoring the migration of large sand dunes, if they exist.

Although aggradation is usually a continuous process, degradation is generally a step function. Bed elevation in the degradation process generally decreases continuously until the bed becomes armored. The armored bed can be destroyed by a large flood, and the bed elevation will decrease suddenly after such a flood until the bed becomes armored again. If the modeling of a short reach of river is required to study the suitability of the reach for fish habitat, detailed knowledge of sediment deposition and erosion is required. This would include the answers to such questions as:

1. What is the length of transition reach required for the flow to pick up enough sediment to reach its transport capability?
2. What variations in sediment transport characteristics exist within this transition scour reach?
3. What is the length of a transition reach over which a flow deposits sediment until it reaches its transport capability?
4. What variations in sediment transport characteristics exist within this transition deposition reach?
5. How does the unsteadiness of the flow affect the sediment transport characteristics?
6. What are the effects of infiltration on sediment transport rate?

A river flow is rarely contained in a straight prismatic channel, and two- or three-dimensional variations of river reach greatly increase the difficulty of estimating the transport capability of a river. Relatively strong secondary currents could be and usually are developed around the river bends. These secondary currents increase the ability of the flow to carry sediment in the longitudinal direction. Toward the inner river bend, the flow is weak and the sediment transport rate in the longitudinal direction is small. The estimation of the sediment transport rate in the longitudinal direction of a river bend is a difficult program and requires much attention. Currently, an uneven river cross section is divided into several lateral sections such as that shown in Fig. 10-2, page 10-5. It is assumed that each section could be treated independently of any other neighboring section in one-dimensional flow analysis. This should be verified. Even in a multi-stream flow approach as described in Fig. 10-24, and a two-dimensional flow model, whether one can use a one-dimensional sediment transport equation to apply it in the dominant resultant flow direction remains to be proven.

Flow Resistance Equation

French engineer Antoine Chezy developed perhaps the first resistance equation for uniform flow in 1769. An Irish engineer, Robert Manning, presented the first version of his famous Manning Equation in 1889. These two formulae have received wide acceptance. Experienced river engineers can visit a river reach and offer their estimation of the Chezy's C value and/or Manning's n value. At first it was thought that these resistance coefficients could be constant for all ranges of flow conditions; however, later it was found that these values may actually change and usually decrease with an increase of flow depth in relatively wide rivers.

Recently it has been found that when fluid flows over a movable river channel bed, the bed surface normally is deformed into various configurations. Some discussion of various bed configurations and resistance is given in Section 5.5 of Chapter 5.

Einstein (1950) was perhaps the first to separate the total resistance into skin resistance and form resistance. Einstein and Barbarossa (1952) proposed the use of the universal logarithmic velocity relationship, developed for a plane rigid boundary, to estimate skin resistance. They also found from laboratory and field data that their form resistance, obtained by subtracting skin roughness from total roughness, varied with a dimensionless parameter called the flow intensity. This flow intensity is actually a ratio between the flow force needed to move sediment and the resistance force of the particle to be moved. Later, many investigators followed Einstein and Barbarossa's (1952) approach and developed their own formulae.

This approach has many inherent difficulties, as discussed in Section 1.6 of Chapter 1. A great deal of research is needed to study, for various flow conditions, the actual skin roughness on dunes, form resistance on dunes, the interference of all neighboring dunes on the skin roughness and the form resistance of a particular dune, the predictability of dune heights and length, the maximum possible heights of dunes, the statistical properties of dunes, and so on.

Infiltration Equation and Other Losses

For rivers with large flows, infiltration and other losses are not important as far as the quantity of flow is concerned. As discussed under Sediment Transport Equation, the effect of infiltration on sediment movement may or may not be important. Infiltration should have an influence on bank stability, also. For flow analysis in an upstream watershed, infiltration, interception and other losses can be of major importance, as discussed in Chapter 9.

The Green-Ampt infiltration model [see Eqs. (9-3) and (9-4)] is probably the most popular approach. Since the variation of soil condition is enormous in almost any watershed, it is questionable whether or not any improvement on this equation is needed because infiltration and other losses are usually determined by rainfall-runoff model calibration.

Section 4.3 of Chapter 4 describes some recent developments on this subject. The author of Chapter 4 has developed an infiltration equation similar to the Green-Ampt equation, but with fewer assumptions. As stated in Chapter 4, perhaps research direction should be aimed at the interflow problem.

20.4 STABILITY OF RIVER CHANNELS

Geomorphologists use field evidence to trace the history of channel developments, as they are particularly interested in long-term effects. Engineers, on the other hand, use a theoretical basis to investigate what would happen to streams under certain conditions of change. From conceptual models, they predict short-term (less than 100 years) effects. Geomorphologists usually try to determine why a stream changes from its manner of change (why from how), whereas engineers try to predict how a stream will change from theoretical considerations (how from why). There is an urgent need for specialists in these two fields to join forces in the investigation of problems related to the stability of river channels.

Basic flow and sediment equations were discussed in the previous section for an idealized one-dimensional flow case. Although these equations can be applied to numerous practical cases, a change of channel regime could cause pronounced changes to rivers. Sections 5.10 through 5.15 of Chapter 5 discuss some of these geomorphic factors.

River Channel Patterns

River channel patterns are the cumulative results of a combination of climatic geological, topographic, hydraulic and human disturbance factors. Basically there are only three types of patterns: straight, meandering, and braided. A straight channel has straight and parallel banks. Flow within the channel is mainly in the longitudinal direction.

A meandering channel consists of many bends separated by short straight reaches ("crossings") between the bends (see Fig. 20-3). The secondary currents in each river bend are significant enough to cause bend modifications. Usually deep scour holes exist at the outer bend and the water near the inside bend is rather shallow. The channel bends may or may not be symmetrical.

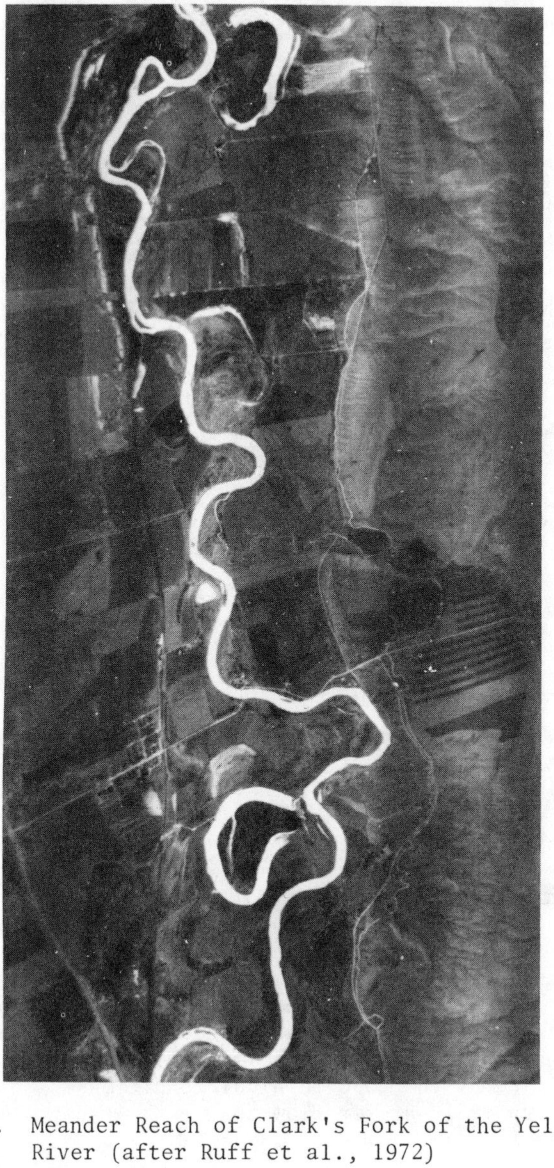

Fig. 20-3. Meander Reach of Clark's Fork of the Yellowstone River (after Ruff et al., 1972)

A braided channel usually has a large width-depth ratio, and there are always many small channels developed within the main channel (see Fig. 20-4). Furthermore, both meandering and braided channels can develop into anabranched channels. Anabranches are simply subsidiary channels that diverge from a stream and eventually rejoin it.

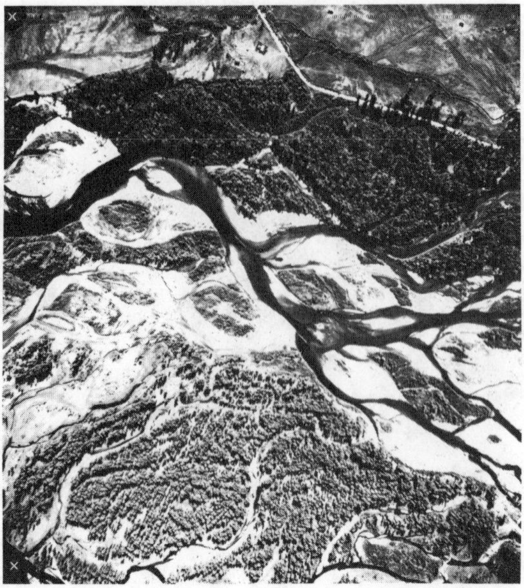

Fig. 20-5. Braided Reach of the Yellowstone River (after Ruff et al., 1972)

Straight, meandering and braided are the basic patterns, and sometimes the distinctions are not too clear. In other words, a particular reach of a river may exhibit both meandering and braided patterns at low flow stages, and yet its overbank flow may be contained within two relatively straight banks. Hence, the river appears to be straight at flood stage.

The basic causes of meandering are still not entirely clear; they are probably numerous. Shen (1971) and Callander (1968) summarized the causes as follows:

1. The development of secondary currents due to the difference of shear stresses between the bed and bank.

2. Misfit between climatic conditions and topographical conditions. Streams have a tendency to reach equilibrium condition. Perhaps the channel slope required to achieve uniform flow for the given precipitation is smaller than the natural slope specified by the topographical conditions. Meandering occurs as a natural way for flow to seek a lesser slope (see Dury, 1964).

3. Dynamic stability of flow. Callander (1977) summarized several studies investigating the possibility that "meandering occurs as a result of unstable response of the bed to a small perturbation. The thesis is that the mobile bed of a channel with straight banks is unstable, so that a perturbation in the set of alternate bars and pools will grow in amplitude. Instability of the bed is seen as the primary cause: erosion of the banks follows at a later stage..."

4. Schumm and Khan (1972) conducted an experiment in a laboratory flume to show that the addition of 3% Kaolinite in suspension with a particular range of flow conditions could cause the development of a meandering channel.

5. Development of secondary currents due to the earth's rotation. Kabelac (1957) and Neu (1967) studied and stressed the possible effect of the earth's rotation on meandering. Many mathematicians, physicists and engineers have investigated this, but until now there was no conclusive evidence that the earth's rotation is a basic cause of meandering. There is still no convincing evidence that streams in the northern hemisphere have more pronounced erosion on their right banks than on their left banks.

6. Meander analysis from the viewpoint of minimum variance. Langbein and Leopold (1966) state that "meanders are the results of erosion-deposition processes tending toward the most stable form in which the variability of certain essential properties is minimized. This minimization involves the adjustment of the planimetric geometry and the hydraulic factors of depth, velocity and local slope. The planimetric geometry of a meander is that of a random walk whose most frequent form minimizes the sum of the squares of the changes in direction in each successive unit length..."

7. Meandering caused by a lateral disturbance. This disturbance can be caused by a tributary, by difference in the soil between the left and right banks, etc. Friedkin (1945) introduced flow at an angle to a straight sand channel in a series of laboratory tests. He found that once the flow had been disturbed and a bend or lateral oscillation had been initiated, there was a marked tendency for the flow to develop a series of bends downstream.

According to Lane (1957), there are two primary causes for braided streams: 1) Overloading; i.e., the stream may be supplied with more sediment than it can carry, and part of it may be deposited. 2) Steep slopes can cause a wide, shallow channel in which bars and islands readily form. Shen and Vedula (1969) presented a basic cause of braided channels based on the considerations of bank erosion and sediment transport. The basic principle is that in a narrow stream, the entire bed can act as a unit to aggrade or degrade according to the difference between the sediment supply and capability of the flow to transport. However, in the case where a stream cross section is too wide (due to excess bank erosion during high flow or weak bank resistance or both), the entire channel cross section cannot act as a single unit, and thus part of the wide channel may be covered by numerous small channels and a braided stream occurs.

Previous Studies

Shen, Schumm and Doehring (1978) reviewed previous studies and presented the synopsis in Tables 20-1 and 20-2.

Table 20-1. Classification of Channel Pattern

Author	Straight	Transitional	Meandering	Braided
Leopold & Wolman (1960)	X		X	X
Lane (1957)			X	Due to steep slope; Due to aggradation
Schumm (1963)	X	X	Regular; Irregular; Tortuous	Straight (island)
Popov (1964)	Alternate bars			Midstream bars
Culbertson et al. (1967)			Uniform width point bars; Point bars, wider at bends	Point bars; Islands
Chitale (1970)	Single channel: X	X	Regular simple; Irregular compound	Multiple channels: X
Garg (1972)			Uniform width; Point bar	Point bar; Bar or island
Kellerhals et al. (1976)	X	Sinuous; Irregular	Irregular; Regular; Tortuous	
Brice et al. (1978)[A]	Equiwidth; Wide-bend	Equiwidth; Wide-bend	Equiwidth; Wide-bend; Braided	Wide bends: Braided; Fully braided

[A]Brice et al. also define another type of channel - anabranched.

Channel Patterns. Kellerhals, Church and Bray (1976) described channel patterns in a different manner, dividing them according to three main headings: 1) Channel types: straight, sinuous, irregular, irregular meander, regular meander, and tortuous meander. 2) Channel islands: occasional, frequent, split, and braided islands. 3) Channel bars: none, side bar, point bar, channel junction bar, midchannel bar, diamond bar, diagonal bar, and sand wave. Table 20-1 shows our arbitrary interpretation of their classification according to the normal divisions of straight, transitional, meandering and braided. However, the braided pattern was actually classified by them as subsets of straight or meandering patterns.

Brice, Blodgett and others (1978) suggested a new criterion classifying the channel into five categories: 1) equiwidth, point-bar stream, 2) wide-bend, point-bar stream, 3) braided, point-bar stream, 4) fully braided stream, and 5) anabranched stream. Their classification is also shown in Table 20-1.

Table 20-2. Investigated Channel Properties

Author	Width, Depth	Meander Wavelength, Amplitude	Sinuosity	Bend Characteristics	Braiding
Ferguson[A]		X			
Jefferson (1902)		X			
Inglis (1949)		X	X	X	
Leopold & Maddock (1953)	X				
Leopold & Wolman (1960)		X		X	
Schumm (1960, 1963, 1968, 1977)	X	X	X		
Dury (1964)		X			
Brice (1964)					X
Carlston (1965)		X			
Speight (1965)		X			
Langbein & Leopold (1966)			X		
Ackers & Charlton (1970)		X	X		
Chitale (1970)			X		
Howard et al. (1970)					X
Daniel (1971)			X		
Krumbein & Orme (1972)					X
Ferguson (1973,1975)		X	X		
Bray (1973)	X		X		

[A]See Inglis (1949)

Shen, Schumm and Doehring (1978) proposed a classification of alluvial stream channel patterns relating to the relative stability of the patterns. Five patterns were selected that include the types of streams common in the U.S. These five types are as follows:

Type 1. Straight channel, equiwidth, with straight thalweg. This channel type is stable.

Type 2. Straight channel, sinuous thalweg. This channel is generally stable, but thalweg shift and bar migration occur.

Type 3. Meandering.

a. Equiwidth - small point bars. Channel is stable; chute cutoffs can occur.

b. Wide bends - large point bars, cut bank outside of meander. Chute and neck cutoffs and meander shift and growth produce a relatively unstable channel.

Type 4. Meander-braid transition, large point bars with frequent chute cutoffs. Unstable channel.

Type 5. Braided channel with multiple thalwegs that shift and numerous bars and islands that shift. Unstable channel.

Obviously, some channel types span a great range of patterns; for example, Type 3 channels can have a great range of sinuosity.

Shen, Schumm and Doehring also presented a useful chart describing the stability of these channel patterns. A great deal of research should be initiated in river channel classification and stability under various conditions.

Other Channel Properties. Numerous investigators have studied different properties of channels using different methods. The authors, together with the particular items studied, are given in Table 20-2. Following is a short discussion of authors who initiated direct approaches.

In 1863 Ferguson (see Inglis, 1949) noticed that meander wavelength bears a relationship to the channel width. Since then various channel properties, including channel width, depth, meander wavelength and amplitude, sinuosity, bend characteristics and braiding, have been investigated by numerous researchers using conventional statistical methods. It was not until 1965 that Speight (1965), and later Chang and Toebes (1970) and Ferguson (1975) started to apply the time series technique to analysis of channel properties. Howard, Keetch and Vincent (1970) and Krumbein and Orme (1972) further used computer simulation techniques to study the process of channel braiding. Table 20-2 summarizes those properties which have been investigated.

20.5 MODELING TECHNIQUES

Physical Modeling

The basic principles in conducting a physical model investigation are explained in Chapter 6 and are reasonably straightforward. However, the application of these principles to the design of a model (especially involving distorted scales) and to the interpretation of model results for application to prototype conditions can be rather complicated. Difficulties may arise in any of the following areas: 1) selection of proper major similarity criteria, 2) selection of model scale or scales, 3) the calibration of model, and 4) the application of model results to prototype conditions including the necessary readjustments.

The major need in physical modeling is to have research engineers present their respective experience and findings on the comparison between model and prototype results. Usually after a model study is completed very little effort will be spent to investigate the accuracies of this model result using future prototype data. We must begin to collect and analyze information on the success and failure of model results in the prediction of prototype conditions. Papers such as that presented by Vries and Zwaard (1975) in summarizing model experience at the large Delft hydraulic laboratory, is rather useful. What we need in this area is not more theory of model construction, but more practical experience on how to make a model work under different conditions for various purposes.

Mathematical Modeling

A mathematical model is defined as a model which can be solved mathematically. Since most of the equations are complex and cannot be solved in a closed form, a mathematical model usually involves the simultaneous solution of several differential equations by numerical techniques with the aid of digital computers.

The major need in this area is the improvement of knowledge on basic equations as described in Section 20.3 and 20.4 of this chapter. It is difficult to visualize how one can obtain accurate results from a computer if the basic equations which build up the computer model involve many uncertainties. Of course, with proper model calibration, some of the uncertainty may be reduced.

It usually takes a great deal of time and effort to improve on our knowledge of basic equations and processes. One cannot and should not wait for the arrival of universally-accepted basic equations to construct mathematical models. Instead, mathematical models should proceed with the best available basic equations to achieve quantitative answers.

Needed research in mathematical models is described in Chapters 7 through 13, and can perhaps be grouped as follows:

1. How can the required computer time be reduced by taking variable space and time increments for different conditions?
2. How can one-dimensional equations be extended to problems involving two-dimensional flow variations?
3. How can the computer time requirement for the solution of two-dimensional flow equations be reduced?
4. When can the unsteady flow terms be neglected by substituting a known discharge at different time intervals?
5. When should the kinematic wave equation be used?
6. How much data is needed for model calibration in order to achieve a certain accuracy?
7. What accuracies can be expected from each model for a given amount of data in a particular situation?
8. How can the applicability and reliability of a particular model in a given situation be defined?
9. How should a model be selected for a given situation?
10. How should a model be evaluated?
11. How can watersheds and river basins be classified for the adoption of the best models?
12. Since the many variables involved (such as flow) are actually random in nature, how can model results be evaluated probabilistically?

Mathematical models of rivers have progressed rapidly in recent years and provide us with much-needed quantitative results. Perhaps we should take some time to evaluate how good these results are under different conditions in a probability context.

20.6 REFERENCES

Ackers, P., and F. G. Charlton, 1970, The geometry of small meandering streams, Proc. Inst. Civil Engineers, Paper 73285, pp. 289-317.

Bray, D. I., 1973, Regime relations for Alberta gravel-bed rivers, *Fluvial Processes and Sedimentation*, National Research Council of Canada, Ottowa, pp. 440-452.

Brice, J. C., 1964, Channel patterns and terraces of the Loup River in Nebraska, U.S. Geol. Surv. Prof. Paper 422-D.

Brice, J. C. and J. C. Blodgett et al., 1978, A study of countermeasures for hydraulic problems at bridges, FHWH draft.

Callander, R. A., 1968, Instability and River Meanders, Ph.D. thesis, University of Auckland, New Zealand, 285 pp.

Carlston, C. W., 1965, The relation of free meander geometry to stream discharge and its geomorphic implications, Amer. J. Sci., 268, pp. 864-885.

Chang, T. P. and G. H. Toebes, 1970, A statistical comparison of meander plane-forms in the Wabash Basin, Water Resources Research 6, pp. 557-578.

Chitale, S. V., 1970, River channel patterns, J. of the Hydr. Div. ASCE, Vol. 96, pp. 201-221.

Chiu, C. L., et al., 1977, Research needs in erosion and sedimentation, Transactions, American Geophysical Union, Vol. 58, No. 12, pp. 1076-1083, Dec.

Colby, B. R., 1964, Discharge of sands and mean-velocity relationships in sand-bed streams, U.S. Geol. Surv. Prof. Paper 462-A.

Culbertson, D. M., L. E. Young, and J. C. Brice, 1967, Scour and fill in alluvial channels, U.S. Geol. Surv. Open-File Report, 58 pp.

Daniel, J. F., 1971, Channel movement of meandering Indiana streams, U.S. Geol. Surv. Prof. Paper 732-A.

Dury, G. H., 1964, Principles of underfit streams, U.S. Geol. Surv. Prof. Paper 452-A.

Einstein, H. A., 1950, The bed load function for sediment transportation in open channel flows, U.S. Dept. of Agriculture Tech. Bull. 1026, 70 pp.

Einstein, H. A., 1961, Needs in Sedimentation, J. of Hydr. Div. ASCE, HY2, Proc. Paper 2756, Mar.

Einstein, H. A., and N. L. Barbarossa, 1952, River channel roughness, Transactions, ASCE, Vol. 117, Paper No. 2528, pp. 1121-1132.

Ferguson, R. A., 1973, Regular meander path models, Water Resources Research, Vol. 9, pp. 1079-1086.

Ferguson, R. I., 1975, Meander irregularity and wavelength estimation, J. Hydrol., Vol. 26, pp. 315-333.

Fisher, R. A. and L. H. C. Tippett, 1928, Limiting forms of the frequency distribution of the largest or smallest member of sample, Proceedings of the Cambridge Philosophical Society, Vol. 24, pg. 180.

Friedkin, J. F., 1945, A laboratory study of the meandering of alluvial rivers, U. S. Waterways Experimental Station, 39 pp.

Garg, M. M., 1972, River classification by photos and map, thesis, Colorado State University, April.

Gumbel, E. J., 1958, *Statistics of Extremes*, Columbia University Press, New York.

Howard, A. D., M. E. Keetch, and C. L. Vincent, 1970, Topological and geometrical properties of braided streams, Water Resources Research, Vol. 6, pp. 1674-1688.

Hoyt, W. G., and W. B. Langbein, 1955, *Floods*, Princeton University Press.

Inglis, C. C., 1949, *The Behavior and Control of Rivers and Canals*, Res. publ. poona (India), No. 13, 2 vols.

Jefferson, M., 1902, The limiting widths of meander belts, Natl. Geogr. Mag., Oct.

Kabelac, O. W., 1957, Rivers under influence of terrestrial rotation, J. of Waterways and Harbors Div. ASCE, Vol. 83, WW1, Proc. Paper 1208.

Kellerhals, R., M. Church, D. I. Bray, 1976, Classification and analysis of river processes, J. Hydr. Div. ASCE, Vol. 102, pp. 813-829.

Krumbein, W. C., and A. R. Orme, 1972, Field mapping and computer simulation of braided stream networks, Geol. Soc. Amer. Bull., Vol. 83, pp. 3369-3380.

Lane, E. W., 1957, A study of the shape of channels formed by natural streams flowing in erodible material, M.R.D. Series No. 9, U.S. Army Eng. Div., Missouri River, Omaha, Nebraska.

Langbein, W. B., and L. B. Leopold, 1966, River meanders--theory of minimum variance, U.S. Geol. Surv. Prof. Paper 422-H, 15 pp.

Leopold, L. B., and T. Maddock Jr., 1953, The hydraulic geometry of stream channels and some physiographic implications, U.S. Geol. Surv. Prof. Paper 242, 57 pp.

Leopold, L. B., and M. G. Wolman, 1960, River meanders, Geol. Soc. Amer. Bull., Vol. 71, pp. 769-794.

National Environmental Research Council, 1975, *Flood Studies Report*, pub. by National Environmental Research Council, London.

Neu, H. A., 1967, Transverse flow in a river due to earth's rotation, J. of Hydr. Div. ASCE, Vol. 93, HY5, Proc. Paper 5434.

Popov, I. V., 1964, Hydromorphological principles of the theory of channel processes and their use in hydrotechnical planning, Soviet Hydrology, pp. 158-195.

Ruff, J. E., M. M. Skinner, J. W. Keys III, and D. Waugaman, 1972, A remote sensing study of sediment in the Clarkes Fork Yellowstone River, Proposed for Bureau of Reclamation, Dept. of the Interior, Nov.

Schumm, S. A., 1960, The effect of sediment type on the shape and stratification of some modern fluvial deposits, Am. J. Sci. Vol. 258, pp. 177-184.

Schumm, S. A., 1963, Sinuosity of alluvial rivers on the Great Plains, Geol. Soc. Amer. Bull., Vol. 74.

Schumm, S. A., 1968, River adjustment to altered hydrologic regimen, Murrumbidgee River and Paleochannels Australia, U.S. Geol. Surv. Prof. Paper 598.

Schumm, S. A., 1977, *The Fluvial System*, John Wiley & Sons, New York.

Schumm, S. A., and H. R. Khan, 1972, Experimental study of channel patterns, Geol. Soc. Amer. Bull. Vol. 83, pp. 1755-1770.

Shen, H. W., 1971, Stability of Alluvial Channels, *River Mechanics*, H.W. Shen (ed.), Vol. I, pp. 16-1 - 16-33.

Shen, H. W., and M. Bryson, 1978, Impact of extremal distributions on analysis of maximum loading, special invited paper, International Symposium on Risk and Reliability in Water Resources, University of Waterloo, Ontario, Canada, June.

Shen, H. W., and R. M. Li, 1976, Watershed sediment yield, *Stochastic Approaches to Water Resources*, Chap. 21, H. W. Shen (ed.), Fort Collins, Colorado.

Shen, H. W., S. A. Schumm, and D. O. Doehring, 1979, Stability of stream channel patterns, paper presented at Transportation Research Board meeting of the National Research Council, Washington, D.C., Jan.

Shen, H. W., and S. Vedula, 1969, A basic cause of a braided channel, Proceedings, 13th Congress of the IAHR, Aug.-Sept., Kyoto, Japan, Vol. 5-1, pp. 201-205.

Simons, D. B., and J. Gessler, 1971, Research needs in fluvial processes, *River Mechanics*, Chap. 32, H. W. Shen (ed.), Fort Collins, Colo.

Speight, J. G., 1965, Meander spectra of the Angabunga River, J. Hydrol., Vol. 3, pp. 1-5.

Toffaletti, F. B., 1969, Definitive computations of sand discharge in rivers, J. of Hydr. Div. ASCE, Vol. 95, HY1, Proc. Paper 6350, Jan.

U.S. Water Resources Council, 1967, A uniform technique for determining flood flow frequencies, Bulletin No. 15.

U.S. Water Resources Council, 1976, Guidelines for determining flood flow frequency, Bulletin No. 17 of the Hydrology Committee, Mar.

Vries, M. de, and J. J. van der Zwaard, 1975, Movable-bed river models, Symposium on Modeling Techniques, ASCE, Vol. 1, pp. 484-500.

Woolhiser, D. A., and J. A. Liggett, 1967, Unsteady, one-dimensional flow over a plane--the rising hydrograph, Water Resources Research, Vol. 3, No. 3, pg. 753.